# The Landscape of the Sierra Nevada

Regino Zamora • Marc Oliva
Editors

# The Landscape of the Sierra Nevada

## A Unique Laboratory of Global Processes in Spain

Springer

*Editors*
Regino Zamora
Department of Ecology
University of Granada
Granada, Spain

Marc Oliva
Department of Geography
University of Barcelona
Barcelona, Spain

ISBN 978-3-030-94218-2 ISBN 978-3-030-94219-9 (eBook)
https://doi.org/10.1007/978-3-030-94219-9



Front cover photo: Andrés Ureña.
Back cover photo: Provided by the SeaWiFS Project, NASA/Goddard Space Flight Center, and ORBIMAGE.

This Springer imprint is published by the registered company Springer Nature Switzerland AG
The registered company address is: Gewerbestrasse 11, 6330 Cham, Switzerland

# Foreword

The mountains of the world cover roughly 12% of the land area outside Antarctica. This includes valleys and immediate forelands, inhabited by roughly 0.5 Billion people. If one includes wider foreland skirts, some mega-urban areas become associated with mountains, raising the potentially influenced global population to several billions. While obviously important to humanity, mountains also host an incredible organismic diversity, which seems like contradicting the often demanding life conditions in mountains. However, this is a human perspective.

Mountains are hotspots of biodiversity for several reasons. The most important one is that they assemble a broad spectrum of climatic conditions over a narrow geographical range. Depending on latitude and elevation, a single mountain range such as the Sierra Nevada may host thermal belts from nearly subtropical to almost polar over a distance of less than 25 km. Since mountains interfere with the weather system, they often exhibit sharp humidity gradients, with more humid frontal ranges and drier interior or continental flanks. Combined with a pronounced seasonality such as the Mediterranean one, this adds to the climatic diversity of life conditions. Once land becomes elevated, topography comes in as an additional driver of habitat diversity through contrasts in exposure to sun and wind. Structured further by the forces of gravity into ridges, gullies, rock fields, fens, etc., the resulting habitat diversity exceeds any other on land. It is the habitat diversity created by all these forces that explains the biological richness of mountains.

This book is unique by assembling information on all abiotic drivers (climate and geology, both current and past), the biological inventory (plants and animals, both aquatic and terrestrial), and the human dimension in mountains. To my best knowledge, this is the first and so far only monographic for a mountain range that takes such an interdisciplinary approach. Covering 2273 km$^2$ of land area, according to the newest global mountain inventory (v2.0) by the Global Mountain Biodiversity Assessment (GMBA), and reaching up to 3479 m at the Mulhacen summit, the Sierra Nevada represents a 'compact unit' that covers all the above-mentioned dimensions of habitat diversity. Not surprisingly, the Sierra Nevada hosts 2348 species of flowering plants, which is similar to the entire Swiss flora, although covering an area 17 times smaller than Switzerland. While the main part of the Alps is covering roughly 188.000 km$^2$, (excluding the lower elevation, most south-western, and the most eastern tails) an area 83-times larger than the Sierra Nevada, the number of flowering species arrives at 4500 (Flora Alpina, 2004), roughly twice that of the Sierra Nevada. This underlines the exceptional biogeographic position of this Mediterranean mountain range.

The book describes in great detail how humans have been shaping the Sierra landscape over millennia from pre-Roman times till more recent centuries when irrigation practices came in use, which changed farming conditions, and thus livelihoods, fundamentally. In an analogy to the dry part of the Swiss central Alps, the Valais, people discovered that the 'water tower' nature of high mountains can facilitate lush life at the foothills, once the mountain waters are channelled to their fields. Water is life. Embedded between the already seasonally dry Mediterranean climate and the almost arid interior of the Iberian Peninsula, the central role of water is the dimension that makes the Sierra Nevada different from most other mountain ranges, those of Europe in particular.

Biogeographers are used to subdivide mountains into bioclimatic belts, addressing the foothills as collines, the naturally commonly forested part as montane, and the world above the low-temperature tree limit as alpine. This temperature-related subdivision of mountains, as already coined by Alexander von Humboldt in his global 'Naturamas', becomes problematic when aridity comes into play. For instance, when an area is too dry for tree growth, it may still host a rich plant and animal life, but there is no treeline that can be employed to define the lower edge of an alpine belt. Hence, the classical definition of 'alpine' must fail under such conditions. The authors adopted a pragmatic approach and identified an area of the size of 242 $km^2$ as 'alpine' based on floristic criteria, although, rated by a global treeline-isotherm approach, there would be hardly any alpine terrain left above such a temperature-only controlled tree limit. As evidenced by current seasonal snow cover and past glaciation, this approach is well justified. With 362 'alpine' species, the Sierra Nevada exceeds by far the alpine plant species diversity of other European mountain ranges of similar size. In Alpine Plant Life (Körner 2021, Springer), I listed species inventories for 10 different, but comparable, mountain ranges from all climatic zones (including the tropics) that average at 250 species per mountain range (210–300), once more underpinning the exceptional position of the Sierra Nevada, although such comparisons are always limited by the uncertain delineation of the lower edge of the alpine belt. I used plant diversity for this comparison, because of the availability of comparable data. Similar comparisons await to be explored using the rich data available for animals and those for aquatic systems in particular (lakes and rivers) as evidenced in several chapters.

Capitalizing on the biological archives in wetlands, the book provides an impressive view of the long-term climatic past of the Sierra Nevada. Once the glaciers melted, there was a maximum temperature and precipitation between 7000 and 10500 years before the present. Thereafter the climate gradually aridified, bringing also Sahara dust to the region, and over the last 3000 years, the human impact became increasingly important. Currently, up to one-third of annual precipitation is stored in the snow at the highest elevation, which extends hydrological discharge into the dry season. In the light of ongoing climatic change, these benefits will decline and the aridity is likely to increase. For a limited period, the still existing rock glaciers above 2500 m elevation may supplement the hydraulic discharge, till they vanish.

This highly diverse and, at the same time, highly utilized mountainous landscape combines elements of the mountain world with the semi-desert world. It thus provides unique possibilities to study the effects of environmental change in all its dimensions. The authors make strong cases for improved inventories, employing remote sensing techniques, building digital data archives, and utilizing local knowledge to facilitate future land development (e.g. reforestation) and projections for the consequences of climatic change.

This impressive assemblage of themes and authors arrived at an almost exhaustive characterization of the Sierra Nevada mountain range. This volume is setting the stage for future attempts at such a synthesis for other mountain ranges. I congratulate the authors for having gone that far. As had been stated many times before, the elevation of mountains above the plains can be viewed as an 'experiment' by nature, where the consequences of environmental contrasts can be studied in a quasi-steady state over short geographical distances. Man-made experiments commonly do not last long enough to establish a new steady state. In contrast, 'experiments' by nature allow for evolutionary adjustment and species selection or replacement. What may look like a hostile world from a human perspective is home to plants, animals, fungi, and microbes that are perfectly adjusted to the local life conditions.

Basel, Switzerland
November 2021

Christian Körner

# Preface

*In Genesis, God blesses human beings and bids us to take dominion over the fish in the sea, the birds in the air, and every other living thing. We are entreated to be fruitful and multiply, to fill the earth, and subdue it (Gen. 1:28).*

Reading these passages of Genesis now from the perspective of the twenty-first century leaves no doubt that the divine mandate has been fulfilled. Our species *Homo sapiens* originated in Africa and spread to Eurasia and the rest of the world, taking dominion over practically all living and non-living things. As hunter-gatherers, humans sought food in the places where it was most abundant and accessible, so groups of *Homo sapiens* first probably occupied the coastline and the lowlands. During glacial periods, mountains covered with ice would have been inhospitable environments for our species. However, the interglacial periods allowed the colonization of these montane environments for hunter-gatherers, for which the high mountains could also have represented sacred landmarks. In the driest regions of Eurasia, our ancestors had to follow water (a resource as well as a reliable sign of plant and animal food), which comes mostly from the mountains. The obligate search for water as a resource for humans as well as for the plants and animals that humans feed on must has pushed our ancestors to occupy the mountains and associated landscapes, especially in dry places, such as the Mediterranean Basin. These mountains in dry areas must have been occupied early on. Evidence has been left by Palaeolithic hunter-gatherer groups as well as by Neolithic settlers implementing agricultural practices, for which mountains guaranteed water during the dry summer. The corollary is simple: mountain ecosystems in the Mediterranean region can be understood only if the geomorphological, biogeographical, and ecological keys are known, but also taking into account the history of human use that has shaped these systems.

Many events have shaped human society and our environment since our species left Africa some 160,000 years ago, but most momentous changes have occurred since the so-called Industrial Revolution. The world's human population reached one billion around 1800. It doubled to two billion around 1930 and doubled again to four billion around 1974. The global population is now approaching 8 billion and is expected to stabilize around 9–11 billion towards the end of this century (UN 2019). The result of this explosion in our population, together with an exponential explosion in our per capita ecological footprint, has caused a series of impacts at the scale of the whole Earth System related to human activities, now known as Global Change.

Over the past 50 years, humans have changed the world's ecosystems more rapidly and extensively than in any other comparable period in human history. Today more than 75% of Earth's ice-free land is directly altered as a result of human activity, with nearly 90% of terrestrial net primary production and 80% of global tree cover under direct human influence. Similarly, in the ocean, no area is unaffected by human influence, and a large fraction is strongly affected by multiple human impacts as a consequence of rising nutrient loads from human actions coupled with warmer temperatures. This situation is new in its speed, its global scale, and its threat to the resilience of the Earth System.

The advent of the Anthropocene, the time interval in which human activities now rival global geophysical processes, suggests that we need to fundamentally alter our relationship with the planet we inhabit. The Holocene, during which complex human societies developed, remained relatively stable in terms of accommodating the environment. As we go further into the Anthropocene, we note major human-related instability, driving the Earth System onto a trajectory towards more hostile states from which we cannot easily return. The need to achieve effective planetary stewardship is urgent. In the twenty-first century, our own future on Earth, as part of the biosphere, is at stake. This new reality has major implications for human wellbeing in the face of climate change, biodiversity loss, and their interplay. The challenge, therefore, is to manage our everchanging planet in a sustainable manner in the face of rapid demographic changes, economic growth, technological innovations, socio-political conditions, and changing social-behavioural patterns.

Mountains are home to a quarter of terrestrial biodiversity and a third of the world's protected areas. These areas store natural water reservoirs on high ground (water towers) and provide vital ecosystem services for the people who live in and around them. Mountain ranges offer extraordinary opportunities to conduct research and to monitor global change, since they can be used as observatories for remote processes. Two types of processes cause changes at a broad scale. On the one hand, certain processes have acquired planetary dimensions by the sum of local impacts (such as land-use changes). On the other hand, regardless of their origin, human impact now propagates at the planet scale through the atmosphere. This second class includes processes considered more genuinely global and are precisely the processes that can be observed more clearly from mountains, which serve as exceptional lookouts. In this sense, mountains are key observatories of the atmosphere and all the aspects related to climate such as energy balance, UV radiation, atmospheric-particle deposition, greenhouse gases, or the transport of resistant biological forms and microorganisms. For example, high-mountain ecosystems are valuable sensors of global pollution because their isolation from human activity makes them receptors, paradoxically, of industrial pollution that circulates through the atmosphere worldwide.

The Sierra Nevada has witnessed glaciations, the emergence of many new species, and the extinction of many others. This high mountain, located between Europe and Africa, also witnessed among the earliest groups of hunter-gatherers arriving from Africa, ascending its valleys, and later establishing early agricultural and livestock nuclei in the mountains. In fact, the first hominid-made utensils found in the Guadix-Baza depression, in the vicinity of the Sierra Nevada, are more than 1,200,000 years old. The Sierra Nevada is a high mountain that has been inhabited by *Homo sapiens* for millennia, but where wilderness landscapes survive to this day. One of the most sweeping and swiftest changes that have occurred in recent history has been the demographic shift of human populations to the cities and the consequent rural abandonment. This human depopulation has led to the abandonment of traditional mountain practices (forest harvesting, livestock rearing, and mountain agriculture) which has had momentous ecological and socio-economic consequences. Rural abandonment during the last century generated a new scenario where the environmental impact (positive or negative) of traditional extensive activities has been diminishing in the Sierra Nevada, at the same time as new intensive exploitation systems have appeared (ski industry, urbanizations, and industrial agriculture), and above all, the global impact undergone by the Sierra Nevada ecosystems has been increasing through climate change and aerosol deposition.

The Sierra Nevada, the mountain of the sun (Sulayr) for the Moors, is also today in the twenty-first century a unique laboratory of global processes because its ecosystems are the simplest and most sophisticated biophysical sensor of global change positioned in one of the most strategic spots on the planet to detect and track regional as well as global changes. The Sierra Nevada is at the same time a climatic, biogeographic, ecological, and historical frontier.

Geographically, the Sierra Nevada stands between Africa and Europe, between the tropical crops of the coast and the semi-arid badlands of the interior of the province of Granada, and between the beaches of the Mediterranean Sea and the sky. Ecologically, it ranges between wild and the humanized, between the pristine and the polluted, and between the exploited and the abandoned.

As Joan Manuel Serrat's song 'Mediterráneo' expressed, the Sierra Nevada was born in the Mediterranean even before this sea existed. This high mountain, standing on one continent (Europe) and facing another across the sea (Africa can be seen from the highest peak of the massif), the Sierra Nevada, stands as a sentinel where the overwhelming implications of the Anthropocene can be tracked through the long history of the mountain, analysed with cutting-edge technology, and scientifically evaluated within a context of Global Change and its impact on the future of humanity.

Granada, Spain Regino Zamora
Barcelona, Spain Marc Oliva

# Introduction

In the South of the Iberian Peninsula, the Sierra Nevada stands out for its historical background and unique geographical characteristics. The massif constitutes a spatial reference in the collective imagination of the southern peninsula, an inherent benchmark of the regional idiosyncrasy. Their highest snow-capped peaks have become an inseparable part of the regional scenery easily seen at long distance and a milestone to be achieved by its inhabitants.

The singularity of the Sierra Nevada lies on the elevation difference between the massif's highlands—with peaks above 3300–3400 m a.s.l.—and the surrounding slopes—at 500–1000 m—together with its location in the southwesternmost corner of the European continent at latitude 37 ºN and the influence of both the Atlantic Ocean and the Mediterranean Sea. All these factors have determined the existence of a wide variety of landscapes resulting from the diversity of climatic and geomorphological settings. In addition, over the last centuries, human activities have modified the natural heritage of the massif introducing new socio-economic practices that have altered the landscape, particularly the middle-low hillsides of the southern slope. As the mountain transformation of the Sierra Nevada accelerated during the last decades, sectors of society raised awareness of the need to preserve its landscapes and traditions, in parallel with sustainable management and exploitation of the natural resources offered by this massif.

Scientific advances during the second half of the twentieth century have advanced our understanding of present and past mountain ecosystem dynamics in the Sierra Nevada from a wide range of perspectives: geology, geomorphology, climate, hydrology, ecology, biology, etc. Indeed, new knowledge generated over the last decades has increasingly focused on the interactions between these research fields. Consequently, like few other mountain regions on Earth, all these progresses have been translated into an exponential increase in our comprehension of natural system dynamics in a massif where the weight of history is also imprinted in its landscape.

## Why Write a Book About the Sierra Nevada?

The Sierra Nevada has traditionally been known for its natural values, which have attracted naturalists from all over Europe. Currently, the Sierra Nevada is also known for the scientific knowledge generated by the study of its ecosystems. The objective of this book is to summarize and update the huge amount of information generated about the Sierra Nevada over the last decades, providing an integrative overview of the most recent scientific advances, science-based conservation and restoration projects, and land-use management practices implemented nowadays in this massif. To this purpose, the book is divided into four main sections that can be read independently but are also interconnected.

This book aims to give visibility to the layer of knowledge that arises from the scientific study of the natural systems of the Sierra Nevada as exceptional research laboratories. For example, since 1970, there are already more than 800 publications (available in Web of Science and SCOPUS) on biodiversity, environment, and ecological aspects of the Sierra Nevada linked to Global Change. The book shows to what extent the uniqueness of the massif (geographical position, high endemicity, climatic variability, secular human impact, strong ecological gradients associated with altitude and orographic gradients, unique ecosystems such as remote process sensors, etc.) makes scientific knowledge generated in the Sierra Nevada of great interest to the international scientific community, and in general to any citizen interested in knowing reliable and multidisciplinary information about the massif. In this sense, the book highlights the relevance of the Sierra Nevada ecosystems as natural laboratories to understand the functioning of ecological systems in scenarios of global change, with their regional and global connections.

## About This Book

This book constitutes an integrated product, much more than the mere sum of the 24 chapters that compose it. To achieve this, the book is structured into four large thematic sections; within each section, the chapters address the different topics following an ordered and complementary storyline. Firstly, the book starts with an introductory section (Chapters "The Geographic Uniqueness of the Sierra Nevada in the Context of the Mid-Latitude Mountains", "Sierra Nevada, a Mediterranean Biodiversity Super Hotspot", "Singular Cultural Landscapes of the Sierra Nevada" and "Scientific Knowledge Generated in Sierra Nevada: Bibliographic Review (1970–2021)") where leading scientists with decades of experience conducting research in the mountain present the geographical, biologic, and landscape features that make the Sierra Nevada a unique mountain in the context of mid-latitude mountains in the Mediterranean region and southern Europe.

The physical setting is widely examined in the next section of the book (Chapters "Geological Setting of Sierra Nevada", "The Impact of Glacial Development on the Landscape of the Sierra Nevada", "Reconstruction of Past Environment and Climate Using Wetland Sediment Records from the Sierra Nevada", "Ancient and Present-Day Periglacial Environments in the Sierra Nevada", "Climate Variability and Trends" and "Snow Dynamics, Hydrology, and Erosion"), where internationally renowned scientists present different particular topics of the massif. The geological setting and how glacial, periglacial, and hydrological processes shaped the mountain building are discussed in the first chapters. This section also includes two contributions about past climate variability and present-day climate trends in order to better frame the past with environmental dynamics prevailing today in the Sierra Nevada. At the end of each section, following an integrative approach, the editors and authors make a brief overview of the main results, proposing some guidelines for future research needs that have not been yet sufficiently addressed in the massif.

The following section (Chapters "Altitudinal Patterns and Changes in the Composition of High Mountain Plant Communities", "Responses of Animal Populations and Communities to Climate Change and Land-Use Shifts", "Forest Dynamics Under Land-Use and Climate Change Scenarios", "Restoration of Mediterranean Forest Ecosystems After Major Disturbances: The Lanjarón Post-fire Experiment Over 15 Years of Succession", "Aquatic Animal Communities of Watercourses from Sierra Nevada", "High Mountain Lakes as Remote Sensors of Global Change", "Paleolimnological Indicators of Global Change" and "Atmospheric Inputs and Biogeochemical Consequences in High-Mountain Lakes") includes eight chapters focusing on the biotic and ecological responses in terrestrial and aquatic systems to global change drivers, particularly to climate change, aerosol deposition, and land-use change. Finally, the last six chapters (Chapters "Remote Sensing in Sierra Nevada: From Abiotic Processes to Biodiversity and Ecosystem Functions and Services", "Managing the Uniqueness of Sierra Nevada Ecosystems Under Global Change: The Value of in situ Scientific Research", "Local Ecological Knowledge and the Sustainable Co-Management of Sierra Nevada's Social-Ecological System", "Data Model, E-Infrastructure Services, and the Virtual Research Environment (VRE)", "Advancing Open Science in Sierra Nevada: Current

Citizen Science Campaigns" and "Filling the Gaps in Research, Monitoring, Management and Social Connection") address a series of new conceptual and methodological approaches that are currently being developed in the Sierra Nevada in close connection with international initiatives. These approaches are also very interdisciplinary and participatory, being in many cases the users themselves (researchers, managers, and citizens) who are responsible for generating the knowledge and the end-users of the same.

Each chapter reviews all the existing information on its subject area, both that published in the scientific literature that can be consulted in large databases and that which exists in formats that are more difficult to access. The book aims to put into value the time series that exist in the Sierra Nevada in order to show how a relevant ecosystem/environmental parameter has changed over time for natural or anthropogenic reasons. This 'travel in time' is complemented by a trip through space visiting other mountains with which to compare the uniqueness of the Sierra Nevada. Another fundamental aspect that we address in the book is to show the interdisciplinarity that has been achieved in research in the Sierra Nevada, both through the combination of complementary scientific approaches and through collaboration with other professionals, particularly with managers responsible for the conservation of the Protected Area of the Sierra Nevada under the auspices of the Observatory of Global Change of the Sierra Nevada. The book also highlights the international projects and collaborations in which it has participated, which reinforce the vision of the Sierra Nevada as an ideal place to work in international networks.

The thematic order of the book allows it to advance through a storyline with abundant cross quotes between chapters to maximize its thematic connection, avoiding repetitions and maximizing the complementarity of themes and approaches. At the end of each section, there is a Concluding Chapter, where an effort is made to integrate and synthesize the most relevant results presented in that section as products that are aimed at: (1) the scientific community, (2) managers, and (3) citizens.

The Sierra Nevada constitutes, thus, a massif encompassing a unique natural and historical heritage where research activities are being performed following an integrative approach in order to maximize time and resources. Indeed, land-use management and social connection are also integrated for an efficient administration of local and regional resources and the protection of the landscape of the summits and surrounding slopes. This book shows a study case, the Sierra Nevada, that can serve as a model for the research, monitoring, management, and environmental protection of other high mountains on the planet.

This book is mainly due to the work, expertise, and willingness of the authors of the different chapters coming from a wide range of disciplines, who have grown upon the scientific atmosphere generated over the previous decades by so many colleagues working in the Sierra Nevada.

This book was created, written, and produced during the COVID-19 pandemic. It is a good example that, even in the most difficult working circumstances, the scientific community is mobilized by an attractive common goal to promote the dissemination of scientific knowledge to society. We would like to acknowledge the authors of the book for their excellent availability, enthusiasm, and work in those very difficult times. They have made very relevant contributions, respecting the strict deadlines imposed by the editors.

This work was conducted in the collaborative framework of the 'Sierra Nevada Global Change Observatory' monitoring program (http://obsnev.es.) under the agreement between the Consejería de Agricultura, Ganadería, Pesca y Desarrollo Sostenible of the Junta de Andalucía, and the Universidad de Granada for the development of activities related to the Sierra Nevada Global Change Observatory. We would also like to acknowledge the institutions that funded the research and monitoring included in this book, which includes the European Union, the Spanish Organismo Autónomo Parques Nacionales (Ministerio de Transición Ecológica), the Spanish Ministerio de Ciencia, the Junta de Andalucía, and the Generalitat de Catalunya among many others public and private funding sources.

We are very grateful to all the staff of the Sierra Nevada National Park (managers, technicians, and park guards) for their encouragement and constant support, especially to Blanca Ramos Losada, and their continued collaboration within the framework of the joint activities that we develop at the Global Change Observatory. Special thanks to both the previous managers of the National Park, Javier Sanchez and Ignacio Henares, as well as the current ones, Francisco de Asís and Jose Luis Martinez.

Very special thanks are due to Manolo Merino Ceballos, our "web magician" for the realization, layout, and artistic revision of a good part of the figures in the book. To Ricardo Moreno Llorca, for his continued help in organizing and editing the book.

Finally, we would like to thank the photographers Luis Ordoñez, José Ignacio Cuenca, and Andrés Ureña for the amazing photos that have selflessly contributed to the edition of this book, providing additional documentary value to the findings presented in this book.

Marc Oliva
Regino Zamora

# Contents

# A Unique Mountain

**Author**: Andrés Ureña

# The Geographic Uniqueness of the Sierra Nevada in the Context of the Mid-Latitude Mountains

Marc Oliva, José M. Fernández-Fernández, and Jordi Martín-Díaz

**Abstract**

Sierra Nevada forms part of the Baetic System, south Iberia. It is located at latitude 37° N and includes the highest peaks of the peninsula above 3300–3400 m a.s.l. in its western fringe. This high semi-arid massif landscape is scarcely vegetated above 2000 m, where geomorphological landforms of glacial and periglacial origin are widespread. Quaternary glaciations shaped the highest lands above 2000 m, whereas periglacial processes remodeled the glaciated domain during interglacial periods. These alternations of glacial/interglacial phases favoured the development of a large number of endemic species; indeed, Sierra Nevada constitutes a refuge of biodiversity in southern Europe. Traditional human activities have historically benefitted from the natural resources provided by the massif that are unique in its regional context (water, pastures, snow, etc.). The geomorphological features existing across the summit areas, together with the richness of its flora and the historical heritage preserved in the massif promoted the designation of the Sierra Nevada as a Biosphere Reserve (1986), Nature Reserve (1989) and National Park (1999).

**Keywords**

Sierra Nevada • Iberian Peninsula • Geography • Quaternary • Glaciers • Biodiversity

## 1 Sierra Nevada Within the Iberian Peninsula

The Iberian Peninsula extends between latitudes 43°47′ to 36° 01′ N and longitudes 9°30′ W to 3°19′ E. It is located in the southwest tip of the Eurasian continent encompassing a surface of 582,925 km$^2$. The peninsula is located in a transitional zone between differing geographical influences: maritime (Atlantic/Mediterranean), climatic (subtropical high-pressure belt/mid-latitude westerly's), and in terms of biomes (Europe/Africa). The interactions among these influences, together with the rough topography, explain the wide range of landscapes existing today in Iberia (Oliva et al. 2018).

The present-day configuration of the Iberian relief results from its very complex geodynamic setting that has resulted in its extraordinary geological diversity. A sequence of tectonic events during the Mesozoic-Cenozoic driven by the collision between the Iberian and European continental plates (i.e. rifting and basin formation, compression and mountain building processes at the margins and inner part of the plate during the Tertiary, followed by Neogene rifting on the Mediterranean side) built the basis of the current geographical setting (Casas-Sainz and de Vicente 2009). Subsequently, during the Quaternary, the highest Iberian mountains have been shaped by both Pleistocene glaciations and post-glacial environmental dynamics, including periglacial, nival, slope and alluvial processes (Oliva et al. 2016b, 2019).

The Iberian Peninsula constitutes a mountainous region, with an average altitude of 660 m. The main mountain ranges in the Iberian Peninsula are distributed in its periphery and generally aligned W–E, surrounding the relatively flat areas of the central part of the peninsula (the "Meseta") and the coastal lowlands of the Atlantic and Mediterranean seas. Six mountain ranges include peaks exceeding 2000 m a.s.l.: the Pyrenees, the Cantabrian Mountains, the NW ranges, the Central Range, the Iberian Range and the Baetic System, where Sierra Nevada is located (Fig. 1).

M. Oliva (✉) · J. Martín-Díaz
Department of Geography, Universitat de Barcelona, Catalonia, Spain
e-mail: marcoliva@ub.edu

J. M. Fernández-Fernández
Centre for Geographical Studies, IGOT, Universidade de Lisboa, Lisbon, Portugal


R. Zamora and M. Oliva (eds.), *The Landscape of the Sierra Nevada*,
https://doi.org/10.1007/978-3-030-94219-9_1

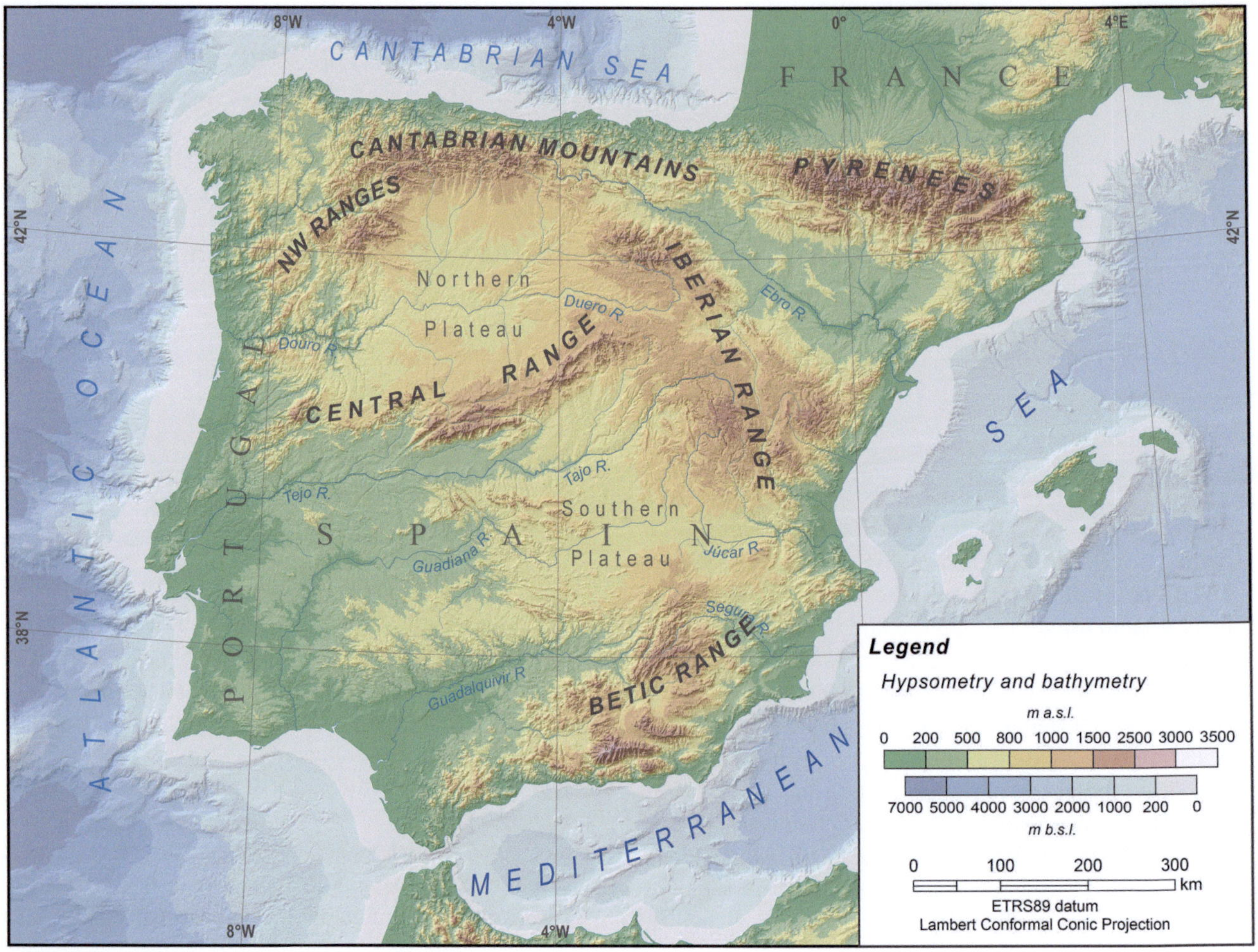


**Fig. 1** Location of the main mountain ranges in Iberia

The prevailing climate of the Iberian Peninsula is affected by both continental and maritime air masses of subtropical, mid-latitude and subpolar origins. This results in a highly seasonal climate, mostly controlled by westerly winds in winter and by the Azores anticyclone in summer. A major proportion of the annual precipitation falls between October and May, when low-pressure systems follow a prevailing zonal trajectory along latitudes 40–50° N (Trigo et al. 2004). These large-scale synoptic patterns, however, lead to substantially different climatic conditions across the peninsula as its rough terrain determines the existence of multiple microclimatic regimes (AEMET-IPMA 2011), particularly in mountain regions (Peña-Angulo et al. 2016). Across Iberia, the mean annual air temperatures and annual precipitation increase and decrease, respectively, towards the S and the E. Currently, negative man annual air temperatures are only recorded in the highest massifs, as the 0 °C isotherm is placed approximately at 2400–2500 m in the Cantabrian Mountains (Muñoz 1982), at 2945 m in the Pyrenees (López-Moreno et al. 2019) and at ~3400 m in the Sierra Nevada (Oliva et al. 2016a).

These climate conditions condition the geoecological processes prevailing in the high lands of the major ranges, although human activities over the last few millennia have significantly modified the natural vertical sequence in some Iberian massifs (Camarero et al. 2015). Human-induced perturbations (e.g. climate change, land use changes) play also a prominent role on the geoecological, geomorphic, hydrological, and edaphic processes prevailing in mountain environments, as well as in the surrounding lowlands. Currently, Iberian mountains do not host extensive glaciers, but rather they only persist as small features in the Pyrenees (Oliva et al. 2021), undergoing a dramatic thinning and recession (López-Moreno et al. 2016). As a result, from the tree-line up to the highest peaks, environmental dynamics in the Iberian mountains is mainly controlled by periglacial activity through a broad variety of processes that result in various landforms and deposits (Oliva et al. 2016b). The periglacial environment integrates two sub-belts constrained by the intensity of cryogenic processes and the degree of vegetation cover. The upper sub-belt is generally rocky and barren, with scarce vegetation cover, which favours the action of ground ice and intense periglacial dynamics; the lower sub-belt has a denser vegetation cover composed mostly of grasslands with weaker periglacial dynamics; since the Neolithic, this environment has been strongly transformed by human deforestation through fire management to favor grazing activities (García-Ruiz et al. 2020). The accelerated mountain depopulation during the second half of the twentieth century favoured the re-colonization of the trees of their natural habitats, with the consequent increase in altitude of the tree-line by several hundreds of meters (Camarero et al. 2015). Forests generally extend, to a larger or lesser extent, below the periglacial domain across the valley bottoms and surrounding slopes. Logically, the spatial distribution of these elevation belts depends on the general climate regime as well as on its topographically-driven modifications within single mountain ranges.

## 2 The Massif of Sierra Nevada

The Baetic System forms a discontinuous mountain region extending across the S–SE of the Iberian Peninsula between the Strait of Gibraltar and the Nao Cape. It extends eastwards along discontinuous submerged mountains, and reappear later forming the Balearic archipelago. This mountain range forms part of the series of Alpine ranges surrounding the Mediterranean basin. Indeed, together with the Atlas Mountains, the Baetic System closes the SW fringe of the Mediterranean Sea. The Baetic System includes three main major units (Fig. 2):

- The Prebaetic System is the northernmost sector that includes several parallel massifs aligned in a SW–NE with elevations exceeding 2000 m (e.g. Cazorla, Segura, la Sagra).
- The Subbaetic System occupies a central position within the Baetic System with several sierras of ca. 2000 m (e.g. Grazalema, Huétor). Several tectonic basins distributed at their foot are drained by the major rivers in S–SE Iberia.
- The Penibaetic System includes the highest peaks of the Baetic System in the Sierra Nevada, which are also some of the highest peaks of the entire Iberian Peninsula (Mulhacén 3479 m; Veleta, 3396 m; Alcazaba, 3369 m). Several other massifs, some of them very close to the coast (at ca. 20–30 km), include elevations above 2000 m (e.g. Gádor, Baza, Filabres).

The massif of the Sierra Nevada stretches W to E along 80 km between latitudes 37°12′ and 36°56′ N and longitudes 2°39′ and 3°32′ W. The highest altitudes are located in the western fringe of the massif, running at ca. 3000 m along 35 km from the Picón de Jeres (3088 m) to the Cerro del Caballo (3011 m). Eastwards of the Cerro de Trevélez (2877 m), altitudes descend gradually and remain above 2000 m along 40 km. The highest summits of the massif are aligned W–E and condition a radial network of valleys on the northern and southern slopes.

The core of the Sierra Nevada where the highest altitudes are located is basically made up of thick layers of graphite schists and mica schists that form the Nevado-Filábride complex, which constitutes the main geological unit of the Baetic System (Fig. 2). A succession of nappes forms a vast anticline that is fractured by the dense network of faults of Alpine origin aligned SW–NE and NW–SE (see chapter "Geological Setting of Sierra Nevada"). The Alpujárrides unit surrounds the western and southern flanks of the massif, with Neogene and Quaternary deposits distributed across the lowlands.

The relief of the Sierra Nevada is organized following the alignment of ridges and remains of ancient erosion surfaces (>2800–3000 m). The valleys shaped by paleoglaciers and rivers are located at their flanks, running radially from the axis of the massif through steep slopes down to the surrounding plains (see chapter "Climate Variability and Trends"). On the northern slope of the Sierra, the rivers draining towards the Genil and Fardes form part of the Guadalquivir River basin, which flows into the Atlantic Ocean. The short and steep river from the south slope of the massif drain to Guadalfeo River, which flows to the Mediterranean Sea.

The schist lithology of the massif was strongly shaped by Quaternary glaciers, which flowed down radially from the axis of the massif to ca. 2000–2500 m during the last glacial cycle (see chapter "The Impact of Glacial Development on the Landscape of the Sierra Nevada"). Unlike in the Alps and to a lesser extent in the northern slope of the Pyrenees, in the Sierra Nevada the glaciers remained confined within the mountains and did not reach the neighbouring plains (Gómez-Ortiz et al. 2002). Glaciers were mostly limited to the western third of the massif, and eastwards the Cerro de Trevélez to the Cerro del Chullo (2612 m), there are only small nivation cirques in the headwaters of the highest valleys. Glacial retreat following the Last Glacial Maximum and the transition towards the Holocene favoured the formation of tens of lakes and lagoons in glacial overdeepened basins. The sediments of these lakes have been used to reconstruct climate oscillations from the onset of deglaciation to present-day (see chapters "Reconstruction of Past Environments and Climate Using Wetland Sediment Records from the Sierra Nevada" and "Paleolimnological Indicators of Global Change"). Geomorphic evidence of periglacial activity is widespread across the massif. Periglacial conditions during cold Quaternary phases were widespread in non-glaciated areas down to 1000–1100 m (Oliva et al. 2014). Currently, the periglacial belt extends from ~2500 m to the summits, as no glaciers exist today in the Sierra Nevada (see chapter "Ancient and Present-Day Periglacial Environments in the Sierra Nevada").

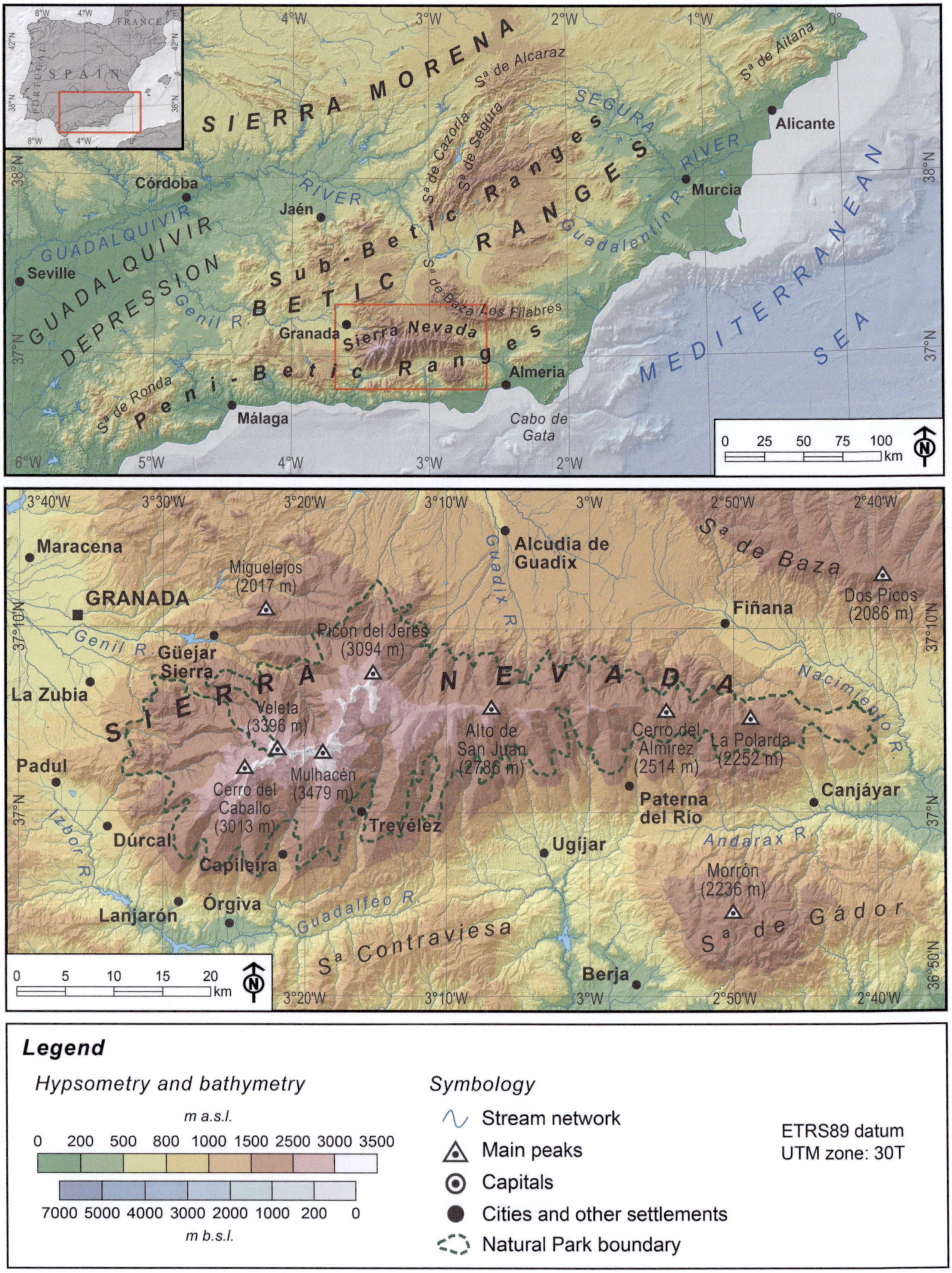


**Fig. 2** **a** Distribution of the main massifs of the Baetic System, including Sierra Nevada (**b**)

The high altitude and latitude of the Sierra Nevada explain the rich biodiversity of the massif. During cold Quaternary phases, the Sierra Nevada—as well as other massifs in central-southern Europe—acted as a genetic refuge for many plant species (Médail and Diadema 2009). Following the deglaciation, these species expanded to higher elevations and adapted to the altitude, cold temperatures and summer aridity. Therefore, geographical conditions and past climate variability explain the large number of species endemic that can be only found in the Sierra Nevada (see chapter "Sierra Nevada, A Mediterranean Biodiversity Super Hotspot").

The present-day climate regime in the massif is characteristic of a Mediterranean semi-arid high-mountain environment (see chapter "Climate Variability and Trends"). It shows a clear contrast between the hot and dry season (June–September) and the cooler and wet period of the year (October–May), with a high interannual and intra-annual variability. From October to May, Sierra Nevada is affected by the humid Atlantic flows associated with the mid-latitude westerlies, whereas from June to September the subtropical high-pressure systems prevail and favour atmospheric stability and high insolation. At 2500 m, the mean annual temperature is 4.4 °C, with negative values from December to April, and the annual precipitation reaches 700 mm, with ca. 80% falling as snow (Oliva and Gómez-Ortiz 2012). The 0 °C isotherm in the western side of the Sierra Nevada is currently placed at ~3400 m (Oliva et al. 2016a). The rough relief imposed by the zonal setting of the massif and the high altitudes existing in its western fringe condition the existence of several topoclimates within the same and in the surrounding lowlands. Generally, temperatures increase southwards and eastwards, whereas precipitation follows the opposite pattern. Indeed, in the rain shadow of the Sierra Nevada, lies the south-eastern coastal region of Almeria, which records the lowest annual precipitation in Europe with only 130 mm $yr^{-1}$ at the Cabo de Gata (Tout 1987).

The combination of high temperatures and very low precipitation during the warm season conditions very high evapotranspiration rates, which is a critical element for development in the massif. The barren and rocky landscape of the periglacial environment enhances the physical weathering of the rocks and limits soil formation and vegetation cover (Fig. 3). Edaphic processes above 2500 m are limited to relatively flat or concave areas in glacial cirques

**Fig. 3** **a** View of the National Park of Sierra Nevada with the Veleta Peak and ski facilities on the left side of the picture; **b** The Mulhacén glacial cirque, including La Mosca Lake; **c** The village of Busquistar, in the southern slope of the massif and **d** Irrigation channels delivering water from high lands to the villages

and valley floors fed by long-lying snow melting waters. These areas are locally known as *borreguiles* in the massif, and the highly organic soils (histosols) favoured the development of wetlands with a large number of endemic species. Slopes at the summit level are mostly vegetation-free, with very incipient soils, including especially cambisols, inceptisols and regosols (Martı et al. 2004). At lower elevations, soils are gradually more developed and support a denser vegetation cover, strongly dependent on topography and moisture supply.

The combination of cold temperatures and high precipitation during the cold semester of the year favoured the presence of snow in the massif, which has promoted the development of socio-economic activities associated with the snow. Indeed, the ski resort of Sierra Nevada is one of the largest in the Iberian Peninsula and constitutes a cornerstone of the local economy, in parallel to recent initiatives promoting a sustainable mountain economy (see chapter "Local Ecological Knowledge and the Sustainable Co-Management of Sierra Nevada's Social-Ecological System").

Historically, the summits of the Sierra Nevada have been only used by local populations during the summer season that have left a wide range of singular landscapes in the slopes of the Sierra Nevada (see chapter "Singular Cultural Landscapes of the Sierra Nevada"). This is particularly evident in the southern slope of the massif, where the villages (their distribution, internal structures and housing typologies), are fully adapted to the topographical setting. Agricultural terraces climbing up the slopes in step-like fashion following the contour lines, the irrigation channels descending from the snow-capped peaks to the villages or the historical ice trade, with the ice first cut and stored in snow caves and cisterns, and later in the summer moved to by pack mules, are some of the traditional uses of the natural resources provided by the Sierra Nevada that were used by local populations (Gómez-Ortiz et al. 2013).

These wide range of historical practices have left an ancestral imprint on the landscape of the Sierra Nevada that, together with the richness of its flora and the diversity of geomorphological landforms, contributed to the designation of the massif as a Biosphere Reserve (1986), Nature Reserve (1989) and National Park (1999).

**Acknowledgements** This work was conducted under the aegis of the research group ANTALP (Antarctic, Arctic, Alpine Environments, 2017-SGR-1102). Marc Oliva is supported by the Ramón y Cajal Program (RYC-2015-17597). José M. Fernández-Fernández is supported by a postdoctoral grant within the NUNANTAR project funded by the Fundação para a Ciência e Tecnologia of Portugal (02/SAICT/2017 - 32002).

## References

AEMET-IPMA (2011) Atlas climático ibérico. Temperatura del aire y precipitación (1971–2000), Agencia Estatal de Meteorología, Ministerio de Medio Ambiente y Medio Rural y Marino

Camarero JJ, García-Ruiz JM, Sangüesa-Barreda G, Galván JD, Alla AQ, Sanjuán Y, Beguería S, Gutiérrez E (2015) Recent and intense dynamics in a formerly static Pyrenean Treeline. Arctic Antarct Alp Res 47:773–783

Casas-Sainz AM, de Vicente G (2009) On the tectonic origin of Iberian topography. Tectonophysics 474:214–235

García-Ruiz JM, Tomás-Faci G, Diarte-Blasco P, Montes L, Domingo R, Sebastián M, Lasanta T, González-Sampériz P, López-Moreno JI, Arnáez J, Beguería S (2020) Transhumance and long-term deforestation in the subalpine belt of the central Spanish Pyrenees: an interdisciplinary approach. Catena 195:104744

Gómez-Ortiz A, Oliva M, Salvà-Catarineu M, Salvador-Franch F (2013) The environmental protection of landscapes in the high semiarid Mediterranean mountain of Sierra Nevada National Park (Spain): Historical evolution and future perspectives. Appl Geogr 42:227–239

Gómez-Ortiz A, Schulte L, Salvador-Franch F, Sánchez S, Simón M (2002) Mapa geomorfológico de Sierra Nevada. Morfología glaciar y periglaciar. Consejería de Medio Ambiente, Junta de Andalucía, Granada

LópezMoreno JI, Alonso-González E, Monserrat O, Del Río LM, Otero J, Lapazaran J, Luzi G, Dematteis N, Serreta A, Rico I, Serrano-Cañadas E, Bartolomé M, Moreno A, Buisan S, Revuelto J (2019) Ground-based remote-sensing techniques for diagnosis of the current state and recent evolution of the Monte Perdido Glacier, Spanish Pyrenees. J Glaciol 65:85–100

López-Moreno JI, Revuelto J, Rico I, Chueca-Cía J, Julián A, Serreta A, Serrano E, Vicente-Serrano SM, Azorin-Molina C, Alonso-González E, García-Ruiz JM (2016) Thinning of the Monte Perdido Glacier in the Spanish Pyrenees since 1981. Cryosphere 10:681–694

Martı JM, Bech J, Delgado R, Martín-García JM, Aranda V, Gámiz E, Bech J, Delgado R (2004) Are Mediterranean mountains Entisols weakly developed? The case of Orthents from Sierra Nevada (Southern Spain). Geoderma 118:115–131

Médail F, Diadema K (2009) Glacial refugia influence plant diversity patterns in the Mediterranean Basin. J Biogeogr 36:1333–1345

Muñoz J (1982) Geografía Física. El relieve, el clima y las aguas, In: Quirós F (ed) Geografía de Asturias. Ayalga Ediciones, pp 1–271

Oliva M, Gómez-Ortiz A (2012) Late-Holocene environmental dynamics and climate variability in a Mediterranean high mountain environment (Sierra Nevada, Spain) inferred from lake sediments and historical sources. Holocene 22:915–927

Oliva M, Gómez-Ortiz A, Palacios D, Salvador-Franch F, Salvà-Catarineu M (2014) Environmental evolution in Sierra Nevada (South Spain) since the Last Glaciation, based on multi-proxy records. Quat Int 353:195–209

Oliva M, Serrano E, Gómez-Ortiz A, González-Amuchastegui MJ, Nieuwendam A, Palacios D, Pérez-Alberti A, Pellitero R, Ruiz-Fernández J, Valcárcel M, Vieira G, Antoniades D (2016) Spatial and temporal variability of periglaciation of the Iberian Peninsula. Quat Sci Rev 137:176–199

Oliva M, Gómez-Ortiz A, Salvador-Franch F, Salvà-Catarineu M, Palacios D, Tanarro L, Ramos M, Pereira P, Ruiz-Fernández J

(2016) Inexistence of permafrost at the top of the Veleta peak (Sierra Nevada, Spain). Sci Total Environ 550:484–494
Oliva M, Ruiz-Fernández J, Barriendos M, Benito G, Cuadrat JM, Domínguez-Castro F, García-Ruiz JM, Giralt S, Gómez-Ortiz A, Hernández A, López-Costas O, López-Moreno JI, López-Sáez JA, Martínez-Cortizas A, Moreno A, Prohom M, Saz MA, Serrano E, Tejedor E, Trigo R, Valero-Garcés B, Vicente-Serrano SM (2018) The little ice age in Iberian mountains. Earth Sci Rev 177:175–208
Oliva M, Palacios D, Fernández-Fernández JM (2021) Iberia, land of glaciers. Elsevier
Oliva M, Palacios D, Fernández-Fernández JM, Rodríguez-Rodríguez L, García-Ruiz JMM, Andrés N, Carrasco RMM, Pedraza J, Pérez-Alberti A, Valcárcel M, Hughes PDD (2019) Late Quaternary glacial phases in the Iberian Peninsula. Earth Sci Rev 192:564–600
Peña-Angulo D, Trigo RM, Cortesi N, González-Hidalgo JC (2016) The influence of weather types on the monthly average maximum and minimum temperatures in the Iberian Peninsula. Atmos Res 178–179:217–230
Tout DG (1987) South-East Almería Province, Spain—the driest region in Europe. Weather 42:242–247
Trigo RM, Pozo-Vázquez D, Osborn TJ, Castro-Díez Y, Gámiz-Fortis S, Esteban-Parra MJ (2004) North Atlantic oscillation influence on precipitation, river flow and water resources in the Iberian Peninsula. Int J Climatol 24:925–944

# Sierra Nevada, a Mediterranean Biodiversity Super Hotspot

Juan Arroyo, Pedro Abellán, Montserrat Arista, María Jesús Ariza, Alejandra de Castro, Marcial Escudero, Juan Lorite, Estefanía Martínez-Borda, José Antonio Mejías, Rafael Molina-Venegas, Juan M. Pleguezuelos, Violeta Simón-Porcar, and Juan Viruel

**Abstract**

The Sierra Nevada massif is a hotspot nested within a biodiversity hotspot in the Western Mediterranean (the Baetic-Rifian range, divided by the Strait of Gibraltar), which in turn is found within one the hottest spots in the World, the Mediterranean Basin. The geographic, geomorphological, geological and climatic history of Sierra Nevada places it at a unique biogeographical crossroads, which explains the high biodiversity levels found there. Here, we review the available literature on the multiple dimensions of biodiversity in Sierra Nevada in different systematic groups. Thus, we go beyond the classical account of endemism as a form of rarity accounting for biodiversity to consider evolutionary and ecological rarities. We focus on successively finer scales, from floristic and faunistic accounts of diversity to community diversity, and then to populations, including phylogenetic, phylogeographic and population genetics information. We also consider biological interactions—mostly mutualisms—that are a fundamental part of biodiversity due to their key role in maintaining species diversity. Finally, we provide some insight into the fate of this biodiversity given the incidence of global change drivers, which may have particularly pernicious consequences in mountainous ecosystems.

J. Arroyo (✉) · M. Arista · A. de Castro · M. Escudero · E. Martínez-Borda · J. A. Mejías · V. Simón-Porcar
Dept. of Plant Biology and Ecology, University of Seville, Sevilla, Spain
e-mail: arroyo@us.es

P. Abellán
Dept of Zoology, University of Seville, Sevilla, Spain

M. J. Ariza
General Research Services for Research, University of Seville Herbarium, Sevilla, Spain

J. Lorite
Dept. of Botany, University of Granada, Granada, Spain

R. Molina-Venegas
Global Change Ecology and Evolution Group (GLOCEE), Dept. of Life Sciences, University of Alcalá, Madrid, Spain

J. M. Pleguezuelos
Dept. of Zoology, University of Granada, Granada, Spain

J. Viruel
Royal Botanic Gardens, Richmond, UK

**Keywords**

Biodiversity • Community • Conservation • Endemism • Flora • Fauna • Hot-spot • Phylogeny • Phylogeography • Population genetics • Rarity • Species interactions

## 1 Biodiversity, a Concept Beyond Species Lists

Biodiversity is a typical complex concept in that it is easy to define but difficult to measure due to its multidimensional nature (Gaston and Spicer 2004). Biodiversity refers to the complexity of life in a given site or region, from genes to ecosystems, thus including all levels of organization as well as interactions among elements within these levels, e.g., interactions among genes in the genome, among species in communities, and so forth (e.g., Valiente-Banuet et al. 2015). Although biodiversity is often used interchangeably with species richness, there is an increasing debate about the appropriate metrics or currency of biodiversity, since aspects beyond the simple sum of species such as historical processes, functional traits, adaptive potential, and hidden patterns of genetic diversity could be critical for a true accounting of the biodiversity value of a place. The fact that many facets of biodiversity are interrelated might make the task of measuring it easier, but it is often difficult to reconcile functional, spatial, historical, and conservation aspects because sometimes the scales of biodiversity components run opposite. Consequently, different sets of variables are being proposed to quantify biodiversity, depending on the specific aim (e.g., Skidmore et al. 2015, 2021), resulting in


R. Zamora and M. Oliva (eds.), *The Landscape of the Sierra Nevada*,
https://doi.org/10.1007/978-3-030-94219-9_2

plenty of literature on the topic (48,714 results for searching "measuring" AND "biodiversity" from Web of Science on July 22nd, 2021).

It is commonly agreed that rare elements and their interactions contribute strongly to biodiversity and serve as a measure of regional conservation value to avoid total or local extinction of rarities. Thus, special attention has been traditionally paid to endemic species as an epitome of rarity. However, rarity and its contribution to biodiversity can be considered not only in a spatial sense (elements present in only a few places), but also in a temporal (formerly common elements becoming more infrequent), functional (particular elements which disproportionately contribute to the maintenance of biological systems) or evolutionary (elements which represent distinct lineages) context. Thus, rarity as a critical component of biodiversity, stemming from the information value of biological systems (Kunin and Gaston 1997), has been also defined as a multidimensional character, even though it is mostly used to refer to local and regional patterns of abundance (Rabinowitz 1981).

Most biodiversity components show biogeographical—that is macroecological and historical—patterns of variation. Different components have different patterns, some of which have been recognized in detail since the seminal work by A. Von Humboldt (1807), which continues to generate debate, particularly in mountains (Morueta-Holme et al. 2015), a favorite theme of the founder of modern biogeography. Ecological and evolutionary dramas play out in highly complex ways in mountains for several reasons. For instance, since mountains represent harsher ecological conditions for many organisms, they tend to harbour fewer species or even lineages at local scales (Körner 2021). However, at the same time, mountain ecosystems can promote higher rates of differentiation and diversification than the surrounding lowlands due to factors such as high rates of environmental change over spatial and temporal scales in aspects such as climate and soils; limited and isolated territories; and geographical positions that promote their role as barriers or corridors. Mountains located at intermediate latitudes and that coincide with transitional biogeographical realms, such as those in the Mediterranean Basin, promote regional biodiversity for both plants and animals (Thompson 2020). Most Mediterranean mountains run along a W-E axis, as in the Iberian Peninsula, North Africa, or Anatolia. On the one hand, they may act as barriers; however, since they harbour strikingly different microclimates on their N versus S slopes, they may act as islands or archipelagos surrounded by lowlands and may therefore ultimately represent refugia during the climatic oscillations throughout the Pleistocene, hampering extinction, due to their milder climates in comparison with northern territories.

## 2 Sierra Nevada: A Unique Scenario for an Intense Evolutionary Drama

The Sierra Nevada massif in SE Spain has been recognized as a biodiversity hotspot for plants (Médail and Quézel 1997, 1999) and animals (Ruano et al. 2013) (Fig. 1), which may be partly due to its environmental and historical conditions. The high elevation—up to 3479 m—promotes the formation of alpine habitats (albeit with Mediterranean conditions of a dry, hot summer in addition to the cold winter). The combination of different geochemical substrates, including limestone, dolomite, mica-schist, and serpentine, among others, and a general arid Mediterranean climate surrounding the mountains, in a location close to African mainland, makes Sierra Nevada the paradigm of a Mediterranean mountain. Furthermore, its historical setting, linked to the Alpine orogeny at about 8 Ma and the intermittent linking and separation of the remains of the Alboran micro-plate between the Iberian (Eurasian) and African plates, explains the dynamic nature of biodiversity accumulation (see chapters "The Geographic Uniqueness of the Sierra Nevada in the Context of the Mid-Lattitude Mountains" and "Geological Setting of Sierra Nevada" for details). Naturalists have long been interested first in describing new endemic species or locating taxonomic outliers that can be found only many kilometers apart in northern or, rarely, southern ranges. More recently, there has been further interest in analyzing ecological and evolutionary processes that may help to explain the complex interactions that sustain current levels of biodiversity. Here, we synthesize the biodiversity of Sierra Nevada from a multilevel perspective, from classical accounts of species richness in different taxonomic groups, to a more biogeographically oriented account (endemics and relicts), then exploring available phylogenetic and phylogeographic information to ascertain the evolutionary singularity of taxa and species assemblages in Sierra Nevada. Finally, we explore the population genetic diversity patterns of selected species from Sierra Nevada to provide insight into microevolutionary processes that may be driving biodiversity.

## 3 The Flora and the Fauna of Sierra Nevada in a Mediterranean Context

### 3.1 Plants

Sierra Nevada harbours 2348 plant taxa (1937 species, 377 subspecies and 34 hybrids), belonging to 756 genera and 146 families (Lorite et al. 2020). Among them, the most prominent families are typically common in the Mediterranean, such as: Asteraceae (274 taxa), Fabaceae (208),

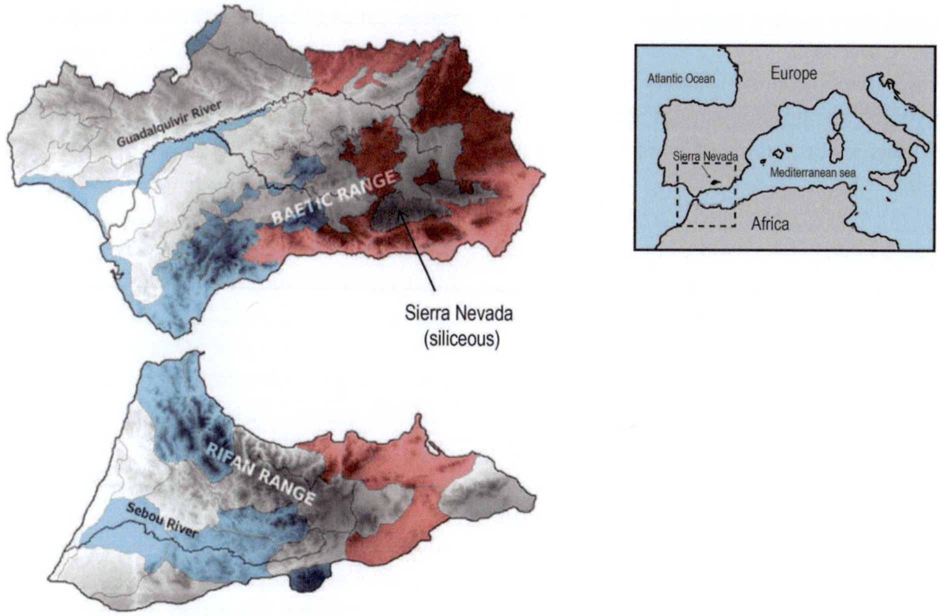


**Fig. 1** Map of Andalusia and northern Morocco showing the position of the Baetic and Rifian ranges and the siliceous Sierra Nevada ecoregion (arrow). The inset shows the position of Sierra Nevada in the Western Mediterranean. Ecoregions coloured in blue represent dominance of palaeoendemism (SES.RPE values above 75th percentile), and in red, neoendemism (SES.RPE values below 25th percentile) (see Molina-Venegas et al. 2017). Note that a region without these colours, like Sierra Nevada, may reflect either lack of endemism or a center of "mixed endemism" that includes both palaeo- and neoendemism. The widely recognized high level of endemism in Sierra Nevada (see text) points to a higher relative incidence of palaeoendemism. Modified from Molina-Venegas et al. (2017) and reproduced with permission from Oxford University Press

Poaceae (207), Brassicaceae (126), Caryophyllaceae (114), and Lamiaceae (112). The richest genera are *Trifolium* (28 taxa), *Silene* (27), *Helianthemum* (26), *Carex* (25), *Galium* (25), *Ranunculus* (25), *Centaurea* (24), *Vicia* (24), *Astragalus* (23), and *Festuca* (21).

Beyond these numbers, there are several reasons that Sierra Nevada is one of the most important plant biodiversity hotspots within the Mediterranean region (Blanca et al. 1998; Médail and Diadema 2009; Cañadas et al. 2014; Lamprecht et al. 2021). First, Sierra Nevada is an isolated high mountain range, separated by 500–700 km from other comparable mountain ranges with elevations above 3000 m (the Pyrenees to the north, the Moroccan Atlas to the south). Secondly, Sierra Nevada marks the southernmost limit of the influence of the Quaternary glaciations in Europe; it was covered with glaciers only in areas above 2500 m, while large areas remained free of glacial ice (Gómez-Ortiz et al. 2013). This geographical setting, together with specific soil types and varied ecological factors, have promoted speciation, resulting in a high number of plant species present today (Lorite 2016), including 95 endemic or subendemic taxa (Lorite et al. 2020).

In addition to the well-known endemism in Sierra Nevada (Blanca et al. 1998), an important element of relict species and lineages deserves special attention. Here, we consider relics in a historical context, including lineages at different time scales. We thus endorse the broad perspective of Vargas et al. (2020) developed for Mediterranean floras across longer climatic periods that have prompted extinction and range retraction, including the most recent at Last Glacial Maximum (Médail and Diadema 2009). This concept relies on detailed dated phylogenetic knowledge of the lineages under scrutiny. Despite dated phylogenetic history is scarce for most putative relict taxa, there are some arctic-alpine species which are outstanding examples within this relict element. Sierra Nevada represents the southernmost limit of twelve cold-adapted species (Abeli et al. 2018), including *Artemisia umbelliformis, Draba dubia, Ranunculus glacialis, Saxifraga oppositifolia, Sibbaldia procumbens* and *Thalictrum alpinum* among others.

## 3.2 Invertebrates

Sierra Nevada also harbours a remarkable diversity of invertebrates, although knowledge of them is still incomplete, especially when compared to plants and vertebrates. About 4000 species of insects have been listed for this massif, of which more than 150 are local endemics (Ruano et al. 2013). For some groups such as ants (Tinaut 2013) and chrysomelid beetles (Vela and Bastazo 2013), one-third of the Iberian taxa are found in Sierra Nevada. These numbers are surely just a small fraction of the actual diversity, as many groups are still understudied (e.g., Acari, Collembola, or some families of Coleoptera, Diptera and Hymenoptera), and new species are added (and even newly described) for this massif every year. The high diversity of invertebrates is a result of the overlap of taxa derived from different biogeographical origins, which arrived during different time periods, including pre-Quaternary relict species as well as Quaternary elements (Tinaut et al. 2008). For instance, there are some alpine elements related to taxa from the Pyrenees or the Alps (e.g., the water beetle *Hydroporus sabaudus* subsp. *sierranevadensis* and the butterfly *Erebia hispania*), elements related to taxa from North Africa (e.g. the millipede *Archipolydesmus altibaeticus*), as well as species related to taxa or populations from Central Asia (e.g., the butterfly *Pseudochazara hippolyte* and the ant *Proformica longiseta*) (see Tinaut et al. 2008; Olivares et al. 2011; Piñero et al. 2011; Millán et al. 2013; Ruano et al. 2013).

### 3.3 Vertebrates

There are 206 species of vertebrates in Sierra Nevada, including three fishes, nine amphibians, 20 reptiles, 131 birds (108 breeding and 23 regular winter visitors), and 43 mammals. Only two of them are endemic to the Baetic mountains (*Alytes dickhilleni*, *Timon nevadensis*; Maia-Carvalho et al. 2014; Ahmadzadeh et al. 2016). Thus, endemicity is apparently not a driver of vertebrate biodiversity in Sierra Nevada, in contrast to other large Mediterranean mountain ranges (e.g., the Middle and High Atlas Mountains). The biogeographical significance of Sierra Nevada, however, is remarkable, as the range limit of many species, particularly congeneric ones, spans both sides of this mountain range. The reasons behind the existence of this crossroad are mostly geological, since the volcanic archipelago that created a bridge across the East Alboran basin permitted the exchange of terrestrial biota during the period ~7–3 Ma (Booth-Rea et al. 2018). Genera involved include *Alytes, Timon, Podarcis, Coronella* (Sánchez-Herráiz et al. 2000; Kaliontzopoulou et al. 2011; Santos et al. 2012), that is, non-flying vertebrates. Sierra Nevada also represents the southernmost limit for some birds (e.g., *Carduelis citrinella*).

## 4 Living Together on a High Mountain: Community-Level Evolutionary and Ecological Correlates of Biodiversity

### 4.1 Phylogenetic Diversity in High Mountain Plant Assemblages

There is clear evidence that current biodiversity patterns are the result of long-standing historical processes. Thus, it is not surprising that biodiversity is deeply rooted in the history of lineages spreading and diversifying in Sierra Nevada. When considering spatial setting, the field of phylogenetic ecology (Swenson 2019; Davies 2021) is particularly promising, since it deals with the history of species assemblages at different spatial scales, including both floristic and faunal assemblages.

Plant communities of Sierra Nevada are unique assemblages in terms of rare species (endemics and relics; see also chapter "Altitudinal Patterns and Changes in the Composition of High Mountain Plant Communities"). Phylogenetic relationships shed light on the evolutionary history of these assemblages and how they came about. There is still little information about phylofloristics of the Mediterranean Basin floras, although that gap is rapidly filling in (Konstantinos et al. 2020; Buira et al. 2021; Cheikh Albassatneh et al. 2021; Monnet et al. 2021). The first account of the woody flora of Sierra Nevada (Simón-Porcar et al. 2018) provides insight into the formation of this floristic super-hotspot. Thus, the phylofloristic information it provides using barcoding approaches from 326 woody taxa sequenced (74% of a total of 440) particularly welcome (Simón-Porcar et al. 2018). The uniqueness of different local plant assemblages and communities within the Baetic range of the hotspot was previously assessed by means of a tree based on a supermatrix obtained with Genbank DNA sequences for virtually the whole flora of the region at the genus level (Molina-Venegas and Roquet 2014).

Any account of plant biodiversity in a region relies on sound floristic data. This is the case of Andalusia and thus Sierra Nevada, for which two complete floras (Valdés et al. 1987; Blanca et al. 2009) and an updated checklist (Cueto et al. 2018) compiled by competent and well-known teams of botanists are available. Sierra Nevada had already been recognized as a super-hotspot based on floristic data, endemism, and threat level, on the Iberian Peninsula (Domínguez-Lozano et al. 2000; Lobo et al. 2001; Moreno-Saiz et al. 2013; Buira et al. 2017) as well as nested within the Andalusian and Baetic ranges (Mota et al. 2002; Peñas et al. 2005; Molina-Venegas et al. 2013). Differences among these studies are due to variation in the completeness of the species pool, the taxonomic treatment, how endemism is defined, and the geographical setting and regional division used. More importantly, these studies do not explicitly consider historical analysis, either of the biotic (evolutionary processes) or the abiotic (dynamic environmental conditions) components.

At the floristic level, available phylogenetic data allow for some inferences about how the current level of biodiversity was reached. Most of the plant lineages inhabiting the Southern Iberian Peninsula (Andalusia) are represented in Sierra Nevada (i.e., the flora of the latter constitutes a large sample of the former; Molina-Venegas et al. 2015a). This pattern recalls the great environmental heterogeneity of the

Sierra Nevada hotspot, which contrasts with that of more phylogenetically clumped neighboring regions such as the semi-arid province of Almería to the East or the Hoya de Baza valley to the North (Molina-Venegas et al. 2015a).

Regarding endemism, it is important to both define what constitutes an endemic species and to determine the evolutionary time required to become endemic. In principle, it is relatively free from assumptions to use the geographic range of species as a continuous variable, rather than using categories of endemic ranges. However, this relies on sound geographical knowledge of the whole flora, which is difficult and time-consuming to compile. Instead, endemism is usually considered within a given range that is frequently determined by political or administrative borders, which precludes any meaningful insight. It is by far more informative to consider species endemic to hotspots, in our case the Baetic-Rifian hotspot (Médail and Quézel 1999) or either of the two main constitutive parts (Baetic and Rifian), and ultimately the Sierra Nevada massif super-hotspot. It is also crucial to consider the possibility that populations that have not yet differentiated into separate taxa might be considered endemic if they have been isolated from the main distribution range of these taxa for a long period. Secondly, evolutionary background can only be determined using dated phylogenetic relationships of endemic lineages. This allows the relative age of endemism to be determined, which is an old pursuit of plant evolutionary biologists (Favarger and Contandriopoulos 1961; Stebbins and Major 1965), who first separated neo- and palaeoendemics. Branch length and extent of geographic range of species can be combined into metrics of timing and endemism for species in regions and sites under scrutiny (Mishler et al. 2014). Molina-Venegas et al. (2017), applying this approach to the whole angiosperm flora of Andalusia, determined that most of the western ecoregions are dominated by palaeoendemism, whereas eastern ecoregions are dominated by neoendemism. Interestingly, Sierra Nevada summits (siliceous) are an exception, likely because palaeondemism is more frequent there than in surrounding ranges, including lower calcareous summits of Sierra Nevada (Fig. 1). Molina-Venegas et al. (2017) suggested that this may reflect the effect of these high mountains acting as cold refugia for northern lineages throughout the late Tertiary and the Ice Ages of the Pleistocene. Phylogenetic and phylogeographic relationships among species fitting this pattern will provide insight into the drivers of palaeoendemism.

It has been long argued that the high diversity and endemism of the Baetic hot-spot is due to elevation and consequent adaptation to high altitudes, to geographic isolation between mountains, and to edaphic constraints. These factors may be disentangled using empirical tests in a number of sites where these factors are well represented. For example, Molina-Venegas et al. (2015a, b) selected seven Baetic mountains up to 2700 m (including in Sierra Nevada) to evaluate the effects of these variables on compositional (taxonomic), phylogenetic and functional beta diversity across elevational belts and mountain systems. Phylogenetic and functional beta diversity across belts within sierras were higher than expected, whereas floristic beta diversity was similar or lower than across sierras within elevational belts. Edaphic characteristics of mountains and their geographic isolation were critical for compositional beta diversity, but not for phylogenetic and functional beta diversity, suggesting that phylogenetic and ecological (functional) processes are similar across mountains, but the speciation processes lead to different species across them, thus explaining their high endemism rate (Mota et al. 2002; Peñas et al. 2005). The highest belt in Sierra Nevada (2700–3479 m) was not included in the full design due to the lack of replication within the Andalusian mountains. However, a specific analysis within Sierra Nevada across its elevational belts showed that the highest belt behaved similarly to the next belt down in phylogenetic beta diversity, showing negligible phylogenetic differentiation (Molina-Venegas et al. 2015a, b).

All former phylofloristic analyses were based on super-matrix phylogeny at the genus level (Molina-Venegas and Roquet 2014). While floristic lists are quite reliable, phylogenetic information suffers from several shortcomings, particularly those derived from loss of specific information on terminal branches and from differences in the DNA sequences of lineages across regions (Simón-Porcar et al. 2018). For the woody flora of Sierra Nevada, a fully resolved, highly congruent phylogenetic tree was assembled using barcoding markers (cpDNA: *rbc*L, *mat*K, *psb*A-*trn*H; nDNA: ITS) from plants collected ad hoc in these mountains (Simón-Porcar et al 2018), which includes 74% of the woody species and 85% of the genera present in Sierra Nevada, according to Lorite (2016). The barcoding approach helped to elucidate some taxonomic controversies. For example, *Juniperus communis* subsp. *hemisphaerica* was found to be clearly differentiated from *J. c.* subsp. *alpina*, and *Sempervivum minutum* and *S. tectorum* were also shown to be different species (see Simón-Porcar et al. 2018 for full list of examples). Given the conservation status of the core area of Sierra Nevada as a National Park and the long list of endemic and threatened taxa, barcoding offers a critical tool for identification of specimens under legal protection or worth conserving. These barcoding data also offer opportunities to delve into elevational and edaphic patterns, which are the two main factors proposed to drive plant differentiation and endemism in Sierra Nevada, albeit only for sequenced woody taxa. The distribution of alpha diversity of woody taxa decreased from the lowlands to summits, both in absolute and relative (i.e., per unit area) terms (Simón-Porcar et al. 2018). However, the distribution of phylogenetic diversity was generally randomly distributed across elevational belts, with no clear clustering or overdispersion. When considering only endemic taxa (Baetic-Rifian or narrower) the pattern was different. Absolute alpha diversity of endemic taxa was highest for the mid-high belt (1900–2800 m), whereas alpha diversity per unit area was highest at the summits (above 2800 m). Forty-three percent and 67% of woody taxa are endemic in the mid-high and summit belts, respectively. Alpha phylogenetic diversity was again random within each elevational belt, meaning that endemic taxa are a random sample of the phylogenetic tree.

Substrates differ in different parts of Sierra Nevada, which may foster differentiation in plant lineages (Mota et al. 2008). Considering the three main substrate groups (mica-schist, limestone, and dolomite), Simón-Porcar et al. (2018) found that, although the vast majority of the area has mica-schist substrate (79% of the area), calcareous limestone (18% of the area) supported the highest absolute alpha diversity of woody taxa. Interestingly, dolomite (3% of the area) had the highest relative alpha diversity, about five times that of limestone, and 26 times that of mica-schist. However, the proportion of woody endemism of the whole woody flora was similar among the three substrates (16–20%). When phylogeny was considered, there was a significant relationship between taxonomic and phylogenetic diversity in the whole flora, but not among endemic species. Most of the distribution of phylogenetic diversity was random across substrates, with only some clustering among taxa inhabiting calcareous substrates (Fig. 2).

The study by Simón-Porcar et al. (2018) on woody flora complements those using the whole flora and super-matrix Genbank sequence data (Molina-Venegas et al. 2015a, 2017). The results from the two approaches are not identical. In fact, Molina-Venegas et al. (2015b) found a gradual decrease in taxonomic and phylogenetic diversity at higher elevations, whereas the endemic woody taxa show the opposite pattern,

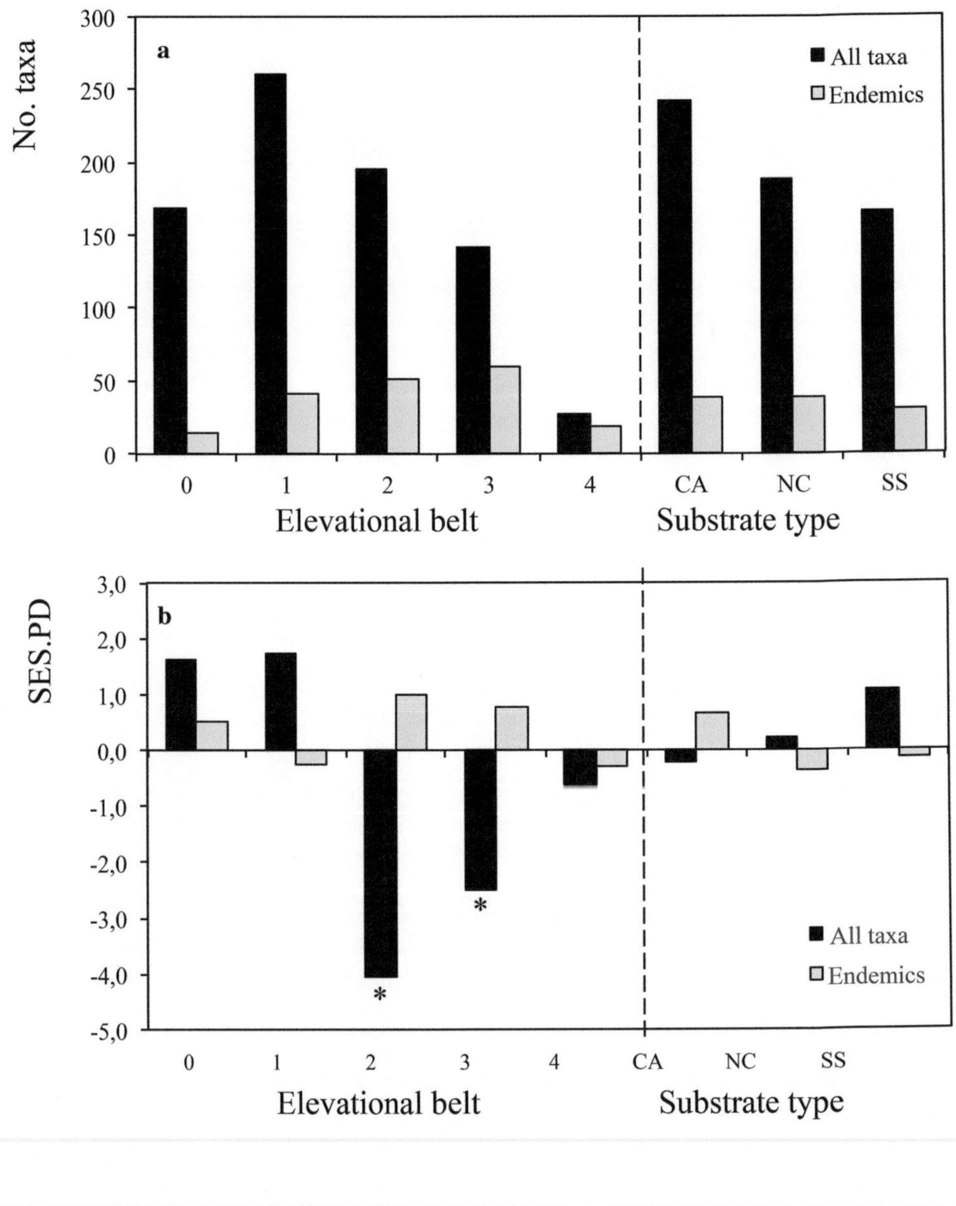


**Fig. 2** a Taxonomic and b phylogenetic alpha diversity of the woody flora of Sierra Nevada in different elevation belts (belt 0, <700 m; belt 1, 700–1500 m; belt 2, 1500–1900 m; belt 3, 1900–2800 m; belt 4, >2800 m) and substrate types (CA: calcareous; NC: non-calcareous; SS: special substrates, mostly dolomites). Positive and negative standardized effect size of phylogenetic diversity (SES.PD) values indicate phylogenetic overdispersion and clustering, respectively. Asterisks indicate significant differences from random expectations. Partially modified from Simón-Porcar et al. (2018) and reproduced with permission from Taylor & Francis

with increasing taxonomic and phylogenetic diversity at the highest elevations. This could be a result of adaptation to high elevation climate and orographic isolation rather than heterogeneous soil conditions, as these high elevations are almost exclusively composed of mica-schist. These differences illustrate the importance of local sequences from properly sampled plants in each target area and show that considering different species pools may yield different results (whole flora vs. woody species). It is also important to properly separate species pools according to ecological conditions of interest (e.g., soils). In fact, the high relative abundance of endemism in dolomites illustrates this point, although lithology does not affect the phylogenetic structure of these species assemblages, as it was mostly random among substrates.

## 4.2 Plant Community Phylogenetics

While the field of phylofloristics (and its counterpart, phylofaunistics; Swenson and Umaña 2014) deals with regional species assemblages where historical factors are critical, the signature of species interactions on phylogenies are better approached at local scales, that is, communities or significant parts of them where species interact. Unfortunately, there are too few studies of community phylogenetics to have a thorough picture of community types across Sierra Nevada. However, the little information that is available has shown some patterns. Given that elevation and soil properties have long been considered key factors for adaptation, differentiation, and speciation in the region (Arroyo and Marañón 1990; Mota et al. 2008), phylogeny may affect in turn community assemblage processes and thus shape their phylogenetic and functional structure and diversity, since elevation and soil discriminate main types of communities. With the goal of elucidating these effects, Molina-Venegas et al. (2016, 2018) designed a study to disentangle the effects of elevation and soils along two elevation transects on the southern slopes of Sierra Nevada (to avoid confounding effects of northern vs. southern microclimates) in each of the three main substrates (dolomite, limestone, and mica-schist). Phylogenetic and functional trait alpha diversities were mostly correlated with elevation, whereas functional beta diversity across communities was more dependent on soil properties. That is, upland communities were phylogenetically overdispersed in comparison with lowland communities. Also, elevation determined functional traits related to plant architecture and phenology, whereas soils were more closely associated with functional traits related to nutrient use, as expected (Fig. 3). The phylogenetic overdispersion at higher elevations could be interpreted as a result of repeated environmental filtering of different clades but could also be due to positive interactions among distantly related species due to environmental stress (Valiente-Banuet and Verdú 2013; Molina-Venegas et al. 2016).

Competitive and facilitative plant-plant interactions are also a critical factor shaping community assemblages. To explore this, Molina-Venegas et al. (2018) used the sampling design described above to compare the effects of elevation and soil on specific leaf area (SLA) of isolated plants in open habitats and overstory and understory plants in clumps. Dolomite had lower SLA values than the other more fertile soil types, likely due to environmental filtering. However, within dolomite, the strong differences in SLA between overstory and open habitat (lower values) and understory plants (higher values) were also indicative of the effect of facilitative interactions. Elevation had a small effect on SLA, at least up to about 2000 m.

Another set of studies (Pistón et al. 2015, 2016) sheds light on the effects of plant-plant interactions on phylogenetic diversity at the highest elevations and between southern and northern slopes of Sierra Nevada, where only mica-schist occurs. Although these studies were limited to the effect of a few nurse plant species on their associated sub-communities, some interesting conclusions can be

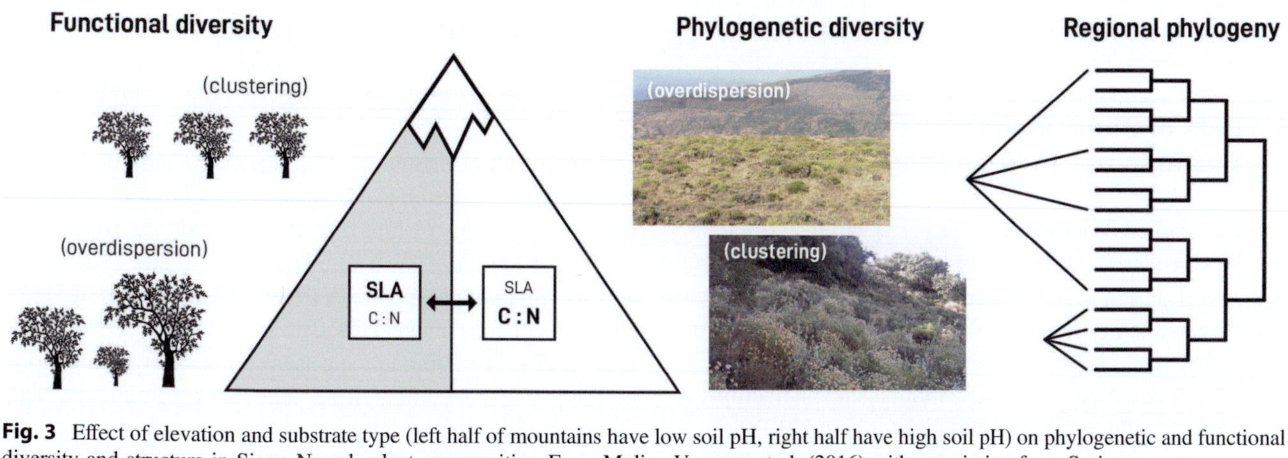


**Fig. 3** Effect of elevation and substrate type (left half of mountains have low soil pH, right half have high soil pH) on phylogenetic and functional diversity and structure in Sierra Nevada plant communities. From Molina-Venegas et al. (2016) with permission from Springer

drawn. At the highest, harshest elevations, facilitation is prevalent and occurs preferentially with species that are phylogenetically related to their nurse plant. This might lead to a clumped phylogenetic structure in those habitats compared to lower elevation communities. However, the community phylogenetic structure remains to be examined in the lower elevational elevational belt. When more nurse species were considered (Pistón et al. 2016), the effects on both species richness and phylogenetic diversity were species-specific. Thus, the composition of the founder community is critical for the results in terms of diversity.

All of these studies on plant community phylogenetics in Sierra Nevada define phylogenetic diversity and structure based on super-matrix or super-tree approaches for assembling local phylogenies. While genus-level and species-level phylogenies show similar results on diversity metrics (Qian and Jin 2021), floristic barcoding will produce increasingly more complete phylogenies at the local species level, which allow a more precise determination of phylogenetic effects (particularly for terminal branches of phylogenies) on phenotypic traits in comparative studies of communities.

### 4.3 Animal Community Phylogenetics

There are virtually no studies assessing the phylogenetic diversity (PD) of invertebrate communities in Sierra Nevada. The only exception is Abellán et al. (2013), in which the PD is quantified for different lineages of water beetles occurring in this massif and other national parks. The PD values in that study were compared to random expectations according to their number of species. Results showed that water beetles in Sierra Nevada tend to display lower PD than expected (i.e., species occurring in SN are more closely related than would be expected at random from the regional species pools). This can be explained by the high number of Iberian endemics occurring in Sierra Nevada, many of which are of recent Pleistocene origin (e.g., Ribera and Vogler 2004), so they added little to the overall PD. We know of no studies at all of vertebrate community phylogenies in Sierra Nevada.

## 5 Role of Biotic Interactions in Plant Communities in Promoting Biodiversity

As stated earlier, biotic interactions per se within a community can be considered a component of biodiversity, aside from their effect on the accumulation and maintenance of species diversity (Valiente-Banuet et al. 2015). In this respect, biotic interactions can be considered a part of functional biodiversity. We focus here on mostly mutualistic interactions as biodiversity builders (Bascompte and Jordano 2013).

### 5.1 Pollination and Seed and Fruit Dispersal

Given that most plant species rely on pollinators to reproduce, zones of high plant diversity harbour a high number of plant–pollinator interactions. Pollinator loss and phenological mismatch decrease plant species diversity mainly when plants depend completely on pollinator services to reproduce (Benadi et al. 2013). Sierra Nevada is a biodiversity hotspot with more than two thousand plant species, so many interactions between plants and pollinators are expected to occur. Both plant and pollinator species vary spatially due to habitat heterogeneity, environmental factors, or gradients, giving rise to spatial variation in plant–pollinator interactions (Burkle and Alarcon 2011). The variability in environmental factors and gradients occurring in Sierra Nevada suggests that plant-pollinator interactions are highly variable.

Most of the published studies on plant–pollinator interaction in Sierra Nevada focus on single plant species occurring along wide elevational gradients. These species are frequently visited by a markedly high number of insect species, indicating that generalization in plant-pollinator interactions is the norm. For example, *Hormathophylla spinosa* and *Erysimum mediohispanicum* are visited and pollinated by more than 70 and 130 insect species respectively, which differ in size, morphology, and behaviour (Gómez and Zamora 1999; Gómez et al. 2009a, b). These interactions showed marked spatial variation even at very small scales due to the mosaic-like distribution of pollinator assemblages (Gómez et al. 2009a, b; Gómez and Perfectti 2010). This leads to pollen limitation in some places (Gómez et al. 2010).

Although plant-pollinator interactions are characterized by generalization, at high elevations floral visitors are usually scarce and unpredictable (Arroyo et al. 1982; Gómez and Zamora 1999) due to the harsh environmental conditions. Studies of plant-pollinator interaction networks at high altitudes in Sierra Nevada showed that only 64% of the plant species studied received efficient pollinator visits (Santamaría and Méndez 2009). Plant–pollinator networks at high elevations have a marked modularity, and plants are visited by about 115 insect species. Both endemic and non-endemic plant species show a high number of interactions, which suggests that endemic plants are not particularly vulnerable to pollinator loss (Santamaría et al. 2014). At high elevations, the pollinator spectrum is dominated by social insects such as *Bombus* spp., *Apis mellifera* and diverse ant species (Santamaría and Méndez 2009). Previous studies had shown that ants are efficient pollinators of diverse plant species in Sierra Nevada, which is unusual (Gómez and Zamora 1999; Gómez et al. 1996). The pollinator spectrum found at high elevations in Sierra Nevada differs from that usually found in high mountain habitats, which tend to be composed mainly of generalist flies (e.g., Arroyo et al. 1982; Elberling and Olesen 1999; Dupont et al. 2003; Santamaría and Méndez 2009).

Lastly, in Sierra Nevada pollinators may also mediate other interaction types, such as those occurring among plants (Losapio et al. 2019). Although there is a lack of information on the mechanisms behind this effect, some studies in Sierra Nevada have clearly shown that some species (e.g., *Arenaria tetraquetra*) facilitate the pollination and establishment of associated plant species (Losapio and Schöb 2020). These nurse plant species act as ecosystem engineers, facilitating pollination networks and increasing insect diversity (Losapio et al. 2021a, b).

Zoochorous seed dispersal determines the abundance and spatial distribution of seed and seedling banks, and this mutualism is key to maintaining and expanding plant communities (Herrera 1985). Most of the available information on seed dispersal in Sierra Nevada focuses on *Quercus ilex*, an abundant evergreen tree species, and *Juniperus communis*, a dominant species in shrublands above the treeline. Both species are mainly dispersed by birds, whose behaviour has a clear demographic effect in both plant species. *Garrulus glandarius*, the main disperser of *Q. ilex* disperses acorns both locally and over long distances (Gómez 2003b) but it moves acorns non-randomly, affecting local regeneration. *Juniperus communis* has berry-like cones that are mainly dispersed by the migratory *Turdus torquatus* and *T. viscivorus* (García 2001), which move more than 600,000 seeds $ha^{-1}$ each year (García et al. 1996). This mutualism is important both quantitatively and qualitatively, since these birds reject fruits damaged by pests, giving rise to a filtered seed bank composed of healthy juniper seeds (García et al. 1999a, b). Both qualitative and quantitative effects of this mutualism are necessary to maintain recruitment given that seedling mortality is high due to the harsh abiotic conditions at high elevations in Sierra Nevada (García et al. 2001).

While small- and medium-sized birds are the main dispersers of fleshy fruits in Sierra Nevada (Zamora and Matías 2014), carnivorous and omnivorous mammal species are also opportunistic dispersers of more than half of the fleshy-fruited species (Matías et al. 2010). These mammals have large home ranges and move seeds over long distances, regardless of habitat type, making them particularly important in the regeneration of degraded habitats in Sierra Nevada (Matías et al. 2010). Studies suggest that seed dispersers in the Sierra Nevada prefer the fruits of shrubs to those of trees and therefore favour the establishment of shrubs over trees among different landscape units (Zamora and Matías 2014, see chapter "Forest Dynamics Under Land-Use and Climate Change Scenarios").

## 5.2 Facilitative Interactions Among Plants

Most available studies have focused on the effect of specific nurse plants on the corresponding facilitated species, rather than on the effect of facilitation-competition on whole plant community diversity. It has been shown that in the harshest habitats at the highest elevations, facilitation is a common mechanism of coexistence, mostly due to modification of micro-abiotic conditions for facilitated species (Callaway 2007; Pistón et al. 2015; Pugnaire et al. 2021). However, the effect is strongly dependent on nurse species identity (in some cases they are responsible for up to one-quarter of species richness in their communities; Schöb et al. 2017) and their phenotypes (Pistón et al. 2018).

At lower elevations, where typical communities are forests and associated shrublands, there is also an effect of nurse shrub species on colonization and early growth of forest species, in this case due to both ameliorated, less dry conditions (Gómez-Aparicio et al. 2004, see chapter "Forest Dynamics Under Land-Use and Climate Change Scenarios") and protection from herbivores (García et al. 2000; Castro et al. 2004; Gómez-Aparicio et al. 2005). It is worth noting that in many cases, both facilitated (*Quercus, Rhamnus, Crataegus, Juniperus*) and nurse species are bird-dispersed (e.g., *Taxus baccata*, see García et al. 2000), implying that the role of both mutualistic and antagonistic interactions should be taken into account in biodiversity accumulation.

These different results illustrate that different interaction networks may also interact among them and affect differently the structure and biodiversity of communities, as it has been shown for pollination interactions of some high elevation nurse plants in Sierra Nevada (Losapio and Schöb 2020). Thus, an interaction between network layers of a community should be expected (Losapio et al. 2021a, b), although the effect is apparently species-specific for nurse species (Losapio et al. 2019). Ultimately, the plant-plant facilitation network is central to the maintenance of other network layers, but could also depend on other layers, such as seed dispersal or soil microbiome (e.g., Cobo-Díaz et al. 2014, 2017; Pascual et al. 2016). It would be interesting to determine to what extent these abiotic and biotic conditions determine patterns of biodiversity across entire communities in different environmental conditions and trait compositions across Sierra Nevada, and then use this information to realistically predict the effects of loss of interactions on loss of biodiversity (Valiente-Banuet et al. 2015; Losapio and Schöb 2017).

## 6 Biotic Interactions in Animal Communities

### 6.1 Invertebrates

Several studies have investigated the biotic interactions between invertebrate and vertebrate species in this massif, focusing especially on the effects of large herbivores such as domestic sheep and Spanish ibex (*Capra pyrenaica*) on arthropod diversity. Large herbivores can affect arthropod diversity directly through unintentional ingestion while foraging (Zamora and Gómez 1993), which can lead to a reduction in the abundance of some arthropod species. Gómez and González-Megías (2007) found contrasting susceptibility to unintentional predation among guilds of phytophagous insects; endophagous insects were often being ingested by ungulates, while ectophagous insects were generally not. Furthermore, grazing by large herbivores modifies vegetation height and structure, providing different abiotic conditions, food resources and predation risk for arthropods. For instance, the removal of aerial structures such as flowers and stems can be detrimental to their consumers, like insect pollinators (Gómez 2003a), and those developing in flowerheads and fruits (Gómez and González-Megías 2007). In addition, herbivorous arthropods compete directly for resources with ungulates, despite the large differences in size and ecology. This competition is highly asymmetrical and likely reduces arthropod diversity through competitive exclusion and decreased population sizes (Zamora and Gómez 1993; Gómez and González-Megías 2002).

### 6.2 Vertebrates

The Sierra Nevada mountain ecosystem is a laboratory for studying species turnover within vertebrate communities. Within the context of global warming, displacement of generalist Mediterranean reptile populations (e.g., *Psammodromus algirus*, *Coronella girondica*, *Malpolon monspessulanus*) up the slopes of Sierra Nevada could lead to negative interactions with high elevation and relict species, such as competitive exclusion and predation (e.g., on *Coronella austriaca*, the only viviparous colubrid in this mountain system; Santos et al. 2009). There are more examples among birds, because of the higher richness of the group and, overall, their moving capacities. Over the last three decades, the Pyrenean oak (*Quercus pyrenaica*) woodland biotope (1700 m) became denser, and there was an increase in breeding bird richness (from 21 to 31 species), along with a conspicuous decline in population densities (from 108.1 to 37.5 birds/10 ha); in the juniper (*Juniperus communis*) scrub biotope (2200 m), there was significant species turnover produced by the arrival of generalist Mediterranean species, and the community density also decreased (from 30.2 to 10.5 birds/10 ha); in the high mountain biotope (3100 m), the density of alpine species slightly declined in parallel with the appearance of newcomers (*Carduelis cannabina, Prunella modularis*; Zamora and Barea-Azcón 2015, see chapter "Responses of Animal Populations and Communities to Climate Change and Land-Use Shifts").

## 7 Paths to Biodiversity: Phylogenetic and Phylogeographical Cues for Biodiversity Accumulation

### 7.1 Plants

Sierra Nevada shares many species or species-subspecies complexes with other high European mountains, such as the Pyrenees or the Alps (see chapter "Altitudinal Patterns and Changes in the Composition of High Mountain Plant Communities"). The current disjunct patterns have been mainly explained under two major hypotheses (see Kropf et al. 2006). The successive vicariance hypothesis suggests that there were successive gene flow disruptions among mountain ranges as the post-glacial period advanced from south to north and cold-adapted species retreated upwards, starting in the Sierra Nevada, and proceeding northwards to the Pyrenees and Alps. Under this hypothesis, populations were generally better connected during the glacial periods than during postglacial periods (like the current one). On the other hand, the long-distance dispersal hypothesis suggests that current species in Sierra Nevada and other European mountains may have also arrived by recent long-distance dispersal, crossing already established barriers. A study (Kropf et al. 2006) on several species (*Kernera saxatilis*, *Silene rupestris* and *Gentiana alpina*) suggests that gene flow was disrupted first between Sierra Nevada and the remaining European high mountain areas as post-glacial periods advanced and plant populations moved to higher altitudes and latitudes. However, recent long-distance dispersals from Sierra Nevada to northern European mountains were also inferred in the same study (*Papaver alpinum*, Kropf et al. 2006). In conclusion, both hypotheses, the successive vicariance and the recent north-to-south long-distance dispersal, may explain the current patterns of species and genetic diversities in high European mountains.

Later studies have shown different patterns. Kropf et al. (2008) detected strong genetic structure among the populations of different mountains in two species of *Saxifraga* (*S. oppositifolia* and *S. stellaris*) but this pattern of genetic structure was incongruent with the successive gene flow restriction hypothesis. The authors associated this pattern with a more complex "vicariance" process. Other studies (*Meum athamanticum*, Huck et al. 2009; *Senecio boissieri*, Peredo et al. 2009) have found also complex patterns that have been associated with the colonization of northern populations not from current southern populations but from periglacial refugia in the northern distribution. The case of *Artemisia umbelliformis* (Sanz et al. 2017) is even more intriguing, since in addition to the northern populations being from periglacial areas, the population in Sierra Nevada is the result of a recent long-distance dispersal from the Alps.

Another question is the mechanism or process that may have shaped the current patterns of differentiation and genetic diversity in species or species complexes within Sierra Nevada. Are the Quaternary glaciations also the most important processes in shaping these patterns? Or are there other equally or even more important processes or mechanisms? A case study in *Armeria* species seems to suggest that Quaternary glaciations may have been the most important factor in plant differentiation and genetic diversity. Gutiérrez-Larena et al. (2002) showed that in Sierra Nevada there are three *Armeria* species that occur at markedly different altitudinal belts and are involved in haplotype sharing; this suggests that altitudinal migrations during the contraction–expansion cycles (i.e., during glacial-interglacial cycles) have shaped the genetic structure and genetic diversity of these species, including the formation of a new hybrid taxon. However, there are other factors that may also explain the current observed patterns. The endemic species of the highest elevations of Sierra Nevada, *Linaria glacialis*, has weak population genetic structure as a consequence of high mutation rates and high levels of gene flow (Blanco-Pastor et al. 2013). The glacial and interglacial stages of the Late Quaternary and climatic oscillations during the last millennium (Medieval Warm Period and Little Ice Age) provoked small changes and moderate changes in the demographic trends, respectively (Blanco-Pastor et al. 2013). Blanco-Pastor et al. (2019) in a subsequent study that included two additional closely related species, *Linaria nevadensis* and *Chaenorrhinum glareosum* in addition to *Linaria glacialis* inferred similar patterns of demographic and niche evolution for the three studied species. However, they also found that genetic diversities were heterogeneous across the landscape of Sierra Nevada for the three species (and uncorrelated among them) but were all correlated with topographic variables. This suggests that topography rather than climatic oscillations seems to be an important factor to explain the distribution of genetic diversity in the alpine belt of Sierra Nevada.

## 7.2 Animals

Invertebrates

Expansions during glacial periods followed by the retreat into higher elevations (Iberian mountain ranges) during postglacial warm phases have been also postulated as an important mechanism of diversification for cold-adapted species of invertebrates. For instance, different lineages within the water beetle genus *Hydraena* diversified independently on the Iberian Peninsula through the fragmentation of a more widely distributed ancestor, with periods of range contraction resulting in the generation of multiple isolated residual species (Ribera et al. 2011). This could be the origin of *Hydraena tatii*, restricted to Sierra Nevada and nearby mountains. Similarly, diversification of the caddisfly genus *Annitella* has been dated to the Pleistocene and associated with southern and extra-Mediterranean refugia located across Europe. Notably, two endemic species inhabiting cold high-elevation headwater streams of Sierra Nevada, *Annitella iglesiasi* and *A. esparraguera*, originated independently from two different lineages by diverging from their respective sister species, likely by long-term isolation between mountain ranges since the Pleistocene (Múrria et al. 2020). Another example is the grasshopper *Omocestus bolivari*, which is restricted to Sierra Nevada and nearby Sierra de Baza-Filabres (Tonzo et al. 2021).

Some studies have also provided evidence of the role of Sierra Nevada as an important refuge for invertebrate populations during the succession of the Quaternary glacial-interglacial cycles. For instance, the populations of the widespread butterfly *Polyommatus icarus*, which is restricted to the high elevations of Sierra Nevada, are included in a distinct lineage that diverged about 0.5 Ma, suggesting that this massif acted as an important refuge that protected ancestral populations from introgressive dispersal events that were common in other parts of the continent (Dincă et al. 2011). Similarly, Todisco et al. (2010) showed very strong phylogeographic structure in the butterfly *Parnassius apollo*, with several distinctive mitochondrial lineages occurring in geographically distinct areas within Eurasia. Among them, a highly divergent lineage is *P. a. nevadensis*, which is restricted to Sierra Nevada.

Vertebrates

Some geological events contribute to understanding the faunal composition of non-flying vertebrates in the southeastern Iberian Peninsula. The existence of a Baetic-Rifian archipelago, and the reopening of the Gibraltar Strait (Barbadillo et al. 1997) promoted diversity by isolation and vicariance within the fragmented archipelago (Kaliontzopoulou et al. 2011) and between continents (Busack 1986). The Messinian Salinity Crisis, on the other hand, favored immigration of other Mediterranean faunas (Booth-Rea et al. 2018). With respect to fishes, the isolated and sedentary populations of the common trout (*Salmo trutta*) in Sierra Nevada belong to the Mediterranean mitochondrial lineage and differentiated during Pleistocene glaciations (Bernatchez 2001); these Iberian meridional populations share a common haplotype (Machordom et al. 2000). Among amphibians, the opening of a strait between the Iberian Peninsula and Northwestern Africa is associated with the split of a common ancestor into the Baetic-endemic *Alytes dickhilleni* and the Rifian *A. maurus*. The estimated time of this phylogenetic process encompasses the timeframe of these geological events (Maia-Carvalho et al. 2014). Among the large Iberian lizards of the genus *Timon*, the split into two species (the western *T. lepidus* and the southeastern *T. nevadensis*) occurred during the late Miocene and it was boosted by the increasing aridity in the southeastern Iberian Peninsula (Ahmadzadeh et al. 2016). Later, when the North Baetic paleo-strait closed, the two species came into secondary contact along a 10 km wide belt that currently crosses the western tip of Sierra Nevada, where some hybrids are found (Miraldo et al. 2012). The Andalusian Wall Lizard, *Podarcis vaucheri*, differentiated within the Baetic-Rifian archipelago, currently spanning northwestern Africa and southern Iberia, including the western half of Sierra Nevada (Busack et al. 2005). Also, the Iberian Wall Lizard, *P. hispanicus*, probably differentiated within the easternmost paleo-islands of this archipelago, now inhabiting the eastern part of Sierra Nevada. It exemplifies a secondary contact with the previous species in this massif and southeastern Iberia (Kaliontzopoulou et al. 2011, 2012).

During the Pleistocene, climatic fluctuations due to alternating glacial periods modified the distribution of boreal vertebrates, and populations of some species with low vagility persisted as refugees in meridional mountains. In Sierra Nevada there is just one case for fishes (*S. trutta*; Almodóvar et al. 2010), one for reptiles (*C. austriaca*; Santos et al. 2009), and one for mammals (*Chionomys nivalis*; Piñeiro 2017). There are no isolated amphibians at high elevations, probably because of excessive solar radiation of this southern mountain massif (see Lizana and Pedraza 1998). The current existence of some northern or mountain bird species isolated within Sierra Nevada must be a very recent process (the cases of *Anthus spinoletta, Prunella collaris, P. modularis, Carduelis citrinella*), since communities of these flying vertebrates show continuous turnover of species composition (Zamora and Barea-Azcón 2015; see also chapter "Responses of Animal Populations and Communities to Climate Change and Land-Use Shifts").

## 8 The Hidden Side of Biodiversity: Population Genetic Structure in Sierra Nevada and Its Drivers

Patterns in population genetic data have been frequently used to ascertain phylogeographical processes hypothesized for isolated taxa and populations in marginal ranges, as apparently occurred in Sierra Nevada during the Pleistocene glacial-interglacial and late Tertiary periods (Nieto-Feliner 2014, see Sect. 7 in this chapter). In most of these studies, inferences about underlying processes are made based on ad hoc correlation analysis between genetic data and historical and current environmental data. Additionally, these data may provide support for some complicated taxonomical issues (e.g., Medrano et al. 2014 for SE Iberian *Narcissus*). However, ascertaining the causes of the current genetic diversity of populations relies on explicit hypotheses and sampling designs, including the hypothesized driving ecological factors and responding biological traits. Here we report a few study cases which may exemplify how natural and artificial conditions (with direct conservation implications) may affect the genetic component of biodiversity at the population level.

### 8.1 Plants

There are several studies on tree species with varied distribution ranges (see chapter "Forest Dynamics Under Land-Use and Climate Change Scenarios"). The widespread Holm oak (*Quercus ilex* s.l.) is one of the ecologically broadest tree species across the Mediterranean Basin. In Sierra Nevada, it is common from lowlands close to the coast up to 2000 m in a variety of soil types, both basic and acidic. An extensive study (Guzmán et al. 2015) across populations in refugial and protected areas on the Iberian Peninsula demonstrated that the genetic diversity included in National Parks populations (including Sierra Nevada) is very high in comparison with refugial and non-protected areas. This is probably due to the highly resilient nature of the species and sheds light on the importance of preserving populations in Sierra Nevada National Park, one of the protected areas where holm oak is currently most frequent. The ecological and biogeographical status as a relic of another oak species (the deciduous *Q. pyrenaica*) in its southernmost range limit poses a different question. Valbuena-Carabaña and Gil (2013) determined that *Q. pyrenaica* populations subjected to intense resource use (coppicing) still conserve a high level of genetic diversity and differentiation, despite intense clonal propagation. The authors argue that for the long time the species has been present in Sierra Nevada (i.e., it is a relict element) and its genetic resilience to perturbations (not only by humans) may have helped to conserve this genetic diversity. Scots pine is an even stronger example of an isolated relic in Sierra Nevada and nearby mountains. Considered a separate taxon, *Pinus sylvestris* var. *nevadensis* has shown particular patterns of neutral genetic variation (Soranzo et al. 2000) and adaptive genetic variation in drought resistance (Alía et al. 2001). Interestingly, in Sierra Nevada, allochthonous *Pinus sylvestris* has been extensively used for afforestation, which poses the question of a possible loss of genetic differentiation and adaptation through introgressions. Unger et al. (2014) determined that the level of gene flow from non-native planted trees is moderate but consistent due to pollen flow, as expected in a wind-pollinated species. It appears that hybrid seedlings are less proficient under severe drought conditions typical of native Sierra Nevada populations, which may help decrease the effects of introgression and may indicate a role of local adaptation in this relict pine (Ramírez-Valiente and Robledo-Arnuncio 2015). However, the long-term consequences on genetic diversity and selected traits of relict *P. sylvestris* var. *nevadensis* are still unknown. In Sierra Nevada, hybridization is a process that has been naturally present in the accumulation of plant biodiversity, although it remains a question if anthropogenic changes such as afforestation, restoration or rapid climatic changes will have detrimental consequences for genetic identity of populations and species (reviewed in Gómez et al. 2015; see chapter "Altitudinal Patterns and Changes in the Composition of High Mountain Plant Communities").

In contrast with trees, small shrubs and perennial herbs are most typical at medium to high elevation belts of Sierra Nevada in different soil types. They are mostly insect-pollinated and have a variety of reproduction modes and life histories which may affect the level of population genetic diversity. In *Helianthemum*, a paradigmatic Mediterranean genus with a diversity center in the south-eastern Iberian Peninsula, narrow-endemic species in dolomites (*H. apenninum* subsp. *estevei* and *H. pannosum*) harbour more genetic diversity than widespread species (*H. apenninum* subsp. *apenninum* and *H. cinereum* subsp. *rotundifolium*) and have traits related to outcrossing (higher pollen production and pollen/ovule ratios). However, these results do not apply to species in subgen. *Plectolobum* (*H.c. rotundifoium, H. pannosum*), suggesting some effect of phylogeny on the distribution of population genetic diversity (Martín-Hernanz et al. 2019).

Herbaceous insect-pollinated species are well suited to the study of the effects of selective regimes on quantitative genetic variation and patterns of differentiation of traits. The well-studied biennial *Erysimum mediohispanicum*, typical to medium to high-elevations in Sierra Nevada, showed high heritabilities in flower traits, which are highly variable and are decoupled from vegetative traits, which have lower variation (Gómez et al. 2009a). This seems to respond to a selective mosaic scenario due to different sets of pollinators across populations in Sierra Nevada (Sect. 5.1; Gómez et al. 2009b, 2010). However, this mosaic is not directly translated into patterns of neutral genetic variation across populations, although their genetic diversity was not explored (Muñoz--Pajares et al. 2017). For this semelparous species, patterns of pollen-mediated gene flow among populations are strongly dependent on flowering overlap among their plants (Muñoz-Pajares et al. 2020), which may vary in changing climatic conditions, thus altering the genetic structure of populations. However, the few case studies apparently do not allow broad generalizations, since results from another insect-pollinated species, *Aquilegia vulgaris* subsp. *nevadensis* and related species in Sierra Nevada and nearby mountains showed more pronounced variation in vegetative traits than in floral traits, despite similar heritabilities among traits. The amount of genetic variation in populations is not a limit for differentiation, and this does not seem to be related to pollinator-mediated selection (Alcántara et al. 2014).

Plants from the highest elevation belt, in the alpine zone (approx. from 2000 to the summits), face the harshest environment in Sierra Nevada, in terms of both climate and rocky substrates. Thus, these species are good candidates for examining the role of these niches in driving patterns of population genetic structure and diversity. In the few cases analyzed, these species harbour high levels of population genetic diversity (Kropf et al. 2006) and weak genetic structure in their limited ranges, which seems to be due to the long-term isolation and coping with past climate changes in different microsites (Blanco-Pastor et al. 2019). This should make them relatively resilient to future climate changes, at least in terms of genetic diversity (Blanco-Pastor et al. 2019; see Sect. 7). It has been shown that some rupicolous relict species have a low demographic stochasticity which may explain their long-term persistence (Picó and Riba 2002; García 2003; García and Zamora 2003), although this kind of study remains to be done in Sierra Nevada (but see García et al. 1999a, b).

## 8.2 Animals

Invertebrates

Intraspecific genetic diversity and structure and population genetics have only been studied in a few invertebrate taxa occurring in Sierra Nevada, usually in the context of habitat fragmentation and often assessing the potential effects of climate change on the genetic patterns. The complex topography of the massif and the steep altitudinal gradient have been identified as key factors explaining the distribution of genetic diversity in some of these species. This is especially the case for low-dispersive insects inhabiting cold high-elevation headwater streams, for which the high and steeply sloped topographic barriers among catchments and internal physical habitat heterogeneity across headwater river sections seem crucial in determining genetic differentiation among populations and gene flow. For instance, Múrria et al. (2020) have studied the intraspecific genetic diversity and structure of two caddisfly species (*Annitella iglesiasi* and *A. esparraguera*) confined to high-elevation headwaters in Sierra Nevada and the nearby Baetic mountains. Evidence of population isolation and limited gene flow was found, especially for *A. esparraguera*, which was explained by the intrinsically limited dispersal of these species and the role of topographic barriers between suitable sites and among catchments. Similarly, Finn et al. (2014) found strong population genetic structure in a lineage (assimilated to the species rank) of the cold-tolerant mayfly *Baetis alpinus*, which is endemic to Sierra Nevada. Interestingly, in addition to the physical isolation among populations occupying the upper tips of mountain headwater streams, they propose a role of reproductive isolation between highly and minimally glacially influenced reaches within the same basin. Another example, in this case a terrestrial species, is the grasshopper *Omocestus bolivari*, which shows strong genetic structure and little genetic admixture among populations located within Sierra Nevada (Tonzo and Ortego 2021).

Furthermore, evidence of isolation by distance has been found to drive the patterns of genetic structure among populations for some insect species. This is the case of the butterfly *Parnassius apollo* subsp. *nevadensis*, restricted to Sierra Nevada, which displays a weak but significant genetic structure and a strong pattern of isolation by distance as one of the main factors driving the differentiation between populations (Mira et al. 2017). The existence of population structure, despite some gene flow, and the asymmetry of gene flow between populations, suggested the existence of source–sink dynamics in this species. Similarly, the populations of the ant *Proformica longiseta* are strongly structured for both nuclear and mitochondrial markers and show strong isolation by distance, which together with the absence of intrapopulation variation in mitochondrial DNA, suggest strong female philopatry and limited male dispersal (Sanllorente et al. 2015).

Vertebrates

Little is known about the population genetic structure of vertebrates in Sierra Nevada. The widespread common trout (*Salmo trutta*) is genetically well structured across basins within the mountain massif (Almodóvar et al. 2010), which may be relevant to conservation management actions involving population translocations. Two kinds of genetic markers (a mitochondrial fragment of the ND4 gene and 20 microsatellite loci) of the endemic Baetic midwife toad (*Alytes dickhilleni*) revealed high genetic diversity and a marked genetic structure among populations. There were also low levels of gene flow, a result that mirrors the presence of the species in fragmented habitats and limited connectivity between Sierra Nevada populations and those in surrounding mountain ranges. The coalescent estimates dating the most recent common ancestor for all haplotypes in the species suggested a Pleistocene origin for the ancestor of the two Sierra Nevada genetic groups (Dias et al. 2015). Sierra Nevada populations of the spiny toad (*Bufo spinosus*) harbour considerable genetic diversity compared to northern populations of the species (Recuero et al. 2012). The populations of the smooth snake (*Coronella austriaca*), a relict in Sierra Nevada evidenced a scenario of refugia within the region; according to a molecular clock, these populations diverged from the other members of their clade approximately 1.2–1.65 Ma, suggesting a fragmentation of the Iberian population of the species during the Pleistocene ice ages (Santos et al. 2008). Moreover, the lineage to which the Sierra Nevada populations belong exhibits higher genetic variability than the two other lineages within the Iberian Peninsula, and the three Iberian lineages together had higher variability than the populations from Central Europe, a region of recent (Holocene) colonization by this ectothermic vertebrate (Santos et al. 2008).

## 9 Current Biodiversity Conservation in Sierra Nevada

The current policy of biodiversity conservation in Sierra Nevada is directly derived from the declaration of an important part of it as a Natural Park (1989) with the core area later designated as a National Park (1999). Additionally, many of the species are legally protected by international, national, or regional rules. The information provided in this review, both evolutionary and ecological, may shed light when establishing priorities for conservation, given that resources are usually limited. Knowledge on the processes underlying rarefaction is critical to choose the right species, communities, and habitats to conserve. However, there is still little information, so most conservation actions are taken at the species level and based mainly on taxonomy and distribution, with some exceptions.

Among the plants, 187 taxa (ca. 8%) have been designated as threatened species (Lorite et al. 2020), of which 15 are Critically Endangered, 16 are Endangered, and 105 are Vulnerable (IUCN categories). In addition, there are 50 plant species lacking data to conduct a proper threat assessment, and thus were classified as Data Deficient. Only 51 (ca. 27%) of the threatened taxa are included in the regional protection catalogue (Peñas and Lorite 2019). However, for most of them some conservation actions have been developed in the last 20 years (Peñas and Lorite 2019). For all threatened species some activities for prospecting, locating and basic monitoring have been carried out. Also, 136 species are currently preserved at local or regional germplasm banks, and 97 are present in ex-situ collections in botanic gardens. There are two in situ botanic gardens in Sierra Nevada including natural habitats, which may facilitate living ex-situ conservation of plants while maintaining their biological interactions. However, in situ conservation activities (reintroductions, reinforcements, habitat management or restoration, etc.), which are the keystone of conservation actions, have been incompletely addressed for only 64 taxa (34%; Lorite et al. 2019). Moreover, in most cases these are isolated actions that do not form part of a coordinated and complete recovery plan focused either on the species, or on the habitats.

Regarding invertebrates, Sierra Nevada harbours 16 threatened taxa according to the Atlas and Red Book of the Threatened Invertebrates of Spain (Verdú and Galante 2009, 2011), most of them restricted to this massif. Among them, five are listed as Endangered and eleven as Vulnerable. The group with the most threatened species is insects (four Lepidoptera, three Coleoptera, three Hymenoptera, one Hemiptera, one Odonata and one Orthoptera), but there are also arachnids (one Araneae and one Opiliones) and one mollusk. Additionally, Millán et al. (2013) identified eight water beetle species occurring in Sierra Nevada as Vulnerable. However, only two (*Baetica ustulata* and *Polyommatus golgus*) of the threatened species listed in these red books are included in both the Regional (Andalusian) and the Spanish Catalogues of Threatened Species. Some of these invertebrate taxa, with very narrow distribution and restricted to the highest elevation zones of the massif, are expected to be especially vulnerable to climate change (Múrria et al. 2020; Pallarés et al. 2020).

Except for the low-diversity fish fauna, the vertebrates of Sierra Nevada are generally not threatened; only 8.2% of the species are Vulnerable, Endangered or Critically Endangered. By groups, 66.7% of fish species, 11.1% of amphibians, 5.0% of reptiles, 2.8% of breeding birds and 18.6% of mammals are threatened. Fish species are mostly threatened by summer drought and hydraulic power stations; amphibians by the loss of breeding ponds; reptiles by the presence of extensive plantations at middle elevation and species-specific life-history traits (dietary specialization and low reproductive frequency); birds from open habitats by the encroachment of plantations, and forest birds have shown an unexplained sharp decrease in density over the last decades (see Sect. 6.2). Among mammals, the most threatened are bats, because of the loss of suitable roosting places. The few populations of northern species isolated in Sierra Nevada (*S. trutta, C. austriaca, C. nivalis*) are consistently threatened.

Management proposals for the conservation of vertebrates in Sierra Nevada must focus on reverting previously cited threat processes, particularly the human-induced ones. Examples include the restoration of natural aquatic habitats and roosting places, thinning of plantations (Azor et al. 2015; Jiménez-Albarral et al. 2020), and bringing the large population of the Iberian ibex (*Capra pyrenaica*) back into balance in the absence of their predators, an imbalance that recently contributed to a dramatic surge in mortality from epizootic mange (*Sarcoptes scabiei*; Pérez et al. 1997). These and other issues represent a challenge for environmental managers of Sierra Nevada. However, there are reasons to hope; during the last decades, the reproduction in Sierra Nevada of some passerines (*Sitta europea, Anthus spinoletta, Prunella modularis, Carduelis citronella*) have been recorded again, and some vultures (*Gyps fulvus, Gypaetus barbatus*) are predicted to reproduce there in the near future.

## 10 Concluding Remarks

The Sierra Nevada massif has long been recognized as a biodiversity hotspot for both animal and plant groups, based mostly on the species richness, endemicity and relictness. However, biodiversity includes many other facets related to functional and historical aspects of biological entities present in hotspots, from genes and genomes to communities and the whole biota. Comprehensive knowledge of these biodiversity components for all species and lineages and also their fine spatial pattern within Sierra Nevada would be ideal. This would allow for the complete mapping of biodiversity and its relationships with environmental conditions, and most importantly, it would help take rapid and accurate conservation actions against threats. Although getting complete information is a hard and time-consuming task, it is feasible given the currently available information and future prospects (see chapters "Data Model, E-Infrastructure Services, and the Virtual Research Environment (VRE)" and "Filling the Gaps in Research, Monitoring, Management and Social Connection"). Here, we have shown that there are some biological groups where the available information evidences high biodiversity value.

Phylogenetic information is available for a relevant floristic pool, the woody flora and some plant communities, and also for some critical plant groups that may serve as model systems to disentangle the critical factors driving current biodiversity. Whereas animal phylogenetics includes studies with insight into keystone species and lineages, studies on animal community phylogenetics are just beginning (a notable exception being the water beetle communities). Population genetic diversity is also a critical factor to study the historical causes of biodiversity in key species, and there are some paradigmatic cases where insight has been deeper with important consequences for conservation, such as the endemic Sierra Nevada Scots pine (*Pinus sylvestris* var. *nevadensis*). Having phylogenetic, phylogeographic and population genetic information for endemic and/or threatened species would help to put conservation efforts in species and populations evolutionarily rare. Also, adapting strategies to cope with climatic change, such as assisted migrations or re-introductions, could greatly benefit with this genetic information.

Lastly, biotic interactions demonstrate the role of mechanisms that promote biodiversity. Fortunately, there are many studies focused mostly on mutualistic interactions, including pollination, seed dispersal and facilitative plant interactions, which demonstrate that interactions are a fundamental part of functional biodiversity. In many cases, these studies bridge the gap between ecological function and evolutionary history by including phylogenetic information on the groups involved in interactions. Other mutualistic interactions such as those occurring in the soil biota, as well as antagonistic interactions, such as herbivory, will help to explain how biodiversity accumulates in Sierra Nevada communities.

**Acknowledgements** The authors are grateful for grant funding by MICINN-FEDER (CGL2013-45037-P, PGC2018 099608 B 100, PID2019-108895G), OAPN (296/2011), FEDER-US (1265280) and PAIDI2020 (P18-RT-3651). Lynna M. Kiere kindly revised the English.

## References

Abeli T, Vamosi JC, Orsenigo S (2018) The importance of marginal population hotspots of cold-adapted species for research on climate change and conservation. J Biogeogr 45:977–985

Abellán P, Sánchez-Fernández D, Picazo F et al (2013) Preserving the evolutionary history of freshwater biota in Iberian National Parks. Biol Conserv 162:116–126

Ahmadzadeh F, Fleck M, Carretero MA et al (2016) Separate histories in both sides of the Mediterranean: phylogeny and niche evolution of ocellated lizards. J Biogeogr 43:122–1253

Alcántara JM, Jaime R, Bastida JM et al (2014) The role of genetic constraints on the diversification of Iberian taxa of the genus Aquilegia (Ranunculaceae). Biol J Linn Soc 11:252–261

Alía R, Moro-Serrano J, Notivol E (2001) Genetic variability of Scots pine (Pinus sylvestris) provenances in Spain: growth traits and survival. Silva Fenn 35:27–38

Almodóvar A, Nicola GG, Leal S et al (2010) Análisis genético de las poblaciones de Trucha Común Salmo trutta en la Comunidad Autónoma de Andalucía. Environmental Council of the Andalusian Government, Sevilla, Spain

Arroyo J, Marañón T (1990) Community ecology and distributional spectra of Mediterranean shrublands and heathlands in Southern Spain. J Biogeogr 173:163–176

Arroyo MTK, Primack R, Armesto J (1982) Community studies in pollination ecology in the high temperate Andes of central Chile. I. Pollination mechanisms and altitudinal variation. Am J Bot 69:82–97

Azor JS, Santos X, Pleguezuelos JM (2015) Conifer-plantation thinning restores reptile biodiversity in Mediterranean landscapes. For Ecol Manag 354:185–189

Barbadillo LJ, García-París M, Sanchiz B (1997) Orígenes y relaciones evolutivas de la herpetofauna ibérica. In: JM Pleguezuelos (ed) Distribución y biogeografía de los anfibios y reptiles en España y Portugal. Monografías de Herpetología 2:47–100

Bascompte J, Jordano P (2013) Mutualistic networks. Princeton University Press, Princeton

Benadi G, Blüthgen N, Hovestadt T et al (2013) When can plant-pollinator interactions promote plant diversity? Am Nat 182:131–146

Bernatchez L (2001) The evolutionary history of brown trout (Salmo trutta, L.) inferred from phylogeographic, nested clade, and mismatch analyses of mitochondrial DNA variation. Evolution 55:351–379

Blanca G, Cueto M, Martínez-Lirola MJ et al (1998) Threatened vascular flora of Sierra Nevada (southern Spain). Biol Conserv 85:269–285

Blanca G, Cabezudo B, Cueto M et al (eds) (2009) Flora Vascular de Andalucía Oriental, 4 vols. Consejería de Medio Ambiente, Junta de Andalucía, Sevilla

Blanco-Pastor JL, Fernández-Mazuecos M, Vargas P (2013) Past and future demographic dynamics of alpine species: limited genetic consequences despite dramatic range contraction in a plant from the Spanish Sierra Nevada. Mol Ecol 22:4177–4195

Blanco-Pastor JL, Fernández-Mazuecos M, Coello AJ et al (2019) Topography explains the distribution of genetic diversity in one of the most fragile European hotspots. Divers Distrib 25:74–89

Booth-Rea G, Ranero CR, Grevemeyer I (2018) The Alboran volcanic-arc modulated the Messinian faunal exchange and salinity crisis. Sci Rep 8:1–14

Buira A, Aedo C, Medina L (2017) Spatial patterns of the Iberian and Balearic endemic vascular flora. Biodivers Conserv 26:479–508

Buira A, Fernández-Mazuecos M, Aedo C et al (2021) The contribution of the edaphic factor as a driver of recent plant diversification in a Mediterranean biodiversity hotspot. J Ecol 109:987–999

Burkle LA, Alarcón R (2011) The future of plant–pollinator diversity: understanding interaction networks across time, space, and global change. Am J Bot 98:528–538

Busack SD (1986) Biogeographic analysis of the herpetofauna separated by the formation of the Strait of Gibraltar. Natl Geogr Res 2:17–36

Busack SD, Lawson R, Arjo WM (2005) Mitochondrial DNA, allozymes, morphology and historical biogeography in the Podarcis vaucheri (Lacertidae) species complex. Amphibia-Reptilia 26:239–256

Callaway RM (2007) Positive interactions and interdependence in plant communities. Springer, New York

Cañadas EM, Fenu G, Peñas J et al (2014) Hotspots within hotspots: endemic plant richness, environmental drivers, and implications for conservation. Biol Conserv 170:282–329

Castro J, Zamora R, Hódar JA et al (2004) Benefits of using shrubs as nurse plants for reforestation in Mediterranean mountains: a 4-year study. Restor Ecol 12:352–358

Cheikh Albassatneh M, Escudero M, Monnet AC et al (2021) Spatial patterns of genus-level phylogenetic endemism in the tree flora of Mediterranean Europe. Divers Distrib 27:913–928

Cobo-Díaz JF, Martínez-Hidalgo P, Fernández-González AJ et al (2014) The endemic Genista versicolor from Sierra Nevada National Park in Spain is nodulated by putative new Bradyrhizobium species and a novel symbiovar (sierranevadense). Syst Appl Microbiol 37:177–185

Cobo-Díaz JF, Fernández-González AJ, Villadas PJ et al (2017) Taxonomic and functional diversity of a Quercus pyrenaica willd.

rhizospheric microbiome in the Mediterranean mountains. Forests 8:390

Cueto M, Melendo M, Giménez E et al (2018) First updated checklist of the vascular flora of Andalusia (S of Spain), one of the main biodiversity centres in the Mediterranean Basin. Phytotaxa 339:1–95

Davies TJ (2021) Ecophylogenetics redux. Ecol Lett 24.1073–1088

Dias G, Beltrán JF, Tejedo M et al (2015) Limited gene flow and high genetic diversity in the threatened Beticmidwife toad (Alytes dickhilleni): evolutionary and conservation implications. Conserv Genet 16:459–476

Dincă V, Dapporto L, Vila R (2011) A combined genetic-morphometric analysis unravels the complex biogeographical history of Polyommatus icarus and Polyommatus celina Common Blue butterflies. Mol Ecol 20:3921–3935

Dupont YL, Hansen DM, Olesen JM (2003) Structure of a plant-flower-visitor network in the high-altitude sub-alpine desert of Tenerife, Canary Islands. Ecography 26:301–310

Elberling H, Olesen J (1999) The structure of a high altitude plant-flower visitor system: the dominance of flies. Ecography 22:314–323

Favarger C, Contandriopoulos J (1961) Essai sur l'endémisme. Travaux de l'Institut de Botanique de l'Université de Neuchâtel 1961. pp 384–408

Finn DS, Zamora-Muñoz C, Múrria C et al (2014) Evidence from recently deglaciated mountain ranges that Baetisalpinus (Ephemeroptera) could lose significant genetic diversity as alpine glaciers disappear. Freshw Sci 33:207–216

García D (2001) Effects of seed dispersal on Juniperus communis recruitment on a Mediterranean mountain. J Veg Sci 12:839–848

García MB (2003) Demographic viability of a relict population of the critically endangered plant Borderea chouardii. Conserv Biol 17:1672–1680

García D, Zamora R (2003) Persistence and multiple demographic strategies in long-lived mediterranean plants. J Veg Sci 14:921–926

García D, Zamora R, Gómez JM et al (1999a) Bird rejection of unhealthy fruits reinforces the mutualism between juniper and its avian dispersers. Oikos 85:536–544

García D, Zamora R, Hódar JA et al (1999b) Age structure of Juniperus communis L. in the Iberian peninsula: conservation of remnant populations in Mediterranean mountains. Biol Conserv 87:215–220

García D, Gómez JM, Hódar JA et al (1996) Ecología reproductiva del enebro Juniperus communis L. en Sierra Nevada: factores que determinan la regeneración natural de las poblaciones. In: 1ª Conferencia Internacional: Sierra Nevada, Granada, pp 441–452

García D, Zamora R, Hódar JA et al (2000) Yew (Taxus baccata L.) regeneration is facilitated by fleshy-fruited shrubs in Mediterranean environments. Biol Conserv 95:31–38

García D, Zamora R, Gómez JM et al (2001) Frugivory at Juniperus communis depends more on population characteristics than on individual attributes. J Ecol 89:639–647

Gaston KJ, Spicer JI (2004) Biodiversity: an introduction, 2nd edn. Wiley, Hoboken

Gómez JM (2003a) Herbivory reduces the strength of pollinator-mediated selection in the Mediterranean herb Erysimum mediohispanicum: consequences for plant specialization. Am Nat 162:242–256

Gómez JM (2003b) Spatial patterns in long-distance dispersal of Quercus ilex acorns by jays in a heterogeneous landscape. Ecography 26:573–584

Gómez JM, GonzálezMegías A (2002) Asymmetrical interactions between ungulates and phytophagous insects: being different matters. Ecology 83:203–211

Gómez JM, GonzálezMegías A (2007) Long-term effects of ungulates on phytophagous insects. Ecol Entomol 32:229–234

Gómez JM, Perfectti F (2010) Evolution of complex traits: the case of Erysimum corolla shape. Int J Plant Sci 171:987–998

Gómez JM, Zamora R (1999) Generalization in the interaction between Hormathophylla spinosa (Cruciferae) and its pollinators. Ecology 80:796–805

Gómez JM, Zamora R, Hódar JA et al (1996) Experimental study of pollination by ants in Mediterranean high mountain and arid habitats. Oecologia 105:236–242

Gómez JM, Abdelaziz M, Muñoz-Pajares J et al (2009a) Heritability and genetic correlation of corolla shape and size in Erysimum mediohispanicum. Evolution 63:1820–1831

Gómez JM, Abdelaziz M, Camacho JPM et al (2009b) Local adaptation and maladaptation to pollinators in a generalist geographic mosaic. Ecol Lett 12:672–682

Gómez JM, Abdelaziz M, Lorite J et al (2010) Changes in pollinator fauna cause spatial variation in pollen limitation. J Ecol 98:1243–1252

Gómez JM, GonzálezMegías A, Lorite J et al (2015) The silent extinction: climate change and the potential hybridization-mediated extinction of endemic high-mountain plants. Biodivers Conserv 24:1843–1857

Gómez Aparicio L, Zamora R, Gómez JM et al (2004) Applying plant facilitation to forest restoration: a meta-analysis of the use of shrubs as nurse plants. Ecol Appl 14:1128–1138

Gómez Aparicio L, Gómez JM, Zamora R et al (2005) Canopy vs. soil effects of shrubs facilitating tree seedlings in Mediterranean montane ecosystems. J Veg Sci 16:191–198

Gómez Ortiz A, Oliva M, Salvà Catarineu M et al (2013) The environmental protection of landscapes in the high semiarid Mediterranean mountain of Sierra Nevada National Park (Spain): historical evolution and future perspectives. Appl Geogr 42:227–239

Gutiérrez Larena B, Fuertes Aguilar J, Nieto Feliner G (2002) Glacial-induced altitudinal migrations in Armeria (Plumbaginaceae) inferred from patterns of chloroplast DNA haplotype sharing. Mol Ecol 11:1965–1974

Guzmán B, López CMR, Forrest A et al (2015) Protected areas of Spain preserve the neutral genetic diversity of Quercus ilex L. irrespective of glacial refugia. Tree Genet Genomes 11:1–18

Herrera CM (1985) Determinants of plant-animal coevolution: the case of mutualistic dispersal of seeds by vertebrates. Oikos 44:132–141

Huck S, Büdel B, Kadereit JW et al (2009) Range-wide phylogeography of the European temperate-montane herbaceous plant Meum athamanticum Jacq.: evidence for periglacial persistence. J Biogeogr 36:1588–1599

Jiménez Albarral JJ, Pleguezuelos JM, Santos X (2020) Positive effect of catastrophic winds on reptile community recovery in pine plantations. Basic Appl Ecol 46:1–9

Kaliontzopoulou A, Pinho C, Harris DJ et al (2011) When cryptic diversity blurs the picture: a cautionary tale from Iberian and North African Podarcis wall lizards. Biol J Linn Soc 103:779–800

Kaliontzopoulou A, Carretero MA, Llorente G (2012) Morphology of Podarcis wall lizards from the Iberian Peninsula and North Africa: patterns of variation in a putative cryptic species complex. Zool J Linn Soc 164:173–193

Konstantinos K, Kokkoris IP, Panitsa M et al (2020) Spatial phylogenetics, biogeographical patterns and conservation implications of the endemic flora of Crete (Aegean, Greece) under climate change scenarios. Biology 8:199

Körner C (2021) Alpine plant life. 3rd ed. Springer, Cham

Kropf M, Comes HP, Kadereit JW (2006) Long-distance dispersal vs vicariance: the origin and genetic diversity of alpine plants in the Spanish Sierra Nevada. New Phytol 172:169–184

Kropf M, Comes HP, Kadereit JW (2008) Causes of the genetic architecture of south-west European high mountain disjuncts. Plant Ecol Divers 1:217–228

Kunin WE, Gaston KJ (eds) (1997) The biology of rarity: causes and consequences of rare—common differences. Springer, Dordrecht

Lamprecht A, Pauli H, Fernández Calzado MR et al (2021) Changes in plant diversity in a water-limited and isolated high-mountain range (Sierra Nevada, Spain). Alp Bot 131:27–39

Lizana M, Pedraza EM (1998) The effects of UV-B radiation on toad mortality in mountainous areas of central Spain. Conserv Biol 12:703–707

Lobo JM, Castro I, Moreno JC (2001) Spatial and environmental determinants of vascular plant species richness distribution in the Iberian Peninsula and Balearic Islands. Biol J Linn Soc 73:233–253

Lorite J (2016) An updated checklist of the vascular flora of Sierra Nevada (SE Spain). Phytotaxa 261:1–57

Lorite J, Ruiz M, Plaza-Arregui L (2019) Conservación ex-situ e in-situ. In: Peñas, J. & Lorite J. (eds.). Biología de la conservación de plantas en Sierra Nevada: Principios y retos para su preservación. pp 247–288. Editorial Universidad de Granada. Granada.

Lorite J, Ros-Candeira A, Alcaraz-Segura D et al (2020) FloraSNevada: a trait database of the vascular flora of Sierra Nevada, southeast Spain. Ecology 101:ecy.03091

Losapio G, Schöb C (2017) Resistance of plant–plant networks to biodiversity loss and secondary extinctions following simulated environmental changes. Funct Ecol 31:1145–1152

Losapio G, Schöb C (2020) Pollination interactions reveal direct costs and indirect benefits of plant–plant facilitation for ecosystem engineers. J Plant Ecol 13:107–113

Losapio G, Norton Hasday E, Espadaler X et al (2021a) Facilitation and biodiversity jointly drive mutualistic networks. J Ecol 109:2029–2037

Losapio G, Fortuna MA, Bascompte J et al (2019) Plant interactions shape pollination networks via nonadditive effects. Ecology 100: e02619

Losapio G, Schmid B, Bascompte J et al (2021b) An experimental approach to assessing the impact of ecosystem engineers on biodiversity and ecosystem functions. Ecology 102:e03243

Lozano FD, Herbada DG, Rivero LM et al (2000) Areas of high floristic endemism in Iberia and the Balearic islands: an approach to biodiversity conservation using narrow endemics. Belg J Entomol 2:171–185

Machordom A, Suarez J, Almodovar A et al (2000) Mitochondrial haplotype variation and phylogeography of Iberian brown trout populations. Mol Ecol 9:1325–1338

Maia-Carvalho B, Gonçalves H, Ferrand N et al (2014) Multilocus assessment of phylogenetic relationships in Alytes (Anura, Alytidae). Mol Phylogenet Evol 79:270–278

Martín-Hernanz S, Martínez-Sánchez S, Albaladejo RG et al (2019) Genetic diversity and differentiation in narrow versus widespread taxa of Helianthemum (Cistaceae) in a hotspot: the role of geographic range, habitat, and reproductive traits. Ecol Evol 9:3016–3029

Matías L, Zamora R, Mendoza I, Hódar JA (2010) Seed dispersal patterns by large frugivorous mammals in a degraded mosaic landscape. Restor Ecol 18(5):619–627

Médail F, Diadema K (2009) Glacial refugia influence plant diversity patterns in the Mediterranean Basin. J Biogeogr 36:1333–1345

Médail F, Quézel P (1997) Hot-spots analysis for conservation of plant biodiversity in the Mediterranean Basin. Ann MO Bot Gard 84:112–127

Médail F, Quézel P (1999) Biodiversity hotspots in the Mediterranean Basin: setting global conservation priorities. Conserv Biol 13:1510–1513

Medrano M, López-Perea E, Herrera CM (2014) Population genetics methods applied to a species delimitation problem: endemic trumpet daffodils (Narcissus section Pseudonarcissi) from the southern Iberian Peninsula. Int J Plant Sci 175:501–517

Millán A, Picazo F, Sánchez-Fernández D et al (2013) Los Coleópteros acuáticos amenazados. In: Ruano F, Tierno de Figueroa M, Tinaut A (eds) Los Insectos de Sierra Nevada. 200 años de historia. Asociación Española de Entomología, Granada, pp 443–456

Mira Ó, Sánchez-Prieto CB, Dawson DA et al (2017) Parnassius apollo nevadensis: identification of recent population structure and source–sink dynamics. Conserv Genet 18:837–851

Miraldo A, Faria C, Hewitt GM et al (2012) Genetic analysis of a contact zone between two lineages of the ocellated lizard (Lacerta lepida Daudin 1802) in south-eastern Iberia reveals a steep and narrow hybrid zone. J Zool Syst Evol Res 51:45–54

Mishler BD, Knerr N, González-Orozco CE et al (2014) Phylogenetic measures of biodiversity and neo-and paleoendemism in Australian Acacia. Nat Commun 5:1–10

Molina-Venegas R, Roquet C (2014) Directional biases in phylogenetic structure quantification: a Mediterranean case study. Ecography 37:572–580

Molina-Venegas R, Aparicio A, Pina FJ et al (2013) Disentangling environmental correlates of vascular plant biodiversity in a Mediterranean hotspot. Ecol Evol 3:3879–3894

Molina-Venegas R, Aparicio A, Lavergne S et al (2015) The building of a biodiversity hotspot across a land-bridge in the Mediterranean. Proc R Soc B Biol Sci 282:20151116

Molina-Venegas R, Aparicio A, Slingsby JA et al (2015) Investigating the evolutionary assembly of a Mediterranean biodiversity hotspot: deep phylogenetic signal in the distribution of eudicots across elevational belts. J Biogeogr 42:507–518

Molina-Venegas R, Aparicio A, Lavergne S et al (2016) How soil and elevation shape local plant biodiversity in a Mediterranean hotspot. Biodivers Conserv 25:1133–1149

Molina-Venegas R, Aparicio A, Lavergne J et al (2017) Climatic and topographical correlates of plant palaeo- and neoendemism in a Mediterranean biodiversity hotspot. Ann Bot 119:229–238

Molina-Venegas R, Aparicio A, Lavergne S et al (2018) Soil conditions drive changes in a key leaf functional trait through environmental filtering and facilitative interactions. Acta Oecol 86:1–8

Monnet AC, Cilleros K, Médail F et al (2021) WOODIV, a database of occurrences, functional traits, and phylogenetic data for all Euro-Mediterranean trees. Sci Data 8:89

Moreno Saiz JC, Donato M, Katinas L et al (2013) New insights into the biogeography of south-western Europe: spatial patterns from vascular plants using cluster analysis and parsimony. J Biogeogr 40:90–104

Morueta-Holme N, Engemann K, Sandoval-Acuña P et al (2015) Strong upslope shifts in Chimborazo's vegetation over two centuries since Humboldt. Proc Nat Acad Sci 112:12741–12745

Mota JF, Pérez-García FJ, Jiménez ML et al (2002) Phytogeographical relationships among high mountain areas in the Baetic Ranges (South Spain). Glob Ecol Biogeogr 11:497–504

Mota JF, Medina-Cazorla JM, Navarro FB et al (2008) Dolomite flora of the baetic ranges glades (South Spain). Flora 203:359–375

Muñoz-Pajares AJ, García C, Abdelaziz M et al (2017) Drivers of genetic differentiation in a generalist insect-pollinated herb across spatial scales. Mol Ecol 26:1576–1585

Muñoz-Pajares AJ, Abdelaziz M, Picó FX (2020) Temporal migration rates affect the genetic structure of populations in the biennial Erysimum mediohispanicum with reproductive asynchrony. AoB Plants 12:plaa037

Múrria C, Sáinz-Bariáin M, Vogler AP et al (2020) Vulnerability to climate change for two endemic high-elevation, low-dispersive

Annitella species (Trichoptera) in Sierra Nevada, the southernmost high mountain in Europe. Insect Conserv Divers 13:283–295
Nieto Feliner G (2014) Patterns and processes in plant phylogeography in the Mediterranean Basin. A review. Perspect Plant Ecol Evol Syst 16:265–278
Olivares FJ, Barca-Azcón JM, Pérez-López FJ et al (2011) Las mariposas diurnas de Sierra Nevada. Consejería de Medio Ambiente, Junta de Andalucía, Granada
Pallarés S, Millán A, Mirón JM et al (2020) Assessing the capacity of endemic alpine water beetles to face climate change. Insect Conserv Divers 13:271–282
Pascual J, Blanco S, García-López M et al (2016) Assessing bacterial diversity in the rhizosphere of Thymus zygis growing in the Sierra Nevada National Park (Spain) through culture-dependent and independent approaches. PLoS One 11(1):e0146558
Peñas J, Lorite J (2019) Biología de la conservación de plantas en Sierra Nevada. Editorial Universidad de Granada, Granada, Principios y retos para su preservación
Peñas J, Pérez-García FJ, Mota JF (2005) Patterns of endemic plants and biogeography of the Baetic high mountains (south Spain). Acta Bot Gallica 152:347–360
Peredo EL, Revilla MÁ, Jiménez-Alfaro B et al (2009) Historical biogeography of a disjunctly distributed, Spanish alpine plant, Senecio boissieri (Asteraceae). Taxon 58:883–892
Pérez JM, Ruiz-Martínez I, Granados JE et al (1997) The dynamics of sarcoptic mange in the ibex population of Sierra Nevada in Spain—influence of climatic factors. J Wildl Res 2:86–89
Picó FX, Riba M (2002) Regional-scale demography of Ramonda myconi: remnant population dynamics in a preglacial relict species. Plant Ecol 161:1–13
Piñero FS, Tinaut A, Aguirre-Segura A et al (2011) Terrestrial arthropod fauna of arid areas of SE Spain: diversity,biogeography, and conservation. J Arid Environ 75:1321–1332
Piñeiro A (2017) Topillo nival—Chionomys nivalis. In: Salvador A, Barja I (eds) Enciclopedia Virtual de los Vertebrados Españoles. Museo Nacional de Ciencias Naturales, Madrid
Pistón N, Armas C, Schöb C et al (2015) Phylogenetic distance among beneficiary species in a cushion plant species explains interaction outcome. Oikos 124:1354–1359
Pistón N, Schöb C, Armas C et al (2016) Contribution of co-occurring shrub species to community richness and phylogenetic diversity along an environmental gradient. Perspect Plant Ecol Evol Syst 19:30–39
Pistón N, Michalet R, Schöb C et al (2018) The balance of canopy and soil effects determines intraspecific differences in foundation species' effects on associated plants. Funct Ecol 32:2253–2263
Pugnaire FI, Losapio G, Schöb C (2021) Species interactions involving cushion plants in high-elevation environments under a changing climate. Ecosistemas 30:2186
Qian H, Jin Y (2021) Are phylogenies resolved at the genus level appropriate for studies on phylogenetic structure of species assemblages? Plant Divers 43:255–263
Rabinowitz D (1981) Seven forms of rarity. In: H Synge (ed) Biological aspects of rare plant conservation. Wiley, New York
Ramírez-Valiente JA, Robledo-Arnuncio JJ (2015) Exotic gene flow affects fitness trait values but not levels of heritable trait variation in the southernmost population of Scots pine (Pinus sylvestris L. var. nevadensis). Biol Conserv 192:331–342
Recuero E, Canestrelli D, Vörös J et al (2012) Multilocus species tree analyses resolve the radiation of the widespread Bufo bufo species group (Anura, Bufonidae). Mol Phylogenet Evol 62:71–86
Ribera I, Castro A, Díaz JA et al (2011) The geography of speciation in narrow-range endemics of the 'Haenydra' lineage (Coleoptera, Hydraenidae, Hydraena). J Biogeogr 38:502–516
Ribera I, Vogler AP (2004) Speciation of Iberian diving beetles in Pleistocene refugia (Coleoptera, Dytiscidae). MolEcol 13:179–193
Ruano F, Tierno de Figueroa J, Tinaut A (2013) Los insectos de Sierra Nevada: 2000 años de historia. Granada, Asociación Española de Entomología
Sánchez-Herráiz MJ, Barbadillo LJ, Machordom A et al (2000) A new species of Pelodytid frog from the Iberian Peninsula. Herpetologica 56:105–118
Sanllorente O, Ruano F, Tinaut A (2015) Large-scale population genetics of the mountain ant Proformica longiseta (Hymenoptera: Formicidae). Popul Ecol 57:637–648
Santamaría S, Galeano J, Pastor JM et al (2014) Robustness of alpine pollination networks: effects of network structure and consequences for endemic plants. Arct Antarct Alp Res 46:568–580
Santamaría S, Méndez M (2009) Cómo conservar las interacciones entre polinizadores y plantas en ecosistemas de alta montaña con ayuda de la teoría de redes complejas. Proyectos de investigación en parques nacionales 2009–2012. OAPN, Madrid, pp 111–122
Santos X, Roca J, Pleguezuelos JM et al (2008) Biogeography and evolution of the smooth snake Coronella austriaca (Serpentes: Colubridae) in the Iberian Peninsula: evidence for Messinian refuges and Pleistocenic range expansions. Amphibia-Reptilia 29:35
Santos X, Brito JC, Caro J et al (2009) Habitat suitability, threats and conservation of isolated populations of the smooth snake (Coronella austriaca) in the southern Iberian Peninsula. Biol Conserv 142:344–352
Santos X, Rato C, Carranza S et al (2012) Complex phylogeography in the Southern Smooth Snake (Coronella girondica) supported by mtDNA sequences. J Zool Syst Evol Res 50:210–219
Sanz M, Schönswetter P, Vallès J et al (2017) Glacial survival in and recent long-distance dispersal to the Iberian Mountains: the phylogeographic history of Artemisia umbelliformis (Asteraceae). Bot J Linn Soc 183:587–599
Schöb C, Macek P, Pistón N et al (2017) A trait-based approach to understand the consequences of specific plant interactions for community structure. J Veg Sci 28:696–704
Simón-Porcar VI, Escudero M, Navarro L et al (2018) Using floristics, modern systematics and phylogenetics for disentangling biodiversity hotspots across scales: a Mediterranean case study. Plant Biosyst 152:1293–1310
Skidmore AK, Pettorelli N, Coops NC et al (2015) Environmental science: agree on biodiversity metrics to track from space. Nat News 523:403
Skidmore AK, Coops NC, Neinavaz E et al (2021) Priority list of biodiversity metrics to observe from space. Nat Ecol Evol 5:896–906
Soranzo N, Alia R, Provan J et al (2000) Patterns of variation at a mitochondrial sequence-tagged-site locus provides new insights into the postglacial history of European Pinus sylvestris populations. Mol Ecol 9:1205–1211
Stebbins GL, Major J (1965) Endemism and speciation in the California flora. Ecol Monogr 35:2–35
Swenson NG (2019) Phylogenetic ecology. University of Chicago Press, Chicago
Swenson NG, Umaña MN (2014) Phylofloristics: an example from the Lesser Antilles. J Plant Ecol 7:166–175
Thompson JD (2020) Plant evolution in the Mediterranean, 2nd edn. Oxford University Press, Oxford
Tinaut A, Ruano F, Ros JL et al (2008) Sierra Nevada. La mayor biodiversidad de Iberia. Nova Ciencia 34:22–25
Tinaut A (2013) Las hormigas. In: Ruano F, Tierno de Figueroa M, Tinaut A (eds) Los Insectos de Sierra Nevada, 200 años de historia. Asociación Española de Entomología, Granada, pp 392–409

Todisco V, Gratton P, Cesaroni D et al (2010) Phylogeography of Parnassius apollo: hints on taxonomy and conservation of a vulnerable glacial butterfly invader. Biol J Linn Soc 101:169–183
Tonzo V, Ortego J (2021) Glacial connectivity and current population fragmentation in sky islands explain the contemporary distribution of genomic variation in two narrow-endemic montane grasshoppers from a biodiversity hotspot. Divers Distrib 27:1619–1633
Tonzo V, Bellvert A, Ortego J (2021) Reticulate evolutionary history in a recent radiation of montane grasshoppers revealed by genomic data. BioRxiv
Unger GM, Vendramin GG, Robledo-Arnuncio JJ (2014) Estimating exotic gene flow into native pine stands: zygotic vs. gametic components. Mol Ecol 23:5435–5447
Valbuena-Carabaña M, Gil L (2013) Genetic resilience in a historically profited root sprouting oak (Quercus pyrenaica Willd.) at its southern boundary. Tree Genet Genomes 9:1129–1142
Valdés B, Talavera S, Fernández-Galiano E (eds) (1987) Flora vascular de Andalucía occidental (3 Vol). Ketres editora, Barcelona
Valiente-Banuet A, Verdú M (2013) Plant facilitation and phylogenetics. Annu Rev Ecol Evol Syst 44:347–366
Valiente-Banuet A, Aizen MA, Alcántara JM et al (2015) Beyond species loss: the extinction of ecological interactions in a changing world. Funct Ecol 29:299–307
Vargas P, Jiménez-Mejías P, Fernández-Mazuecos M (2020) 'Endangered living fossils'(ELFs): long-term survivors through periods of dramatic climate change. Environ Exp Bot 170:103892
Vela JM Bastazo G (2013) Los Crisomélidos. In: Ruano F, Tierno de Figueroa M, Tinaut A (eds) Los insectos de Sierra Nevada, 200 años de historia. Asociación Española de Entomología, Granada, pp 118–145
Verdú JR, Numa C, Galante E (2011) Atlas y libro rojo de los invertebrados amenazados de España (especies vulnerables). Dirección General de Medio Natural y Polıtica Forestal, Ministerio de Medio Ambiente, Medio Rural y Marino, Madrid
Verdú JR, Galante E (eds) (2009) Atlas de los invertebrados amenazados de España (especies en peligro crítico y en peligro). Organismo Autónomo Parques Nacionales, Madrid
Von Humboldt A (1807) Essai sur la géographie des plantes: accompagné d'un tableau physique des régions équinoxiales, fondé sur des mesures exécutées, depuis le dixième degré de latitude boréale jusqu'au dixième degré de latitude australe, pendant les années 1799, 1800, 1801, 1802 et 1803. Schoell, Paris
Zamora R, Barea-Azcón JM (2015) Long-term changes in mountain passerine bird communities in the Sierra Nevada (southern Spain): a 30-year case study. Ardeola 62:3–18
Zamora R, Gómez JM (1993) Vertebrate herbivores as predators of insect herbivores: an asymmetrical interaction mediated by size differences. Oikos 66:223–228
Zamora R, Matías L (2014) Seed dispersers, seed predators, and browsers act synergistically as biotic filters in a mosaic landscape. PLoS One 9(9):e107385

# Singular Cultural Landscapes of the Sierra Nevada

José Mª Martín Civantos, Jorge Rouco Collazo, José Abellán Santisteban, Blas Ramos Rodríguez, Agustín Sánchez García, Sergio Martos Rosillo, and Antonio González Ramón

**Abstract**

Human activity has been fundamental in shaping the natural landscapes of Sierra Nevada. This chapter offers an overview of the evolution of the occupation of this mountain range and of the changes in the ways of managing and exploiting its resources. The study spans the still poorly known Metal Ages and Roman period to medieval times, a period characterized by a great transformation that marked Sierra Nevada cultural landscape due in large part to the introduction of irrigation systems and a complex socio-ecosystem that conditioned most economic activities. The current analysis also delves into the changes provoked by the Castilian conquest, the arrival of the new settlers, and the later Industrial Revolution and application of liberal policies. The irrigation systems reveal a great resilience, surviving even today and playing a key role in Sierra Nevada ecosystems and landscapes.

**Keywords**

History • Archaeology • Cultural landscape • Traditional territorial and resource management • Irrigation systems • Middle ages • Modern era

J. Mª. Martín Civantos (✉) · J. Rouco Collazo · J. Abellán Santisteban · B. Ramos Rodríguez
MEMOLab, Biocultural Archaeology Laboratory. Medieval History Department, Universidad de Granada, Granada, Spain
e-mail: josemaria.martin@civantos.com

A. Sánchez García
ERC Project ANTIGONE, Università Degli Studi Di Genova, Genova, Italy

S. Martos Rosillo · A. González Ramón
Instituto Geológico y Minero de España, Madrid, Spain

## 1 Sierra Nevada as a Cultural Landscape: A Coevolutive Process, a Transformed Mountain Range

Sierra Nevada landscape has its roots in both natural and historical processes. It is the result of the interaction between human beings and nature over hundreds and thousands of years. It is a co-evolutionary process in which humans adapted to the conditions of the mountains that can undoubtedly be classified as unfavourable, especially to the development of certain productive activities, and the establishment of stable settlements and community and communication networks. Even today these pitfalls are experienced although our perception, obviously, does not equate with that of Pre-industrial societies or the poorly developed modernization processes still present in the Sierra Nevada until at least the 1970s. And yet, at the same time, humans were able to introduce great modifications throughout history. Certain were radical: water and irrigation management systems, deforestation processes and alterations of land use that often become more difficult to identify as one goes farther back in time. This caused nature itself to progressively adapt to the changes, generating new balances and relationships between the different abiotic, biotic and cultural elements. This led to a process of mutual adaptation, a co-evolution that transformed the original ecosystems and natural conditions generating new socio-ecosystems that can often be discerned even today regardless of the more recent transformations (see chapter "Local Ecological Knowledge and the Sustainable Co-Management of Sierra Nevada's Social-Ecological System"). They correspond to cultural landscapes which have to a great extent paved the way to the uniqueness of Sierra Nevada and a good part of its values.

Not all societies historically share the same relationship with the environment. They have not only depended on their capacity of intervention or transformation due to the development of productive forces or technology. They actually rhyme with different strategies and preferences when



R. Zamora and M. Oliva (eds.), *The Landscape of the Sierra Nevada*,
https://doi.org/10.1007/978-3-030-94219-9_3

exploiting territories and their resources and experienced changes of density of occupation and intensity of use, as well as productive options and practices. As has been noted, certain stages left indelible marks that greatly conditioned subsequent developments and the very evolution of ecosystems in this co-evolutionary process. Other periods experienced processes of abandonment and renaturation, conditioned in any case by previous phases where certain spaces or resources saw an intense exploitation.

It is thus necessary to recognize that landscapes visible today stem from a historical process resulting from a (not always positive) cumulative process. It is, in short, a stratified landscape like that of an archaeological site that is nonetheless very extensive in space and time. It is a process marked by a recurrence of contrasting landscapes built not only by the addition of new elements, but by destroying and cancelling features of earlier landscapes in function of the strategies, needs or capacities of each new society or period (Martín Civantos 2006; Brogiolo 2007).

However, the most significant element with the greatest influence throughout all of Sierra Nevada and its surroundings is water and its management. This resource served to structure the traditional landscape and organization of the territory. It served as the main fundament of the socio-ecosystems that shaped the current image of these mountains and, without a doubt, serves to attract the greatest notice of any minimally attentive observer. It is also a clear example of the different social and productive options referred to above. It is not the only factor to take into account when analysing the relationship of history with the environment. Although the most obvious element is altitude level, there is also the factor of humidity marked by a gradual eastward decrease in rainfall (both rain and snow) and therefore water availability. A notable difference also exists between the northern and southern watersheds (to which can be added the westward slopes leaning towards Granada and its plain and the Padul-Lecrín Valley lowlands) that directly influenced the occupation of the slopes and foothills, and structured the settlement patterns and exploitation of the territory mainly through agricultural activities. Moreover, key determinants at the sub-basin scale are sun exposure and the variations between shaded and sunlit areas in the valleys, as well as geomorphology that conditioned the slopes and soil formation and steep rocky outcrops. These variables fall in line with differences of resource use, the creation of productive spaces and territorial organization which include settlement patterns and the scope of their boundaries. All form part of the local ecological knowledge assembled by the communities inhabiting and toiling in Sierra Nevada, shaping the landscapes and yielding new socio-ecosystems with extraordinary environmental, cultural and social values.

## 2 The Historical Evolution of the Occupation of Sierra Nevada

### 2.1 From Prehistory to the Fall of the Roman Empire: Little Data or Little Presence?

There is little evidence of the periods preceding the Middle Ages in the Sierra Nevada. Furthermore, our grasp of these periods from area to area remains uneven. As noted above, the Sierra Nevada is a humanized mountainous space which was never hostile or represented an insurmountable barrier. It, on the contrary, was a sector whose resources were exploited to a greater or lesser extent depending on the interests or strategies of each society and each period. Thus, it is possible to refer to a use, rather than an effective occupation, since at least the Bronze Age, and most probably already during the Copper Age. Phases of occupation of this earlier time frame, still poorly recorded and characterized in detail, have been identified at the sites of Zamarriche (Dólar) and Puntal de Alquife.

A more widespread presence from the subsequent Bronze Age is evidenced by settlements from the Argaric Culture in the foothills, and at least along their northern and western faces. These include sites such as Cerro de la Encina (Monachil), Zamarriche (Dólar) and Alrután (Jérez del Marquesado). The southern face, in turn, is practically devoid of occupations probably due to the lack of systematic fieldwork (Fig. 1). It is nonetheless possible to speculate as to a lower intensity of occupation in this sector, at least in specific areas such as Tajo del Águila (Juviles), Peñón de la Reina (Alboloduy), Peñón Hundido (Tímar) and Castillejo de Escariantes (Ugíjar). These settlements were presumably linked to modest copper and silver mining operations such as at Lanteira and Jérez del Marquesado, and most likely also took part in raising mountain livestock. The only features evidencing livestock activity are the numerous petroglyphs, arduous to interpret, dotting the Sierra Nevada. Although research on them is scant, and their chronology still ill-defined, these engravings of the faces of schists appear to be linked to livestock transfer. This could explain the presence of *cocones*, rounded hollows carved into the rock that are traditionally interpreted as serving to collect dew water for shepherds. However, the significance and function of these rock cuttings are probably much more complex. The best example is in Dílar at the Peñón de las Letras (Peña Madura) (Redondo Ortega and Martínez Rodríguez 2021), initially interpreted as a petroglyph of 'Atlantic style'.

If evidence of Prehistoric times is scarce in the Sierra Nevada, that of the Iron Age is even more so. Traces of the Iberian Culture are practically absent except for a few finds associated with settlements in the foothills of the northern slopes (Martín Civantos 2007: 619–623; Adroher Auroux

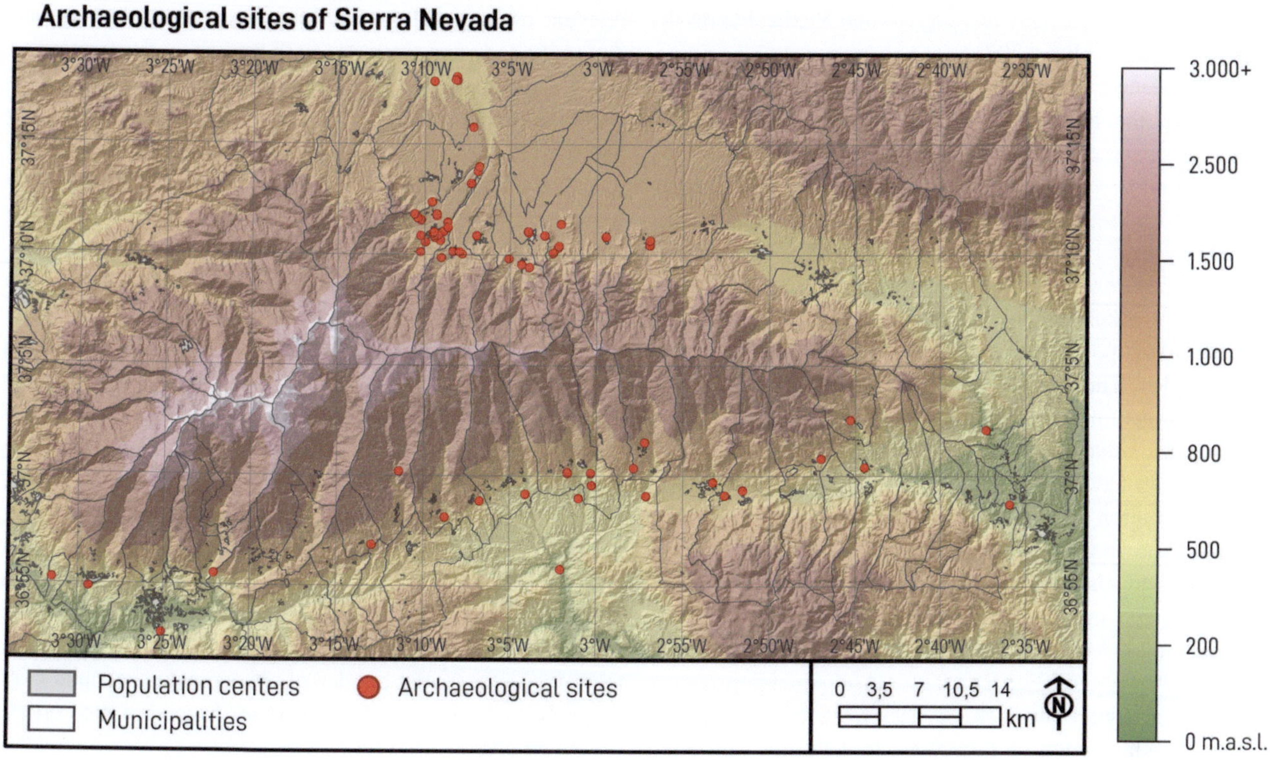


**Fig. 1** Archaeological sites of Sierra Nevada

1990) at Las Juntas (Abla), La Calera (Dólar), El Cardal (Ferreira), Campo de Marte (Lanteira) and Arruta (Jérez del Marquesado). Most reveal clear links with mining and iron, silver and copper working. However, it is still impossible to advance ideas as to the occupation or use of the high mountain areas during this period, although raising livestock and grazing were certainly key.

Many of these settlements persisted throughout the Roman Republican period at the end of the first century BC. They coincided with the creation of the Roman colony of Guadix which provoked organizational changes to the Sierra's northern area and the abandonment of mining operations in favour of other exploitations elsewhere in the Iberian Peninsula (Martín Civantos 2005, 2007; Bertrand and Sánchez Viciana 2008). The southern slope offers at least partial data from subsequent Roman Imperial times based on pottery finds at Castillejo (Órgiva) and the Peñón Hundido (Tímar), as well as sporadic finds at later medieval fortified features such as the Fuerte de Juviles, Piedra Fuerte de Yegen (Alpujarra de la Sierra), Escariantes (Ugíjar), Guarros (Paterna del Río), Ohanes and Marchena (Terque) (Rouco Collazo 2021). Yet again, the lack of systematic explorations of many areas of the Sierra Nevada impedes gaining a grasp of these periods. It nonetheless is obvious that the entire region did not arouse any special interest during the Roman Empire (with certain exceptions, particularly the Vega of Granada as evidenced by recent finds). In fact, new research reveals that the mountainous and semi-arid areas of Eastern Andalusia remained marginal when compared to those of the Guadalquivir Valley (Román Punzón and Martín Civantos 2014).

## 2.2 The 'Assault' of the Sierra Nevada During Late Antiquity

Clear evidence of a real occupation of Sierra Nevada is not evident until the definitive crisis of the Roman Empire in the fifth century. It is possible to view this period in many ways as the veritable 'assault of the mountain' as the crisis at this time explains what is apparently a move into its mountainous sectors. This phenomenon, visible already in the fifth century with the abandonment of many rural Roman settlements in the flat lands and along the Mediterranean coast itself, became totally manifest later in the sixth century. New settlements on the higher and more isolated slopes are more difficult to identify in part due to their topography, but mainly due to their scarcer material culture. Moreover, there is a dearth of systematic fieldwork throughout many of these areas, especially in the Alpujarra region, that is the southern slopes, where finds from Late Antiquity are scant. This period is only represented by specific sites such as the cemetery of Ana Cortijo of Órgiva (Trillo San José et al. 1999) and surface potsherds at several fortified features

(Escariantes, Guarros, Juviles, Órgiva and Piedra Fuerte de Yegen) (Rouco Collazo 2021).

Yet evidence of occupations from Late Antiquity are clear along the northern face of the Sierra Nevada. The Zenete area, for example, is dotted by numerous sites that sprang up either ex novo or reoccupied earlier protohistoric or prehistoric settlements. Among the ex-novo sites are Alcázar (Jérez del Marquesado), Loma Bermite (La Peza), El Castillejo (La Peza) and Castillejo Abrucena. Those capping earlier occupations include Zamarriche (Dólar), El Cardal (Ferreira), Campo de Marte (Lanteira) and Arruta (Jérez del Marquesado). They represent a clear change in settlement patterns, largely breaking with prior phases in spite of cases occasionally stretching over time as late as the Arab conquest (Martín Civantos 2007).

This evolution is likewise linked to changes in productive systems. The most obvious is the reactivation of mining which led to the reopening of numerous metallurgical workshops and mines that had been abandoned in Roman Imperial times. Many are small and directly related to settlements cited above (Fig. 2).

Again, although the lack of data from other areas of Sierra Nevada renders it hazardous to interpret this phenomenon throughout the whole of the massif, a general pattern does progressively emerge. This pattern is also in fact evidenced by toponymy, which in this case is particularly relevant as many of the current place names date to Late Antiquity and stem from the sixth century in the framework of the 'assault of the mountain'. The names Ferreira, Ferreirola and Lanteira, for example, can be linked to mining and metallurgy. The last derives from '*argentaira*' literally meaning 'silverwork'. They likewise derive from raising livestock (Poqueira in its Arab form is '*Buqayra*' meaning 'dairy') or Christianization processes (Capileira and Capilerilla are derivatives of 'chapel' and Monachil comes from '*Monastil*' meaning 'monastery').

In many cases they are derivations of late vulgar Latin used at the time. This is the case of famous place names ending in '*-eira*', as well as many others ending in '*-ar*' identified originally as having a Mozarabic origin (Martínez Ruiz 1987), but in reality, stem from Late Antiquity. Examples are presumably Cáñar, Soportújar, Almegíjar, Busquistar, Cádiar, Tímar, Ugíjar, Canjáyar, Laujar, Íllar and Bacares and probably other derivations that underwent a number of transformations (e.g., Bayacas, Carataunas and Cástaras). All of these (and many others evidenced by microtoponymy) offer an idea of the extensive occupation of Sierra Nevada since the sixth century framed in the context described above.

To these places names can be added a lengthy series of Latin anthroponyms stemming from the names of Roman landowners that have survived even if the settlements

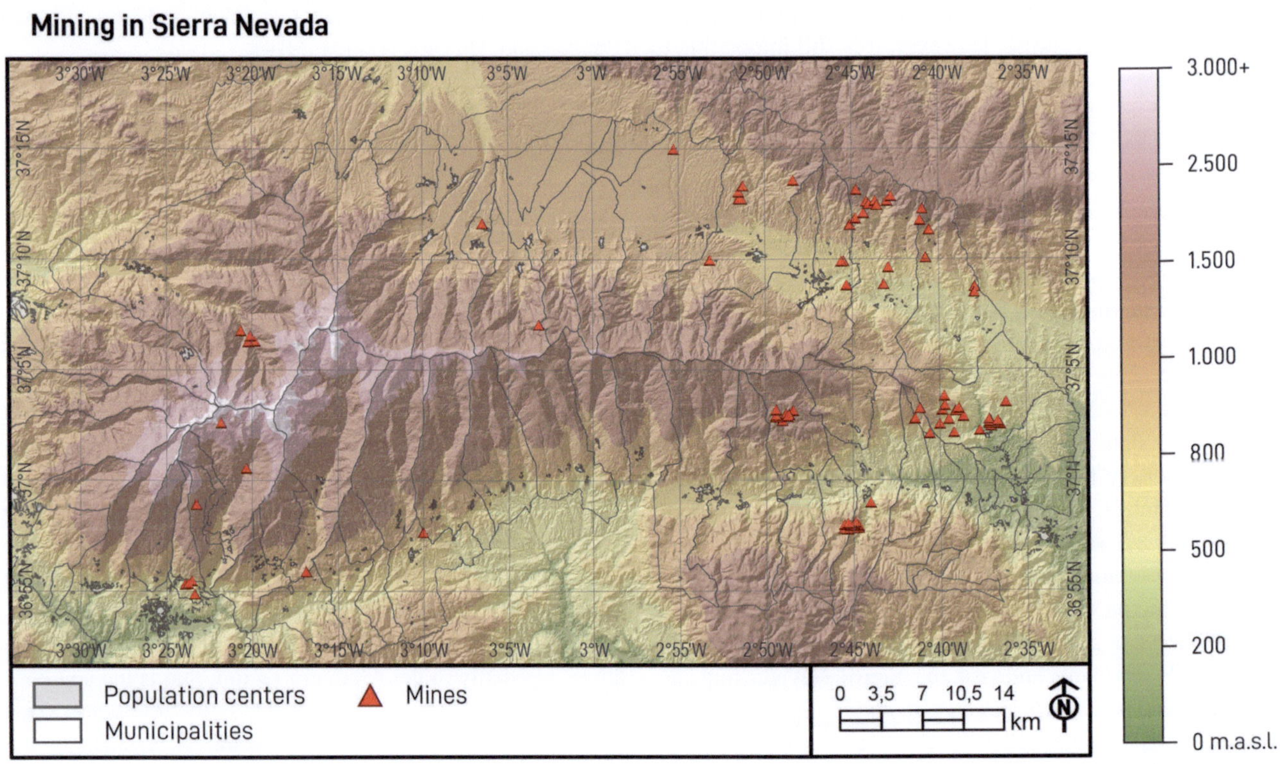


**Fig. 2** Mining sites in Sierra Nevada

changed their location. Examples are Abrucena, from the lands of '*Laurusana*' (related to a certain *Laurus*). Other cases are Fiñana (*Finius*), Bartillana in Jérez del Marquesado, Picena in Nevada, Juliana in Murtas or Marchena. It is arduous to determine if these place names were preserved due to a partial presence of the old aristocratic landowning families. Mining is again certainly a key factor in identifying these elites, at least in cases where the exploitations include silver and copper such as at Lanteira, Jérez del Marquesado and La Peza. It is also compelling to link these place names to a subsequent participation during the *fitna* of the ninth century by a population of Hispano-Roman origin who rebelled against the Umayyad emirs in an attempt to maintain their autonomy and control over the territory in the face of an attempt by the State at centralization. It is here that the aristocracies with the capacity to build fortifications obliged the Emirs of Cordoba to displace themselves and lay siege to quell the revolt. This is the case of Fiñana, Marchena and Juviles and can potentially be connected to many of the other sites identified as fortifications (Martín Civantos 2007; Rouco Collazo 2021) (Fig. 3).

Furthermore, of special interest are the toponyms associated with the Christianization process in mountainous rural areas. Mentioned above are place names linked to chapels (Capileira and Capilerilla) and monasteries (Monachil). This process can also be traced to the proliferation of names such as Aldeire stemming from the Arabic *al-dayr* which precisely designates a monastery (Martínez Ruiz 1987). Adjacent to this town of the Zenete area are *cortijos* (country estates), *pagos* (farming districts) and *eras* (threshing floors) bearing identical names to those of Fiñana, Güéjar Sierra, Capileira and Bérchules. To these can be added other microtoponyms that have since disappeared such as *Kanisa/Quiniçia* ('church') or *al-Qasis* ('priest') which appear as *pagos* and territorial districts cited in al-Andalus and Morisco written records.

**Fig. 3** Castle of Beires

## 2.3 Islamic Conquest and Territorial Transformation During the Al-Andalus Period

The conditions described above served as foundations for the Arab conquest and colonization that, accompanied by Berber contingents, took place from the outset of the eighth century. Productive strategies with the arrival of the Arabs were radically altered by the development of hydraulic systems and the extension of irrigated agriculture, as well as by the design of *alquerías* (villages) and their surrounding domains, including communal lands (Fig. 4).

The real transformation of Sierra Nevada's environment undoubtedly was initiated at this time. The Arab conquest therefore represents the outset of a new phase marked by radically different forms of territorial and resources management based on an intensification of agricultural production. This was implemented through an extension of hydraulic features, the introduction of new crops, systems of recharging aquifers and new strategies of communal land and resource management. The most vital of these was undoubtedly that related to water. Mining played a secondary role, especially since the quelling of the *fitna*, a general revolt in the ninth century, and the proclamation of the Umayyad Caliphate in the early tenth century (Martín Civantos 2005, 2007). But as noted, the result was a heretofore unknown complex system of exploitation that radically altered the mountain range and peasant way of life.

There is evidence at this time of the installation of numerous Arab tribal groups of Qaisi and Yemeni origin. Berbers also most likely arrived albeit left no trace. The revolt of the ninth century cited above offers an image of a territory populated by different social groups coexisting more or less peacefully. The process of integration subsequent to the final triumph of the Umayyad State in the tenth century was fully established not only from the political standpoint, but above all from social and territorial points of view. Its main expression was consolidation of territories organized through *alquerías*, that is, rural communities adopting a communal management of the territory that apart from being the source of units of population also formed fiscal units. The development of workspaces (especially irrigation systems) and communal management of resources is one of the keys to grasping the blending, and especially the acculturation processes, that took place throughout the first centuries of al-Andalus rule.

**Fig. 4** Laroles, old *alquería*, and its agricultural terraces

Ahmad al-Rāzī in the tenth century offers a developed definition of the agriculture production and population of Sierra Nevada. The crops listed in his writings include hazelnuts, walnuts, citrus fruits, pomegranates and figs (Al-Rāzī 1974: 23–24, 291). Ibn Gālib, a contemporary author, also records that linen production was key in this mountain range (Ibn Gālib 1955, 285). As we have stated elsewhere, the hydraulic irrigation systems of the Sierra Nevada in the tenth century had fully come into being. Moreover, other evidence of this development stems from the vast silk production and other goods transiting through the Port of Pechina (Almería) since the ninth century (Martín Civantos 2007: 483–484).

The *alquerías* were organized into *aljamas*, that is, communities of residents managing the land which was, according to Islamic law, divided into several parts, notably *mamlūka* or privately owned lands and *mubāha*, lands of communal character (Trillo San José 1999, 12–13). Despite this theoretical communal approach, there is evidence that the use of land by residents of other *alquerías* required the permission of the *aljama*. This is the case of a fifteenth-century legal dispute between Jérez and Lanteira clearly stating that the inhabitants '*… of Al-yantaira asked those of Jérez and Alcázar permission to access and graze the highland pastures of two mountains. The residents of Jérez and Alcázar consented to this request to all who asked for access to these pastures by reason of their generosity, and good neighbourship and proximity. The sole condition was that the* qādī *certify that this was not a legal obligation, but due to their generosity, as noted before, and that they, the residents of Jérez and Alcázar, could block access to the pastures at any moment*' (González Palencia 1940: 348). The significance of a control of these mountains derives, as indicated by written records, from the need to manage the watersheds and, therefore, the headwaters of rivers serving to irrigate crops. The hydraulic systems at this time were nonetheless very complex and not limited exclusively to areas of intensive irrigation, although these represented the heart of productive spaces.

## 2.4 Changes After the Castilian Conquest: The Expulsion of the Moriscos and the Stagnation of the Region

The conquest by Castilian troops of Granada's Nasrid kingdom in 1492 marked the outset of a gradual and widespread transformation stemming from the arrival of new settlers with different perspectives as to the territory and the exploitation of agricultural resources. The transformation in general throughout the first years after the conquest was nonetheless marginal with the arrival of few Christian settlers.

The migration of Moriscos was then accelerated after the policy of forced conversion subsequent to the revolt of 1500 which culminated in a later great rupture sparked by the Morisco Rebellion (1568–1571). The main scene of this event, initially headed by Aben Humeya, was in Alpujarra. The repression after the Morisco defeat and expulsion culminated in a forced displacement of the vast majority of the population followed by an intense process of repopulation that did not, in fact, affect all the old nuclei.

New actors sprung up with the arrival of the new Christian settlers who filled the demographic void provoked by the forced emigration. This new population likewise led to cultural, hence socioeconomic, changes. Many of the al-Andalus management and exploitation practices throughout the territory were either modified or directly abandoned by the new inhabitants who, in spite of everything, attempted as best they could to settle these inhospitable and unknown lands. Evidence of a transfer of knowledge between the earlier and the new cultures is evidenced by the survival of Morisco families in many *alquerías* who served as the link ensuring the passing on of unknown techniques from the original to the new residents (Estévez Callejón 1995: 107–110).

The many preserved local *Libros de Apeo y Repartimiento*, record the transfer of properties, indicate that the redistributions were carried out along relatively egalitarian parameters, with the settlers receiving plots of land of diverse quality. These small highly dispersed properties forced the new farmers to adopt a wide range of livestock or agriculture ventures to parcels of different nature, that is, irrigated or dry lands, vineyards, etc. These private properties were not the only means to glean resources from nature as there were also communal lands (forests, meadows, etc.) that complemented the local subsistence economies (González de Molina 1996: 10).

Later records from the mid-eighteenth century reveal agro-pastoral societies still based on subsistence. It is in this period that there are references, often for the first time, of *propios* (common lands), that is, communal lands of the councils and residents. *Propios*, as noted above, served as a fundamental pillar among local economies as spaces to raise livestock, exploit forests and for agriculture. They could not be appropriated by private individuals as they belonged to local councils and every resident had the right to a rational exploitation of them following various regulatory mechanisms designed specifically for them.

But this eighteenth-century data stems from an interest by the State to pervade and identify the local realities of unfamiliar places, an interference by an external agent that generated changes in the forms of territorial management. Visits by the *Intendencia de la Marina* (Logistics Corps of the Navy) during the second half of 1740 to determine the state of the various local forests in order to exploit them for different purposes (not all necessarily naval) serve as proof of this interference. The southern sector of the Sierra Nevada retains ample records reporting the visits of diverse intendants to describe and study the forests, particularly those of holm oak, chaparral, chestnut and oak, with the intention of exploiting them mostly for lumber. These records reveal an interest by the State itself of a sustainable exploitation of the environment, as it urged—and threatened to levy fines—on all the towns and their mayors who did not guarantee tree repopulating after felling them.

The interference by an external agent as a new manager of ancient communal spaces led to drastic changes in the nature of exploitations. It was now the State, and not the community, that kept watch on certain of these communal spaces. It reached the point that the State zealously managed these forests to avoid potential abuse and deterioration as they served as a source of wealth. Evidence of this is the appearance of a new protagonist hitherto unknown in the Sierra Nevada, the guard or caretaker responding directly to the State that ensured a proper forest management. The influence of the communities remained nonetheless strong as, for example, it was the residents who chose the guard, as noted in a record generated by the visits of a guard to Barxa (Barjas) (AHDG, File 402, no. 6 and File 402, no. 9).

## 2.5 The Attack on the Communal and Liberal Policies

While it is true that the expulsion of the Moriscos in the sixteenth century led to a change in the management of Sierra Nevada, the most abrupt transformation is rooted in the introduction of liberal and capitalist policies, especially those of the nineteenth century. It is clear, based on the cadastres, that the local communities until the middle of the eighteenth century continued to play a key and relatively equitable role in land management. But new economic and political trends and their penetration into these mountains led to medium-term enormous social and economic imbalances altering the existing agro-forestry-pastoral world.

Many of the spaces that were not private, including those known as *propios*, did suffer an aggressive campaign of exploitation and appropriation by various local and external agents as they were deemed by the new economic doctrines as unproductive and remnants of the Old Regime. This situation led to prizing their direct exploitation and, in many cases, led to their destruction and exhaustion.

Certain of these processes of change can be detected through the setting up of mining companies and metallurgical areas. These industries were mainly fueled by coal and the most convenient and profitable option was to manufacture *picón*, a charcoal produced by burning specific plants that grew in the surroundings of the factories. This appears to have done away with much of the communal brush and forests belonging to the local councils, who presumably did not hinder the arrival of the new industries in their territory despite their much more long-term harm to both the population and the environment. A text from 1804 by the scientist Simón de Rojas Clemente referencing the town of Lugros describes a space dominated by kermes oak between Güéjar Sierra and Lugros whose sole purpose was to supply coal to the local smithy and factory (Titos Martínez 1991: 168–170). The growing demand for charcoal led to disproportionate abuse of the forests, adversely affecting mainly the local communities. Examples of intense grievances arose shortly thereafter in the 1830s.

Another consequence of these changes was the disproportionate increase in local demographics. The decline in mortality due to hygienic-sanitary advances among all age groups obviously led to a greater population. From the middle of the nineteenth century—since the existence of the first state-sponsored population censuses—and until the middle of the twentieth century, the population continued to rise. This became a major issue as family units maintained a relative balance with their land, without the need to expand their property. But the disproportionate rise of the number of mouths to feed with regard to food production, resulted in a need for more land. This provoked a clearing and ploughing of forests and wastelands that were often communal or had never been directly exploited because of their marginal and apparently unproductive nature. The aim was none other than to exploit these new lands by applying private productive criteria. Hence many spaces that had been communally managed and exploited now became closed and private spaces intended for the sole purpose of feeding a family unit or to produce a good for the market.

Gradually the head of the family, to earn money, changed from playing the role of *director* of an agro-livestock farm to becoming an employee of a trade, leaving the responsibility of management of the subsistence farm to other family members (the wife or the oldest children). In certain cases, these activities took place in cities away from the localities giving rise to the phenomenon of temporary migrations, either to cities in the region itself or to other more distant foreign industrial centres in the Americas or Europe. This trend, known as part-time agriculture (Naredo 1974: 93–104), became an indicator of changes in agro-livestock management. It likewise represented a step into a new phase of the history of agriculture and livestock among European industrialized societies that often became definitive as it marked a point of no return to the primary sector, thus leading to its decline, marginalization and abandonment.

But in the Sierra Nevada, as in many other areas of Spain, the real milestone leading to the end of the primary sector as a hegemonic productive order was the adoption in the 1950s of developmental policies sponsored by the dictatorship, specifically the *Stabilisation Plan* of 1959 (Naredo 1974: 37–38), which provoked the greatest demographic movement from the countryside to cities in Andalusia, Spain and abroad. The mechanization of agriculture from the middle of the twentieth century contributed to the obsolescence of a large number of farmhands that saw their fate linked to an emerging industrial fabric bolstered by the new economic policy of Spanish cities. The farms of Sierra Nevada thus began to be abandoned after the 1960s. This exodus did not only affect the younger generation who left in search of a 'better future', but entire families who picked up roots as they could no longer survive due to the suffocating land management transformations experienced throughout the previous decades. Many of the communities throughout this decade lost up to half of their population, an abandonment that half a century later, though slowed, has not halted.

Thus small farms at this time no longer represented the main means of survival and the population initiated an ageing process which led to a neglect of local ecological knowledge that had evolved over time by generations of farmers. It is at this time that new activities and industries emerged throughout the Sierra Nevada designed for those who either chose to remain or were not able to emigrate. The new pursuits were mostly in construction and tourism, representing a definitive break with traditional small farming.

## 3 The Historical Management of Water as the Mainstay of the Territory

### 3.1 *Careo* Ditches: 'Sowing' and 'Harvesting' Water

Glacial and periglacial sediments characterized by high levels of permeability are pervasive towards the summits of the Sierra Nevada. Specifically, the highest zones of these mountains marked by great precipitation and low levels of evapotranspiration correlate with extensions of very permeable glacial and periglacial deposits susceptible to water infiltration and circulation (Castillo 2009; Gómez-Ortiz et al. 2013; Martos Rosillo et al. 2019b). The yearly rise of temperatures and melting of snow yield water that gradually seeps and circulates through the ground, feeding springs and numerous *borreguiles* (high pastures) characteristic of the Sierra Nevada highlands, serving since ancient times for grazing. Today, unfortunately, few have grasped that this slow seepage of groundwater is what contributes to maintaining the water tables of the lagoons of the Sierra Nevada and that it feeds the headwaters of the rivers during summer when the snow disappears and when there is practically no rain.

The hydraulic properties of these sediments, their capacity of infiltration and ability to retain water due to low rates of underground circulation, were wisely exploited by the local population over the years, especially at altitudes ranging between 2500 and 1000 m a.s.l. (Martos-Rosillo et al. 2019a). The processes of chemical weathering and rock edaphology predominated below 2500 m due to increases in average annual temperatures. This generates an alteration of the outcropping rocks marked by higher clay content than the rocks weathered by glacial and periglacial actions that have resulted in a vast aquifer attaining at times up to 50 m in thickness stretching throughout most of the slopes (Castillo et al. 1996, 1999; Castillo and Fedeli 2002; Fedeli and Castillo 1998; González-Ramón et al. 2015; Martos-Rosillo et al. 2017, 2019a).

To increase the volume of water and expand the flow of springs serving for irrigation and for humans and livestock, the al-Andalus populace designed a means of water management where *careo* ditches were essential (Martos-Rosillo et al. 2019b). These consisted of channels dug in the ground in the higher sectors of the slopes which served to capture water from rivers during the melting of the snow. The water circulating through them infiltrated both along their course and in specific zones known by the mountain inhabitants where the land had a higher capacity of infiltration. These features are known as *simas*, *calaeros* and *matas,* sorts of chasms that are most often upstream from the farmlands near settlements or upstream from springs that are key in supplying them.

A premeditated derivation and infiltration of melted waters from the summits of Sierra Nevada yielded an optimal means of exploiting the hydrogeological characteristics of the environment stemming from a rock type, schist, traditionally considered impermeable. These small hillside aquifers are known for their capacity to capture and temporarily store large volumes of water. Certain of these chasms can capture flows exceeding far beyond 100 L per second. *Careo* ditches are thus one of the most efficient and resilient ancestral features of water management, allowing them to overcome major climatic and social crises. They are thus tools serving to adapt to climate change and how local ecological knowledge represents an exemplary means of integrating humans into conditions that, in principle, are thought to be hostile to settling. Although this complex system has its origins in Islamic times, its accurate dating remains very difficult to pinpoint. *Careo* ditches, as indicated in a prior study (Martín Civantos 2011), must have existed already in the eleventh century in the higher zones of the Sierra Nevada based on the records of a legal dispute from the subsequent century between the *alquerías* of Lubros (Lugros) and Bartillana (González Palencia 1940: 321–328).

These ditches were likewise linked to the creation of pastures and fields to cultivate the higher sectors of the Sierra. In fact, the term *careo*, according to the *Diccionario de Uso del Espanol*, denotes areas for 'grazing livestock' and *carear* designates 'directing livestock to graze somewhere' or 'grazing livestock', a traditional system maintained partly by shepherds. This thus represents an almost perfect symbiosis where livestock provided abundant organic matter required to maintain an intensive agricultural activity in return for the most of the energy required for transport and draught.

## 3.2 Historical Irrigation Systems as the Centre of a Complex Socioecosystem

It is hardly possible to distinguish between livestock and agricultural activities in rural communities, a notion that emerges from records dating to the end of the Nasrid and Morisco periods (Figs. 5 and 6). The *alquerías* of the Sierra Nevada for the most part possessed some livestock. They also had access to irrigated lands that could have served indiscriminately for pasture or to plant crops. There were also areas, including certain intensive farms, intended for animal fodder. It is noteworthy that certain millstones driven by hydraulic mills from modern times were dedicated exclusively to grinding grains or legumes for fodder as opposed to other stones intended solely to grind products for human consumption.

It is clear that there were zones intended uniquely for livestock such as the so-called *borreguiles*, pastures around high mountain lagoons (Castillo Martín 2009). There were also until a few decades back seasonal high mountain habitats intended for the summer months for livestock or to grow mountain crops. Although they have yet to be studied, these habitats once again reveal the versatility and assorted uses of these spaces.

It is certainly necessary to view this sort of diversity, or mixture of concepts, uses and spaces, as elements of a complex system of exploiting the environment that also probably has an important capacity of adaptation or, rather, resilience, depending on, for example, annual climatic conditions or needs of supply. It is our belief that the 'intermediary' spaces of this complex system played a decisive role in spite of traditionally suffering from little research and that in certain instances have not received any attention from academia. These 'intermediary' spaces refer, on the one hand, to communal lands and, on the other, to occasionally irrigated lands.

There are records describing occasionally irrigated lands which received water only during winter and spring when it was abundant and the needs of crops were lower. Their extension was always variable depending on climatic conditions, both rain and snowfall, as well as insolation and temperature. The *careo* ditches themselves played a role in irrigating the highland wheats, ryes, potatoes and beans. Other ditches yielding spaces of intensive irrigation called *pagos* were found most often at high altitudes. They were labelled by the inhabitants as *sierra* to distinguish them from

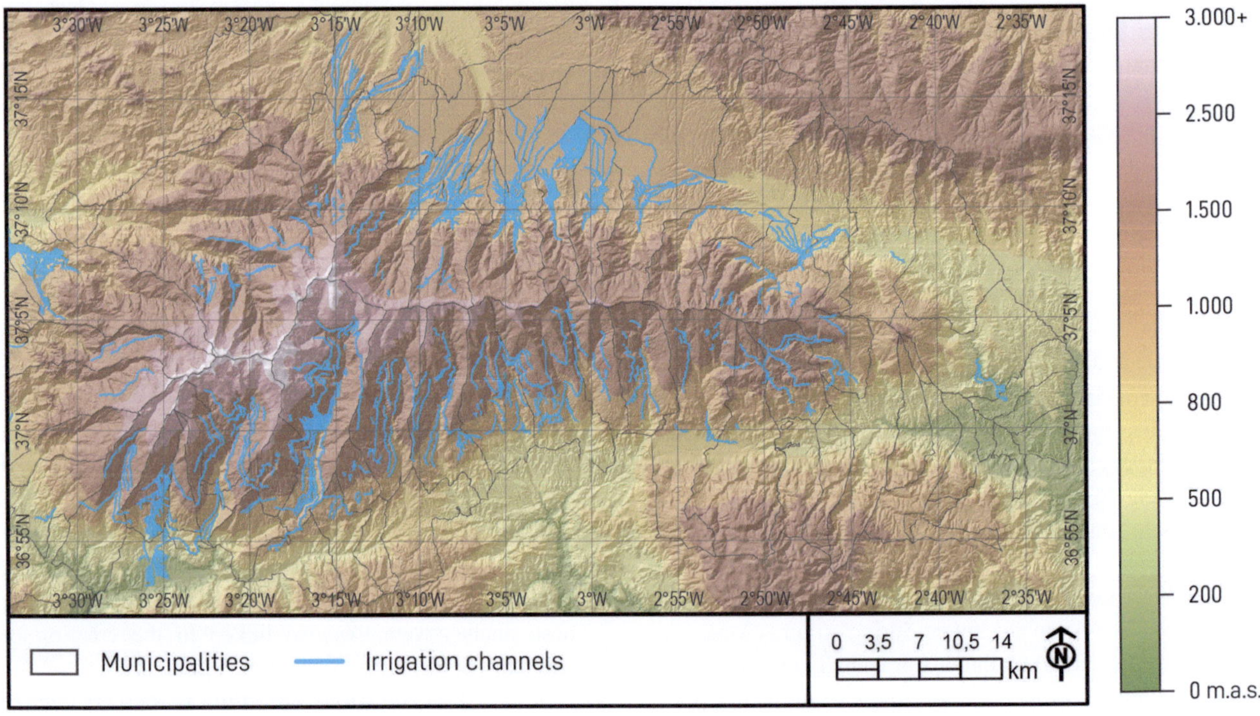


**Fig. 5** Irrigation systems of Sierra Nevada

**Fig. 6** Irrigation canal of Alcázar (Jérez del Marquesado)

those of the lowlands called *vega*. Although the only examples identified to date are at Trevélez and Cáñar, the phenomenon was certainly widespread.

Evidence from Trevélez points to a cultivated zone of the Alta irrigation ditch at the furthest point from the village which possessed no right to water during the summer. Thus when the Culo de Perro River which supplied its water did not suffice for the entire irrigable surface, only its lowest sector closest to the urban core received water. The difference is linked to the *balsa común* (common pond or tank) which served to store water at night to irrigate the following day. The lands above this pool, as noted, were labelled '*sierra*' while those below took on the name '*vega*' (plain) as they did possess the right to water during the summer. The *sierra* plots that were only occasionally irrigated usually served to sow cereals. Numerous farmhouses dotting these highlands were occupied by a great proportion of the population during the spring and summer to take advantage of the highland crops and pastures. Moreover, each of these farmhouses featured a threshing floor. These dry and occasionally irrigated lands were fundamental to the al-Andalus peasant communities and most likely remained so until the outset of the liberalizing and confiscating processes of the nineteenth century (Herrera González de Molina et al. 2002).

These elements offer the possibility to begin to identify real physical features of these structures of communal lands and their uses. They serve to gain a better grasp of the organizational patterns of the territories of the peasant communities since the al-Andalus period of the lands that did not benefit from intensive irrigation. The possibility of being able to count on resources, at least in the case of Sierra Nevada, which include not only occasionally irrigated lands but areas of pasture, mountain or forest, guaranteed the survival of the local residents and can be added to those of water management and intensive irrigated agriculture (Martín Civantos 2011).

These areas comprised a number of activities and uses that were at times, as noted, of mixed character. There is evidence, for example, of beehives and beekeeping which appear associated with *pagos* and *vegas* such as those of Juviles and Cástaras (Trillo San José 1994: 263). Moreover, a total of 378 beehives owned by Moriscos in the mid-sixteenth century are recorded in six of the villages of the Zenete region (Martín Civantos 2007: 341).

These were likewise lands serving to gather aromatic and medicinal plants. Al-ʻUmarī in the fourteenth century emphasized that it is possible in the Sierra Nevada to collect all kinds of native medicinal plants as well as 'drugs of the types produced in India' (Torres Palomo 1967–1968: 76).

## 3.3 Livestock as a Necessary Complement to Subsistence and Soil Fertilization

As noted above, livestock played a key role in the Sierra Nevada. It was likewise essential in developing irrigation systems precisely because a fundamental requirement for intensive cultivation was generating and maintaining soil fertility. In fact, it was not possible to develop intensive agriculture without a continuous supply of organic animal fertilizer. Moreover, it was traditional that part of the irrigated lands serves to produce fodder for livestock (Fig. 7).

Yet the question of livestock remains one of the great unknowns in the history of the Sierra Nevada. A more or less widespread belief exists that raising livestock during the al-Andalus period was a secondary activity (Trillo San José 1994: 264–265). This view can nonetheless be contrasted by the fact that livestock was in fact subjected to at least seven types of taxes in 1496 (Trillo San José 1994), a significant level of fiscal sophistication suggesting it played a key role.

The earliest evidence of livestock appears indirectly in the account of the life of *Shaikh* Abū Marwān al-Yuḥānisī, a thirteenth-century Sufi master from Ohanes (Qashtālī 2010: 271) who cites the presence of shepherds and hunters along the summits of the Sierra Nevada. What is particularly striking throughout his book is the apparent ease with which individuals crossed the massif from the southern Alpujarra to the northern Granada, Guadix and Zenete regions through numerous roads and passes. These old crossings were a common phenomenon until relatively recently when motorized transport and new roads replaced the ancient paths serving muleteers and pack-horse drivers. Most are still recognizable and can be transited. They form part of local oral memory, especially among the older generations who recall travelling.

Al-Yuḥānisī's biographical text, apart from its data on livestock, relates a hunting episode in Abrucena. This episode, the oldest instance of this type, pertains to hunting deer, an animal extinct today in the Sierra Nevada. What is nonetheless noteworthy is that the hunt did not take place in the mountains but among the orchards of irrigated spaces.

**Fig. 7** Careo canal of El Jaral (Jérez del Marquesado)

Along these lines, Hieronymus Münzer, a German visitor to the Kingdom of Granada just after the Castilian conquest, noted that '… *Granada also has high mountains, plains, valleys, which due to the scarcity of water cannot be irrigated or inhabited. It has endless herds of goats, sheep, big and stout oxen. It is almost unbelievable that its mountain forests have so many deer, bears, fallow deer, rabbits and especially wild boars*' (Münzer 1991: 107).

Furthermore, the sixteenth-century writer L. del Mármol Carvajal in his *Historia de la rebelión y castigo de los moriscos* cites the existence of '… *beautiful pastures for livestock…*' in the *Taha* (district) of Órgiva. He also states, in reference to the surroundings of Ferreira and Poqueira, '*In summer there are in these mountain ranges very beautiful pastures for livestock; and in winter, because it is a very cold land, the livestock is taken to Dalías, or to Motril and Salobreña, which are warmer and more template due to the sea air*' (Del Mármol Carvajal 1946: 189, 191). The same can be said of other localities he cites of the Sierra Nevada that offer a rather different image in terms of the richness of their pastures and livestock.

This information corroborates the traditional view that livestock was subjected to transhumance, and that this became more common as a result of the privatization processes of communal spaces in the nineteenth century which yielded large *dehesas* (spaces for pasture) such as that of Camarate leading to an increase in many areas of the number of livestock. However, the same sources signal that a great part of the livestock that grazed in the Sierra Nevada belonged to farmers living in the mountain villages and practicing transtermitance (short-distance transhumance), a system persisting partly until today which included *dulas* or shifts to care for the animals (Martín Civantos 2007).

## 3.4 Communal Governance Systems and Land Management

The written references cited above corroborate what we noted as to the nature of communal lands and their uses. Indeed, residents of *alquerías* had the right to exploit the pastures and forests and to hunt in common lands. Although this certainly led to livestock development, it is impossible for the moment to precisely identify what type and, especially, to what scale.

Ownership and use of communal lands are likewise evidenced by the fifteenth-century legal dispute between the *alquerías* of Jérez-Alcázar and Lanteira on the northern face of the Sierra Nevada. This procedure, cited above (González Palencia 1940: 345–362, 366–368), pertained to the use of the forests and pastures of Jérez-Alcázar. A declaration from February 22, 1422, among other records, alludes to the declaration by 145 residents described above.

As noted, many of these communal spaces and their means of exploitation persisted into the Modern Age, even after the expulsion of the Moriscos. Communal uses of forests and pastures were for the most part conserved while the majority of the dry lands and occasionally irrigated lands became part of the '*suertes*' ('prizes') distributed among the new settlers. These systems were maintained until, as explained above, the introduction by the State in Modern times of new forms of management and intervention, first as interests in forests and later through liberal policies and an attack on the *propios* and communal types of goods by means of privatization. This was behind the origin of many zones of public forest which became the property of municipalities. It was a process, yet to be studied in depth, in which peasant resistance itself played a key role. We will not delve deeper into this issue as those forms of management today are only memories, as is the case of the oak grove of Cáñar. Perhaps one of the few exceptions of communal management of the territory still remaining is that of the *Junta de Propiedad Particular Colectiva* of the town of Güéjar Sierra which today remains a compelling case that is largely transformed and adapted to the new socioeconomic contexts (Ortega Santos 2012).

But if the Sierra Nevada massif stands out for anything, it is undoubtedly its management of water. As we have highlighted above, water to a great extent articulated the productive strategies and the forms of territorial organization since the al-Andalus period. This is where it is possible to identify forms of communal management that remain active despite the processes of change and, above all, the abandonment of the region in recent decades. *Comunidades de regantes* (irrigation associations) are the institutions responsible for operating and preserving irrigation systems. Their tasks are based on two fundamental pillars. The first is the preservation of a communal organization serving to harness a scarce resource. The second is its task in preserving the ethnoecological knowledge vital to understanding the environment. From the legal point of view, these associations are entities of public law since their recognition and the drafting in the first Laws of Water of 1866. Although labelled under different names (*Hermandad de Labradores*—Brotherhood of Farmhands and *Sindicatos de Usuarios*—Unions of Users), their focus has remained the same, formed by all landowners with the right to water and who organized themselves collectively to maintain and manage the hydraulic features. Certain authors have emphasized the ability of the irrigation associations to control access to the resource as one of the keys to its long-term operation (Aguilera 1987; Ostrom 1990). The right to water, in fact, is linked to the land and not of its owner. Hence a parcel cannot be sold without its respective water. This is one of the basic principles that has allowed the system to survive over the centuries.

There are no records shedding light on the original forms of organization dating to the implementation of these irrigation systems. There are likewise no records from the Modern Castilian era apart from a few legal actions narrating water distribution and user conflicts. In fact, the organization of water management is gleaned fundamentally and almost exclusively from oral tradition. It is one of the defence mechanisms of the communities themselves, serving to prevent interference of external powers, either public or private.

As noted above, water was communal in spite of the fact that since the adoption of the Law of Waters it became a public good managed by the communities in the form of concessions based on the recognition of historical rights. However, despite these apparent changes, water management systems remained for the most part by adapting to new social, political and environmental circumstances, demonstrating an enormous resilience. 'The concept of resilience is of great analytical utility to contend with the nature of irrigation systems themselves. These spaces are subject to great changes and pressures, both internal and external, which suppose social or environmental risks to their survival. A growing scarcity of water in certain cases puts the continuity of the system at risk and forces a revision of the forms of traditional water management. In other cases, it is urban pressure that endangers areas of cultivation and the layouts of water distribution networks. Even the adoption of new European and national legislation threatens the centuries-old systems of communal use of water. It is also noteworthy that one of the main challenges to these traditions is the progressive ageing of farmers and the lack of generational replacements to preserve their systems and infrastructures. In any case, these systems, from their creation until today, had to overcome many environmental and social challenges. Hence any static and balanced image of them is not in line with reality' (Ruiz Ruiz and Martín Civantos 2017: 83–84). The approval in 2000 of the EU Water Framework Directive, as well as its reflection in national legislation in the form of the Consolidated Text of the Water Law, promote new, reputedly more efficient, uses of water, have forced irrigation associations to replace traditional infrastructures by pipes and blanket methods (sprinkler or drip irrigation).

For the different irrigation associations to operate, they were provided with traditional norms to regulate the rights and distribution of water throughout society according to volume and duration, means of distribution and control mechanisms. Regulations also defined the obligations of participants in preserving irrigation infrastructures, the ways in which decisions were reached, the governing bodies and, specifically, the mechanisms to resolve conflicts and even impose sanctions. These systems of governance were designed as social institutions and defined as sets of rules determining who possessed the right and how decisions were undertaken, rules of affiliation or membership, activities allowed or prohibited, procedures, if information can be divulged or not, and the existence or not of remuneration (Ostrom 1990: 94).

This study avoids delving into these types of institutions as they have been extensively examined by E. Ostrom and many others (Anderies et al. 2004; Janssen and Ostrom 2006). We nonetheless insist that, '*... as certain authors* (Steins and Edwards 1999; Steins et al. 2000) *have pointed out, the design of institutions in itself is not a guarantee of the success of managing a resource. It is necessary, beyond this issue, to take into account historical, contextual, social, etc. factors which are undoubtedly fundamental to their function and particularly the perceptions and motivations of the users that make them up* (McCay 2002). *In this sense, institutions become more than just a set of operating rules and assume a much more social dimension. Not only do they define how to make collective use of the resource, but also act as recognised centres for the conservation and dissemination of ethnoecological knowledge accumulated over time that bolster the management*' (Ruiz Ruiz and Martín Civantos 2017: 84).

In the case of the Sierra Nevada there is currently an overlap between municipalities and irrigation associations. Yet this was not always the case. Certain systems were shared by different municipalities (not only towns) with a single irrigation association or municipalities with several irrigation associations, one for each of the types of irrigation. This therefore represents a distinction between the system, the technological unit consisting of the catchment infrastructure, the main or 'mother' irrigation ditch and its derivations, and the social unit, the community that manages the infrastructure and is responsible for its maintenance and the social distribution of water and its governance. Due to this distinction, it is often possible to observe an evolution of territorial and settlement patterns since medieval times. There are cases of disaggregation, such as between Aldeire-La Calahorra and Lanteira-Alquife localities along Sierra Nevadas northern face (Martín Civantos 2007). An opposite aggregation process is identified elsewhere where old *alquerías* or towns began to disappear while their spaces of irrigation, water rights and limits were absorbed by other neighbouring bodies. This is the case of Jérez del Marquesado which now groups five different irrigation areas that gradually disappeared between the twelfth century and the moment of the expulsion of the Moriscos (Martín Civantos 2007). It is also the case of the Municipality of Cáñar where both the small towns of El Fex and Barjas disappeared respectively following the Morisco rebellion at the end of the sixteenth century and a landslide towards the middle of the nineteenth century.

Although the different communities arrange themselves into practically identical institutional schemes, each possesses its own traditional regulations to manage and distribute water. This affects the way of measuring both the land and the water, the variations between '*turnos*' (shifts) and '*tandas*' (rounds), the presence or not of ponds or tanks to regulate water flow, the existence or not of occasional irrigation areas, etc. In this way, the systems become more complex, often rendering the forms of water distribution almost unintelligible, especially when the mechanisms of adaptation to scarcity come into play and the rights of each parcel are diminished proportionally to the flow of water available at any given time, or among the systems where water can be changed from one parcel to another at the request of the irrigator, depending on cultivation strategies and the presence or not of trees.

## 4 Concluding Comments

The Sierra Nevada is an anthropized range from its highest picks to the foothill. Is the result of a co-evolutive process that started at least in the Copper Age (5.100 B.P) and mainly in the Bronze Age (4.000 B.P). This historical process includes radical modifications such as deforestation, pasture creation, terracing, mining or water management. We know little about the prehistoric and classical periods due not only to the lack of research, but because it seems that settlement was already established on the foothill. For sure, the main change started during the Late Roman period, from the fifth and for sure the sixth century, when a new stable settlement network appeared for the first time in the framework of the Roman Empire crisis and a deep social transformation. Indeed, it can be considered as a real assault on the mountain. The Islamic conquest at the beginning of the eighth century consolidated and enlarged that settlement by the creation of a new productive system based on water management and irrigated agriculture, including meadows in the upper lands linked to complex artificial aquifer recharge (*careos*). That process completely changed the relationship between humans and the Sierra Nevada, including not only water uses, but pastures and cattle, soil creation and management, terracing and territorial organization. These are the bases of the later development and the current landscapes and socioecosystems already recognized and protected from an environmental and cultural perspective. Nevertheless, these landscapes are now threatened by modernization, mainly translated into migration and abandonment of traditional practices and knowledge, but also into new pressures and tensions coming from the surrounding processes of agrarian intensification and industrialization and by conflicts provoked by the conservation measures not always well implemented and understood.

**Acknowledgements** This paper has received funding under the projects 'Thematic Center on Mountain Ecosystem & Remote sensing, Deep learning-AI e-Services University of Granada-Sierra Nevada'. LifeWatch-ERIC infrastructure (Convocatoria de Operaciones cofinanciadas FEDER-Programa Operativo Pluriregional de España); 'Inspiring Rural Heritage: Sustainable Practices to Protect and Conserve Upland Landscapes and Memories – IRIS'. JPI Cultural Heritage Project funded by the EU's H2020 research and innovation programme under grant agreement No 699523; ERC Stg 2019 'ANTIGONE - Archaeology of shariNg pracTIces: the material evidence of mountain marGinalisatiON in Europe (18th-21st c. AD)' (GA n. 853539).

## References

Adroher Auroux A (1990) Prospección superficial en el pasillo de Fiñana, Sierra de Baza y Sierra Nevada. In: Anuario Arqueológico de Andalucía 1987. Junta de Andalucía, Sevilla, pp 77–80

Aguilera Klink F (1987) Los recursos naturales de propiedad común. Una Introducción. Hacienda Pública Española 107:121–127

al-Qashtalı A (2010) Tuhfat al-mugtarıb Prodigios del maestro sufí Abu Marwan al-Yuhanisi de Almería (ed and trans: Boloix Gallardo B) Mandala, Madrid

al-Rāzī (1974) Ajbar muluk al-Andalus. Crónica del Moro Rasis. Versión del ajbār mulūk al-Andalus de Ahmad ibn Muhammad ibn Musa al-Razi, 889–995; romanzada para el rey don Dionís de Portugal hacia 1300 por Mahomad, alarife, y Gil Pérez, clérigo de don Perianis Porçel (trans: Catalán D, De Andrés MS) Madrid

Anderies JM, Janssen M, Ostrom E (2004) A framework to analyze the robustness of social-ecological systems from an institutional perspective. Ecol Soc 9(1):1–18

Bertrand M, Sánchez Viciana J (2008) Production du fer et peuplement de la region de Guadix (Grenade) au cours de l'Antiquité tardive et du haut Moyen Âge. In: Canto García A, Cressier P (eds) Minas y metalurgia en al-Andalus y Magreb occidental. Explotación y poblamiento. Casa de Velázquez, Madrid, pp 127–147

Brogiolo GP (2007) Dall'archeologia Dell'architettura All'archeologia Della Complessità. Pyrenae 38(1):7–38

Castillo A, Fedeli B (2002) Algunas pautas del comportamiento hidrogeológico de rocas duras afectadas por el glaciarismo y periglaciarismo en Sierra Nevada. Geogaceta 32:189–191

Castillo A, Cruz-Sanjulián JJ, Benavente J (1999) Aguas de Sierra Nevada; Aguas de Lanjarón. Lanjarón, paisajes del agua. Granada, Balneario de Lanjarón, pp 35–64

Castillo A, del Valle M, Rubio-Campos JC et al (1996) Síntesis hidrogeológica del macizo de Sierra Nevada (Granada y Almería). In: 1ª Conferencia Internacional de Sierra Nevada, vol 1, pp 389–417

Castillo A (2009) Lagunas de Sierra Nevada. Todas las lagunas, valle a valle. Universidad de Granada, Granada

Estévez Callejón M (1995) Capileira de Poqueira. Granada, Ayuntamiento de Capileira

Fedeli B, Castillo A (1998) Condiciones de surgencia en una cuenca esquistosa de alta montaña (Sierra Nevada; Granada, España). Geogaceta 23:47–50

Ibn Gālib (1955) Nassāndalusīya dīdqit`a min Kitāb Farḥatal-ānfus li Ibn Gālib `ankūwar al-Andalus wamuduniha. In: Lutfī 'Abd Albadi` (ed) El Cairo

Gómez-Ortiz A, Oliva M, Salvador F (2013) The environmental protection of landscapes in the high semiarid Mediterranean mountain of Sierra Nevada National Park (Spain): historical evolution and future perspectives. Appl Geogr 42:227–239

González de Molina M (1996) Hombre y naturaleza en el siglo XIX. Las raíces de la crisis ecológica. Santillana, Madrid

Herrera González de Molina A, González de Molina M, Ortega Santos A (2002) Bienes comunales desde la perspectiva socioambiental. In: de Dios de Dios S et al (eds) Historia de la propiedad en España. Bienes comunales, pasado y presente: II encuentro interdisciplinar. Salamanca, pp 493–532

González Palencia Á (1940) Documentos árabes del Cenete (siglos XII-XV). Al-Andalus 5:301–382

González-Ramón A, Moral-Martos F, Marín-Lechado C et al (2015) Factores geomorfológicos condicionantes de la hidrogeología de la cuenca alta del río Bérchules (Sierra Nevada, Granada). In: Andreo Navarro B et al (eds) El Agua en Andalucía. El agua clave medioambiental y socioeconómica. IGME 1:283–294

Janssen MA, Ostrom E (2006) Governing social-ecological systems. In: Tesfastion L, Judd KL (eds) Handbook of computational Economics, vol 2. Elsevier, Amsterdam, pp 1465–1509

del Mármol Carvajal L (1946) Rebelión y castigo de los moriscos de Granada In: Rossell C (ed) Atlas, Madrid
Martín Civantos JMª (2005) La minería altomedieval en la kura de Ilbira (provincias de Granada y Almería, España). Archeol Mediev 32:35–49
Martín Civantos JMª (2006) Il territorio stratificato: proposte dall'archeologia del paesaggio. In: Francovich R, Valenti M (eds) IV Congresso Nazionale di Archeologia Medievale. Florencia, pp 12–16
Martín Civantos JMª (2007) Poblamiento y territorio medieval en el Zenete (Granada). Editorial Universidad de Granada, Granada
Martín Civantos JMª (2011) Working in landscape archaeology: the social and territorial significance of the agricultural revolution in al-Andalus. Early Mediev Eur 19(4):385–410
Martínez Ruiz J (1987) Huellas de las tres religiones en la toponimia medieval granadina. In: Homenaje al profesor Darío Cabanelas Rodríguez, O.F.M. Granada, pp 53–67
Martos-Rosillo S, González-Ramón A, Marín-Lechado C et al (2017). Las acequias de careo de Sierra Nevada (Sur de España), un sistema de recarga ancestral en acuíferos de alta montaña. In: Escolero Ó, Carlos Gutiérrez C, Mendoza E (eds) Manejo de la recarga de acuíferos un enfoque hacia Latinoamérica. Instituto Mexicano de Tecnología del Agua, México, pp 527–563
Martos-Rosillo S, González-Ramón A, Ruiz-Constán A et al (2019a) El manejo del agua en las cuencas de alta montaña del Parque Nacional de Sierra Nevada (Sur de España). Un ejemplo ancestral de Gestión Integral del Agua. Boletín Geológico y Minero 130(4):729–742
Martos-Rosillo S, Ruiz-Constán A, González-Ramón A et al (2019b) The oldest managed aquifer recharge system in Europe: new insights from the Espino recharge channel (Sierra Nevada, southern Spain). J Hydrol 578:124047
McCay B (2002) Emergence of institutions for the commons: contexts, situations and events. In: Ostrom E (ed) Drama of the commons. National Academies Press, Washington D.C., pp 361–402
Münzer J (1991) Viaje por España y Portugal (1494-1495). Madrid
Naredo JM (1974) La evolución de la agricultura en España. Editorial Laia, Barcelona
Ortega Santos A (2012) El comunal imaginado. De la transición en los usos de la propiedad comunal en el siglo XX. Güejar Sierra. Historia Agraria 58:73–112
Ostrom E (1990) Governing the commons: the evolution of institutions for collective action. Cambridge University Press, Cambridge
Redondo Ortega JJ, Martínez Rodríguez F (2021) Primeros datos sobre un petroglifo prehistórico de estilo atlántico hallado en Peña Madura, Cuenca Alta del Río Dílar, Sierra Nevada (Granada). Cuadernos De Arte Prehistórico 11:124–160
Román Punzón JM, Martín Civantos JMª (2014) Aproximación al poblamiento tardoantiguo en Andalucía. In: Catalán R, Fuentes P, Sastre JC (eds) Las fortificaciones en la Tardoantigüedad. Élites y articulación del territorio (siglos V-VIII). La Ergástula, Madrid, pp 57–78
Rouco Collazo J (2021) Las fortificaciones medievales de las Alpujarra Alta desde la Arqueología de la Arquitectura y del Paisaje. Universidad de Granada
Ruiz Ruiz JF, Martín Civantos JM (2017) La gestión comunitaria del agua en la cara norte de Sierra Nevada: acción colectiva y saberes etnoecológicos en los sistemas de riego de origen andalusí. e-rph 20:76–103
Steins NA, Edwards VM (1999) Collective action in common-pool resource management: the contribution of a social constructivist perspective to existing theory. Soc Nat Resour Int J 12(6):539–557
Steins N, Röling NG, Edwards VM (2000) Re-"designing" the principles: an interactive perspective to CPR theory. In: 8th conference of the international association for the study of common property. Bloomington, Indiana, USA
Titos Martínez M (1991) Textos primitivos sobre Sierra Nevada (1754/1838). Granada, Caja General de Ahorros de Granada
Torres Palomo MP (1967–1968) Sierra Nevada en los escritos árabes. Miscelánea de Estudios Árabes y Hebraicos 16–17:57–88
Trillo San José MC (1994) La Alpujarra antes y después de la conquista castellana. Editorial Universidad de Granada, Granada
Trillo San José MC (1999) El paisaje vegetal en la Granada Islámica y sus transformaciones tras la conquista castellana. Historia Agraria 17:3–24
Trillo San José MC, Álvarez García JJ, Jiménez Lozano E (1999) Informe de la excavación de urgencia realizada en el cortijo de Ana, término municipal de Órgiva (Granada). In: Anuario Arqueológico de Andalucía 1994, vol 3. Junta de Andalucía, Sevilla, pp 172–175

# Scientific Knowledge Generated in Sierra Nevada: Bibliographic Review (1970–2021)

Ricardo Moreno-Llorca and Regino Zamora

**Abstract**

The high-mountain range Sierra Nevada constitutes a hot spot of biodiversity in the western Mediterranean. This massif also represents an excellent mountain landscape to analyse the effects of global change on ecosystems due to a considerable altitudinal gradient, its proximity to the Mediterranean Sea, its geographical position and its complex and extensive natural as well as cultural landscape history. The literature review has been conducted in accordance with the principles and protocol established by the PRISMA declaration, collecting all the scientific research that has been generated in this protected area from 1970 to 2021. The results identify the institutions that have generated this scientific knowledge as well as the development of these topics over time, mostly in relation to the thematic areas of the Global-Change Observatory. Finally, the study focuses on analysing specific research topics of prominence such as recent geomorphological changes, the reconstruction of the past (historical scale), aquatic systems (ecosystem scale), community ecology in terrestrial ecosystems (ecological interactions) and Iberian ibex as a representative species.

**Keywords**

Bibliographic review • Web of science • Scopus • Scientific knowledge • Research areas • Decision-making • Bibliometrix

R. Moreno-Llorca (✉) · R. Zamora
Andalusian Institute for Earth System Research (IISTA), University of Granada, Granada, Spain

R. Zamora
Department of Ecology, University of Granada, Granada, Spain

## 1 Introduction

Sierra Nevada, a high-mountain range in southern Spain, has unique ecosystems that for millennia have been co-evolving with humans inhabiting the area. During the last centuries, national as well as international explorers, travellers, artists and scientists have explored, described and studied diverse aspects of this Mediterranean massif, which was once called the Alps of Andalusia (Titos Martínez 1997), generating publications dating back to, at least, the eighteenth century.

Because of its location (the southernmost high-mountain in Western Europe), elevational range and geomorphology, Sierra Nevada has biological and ecological uniqueness that has attracted botanists, zoologists, ecologists, geologists, geographers and historians worldwide. Sierra Nevada is the richest centre of plant diversity in the western Mediterranean and one of the most outstanding in Europe, where 2353 species of vascular plants have been registered, some 20,000 species of invertebrates, and 290 species of vertebrates (43 mammals, 123 birds, 20 reptiles, 7 amphibians and 6 fish, among others). Sierra Nevada (Andalusia, SE Spain) is a mountainous region covering more than 2000 km$^2$ with an elevation range of between 860 m and 3479 m a.s.l. This mountain area harbours 27 habitat types of the EU Habitat Directive (92/43/EEC), and it is considered one of the most important biodiversity hotspots in the Mediterranean region. Sierra Nevada receives legal protection in multiple ways: it is a MAB Biosphere Reserve, Special Area of Conservation (Natura 2000 network), Natural Park and National Park, and Important Bird Area. Sierra Nevada is included in the World Green List of Protected Areas (IUCN) and is part of the Spanish Long-Term Ecological Research. The main economic activities in this mountain region are agriculture, tourism, livestock raising, beekeeping, mining and skiing.

Furthermore, Sierra Nevada offers advantageous sites for astronomy, meteorology and the study of atmospheric remote processes. While in past centuries, its natural values have been the subject of sectoral studies by different



R. Zamora and M. Oliva (eds.), *The Landscape of the Sierra Nevada*,
https://doi.org/10.1007/978-3-030-94219-9_4

disciplines, today Sierra Nevada is the focus of interdisciplinary studies, serving as a European laboratory of global change.

## 2 Research in Sierra Nevada: Initial Scientific Contributions

With regard to recovering information about Sierra Nevada and making it known, the work of Manuel Titos Martínez (professor and historian at the University of Granada) should be highlighted for compiling the scientific, naturalistic, economic and social history (Titos Martínez 1997). Also noteworthy are the bibliographical reviews of specific topics, such as those performed by a large group of zoologists directed by Alberto Tinaut (Ruano et al. 2013). Likewise, scientists such as Gabriel Blanca (Blanca López 2002) and colleagues have examined the great botanical biodiversity of this mountain (Fig. 1).

### 2.1 First References to Sierra Nevada

Sierra Nevada was originally called the Mountain of the Sun, the Mountain of the Sun and the Air, the Mountain of Snow or the Frozen Mountain by the different cultures that have inhabited it (Romans, Goths, Arabs, etc.). However, the earliest texts that mention the mountain date from after the conquest of Granada by the Christians (1492). The Italian historians Andrea Navagero and Lucio Marineo, and the Spaniards Diego Hurtado de Mendoza and Luis Mármol y Carvajal (sixteenth century) were the first to describe Sierra Nevada. Other later historians (seventeenth century) include Bermúdez de Pedraza and Henríquez de Jorquera, who made a more in-depth analysis of the hydrology, geography and the human population that settled there (Titos Martínez 1990).

However, the first naturalist to undertake an expedition expressly to study Sierra Nevada with the intention of publishing and disseminating knowledge was Antonio Ponz. His expedition in 1754, promoted by the "Marqués de la Ensenada", focused on studying the general relief, high peaks, rivers and small lakes, emphasizing the need to broaden scientific knowledge concerning botanical discoveries he made (Titos Martínez 1991). This naturalist was the precursor of a long tradition of scholars who gradually began to reveal the natural richness of Sierra Nevada.

### 2.2 Primary Biological Research

The extraordinary biodiversity of Sierra Nevada makes it a focal point for national and international naturalists. The first botanist to study the massif was Fernández Navarrete, born in Granada (Spain), who published a nationwide essay in 1742, with 82 pages describing species and their location on the mountain (Titos Martínez 1990). Other noteworthy botanists include José Quer, Guillermo Bowles and Antonio Palau. In 1760, a collection was made of species that Linnaeus included in his work *Species Plantarum* (Titos Martínez 1997). The first botanist to publish texts specifically on Sierra Nevada was William Thalacker, who travelled across the mountains for six days collecting botanical specimens. The result of his work was later published under the name

**Fig. 1** Main compilations on the Sierra Nevada edited by Manuel Titos, Alberto Tinaut and Gabriel Blanca, researchers at the University of Granada

*Anales de las Ciencias Naturales*. His works, which also included mineralogy, were completed by the naturalist Simón de Rojas Clemente y Rubio, who focused on the study of zoology and botany with emphasis on lichens. These works were later taken up by the renowned Charles Edmund Boissier (Titos Martínez 1991).

At the end of the nineteenth century, Boissier's work was considered to be the most extensive and meticulous study that had been published on the botany of Sierra Nevada to date. Boissier made three trips, visiting many of its peaks, lakes, ravines and villages. Boissier described its vegetation and flora in detail, and made the first barometric measurement on Veleta Peak (3396 m a.s.l.). He collected plants and seeds and made detailed illustrations of the most notable species of the massif. Later Moritz Willkomm (in the mid-nineteenth century) visited Sierra Nevada and focused his study on the vegetation, collecting numerous plant species. This work of documenting plant species was continued by authors such as Funk, Alioth and Rossmaesler. Many of these works initially went unnoticed, until later Spanish botanists, such as Mariano Lagasca, Elías Pelayo, Mariano del Amo y Mora and Pedro del Campo (Titos Martínez 1997) revived and highlighted these studies.

Regarding research on fauna in Sierra Nevada, early studies were not as prolific as they were for flora. The first publication related to the animals of Sierra Nevada was written by Hübner in 1813. Further efforts were made in the mid-nineteenth century by Rambur (1837) and Rosenhauer (1856) and in the late same century by the Spanish biologist Sánchez García (1885) (Ruano et al. 2013). The Spanish ibex (*Capra pyrenaica hispanica*), as one of the symbols of this massif, is one of the first animal species to arouse the interest of naturalists, as in the case of D'Ottensheim and Schimper during the same period. The detection of the ibex, and its definition as the same species as the one found in the Pyrenees, but different from those found in the rest of Europe, attracted the attention of European zoologists, including Willhelm Rossenhauer. He published a book in 1856 with 189 new species in Andalusia, including the golden eagle, the bearded vulture and the wolf. The latter, considered to be a predator to be exterminated, is mentioned by Johannes Rein in 1892, in which he chronicles the beginning of its extinction in Sierra Nevada. In 1879, an entomological sampling campaign was led by Laureano Pérez Arcas. Other contributions defining new species were made by José Sánchez, Antonio Rubio and Máximo Hertting, who wrote about the brown vulture, the monk vulture, the purple vulture, the bearded vulture and the lynx.

Finally, biological research on the lakes began considerably later with the study of algae by González Guerrero and Martínez and the limnology research by Rosa Martínez-Silvestre in 1975 (Sánchez Castillo 1988).

### 2.3 The Physical Environment

The beginning of research on Sierra Nevada in the physical sphere focuses on glaciers and geology. The first references to glaciers that shaped the mountain landscape in the past were carried out by Antonio Ponz, William Thalacker, Simón de Rojas, Simón de Argote, Charles Rochfort Scott, Richard Ford, Willian George and Pacual Madoz. The controversy over glaciers on Sierra Nevada was clarified in the publication in 1916 by Hugo Obermaier and Juan Carendel, in their work "Los glaciares cuaternarios de Sierra Nevada" (The Quaternary Glaciers of Sierra Nevada) (López Ontiveros and Naranjo Ramírez 2000). This work was the result of an expedition the previous year, based at the lodge of Sierra Nevada Society, and taking as a reference the cartographic work by Bertuchi, Bide, Prudent and Rein, who mapped the entire mountain in detail. In this work, Obermaier produced perhaps the first glaciological bibliography on Sierra Nevada, citing authors such as Schimper, Casiano de Prado, MacPherson, Geikie, Von Drasche, Hellman, Rein, Richter and Reger (Titos Martínez 1997). Following Obermaier, the works by Quelle confirmed the existence of glaciers on the summits of Sierra Nevada, using the scientific method and terminology to analyse and describe them (Gómez Ortiz and Plana Castellví 2004).

Thalacker, in 1802, was the first expert to study the geology of Sierra Nevada (Titos Martínez 1991). In 1842, the German geologist Hausmann published a book on Sierra Nevada mountain system, a work which was continued by the Austrian Richard von Drasche, who published the first geological analysis of the massif in 1878 (Titos Martínez 1990), a study that analysed both the mineralogical composition of the rocks and the stratigraphy. Other works on more specific areas are those of Arévalo and Baca in Lanjarón and Barrois and Offret in the Alpujarra at the end of the nineteenth century.

Regarding astronomical research, in 1966 the observatory called "Mojón del Trigo" was built and equipped with a telescope having a lens 32 cm in diameter. In 1978 a new astronomical observatory with two domes was built by the Astrophysics Institute of Andalusia. Finally, in 1981, close to the previous observatory, construction began on one of the three largest radio telescopes in the world at that time (Titos Martínez 1997).

## 3 Bibliographic Review of the Scientific Literature in Sierra Nevada

### 3.1 Introduction

After analysing the first historical naturalist contributions on Sierra Nevada, a bibliographic revision of the scientific studies in the last decades has been carried out. Publications about Sierra Nevada in indexed journals and internationally recognized databases began in the 1970s. Hence this is the starting point for our study. The method chosen for the collection is the literature review. A literature review (as opposed to a systematic review or meta-analysis) summarizes a wide-ranging topic (Higgins et al. 2019). It therefore does not define a priori which studies should be included or excluded. The literature review of Sierra Nevada adopted the definitions used by Cochrane Collaboration (Higgins et al. 2019).

## 3.2 Methods and Review Protocol

*Objective of the review*

The area covered by the present review is Sierra Nevada Biosphere Reserve, which is part of the first Global Green List of Well-Managed Protected Areas (Gilbert 2014). The review compiles all the scientific research that has been generated in relation to global change in this protected area.

The study spans 1970–2021. The beginning of the period is when major publications on Sierra Nevada started to appear in the selected databases. All searches were conducted in June 2021.

*Protocol*

The study follows the principles and protocol established by the PRISMA declaration (Moher et al. 2009; Urrútia and Bonfill 2010), being adapted to an iterative model in which the different databases are sequentially included (Fig. 2). The stages of the protocol involve the identification of records, screening, assessment of suitability and inclusion in the final database for analysis.

1. Identification

Within the identification phase, the search criteria were based on a population-intervention-comparison of results (PICO) strategy (Higgins et al. 2019), in which the terms "Sierra Nevada" was defined as Population, "Spain" as Intervention, and "Global Change" as Outcome.

A search for scientific publications concerning Sierra Nevada (Spain) was made with the ISI Web of Science (Wos) search engines (http://webofknowledge.com/), on Scopus (https://www.scopus.com/) and subsequently with complementary sources. The search string on Wos was "TS = (Sierra Nevada* AND [Spain* OR España*])", returning a total of 720 records. The search field "TS" included the scanning in the title, abstract, keywords of the authors and the keywords created by Wos. The search on Scopus was made using the TITLE-ABS-KEY ("Sierra Nevada") AND

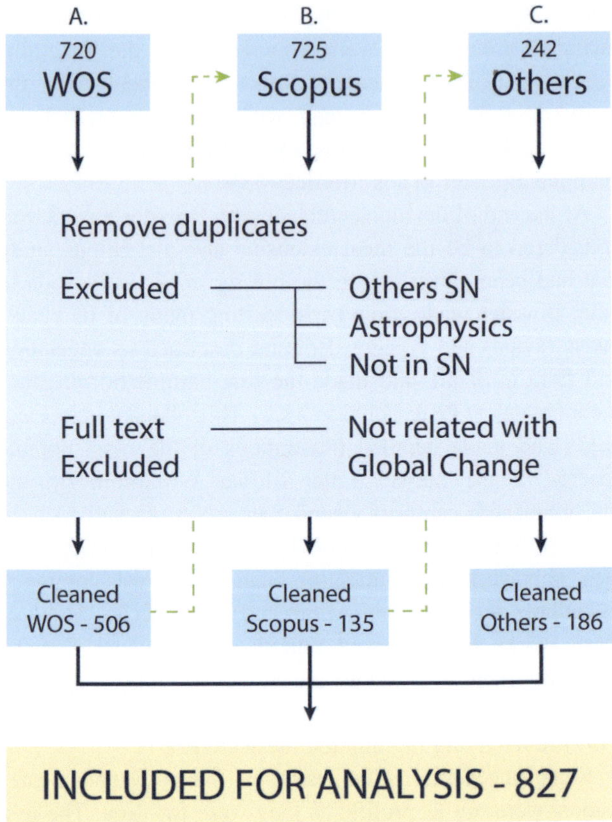


**Fig. 2** Adaptation of the Prisma protocol to an iterative model of recurrent incorporation of different data sources

TITLE-ABS-KEY ("España" OR "Spain") query, looking up the title, summary and keywords of the authors. This request gave rise to 725 records. The resulting database of bibliographic references was analysed in a participatory process involving a team of researchers with experience in the scientific knowledge of Sierra Nevada. In the first few decades of the study, the lack of several leading journals that were not indexed, either totally or partially, was detected. In addition, it was found that some widely cited and scientifically relevant works were missing. Therefore, a complementary search was carried out by research domains and relevant citations in some other repositories such as Dialnet, Digital CSIC, Agris FAO, Research Gate or specific journals (*Amphibia-Reptilia*, *Aquatic Insects*, *Ardeola*, *Oikos*, *Donana Acta Vertebrata*, among others). The last identification added 186 records to the global database.

2. Screening and eligibility

The procedure started with the ISI Web of Science database. Once debugged, it was used as a reference to clean the Scopus database dataset and add it to the combined database for the study. In addition, this database was used for the debugging and the incorporation of the references identified through other sources, in an iterative process (Fig. 2).

This was undertaken by screening the Wos database, eliminating duplicate citations. After screening of the database, the suitability of the records was assessed and references were excluded based on the following criteria:

- The focus of the study was on Astrophysics
- Studies referring to other spaces called Sierra Nevada (in the USA or Colombia)
- Studies in which Sierra Nevada appeared in the search fields but was not the focus of their analysis.

After this exclusion, the full text of the references was analysed to screen works that were not related to global change. In this sense, articles included those related to biodiversity, environmental sciences and ecological aspects of global change in Sierra Nevada (publications dealing with the composition, structure and functioning of natural systems in interaction with human activities). Publications related to geological topics (mineralogy or internal geodynamics), non-environmental historical matters or astronomy were excluded from the dataset. As a result, a refined version of the database of 506 records was drawn from the Wos repository.

Following the iterative model, the result of the reference identification in Scopus was incorporated into the process. The first step was to eliminate duplicate records with respect to the previous database (Wos). Afterwards, items were excluded based on screening and eligibility criteria, leaving a total of 135 records. These were added to the bibliography repository, which resulted in a total of 641 records.

A preliminary exploratory analysis of the joint data of Wos and Scopus revealed the absence of relevant works, mostly belonging to the 70s, 80s and 90s. These works were not indexed for different reasons (e.g. journals that had changed names, journals and references whose metadata were not indexed due to their age). For this reason, new records were identified in other repositories and specific publications. This was a more complex and time-consuming search, in which the references had to be selected one by one. It resulted in the addition of 186 new records, leaving a total of 827 items in a final dataset for analysis.

3. Database integration

Combining all records into a single database for bibliometric analysis has several challenges. The first is the refinement of terminology. A single database has several ways of addressing the same author, institution or other terms. When items from different repositories are added, this complexity becomes greater. When results from a manual search are added to this dataset, where each item comes from a different source, the variety of these terms becomes very high.

For this purpose, a thesaurus of these terms was needed to refine these fields from the three main datasets (Wos, Scopus and other sources). The second step in data integration concerns the tool used for the analysis of the bibliographic data: Bibliometrix (Aria and Cuccurullo 2017). Bibliometrix is a tool for bibliometric analysis that is developed in the R language, with a graphic interface (Biblioshiny). Bibliometrix operates with a variety of data input and output formats and combines the classic functions of bibliographic analysis with more complex techniques such as the analysis of the conceptual, intellectual and social structure of the bibliographic references. However, when it is intended to analyse multi source datasets (Wos, Scopus, Dimensions, Pubmed or Cochrane), the challenge is to combine all the records in a format that Bibliometrix is capable of handling. This is an intricate process, given that each repository returns data with a different structure that needs to be standardized. Moreover, searched references must be added manually, presenting a broad diversity of data. This process involves receiving the original data, importing it for debugging in "Tableau" (Stolte 2003), getting the databases debugged and integrating them into a dataset in "Bibliometrix" format for analysis.

4. Data analysis

Firstly, the database was analysed to provide a preliminary description of the set of scientific publications: total volume of publications, number of documentary sources, bibliographic citations, type of publications and other parameters related to keywords and authorship. Subsequently, all references were reviewed and classified based on the research categories established by the repositories. This indicated the most relevant areas of scientific research regarding Sierra Nevada. For this analysis, the main category of the reference was taken into account.

Afterwards, the scientific journals that have published the largest number of works involving global change were analysed. In addition, the data set was based on the key words used by researchers to define the main areas of their work. Likewise, the institutions most active in the research and publication of scientific knowledge on Sierra Nevada were identified. The indicator used for this analysis was the institution to which the first author of each of the works belonged.

Finally, the studies in relation to Sierra Nevada Global-Change Observatory were analysed. The Observatory operationalizes the research and monitoring of global change through a set of thematic areas (Table 1). For each of the thematic areas of our research procedure, methods were

**Table 1** Sierra Nevada global-change observatory thematic areas

| |
|---|
| Climatology |
| Temporal change in the cryosphere |
| Palaeo perspective |
| Land-use and land-cover changes |
| Atmospheric physicochemistry |
| Population trends and community changes |
| Biodiversity changes |
| Phenological changes |
| Emerging diseases |
| Biogeochemical changes in aquatic systems |
| Primary productivity and carbon fluxes |
| Assessments of ecosystem management activities |
| Ecosystem services and socio-economy |
| Extreme events |

defined to evaluate both the state of key ecological functions, as well as the possible impacts of global change on ecosystems of Sierra Nevada. The thematic areas were organized according to our understanding of the causes and consequences of global change, their ecological consequences and the corresponding biotic and socioeconomic responses to changes (Zamora et al. 2017).

Following this approach, each bibliographical reference was analysed and linked to one or more of these thematic areas. The result enabled us to highlight the research effort related to Sierra Nevada in each thematic area throughout the overall period of publications.

## 3.3 General Overview of Scientific Publications

Sierra Nevada is a widely studied and published mountain system over a diverse range of disciplines (Titos Martínez 1991). In addition to the broad variety of information generated in different types of documents (texts, maps, illustrations, photographs, etc.), the scientific knowledge generated, disseminated and compiled is highly valuable. This overall compilation of research generated on Sierra Nevada is focused on issues related to the processes of global change. In this way, bringing together all the scientific knowledge on Sierra Nevada massif should facilitate access to such information in different areas to be used in management decision-making. As added value, it also helps the entire scientific community to find areas of work and results from colleagues in order to promote joint research and interdisciplinary collaboration.

After eliminating duplicate references, studies carried out elsewhere, those monitoring remote processes, and those not related to global change, the final dataset was left with 827 records from the total of 1632 records gathered from the different search engines (see Sect. 2 Screening and eligibility). Thus, the 827 documents appeared in a total of 407 different sources (journals, books), for a total of 22,624 bibliographic references with an average of 14.1 citations per document.

The main vehicle chosen by the scientific community to disseminate their results among these search engines proved to be the scientific article, which accounted for 86.98% of the total number of entries. Analysed throughout the entire series, data articles accounted for a very low percentage (0.006%). However, given the new trend of standardization and making research data available to the public, the percentage of this type of article is expected to increase in the future. The second channel for the dissemination of scientific findings was the conference article (7.87%), followed by the book chapter (2.72%).

## 3.4 Thematic Analysis of the Dataset

A synthetic method to characterize a scientific work is through their keywords. Search engines provide two different datasets of keywords: those specified by the authors of the publication and those generated by algorithms. A total of 5029 different keywords, used by the authors or generated by the repositories to define the main lines of their study are also used to characterize the most heavily researched topics in Sierra Nevada.

A total of 1603 different authors have studied Sierra Nevada, reflecting the attention accorded this mountain in the scientific sphere. A high proportion (97%) of the works were multi-authored publications. The average co-authorship per paper was 4.3, with the collaboration index value (total authors of multi-authored papers/total multi-authored papers) being 2.1. Single-authored papers accounted for only 9.5% of the total.

Although some publications from the 1970s contained references to Sierra Nevada, it was in the 1980s that a considerable amount of scientific knowledge about Sierra Nevada appeared in publications of international relevance, increasing from 6 to 83 publications (Fig. 3). The number of publications has been increasing decade by decade (168 in the 1990s and 221 in the 2000s), reaching its highest rate between 2010 and 2020 with a total of 412 references.

Figure 4a shows the volume of scientific information generated over these five decades in relation to each of the research categories (established by ISI web of Sciences and manually assigned in missing data). This indicates the prominence of topics related to mountain ecosystems and biodiversity (e.g. "Ecology", "Plant sciences", "Zoology",

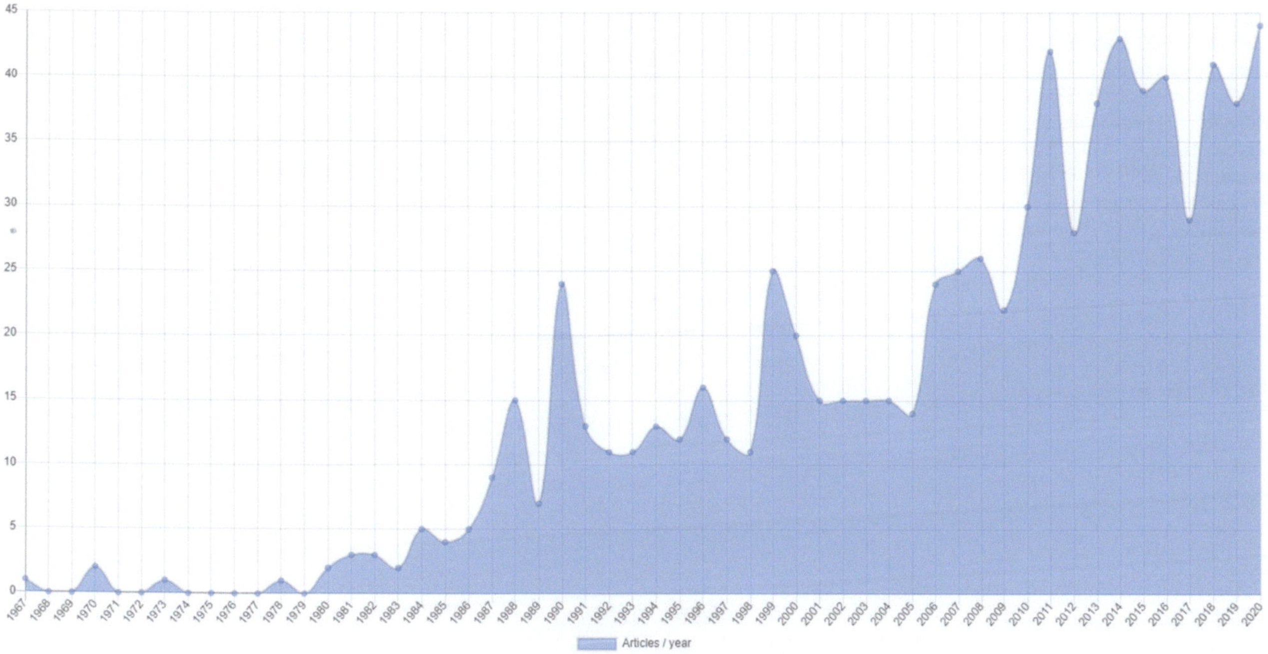


**Fig. 3** Evolution of the number of scientific publications over the study period

"Biodiversity conservation" and "Forestry") with a total of 605 records. Similarly, aspects of aquatic systems ("Limnology" and "Water resources") represented a large volume of research in Sierra Nevada (118 records), exerting a major impact on the study of climate change in fragile Mediterranean mountain ecosystems. Other outstanding areas were multidisciplinary or were related to the biophysical support of ecosystems (e.g. "Environmental Sciences", "Meteorology & Atmospheric Sciences" or "Geosciences").

The 10 leading bibliographic sources of scientific articles (Fig. 4b) include a set of journals related to aquatic systems research: *Hydrobiologia*, *Limnetica* and *Journal of Plankton Research*. In second position is a journal covering issues closely related to global change, i.e. *Science of Total Environment*, followed by a number of publications related to the study of biodiversity, ecology and conservation: *Biological Conservation*, *Forest Ecology and Management*, *Biodiversity and Conservation*, *Ecology* and *Oikos*. A strong position is also held by journals dedicated to taxonomic and morphologic studies of organisms, such as *Aquatic Insects* and *Nova Hedwigia*, this subject being closely related to biodiversity and assiduously researched in Sierra Nevada.

Among the institutions, the University of Granada (Spain), geographically closest to Sierra Nevada, is the research centre that has dedicated by far the most effort to the study of this mountain. In the study period 1970–2021, it carried out nearly 49% of all research activity in Sierra Nevada. Nevertheless, a large number of institutions that have intensively studied the massif, with a total of 168 research institutions that have addressed issues related to global change in Sierra Nevada. Given that the analysis is made only by the institution to which the first author belongs, this is a remarkably large number of institutions involved, reflecting the significant role that Mediterranean mountains play as laboratories for analysing the consequences of global change on ecosystems and the services derived from them. It also offers a valuable scenario to aid decision making in the management of protected natural areas based on scientific knowledge.

### 3.5 Sierra Nevada Global-Change Observatory and Scientific Knowledge of Bibliographic Review

The publications related to the thematic areas of the Global-Change Observatory were evaluated and classified. Each publication was matched to one or more of the observatory's thematic areas. As a result, we were able to identify the areas of global-change monitoring where the most scientific knowledge has been produced and trace its evolution over time.

In a snapshot of the total data from 1970 to date (Fig. 5), tracking Population trends and Changes in biological communities is by far the most published area with 433 matches (32.4%). It should be noted that the study of Biodiversity (215 matches, 15.7%) and Aquatic systems (mainly lakes and rivers, with 148 matches, 11%) are the next two most studied areas. Further in descending order of magnitude, Climatology proved to be a major area, as expected (139

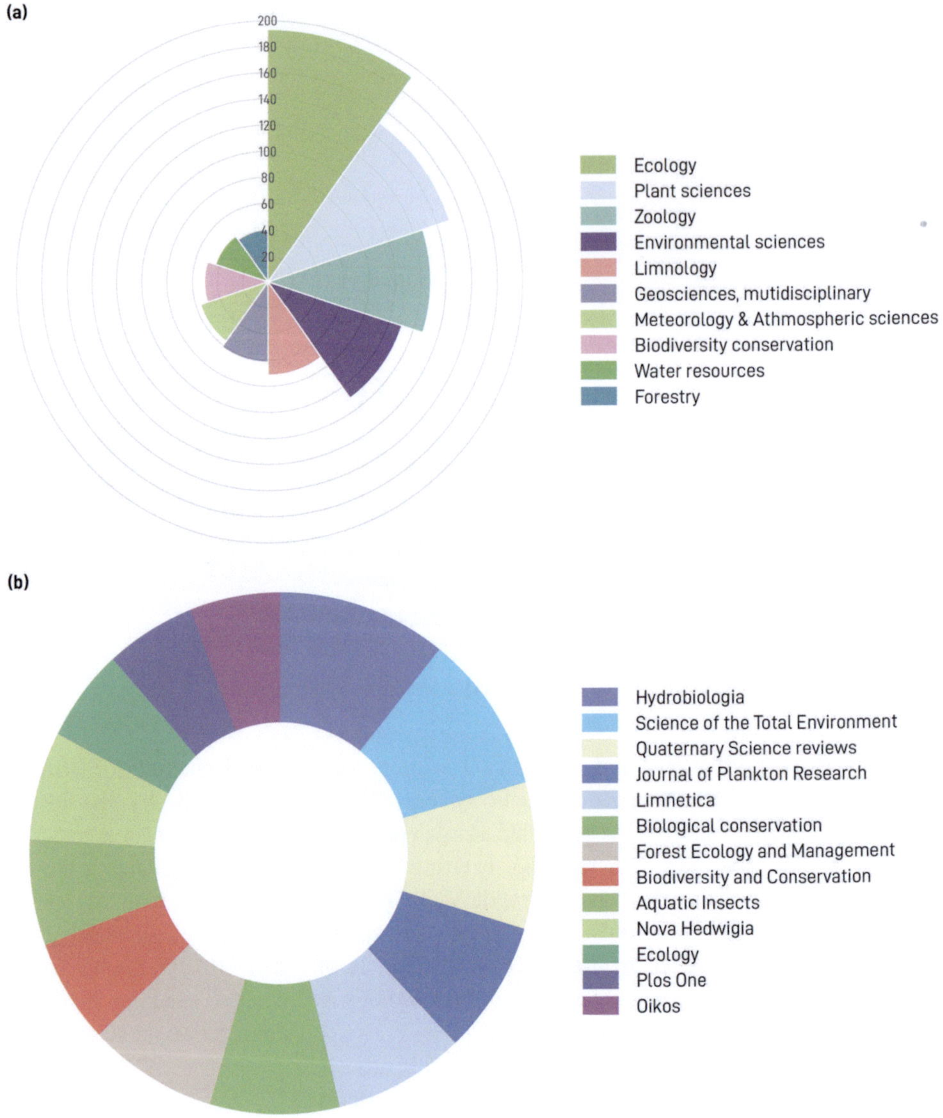


**Fig. 4** **a** Scientific categories (defined by ISI web of Science) most present in the scientific studies carried out in Sierra Nevada. The size of the pie represents the amount of scientific production published in each of the research categories. **b** Most relevant sources of scientific articles in Sierra Nevada

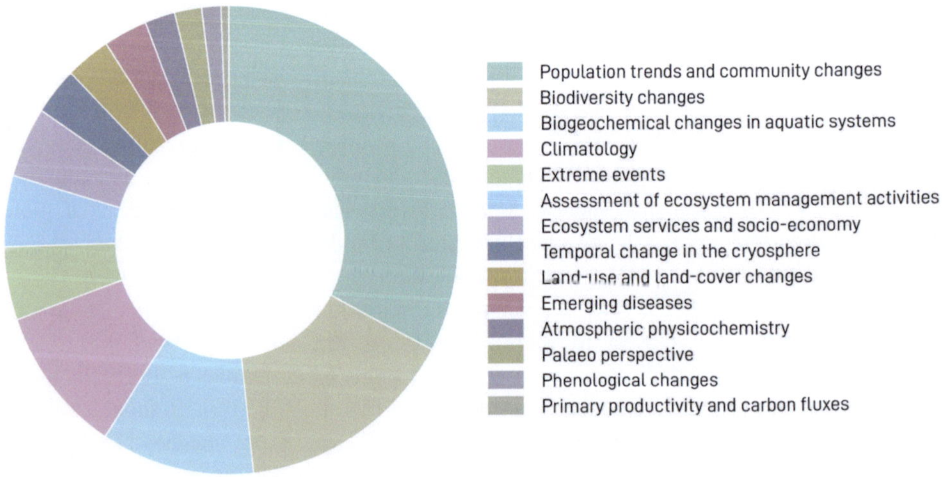


**Fig. 5** Most studied observatory thematic areas in Sierra Nevada (pie size corresponds to the amount of records related with each thematic area of the Global Change Observatory of Sierra Nevada)

matches, 10.1%). Other categories, such as Ecosystem services and socio-economy, Extreme events and Assessment of ecosystem management activities have similar proportions (71, 66 and 66 matches, respectively, representing approximately 5% each), followed by Land use and land cover, Temporal change in the cryosphere and Emerging diseases (45, 45 and 44 matches, respectively, for some 3% each).

In terms of the time course of research in the thematic areas of the Observatory, in the first two decades of the study (the 1970s and 1980s), publications related to Biodiversity and Ecology accounted for 90% of the total number of publications. In this period, other articles were related mainly to the Climate and Palaeo perspective. Over the following decades, articles related to the other thematic areas, such as Assessments of ecosystem management activities, Ecosystem services and socio-economy or Land-use and land-cover changes, gradually began to emerge. Nevertheless, during the period 1990–2020, the categories related to Biodiversity and Ecology rose above 57%, with Climate gradually attracting more attention than less represented areas.

## 4 Development of Representative Case Studies

Having characterized the overall research on Sierra Nevada, we now focus on analysing specific research topics of prominence. Firstly, we highlight the reconstruction of the past (glaciers and palaeoclimatology), with a total of 72 references. We then focus on the ecosystem scale, taking as a reference the aquatic systems (small lakes) that have generated extensive scientific knowledge in this period (148 references). Then, we stress the importance of community ecology in terrestrial ecosystems, focusing on plant-animal interactions that have been extensively studied in Sierra Nevada (178 records). Finally, as a representative of the ecological scale of species, the Iberian ibex (*Capra pyrenaica hispanica)* is examined as one of the emblematic species of this Mediterranean mountain.

### 4.1 Glacial Geomorphological Dynamics

Cold-climate regions have been undergoing large geoecological changes in response to climate warming that occurred over recent decades. Terrestrial and aquatic ecosystems in polar regions and high mountains are among the most heavily affected environments due to their high sensitivity and accelerating climate feedback processes in these regions. The ANTALP group conducts research in the Antarctica, the Arctic and Alpine regions, seeking a fuller understanding of present environmental dynamics in these areas from a multidisciplinary perspective, with a focus on geomorphology, permafrost and geoecology. One of the goals of the team is to reconstruct past environmental and climate conditions using a wide range of natural records including glacial and periglacial landforms, lake sediments, peat bogs and other environmental cues. The main objective is to bridge the gap between past and present processes in order to evaluate whether recent climate and environmental changes are unprecedented or whether analogous phenomena have left imprints in the geological record.

ANTALP research in mountain regions focuses mostly on the Iberian mountains, namely in the Pyrenees, Cantabrian Mountains and Sierra Nevada. Here, in the highest massif in the southern Iberian Peninsula, over the last three decades, members of the team have examined different natural records (glacial, periglacial, lakes, peat bogs, alluvial deposits), combining the study of present-day processes and past environments through cutting-edge field technologies, advanced analytical methodologies and the latest statistical techniques. Present-day processes have been characterized through a wide range of complementary methods (remote-sensing analysis, geophysical measurements, monitoring geomorphic processes and soil thermal regime, etc.), while past environments have been inferred by combining absolute and relative dating techniques (cosmogenic dating, 14C, OSL, Schmidt hammer, etc.) with a multi-proxy approach on the sediments collected from lakes and peat bogs (geochemical, biological and physical indicators).

As a result, Sierra Nevada is one of the most thoroughly studied massifs in Europe in terms of its past environmental evolution. The team has reconstructed the presence of glaciers since the glaciation occurred prior to the last glacial cycle (Palacios et al. 2019) to the maximum ice expansion of the last glacial cycle (Gómez-Ortiz et al. 2012; Palacios et al. 2020, 2016) until the present (Oliva et al. 2014a). Moreover, the team has drawn a detailed picture of the landscape evolution during the Holocene (Oliva et al. 2019, 2011), particularly during the Little Ice Age (Gómez-Ortiz et al. 2018; Oliva and Gómez-Ortiz 2012).

In addition, ANTALP members have monitored present-day geomorphological processes prevailing in the highest sectors of Sierra Nevada. Research has documented periglacial processes that are currently active above 2500 m and are driven primarily by seasonal frost (Oliva et al. 2014b, c, 2009). Permafrost conditions have not been detected at the summit level (Oliva et al. 2016), and only some isolated permafrost patches have been found in the highest northern cirques that were glaciated during the Little Ice Age, such as the Veleta cirque (Gómez-Ortiz et al. 2019). The monitoring of permafrost-related landforms in these areas shows evidence of the degradation of these buried

**Fig. 6** The Veleta Cirque encompassed the last glacier existing in the Sierra Nevada, which disappeared during the mid-twentieth century. The area covered by the snow field that persisted in August 2011 (as a result of a very snowy and cold year) was occupied by a southernmost glacier in Europe during the Little Ice Age (1300–1850 AD) (Photograph by Marc Oliva)

frozen masses in response to the ongoing warming trend (Gómez-Ortiz et al. 2014) (Fig. 6).

## 4.2 Palaeoenvironment and Palaeoclimate in Sierra Nevada

The footprint of global change and disturbances in the environment (e.g. changes in vegetation, fires and erosion) becomes recorded in lake sediments through, for example, the accumulation of fossil remains, charcoal fragments, presence of specific biomarkers, chemical elemental composition and sediment granulometry. In this respect, a significant effort has been made in the last decade with the multidisciplinary study of sedimentary records from several alpine lakes and bogs in Sierra Nevada to understand palaeoenvironmental and palaeoclimatic change throughout the Holocene (last 11,500 years). Palynological studies, with the goal of deciphering vegetation dynamics, were undertaken in Laguna de Río Seco (Anderson et al. 2011), Borreguil de los Tajos la Virgen (Jiménez-Moreno and Anderson 2012), Laguna de la Mula (Jiménez-Moreno et al. 2013), Borreguil de la Caldera (Ramos-Román et al. 2016), Laguna Hondera (Mesa-Fernández et al. 2018) and Laguna de la Mosca (Manzano et al. 2019). Two synthetic works on *Olea* and *Cedrus* regional vegetation dynamics in the last millennia in those Sierra Nevada wetland sites have also been recently published (Ramos-Román et al. 2019; Jiménez-Moreno et al. 2020). Geochemical studies have also been made with the same palaeoenvironmental and palaeoclimatic goals in those alpine wetland sedimentary records (García-Alix et al. 2012, 2013, 2017, 2018, 2020a; Jiménez-Espejo et al. 2014; Toney et al. 2020). More recently, these multidisciplinary palaeoecological analyses have concentrated on a longer and older sedimentary sequence located in Padul, at a foothill elevation of Sierra Nevada (Ramos-Román et al. 2018a, b; Camuera

**Fig. 7** Recovery of a sedimentary sequence from the peat bog of the "Caldera" alpine lake, western "Sierra Nevada" (Photograph by Gonzalo Jiménez-Moreno)

et al. 2018, 2019, 2021). These studies have offered new insight into environmental (vegetation, sedimentation) change related with climate variability and human impact in Sierra Nevada area over the last 200,000 years (Fig. 7).

### 4.3 Limnological Studies

The geographic location and geological history of Sierra Nevada cause high-mountain lakes to the simultaneous exposition of several environmental stressors: climatic anomalies (temperature and precipitation), UV radiation, aerosol deposition and allochthonous nutrient input. High transparency, low nutrient content, narrow temperature ranges found in high-mountain lakes and great simplicity of their biological communities qualifies them as sentinels of global change (Morales-Baquero et al. 2006a; Medina-Sánchez et al. 2016; Villar-Argaiz and Bullejos 2016). These lakes receive mineral nutrients, for instance, phosphorus (Morales-Baquero et al. 2006b), calcium (Pulido-Villena et al. 2006) and iron (Bhattachan et al. 2016), among others elements, in concentrations that depends on the type of atmospheric deposition (wet vs. dry) and its Saharan dust influence (Morales-Baquero et al. 2013). Saharan dust deposition also includes organic matter compounds of humic nature affecting water light transparency and microbial organic substrates (Mladenov et al. 2008, 2009, 2011). In these pristine lakes, the nutrients carried out by the atmospheric deposition boost phytoplankton and bacterioplankton activity (Morales-Baquero et al. 2006b; Pulido-Villena et al. 2008; Reche et al. 2009). Aerosols contain viable bacteria and viruses (Reche et al. 2009, 2018) that can be considered a continuous microbial seed bank stored in the sediments of these lakes. The bioaerosol composition collected in these mountains is similar to that found in the free troposphere microbiome (Triadó-Margarit et al. 2019). Therefore, these alpine lakes above the planetary boundary layer are very useful sites to monitor long-term changes in the atmospheric microbiome.

The altitude and remoteness of Sierra Nevada lakes also make them excellent sites to study long-term effects of climate change. These lakes and their catchments have undergone significant changes throughout the twentieth century. Their lake algal biomass has significantly increased due to warmer and drier conditions and Saharan P deposition since the 1970s (Jiménez et al. 2018). Compositional changes in primary producers and consumer's assemblages are also observed in concordance with warmer conditions and Saharan Ca and P inputs (Jiménez et al. 2015; Burillo et al. 2019; Pérez-Martínez et al. 2020a; Conde-Porcuna et al. 2021). Hydrological and chemical lake regimes have experienced an increase of lake water temperature, reductions of lake water turbulence associated with reduced water levels and water inflows into the lakes (Jiménez et al. 2019; Pérez-Martínez et al. 2020b) and the alkalinization of lake water and evapoconcentration of solutes (Jiménez et al. 2018; Pérez-Martínez et al. 2020a, b).

Some of the scientific achievements in the high-mountain lakes of Sierra Nevada include theoretical-conceptual advances in models of microbial food webs and biogeography (Reche et al. 1997; Medina-Sánchez et al. 1999, 2004, 2013; Villar-Argaiz et al. 2002a; Reche et al. 2005, 2007; Carrillo et al. 2006, 2008a; Biddanda et al. 2021), patterns of microbial limitation across trophic gradients (Medina-Sánchez et al. 2010), life history, ontogeny and homeostasis of zooplankton from ecological stoichiometric theory (Barea et al. 2001; Carrillo et al. 2001; Villar-Argaiz et al. 2002b, c; Pérez-Martinez et al. 2007, 2013; Conde-Porcuna et al. 2011, 2014; Bullejos et al. 2014a), patterns of herbivorous consumer growth depending on food quantity and quality (Souza et al. 2010; Villar-Argaiz et al. 2012, 2018a; Bullejos et al. 2014b) and loss of functional biodiversity due to global change (Delgado-Molina et al. 2009; Carrillo et al. 2017). An important issue addressed in the research on high-mountain lakes of Sierra Nevada is the impact of multiple global-change stressors on phytoplankton, phytobenthos, zooplankton, bacterioplankton and the ecological relationships among them (Carrillo et al. 2002, 2008b, 2015; Medina-Sánchez et al. 2002, 2006, 2013; Villar-Argaiz et al. 2001, 2009, 2016, 2018b; Figueroa et al. 2009; Bullejos et al. 2010; Korbee et al. 2012; Helbling et al. 2013; Durán et al. 2016, 2020; Cabrerizo et al. 2017; González-Olalla et al. 2018) (Fig. 8).

## 4.4 Plant–Animal Interactions: Ecological and Evolutionary Consequences

Plant-animal interactions that have been extensively studied in Sierra Nevada. Probably the most successful case study has been *Hormathophylla spinosa* and its diverse network of biotic interactions. These investigations have given rise to some highly cited papers, being also included as an example in some major ecology textbooks. *H. spinosa* is a stunted shrub distributed throughout most of the mountains of the western Mediterranean (Salmerón-Sánchez et al. 2018) and is common in the highland areas of Sierra Nevada (Lamprecht et al. 2021). The plant is widely covered with thorny outgrowths that consist in most cases of the stems of inflorescences from previous years that have hardened (Zamora and Gómez 1993). In Sierra Nevada, the individuals of this species bloom every year (Gómez 1993b) in a mass flowering habit, from June to the end of August (Gómez 1993a). The flowers are hermaphroditic, with four ovules and six stamens each (Gómez 1993b), although this species produces flowers with a variable number of stamens (Méndez and Gómez 2006).

This scrub species plays a central role in the food webs of Sierra Nevada range by maintaining a diverse network of interactions with innumerable mutualistic and antagonistic organisms and behaving as an ecosystem engineer (Gómez and Zamora 1999, 2000a, b, 2002, 2003, 2006; Zamora 2000). For example, *H. spinosa* flowers are pollinated by more of 70 species of insects, notably the ant *Proformica longiseta* Collingwood (Formicidae), which makes up more than 80% of the floral visits that this plant receives (Gómez and Zamora 1992; Gómez et al. 1995). *H. spinosa* can be pollinated not only by insects but even by wind in populations located at higher altitudes (Gómez and Zamora 1996; Gómez et al. 1996). However, only 20% of the flowers produce fruits, each with one or two seeds (Gómez 1993a). Many of these seeds are eaten inside the fruit by a species of weevil, *Ceutorhynchus* sp. nova (Curculionidae), which considerably reduces the final number of seeds that can be dispersed (Gómez 1993a; Gómez and Zamora 1994). The effect of these insects is not only quantitative, but also qualitative, since they selectively prey on the largest seeds. The populations of these predatory insects are controlled by

**Fig. 8** "Río Seco" alpine lake. Poqueira´s valley in Sierra Nevada (Photograph by Juan Manuel Medina)

several species of parasitoids, this relation exerting a positive indirect effect for the plant, which can disperse more and larger seeds when the parasitoid is present (Gómez and Zamora 1994). *H. spinosa* is also attacked by a parasitic plant, *Cuscuta epithymum* L. (Cuscutaceae), which reduces the reproductive potential of the host plant (Gómez 1994). The seed production of *H. spinosa* also depends partially on flower and leaf consumption by a species of endemic beetle of Sierra Nevada, *Timarcha lugens* (Chrysomelidae) (Gómez and González-Megías 2002). This species is strictly monophagous and feeds exclusively on *H. spinosa* during both its larval and adult stages (González-Megías and Gómez 2001). This plant also maintains a network of antagonistic invertebrate interactors, including leaf miners, stem miners, sap suckers and folivores (González-Megías and Gómez 2003). However, all these interactions are conditioned by the action of the ibex (*Capra pyrenaica* Schinz, Bovidae) and domestic cattle, which become the most influential factor, determining both the direct and indirect interactions between the plant and its constellation of organisms that govern the final reproductive success of the plant (Zamora and Gómez 1993; Gómez 1993a, 1994; Gómez and González Megías 2002; González-Megías and Gómez 2003). Despite these reproductive losses, each plant disperses, on average, between 1000 and 2000 seeds at the end of each reproductive event, although the inter-individual variation is very large. More than 75% of these seeds fall below the canopy of the mother plant while less than 3% are dispersed to distances greater than two meters away (Gómez 1991). *H. spinosa*, due to its great abundance in the high-mountains and its pulvinular growth pattern, frequently interacts with other plant species. Thus, it is a competitor of some herbaceous species (Macek et al. 2016). On the other hand, with its ability to retain soil moisture and reduce insolation, it can also act as a facilitator for many other high-mountain plant species (Losapio et al. 2019; Losapio and Schöb 2020; Schöb et al. 2012, 2013). This effect sometimes transcends the merely ecological effect, and implies that some species are locally adapted to grow, facilitated by *H. spinosa* (O'Brien et al. 2021). In many cases, this facilitating role is fulfilled at a cost in terms of growth and interaction with pollinators (Fig. 9).

**Fig. 9** *Hormathophylla spinosa* is a mass-flowering plant of Sierra Nevada that interacts with many organisms (Photograph by Andrés Ureña)

## 4.5 Biology and Ecology of *Capra pyrenaica*

Sierra Nevada is home to the largest *Capra pyrenaica* population of all the existing ones (Pérez et al. 2002) and the one with the greatest genetic diversity (Márquez et al. 2020). The parameters that define the population are monitored annually (Granados et al. 2020). Parallel to the monitoring of the population parameters, the individuals that make up the population have been characterized in detail, addressing the biometric characteristics and growth of the individuals (Granados et al. 1997; Sarasa et al. 2012), and determining haematological and physiological parameters (Pérez et al. 1999; Serrano et al. 2008) or body condition of the individuals (Sarasa et al. 2010; Serrano et al. 2011). The

**Fig. 10** Male of Iberian ibex *Capra pyrenaica* (Photograph by Andrés Ureña)

movement of specimens and habitat selection has also been studied (Viana et al. 2018).

In addition to population monitoring, studies have been made of the main infectious diseases that affect this species. Sarcoptidosis is perhaps the most thoroughly studied parasitosis since its seasonal dynamics are known (Pérez et al. 1997) and advances have been made in the diagnosis of the disease (Ráez Bravo et al. 2016; Valldeperes et al. 2019), as well as effects in the host (Ráez-Bravo et al. 2015; López Olvera et al. 2015; Espinosa et al. 2017a, b). Other parasites have also been studied: oestridae (Pérez et al. 1996), gastrointestinal nematodes (Pérez et al. 2003), bronchopulmonary nematodes (Alasaad et al. 2009) and trematodes (Alasaad et al. 2008).

The above cumulative experience has made it possible to draft a management protocol for the ibex in captivity (Espinosa et al. 2017c) as well as to establish recommendations on the management of domestic goat populations affected by sarcoptidosis (Espinosa et al. 2020; Pérez et al. 2021) (Fig. 10).

## 5 Conclusions

Sierra Nevada has long been the focal topic of scientific interest due to its natural richness, its ecological values and the interaction of society with the ecosystems. Very relevant scientific knowledge has been generated in recent decades, but it still does not have the necessary visibility for the general public. This bibliographic compilation of scientific knowledge about global change in Sierra Nevada aims to make accessible the research work of hundreds of researchers. Thus, the purpose is to give added value to the scientific information, being used by managers and the general public. To achieve this, all the scientific publications in the Wos and Scopus databases have been compiled, as well as other additional repositories such as Dialnet, Digital CSIC, Agris FAO, ResearchGate or specific journals, less accessible to society. A unique repository of scientific information from such diverse sources requires an important effort of comparison, removal of duplicates, standardization of the databases and classification of the information obtained in order to increase its usability by different social profiles (scientists, managers and citizens in general). The result is an application called "Biblionevada" (https://biblionevada.obsnev.es/), which allows the search of bibliographic references (alphanumeric and graphic) based on fields such as "title of the publication", "author/s", "abstract", "date", "institution", "name of the journal or book", "keywords", "research areas", "thematic areas of Sierra Nevada Global Change Observatory" as well as "traditional categories".

Biblionevada is aligned with the European philosophy of proper information management, following international standards, which enables free access to scientific information for the whole society. In the same way, the University of Granada participates as a research mountain node in the Lifewatch European Research Infrastructure, with the development of virtual research environments (VREs). VREs are designed to transform scientific knowledge into useful information for the resolution of society's environmental challenges (see chapter "Data Model, E-Infrastructure Services, and the Virtual Research Environment (VRE)").

**Acknowledgements** Thanks to Manuel Titos, Daniel Torres, José María Gómez-Reyes, Marc Oliva, Gonzalo Giménez, Juan Manuel Medina and José Enrique Granados for providing a summary text of their group's research. Thanks to Pedro Magaña for his work on the integration of the databases. This work has been carried out within the framework of the Collaboration Agreement between the "Consejería de Medio Ambiente y Ordenación del Territorio de la Junta de Andalucía" and the University of Granada for the development of activities related to the Global Change Observatory of Sierra Nevada. We are also grateful for the partial support received from the LIFE-ADAPTAMED projects (LIFE14 CCA/ES/ 000612): Protection of Key Ecosystem Services Threatened by Climate Change through Adaptive Management of Climate Change (LIFE14 CCA/ES/ 000612). by climate change through adaptive management of Mediterranean socio-ecosystems and the H2020 project European Long-Term Ecosystem and socio-ecological Research Infrastructure (eLTER). Likewise, we are grateful for the partial support received from Thematic Center on Mountain Ecosystem & Remote sensing, Deep learning-AI e-Services University of Granada-Sierra Nevada (LifeWatch-2019-10-UGR-01). We are grateful to the management and technical team of the Sierra Nevada Natural Area for their continued collaboration in the activities that we develop in the Global Change Observatory.

## References

Aria M, Cuccurullo C (2017) bibliometrix: an R-tool for comprehensive science mapping analysis. J Inf 11:959–975.

Blanca López G (2002) Flora amenazada y endémica de Sierra Nevada. Universidad de Granada, Granada

Gilbert N (2014) Green List promotes best conservation areas. Nature 515:322.

Gómez Ortiz A, Plana Castellví JA (2004) El paisaje glaciar de Sierra Nevada a través de los escritos de época (ss. XVIII-XIX). Contribución al conocimiento geográfico español/Antonio Gómez Ortiz y José Antonio Plana Castellví

Higgins JPT, Thomas J, Chandler J et al (2019) Cochrane handbook for systematic reviews of interventions. Wiley

López Ontiveros A, Naranjo Ramírez J (2000) Juan Carandell Pericay (1893–1937) y Sierra Nevada. Cuad Geogr 281–324

Moher D, Liberati A, Tetzlaff J et al (2009) Preferred reporting items for systematic reviews and meta-analyses: the PRISMA statement. PLoS Med 6

Ruano F, de Figueroa JM, Tinaut Ranera A (2013) Los insectos de Sierra Nevada: 200 años de historia. Asociación Española de Entomología

Sánchez Castillo PM (1988) Algas de las lagunas de alta montaña de Sierra Nevada. ABM 13:21–52

Stolte C (2003) Tableau desktop. Tableau Software, Seattle, Washington (Estados Unidos)

Titos M (1997) Sierra Nevada: una gran, historia. Universidad de Granada, Granada

Titos M (1990) La aventura de Sierra Nevada: 1717–1915. Universidad de Granada, Granada

Titos Martínez M (1991) Textos primitivos sobre Sierra Nevada (1754–1838). Caja General de Ahorros y Monte de Piedad de Granada, Granada

Urrútia G, Bonfill X (2010) Declaración PRISMA: una propuesta para mejorar la publicación de revisiones sistemáticas y metaanálisis. Med Clin (Barc) 135:507–511

Zamora R, Pérez-Luque AJ, Bonet FJ (2017) Monitoring global change in high mountains. In: Catalan J, Ninot JM, Aniz MM (eds) High mountain conservation in a changing world. Springer International Publishing, Cham, pp 385–413

## Case Study 4.1. Antarctic, Arctic and Alpine Environments (ANTALP)—Sierra Nevada

Gómez-Ortiz A, Oliva M, Salvador-Franch F et al (2019) Monitoring permafrost and periglacial processes in Sierra Nevada (Spain) from 2001 to 2016. Permafr Periglac Process

Gómez-Ortiz A, Oliva M, Salvador-Franch F et al (2014) Degradation of buried ice and permafrost in the Veleta cirque (Sierra Nevada, Spain) from 2006 to 2013 as a response to recent climate trends. Solid Earth 5:979–993

Gómez-Ortiz A, Oliva M, Salvador-Franch F et al (2018) The geographical interest of historical documents to interpret the scientific evolution of the glacier existing in the Veleta cirque (Sierra Nevada, Spain) during the Little Ice Age. 44:267–292

Gómez-Ortiz A, Palacios D, Palade B et al (2012) The deglaciation of the Sierra Nevada (Southern Spain). Geomorphology 159–160:93–105

Oliva M, Gómez-Ortiz A (2012) Late-Holocene environmental dynamics and climate variability in a Mediterranean high mountain environment (Sierra Nevada, Spain) inferred from lake sediments and historical sources. Holocene 22:915–927

Oliva M, Gómez-Ortiz A, Palacios D et al (2019) Multiproxy reconstruction of Holocene glaciers in Sierra Nevada (south Spain). Mediterr Geosci Rev

Oliva M, Gómez-Ortiz A, Salvador-Franch F et al (2016) Inexistence of permafrost at the top of the Veleta peak (Sierra Nevada, Spain). Sci Total Environ 550:484–494

Oliva M, Gómez Ortiz A, Palacios D et al (2014a) Environmental evolution in Sierra Nevada (South Spain) since the Last Glaciation, based on multi-proxy records. Quat Int 353:195–209

Oliva M, Gómez Ortiz A, Salvador F et al (2014b) Long-term soil temperature dynamics in the Sierra Nevada, Spain. Geoderma 235–236:170–181

Oliva M, Ortiz AG, Franch FS et al (2014) Present-day solifluction processes in the semi-arid range of Sierra Nevada (Spain). Arctic Antarct Alp Res 46:365–370
Oliva M, Schulte L, Gómez-Ortiz A (2009) Morphometry and Late Holocene activity of solifluction landforms in the Sierra Nevada, southern Spain. Permafr Periglac Process 20:369–382
Oliva M, Schulte L, Ortiz AG (2011) The role of aridification in constraining the elevation range of Holocene solifluction processes and associated landforms in the periglacial belt of the Sierra Nevada (southern Spain). Earth Surf Process Landf 36:1279–1291
Palacios D, Gómez-Ortiz A, Alcalá-Reygosa J et al (2019) The challenging application of cosmogenic dating methods in residual glacial landforms: The case of Sierra Nevada (Spain). Geomorphology 325:103–118
Palacios D, Gómez-Ortiz A, Andrés N et al (2016) Timing and new geomorphologic evidence of the last deglaciation stages in Sierra Nevada (southern Spain). Quat Sci Rev 150:110–129
Palacios D, Oliva M, Gómez-Ortiz A et al (2020) Record of several glacial advances from Late Pleistocene to Late Holocene in the Mulhacén cirque, Sierra Nevada, Southern Spain. The Holocene

## Case Study 4.2. Paleoenvironmental and Paleoclimate—Sierra Nevada

Anderson RS, Jiménez-Moreno G, Carrión JS et al (2011) Holocene vegetation history from Laguna de Río Seco, Sierra Nevada, southern Spain. Quat Sci Rev 30:1615–1629
Camuera J, Jiménez-Moreno G, Ramos-Román MJ et al (2018) Orbital-scale environmental and climatic changes recorded in a new ~200,000-year-long multiproxy sedimentary record from Padul, southern Iberian Peninsula. Quat Sci Rev 198:91–114
Camuera J, Jiménez-Moreno G, Ramos-Román MJ et al (2019) Vegetation and climate changes during the last two glacial-interglacial cycles in the western Mediterranean: a new long pollen record from Padul (southern Iberian Peninsula). Quat Sci Rev 205:86–105
Camuera J, Jiménez-Moreno G, Ramos-Román MJ et al (2021) Chronological control and centennial-scale climatic subdivisions of the Last Glacial Termination in the western Mediterranean region. Quat Sci Rev 255:106814
García-Alix A, Jiménez-Moreno G, Anderson RS et al (2012) Holocene paleoenvironmental evolution of a high-elevation wetland in Sierra Nevada, southern Spain, deduced from an isotopic record. J Paleolimnol 48:471–484
García-Alix A, Jimenez-Espejo FJ, Lozano JA et al (2013) Anthropogenic impact and lead pollution throughout the Holocene in southern Iberia. Sci Total Environ 449:451–460
García-Alix A, Jiménez-Espejo FJ, Toney JL et al (2017) Alpine bogs of southern Spain show human-induced environmental change superimposed on long-term natural variations. Sci Rep 7:7439
García-Alix A, Jiménez-Espejo FJ, Jiménez-Moreno G et al (2018) Holocene geochemical footprint from Semi-arid alpine wetlands in southern Spain. Sci Data 5:180024
García-Alix A, Toney JL, Jiménez-Moreno G et al (2020) Algal lipids reveal unprecedent warming rates in alpine areas of SW Europe during the industrial period. Clim Past 16:245–263
Jiménez-Espejo FJ, García-Alix A, Jiménez-Moreno G et al (2014) Saharan aeolian input and effective humidity variations over Western Europe during the Holocene. Chem Geol 374–375:1–12
Jiménez-Moreno G, Anderson RS (2012) Holocene vegetation and climate change recorded in alpine bog sediments from the Borreguiles de la Virgen, Sierra Nevada, southern Spain. Quat Res 77:44–53
Jiménez-Moreno G, García-Alix A, Hernández-Corbalán MD et al (2013) Vegetation, fire, climate and human disturbance history in the southwestern Mediterranean area during the late Holocene. Quat Res 79:110–122
Jiménez-Moreno G, Anderson RS, Ramos-Román MJ et al (2020) The Holocene Cedrus pollen record from Sierra Nevada (S Spain), a proxy for climate change in N Africa. Quat Sci Rev 242:106468
Manzano S, Carrión JS, López-Merino L et al (2019) A palaeoecological approach to understanding the past and present of Sierra Nevada, a Southwestern European biodiversity hotspot. Glob Planet Chang 175:238–250
Mesa-Fernández JM, Jiménez-Moreno G, Rodrigo-Gámiz M et al (2018) Vegetation and geochemical response to Holocene rapid climate change in the Sierra Nevada (southeastern Iberia): the Laguna Hondera record. Clim Past 14:1687–1706
Ramos-Román MJ, Jiménez-Moreno G, Anderson RS et al (2016) Centennial-scale vegetation and North Atlantic Oscillation changes during the late Holocene in the western Mediterranean. Quat Sci Rev 143:84–95
Ramos-Román MJ, Jiménez-Moreno G, Camuera J et al (2018) Holocene climate aridification trend and human impact interrupted by millennial and centennial-scale climate fluctuations from a new sedimentary record from Padul (Sierra Nevada, southern Iberian Peninsula). Clim Past 14:117–137
Ramos-Román MJ, Jiménez-Moreno G, Camuera J et al (2018) Millennial-scale cyclical environment and climate variability during the Holocene in the western Mediterranean region deduced from a new multiproxy analysis from the Padul record (Sierra Nevada, Spain). Glob Planet Chang 168:35–53
Ramos-Román MJ, Jiménez-Moreno G, Anderson RS et al (2019) Climate controlled historic olive tree occurrences and olive oil production in southern Spain. Glob Planet Chang 182:102996
Toney JL, García-Alix A, Jiménez-Moreno G et al (2020) New insights into Holocene hydrology and temperature from lipid biomarkers in western Mediterranean alpine wetlands. Quat Sci Rev 240:106395

## Case Study 4.3. Limnological Studies

Barea-Arco J, Pérez-Martínez C, Morales-Baquero R (2001) Evidence of a mutualistic relationship between an algal epibiont and its host, Daphnia pulicaria. Limnol Ocean 46(871):871–881
Bhattachan A, Reche I, D'Odorico P (2016) Soluble ferrous iron (Fe (II)) enrichment in airborne dust. J Geophys Res Atmos 121 (17):10–153
Biddanda B, Dila D, Weinke A et al (2021) Housekeeping in the hydrosphere: microbial cooking, cleaning, and control under stress. Life 11:152
Bullejos FJ, Carrillo P, Gorokhova E et al (2014a) Shifts in food quality for herbivorous consumer growth: multiple golden means in the life history. Ecology 95:1272–1284
Bullejos FJ, Carrillo P, Gorokhova E (2014b) Nucleic acid content in crustacean zooplankton: bridging metabolic and stoichiometric Predictions. Plos One 9(1):e86493
Bullejos FJ, Carrillo P, Villar-Argaiz M et al (2010) Roles of phosphorus and ultraviolet radiation in the strength of

phytoplankton-zooplankton coupling in a Mediterranean high mountain lake. Limnol Ocean 55:2549–2562

Burillo JP, Jiménez L, Pérez-Martínez C (2019) Identifying invasive Daphnia species by morphological analysis of postabdominal claws in Sierra Nevada alpine lakes. J Paleolimnol 62:121–135

Cabrerizo MJ, Medina-Sánchez JM, Dorado I et al (2017) Rising frequency of resource pulses under solar UVR strengthen microbial interactions. Sci Rep 7:43615

Carrillo P, Delgado-Molina JA, Medina-Sánchez JM et al (2008a) Phosphorus inputs unmask negative effects of UV radiation on algae in a high mountain lake. Glob Chang Biol 14:423–439

Carrillo P, Medina-Sánchez JM, Duran C et al (2015) Synergistic effects of UVR and simulated stratification on commensalistic algal-bacterial relationship in two optically contrasting oligotrophic Mediterranean lakes. Biogeosciences 12:697–712

Carrillo P, Medina-Sánchez JM, Villar-Argaiz M (2002) The interaction of phytoplankton and bacteria in a high mountain lake: importance of the spectral composition of solar radiation. Limnol Ocean 47:1294–1306

Carrillo P, Medina-Sánchez JM, Villar-Argaiz M et al (2006) Complex interactions in microbial food webs: stoichiometric and functional approaches. In: Armengol J, Muñoz I (eds) The ecology of the Iberian inland waters: homage to Ramón Margalef. Páginas, pp 189–204. Asociación Española de Limnología, Madrid. ISSN 0213-8409

Carrillo P, Villar-Argaiz M, Medina-Sánchez JM (2001) Relationship between N:P ratio and growth rate during the life cycle of calanoid copepods: an in situ measurement. J Plankton Res 23:537–547

Carrillo P, Villar-Argaiz M, Medina-Sánchez JM (2008) Does microorganism stoichiometry predict microbial food web interactions after a phosphorus pulse? Microb Ecol 56:350–363

Carrillo P, Medina-Sanchez JM, Villar-Argaiz M et al (2017) Vulnerability of mixotrophic algae to nutrient pulses and UVR in an oligotrophic Southern and Northern Hemisphere lake. Sci Rep 7:6333

Conde-Porcuna JM, Valdés-Cano J, Romo S et al (2011) Ephippial and subitaneous egg abortion: relevance for an obligate parthenogenetic Daphnia population. J Limnol 70(1):69–75

Conde-Porcuna JM, Ramos-Rodríguez E, Pérez-Martínez C (2014) In situ production of empty ephippia and resting eggs by Daphnia: relationships with photoperiod and food conditions. J Plankton Res 36:157–169

Conde-Porcuna JM, Veiga J, Moreno E et al (2021) Phylogeny and spatiotemporal genetic patterns in the Daphnia pulex complex from Sierra Nevada lakes (Spain): first record of North American D. pulex in a European high mountain lake. J Plankton Res 43(3):380–395

Delgado-Molina JA, Carrillo P, Medina-Sánchez JM, Villar-Argaiz M, Bullejos FJ (2009) Interactive effects of phosphorus loads and ambient ultraviolet radiation on the algal community in a high-mountain lake. J Plankton Res 31:619–634

Durán C, Medina-Sánchez JM, Carrillo P (2020) Uncoupled phytoplankton-bacterioplankton relationship by multiple drivers interacting at different temporal scales in a high-mountain Mediterranean lake. Sci Rep 8:350

Durán C, Medina-Sánchez JM, Herrera G et al (2016) Changes in the phytoplankton-bacteria coupling triggered by joint action of UVR, nutrients, and warming in Mediterranean high-mountain lakes. Limnol Ocean 61:413–429

Figueroa FL, Korbee N, Carrillo P et al (2009) The effects of UV radiation on photosynthesis estimated as chlorophyll fluorescence in Zygnemopsis decussata (Chlorophyta) growing in a high mountain lake (Sierra Nevada, Southern Spain). J Limnol 68:206–216

González-Olalla JM, Medina-Sánchez JM, Lozano IL et al (2018) Climate-driven shifts in algal-bacterial interaction of high-mountain lakes in two years spanning a decade. Sci Rep 8:10278

Helbling EW, Carrillo P, Medina-Sánchez JM et al (2013) Interactive effects of vertical mixing, nutrients and ultraviolet radiation: In situ photosynthetic responses of phytoplankton from high mountain lakes of Southern Europe. Biogeosciences 10:1037–1050

Jiménez L, Rühland KM, Jeziorski A et al (2018) Climate change and Saharan dust drive recent cladoceran and primary production changes in remote alpine lakes of Sierra Nevada, Spain. Glob Chang Biol 24:e139–e158

Jiménez L, Conde-Porcuna JM, Heiri O et al (2019) Ecosystem responses to climate-related changes in a Mediterranean alpine environment over the last ~180 years. Ecosystems 22:563–577

Jiménez L, Romero-Viana L, Conde-Porcuna et al (2015) Sedimentary photosynthetic pigments as indicators of climate and watershed perturbations in an alpine lake in southern Spain. Limnetica 34 (2):439–454

Korbee N, Carrillo P, Mata MT et al (2012) Effects of ultraviolet radiation and nutrients on the structure-function of phytoplankton in a high mountain lake. Photochem Photobiol Sci 11:1087–1098

Medina-Sánchez JM, Carrillo P, Delgado-Molina JA et al (2010) Patterns of resource limitation of bacterial on a trophic gradient in Mediterranean inland waters. FEMS Microbiol Ecol 74:554–565

Medina-Sánchez JM, Delgado-Molina JA, Carrillo P (2016) Sentinels of global change (I): mixotrophic algae in La Caldera Lake. In: Zamora R, Pérez-Luque AJ, Bonet FJ, Barea-Azcón JM and Aspizua R (editors) Global Change Impacts in Sierra Nevada: Challenges for Conservation. Consejería de Medio Ambiente y Ordenación del Territorio. Junta de Andalucía

Medina-Sánchez JM, Delgado-Molina JA, Bratbak G et al (2013) Maximum in the middle: nonlinear response of microbial plankton to ultraviolet radiation and phosphorus. Plos One 8(4):e60223

Medina-Sánchez JM, Villar-Argaiz M, Carrillo P (2002) Modulation of the bacterial response to spectral solar radiation by algae and limiting nutrients. Freshw Biol 47:2191–2204

Medina-Sánchez JM, Villar-Argaiz M, Carrillo P (2004) Neither with nor without you: a complex algal control on bacterioplankton in a high mountain lake. Limnol Ocean 49:1722–1733

Medina-Sánchez JM, Villar-Argaiz M, Carrillo P (2006) Solar radiation-nutrient interaction enhances the resource and predation algal control on bacterioplankton: a short-term experimental study. Limnol Ocean 51:913–924

Medina-Sánchez JM, Villar-Argaiz M, Cruz-Pizarro L et al (1999) Structure changes in a planktonic food web: biotic and abiotic controls. J Limnol 58:213–222

Mladenov N, Lípez-Ramos J, McKnight DM et al (2009) Alpine lake optical properties as sentinels of dust deposition and global change. Limnol Ocean 54(6part2):2386–2400

Mladenov N, Pulido-Villena E, Morales-Baquero R et al (2008) Spatiotemporal drivers of dissolved organic matter in high alpine lakes: role of Saharan dust inputs and bacterial activity. J Geophys Res Biogeosci 113(G2)

Mladenov N, Sommaruga R, Morales-Baquero R et al (2011) Dust inputs and bacteria influence dissolved organic matter in clear alpine lakes. Nat Commun 2(1):1–7

Morales-Baquero R, Pulido-Villena E, Reche I (2006a) Atmospheric inputs of phosphorus and nitrogen to the southwest Mediterranean region: biogeochemical responses of high mountain lakes. Limnol Ocean 51(2):830–837

Morales-Baquero R, Pulido-Villena E, Reche I (2013) Chemical signature of Saharan dust on dry and wet atmospheric deposition in the south-western Mediterranean region. Tellus B Chem Phys Meteorol 65(1):18720

Morales-Baquero R, Pulido-Villena E, Romera O et al (2006) Significance of atmospheric deposition to freshwater ecosystems in the southern Iberian Peninsula. Limnetica 25(1–2):171–180

Pérez-Martínez C, Barea-Arco J, Conde-Porcuna JM (2007) Reproduction strategies of Daphnia pulicaria population in a high mountain lake of Southern Spain. Hydrobiologia 594(1):75–82

Pérez-Martínez C, Conde-Porcuna JM, Moreno E (2020a) Cladoceran assemblage distribution in shallow alpine lakes of Sierra Nevada (Spain) and its relationship with environmental variables. Aquat Sci 82:4

Pérez-Martínez C, Jiménez L, Moreno E et al (2013) Emergence pattern and hatching cues of Daphnia pulicaria (Crustacea, Cladocera) in an alpine lake. Hydrobiologia 707(1):47–57

Pérez-Martínez C, Rühland KM, Smol JP et al (2020b) Long-term ecological changes in Mediterranean mountain lakes linked to recent climate change and Saharan dust deposition revealed by diatom analyses. Science of The Total Environment, 138519

Pulido-Villena E, Reche I, Morales-Baquero R (2006) Significance of atmospheric inputs of calcium over the southwestern Mediterranean region: High mountain lakes as tools for detection. Glob Biogeochem Cycles 20(2)

Pulido-Villena E, Reche I, Morales-Baquero R (2008) Evidence of an atmospheric forcing on bacterioplankton and phytoplankton dynamics in a high mountain lake. Aquat sci 70(1):1–9

Reche I, D'Orta G, Mladenov N et al (2018) Deposition rates of viruses and bacteria above the atmospheric boundary layer. ISME J 12 (4):1154–1162

Reche I, Ortega-Retuerta E, Romera O et al (2009) Effect of Saharan dust inputs on bacterial activity and community composition in Mediterranean lakes and reservoirs. Limnol Ocean 54(3):869–879

Reche I, Pulido-Villena E, Morales-Baquero R et al (2005) Does ecosystem size determine aquatic bacterial richness? Ecology 86 (7):1715–1722

Reche I, Pulido-Villena E, Morales-Baquero R et al (2007) Does ecosystem size determine aquatic bacteria richness? Reply. Ecology 88(1):253–255

Reche I, Carrillo P, Cruz-Pizarro L (1997) Influence of metazooplankton on interactions of bacteria and phytoplankton in an oligotrophic lake. J Plankton Res 19:631–646

Souza MS, Modenutti BE, Carrillo P et al (2010) Stoichiometric dietary constraints influence the response of copepods to ultraviolet radiation-induced oxidative stress. Limnol Ocean 55:1024–1032

Triadó-Margarit X, Caliz J, Reche I et al (2019) High similarity in bacterial bioaerosol compositions between the free troposphere and atmospheric depositions collected at high-elevation mountains. Atmos Environ 203:79–86

Villar-Argaiz M, Balseiro EG, Modenutti BE et al (2018a) Resource versus consumer regulation of phytoplankton: testing the role of UVR in a Southern and Northern hemisphere lake. Hydrobiologia 816:107–120

Villar-Argaiz M, Bullejos FJ, Medina-Sánchez JM et al (2012) Disentangling food quantity and quality effects in zooplankton response to P-enrichment and UV radiation. Limnol Ocean 57:235–250

Villar-Argaiz M, Medina-Sánchez JM, Biddanda BA et al (2018) Predominant non-additive effects of multiple stressors on autotroph C:N:P ratios propagate in freshwater and marine food webs. Front Microbiol 9:69

Villar-Argaiz M, Medina-Sánchez JM, Carrillo P (2002a) Microbial plankton response to contrasting climatic conditions: insights from community structure, productivity and fraction stoichiometry. Aquat Microb Ecol 29:253–266

Villar-Argaiz M, Medina-Sánchez JM, Carrillo P (2002b) Linking life history strategies and ontogeny in crustacean zooplankton: implications for homeostasis. Ecology 83:1899–1914

Villar-Argaiz M, Medina-Sánchez JM, Carrillo P (2002c) Interannual changes in the C:N:P ratios of seston and zooplankton of a high mountain lake in Sierra Nevada, Spain. Water Air Soil Pollut Focus 2:359–378

Villar-Argaiz M, Medina-Sánchez JM, Carrillo P (2016) Microbial carbon production and transfer across trophic levels is affected by solar UVA and phosphorus. Hydrobiologia 776:221–235

Villar-Argaiz M, Medina-Sánchez JM, Cruz-Pizarro L et al (2001) Inter- and intra-annual variability in the phytoplankton community of a high mountain lake: the influence of external (atmospheric) and internal (recycled) sources of phosphorus. Freshw Biol 46:1017–1034

Villar-Argaiz M, Medina-Sánchez JM, Bullejos FJ et al (2009) UV radiation and phosphorus interact to influence the biochemical composition of phytoplankton. Freshw Biol 54:1233–1245

Villar-Argaiz M, Bullejos FJ (2016) Sentinels of global change (II): herbivorous consumers in La Caldera Lake. In: Zamora R, Pérez-Luque AJ, Bonet FJ, Barea-Azcón JM and Aspizua R (editors) Global Change Impacts in Sierra Nevada: Challenges for Conservation. Consejería de Medio Ambiente y Ordenación del Territorio. Junta de Andalucía

## Case Study 4.4. Plant-Animal Interactions: *Hormathophylla spinosa* Study Case

Gómez JM, Zamora R (1992) Pollination by ants: consequences of the quantitative effects on a mutualistic system. Oecologia 91:410–418

Gómez JM (1993) Phenotypic selection on flowering synchrony in a high mountain plant, *Hormathophylla spinosa* (Cruciferae). J Ecol 81:605–613

Gómez JM (1993b) Ecología predispersiva de. una planta de alta montaña: efecto de la herbivoría en la' producción de semillas de *Hormathophylla spinosa* (Cruciferae). Tesis doctoral, Universidad de Granada, Granada

Gómez JM (1994) Importance of direct and indirect effects in the interaction between a parasitic plant (*Cuscuta epithymum*) and its host plant (*Hormathophylla spinosa*). Oikos 71:97–106

Gómez JM, Zamora R (1994) Top-down effects in a tritrophic system: parasitoids enhance plant fitness. Ecology 75:1023–1030

Gómez JM, Zamora R (1996) Wind pollination in high-mountain populations of *Hormathophylla spinosa* (Cruciferáe). Am J Bot 83 (5):580–585

Gómez JM, Zamora R, Hódar JA et al (1996) Experimental study of pollination by ants in Mediterranean high mountain and arid habitats. Oecologia 105(2):236–242s

Gómez JM, Zamora R (1999) Generalization vs. Specialization in the pollination system of *Hormathophylla spinosa* (cruciferae). Ecology 80(3):796–805

Gómez JM, Zamora R (2000) Spatial variation in the selective scenarios of *Hormathophylla spinosa* (Cruciferae). Am Nat 155 (5):657–668

Gómez JM, Zamora R (2000) Differential impact of vertebrate and invertebrate herbivores on the reproductive output of *Hormathophylla spinosa*. EcoScience 7(3):299–306

Gómez JM, González-Megías A (2002) Asymmetrical interactions between ungulates and phytophagous insects: being different matters. Ecology 83(1):203–211

González-Megías A, Gómez JM (2001) Adult and larval plant range and preference in *Timarcha lugens* (Coleoptera: Chrysomelidae): strict monophagy on an atypical host. Ann Entomol Soc Am 94 (1):110–115

Gómez JM, Zamora R (2002) Thorns as induced mechanical defense in a long-lived shrub (Hormathophylla spinosa, Cruciferae). Ecology 83(4):885–890

Gómez JM, Zamora R (2003) Factors affecting intrafruit pattern of ovule abortion and seed production in *Hormathophylla spinosa* (Cruciferae). Plant Syst Evol 239(3):215–229

Gómez JM, Zamora R (2006) Ecological factors that promote the evolution of generalization in pollination systems. In: Plant–pollinator interactions: from specialization to generalization. In: Waser NM, Ollerton J (eds) University of Chicago Press, pp 145–166

González-Megías A, Gómez JM (2003) Consequences of removing a keystone herbivore for the abundance and diversity of arthropods associated with a cruciferous shrub. Ecol Entomol 28(3):299–308

Lamprecht A, Pauli H, Calzado MRF et al (2021) Changes in plant diversity in a water-limited and isolated high-mountain range (Sierra Nevada, Spain). Alp Bot 131(1):27–39

Losapio G, Fortuna MA, Bascompte J et al (2019) Plant interactions shape pollination networks via nonadditive effects. Ecology 100(3): e02619

Losapio G, Schöb C (2020) Pollination interactions reveal direct costs and indirect benefits of plant–plant facilitation for ecosystem engineers. J Plant Ecol 13(1):107–113

Macek P, Prieto I, Macková J et al (2016) Functional plant types drive plant interactions in a Mediterranean mountain range. Front Plant Sci 7:662

Méndez M, Gómez JM (2006) Phenotypic gender in *Hormathophylla spinosa* (Brassicaceae), a perfect hermaphrodite with tetradynamous flowers, is variable. Plant Syst Evol 262(3):225–237

O'Brien MJ, Carbonell EP, Losapio G et al (2021) Foundation species promote local adaptation and fine-scale distribution of herbaceous plants. J Ecol 109(1):191–203

Salmerón-Sánchez E, Fuertes-Aguilar J, Španiel S et al (2018) Plant evolution in alkaline magnesium-rich soils: a phylogenetic study of the Mediterranean genus *Hormathophylla* (Cruciferae: Alysseae) based on nuclear and plastid sequences. PloS one 13(12):e0208307

Schöb C, Butterfield BJ, Pugnaire F (2012) Foundation species influence trait-based community assembly. New Phytol 196:824–834

Schöb C, Armas C, Guler M et al (2013) Variability in functional traits mediates plant interactions along stress gradients. J Ecol 101:753–762

Zamora R, Gómez JM (1993) Vertebrate herbivores as predators of insect herbivores: an asymmetric.al interaction mediated by size differences. OikOs 66:223–228

Zamora R (2000) Functional equivalence in plant-animal interactions: ecological and evolutionary consequences. Oikos 88(2):442–447

## Case Study 4.5. Biology and Ecology of *Capra Pyrenaica*

Alasaad S, Granados Torres JE, Cano-Manuel León FJ et al (2008) Epidemiology of fasciolosis affecting Iberian ibex (Capra pyrenaica) in southern Spain. Parasitol Res 102(4):751–755

Alasaad S, Morrondo P, Dacal V et al (2009) Bronchopulmonary nematode infection of Capra pyrenaica in the Sierra Nevada massif, Spain. Vet Parasitol 162(2–4):340–343

Espinosa J, Ráez Bravo A, López Olvera JR et al (2017) Histopathology, microbiology and the inflammatory process associated with Sarcoptes scabiei infection in the Iberian ibex, Capra pyrenaica. Parasites Vectors 10:596

Espinosa J, Pérez-Jiménez JM, López Olvera JR et al (2017b) Evaluation of oxidant/antioxidant balance in Iberian ibex (Capra pyrenaica) experimentally infested with Sarcoptes scabiei. Vet Parasitol 242(1):63–70

Espinosa J, López Olvera JR, Cano-Manuel León FJ et al (2017) Guidelines for managing captive Iberian ibex herds for conservation purposes. J Nat Conserv 40(1):24–32

Espinosa J, Pérez Jiménez JM, Ráez Bravo A et al (2020) Recommendations for the management of sarcoptic mange in free ranging Iberian ibex populations. Anim Biodivers Conserv 43(1):137–149

Granados Torres JE, Pérez Jiménez JM, Soriguer Escofet RC et al (1997) On the biometry of the Spanish ibex, Capra pyrenaica, from Sierra Nevada (Southern Spain). Folia Zool 46(1):9–14

Granados Torres JE, Ros Candeira A, Pérez Luque AJ et al (2020) Long-term monitoring of the Iberian ibex population in the Sierra Nevada of the southeast Iberian Peninsula. Sci Data 7(203):1–10

López Olvera JR, Serrano Ferrón E, Armenteros A et al (2015) Sex-biased severity of sarcoptic mange at the same biological cost in a sexually dimorphic ungulate. Parasites Vectors 8(583)

Márquez Jiménez FJ, Granados Torres JE, Caruz A et al (2020) Genetic diversity of cytochrome b in Iberian ibex from Andalusia. Mamm Biol 100:675–684

Pérez Jiménez JM, Granados Torres JE, Soriguer Escofet RC et al (1996) Prevalence and seasonality of Oestrus caucasicus Grunin, 1948 (diptera: Oestridae) parasitizing the Spanish ibex, Capra pyrenaica (Mammalia: Artiodactyla). J Parasitol 82(2):233–236

Pérez Jiménez JM, Ruiz-Martínez I, Granados Torres JE et al (1997) The dynamics of sarcoptic mange in the ibex population of Sierra Nevada in Spain. Influence of climatic factors. J Wildl Res 2(1):86–89

Pérez Jiménez JM, Granados Torres JE, González Rodríguez FJ et al (1999) Hematologic parameters of the Spanish ibex, (Capra pyrenaica). J Zoo Wildl Med 30(4):550–554 PMID: 10749444

Pérez Jiménez JM, Granados Torres JE, Soriguer Escofet RC et al (2002) Distribution, status and conservation problems of the Spanish ibex, Capra pyrenaica (Mammalia: Artiodactyla). Mammal Rev 32(1):26–39

Pérez Jiménez JM, Granados Torres JE, Pérez Alvarado MC et al (2003) A survey of the gastrointestinal nematodes of Spanish ibex

(Capra pyrenaica) in a hish mountain habitat. J Parasitol 89(2):315–318

Pérez Jiménez JM, Granados Torres JE, Espinosa J et al (2021) Biology and management of sarcoptic mange in wild Caprinae populations. Mammal Rev 51:82–94

Raéz Bravo A, Granados Torres JE, Cerón JJ et al (2015) Acute phase proteins increase with sarcoptic mange status and severity in Iberian ibex (Capra pyrenaica, Schinz 1838). Parasitol Res 114:4005–4010

Raéz Bravo A, Granados Torres JE, Serrano Ferrón E et al (2016) Evaluation of three enzyme-linked immunosorbet assays for sarcoptica mange diagnosis and assessment in the Iberian ibex, Capra pyrenaica. Parasite Vectors 8:558

Sarasa M, Serrano Ferrón E, Pérez Jiménez JM et al (2010) Effect of season, age and body condition on allocation to testis weight in Iberian ibex. J Zool 281(2):125–131

Sarasa M, Soriguer Escofet RC, Granados Torres JE et al (2012) Mismeasure of secondary sexual traits: an example with horn growth in the Iberian ibex. J Zool 288:170–176

Serrano Ferrón E, González Rodríguez FJ, Granados Torres JE et al (2008) The use of total serum proteins and triglycerides for monitoring body condition in the Iberian wild goat (Capra pyrenaica). J Zoo Wildl Med 39(4):646–649

Serrano Ferrón E, Granados Torres JE, González Rodríguez FJ et al (2011) The effects of winter severity and population density on body stores in the Iberian wild goat (Capra pyrenaica) in a highly seasonal mountain environment. Eur J Wildl Res 57:45–55

Valldeperes M, Granados Torres JE, Pérez Jiménez JM et al (2019) How sensitive and specifc is the visual diagnosis of sarcoptic mange in free-ranging Iberian ibexes? Parasites Vectors 12(405)

Viana D, Granados Torres JE, Fandos Paris P et al (2018) Linking seasonal home range size with habitat selection and movement in a mountain ungulate. Mov Ecol 6:1

**Author**: Jose Ignacio Cuenca

# The Physical Setting

**Author**: Andrés Ureña

# Geological Setting of Sierra Nevada

Ángel Santamaría-López, Lourdes González-Castillo, Jesús Galindo-Zaldívar, and Carlos Sanz de Galdeano

**Abstract**

Sierra Nevada is the highest mountain range of the Iberian Peninsula, and forms part of the Betic Cordillera, which is connected with the Rif Mountains through the Gibraltar Arc. This cordillera is divided into the External and the Internal zones, and also the so-called Flysch units, and Neogene basins are present. The Internal zone is formed by three main tectonic superimposed complexes that from bottom to top are named Nevado-Filabride, Alpujarride and Malaguide. The core of Sierra Nevada is mainly formed by the Nevado-Filabride complex, whereas the Alpujarride complex is particularly well exposed in the Alpujarra and in the western part of Sierra Nevada. During the Alpine orogeny, the rocks of the Nevado-Filabride and Alpujarride complexes were subducted, metamorphosed and later exhumed and uplifted to their actual emplacement. These two complexes show a penetrative main planar or planar-linear fabric, affected by later folds. Brittle deformation consists of high-angle and low-angle faults and joints formed during exhumation of the metamorphic complexes. After these deformations, a new stage of compression triggered the development of broad late E-W and NE-SW folds throughout the Internal zone, and the elevation of the area (starting at $\sim$8 Ma). Contemporaneously, marine sediments were deposited in the surrounding basins situated between the higher reliefs, forming the Neogene basins observed nowadays. The relief of Sierra Nevada is mainly controlled by its great E-W antiform and by the schistosity, fundamentally that of its dark schists. Moreover, the transversal incision of the rivers is very important, together with the glacial and periglacial reliefs in the higher peaks.

**Keywords**

Sierra Nevada • Betic-Rif Cordillera • Nevado-Filabride complex • Alpujarride complex

## 1 The Geological Setting

Sierra Nevada, the highest mountain range of the Iberian Peninsula (reaching 3482 m a.s.l., Mulhacén Peak), is formed by the two deepest tectonic superposed complexes of the Betic Cordillera (Fig. 1), the Nevado-Filabride and the Alpujarride complexes.

Our present-day geological knowledge of Sierra Nevada is the result of over 150 years of research. In this sense can be cited Ezquerra de Bayo (1856) who described the presence of crystalline rocks, more precisely identified by the earliest lithological maps from Von Drasche (1879), Barrois and Offret (1889) and Brouwer (1926). This last author distinguished in the Nevado-Filabride complex two groups of rocks. To the deepest one he gave the name of 'Crystalline Rocks of Sierra Nevada', and to the upper one that of the 'Mischungszone'. The first group comprises a monotonous metamorphic series of dark schists and quartzites, while the second one is formed by feldspathic schists, marbles, gneisses and metabasite intercalations. In a regional overview, Sierra Nevada forms part of the Betic Cordillera, in which nappe structures can be clearly identified (Van Bemmelen 1927). At present, there is debate on the division in tectonic units of the Nevado-Filábride complex, while it is generally accepted that the Alpujarride complex is divided in three main groups of tectonic units.

Á. Santamaría-López
Departamento de Geologia, Universidad de Salamanca, Salamanca, Spain

L. González-Castillo · J. Galindo-Zaldívar (✉)
Departamento de Geodinámica, Universidad de Granada, Granada, Spain
e-mail: jgalindo@ugr.es

J. Galindo-Zaldívar · C. Sanz de Galdeano
Instituto Andaluz de Ciencias de la Tierra, CSIC-Universidad de Granada, Granada, Spain


R. Zamora and M. Oliva (eds.), *The Landscape of the Sierra Nevada*,
https://doi.org/10.1007/978-3-030-94219-9_5

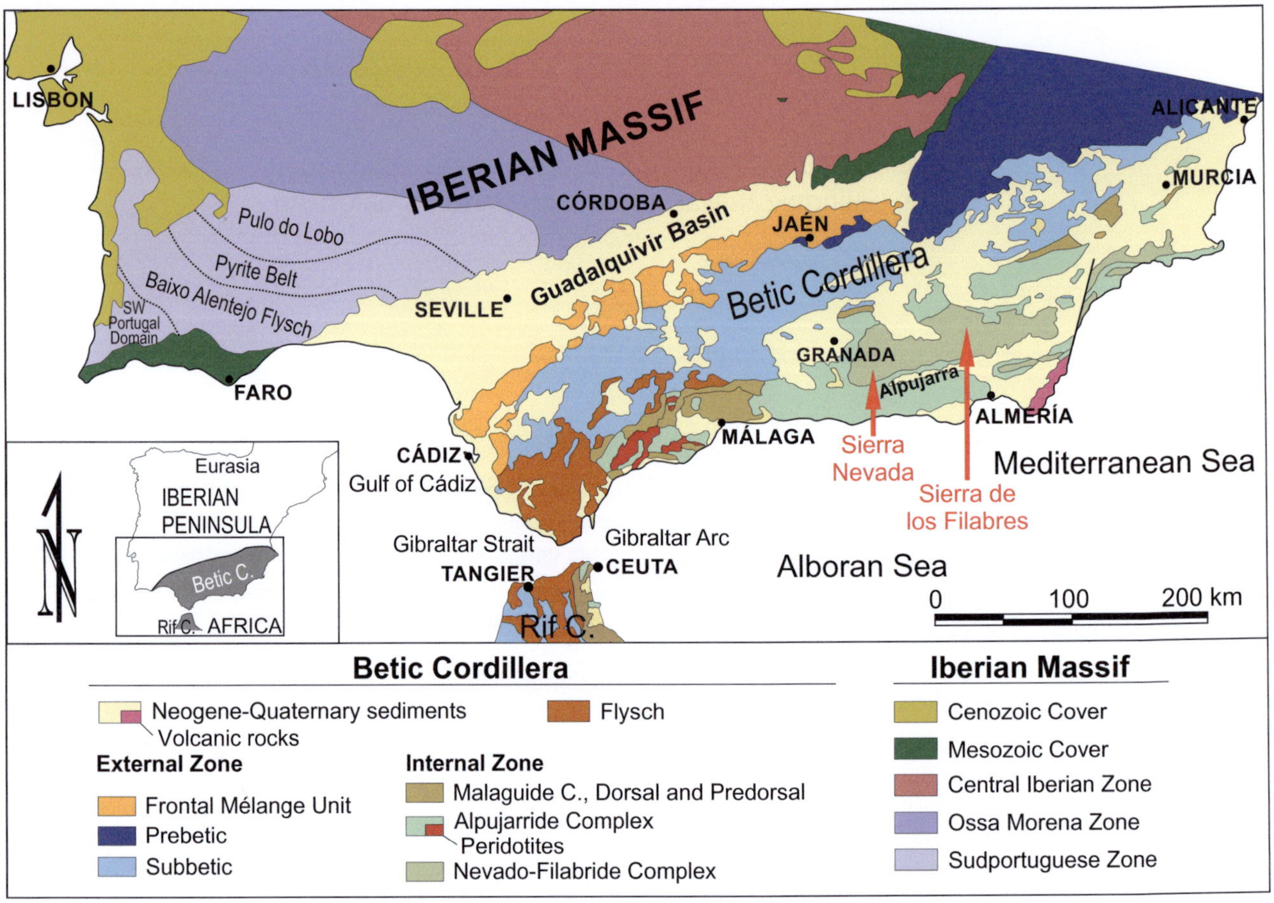


**Fig. 1** Geological map of the Betic Cordillera and the southwestern Iberian Massif (Adapted from González-Castillo 2015)

## 2 Sierra Nevada in the Framework of the Betic-Rif Cordillera

The Betic Cordillera can be divided into External and Internal zones, moreover the so-called Flysch units (or Campo de Gibraltar units), and several Neogene basins (Fig. 1). The greatest of these basins is the Guadalquivir foreland basin, separating the cordillera from the Iberian Paleozoic Massif. The External Zone—constituted by Mesozoic and Cenozoic sedimentary formations deposited on the Paleozoic S and SE margins of the Iberian Massif—now corresponds to a series of fold and thrust belts, unaffected by alpine metamorphism.

The Internal Zone comprises three main tectonostratigraphic superimposed complexes, from bottom to top: the Nevado-Filabride, the Alpujarride and the Malaguide. The two lower complexes have undergone different degrees of metamorphism during the Alpine Orogeny (Fig. 2). The nucleus of Sierra Nevada and the Sierra de Los Filabres (Figs. 1 and 3) are formed by the Nevado-Filabride complex. The overlying Alpujarride complex is particularly well exposed in the Alpujarra. The occurrence of the uppermost Malaguide complex in Sierra Nevada is restricted to scarce outcrops along the northwestern slopes of Sierra Nevada and the adjacent sierras (Figs. 1 and 3).

Since the late Miocene ($\sim$10 Ma) new deformations affected the entire cordillera. Then Sierra Nevada was formed by a fold that raises the central part (the antiform of Sierra Nevada), similar to others, as those of Sierra de Los Filabres, Sierra de Gádor, Sierra Contraviesa and Sierra de Lujar (Galindo-Zaldívar et al. 2003). Synforms, located between these antiforms, along the elongated valleys, are affected by faulting (Figs. 1 and 3).

## 3 Lithology, Tectonic Units and Deep Structure of Sierra Nevada

### 3.1 The Nevado-Filabride Complex

The Nevado-Filabride metamorphic complex appeared during the uplift of the sierra, after the erosion of the overlying rocks. For this reason, the higher reliefs (including Mulhacén, Veleta and Alcazaba peaks) are formed by rocks of this complex.

Several lines of evidence suggest that the Nevado-Filabride complex recorded an Alpine metamorphic event reaching high pressure (P) and low temperature (T) conditions during subduction. Thermobarometric data moreover show higher P–T values in the upper part of the complex than in the lower one (e.g. Puga et al. 2002; Augier et al. 2005; Behr and Platt 2012; Booth-Rea et al. 2015; Ruiz-Cruz et al. 2015; Li and Massonne 2018; Santamaría-López et al. 2019).

The lithological succession of the Nevado-Filabride complex in Sierra Nevada is the next: the lowest part of the complex is formed by a monotonous lithological sequence (Fig. 2). It includes graphite-bearing black, dark grey and brown schists, as well as (locally) dark slates (Fig. 4A). The thickness of this lithological succession is estimated at 4000 m (Fallot et al. 1960), even though the bottom of the sequence does not crop out in the sierra. The schists are interbedded with quartzite layers, in some points up to tens of meters thick (Fig. 2).

Atop the former rocks, a change in the lithological sequence marks the beginning of Mischungszone group of Brouwer (1926). The lowest part of this sequence is dominated by alternating dark schists and light quartzite levels, whereas its upper part shows a progressive increase—in both occurrence and thickness—of light schists (Fig. 4b) and occasional marbles (Fig. 2). These rocks often alternate with discontinuous bodies of metabasites (locally reaching kilometric length) (Fig. 4c), levels of gneisses (from several centimetres to tens of meters thick) (Fig. 4d) and scarce meta-evaporitic levels.

In the upper position there is a sequence of alternating light/dark schists and marbles (Fig. 4g), and occasional calc-schists (Fig. 2), particularly well preserved in the western part of Sierra Nevada. Their thickness ranges between 200 and 1300 m.

Radiometric dating carried out on both metasediments and meta-igneous rocks gave ages ranging from upper Carboniferous to Triassic (e.g. Martínez-Martínez et al. 2010; Gómez-Pugnaire et al. 2012; Ruiz-Cruz and Sanz de Galdeano 2017; Santamaría-López and Sanz de Galdeano 2018; Jabaloy et al. 2021).

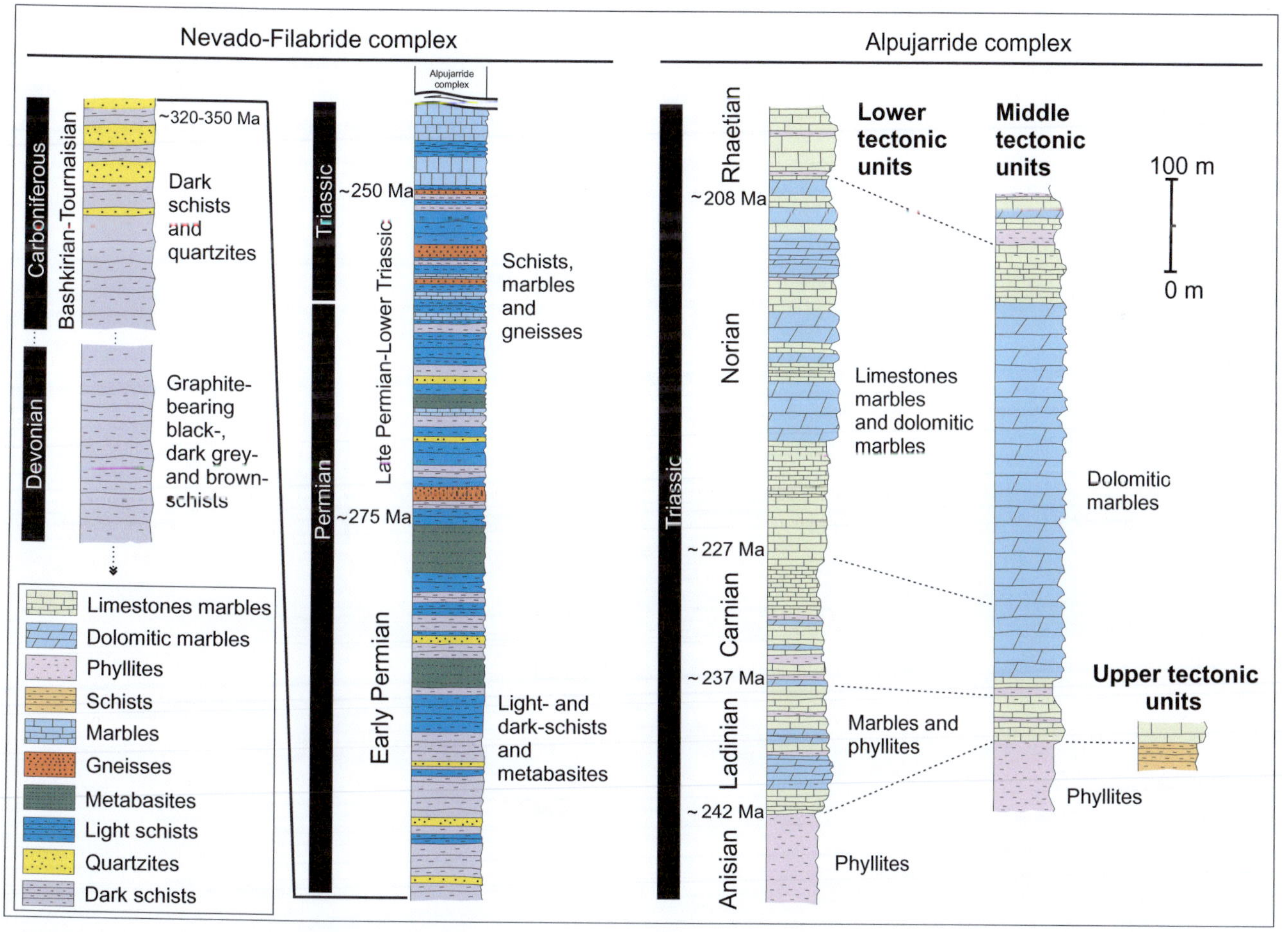


**Fig. 2** Lithological succession of the Nevado-Filabride and Alpujarride complexes 1999b (Adapted from Sanz de Galdeano and López-Garrido; and Santamaría-López 2019)

## 3.2 The Alpujarride Complex

Originally, the Alpujarride complex completely covered the Nevado-Filabride complex in the Sierra Nevada. However, subsequent uplift and erosion determined its disappearance in the upper parts of the sierra. At present, it forms a rim around the outcrops of the Nevado-Filabride complex (Fig. 3), being absent only in an area of connection with the eastern Sierra de los Filabres, where the Nevado-Filabride complex passes from one massif to the other.

Regionally considered, the Alpujarride complex harbors several superimposed tectonic units pertaining to three groups (Aldaya et al. 1979), whose metamorphic grade increases from bottom to top. In Sierra Nevada this complex is distributed around the Nevado-Filabride complex. Here, its best represented units are those of the lowest group, the one directly overlying the Nevado-Filabride complex.

*Lithological series of the Alpujarride complex in Sierra Nevada*

The preserved lithological sequences of the lower and middle units are similar, mainly differentiated by the metamorphic grade (Fig. 2).

In both types of units, the visible lower terms are phyllites (Fig. 4h), and to a lesser extent, quartzites (Fig. 2). The phyllites are generally greyish to bluish in colour; but in some places they become reddish-purple or even yellowish-green. Exudation quartz may be locally abundant. The quartzites are predominantly white to beige. These rocks, attributed to the Middle Triassic, can reach a thickness of several hundreds of metres, but overall, the preserved thickness is much lesser due to erosion and tectonic deformations. In addition, gypsum of a predominantly white colour can be found, its maximum thickness up to several metres.

Above the previous rocks, a thick carbonate series (Fig. 2) corresponds to over 700 m of marbles (Fig. 4i)—elsewhere reaching thickness over 1000 m (Sanz de Galdeano, 2019). Variations in their composition are common: i.e. some are more calcareous and others more dolomitic, the thickness of their layers and their colour may vary, and there may even be intercalations of detritic rocks, currently calc-schists, phyllites and quartzites. Those marbles contain fossils (Fig. 4j)—algae, bivalves, gastropods—that point to ages from Middle to Upper Triassic, and possibly reach Jurassic ages (Delgado et al. 1981; Martín and Braga 1987). They formed in a very shallow and tepid sea, over a vast marine platform that even extended into the depositional area of the future Nevado-Filabride complex. The clearest difference between the lower tectonic units and the intermediate ones is that the carbonates are poorly recrystallized in the former, essentially limited to limestones, or dolomites in some cases; yet in the latter they are already marbles. The Alpujarride complex contains basic volcanic rocks

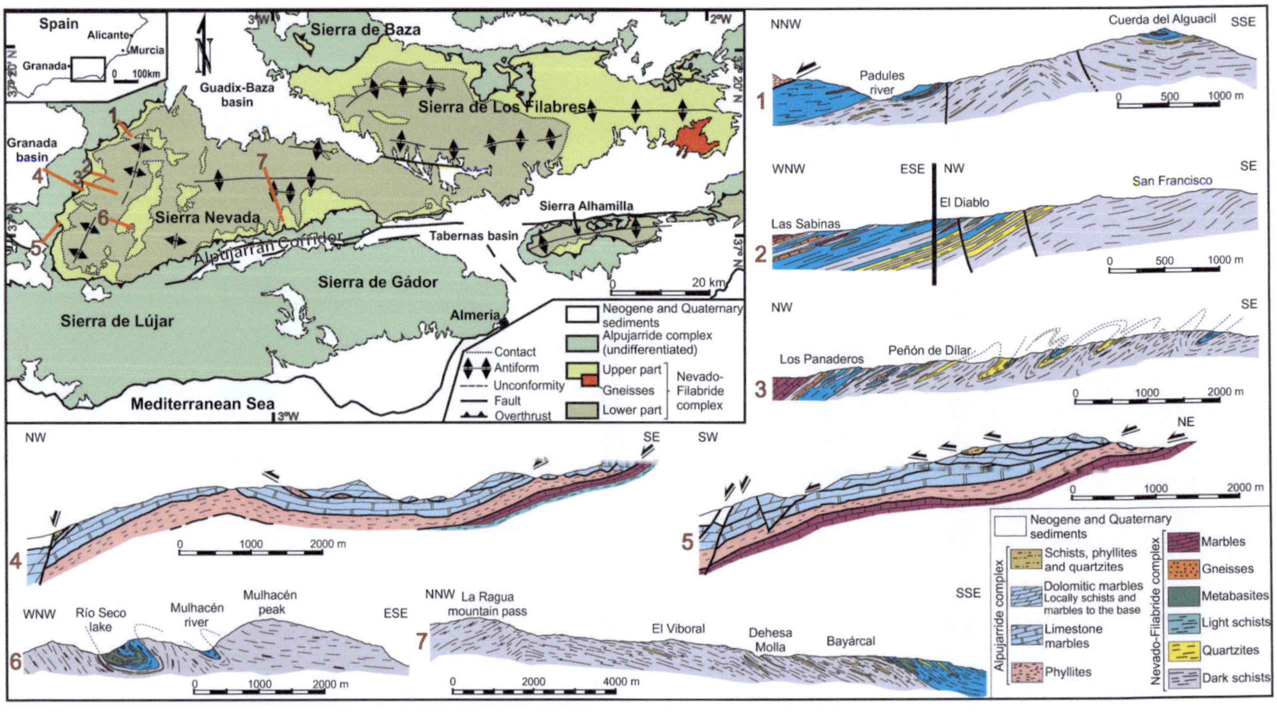


**Fig. 3** Geological map of the Betic Cordillera and cross sections of the western and central areas of Sierra Nevada 1999a (Adapted from Sanz de Galdeano and López-Garrido; Sanz de Galdeano and López-Garrido 2016; Santamaría-López 2019)

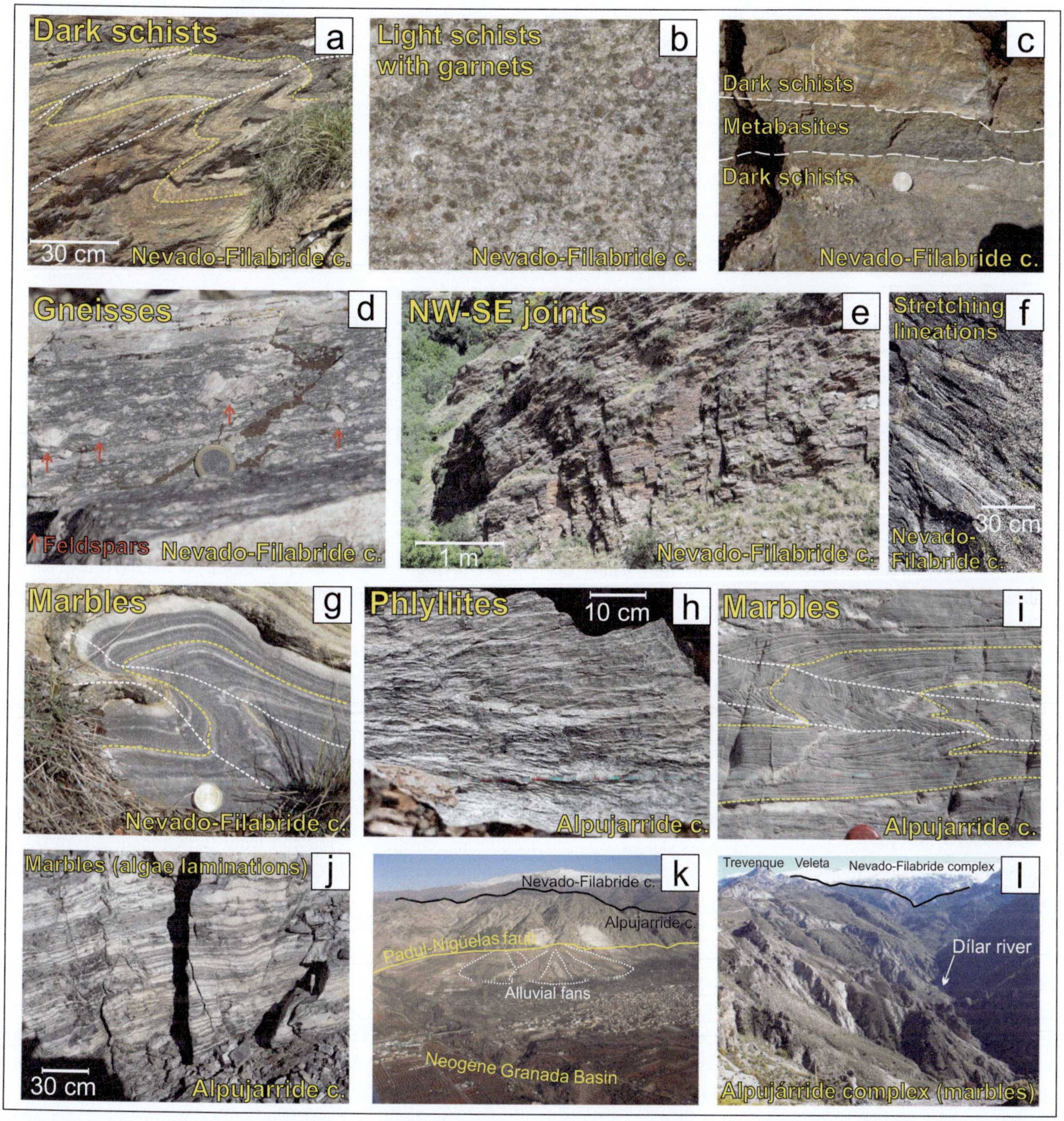


**Fig. 4** Field images of distinctive lithologies and structures of the Nevado-Filabride and Alpujarride complexes. Nevado-Filabride complex: **a** dark schists showing the original schistosity (yellow lines) overprinted by a second one (white lines), Güéjar-Sierra area; **b** light schists with garnets, Güéjar-Sierra area; **c** succession of dark schists and metabasites, Güéjar-Sierra area; **d** Gneisses, Loma Panaderos area; **e** NW–SE joints, upper basin of the Genil river; **f** stretching lineations, upper basin of the Genil river; **g** marbles, Güéjar-Sierra area. Alpujarride complex: **h** Phyllites, Güéjar-Sierra area; **i** Marbles, Collado de las Sabinas area; **j** Marbles showing algae laminations, Trevenque area. **k** Panoramic view from the Dúrcal town showing the Padul-Nigüelas fault (yellow line) and associated alluvial fans (photo: Javier Sanz de Galdeano); **l** panoramic view from the Trevenque area showing the Triassic marbles pertaining the Alpujarride complex

intercalated between the carbonates that appear as amphibolites, although they are very scarce in the Sierra Nevada rim units.

The upper Alpujarride group of units (Fig. 2) is poorly preserved around Sierra Nevada. It is mainly seen to the west of Sierra Nevada, now showing only dark schists and, locally, some marbles. The schists largely correspond to the Paleozoic, perhaps reaching the Triassic, while the marbles are of Triassic age.

The marbles form abrupt reliefs locally. In western Sierra Nevada, Trevenque (Fig. 4l) and nearby areas stand out, though there are many others. In some cases, they are highly fractured, forming crumbly, so-called kakiritas; therefore, their accumulated remains are sometimes called *arenales* (sandy grounds), which is incorrect from a geological point of view.

On the edges of Sierra Nevada, and within the Alpujarride complex, in addition to quarries for aggregates (generally located in the kakiritas), there are old mines. Deserving mention here are the El Conjuro iron mines at the base of the carbonates. Further mines extracted, for instance, mercury at Tímar (Juviles) in the Alpujarra, or from the Minas del Águila (Martín and Arana 1976).

In short, the Alpujarride complex, which originally covered the whole of the Nevado-Filabride, now occupies only part of the slopes of the Sierra Nevada. Its marbles sometimes form abrupt reliefs and trace a habitat far different from the one formed by Nevado-Filabride rocks.

## 3.3 The Neogene-Quaternary Basins

Sedimentary post-metamorphic deposits are found in and around the edges of Sierra Nevada. We discern three sectors (Fig. 3):

a. The Southern sector corresponds to the Alpujarran corridor (Fig. 3), an E-W valley extending S of the Sierra Nevada. In the central part of the corridor there are sediments from the Middle Miocene to the Quaternary (Aldaya et al. 1983; Rodríguez-Fernández et al. 1990). On the whole, they are made up of detrital deposits, clays, silts, sands and conglomerates, although there are also some carbonate sediments (Tortonian, Upper Miocene). The thickness of each formation is highly variable owing to erosion and substantial tectonic deformations—the Alpujarran corridor is controlled by important faults of approximately E-W direction, in addition to the effects produced by the development of the great antiform of Sierra Nevada. Yet in no case do the sedimentary formations exceed 200 m in thickness, the value generally being much less.

These sediments collect the erosion remnants from nearby sectors. It is noteworthy that the Middle Miocene conglomerates contain pebbles inherited from the Malaguide complex, along with some from the Alpujarrides, but none from the Nevado-Filabride complex. This means that, at that age, the latter complex had not yet reached the surface and was therefore not yet subjected to erosion. Hence the Sierra Nevada did not exist at that time, even though some early reliefs began to form, exhibiting a very different geography picture from the present one. Later, from the Upper Miocene onward, sediments containing pebbles and boulders inherited from the Nevado-Filabride complex were deposited in the near Neogene basins.

b. In the western sector of Sierra Nevada appears the contact of basement rocks with the Neogene Granada basin (Figs. 3 and 4j). The age of their sediments ranges from the Middle Miocene to the Pliocene. Calcarenites of Tortonian age (lowermost Upper Miocene) are well developed, reaching almost 100 m in thicknesses at some places. It is noteworthy that on top of them, the new sediments contain pebbles and blocks inherited from the Nevado-Filabride complex (hence its original denomination, "Blockformation", though it was later re-baptized as the Pinos-Genil formation). This indicates that the Nevado-Filabride complex was exposed in these times and underwent intense erosion simultaneous to the uplift of Sierra Nevada.

In western Sierra Nevada, the Neogene sediments show a clearly progressive angular unconformity with the oldest sediments with a greater dip (inclination of their layers), thus recording the progressive uplift of the adjacent sierra (Sanz de Galdeano and Alfaro 2004). Additional data of the uplift process are provided by several small, isolated Neogene outcrops, currently separated from the Granada Basin (Díaz de Federico et al. 1980), where shallow marine sediments from the Upper Miocene are found at an altitude of up to 1830 m above sea level, yet there was no doubt that others at higher altitudes were destroyed by erosion.

In the Trevenque area (Fig. 4L) the Dílar river, incised at roughly 550 m approximately in 0.5 Ma, evidences a fast erosion in response to the rapid uplift taking place in the western sector of Sierra Nevada. In fact, the characteristics of numerous nearby faults would indicate they were active at this time.

c.- The sediments existing in the northern Guadix-Baza Basin correspond to the Miocene, Pliocene and Quaternary. Near Sierra Nevada, the most abundant sediments are Upper Miocene lacustrine silts, although there are also calcarenites showing angular unconformities. The Pliocene and Quaternary sediments correspond to sands, conglomerates and clays. On top, an important glacis surface covered this basin, at present very incised by the rivers in many parts of the basin, although, in the northern border of Sierra Nevada part of this Quaternary glacis (García-Tortosa et al. 2011) is preserved.

No eastern sector is mentioned, as the N and S edges of the Sierra Nevada progressively converge towards the E, near Tabernas basin.

In summary, the Neogene sediments, mainly distributed along the margins of the Sierra Nevada (Fig. 4k), offer clear evidence of the uplift process undergone, according to the different types of inherited rocks they contain, as well as the heights currently reached on their slopes.

## 4 Tectonic Deformation and Evolution

### 4.1 Ductile Structures (Foliations, Lineations and Folds)

The Nevado-Filabride and the Alpujarride complexes were affected by several stages of ductile and brittle deformations of variable intensity and features. The two complexes share the presence of a penetrative main planar or planar-linear fabric, generally low dipping, which is their main reference foliation.

In the Nevado-Filabride complex, the main foliation is generally linked to stretching lineations that increase in intensity upwards and westwards. Foliation is determined by the flattening of most planar minerals (Fig. 4a, c, d, g). Meanwhile, stretching lineation (Fig. 4f) is indicated by the rotation of elongated minerals or by crystal deformation. This penetrative foliation is overprinted upon previous foliations. The main foliation coincides with the axial planes of the corresponding folds—sometimes tight to isoclinal—that can reach kilometric dimensions (Fig. 3); they may include sheath folds whose axes are usually parallel to the stretching lineations. The ductile microstructures suggest that all these structures developed in a ductile shear zone with regional top-to-the-W-SW simple shear.

The main foliation is affected by later folds (Fig. 4a, g), open to tight, that develop crenulation cleavage lineation, and allow for observation of fold interference patterns (Fig. 4a). The folded structures evidence ductile–brittle extensional crenulation cleavages (Platt and Vissers 1980) indicating a dominance of areas with top-to-the-W-SW kinematics in addition to areas of pure shear and flattening.

The Alpujarride complex is furthermore affected by penetrative foliation in phyllites and schists, and the stretching lineation is very poorly developed. The main foliation is affected by subsequent open to tight folds (Fig. 4i).

### 4.2 Brittle Deformations (Joints and Main Low-Angle and High-Angle Faults)

Brittle deformation consists of high-angle and low-angle faults and joints formed in the latest stages of exhumation of the metamorphic complexes. The age and brittle or ductile character of thrusting of the Alpujarride upon the Nevado-Filabride complex, and the stacking of the different thrust sheets or recumbent folds of the Alpujarride complex, are two matters under discussion in the last half century.

Simancas (2018) suggests a first stage of top-to-the-N thrusting in the Alpujarrides, which may have determined its superposition on the Nevado-Filabride complex, followed by the activity of top-to-the-N low-angle normal faults of probable Burdigalian age. This stage would have been followed by top-to-the-SW low-angle normal faulting since the Serravallian. In this setting, the top-to-the-W-SW kinematics of the Alpujarride-Nevado Filabride contact in Sierra Nevada was probably overprinted on the initial thrusting, developing a wide ductile–brittle shear zone, also affecting most of the Alpujarrides (Galindo-Zaldívar et al. 1989).

South of Sierra Nevada, a wide dextral strike-slip zone developed along the Alpujarran corridor (Figs. 3 and 5) (Sanz de Galdeano et al. 1985). This zone can be characterized by occasionally active extensional faults (Martínez-Martos et al. 2017), though transpressional thrust structures are also observed (Galindo-Zaldívar 1986).

Finally, the E-W to NE-SW regional extension, orthogonal to the NW–SE main compression, developed a penetrative set of N-S to NW–SE tensional joints (Fig. 4e) affecting most of Sierra Nevada (Galindo-Zaldívar and González-Lodeiro 1990), in addition to a conspicuous system of extensional faults. The faults located to the west of Sierra Nevada, in the Granada basin, present high seismic activity (Sanz de Galdeano et al. 2003). The Padul-Nigüelas

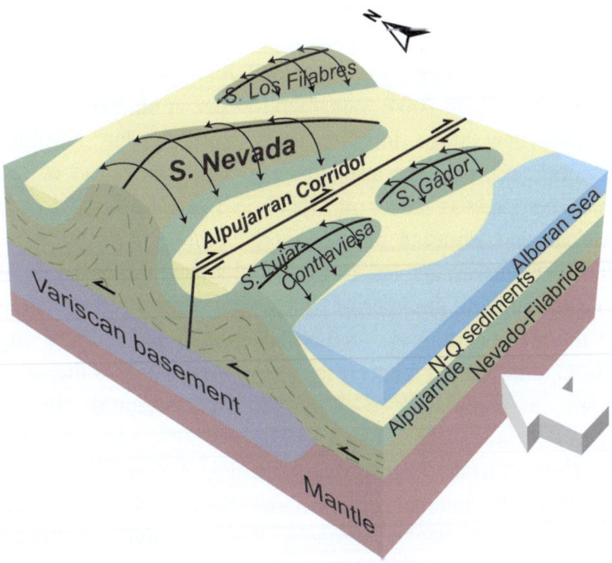


**Fig. 5** 3D model of the SE area of Iberian Peninsula

fault (Fig. 4k) is the most significant, having an average extensional rate of 0.5 mm $yr^{-1}$ (Gil et al. 2017).

## 4.3 Evolution of Sierra Nevada

The Betic Cordillera is connected with the Rif Mountains through the Gibraltar Arc, forming a strongly curved and arched cordillera in the westernmost part of the Mediterranean Alpine Belt (Fig. 1). The formation of the Betic-Rif Cordillera implied the convergence of two main tectonic plates, the African and Eurasian plates (the south-west end of the latter, the Iberian Massif). Due to this convergence, the Betic-Rif Internal Zone (the three previously cited complexes) was strongly affected. In this process of convergence, this Internal Zone was partially metamorphosed and structured as superposed tectonic complexes. Later, owing to the opening of important new oceanic areas in the western Mediterranean, the Internal Zone began to be westwards expulsed.

The drift of the Internal Zone to the west and the eventual collision with the south margin of the Iberian Massif was very important because it produced deformations in the Internal Zone, also in the external zones of the Betics and the Rif, and formed the arc of Gibraltar.

A late stage of deformation caused the formation of large kilometer-size folds that gave rise to the main shape of Sierra Nevada (Figs. 3 and 5). Whereas in the eastern part the fold is roughly E-W in orientation, towards the western Sierra Nevada it becomes NE-SW, with a southwestwards termination. These important folds facilitated the erosion of the Alpujarride complex and the uplift of the Nevado-Filabride complex.

The exhumation of the Nevado-Filabride (López Sánchez-Vizcaíno et al. 2001; de Jong 2003; Platt et al. 2006; Behr and Platt 2012; Kirchner et al. 2016; Santamaría-López et al. 2019), and Alpujarride complexes (Platt et al. 2005 2013) would have been rapid and accompanied by high cooling rates (Johnson et al. 1997; Lonergan and Johnson 1998; Esteban et al. 2004; Zeck 2004; Santamaría-López et al. 2019).

This new stage forming folds, uplifting the Nevado-Filabride complex, corresponds with the reactivation of the main compressive stress of NNW-SSE direction due to the African and Eurasian plate convergence (e.g. Jabaloy et al. 2003). Compression led to the development of broad E-W folds throughout the Internal zone, and a generalized elevation of the area (starting at ~8 Ma; Johnson et al. 1997), giving rise to today´s high reliefs (Fig. 5). Approximately at the same point, a remarkable rise in sea-level occurred (Rodríguez-Fernández and Sanz de Galdeano 2006), which yielded the formation of straight marine corridors between the higher reliefs of the Internal zone terrains. Sanz de Galdeano and López-Garrido (1999a) propose that the uplift of Sierra Nevada could have been even higher (up to 5000 m) considering the progressive erosion that affected the Tortonian rocks.

What is clear is that the maximum vertical movement was focused on the sierras Nevada and Filabres (e.g. Sanz de Galdeano et al. 2007). In this sense, Azañón et al (2015) argue that the western area of Sierra Nevada concentrates the main elevation. Using the Tortonian rocks as a marker, some authors (Braga et al. 2003; Sanz de Galdeano and Alfaro, 2004) have estimated that regional elevation currently attains a maximum value of 0.5 mm $yr^{-1}$ in the central sector of the cordillera, corresponding to Sierra Nevada.

# 5 The Role of Tectonics Shaping the Landscape of the Sierra Nevada

The landforms of Sierra Nevada are determined by the interaction of lithology, recent tectonics and climate. The presence of metamorphic rocks with well-defined foliations, together with the nappe structure, influences its heterogeneous behaviour during exposure to the external geodynamic factors.

The elevation of Sierra Nevada is a consequence of the activity of wide kilometric-scale antiforms, with E-W orientation in eastern Sierra Nevada that becomes NE-SW oriented in the westernmost part (Fig. 5). The western sector includes the maximum elevations, orthogonal to plate convergence. It is surrounded by the synforms and faults that enhance Sierra Nevada´s uplift with respect to the surrounding regions.

The core of the Sierra is remarkable, then, for a number of reasons. In addition to its outstanding altitude and wide, elongated dome-shape, incised by the fluvial network and affected by glacial and periglacial environments during the last glacial cycle (Gómez-Ortiz et al. 2015, see Chaps. "The Impact of Glacial Developmenton the Landscape of the Sierra Nevada" and "Ancient and Present-Day Periglacial Environments in the Sierra Nevada"), in its uppermost part, glaciers separated by impressive arêtes and moraines are still easily recognizable. The reliefs of this sector of Sierra Nevada are generally asymmetrical. The sharp northern slopes correspond to several well-developed glacial cirques, currently occupied by several lakes. In contrast, the southern slopes are smoother, owing to the low-dipping foliation of the metamorphic rocks of the Nevado-Filabride complex on the southern flank of the antiform. The Alpujarride marbles produce very marked irregular 'alpine' reliefs surrounding the western and the southern slopes of Sierra Nevada (Fig. 4l). This setting is especially relevant to the southwest, where incision is a consequence of the tectonic uplift and the close proximity of the Mediterranean base level (Pérez-Peña et al. 2010). The morphology of the western slopes is further determined by a set of active normal faults, generally with

NW–SE orientation (e.g. Padul-Nigüelas fault; Fig. 4k); they also produce uplift of the Sierra, together with basin formation, and determine the development of large alluvial fans (Fig. 4k).

## 6 Conclusions

Sierra Nevada is formed by rocks belonging to the two lower tectonic complexes of the Betic Internal Zone: the Nevado-Filabride, situated in the bottom, and the Alpujarride, above. Both complexes underwent Alpine deformations and metamorphism. The lower part of the Nevado-Filabride complex is formed by dark schists with quartzites, whereas the upper part is formed by lighter schists, quartzites and marbles. In the Alpujarride complex appears phyllites in the visible bottom, followed by marbles. In addition to the tectonic superposition of the Alpujarride complex over the Nevado-Filabride, the main structural feature of Sierra Nevada is the big E-W antiform, cut by several transversal normal faults in its western side. Consequently, the relief of Sierra Nevada is mainly controlled by this great antiform. In detail, the disposition of the Nevado-Filabride´s dark schists, whose schistosity, dipping to the south in the southern slope and to the north in the northern slope, determine at a small scale a great part of the surfaces of this sierra. Also, the Alpujarride marbles locally form strong reliefs. To this general disposition, it is necessary to add the active incision of the rivers, which form many important transversal valleys, and the glacial and periglacial reliefs existing in the higher parts of the sierra.

**Acknowledgements** This study was supported by projects CGL2016-80687-R AEI/FEDER, P18-RT-3275, B-RNM-301-UGR18, PID2019-108880RJ-I00/AEI/https://doi.org/10.13039/501100011033, and RNM148 (Junta de Andalucía/FEDER).

## References

Aldaya F, Baena J, Ewert K (1983) Mapa Geológico de España, e:1:50.000, number 1043 (Berja). I.G.M.E. 60 p

Aldaya F, García-Dueñas V, Navarro Vila F (1979) Los Mantos Alpujárrides del tercio central de las Cordilleras Béticas. Ensayo de correlación tectónica de los Alpujárrides. Acta Geol Hisp. Homenatge a Lluis Solé i Sabarís 14:154–166

Augier R, Agard P, Monié P, Jolivet L et al (2005) Exhumation, doming and slab retreat in the Betic Cordillera (SE Spain): in situ Ar-40/Ar-39 ages and P-T-d-t paths for the Nevado-Filabride complex. J Metamorph Geol 23:357–381

Azañón JM, Galve JP, Pérez-Peña JV, Giaconia F et al (2015) Relief and drainage evolution during the exhumation of the Sierra Nevada (SE Spain): is denudation keeping pace with uplift? Tectonophysics 663:19–32

Barrois C, Offret A (1889) Mémoire sur la constitution géologique du Sud de l'Andalousie; de la Sierra Tejeda à la Sierra Nevada. In Mission D'andalousie. Mém Ac Sc 30:79–169

Behr WM, Platt JP (2012) Kinematic and thermal evolution during two-stage exhumation of a Mediterranean subduction complex. Tectonics 31:TC4025

Booth-Rea G, Martínez-Martínez JM, Giaconia F (2015) Continental subduction, intracrustal shortening and coeval upper-crustal extension: P-T evolution of subducted south-Iberian paleomargin metapelites. Tectonophysics 663:122–139

Braga JC, Martín JM, Quesada C (2003) Patterns and average rates of late Neogene-recent uplift of the Betic Cordillera, SE Spain. Geomorphology 50:3–26

Brouwer HA (1926) Zur Tektonik der betischen Kordilleren. Geol Rundsch 17:332–336

de Jong K (2003) Very fast exhumation of high-pressure metamorphic rocks with excess 40Ar and inherited 87Sr, Betic Cordilleras, southern Spain. Lithos 70:91–110

Delgado F, Estévez A, Martín JM, Martín Algarra A (1981) Observaciones sobre la estratigrafía de la formación carbonatada de los mantos alpujárrides (Cordilleras béticas). Estud Geol 37:45–57

Díaz de Federico A, Puga E, Burgos J, Gallegos JA, et al (1980) Mapa Geológico de España, e: 1:50.000, number 1027 (Güéjar-Sierra). I. G.M.E. 53 p

Esteban JJ, Sánchez-Rodríguez L, Seward D, Cuevas J et al (2004) The late thermal history of the Ronda area, southern Spain. Tectonophysics 389(1):81–92

Ezquerra del Bayo J (1856) Ensayo de una descripción general de la estructura de España en la Península. Mem R Ac Ciencias Madrid 1

Fallot P, Faure-Muret A, Fontboté JM, Solé Sabarís L (1960) Estudio sobre las series de Sierra Nevada y de la llamada Mischungzone. Boletín Del Instituto Geológico y Minero España 81:347–557

Galindo-Zaldívar J (1986) Etapas de fallamiento neógenas en la mitad occidental de la depresión de Ugijar (Cordilleras Béticas). Estud Geol 42:1–10

Galindo-Zaldívar J, González-Lodeiro F (1990) Diaclasas de tensión regionales en el Complejo Nevado-Filábride y su relación con el contacto Alpujárride/Nevado-Filábride (Sierra Nevada, Cordilleras Béticas). Sociedad Geológica De España 7:9–11

Galindo-Zaldívar J, González-Lodeiro F, Jabaloy A (1989) Progressive extensional shear structures in a detachment contact in the Western Sierra Nevada (Betic Cordilleras, Spain). Geodin Acta 3(1):73–85

Galindo-Zaldívar J, Gil AJ, Borque M, González-Loderiro F, et al (2003) Active faulting in the internal zones of the central Betic Cordilleras (SE, Spain). J Geodyn 36:239–250

García-Tortosa FJ, Alfaro P, Sanz de Galdeano C, Galindo-Zaldívar J (2011) Glacis geometry as a geomorphic marker of recent tectonics: the Guadix-Baza basin (South Spain). Geomorphology 125:517–529

Gil AJ, Galindo-Zaldivar J, Sanz de Galdeano C, Borque M et al (2017) The Padul normal fault activity constrained by GPS data: Brittle extension orthogonal to folding in the central Betic Cordillera. Tectonophysics 712–713:64–71

Gómez-Ortiz A, Oliva M, Palacios D et al (2015) The deglaciation of Sierra Nevada (Spain), synthesis of the knowledge and new contributions. Cuad Investig Geográfica 41:409

Gómez-Pugnaire MT, Rubatto D, Fernández-Soler JM, Jabaloy A et al (2012) Late Variscan magmatism in the Nevado-Filábride complex: U-Pb geochronologic evidence for the pre-Mesozoic nature of the deepest Betic complex (SE Spain). Lithos 146–147:93–111

González-Castillo L (2015) Crustal structure and active deformation in the westernmost Betic Cordillera and its foreland. Thesis Univ Granada 190 pp. ISBN 9788491251217

Jabaloy A, Galindo-Zaldívar J, González-Lodeiro F (2003) Evolución de los paleoesfuerzos en Iberia desde el Carbonífero superior a la actualidad. Enseñanza de las Ciencias de la Tierra 1(1):10–21. ISSN 1132-9157

Jabaloy A, Talavera C, Rodríguez-Peces M, Vázquez-Vílchez M et al (2021) U-Pb geochronology of detrital and igneous zircon grains from the Águilas Arc in the Internal Betics (SE Spain): implications for Carboniferous-Permian paleogeography of Pangea. Gondwana Res 90:135–158

Johnson C, Harbury N, Hurford AJ (1997) The role of extension in the Miocene denudation of the Nevado-Filábride Complex, Betic Cordillera (SE Spain). Tectonics 16:189–204

Kirchner KL, Behr WM, Loewy S, Stockli DF (2016) Early Miocene subduction in the western Mediterranean: constraints from Rb-Sr multimineral isochron geochronology. Geochem Geophys Geosyst 17:1842–1860

Li B, Massonne HJ (2018) Two Tertiary metamorphic events recognized in high-pressure metapelites of the Nevado-Filábride complex (Betic Cordillera, S Spain). J Metamorph Geol 36:603–630

Lonergan L, Johnson C (1998) Reconstructing orogenic exhumation histories using synorogenic detrital zircons and apatites: an example from the Betic Cordillera, SE Spain. Basin Res 10(3):353–364

López Sánchez-Vizcaíno V, Rubatto D, Gómez-Pugnaire MT, Trommsdorff V et al (2001) Middle Miocene high-pressure metamorphism and fast exhumation of the Nevado-Filábride complex, SE Spain. Terra Nova 13:327–332

Martín JM, Arana R (1976) Estudio geológico de las Minas del Aguila (Sierra Nevada, Cordilleras Béticas). Estud Geol 32:385–390

Martín JM, Braga JC (1987) Alpujárride carbonate deposits (Southern Spain)—marine sedimentation in a Triassic Atlantic. Palaeogeogr Palaeoclimatol Palaeoecol 59:243–260

Martínez-Martínez JM, Torres-Ruiz J, Pesquera A, Gil-Crespo PP (2010) Geological relationships and U-Pb zircon and 40 Ar/39Ar tourmaline geochronology of gneisses and tourmalinites from the Nevado-Filabride complex (western Sierra Nevada, Spain): tectonic implications. Lithos 119:238–250

Martínez-Martos M, Galindo-Zaldívar J, Sanz de Galdeano C, García-Tortosa FJ et al (2017) Latest extension of the Laujar fault in a convergence setting (Sierra Nevada, Betic Cordillera). J Geodyn 104:15–26

Pérez-Peña A, Martín-Dávila J, Garate J, Berrocoso M et al (2010) Velocity field and tectonic strain in Southern Spain and surrounding areas derived from GPS episodic measurements. J Geodyn 49:232–240

Platt JP, Anczkiewicz R, Soto JI, Kelley SP et al (2006) Early Miocene continental subduction and rapid exhumation in the western Mediterranean. Geology 34:981–984

Platt JP, Behr WM, Johansen K, Williams R (2013) The Betic-Rif arc and its orogenic hinterland: a review. Annu Rev Earth Planet Sci 41:313–357

Platt JP, Kelley SP, Carter A, Orozco M (2005) Timing of tectonic events in the Alpujárride Complex, Betic Cordillera, S. Spain. J Geol Soc London 162:451–462

Platt JP, Vissers RLM (1980) Extensional structures in anisotropic rocks. J Struct Geol 2:397–410

Puga E, De Federico AD, Nieto JM (2002) Tectonostratigraphic subdivision and petrological characterisation of the deepest complexes of the Betic zone: a review. Geodin Acta 15(1):23–43

Rodríguez-Fernández J, Sanz de Galdeano C (2006) Late orogenic intramontane basin development: the Granada basin, Betics (southern Spain). Basin Res 18:85–102

Rodríguez-Fernández J, Sanz de Galdeano C, Serrano F (1990) Le Couloir des Alpujarras. Documents Et Travaux De l'IGA 12–13:87–100

Ruiz-Cruz MD, Sanz de Galdeano C (2017) Genetic significance of zircon in orthogneisses from Sierra Nevada (Betic Cordillera, Spain). Mineral Mag 81(1):77–101

Ruiz-Cruz MD, Sanz de Galdeano CS, Santamaría A (2015) Petrology and thermobarometric estimates for metasediments, orthogneisses, and eclogites from the Nevado-Filábride complex in the western Sierra Nevada (Betic Cordillera, Spain). Can Mineral 53(6):1083–1107

Santamaría-López Á (2019) Discussion on the division of the Nevado-Filábride complex (Betic Cordillera) based on tectono-stratigraphic, geochronological and petrochronological data. Thesis Univ Granada 150 pp. ISBN: 9788413061719

Santamaría-López Á, Lanari P, Sanz de Galdeano C (2019) Deciphering the tectono-metamorphic evolution of the Nevado-Filábride complex (Betic Cordillera, Spain)—a petrochronological study. Tectonophysics 767:128158

Santamaría-López Á, Sanz de Galdeano CS (2018) SHRIMP U-Pb detrital zircon dating to check subdivisions in metamorphic complexes: a case of study in the Nevado-Filábride complex (Betic Cordillera, Spain). Int J Earth Sci 107(7):2539–2552

Sanz de Galdeano C (2019) Paleogrographic reconstruction of the Betic-Rif Internal zone. Revista de la Sociedad Geológica de España 32(2). ISSN: 0214-2708

Sanz de Galdeano C, Alfaro P (2004) Tectonic significance of the present relief of the Betic Cordillera. Geomorphology 63:175–190

Sanz de Galdeano C, Galindo-Zaldívar J, Alfaro P, Ruano P (2007) El relieve de la Cordillera Bética. Enseñanza de las Ciencias de la Tierra [en línia] 15(2):185–195

Sanz de Galdeano C, López-Garrido AC (1999a) Nature and impact of the Neotectonic deformation in the western Sierra Nevada (Spain). Geomorphology 30:259–272

Sanz de Galdeano C, López-Garrido AC (1999b) Estratigrafía y estructura de las unidades alpujárrides en el borde occidental de Sierra Nevada (Granada, España). Revista de la Sociedad Geológica de España 12(2):187–198. ISSN: 0214-2708

Sanz de Galdeano C, López-Garrido AC (2016) The Nevado-Filábride Complex in the western part of Sierra de los Filabres (Betic Internal Zone). Structure and lithological succession. Bol Geol Min 127:823–836

Sanz de Galdeano C, Peláez Montilla J, López Casado C (2003) Seismic potential of the main active faults in the Granada Basin (Southern Spain). Pure Appl Geophys 160:1537–1556

Sanz de Galdeano C, Rodríguez-Fernández J, López-Garrido AC (1985) A strike-slip fault corridor within the Alpujarra Mountains (Betic Cordilleras, Spain). Geol Rundsch 74:641–655

Simancas FJ (2018) A reappraisal of the Alpine structure of the Alpujárride Complex in the Betic Cordillera: interplay of shortening and extension in the westernmost Mediterranean. J Struct Geol 115:231–242

Van Bemmelen RW (1927) Bijdrage tot de geologie der Betische ketens in de provincie Granada. Thesis Univ Delft 176 pp

Von Drasche R (1879) Geologische Skizze des Hochgebirgsteiles der Sierra Nevada in Spanien. Jahrbuch der K.K. Geol Reischsanstalt, Wien 29:93–122

Zeck HP (2004) Rapid exhumation in the Alpine Belt of the Betic-Rif (W Mediterranean): tectonic extrusion. Pure Appl Geophys 161:477–487

# The Impact of Glacial Development on the Landscape of the Sierra Nevada

Antonio Gómez-Ortiz, Marc Oliva, David Palacios, Ferran Salvador Franch, and José M. Fernández-Fernández

**Abstract**

The Sierra Nevada was the southernmost massif in Europe hosting glaciers during the Quaternary. The remains of those ice masses disappeared in the middle of the twentieth century. The Pleistocene glaciers in the Sierra Nevada remained confined to the high mountains, occupying the headwaters of ravines, with maximum lengths of 6–9 km. During the last decade, Cosmic-Ray Exposure (CRE) dating has complemented previously known geomorphological evidence, shedding new light on the chronology of various phases of glacial expansion and retreat. A moraine development phase occurred prior to the Last Glacial Cycle, around 130 ka, although most depositional and erosive records of glacial origin have been dated to the Last Glacial Cycle. This phase recorded two maximum advance pulses, the first around 30 ka and another of very similar extent at 20–19 ka. Later, the glaciers retreated significantly until they advanced again during the Oldest Dryas (17–16 ka). Then, after the almost total disappearance of the ice masses in the Sierra Nevada during the Bølling–Allerød (15–14 ka), small glaciers were formed during the Younger Dryas (12–11 ka). These glaciers disappeared at the beginning of the Holocene (10–9 ka), generating rock glaciers in the recently deglaciated cirques. At the end of the Holocene, including the Little Ice Age (LIA, 1300–1850 AD), very small glaciers were formed in the shelter of the northern walls of the highest peaks (Mulhacén and Veleta). They disappeared at the end of the LIA, giving rise to incipient rock glaciers.

**Keywords**

Sierra Nevada • Glaciers • Last Glacial cycle • Holocene • Little Ice Age

## 1 The Discovery of Glacial Landforms

The imprint of the glaciers on the landscape of the Sierra Nevada is one of the greatest attractions of this massif. Its peaks, cirques, and high valleys include a wide variety of landforms carved out by the ice during the Quaternary. In the regional context, in the extreme south of the Iberian Peninsula, the Sierra Nevada is the only mountain to have been extensively glaciated during the Pleistocene cold phases, the last of which reached its zenith some 20,000 years ago.

Similar to the case of the other Iberian mountains that were home to glaciers during the Quaternary, the discovery about glacial development modifying the landscape of the Sierra Nevada took place in the third decade of the nineteenth century. Boissier's (1839) research stands out among the first references to glacial development with its description of a small glacier enclosed in the Corral del Veleta. A decade later, Schimper (1849) was the first to cite the existence of a regional Quaternary glaciation in the Sierra Nevada and describe the small glacier that (even then) was located in the Corral del Veleta (Gómez-Ortiz et al. 2018; Titos Martínez 2019). Throughout the twentieth century and until the 1960s, numerous researchers published their findings regarding the delimitations of glacial systems, their descriptions, and morphological analyses as well as the chronologies of glacial events (Quelle 1908; Obermaier 1916; Dresch 1937; Casas-Morales 1943; García-Sainz 1947; Hempel 1960; Messerli 1965).

A. Gómez-Ortiz · M. Oliva (✉) · F. S. Franch
Department of Geography, University of Barcelona, Barcelona, Spain
e-mail: marcoliva@ub.edu

D. Palacios
Department of Geography, Universidad Complutense de Madrid, Madrid, Spain

J. M. Fernández-Fernández
Centre for Geographical Studies, IGOT, Universidade de Lisboa, Lisbon, Portugal


R. Zamora and M. Oliva (eds.), *The Landscape of the Sierra Nevada*,
https://doi.org/10.1007/978-3-030-94219-9_6

Obermaier (1916) highlighted the existence of glaciers circumscribed at the summit level and located the maximum advances of the glacier tongues. He also proposed a tentative glacial chronology, locating evidence of a last glaciation and probably other previous ones, as did García-Sainz (1947) and Hempel (1960) decades later. Half a century after, Messerli (1965) described the Sierra Nevada glaciation as typical of an arid high mountain, where the glaciers were always enclosed in a network of ravines conditioned by the geological structure and pre-glacial relief. The author presented a geomorphological sketch of the glacial complex, differentiating the various moraine systems and recognizing deposits of the Riss glaciation, which was the penultimate glacial cycle according to the alpine glacial terminology prevailing during that time. This glaciation was also the most extensive one, as proposed by Penck (1883) and discussed years later by Lhenaff (1977). The research also located the moraines of the Last Glacial Cycle (LGC; Würm) in the valley bottoms, between the altitudes of 1450 and 2000 m. Finally, he placed the present-day periglacial environment above 2500–2600 m, where solifluction processes are active; during the cold Pleistocene phases, these processes were active up to 700 m on the northern face.

The last decades of the twentieth century were characterized by the implementation of new methodologies introduced to examine the geomorphological implications of the Quaternary cold periods. These methods were accompanied by notable advances in detailed geomorphological mapping (Gómez-Ortiz 1987). During the twenty-first century, glacial and periglacial studies in the Sierra Nevada experienced a notable boost. A significant contribution was the publication of a geomorphological map at a scale of 1:25,000 (Gómez-Ortiz 2002), which was later complemented with detailed quantitative information about the glacial domain and its morphologies (Palma et al. 2017). Likewise, in parallel to the rest of the Iberian mountains (Oliva et al. 2019), remarkable advances were made with regard to the absolute chronology of the erosive and depositional glacial records of the Sierra Nevada thanks to CRE dating through $^{36}$Cl and $^{10}$Be isotopes. The results obtained confirmed the existence of glacial phases prior to the LGC in the headwaters of the Poqueira River, as well as the occurrence of glacial advances in different valleys during the global Last Glacial Maximum (LGM), Oldest Dryas, and Younger Dryas (Gómez-Ortiz et al. 2012a; Palacios et al. 2016, 2019). In the last decade, researchers have also made strides in the Holocene paleoenvironmental evolution in the massif, with particular interest in the reconstruction of the historical glaciers that Sierra Nevada hosted during the Little Ice Age (LIA, 1300–1850 AD) (Oliva and Gómez-Ortiz 2012; Oliva et al. 2014a, 2018; Gómez-Ortiz et al. 2018; Palacios et al. 2019).

## 2 The Geographical Context

Located in the southeast of the Iberian Peninsula (37°N) and in the southwest of the European continent, the Sierra Nevada is part of the Betic Range and is included in the so-called Internal Zone. The tectonic structure and structural units that make up the massif are described in detail in chapter "Geological Setting of Sierra Nevada."

The western sector of the Sierra Nevada's summits includes the highest peaks of the Iberian Peninsula (Mulhacén, 3479 m; Veleta, 3396 m; and Alcazaba, 3369 m; Fig. 1). Its current landscape is the result of the erosive action from the Quaternary glaciers, the southernmost in Europe and unique in the Betic Range. The configuration of the relief is structured according to the alignments of ridges and remains of ancient erosion surfaces, mostly located above 2800–3000 m. The headwaters of the rivers are located at its flanks, which run radially from the axis of the massif through steep slopes until they reach the surrounding plains. The rivers on the northern slope are tributaries of the Genil and Fardes, both of which are part of the Guadalquivir Basin, which flows into the Atlantic Ocean. Those located on the south are tributaries of the Guadalfeo River, which drains directly into the Mediterranean Sea.

The high peaks above 2000 m extend for more than 65 km, with a west–east arrangement only 30–40 km from the Mediterranean Sea. The volume and high altitudes of the Sierra Nevada distinguish it from the adjacent flat areas and determine a topoclimatic mosaic in and around the massif itself. The climate of the Sierra Nevada presents the characteristics of a semi-arid high mountain with a Mediterranean influence and notable contrasts depending on its altitude and orientation. At an altitude of 2500 m, the mean annual air temperature is 4.4 °C, and the annual precipitation is 710 mm, which occurs in the form of snow during the cold season (Oliva et al. 2014b). At 3300–3400 m, namely at the summit level, the mean annual air temperature reaches 0 °C (Oliva et al. 2016), showing negative average monthly values from November to April. The snow on the summits of the massif lasts from November to June, and it can persist throughout the year in the higher northern cirques. There is little or no precipitation during the hot season, from the end of June to August, and the average temperature ranges from ~15 °C at 2500 m to 10 °C at 3400 m (Salvador-Franch et al. 2011; Gómez-Ortiz et al. 2019, see chapter "Climate Variability and Trends"). These climatic conditions favor the predominance of a periglacial environment above 2500–2600 m. Above these altitudes, there is scarce vegetation, with the only presence of hygrophilous grasslands (locally named *borreguiles*) in the valley bottoms.

The singularity of the glacial landscape of the Sierra Nevada, as well as the uniqueness of the vegetation that

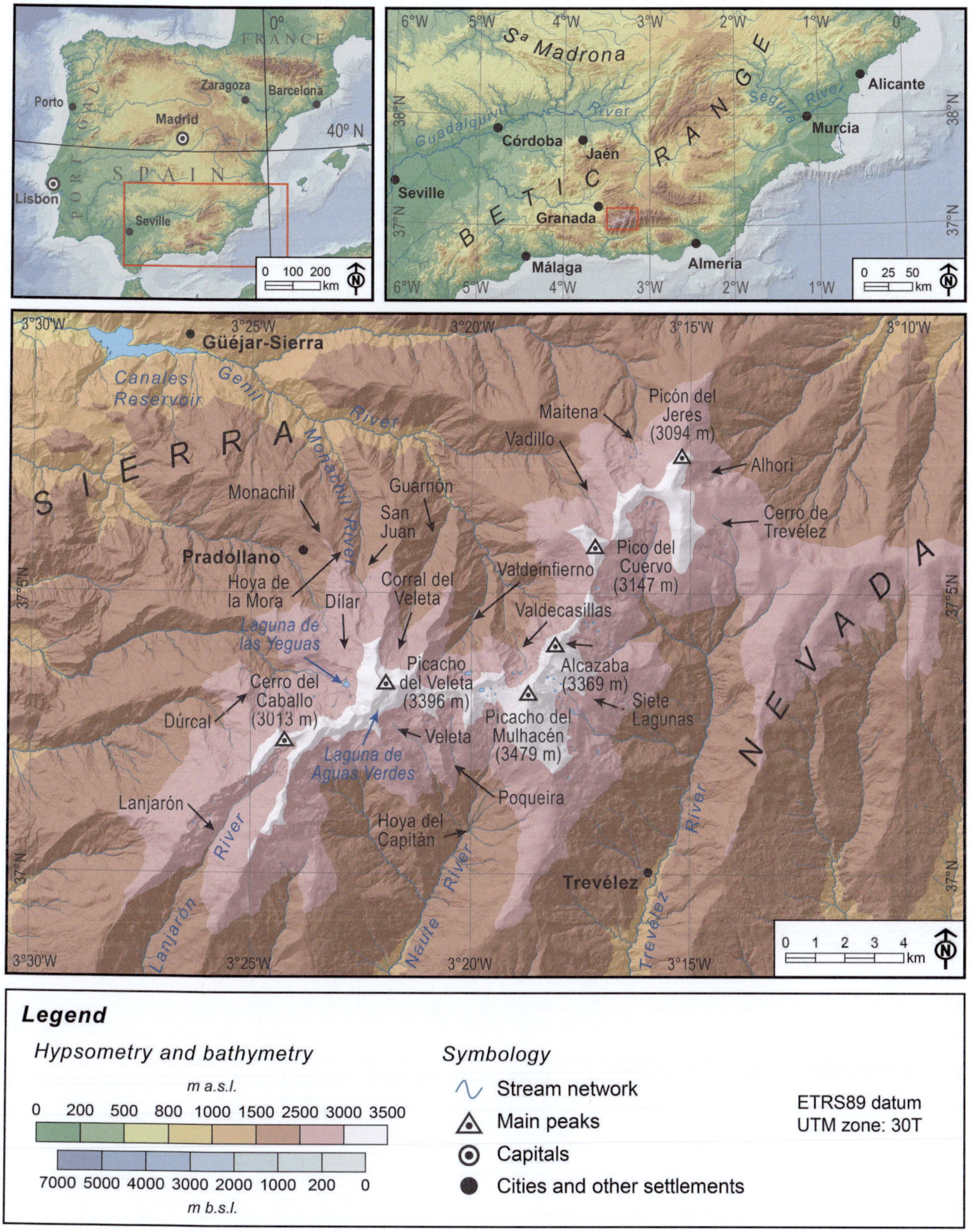


**Fig. 1** Locations of the places cited in this chapter

colonizes its peaks (see chapter "Altitudinal Patterns and Changes in the Composition of High Mountain Plant Communities"), were decisive for its declaration as a National Park in 1999, the aim being to preserve the inherent natural value of its landscapes and the ecosystems of the high Mediterranean mountains.

## 3 The Distribution of Paleoglaciers and the Glacial Landscape

The Quaternary glaciers hosted in the Sierra Nevada were the only ones developed in the whole Betic Range. They had limited dimensions, and their development was conditioned by the morphostructure of the massif and the pre-glacial relief organizing the network of the main ravines (Fig. 2). They were formed at the base of the highest peaks, showing a glaciation style typical of high semi-arid mountains.

In the Sierra Nevada, glaciers developed in the western third of the massif along 35 km, between the Picón de Jeres (3094 m) and the Cerro del Caballo (3013 m), where the summits reach the highest altitudes in the Iberian Peninsula, with many sectors above 3300 m. From Cerro de Trevélez (2877 m) in the easterly direction to Cerro del Chullo (2609 m), the impact of glacial development on the landscape tended to diminish progressively, with masses of ice composed of small glaciers enclosed in the high headwaters of the ravines of the Marquesado del Zenete (Casas-Morales 1943). Eastwards Cerro del Chullo, there is no trace of glacial footprints on the summits, but there is evidence of periglacial processes shaping the slopes.

The distribution of the different glacial systems followed a radial pattern from the main axis of the massif. On the north face, the units of the Hoya de la Mora, San Juan, and Genil systems (i.e., Guarnón, Valdeinfierno, Valdecasillas, and Vacares) and those of Vadillo, Maitena, and Alhorí stood out. On the southern side, the Lanjarón system (the Cornavaca, Lagunillos, and Nevero units), the Poqueira system (the Veleta, Río Seco, and Mulhacén units), and the Trevélez system (the Goterón, Buitrera, Bacares, Juntillas, Peñón del Puerto, and Las Albardas units) include remarkable glacial features. The western side is notable for the units of Dúrcal, the Dílar system (the Cartujo and Fraile de Capileira units), and the Monachil unit. The eastern side was home to the Siete Lagunas system. In total, up to 27 glaciers of different dimensions were formed.

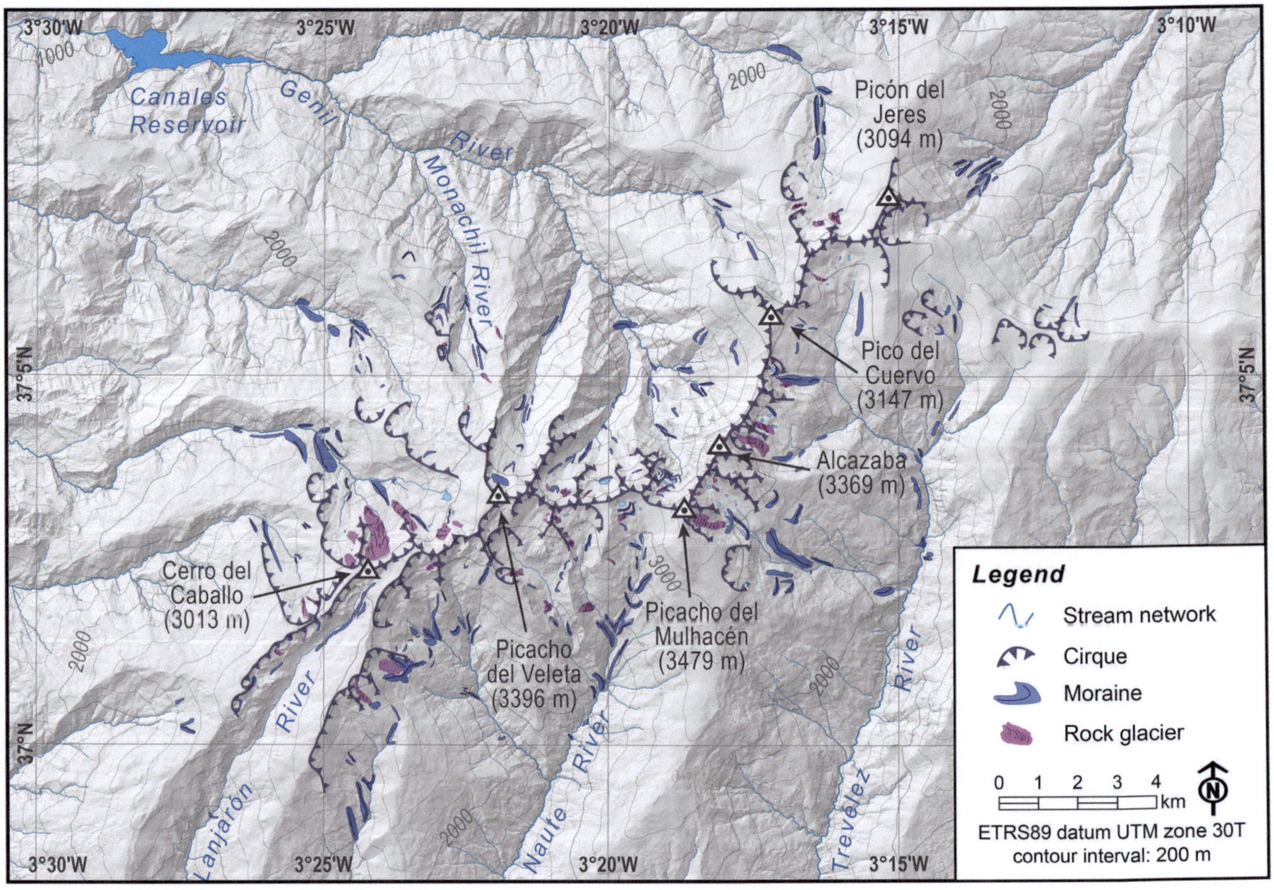


**Fig. 2** Geomorphological map of the Sierra Nevada showing the distribution of the main landforms of glacial origin (Adapted from Palma et al. 2017)

### 3.1 Glaciers Confined to the Valley Heads

The glacial morphology of the Sierra Nevada shows the greatest geomorphological development in the cirques and high valleys shaped on the main peaks of the massif (Fig. 3). The cirques are very varied in size and morphology, in line with the morphostructure and prevailing lithology of the massif (Messerli 1965). This aspect can be particularly observed on the northern and southern slopes between the Mulhacén and Veleta peaks. The cirques end in concave depressions open to the valley, limited by steep slopes filled by frost shattered rock fragments in their lower areas. The surfaces of the cirques exhibit glacial polished surfaces and house over-deepened basins, some of which are filled by lakes, such as in the areas of the Las Yeguas, Laguna Grande de Río Seco, Aguas Verdes, Las Cabras, Juntillas, La Mosca, and Larga, among others. The structure of the substrate has also led to the existence of thresholds and rock steps at the base of the cirques (*basares*), with traces of glacial abrasion, as is the case in El Goterón, Basares del Veleta, Corral del Veleta, Juntillas, etc. As far as the edges between the cirques are concerned, it is necessary to emphasize the succession of the dividing crests (*raspones*) that delimit their spatial domains, as in the case of the prolonged ridge between the peak of the Cuervo and the Picacho del Veleta. The crests include *horn*-shaped peaks (*puntales*), as observed in the Puntal de Vacares, Alcazaba peak, and Picacho del Veleta, among others. To a much lesser extent, mountain passes exist between the highest peaks, which acted as transfluence or glacial diffluence sites during the Quaternary glacial phases (e.g., Collado del Ciervo, Cuneta de Vacares, Portillo de la Carihuela, etc.). This last fact explains that in a few cases, the retreat of the walls of the cirques caused them to coalesce and form individualized units.

In terms of the erosive landforms in the valleys, the most significant type is presented by the U-shaped valleys existing in their middle and lower sections. The best examples are located on the southern slope, where the paths of the valleys and hillsides follow gentler slopes with few abrupt changes in level. The most significant case is that of the Lanjarón Valley, which extends more than 4 km from its headwaters to beyond the water intake from the del Caballo sector, where the slope substrate includes a glacial shoulder and long stretches of glacial polishing crowned by small aligned cirques. The U shape of the Veleta Valley is also remarkable, the clearest in the Sierra, with a 2-km-long development between the Aguas Verdes Lake (3050 m) and the shoulder hanging from the surroundings of the Las Cabras Lake (3020 m). This pattern is less repeated on the northern slope, although it is presented in short sections due to the steep slopes, as in the Guarnón, in its middle section, from the Chorreras. Due to their smaller size, the glacier systems hanging on the slopes (e.g., Hoya de la Mora, Cornavaca,

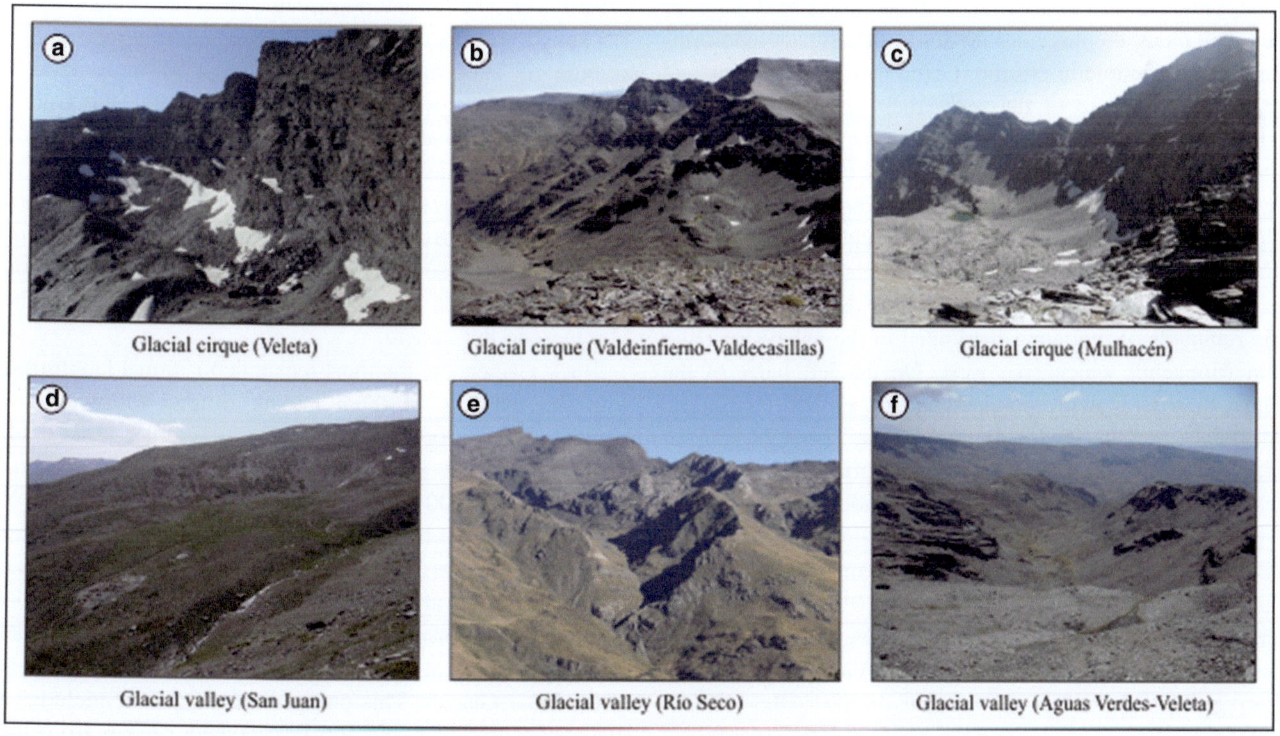


**Fig. 3** Examples of cirques and glacial valleys in the Sierra Nevada

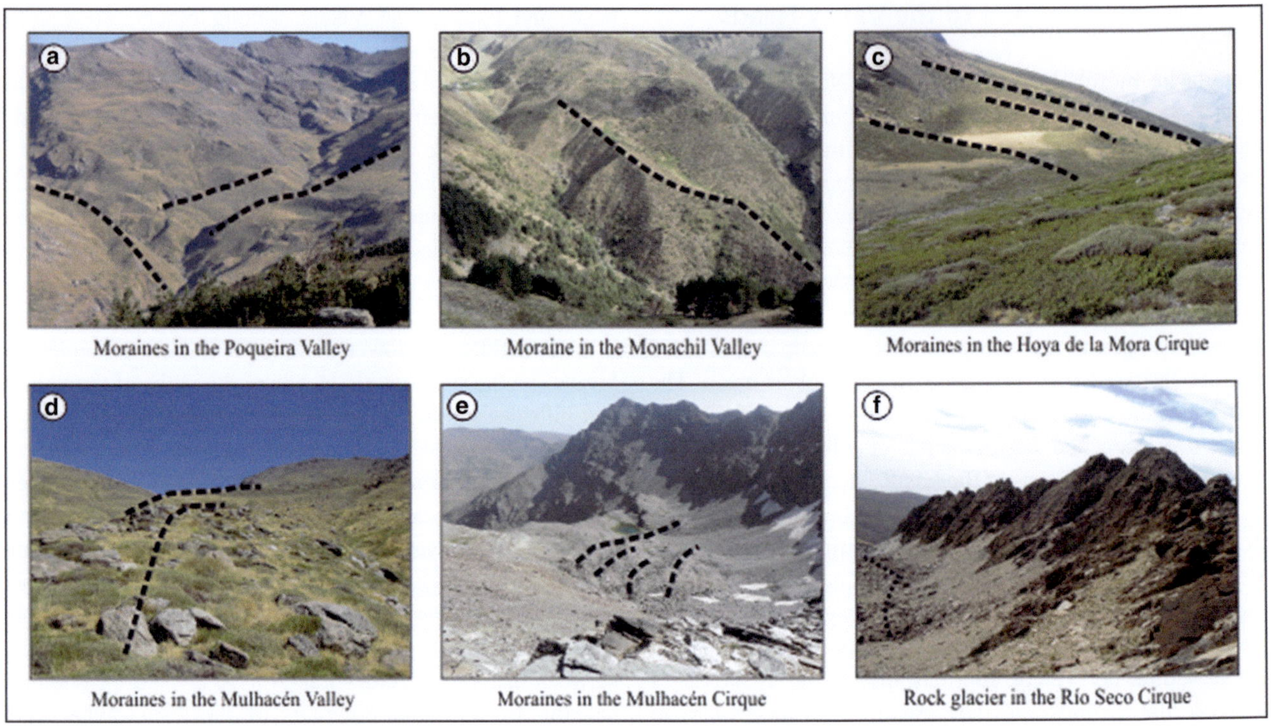


**Fig. 4** Systems of moraine deposits and rock glaciers developed during the different periods described in the chapter

Siete Lagunas, and Alhorí) do not include U-shaped glacial valleys (Figs. 4 and 5).

Depositional landforms, mainly moraines and rock glaciers, are also abundant geomorphological features. Due to their unconsolidated material, without developed soils to stabilize them, the moraines do not form compact landforms and are susceptible to erosion. Given their younger age and the compact nature of the rock, rock glaciers tend to be spatially distributed more frequently in the cirques and headwalls. The most remarkable moraine complex is located in the Poqueira Basin and includes the southern valleys of Veleta, Río Seco, and Mulhacén. The segments to the right and left of the Mulhacén and Río Seco glaciers, which must have entered the confluence, stand out. Also noteworthy are the remnants of moraines distributed in the valleys of Dílar and Monachil, which face west. On the left bank of the Dílar, there is a lateral moraine that extends between the La Mula lake and the Prados de las Monjas. The right bank of the Monachil Valley is home to the remains of moraines, which are located downstream of the Pradollano resort, over a 620-m-long segment at an altitude of 1950 m, which tends to close off the riverbed. Its stabilization must have occurred around 19.6 ka (Gómez-Ortiz et al. 2012a). Finally, we should also highlight the sequence of the different generations of lateral moraines in the San Juan Valley that close the *borreguiles* installed at the bottom of the valley (2200–2500 m).

The Sierra Nevada also hosts the remains of moraines hanging from slopes or hills of cirques that were confined to summits. This is the case of the Siete Lagunas Glacier, which was oriented to the east over the course of the Trevélez River. The left lateral moraine alignment extends over 1.5 km along the Culo de Perro Ravine, its front reaching an altitude of 2200 m. Likewise, although of more modest dimensions, it is worth mentioning the lateral and frontal moraines of the small cirque of the Hoya de la Mora, which belonged to the lowest glacier of the northern slope. Also remarkable are the sequence of east-facing arched moraines of the Cornavaca Basin at the foot of the Tajo de los Machos (3081 m) and those of the small Alhorí Glacier installed on the northern slope of the Picón de Jeres and facing the Marquesado del Zenete.

Rock glaciers are the most recent depositional landforms, with the majority of them having formed following ice retreat at the end of the Younger Dryas (Gómez-Ortiz et al. 2015; Palacios et al. 2016). Almost all of them are distributed above 2500 m, with some located as high as 3200 m and occupying the headwaters of valleys and cirque cavities (Río Seco, Dílar, Valdecasillas, Corral del Veleta, etc.) (Gómez-Ortiz et al. 2013; Palma et al. 2017). The most developed landform corresponds to that of Cascajar del Cartujo, installed at the westernmost end of the headwaters of the Dílar Glacier. This polygenic rock glacier includes a succession of ridges and furrows located 1.6 km from the

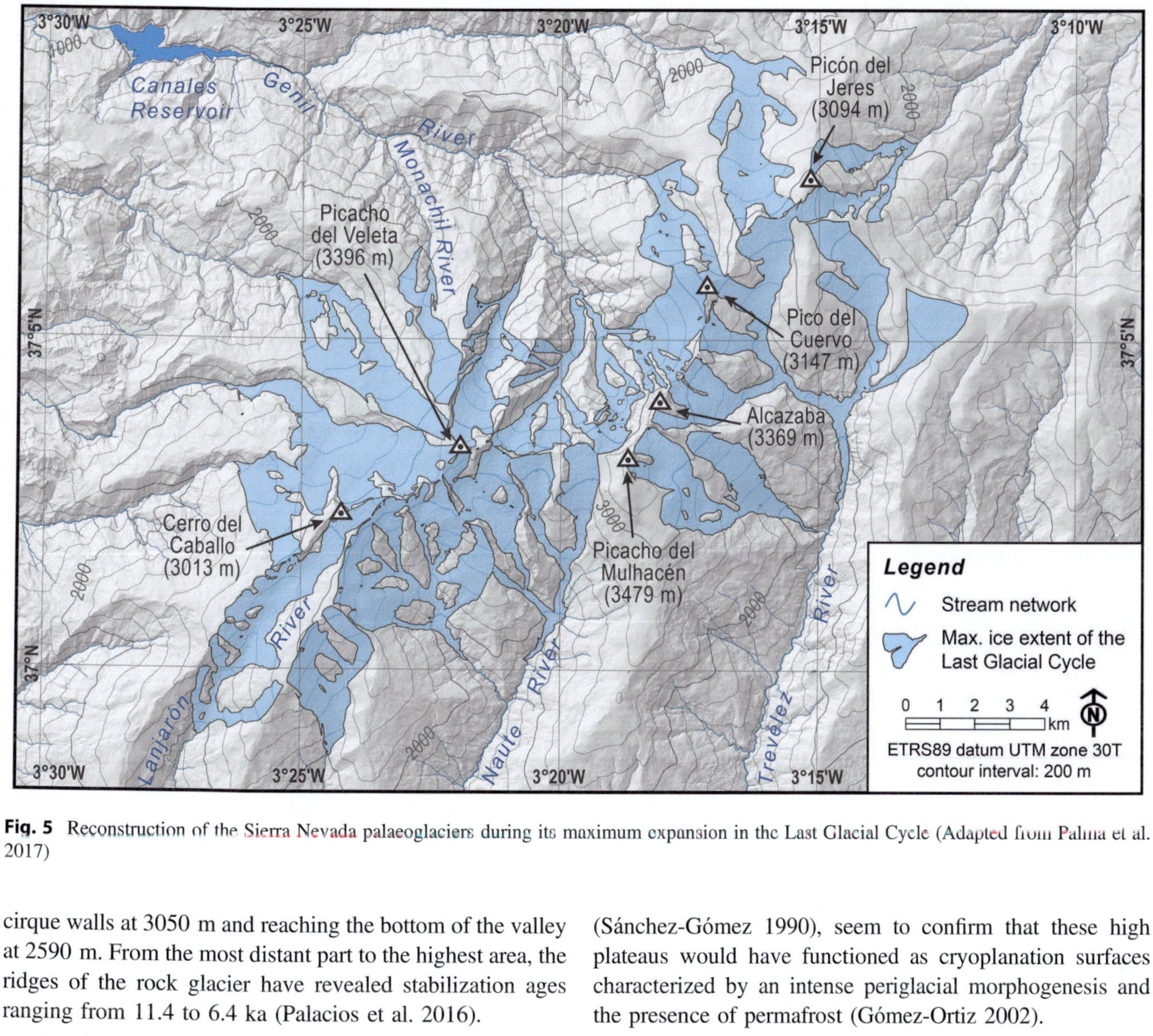


**Fig. 5** Reconstruction of the Sierra Nevada palaeoglaciers during its maximum expansion in the Last Glacial Cycle (Adapted from Palma et al. 2017)

cirque walls at 3050 m and reaching the bottom of the valley at 2590 m. From the most distant part to the highest area, the ridges of the rock glacier have revealed stabilization ages ranging from 11.4 to 6.4 ka (Palacios et al. 2016).

## 3.2 Weakly Glaciated Plateaus

The imbricated orography of the Sierra Nevada must not have been favorable for the existence of large ice caps during the LGC. The plateaus of the highest summit surfaces present highly developed periglacial landforms, namely patterned ground features (e.g., stone cirques of metric dimensions), suggesting the prevalence of very intense periglacial conditions without the presence of ice fields during the LGC. This is the case for the summit surfaces of Picón de Jeres-Cerro Pelao (3144 m), Collado de los Machos (3329 m), and Allanada del Mulhacén (3450 m). The absence of erratic boulders or polished surfaces as well as the presence of subsurface soil levels with a degree of development similar to those of the moraine deposits identified in the valley bottoms, such as in the Lanjarón Valley (Sánchez-Gómez 1990), seem to confirm that these high plateaus would have functioned as cryoplanation surfaces characterized by an intense periglacial morphogenesis and the presence of permafrost (Gómez-Ortiz 2002).

## 4 The Chronology of Glacial Landforms

Until the last decades of the twentieth century, the chronology of the deglaciation in the Sierra Nevada was based especially on the altitudinal and morphostratigraphic distribution of the moraines as well as the degree of preservation of their sediments and the development of the glacial cirques. In this phase, the chronological determination of glacial events was always relative and identified in concurrence with the alpine terminology prevailing at the time (Penck 1883; García-Sainz 1947; Hempel 1960; Messerli 1965; Lhenaff 1977). Subsequently, $^{14}C$ dating provided information on the paleoenvironmental evolution once the ice disappeared from the Sierra Nevada, confirming that the southern slope of the massif was already deglaciated around 12.9 ka (Oliva 2009; Anderson et al. 2011; Oliva et al. 2014a). It was not until the second decade of the new millennium that absolute data on glacial chronology began to be provided via Cosmic Ray Exposure (CRE) dating, particularly $^{36}$Cl and $^{10}$Be dating (Gómez-Ortiz et al. 2012a; Palacios et al. 2016, 2019; 2020). The results obtained for the valleys of Mulhacén, Río Seco, Lanjarón, Dílar, Hoya de la Mora, Guarnón, San Juan, and Valdecasillas confirmed that glacial processes in the Sierra Nevada were active during the periods prior to the LGC, the LGC, and the Oldest Dryas and Younger Dryas. Thus, similar to the Pyrenees and Picos de Europa among the Iberian mountains, the Sierra Nevada hosted glaciers during the Holocene and the LIA (Oliva et al. 2019).

## Glacial phases before the Last Glacial Cycle

The oldest dates obtained in the Sierra Nevada correspond to the southern slope of the massif, specifically at the head of the Poqueira Valley, in the geomorphological unit of the Mulhacén (Table 1). This evidence corresponds to two blocks anchored in a right lateral moraine at 2164 m and 2263 m. The samples dated by the Cosmic Ray Exposure (CRE) dating reported ages of 134.7 and 129.2 ka (Palacios et al. 2019). These ages suggest the existence of a glaciation older than the LGC, as already proposed by Obermaier (1916), García-Sainz (1947), and Messerli (1965). These authors attributed moraine deposits located at very high levels with respect to the valley floor or fluvio-glacial deposits downstream of the main moraine ranges to the so-called Riss Alpine glaciation. This phase has hardly been detected in the Pyrenees and the mountains of Northwest Iberia, where a glacial expansion phase may have taken place around 170–130 ka (Oliva et al. 2019).

## Last Glacial Cycle

During the LGC, the Sierra Nevada hosted a significant development of glaciers in the cirques and valleys, in some cases with transfluences in the headwaters and coalescences of fronts in the valleys. Due to the narrowness and steep slope of its valleys, the glacial domain extended over 105 km$^2$ to heights slightly below 2000 m in the northern valleys, the maximum ice thickness being 135 m (Palma et al. 2017). The most extensive glacial systems during this phase reached lengths of 9 km in some valleys (Lanjarón, Dílar, Monachil, Guarnón, etc.). A moraine ridge of the maximum local extent of the LGC located on the left slope in the Mulhacén Valley in the Poqueira basin at 2450 m yielded an age of 29.9 ka; a very similar age (29.7 ka) was also revealed by a sample of the horn substrate at 3210 m of the Picacho del Veleta (Palacios et al. 2016). After these maximum advances of the ice around 20 ka, the glaciers in the

**Table 1** Chronologies of the glacial phases in the Sierra Nevada and their geomorphological evidence

| Period | Chronology | Relief landforms and environments |
|---|---|---|
| Previous glacial cycles (<100 ka) | Expansion of the glacier dated around 135–130 ka | • Presence of moraines very degraded by the passage of time and fluvio-glacial sediments at lower altitudes than those of moraines from the Last Glacial Cycle (LGC) |
| LGC (100–20 ka) | Maximum expansion of the ice in the LGC during 32–30 ka and a later retreat of almost the same extent during 20–19 ka | • Tongues of valley glaciers that occupied the headwaters of the valleys reaching heights slightly below ~2000 m (north face) and ~2500 m (south face), always lower on the western face of the massif<br>• Glaciated area during the maximum glacial advance (105 km$^2$) almost entirely centered in the western third of the Sierra Nevada |
| Deglaciation 201–11.7 ka) | • Beginning of deglaciation in ~19 ka<br>• Glacier expansion in ~17 ka (Oldest Dryas)<br>• Glacial retreat during ~14.5–14 ka (Bølling–Allerød)<br>• Glacial expansion during ~13–12 ka (Younger Dryas) | • Rapid deglaciation<br>• Glacial advance with moraine formation near the limits of the LGC<br>• Rapid glacial retreat toward the headwaters<br>• Expansion of glaciers in the highest cirques, with few glaciers on the southern slope |
| Holocene (<11.7 ka) | • Early Holocene during ~10–9 ka<br>• Late Holocene, glacier formation from ~2.8–2.7 and 1.4–1.2 ka cal BP, and during the LIA | • Glacial retreat and development of rock glaciers, which stabilized at ~7–6 ka<br>• Formation of small glaciers in the Mulhacén and Veleta cirques |
| LIA (1300–1850 AD) | Glacier formation in the Hoya del Mulhacén (1440–1710 AD) and the Corral del Veleta (1350–1950 AD) | Small glaciers covering only a few hectares in the highest northern cirques above 2900 m between the Mulhacén and Veleta peaks |
| Current | Disappearance of the Corral del Veleta Glacier in the mid-twentieth century | • No glaciers in the Sierra Nevada today<br>• Only masses of ice buried in the enclaves of the glacial cirques during the LIA |

Sierra Nevada retreated toward the headwaters before re-advancing again with the formation of new moraine systems during the global LGM (26–19 ka; Clark et al. 2009). Geomorphological evidence from this phase is scarce; the moraines distributed in the valleys of Monachil and Hoya de la Mora stand out. These complexes were generated a few hundred meters behind and at a higher altitude than those of the maximum glacial expansion phase (Gómez-Ortiz et al. 2012a). Other examples of this phase are the fronto-lateral moraine of the Monachil, downstream from Pradollano (1975 m, 19.6 ka), and the outermost arches of the Hoya de la Mora Cirque (2266 m, 21.3 ka). This advance synchronous with the global LGM was also detected in most Iberian mountains, although in some cases, the maximum expansion of the ice occurred even before the advance of 30 ka recorded in the Sierra Nevada, as was the case in the northern and northwestern mountains, where it occurred between 45 and 35 ka (Oliva et al. 2019). As mentioned above, the summit surfaces of the Sierra Nevada during this phase would have been predominantly ice-free, with average annual air temperatures of $\sim -7$ or $-8$ °C favoring very active periglacial processes in a permafrost environment. In turn, the periglacial dynamics would have extended to relatively low levels (1200–1300 m), particularly at its western end.

## Deglaciation and pre-Holocene glacial retreats

The notable increase in temperatures recorded on a planetary scale at $\sim$20–19 ka and following the LGM (Clark et al. 2009) also led to the recession of the Sierra Nevada glaciers, which retreated toward the headwaters of the valleys at very high altitudes. Subsequently, glaciers advanced again sometime around 17 ka during the Oldest Dryas due to the sudden drop of 4–5 °C in global temperatures. Evidence of these oscillations are found in the rock surfaces polished by the erosive action of the glacial ice as well as the presence of moraines located in many valleys of the massif. The case of the head of the Lanjarón Valley is significant because of the abundance of polished surfaces that suggest the disappearance of the ice in 18.3 ka (2935 m). There is also evidence of glacial advances during this period in other Iberian mountain massifs (Oliva et al. 2019), such as the Pyrenees (17 to 16 ka), Cantabrian Mountains (17.5 ka), Central Range (17 ka), and the Iberian Range (17 ka).

However, this phase of glacial advance within the deglaciation process was short. Around 14.5–14 ka (Bølling–Allerød), temperatures increased significantly, which promoted the disappearance of the glaciers from a large part of the massif (Palacios et al. 2016, 2020) as well as the rest of the Iberian mountains (Oliva et al. 2019). The CRE data show the accelerated retreat of ice between 15 and 14 ka in many valleys of the massif, such as Monachil, Río Seco, Lanjarón, Dílar, San Juan, and Hoya de la Mora. Paraglacial dynamics in these recently deglaciated enclaves as well as the presence of permafrost favored the formation of rock glaciers (Gómez-Ortiz et al. 2013; Palacios et al. 2015).

Around 12.9 ka (Younger Dryas), global temperatures dropped sharply, once again favoring the expansion of the high-latitude ice sheets as well as the advance of the mountain glaciers in the mid-latitudes and, of course, in Iberian mountains (Oliva et al. 2019), such as the Pyrenees (13–11.5 ka), possibly in the Cantabrian Mountains (13.2–12.1 ka) and the Central Range (13–12 ka). Small cirque glaciers formed again in the Sierra Nevada on both sides of the massif, such as in San Juan, Hoya de la Mora, Hoya del Mulhacén, and Lanjarón, where the glacier tongues extended less than 2 km from the headwaters (Palacios et al. 2016). This phase lasted for $\sim$1500 years until the glaciers disappeared around 10–9 ka with the beginning of the present interglacial (Gómez-Ortiz et al. 2012b, 2015). As in the case at the end of the Younger Dryas, new rock glaciers were formed in the headwaters of cirques with the melting of these last glaciers. Else, other existing rock glaciers were reactivated, but in all cases their fronts soon stabilized. These events occurred at the head of the Dílar (as attested by the sequence of arches dated at 11.4, 8.6, and 6.4 ka) and in the Río Seco Cirque (9.6 ka) (Gómez-Ortiz et al. 2012b; Palacios et al. 2016).

## The Holocene and LIA

In the Sierra Nevada, as in the rest of the Iberian mountains, the temperatures prevailing during the present interglacial, namely the Holocene, have not favored glacial development. In fact, other Iberian mountains include very little evidence of the persistence of glaciers during this period, which was primarily restricted to some Pyrenean cirques, and small shrinking ice masses in the Iberian Range. Meanwhile, the glaciers of the Central Range disappeared (Oliva et al. 2019). In the Sierra Nevada, the existence of glaciers during the Holocene has been short-lived and only restricted to the highest northern cirques of the massif. Small Holocene glaciers have been detected in two enclaves only: the Hoya del Mulhacén (2900 m) and the Corral del Veleta (3150 m). According to the analysis of the sediments contained at the bottom of the La Mosca Lake in the Hoya del Mulhacén, a small glacier must have existed between 2.8 and 2.7 ka cal BP, 1.4 and 1.2 ka cal BP, as well as during the LIA, between 510 and 240 cal yr BP (Oliva and Gómez-Ortiz 2012). As per the CRE dating and written evidence by historical documents, a glacier existed in the Corral del Veleta (southernmost Europe) from the fourteenth to the mid-twentieth century (Gómez-Ortiz et al. 2018; Palacios et al. 2019). The degradation and shrinkage of this glacier led rise to a small rock glacier that has been monitored since the end of the 1990s and testifies to the degradation of permafrost conditions in the Sierra Nevada (chapter "Ancient and Present-Day Periglacial Environments in the Sierra Nevada").

## 5 Concluding Comments: The Uniqueness of Glacial Development in the Sierra Nevada and in the Context of the Iberian Region as a Whole

The Quaternary glaciers in the Sierra Nevada were the southernmost in Europe. The latitudinal position of the massif at 37° N, its geographical location on the southwest edge of the continent, and its morphological characteristics conditioned a glacial development characteristic of semi-arid mountains, closer to those that developed in the Atlas compared to the glaciers formed in the Pyrenees or Alps. In the Sierra Nevada, the different glacial systems remained confined to the summit level and enclosed in the headwaters of the ravines according to the morphostructure and pre-glacial relief. Although the glaciated domain was modest in comparison with the glacial development recorded in other Iberian mountains of lower altitudes, its cirques and valleys hosted glaciers that were very sensitive to the climatic changes that occurred in the transition from the Pleistocene to the Holocene. Unlike other mountains of Northwest Iberia, the glaciers in the Sierra Nevada did not cover the summit plains, which acted as cryoplanation surfaces under very intense periglacial dynamics and the presence of permafrost.

Glacial dynamics during the LGC in the Sierra Nevada must have erased or masked the traces of a previous glaciation, as was the case in the majority of the glaciated Iberian mountains. However, recently, remains of very old and degraded moraines have been detected and dated, corresponding to a glacial advance that occurred around 130 ka (Palacios et al. 2019). This recent finding reinforces the hypothesis of a glaciation prior to the LGC in the Sierra Nevada and provides new perspectives on the paleoenvironmental evolution of this part of the Western Mediterranean and the glaciated mountains of the Iberian Peninsula.

The maximum advance of the LGC in the Sierra Nevada must have occurred at ~30 ka as suggested by the left lateral moraine of the Veleta Valley, coinciding with the maximum development of the glacial tongues in valleys and, in some cases, a confluence of tongues. In this phase, the Equilibrium Line Altitude would have been located between 2300 and 2500 m, depending on the orientations and the topographical characteristics of each valley (Paschinger 1957; Messerli 1965; Oliva et al. 2014a; Palma et al. 2017). An advance of similar scope took place at ~20–19 ka (Gómez-Ortiz et al. 2012b; Palacios et al. 2016, 2019) and later began a process of accelerated deglaciation, as in most mid-latitude mountain ranges. The glaciers of the Sierra Nevada tended to retreat toward the headwaters of the valleys, although this occurred in recurrent episodes of advance–retreat along the Oldest and Younger Dryas, when the glaciers once again occupied the valley bottoms and the highest cirques (Gómez-Ortiz et al. 2012b; Palacios et al. 2016, 2019). The temperature rise recorded during the Holocene conditioned a very ephemeral presence of glaciers during the current interglacial.

The periods of glacial retreat also had a significant impact on the development of the current landscape of the Sierra via the formation of rock glaciers. Some of these periglacial landforms, which are indicative of the presence of permafrost, were active until well into the Holocene, as was the case of the rock glaciers of Río Seco (9 ka) and the headwaters of the Dílar (8.6 and 6.4 ka), where the definitive disappearance of permafrost favored their stabilization (Palacios et al. 2016). Ephemeral cold phases during the Late Holocene also favored a punctual glacial development in the peaks of the Sierra Nevada; the appearance of glaciers in the highest northern cirques, similar to the case in the Hoya del Mulhacén, attests to this point (Oliva and Gómez-Ortiz 2012; Oliva et al. 2019).

The general decline in temperature and increase in precipitation recorded during the LIA intensified periglacial dynamics on the summits of the massif, as well as the generation of new small glaciers within some north-facing cirques under the vertical walls of the main peaks. Among them, the Corral del Veleta stands out as it "*presents in miniature all the characters of the Alpine glaciers*" (Boissier 1839). This glacier developed in the shade of the Guarnón Cirque at an altitude of 3150 m and under the shelter of the Picacho del Veleta. The remains of its ice were visible until the middle of the twentieth century, after which they were buried by a thick debris cover from the rock wall triggered by post-LIA paraglacial dynamics. The monitoring carried out since 1999 confirms that the ice, a legacy of the LIA, is undergoing degradation (Gómez-Ortiz et al. 2014, 2019). The main cause for the degradation has been attributed to the increase in post-LIA temperatures and the decrease in the duration of the snow cover on the ground, which affect the dynamics and evolution of the Sierra Nevada summit ecosystems.

**Acknowledgements** This work was conducted under the aegis of the research group ANTALP (Antarctic, Arctic, Alpine Environments, 2017-SGR-1102). Marc Oliva is supported by the Ramón y Cajal Program (RYC-2015-17597).

## References


Anderson RS, Jiménez-Moreno G, Carrión JS, Pérez-Martínez C (2011) Postglacial history of alpine vegetation, fire, and climate from Laguna de Río Seco, Sierra Nevada, southern Spain. Quat Sci Rev 30:1615–1629

Boissier CE (1839) Viaje botánico al sur de España durante el año 1837. Granada, Fundación

Casas-Morales A (1943) Contribución al estudio del glaciarismo cuaternario en Sierra Nevada. Boletín la Real Soc Española Hist Nat Sección Geológica XLI:543–567

Clark PU, Dyke AS, Shakun JD et al (2009) The Last Glacial Maximum. Science (80- ) 325:710–714

Dresch J (1937) De la Sierra Nevada au Grand Atlas, formes glaciaires et de nivation. Mélanges Géographie d'orientalisme Offer à EF Gautiers 194–212

García-Sainz L (1947) El clima de la España cuaternaria y los factores de su formación. Secretariado de Publicaciones Universidad de Valencia, Valencia

Gómez-Ortiz A (1987) Morfología glaciar en la vertiente meridional de Sierra Nevada. Estud Geográficos 527–558

Gómez-Ortiz A (2002) Mapa geomorfológico de Sierra Nevada. Consejería de Medio Ambiente, Junta de Andalucía, Granada

Gómez-Ortiz A, Oliva M, Palacios D et al (2015) The deglaciation of Sierra Nevada (Spain), synthesis of the knowledge and new contributions. Cuad Investig Geográfica 41:409

Gómez-Ortiz A, Oliva M, Salvador-Franch F et al (2018) The geographical interest of historical documents to interpret the scientific evolution of the glacier existing in the Veleta cirque (Sierra Nevada, Spain) during the Little Ice Age. Cuad Investig Geográfica 44:267–292

Gómez-Ortiz A, Oliva M, Salvador-Franch F et al (2019) Monitoring permafrost and periglacial processes in Sierra Nevada (Spain) from 2001 to 2016. Permafr Periglac Process

Gómez-Ortiz A, Oliva M, Salvador-Franch F et al (2014) Degradation of buried ice and permafrost in the Veleta cirque (Sierra Nevada, Spain) from 2006 to 2013 as a response to recent climate trends. Solid Earth 5:979–993

Gómez-Ortiz A, Palacios D, Palade B et al (2012a) Geomorphology The deglaciation of the Sierra Nevada (Southern Spain). Geomorphology

Gómez-Ortiz A, Palacios D, Palade B et al (2013) La evolución glaciar de Sierra Nevada y la formación de glaciares rocosos. Bol La Asoc Geogr Esp 61:139–162

Gómez-Ortiz A, Palacios D, Palade B et al (2012b) The deglaciation of the Sierra Nevada (Southern Spain). Geomorphology 159–160:93–105

Hempel L (1960) Límites geomorfológicos altitudinales en Sierra Nevada. Estudios Geográficos 78:81–93

Lhenaff R (1977) Recherches géomorphologiques sur les Cordillères Bétiques Centro-Occidentales (Espagne). Université de Lille

Messerli B (1965) Beiträge zur Geomorphologie der Sierra Nevada (Andalusien). Juris Verlag, Zürich

Obermaier H (1916) Los glaciares cuaternarios de Sierra Nevada. Trab Del Mus Nac Ciencias Nat 17:1–68

Oliva M (2009) Holocene alpine environments in Sierra Nevada (southern Spain). University of Barcelona

Oliva M, Gómez-Ortiz A (2012) Late-Holocene environmental dynamics and climate variability in a Mediterranean high mountain environment (Sierra Nevada, Spain) inferred from lake sediments and historical sources. Holocene 22:915–927

Oliva M, Gómez-Ortiz A, Salvador-Franch F et al (2016) Inexistence of permafrost at the top of the Veleta peak (Sierra Nevada, Spain). Sci Total Environ 550:484–494

Oliva M, Gómez Ortiz A, Palacios D et al (2014a) Environmental evolution in Sierra Nevada (South Spain) since the Last Glaciation, based on multi-proxy records. Quat Int 353:195–209

Oliva M, Gómez Ortiz A, Salvador F et al (2014b) Long-term soil temperature dynamics in the Sierra Nevada, Spain. Geoderma 235–236:170–181

Oliva M, Palacios D, Fernández-Fernández JM et al (2019) Late Quaternary glacial phases in the Iberian Peninsula. Earth Sci Rev 192:564–600

Oliva M, Ruiz-Fernández J, Barriendos M et al (2018) The Little Ice Age in Iberian mountains. Earth Sci Rev 177:175–208

Palacios D, Gómez-Ortiz A, Alcalá-Reygosa J et al (2019) The challenging application of cosmogenic dating methods in residual glacial landforms: the case of Sierra Nevada (Spain). Geomorphology 325:103–118

Palacios D, Gómez-Ortiz A, Andrés N et al (2016) Timing and new geomorphologic evidence of the last deglaciation stages in Sierra Nevada (southern Spain). Quat Sci Rev 150:110–129

Palacios D, Gómez-Ortiz A, Andrés N et al (2015) Maximum extent of Late Pleistocene glaciers and last deglaciation of La Cerdanya mountains, Southeastern Pyrenees. Geomorphology 231:116–129

Palacios D, Oliva M, Gómez-Ortiz A, Andrés N, Fernández-Fernández JM, Schimmelpfennig I Léannid L, ASTER Team (2020) Climate sensitivity and geomorphologic response of cirque glaciers from the late glacial to the Holocene, Sierra Nevada, Spain. Quat Sci Rev 248:106617

Palma P, Oliva M, García-Hernández C et al (2017) Spatial characterization of glacial and periglacial landforms in the highlands of Sierra Nevada (Spain). Sci Total Environ 584–585:1256–1267

Paschinger H (1957) Las formas glaciares de Sierra Nevada. España. Memorias y Comun Inst Geológico Prov 81–94

Penck A (1883) La période glaciaire dans les Pyrénées. etin la Société d'Histoire Nat Toulouse 19:200

Quelle O (1908) Beiträge zur Kenntnis der spanischen Sierra Nevada. University of Friedrich-Wilhelm, Germany

Salvador-Franch F, Gómez Ortiz A, Salvà Catarineu M, Palacios Estremera D (2011) Caracterización térmica de la capa activa de un glaciar rocoso en medio periglaciar de alta montaña mediterránea: El ejemplo del Corral del Veleta (Sierra Nevada, España). Cuad Investig Geográfica 37:25

Sánchez-Gómez S (1990) Aplicación del estudio de suelos a la dinámica de la cuenca del río Lanjarón: relación suelos-geomorfología. University of Granada

Schimper WP (1849) Sur la géologie, la botanique et la zoologie du midi de l'Espagne. J Univers Des Sci 806:189–192

Titos Martínez M (2019) La expedición del naturalista alsaciano Guillaume Philippe Schimper a Sierra Nevada en 1847. Ería 2:207

# Reconstruction of Past Environment and Climate Using Wetland Sediment Records from the Sierra Nevada

Gonzalo Jiménez-Moreno, Antonio García-Alix, R. Scott Anderson, María J. Ramos-Román, Jon Camuera, José Manuel Mesa-Fernández, Jaime L. Toney, Francisco J. Jiménez-Espejo, José S. Carrión, Alejandro López-Avilés, Marta Rodrigo-Gámiz, and Cole E. Webster

**Abstract**

Understanding the effects of climate change and human activities on fragile mountain ecosystems is necessary to successfully managing these environments under future climate scenarios (e.g., global warming, enhanced aridity). This can be done through the study of paleoecological records, which can provide long paleoenvironmental databases containing information on how ecosystems reacted to climate change and human disturbances before the historical record. These studies can be particularly interesting when focusing on especially warm and/or dry past climatic phases. Biotic (pollen, charcoal) and abiotic (physical, geochemistry) analyses from wetland sediment records from the Sierra Nevada, southern Spain record changes in vegetation, fire history and lake sedimentation since ~11,700 years (cal yr BP). This multiproxy paleoecological study indicates that maxima in temperature and humidity occurred in the area in the Early and Middle Holocene, with a peak in precipitation between ~10,500 and 7000 cal yr BP. This is deduced by maxima in water runoff, the highest abundance of tree species and algae and high total organic carbon values recorded in the alpine wetland's sedimentary records of the Sierra Nevada during that time period. In the last 7000 cal yr BP, and especially after a transition period between ~7000 and 5000 cal yr BP, a progressive aridification process took place, indicated by the decrease in tree species and the increase in xerophytic herbs in this region and a reduction in water runoff evidenced by the decrease in detritic input in the wetland sedimentary records. An increasing trend in Saharan dust deposition in the Sierra Nevada wetlands is also recorded through inorganic geochemical proxies, probably due to a coetaneous loss of vegetation cover in North Africa. The process of progressive aridification during the Middle and Late Holocene was interrupted by millennial-scale climatic oscillations and several periods of relative humid/droughty conditions and warm/cold periods have been identified in different temperature and/or precipitation proxies. Enhanced human impact has been observed in the Sierra Nevada in the last ~3000 cal yr BP through the increase in fires, grazing, cultivation, atmospheric pollution as well as reforestation by *Pinus* and the massive cultivation of *Olea* at lower altitudes.

G. Jiménez-Moreno (✉) · A. García-Alix · A. López-Avilés · M. Rodrigo-Gámiz
Departamento de Estratigrafía y Paleontología, Universidad de Granada, Granada, Spain
e-mail: gonzaloj@ugr.es

A. García-Alix · J. M. Mesa-Fernández · F. J. Jiménez-Espejo
Instituto Andaluz de Ciencias de la Tierra (IACT), CSIC-UGR, Armilla, Spain

R. S. Anderson · C. E. Webster
School of Earth and Sustainability, Northern Arizona University, Flagstaff, USA

M. J. Ramos-Román · J. Camuera
Department of Geosciences and Geography, University of Helsinki, Helsinki, Finland

J. L. Toney
School of Geographical and Earth Sciences, University of Glasgow, Glasgow, UK

J. S. Carrión
Department of Plant Biology, Faculty of Biology, University of Murcia, Murcia, Spain

**Keywords**

Holocene • Environmental change • Climate change • Human impact • Alpine wetlands • Sierra Nevada • Southern Spain

## 1 Introduction

Alpine environments are especially fragile and sensitive to recent climate change, which is causing altitudinal displacements of plant and animal species, and loss of


R. Zamora and M. Oliva (eds.), *The Landscape of the Sierra Nevada*,
https://doi.org/10.1007/978-3-030-94219-9_7

biodiversity (Thuiller et al. 2005; Gehrig-Fasel et al. 2008; Malanson et al. 2019). The stress on the biotic component of ecosystems is intensified when adding the effect of recent and projected drought in the Mediterranean region (Páscoa et al. 2017; Sousa et al. 2019), which will also cause enhanced fire activity (Pausas and Fernández-Muñoz 2012; IPCC 2013; Sousa et al. 2015).

The study of past environment and climate changes is necessary to improve our insight on future climate scenarios by providing information on how mountain ecosystems responded to long- and short-term variations in climate during the Holocene (last 11,700 years), particularly during warmer and drier climates. In this respect, wetlands in the Sierra Nevada are a model system for such studies, as they are remote and relatively pristine areas of high biodiversity, not significantly modified by human activities and preserve a high-quality signal of past Holocene natural and anthropogenic environmental and climate change (see references below).

The footprint of climate change and human impact and the generated response in the environment (e.g., changes in vegetation, fires, erosion, pollution) is recorded in lake sediments, for example, through the accumulation of fossil remains, charcoal fragments, presence of specific biomarkers, sediment granulometry and elemental chemical composition. In this respect, a significant effort has been made in the last decade by our research group to further multidisciplinary study of sediment records from seven alpine lakes and bogs in the Sierra Nevada to understand paleoenvironmental and paleoclimate change (Anderson et al. 2011; Jiménez-Moreno and Anderson 2012; García-Alix et al. 2012, 2013, 2017, 2018, 2020; Jiménez-Espejo et al. 2014; Jiménez-Moreno et al. 2013, 2020; Ramos-Román et al. 2016, 2018a, b, 2019, 2021; Mesa-Fernández et al. 2018; Camuera et al. 2018, 2019; Manzano et al. 2019; Toney et al. 2020). Here we provide a review of the research in alpine wetland sedimentary records from the Sierra Nevada with the main goal of linking biotic changes with climate and human impacts on these fragile environments.

## 2 Materials and Methods

### 2.1 Sierra Nevada Sites

In this study, we used multi-proxy (lithological, sedimentological, palynological, anthracological, geochemical) data from seven radiometrically well-dated Holocene sedimentary records from wetland sites located at different elevation and orientations in the Sierra Nevada area (Fig. 1; Table 1). These sedimentary lake and bog archives were recovered by the authors between 2006 and 2015 (Table 1). Six of the studied sites are situated above treeline in the cryoromediterranean vegetation belt where typical vegetation is alpine tundra, while one site occurs in the mesomediterranean vegetation belt, where natural potential vegetation is mostly characterized by evergreen sclerophyllous oak forests. Three sites—Laguna de la Mula (LdlM) (Jiménez-Moreno et al. 2013), Laguna de la Mosca (LdlMo) (Manzano et al. 2019) and Borreguil de la Virgen (BdlV) (García-Alix et al. 2012; Jiménez-Moreno and Anderson, 2012)—are situated at the north-facing side of the Sierra Nevada, while three others—Laguna de Río Seco (LdRS) (Anderson et al. 2011; García-Alix et al. 2013, 2018; Jiménez-Espejo et al. 2014; Toney et al. 2020), Borreguil de la Caldera (BdlC) (Ramos-Román et al. 2016; García-Alix et al. 2017) and Laguna Hondera (LH) (Mesa-Fernández et al. 2018)—are located in the south face. The Padul wetland site is located at the southwest foothills of the Sierra Nevada (Camuera et al. 2018, 2019, 2021; Ramos-Román et al. 2018a, b; Webster 2018). These sites are thus situated between ~725 m in the mesomediterranean to ~3020 m in the cryoromediterranean vegetation belt (Table 1).

### 2.2 Sedimentary Records: Sediment Collection, Chronology, Lithology and Magnetic Susceptibility

Sediment core collection in the alpine Sierra Nevada sites was done using a Livingstone square-rod piston corer in the visual depocenter of the studied wetlands. Coring in lakes was done from a floating platform anchored to shore and short cores with the unconsolidated 10–20 cm of the lake sediment records were retrieved using a universal corer (Aquatic Research Inc). Padul-15–05 core was collected using a Rolatec RL-48-L drilling machine equipped with a hydraulic piston corer from the Scientific Instrumentation Centre of the University of Granada (CIC-UGR). The sediment cores were wrapped in plastic wrap and alu minum foil in the field, put in core boxes, and transported back to the laboratory where they were stored in a cool room at 4 °C.

The chronology for the different wetland sedimentary records was constrained using accelerator mass spectrometry (AMS) radiocarbon dates from plant remains and organic bulk samples taken from the cores. ICP-MS Plutonium, $^{210}$Pb and $^{137}$Cs profiles were also built on the uppermost part of the lake sedimentary cores (Table 1). Radiocarbon dates were calibrated with the IntCal13.14C calibration curve (Reimer et al. 2013), and used together with the Pb, Cs and Pu dates and the year of core sampling as the top of the sediment record in order to build accurate and robust age-depth models (Fig. 2). Age models for the individual sites were built using different approaches, including linear interpolation (BldV, LdlMo) and smooth spline option

**Fig. 1** Sierra Nevada wetland site location and photographs. **a** Map of the Sierra Nevada mountain range, southern Iberian Peninsula. The rectangle shows the magnified area shown in **b**. **b** Site location of the Sierra Nevada mutiproxy records used in this study: Laguna de Río Seco (LdRS), Borreguil de la Caldera (BdlC), Borreguil de la Virgen (BdlV); Laguna de la Mosca (LdlMo), Laguna Hondera (LH), Laguna de la Mula (LdlM), and Padul-15–05 (Padul). See Table 1 and text for explanation

(LdRS, BdlC, LH, LdlM, Padul) using the "clam" software (Blaauw 2010; Fig. 2) or Heegaard et al. (2005) approach to calculate the age-depth (see original studies for more details on the age model construction; Table 1).

The lithology of the Sierra Nevada sediment cores (Fig. 2) was described from split core segments in the laboratory (Table 1). Magnetic susceptibility (MS), a measure of the tendency of sediment to carry a magnetic charge

**Table 1** Site description of the Sierra Nevada records synthetized in this study. Sites from the Sierra Nevada are: BdlV: Borreguil de la Virgen; LdlMo: Laguna de la Mosca; LdlM: Laguna de la Mula; LdRS: Laguna de Río Seco; BdlC: Borreguil de la Caldera; LH: Laguna Hondera; Padul: Padul peat bog. Asterisks (*) indicate information for the Holocene part of those sedimentary records

| Site | Coordinates and altitude (m asl) | Vegetation belt | Site environment | Core length (cm) | Period covered (cal yr BP) | Dating method | References |
|---|---|---|---|---|---|---|---|
| BdlV | 37°03′15″N, 3°22′40″ W; 2945 m | Cryoromediterranean | Peatland | 169 | 0–8200 | 9 AMS $^{14}$C dates | Jiménez-Moreno and Anderson (2012), García-Alix et al. (2012) |
| LdlMo | 37°03′ 34.88″N, 3° 18′52.98″W; 2889 m | Cryoromediterranean | Lake | 190 | 0–8300 | 10 AMS $^{14}$C dates, 1 Pu profile | Manzano et al. (2019) |
| LdlM | 37°3.583′N, 3°25.017′W; 2497 m | Cryoromediterranean | Lake | 32.5 | 0–4100 | 6 AMS $^{14}$C dates | Jiménez-Moreno et al. (2013) |
| LdRS | 37°02.43″N, 3°20.57′W; 3020 m | Cryoromediterranean | Lake | 150 | 0–11,000 | 9 AMS $^{14}$C dates, 1 $^{137}$Cs, 1 $^{210}$Pb profiles | Anderson et al. (2011), García-Alix et al. (2013, 2020), Jiménez-Espejo et al. (2014), Toney et al. (2020) |
| BdlC | 37°03′02″ N, 3°19′24″ W; 2992 m | Cryoromediterranean | Peatland | 56 | 0–4400 | 5 AMS $^{14}$C dates | Ramos-Román et al. (2016), García-Alix et al. (2017) |
| LH | 37°02.88′N, 3°17.66′W; 2899 m | Cryoromediterranean | Lake | 83 | 0–11,000 | 7 AMS $^{14}$C dates | Mesa-Fernández et al. (2018) |
| Padul | 37°00′ 39.77″N, 3° 36′14.06″W; 725 m | Mesomediterranean | Lake/Peatland | 327* | 0–11,000 | 14 AMS $^{14}$C dates* | Ramos-Román et al. (2018a, b), Camuera et al. (2018, 2019) |

(Snowball and Sandgren 2001), was measured with a Bartington MS2E meter in dimensionless SI units (Fig. 2). Measurements were taken directly from the core surface every 0.5 cm for the entire length of the cores.

## 2.3 Inorganic Geochemistry

Inorganic geochemical analyses in sediment samples from cores LdRS and LH were performed by means of (1) inductively coupled plasma mass spectrometry (ICP-MS for minor and trace elements in both cores), (2) flame Atomic Absorption (AAS for major elements in LdRS core) and (3) inductively coupled plasma–optical emission spectrometry (ICP-OES for major elements in LH core) (Table 1) at the Centro de Instrumentación Científica of the University of Granada (CIC-UGR). See details in Jiménez-Espejo et al. (2014) and Mesa-Fernández et al. (2018).

Additionally, high-resolution and continuous elemental geochemical analyses from the BdlC, LH and Padul Sierra Nevada cores were obtained using an Avaatech X-Ray fluorescence (XRF) core Scanner at the XRF-Core Scanner Laboratory (University of Barcelona, Spain). The cores were scanned two times with a point sensor: one at 10 s count time (10 kV X-ray voltage and 650 mA X-ray current for light elements), and another one at 35 s count time (30 kV X-ray voltage and 1700 mA X-ray current for heavy elements). See details in García-Alix et al. (2017), Mesa-Fernández et al. (2018), Camuera et al. (2019).

## 2.4 Organic Geochemistry—Bulk Sediments and Specific Organic Lipids

The elemental composition (C and N) of bulk sediment samples was measured (after acid digestion to remove potential carbonates) using a Thermo Scientific Flash 2000 elemental analyzer at CIC-UGR. C isotopes were measured in an aliquot of the same samples by means of isotope-ratio mass spectrometry (IRMS) with a coupled elemental analyzer (EA). We used two different configurations: a Carlo Erba Ba 1500 series 2 EA attached to a Thermo Finnigan Delta plus XL IRMS (Instituto Andaluz de Ciencias de la Tierra CSIC-UGR, Spain) in samples from LdlM, BdlV and

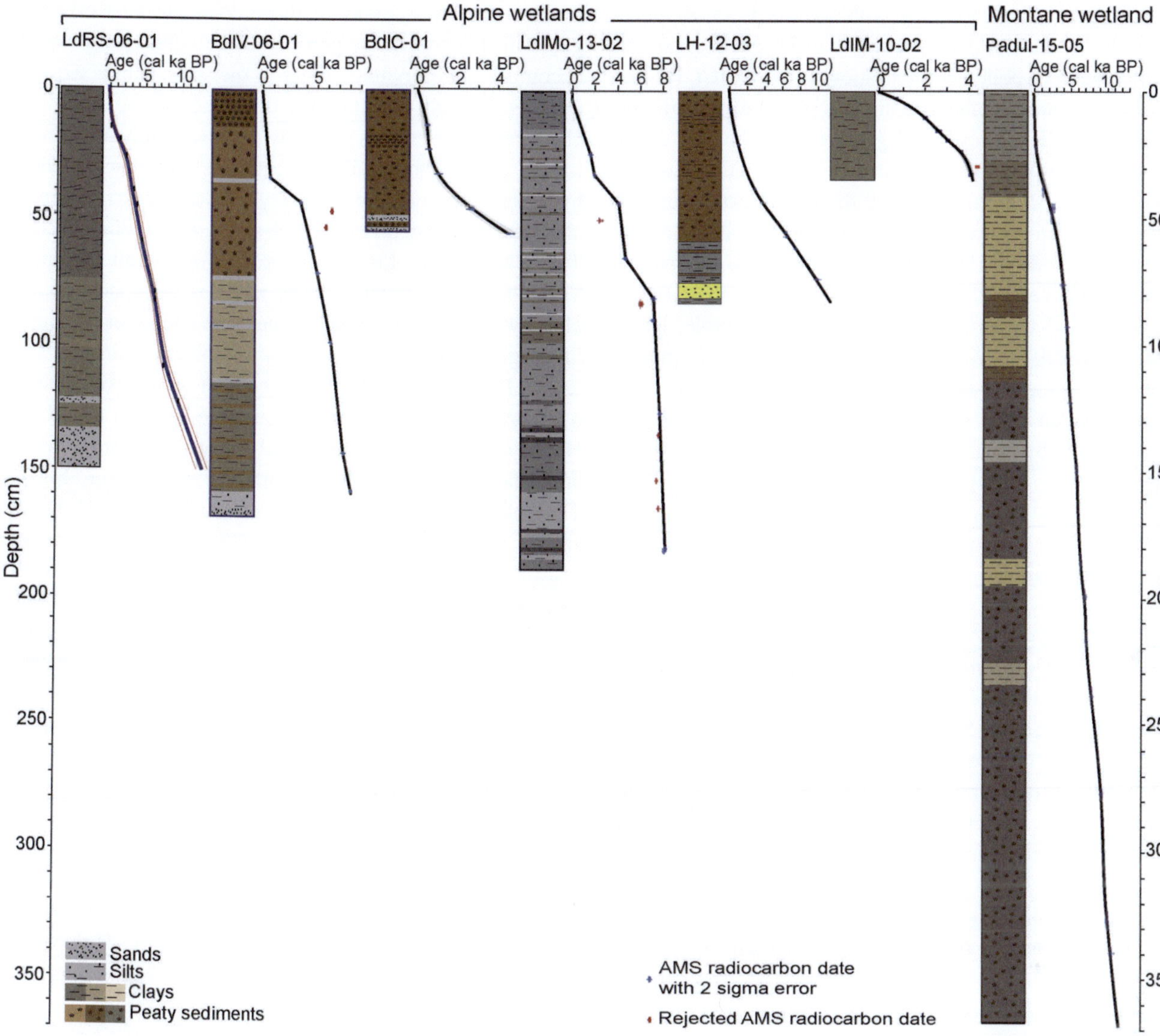


**Fig. 2** Sediment core lithology and age–depth models for the studied Sierra Nevada records. Red squares are radiocarbon dates that were not used in the age models. The studied sites are Laguna de Río Seco (LdRS-06–01: LdRS), Borreguil de la Caldera (BdlC-01: BdlC), Borreguil de la Virgen (BdlV-06–01: BldV); Laguna de la Mosca (LdlMo-10–02: LdlMo), Laguna Hondera (LH-12–03: LH), Laguna de la Mula (LdlM-10–02: LdlM), and Padul (Padul-15–05: Padul)

LdRS, and a Euro EA 300 EA attached to an Isoprime 50 V IRMS (CIC-UGR) in samples from BdlC. The isotopic measurements were calibrated using internal and international standards, and expressed using the δ notation and the reference V-PDB. See details in García-Alix et al. (2012, 2017), Jiménez Moreno et al. (2013), Jiménez-Espejo et al. (2014), Camuera et al. (2019).

Lipid extractions were performed by means of silica gel chromatography using different solvents in order to separate the organic compounds depending on their polarity. The *n*-alkanes were analyzed at the BECS laboratory (University of Glasgow, UK) by means of a Shimadzu 2010 GC-FID in order to quantify them. Long chain diols were analyzed after derivatization by bis- (trimethylsilyl) trifluoroacetamide (BSTFA) at the BECS laboratory (University of Glasgow, UK) in the Shimadzu QP2010 Plus mass spectrometer interfaced with a Shimadzu 2010 GC using a scan mode. Afterwards, an ion monitoring mode (SIM) was selected specifying the characteristic fragment ions of the most important long-chain diols ($C_{28}$, $C_{30}$ and $C_{32}$ diols). See more details in Garcia-Alix et al. (2020) and Toney et al. (2020).

## 2.5 Palynological Analysis

Sample processing for pollen analysis in all the alpine Sierra Nevada records was done following a modified Faegri and Iversen (1989) methodology, using 1 cm$^3$ of sediment.

Processing included treatment with sodium hexametaphosphate for clay deflocculation, addition of *Lycopodium* spores for calculation of pollen concentration, sieving, HCl, HF and acetolysis solution. Pollen residue was suspended in silicone oil (for the LdRS record) and glycerol (for the rest of the studied records), and analyzed at 400 magnifications using a light microscope with a goal of the identification of 300 terrestrial pollen grains. Pollen identification was done with the help of modern pollen collections at the University of Granada, University of Murcia and Northern Arizona University and modern pollen atlases (for example, Reille 1992; Beug 2004). Raw pollen counts were transformed to pollen percentages based on the terrestrial pollen sum. Arboreal pollen (AP) was calculated with the sum of tree pollen divided by the total terrestrial (tree, herbs and shrubs) pollen.

### 2.6 Charcoal

Samples for macroscopic charcoal analysis (1 $cm^3$) were taken every 0.5 cm throughout selected Sierra Nevada sediment cores (LdRS, LdlMo, LdlM and BdlC). Charcoal analysis followed the protocol described in Whitlock and Anderson (2003). Processing included pretreatment with sodium hexametaphosphate to deflocculate clays and sieving with mesh sizes of 250 and 125 μm. Counting of charcoal particles was performed with a stereomicroscope at 10–70 magnifications. Charcoal counts for each sample were converted to charcoal concentrations [CHAC; number of charcoal particles per $cm^3$ or cubic centimeter (cc)].

## 3 Results

### 3.1 Age Control of the Sedimentary Sequences

The age-depth models for the studied Sierra Nevada records are constrained by a total of 60 AMS $^{14}C$ radiocarbon dates and 3 ICP-MS Plutonium, $^{210}Pb$ and $^{137}Cs$ profiles for the upper part of the LdRS and LdlMo records (Fig. 2; Table 1). All the absolute age data fall within the Holocene period (11,700–0 cal yr BP). The age of wetland formation and beginning of sediment accumulation in the studied basins shows a three-step chronological pattern at slightly older than ~11,000, ~8200 and ~4200 cal yr BP.

### 3.2 Lithology and Magnetic Susceptibility (MS)

Sediment coring in all the studied sites from alpine Sierra Nevada ended when coarse sediments were reached and the human-powered Livingstone corer was unable to deepen any further. Therefore, many of those cores are relatively inorganic in the lowermost portion of the cores, progressively becoming more organic higher in the sedimentary sequence (Fig. 2). Overall, the lithology of the lake records is mostly characterized by clays and silts, whereas lithology in the locally called “borreguiles” (peat bogs) cores is dominated by peat. Very organic-rich clay (called gyttja) sediments characterize most of the LdRS and LdlM sediment cores. The 190-cm-long LdlMo is the longest of all the alpine Sierra Nevada sediment records and it is dominated by silty clays, which vary greatly in grain size and color. A similar lithological sedimentary sequence is observed in the BdlV and LH cores, with high sand content at the core bottom, overlain by clays and transitioning into peats in the Middle Holocene, which dominate until Present (Jiménez-Moreno and Anderson 2012; Mesa-Fernández et al. 2018; Fig. 2). The Late Holocene BdlC core, also located in a peat bog, is shorter and shows sand at the bottom and then peat for the rest of the record. Lithology of the Padul record is characterized by dark organic rich peat sediments from ~11,600 to ~7600 cal yr BP. Dark organic-rich peats with intercalations of three grey or brown clay layers occurred between ~7600 and 4700 cal yr BP. The uppermost 4700 cal yr BP of the record are characterized by clays and clayey-carbonates (Ramos-Román et al. 2018a, b).

Holocene MS records from the Sierra Nevada show unique individual patterns but share long- and short-scale variations (Fig. 3). The Early Holocene in the alpine Sierra Nevada environments (i.e., LdRS and LH records) is characterized by overall high MS values. This contrasts with very low MS values recorded during the Early Holocene in the lower elevation sedimentary record from Padul. MS values decreased during the Middle Holocene reaching minima in some of the records right at the transition with the Late Holocene (between ~5000 and 4400 cal yr BP). MS depicts an increasing trend in the Late Holocene (clear in Padul, LH, LdlMo and LdlM).

### 3.3 Inorganic Geochemistry

In order to simplify, in this section we show a synthesis of the most environmentally significant variations in major and trace element concentrations from the studied Sierra Nevada sites (Fig. 4; check specific publications for more detailed information). The Early Holocene and early-Middle Holocene (from ~11,000 until ~7000 cal yr BP) in LH is characterized by maximum values of K/Ti ratios, coinciding with the lowest values in Ca/Al, Ca/Ti and Zr/Al ratios (Mesa-Fernández et al. 2018). On the other hand, an opposite pattern is observed in Zr/Th data from LdRS, LH and BdlC, which show minima at that time and an increasing trend in the last ~7000 cal yr BP (Jiménez-Espejo et al.

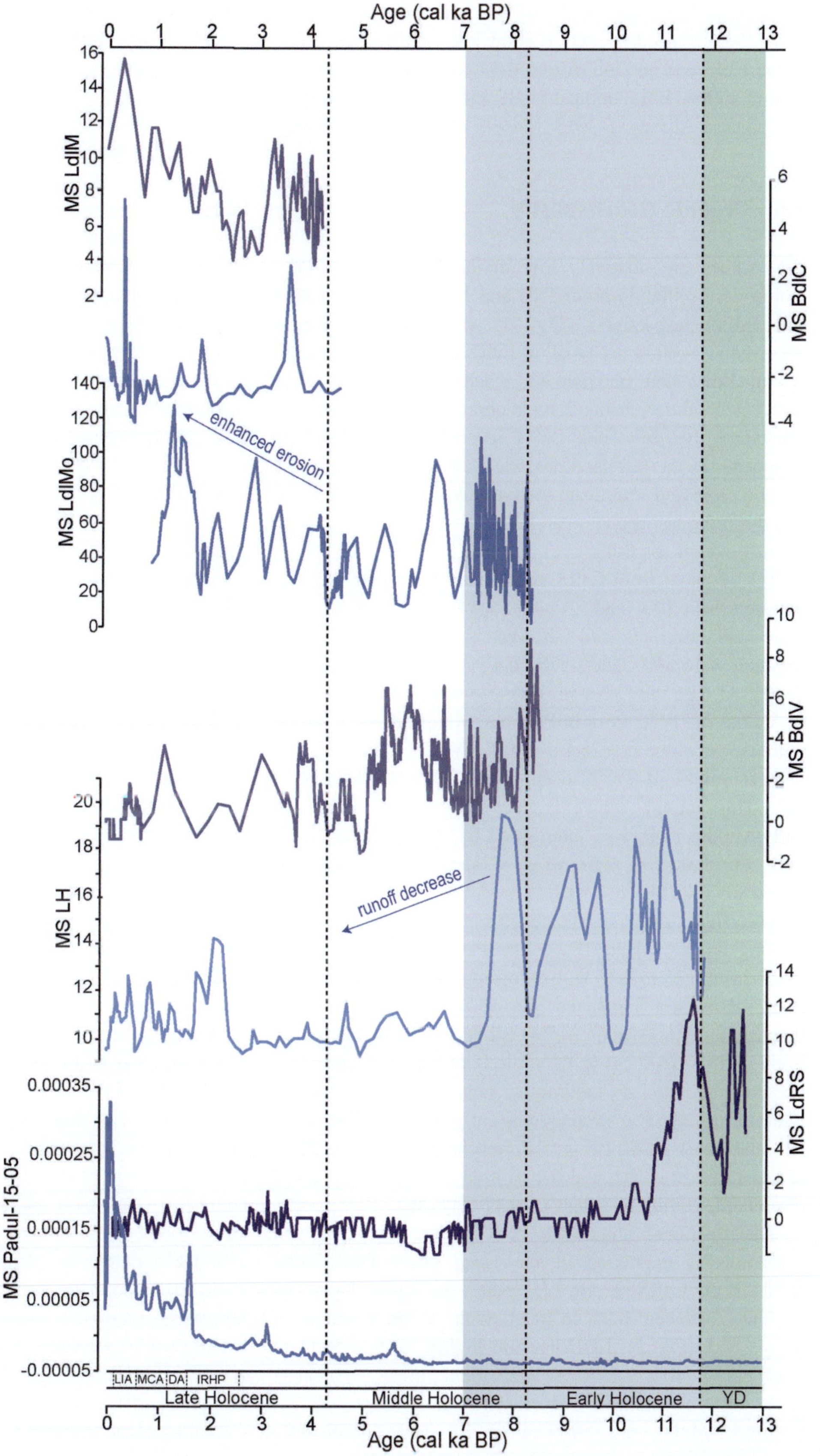


**Fig. 3** Magnetic susceptibility (MS) records from the Sierra Nevada sediment cores for the last 13,000 cal yr BP. MS data is shown in SI units. YD stands for Younger Dryas (in green shading color). In grey shading is the warmest-wettest period deduced by other climatic proxies discussed in the text and overall highest MS data. IRHP, DA, MCA and LIA stand for Iberian Roman Humid Period, Dark Ages, Medieval Climate Anomaly and Little Ice Age, respectively

2014; Mesa-Fernández et al. 2018; García-Alix et al. 2017). Pb/Al data from LdRS and LH show low values during most of the Holocene and increase in the Late Holocene, peaking around 3000–2500, 2000 cal yr BP and between 1950–1970 AD (García-Alix et al. 2013; Mesa-Fernández et al. 2018).

## 3.4 Organic Geochemistry

C/N values are relatively low in the Early and Middle Holocene, oscillating around 10 and 16, recorded in LdRS and BdlV, respectively (Fig. 5). A significant increase occurred at ~6000 cal yr BP in LdRS and at ~5000 cal yr BP in BdlV and maxima are reached until 4000 cal yr BP. A general decreasing trend is observed in all the records (including the BdlC and LdlM records) during the Late Holocene. In the last centuries, the BdlC and BdlV records show a rapid and significant increase. A significant decrease in C/N in the last decades can be observed in all the studied records except for BdlV.

TOC content from LdRS shows very low values (2.1%) between ~11,100 and 10,500 cal yr BP. TOC values increased later on and showed a maximum (11.2%) between ~10,500 and 5700 cal yr BP. TOC values decreased then until present (down to 4.8%).

Values of ™$^{13}$C from BdlV range from −28.1 to −23.1% with a mean value of −26.4 ± 1.0 ‰ (V-PDB) (Fig. 5). The highest values of ™$^{13}$C occur from 8200 to 5100 cal yr BP. This record shows a decreasing trend since the early-Middle Holocene, interrupted by several oscillations, until the last 300 years when a significant increase is observed (Fig. 5).

The range of $nC_{21}$-$_{25}$-alkanes is related to submerged and floating aquatic plants ($P_{aq}$). The Paq values from LdRS range from 0.25 to 0.48 with an average value of 0.37. High Paq values are registered during the Early Holocene between ~11,800 and 10,500 cal yr BP. An increasing trend until ~6300 cal yr BP up to 0.47 values is observed, subsequently to a minimum value of 0.34 around 9300 cal yr BP. Paq decreased linearly after ~6300 cal yr BP and until ~260 cal yr BP, reaching values of 0.29. A fast increase in Paq is detected in the last ~160 years of the record, getting to values of 0.47 (Fig. 5).

The fractional abundance of $C_{28}$ and $C_{30}$ 1,13 and 1,15-diols is expressed in the Long chain Diol Index (LDI) in the LdRS record. LDI values during the Holocene oscillate between 0.05 to 0.31 with a mean value of 0.18 ± 0.5 (Fig. 5). LDI data are higher from ~5000 to 4200 cal yr BP, ~2540, ~1020, and after −10 cal yr BP. LDI values decreased at ~6560, ~6170, between ~4100 and 3900 cal yr BP and significantly between ~450 and 150 cal yr BP and between ~40 and 30 cal yr BP. Highest LDI values are reached at around 4800–4650 cal yr BP and at present.

## 3.5 Palynological Analysis

Pollen records from the Sierra Nevada show highest AP abundances during the Early and early-Middle Holocene between ~11,500–7000 cal yr BP (LdRS, BdlV and LdlMo at high elevations and Padul-15–05 at lower elevation; Fig. 6). In the alpine environments, the highest abundance of tree species is reached at that time, especially in *Pinus* (most likely *P. sylvestris* or *P. nigra*) and *Quercus* (deciduous), but also in *Betula*, *Alnus* and *Salix* (Anderson et al. 2011; Jiménez-Moreno and Anderson 2012; Manzano et al. 2019). On the other hand, *Artemisia*, *Juniperus* and Amaranthaceae show their minima. Pelagic algae, such as *Botryococcus* and *Pediastrum*, reach maxima during this period (Anderson et al. 2011; Jiménez-Moreno and Anderson 2012). At lower elevation, the Padul record shows high AP (mostly Mediterranean forest with evergreen and deciduous *Quercus*) from ~11,600 to 7600 cal yr BP (Ramos-Román et al. 2018a, b).

The Sierra Nevada pollen records show that between ~7000 and 5000 cal yr BP AP (mostly *Pinus*) remains abundant, but starts showing a decreasing trend (Fig. 6), together with other forest species such as deciduous *Quercus* and *Betula* (Anderson et al. 2011; Jiménez-Moreno and Anderson 2012; Mesa-Fernández et al. 2018; Manzano et al. 2019). Aquatic species also declined considerably at that time (see previous references). A more pronounced decline in forest species is observed in all the Sierra Nevada records from 5000 cal yr BP until the last centuries (Fig. 6). *Pinus* and other forest species (such as *Quercus*) decreased considerably while *Artemisia*, Amaranthaceae or Caryophyllaceae increased (Anderson et al. 2011; Jiménez-Moreno and Anderson 2012; Ramos-Román et al. 2016, 2018b; Mesa-Fernández et al. 2018; Manzano et al. 2019).

The high-elevation (alpine) Sierra Nevada records also show that *Olea* expanded rapidly in the last ~950 cal yr BP (Ramos-Román et al. 2019) and *Pinus* increased in the last centuries (Anderson et al. 2011; Jiménez-Moreno and Anderson 2012; Mesa-Fernández et al. 2018; Manzano et al. 2019) and particularly in the last decades (Ramos-Román et al. 2016). *Sporormiella*, a fungus spore type indicating herbivory, increased considerably in the Late Holocene in LdRS (last 3000 years) and BdlV and BdlC (last 200 years) records (Anderson et al. 2011; Jiménez-Moreno and Anderson 2012; Ramos-Román et al. 2016). Increases in Cerealia pollen, together with other nitrophilous and ruderal

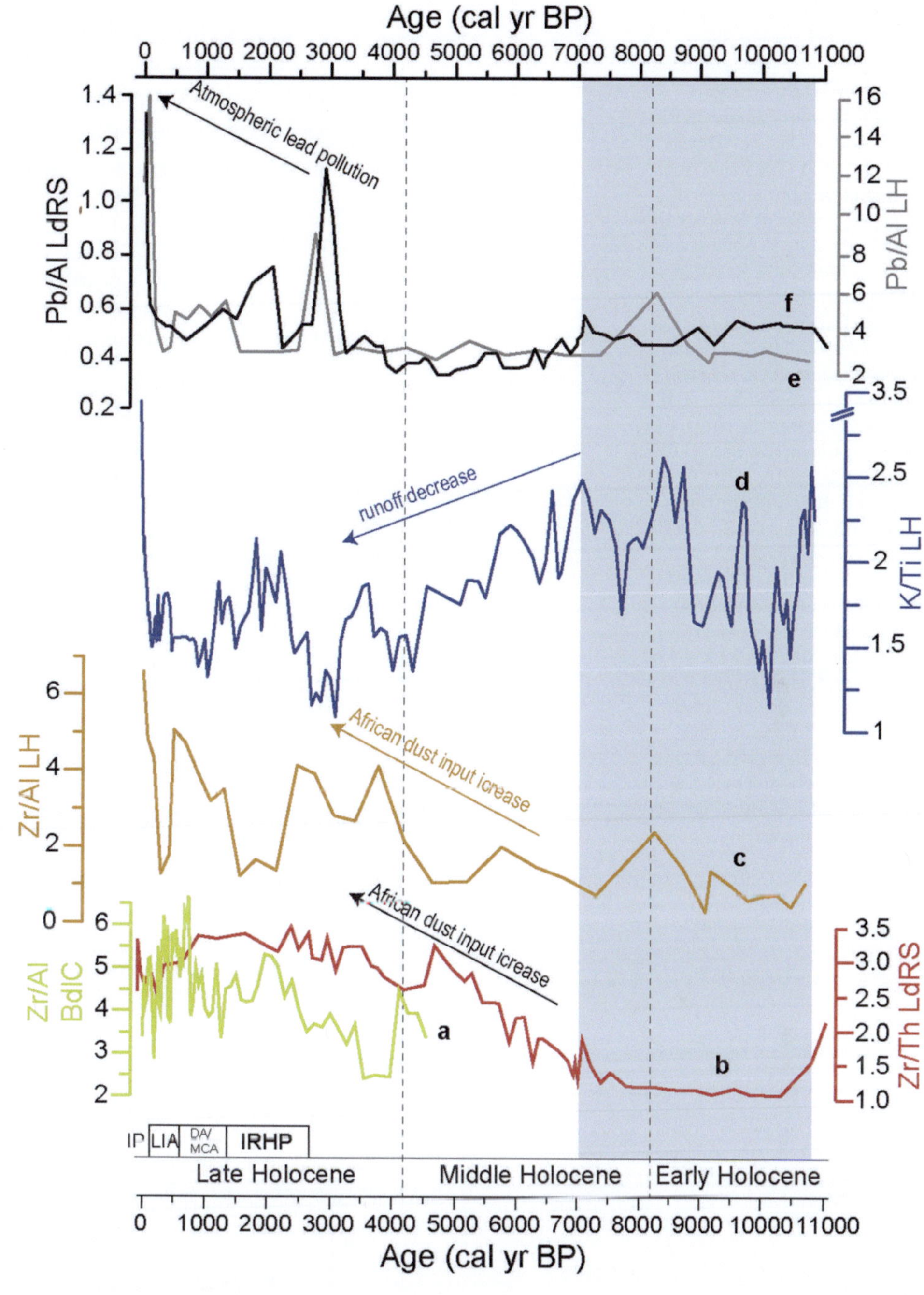


**Fig. 4** Inorganic geochemistry data from the studied Sierra Nevada wetland sediment records for the last 11,000 cal yr BP. Runoff, lake level and eolian dust proxies. **a** Zr/Al from Borreguil de la Caldera (BdlC; García-Alix et al. 2017). **b** Zr/Th from Laguna de Río Seco (LdRS) (Jiménez-Espejo et al. 2014). **c** Zr/Al from Laguna Hondera (LH) (Mesa-Fernández et al. 2018). **d** K/Ti from Laguna Hondera (LH; Mesa-Fernández et al. 2018). **e** Pb/Al from Laguna Hondera (LH; Mesa-Fernández et al. 2018). **f** Pb/Al from Laguna de Río Seco (LdRS; García-Alix et al. 2013). IRHP, DA, MCA, LIA and IP stand for Iberian Roman Humid Period, Dark Ages, Medieval Climate Anomaly, Little Ice Age and Industrial Period, respectively. In grey shading is the wettest period deduced by these and other climatic proxies discussed in the text

pollen taxa such as *Rumex*, *Plantago*, Urticaceae or Convolvulaceae are observed in the last ~3000 cal yr BP and are particularly abundant in the last ~1500 cal yr BP.

In addition to the general decreasing trend in tree species after ~7000 cal yr BP, important cyclical variations in pollen abundances are observed, especially in *Pinus* and *Quercus* and thus in AP (Fig. 6). In most of the Sierra Nevada pollen sequences, the AP minima are observed at ~7500, 6500, 5000 and 4200, 3000 and 1200 cal yr BP.

## 3.6 Charcoal

The occurrence of charcoal particles in the alpine Sierra Nevada Holocene sedimentary records is low, with values ranging between 0 and 20 particles per cubic centimeter (cc) and a maximum value of 60 particles per cc reached in LdRS record (Anderson et al. 2011; Fig. 7). Only one charcoal record, LdRS, is available for the Early Holocene, which suggests relatively low fire occurrence. The Middle Holocene is characterized in the LdRS record by an increase

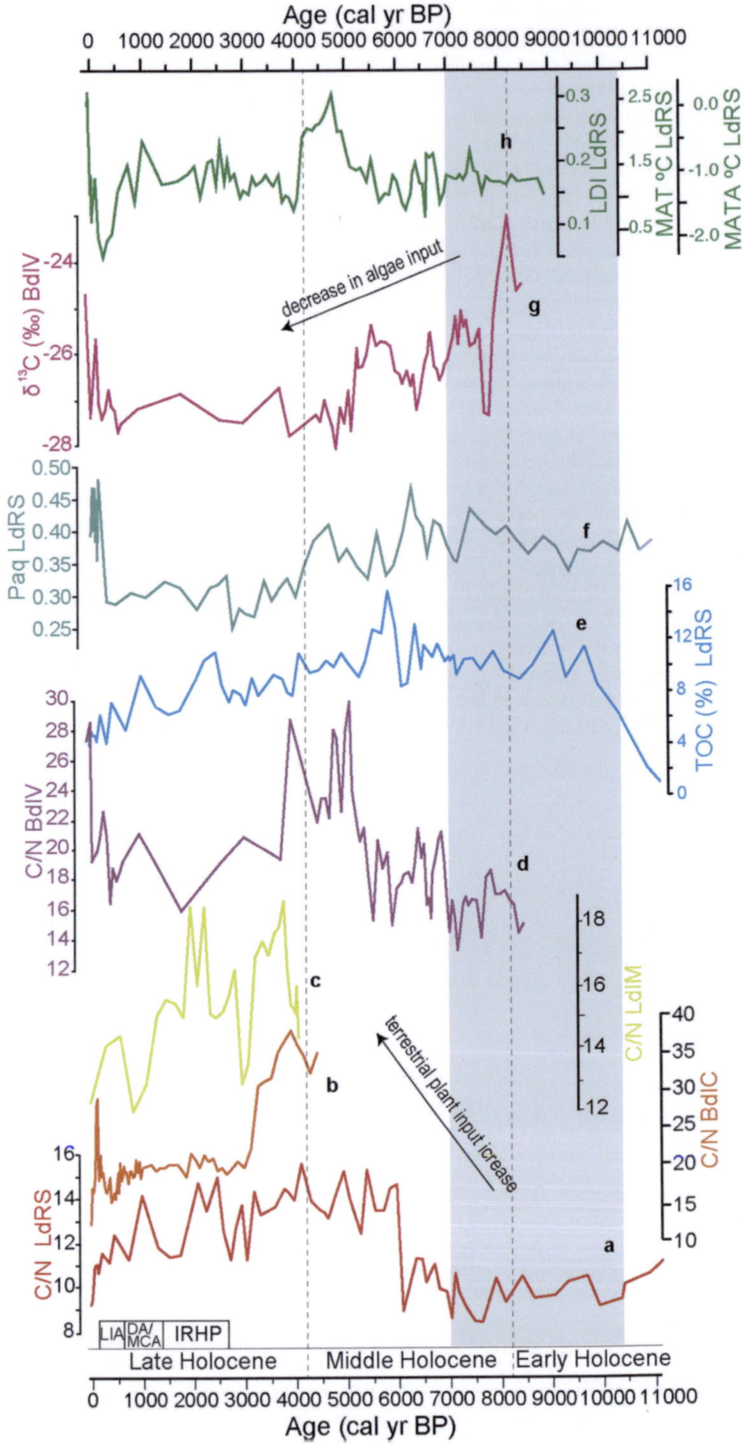


**Fig. 5** Organic geochemistry data from the studied Sierra Nevada wetland sediment records for the last 11,000 cal yr BP. **a** C/N from Laguna de Río Seco (LdRS) (Jiménez-Espejo et al. 2014). **b** C/N ratios from Borreguil de la Caldera (BdlC; Ramos-Román et al. 2016). **c** C/N from Laguna de la Mula (LdlM; Jiménez-Moreno et al. 2013). **d** C/N from Borreguil de la Virgen (BdlV; Jiménez-Moreno and Anderson, 2012). **e** Total organic carbon (TOC) content from Laguna de Río Seco (LdRS; Jiménez-Espejo et al. 2014). **f** $P_{aq}$ from Laguna de Río Seco (LdRS; Toney et al. 2020). **g** ™$^{13}$C from Borreguil de la Virgen (BdlV; García-Alix et al. 2012). **h** LDI record and temperature reconstruction from Laguna de Río Seco (LdRS; García-Alix et al. 2020; Toney et al. 2020). IRHP, DA, MCA and LIA stand for Iberian Roman Humid Period, Dark Ages, Medieval Climate Anomaly and Little Ice Age, respectively. In grey shading is the wettest period deduced by these and other climatic proxies discussed in the text

in charcoal particles, reaching a peak at ~6700 cal yr BP and a secondary lower peak at ~5500 cal yr BP. High relative values are reached in the LdlMo record during the Middle Holocene, with peaks around 7500, 6700 and 5500 cal yr BP. The Late Holocene is characterized in LdRS by an increase in fire activity between 3000 cal yr BP and present and a significant peak between ~2700 and 1600 cal yr BP in LdlM and BdlC records (Fig. 7).

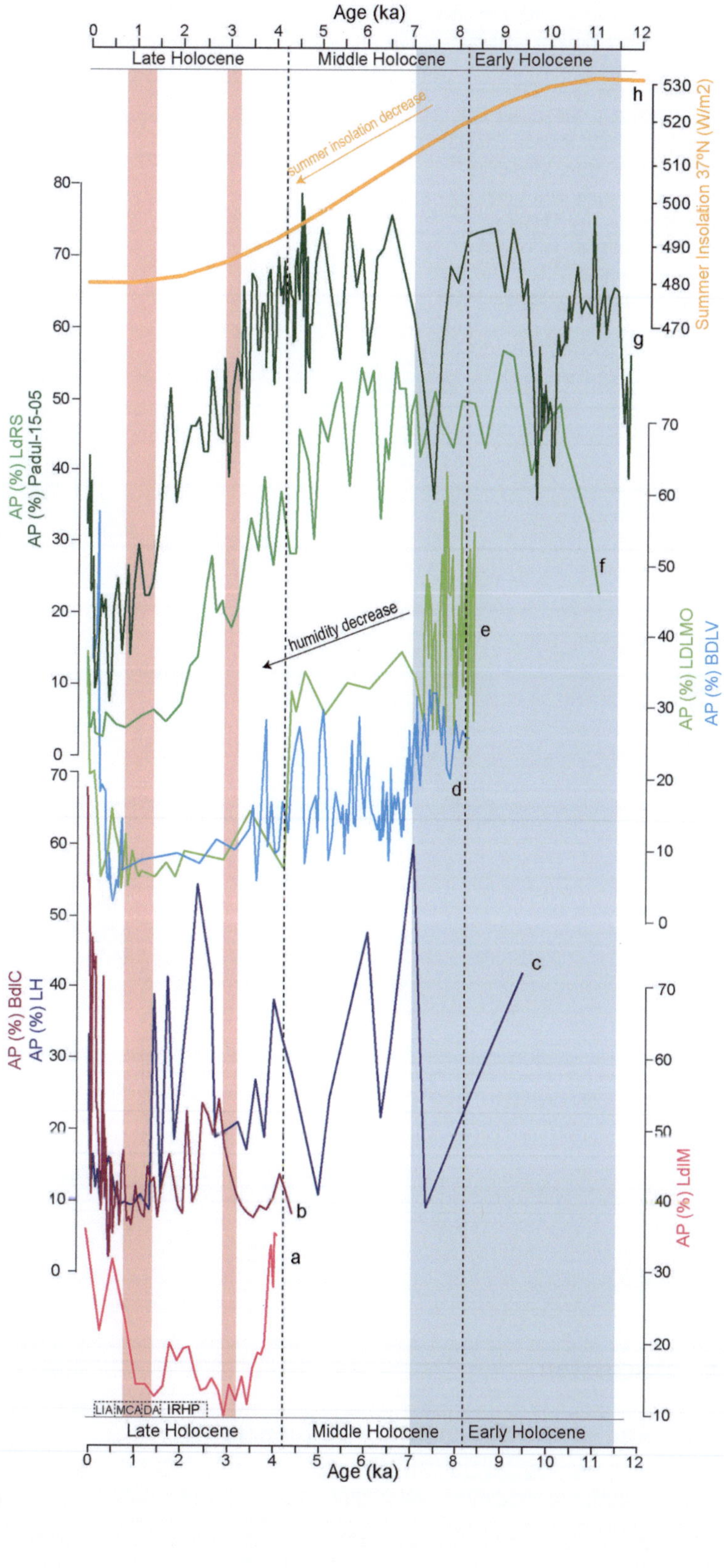


**Fig. 6** Arboreal pollen (AP) records from the studied Sierra Nevada wetland sediment cores and summer insolation for the last 12,000 cal yr BP. **a** AP from Laguna de la Mula (LdlM; Jiménez-Moreno et al. 2013). **b** AP from Borreguil de la Caldera (BdlC; Ramos-Román et al. 2016). **c** AP from Laguna Hondera (LH; Mesa-Fernández et al. 2018). **d** AP from Borreguil de la Virgen (BdlV; Jiménez-Moreno and Anderson, 2012). **e** AP from Laguna de la Mosca (Manzano et al. 2019). **f** AP from Laguna de Río Seco (LdRS; Anderson et al. 2011). **g** AP from Padul (Padul-15–05; Ramos-Román et al. 2018a, b). **h** Summer insolation for 37 °N (Laskar et al. 2004). IRHP, DA, MCA and LIA stand for Iberian Roman Humid Period, Dark Ages, Medieval Climate Anomaly and Little Ice Age, respectively. In grey shading is the deduced warmest and wettest period. In red shading are especially arid periods of the Late Holocene that triggered forest reductions

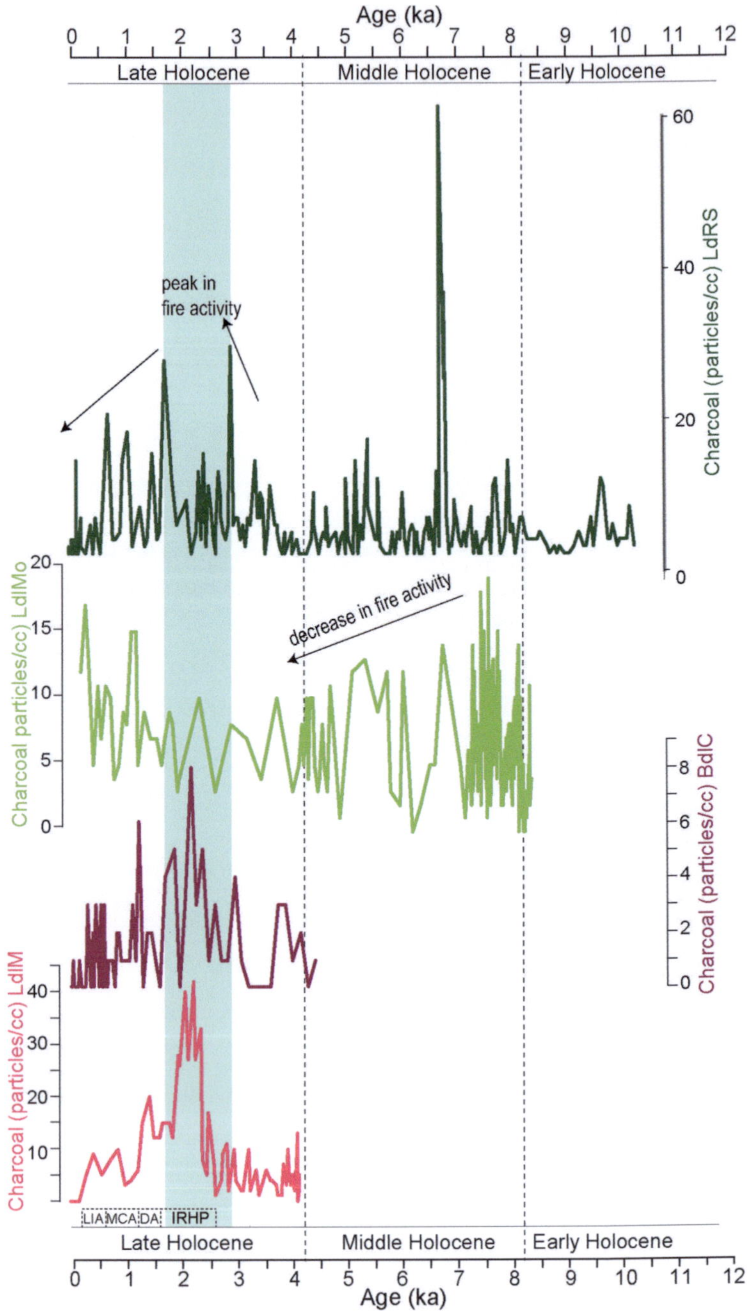


**Fig. 7** Charcoal records from the studied Sierra Nevada wetland sediment cores indicate fire activity for the last 12,000 cal yr BP. IRHP, DA, MCA and LIA stand for Iberian Roman Humid Period, Dark Ages, Medieval Climate Anomaly and Little Ice Age, respectively. The blue shading highlights an overall increase in charcoal and thus fire activity coinciding with the IRHP

## 4 Discussion

### 4.1 Age of the Sedimentary Sequences

Geomorphologic evidences show that alpine environments above 2000–2200 m in the north face and 2300–2400 m in the south face in the Sierra Nevada were glaciated during the Last Glacial Cycle and valley and cirque glaciers occupied and eroded the basins where the lakes and bogs occur at present (Palacios et al. 2016). Radiocarbon datings from the studied alpine paleoecological records indicate that wetlands formed in the area above ~2500 m after deglaciation (Fig. 2; Table 1). The different absolute ages for the beginning of sedimentation in different basins seems to be related to climate events and/or orientation of the studied sites (Manzano et al. 2019). Two south-facing sites, LdRS (3020 m) and LH (2899 m), accumulated sediments for the entire (or most of the) Holocene with estimated ages for the beginning of sedimentation of about 11,000 cal yr BP. Sediments in those two basins started accumulating right after deglaciation due to climate warming. Sediment accumulation began somewhat later, at ~8200 cal yr BP, at two other sites located in the north face, BdlV (2945 m) and LdlMo (2889 m). Two alternative hypotheses should be considered to explain this. Firstly, it is possible that any sediments that accumulated during the Early Holocene in those basins were eroded by a short glaciation during the well-known 8.2 ka cold-arid event. However, there is no direct evidence of glaciation at that time in the Sierra Nevada (See Palacios et al. 2016). Secondly, coarse sands-conglomerates could have been deposited in the studied basins due to enhanced erosion caused by especially cold-dry conditions by the ~8200 cal yr BP event, and sediments below that coarse layer could not be retrieved by our coring methods. Two additional sediment records from LdlM (2497 m) and BdlC (2992 m), located in the north and south face, respectively, have basal sediment accumulation ages between ~4400 and 4100 cal yr BP. Interestingly, this coincides in time with the well-known 4.2 ka arid climatic event and the Middle-Late Holocene transition (Zielhofer et al. 2019). Perhaps, as mentioned above, older sediments were eroded from the basins due to particularly arid conditions at that time (see next section below) and/or our coring technique precluded us from recovering the older part of the sedimentary sequence from those two basins. The studied records show that sediment accumulation seems to be continuous after the formation of the wetlands in the alpine sedimentary basins (Fig. 2).

## 4.2 Changes in Lithology and MS in Relation with Climate and Human Impact

MS, a measure of the tendency of sediment to carry a magnetic charge (Snowball and Sandgren 2001), is in alpine lakes mostly related to the relative abundance of detritics (thus magnetic minerals) and organic matter in the sediments. MS is relatively lower when organic matter is abundant, as it is a diamagnetic material (Dearing 1999).

MS maxima during the Early Holocene could be related with maxima in winter precipitation and thus maxima in snowpack and highest summer insolation, melting the snow and producing a large amount runoff, fluvial erosion and sedimentation of detritics into the alpine sedimentary basins. Highest summer insolation at this time generated highest evaporation in summer at lower elevations and low lake level in Padul, with the development of marshy vegetation and sedimentation of peat, depicting low MS in this record (Ramos-Román et al. 2018b; Camuera et al. 2018),

Middle Holocene is characterized by minima in MS in the alpine sedimentary basins, warm but decreasing summer insolation and still quite humid conditions most likely benefited vegetation and algae to flourish in the Sierra Nevada environments, generating a major content in organics in the sediments deposited in the basins with respect to detritics. Decreasing precipitation, particularly in winter snowpack, at this time (Ramos-Román et al. 2018b) might have also produced a reduction in erosion and sedimentation in the studied Sierra Nevada basins. Perhaps both factors were in play at that time.

MS shows an overall increasing trend in the Late Holocene. This increase in MS has previously been interpreted as produced by enhanced erosion and increase in siliciclastic sediments due to a decrease in winter precipitation, enhanced aridification and decrease in forested vegetation in the area related to a decrease in summer insolation (Jiménez-Moreno and Anderson 2012; Fletcher et al. 2013). The occurrence of small glaciers in the highest catchment areas during several phases of the Late Holocene at ~2800–2700, 1400–1200 cal yr BP and during the Little Ice Age (1300–1850 CE) could have also generated enhanced erosion in the study area (Oliva et al. 2020). Finally, a clear increase in land use is also observed during the last ~1550 cal yr BP in Padul (Ramos-Román et al. 2018a), suggesting that humans could have been partially responsible for this sedimentary change.

## 4.3 Inorganic Geochemistry—Holocene Evolution of Lake-Level Changes, Runoff and Dust Deposition in Relation with Climate Change and Human Impact

Inorganic geochemistry data from the Sierra Nevada wetland sedimentary records inform us about variations in the lithology related to changes in the local depositional environment as well as with allochthonous sediment inputs into the basins (e.g., aeolian dust deposition) (Jiménez-Espejo et al. 2014; García-Alix et al. 2018). The rocks that occur in the Sierra Nevada alpine wetland drainage basins are metamorphic mica schists (rich in mica and feldspars with abundant Si, K, Al) (Camuera et al. 2018). Therefore, the K/Al and K/Ti ratios can be interpreted as detrital inputs, and thus a proxy of surface runoff. In this respect, the highest K/Ti ratio reached from ~11,000 until ~7000 cal yr BP in LH points to the wettest climate conditions recorded in the Sierra Nevada during the Early and Middle Holocene (Fig. 3). This, together with lowest values in Ca/Al, Ca/Ti and Zr/Al ratios also recorded in LH, suggests that runoff dominated over eolian inputs at that time (Mesa-Fernández et al. 2018; see explanation about eolian proxies below).

The inorganic geochemistry data show a trend toward more arid conditions in the Middle and Late Holocene in the Sierra Nevada. This is deduced by decreasing K/Ti ratios in LH in the last ~7000 cal yr BP (Mesa-Fernández et al. 2018; Fig. 3). Further support of enhanced aridity comes from the patterns observed in typical Saharan eolian dust input proxies, such as the Zr/Th (Zr/Al, Ca/Ti, or Ca/Al) data from LdRS, LH and BdlC, which show minima in the Early and early-Middle Holocene and an increasing trend in the last ~7000 cal yr BP (Jiménez-Espejo et al. 2014; Mesa-Fernández et al. 2018; García-Alix et al. 2017). Zr/Th (Zr/Al) and Ca/Al (Ca/Ti) ratios have been used as eolian input proxies because Saharan dust is enriched in heavy minerals such as zircons, rutiles, as well as carbonates, which are absent in the alpine Sierra Nevada drainage basins (Mesa-Fernández et al. 2018). These ratios have been extensively used as a proxy of aeolian input in the Mediterranean or Atlantic regions (e.g., Rodrigo-Gámiz et al. 2011; Martínez-Ruiz et al. 2015). Enhanced eolian dust emissions from the Sahara were probably due to a coetaneous loss of vegetation cover in North Africa (Jiménez-Espejo et al. 2014).

Anthropogenic heavy metal atmospheric pollution at these high elevation Sierra Nevada wetlands was tracked using the Pb and Pb/Al data from selected sedimentary records (García-Alix et al. 2013; Mesa-Fernández et al. 2018). Pb/Al data from LdRS and LH show low values during most of the Holocene but increased in the Late Holocene, with peaks around 3000–2500, 2000 cal yr BP and between 1950–1970 AD (García-Alix et al. 2013; Mesa-Fernández et al. 2018). Lead was intensively extracted during the Late Bronze and Early Iron ages (~3500–2500 cal yr BP/~1550–550 cal BC) for metallurgical activities. It was used for ternary alloys (lead, tin and copper) and in cupellation processes. The development of cupellation processes in the southwestern Iberia pyritic belt for silver extraction led to an intensive exploitation of southeastern Iberia lead outcrops. High lead pollution is observed during the Roman Empire (between ~2100 and ~1700 cal yr BP), which is coherent with other regional records and data from historical sources (García-Alix et al. 2013). Maximum values of lead pollution in the Sierra Nevada are recorded in the last decades, between the 1950–1970 AD, related to the use of leaded-fuel in vehicles (García-Alix et al. 2013). Subsequently, the increase in unleaded-fuels, and the environmental alertness about lead emissions generated a decrease in lead pollution, which is clearly documented with a reduction in Pb/Al in the Sierra Nevada records (García-Alix et al. 2013; Mesa-Fernández et al. 2018; Fig. 3).

## 4.4 Organic Geochemistry—Total Organic Carbon Content and C/N Ratio in Relation with Climate, Temperature Estimations and Precipitation Patterns Derived by Specific Organic Lipids

Organic geochemistry data from bulk sediments such as total organic carbon (TOC) content, carbon to nitrogen atomic ratio (C/N), and stable carbon isotopes ($\delta^{13}C$), together with specific lipid biomarkers (*n*-alkanes: $P_{aq}$, long chain diols: LDI), give us insight about the biological sources of the organic matter and past biogeochemical cycles in the Sierra Nevada wetlands.

The C/N ratio can be used as a proxy for organic matter source in lacustrine sediments (Meyers 1994; Meyers and Teranes 2001). Protein-rich and cellulose-poor organic matter from lake algae is generally characterized by C/N values around 10, whereas values for protein-poor and cellulose-rich vascular land plants regularly exceed 20. A mix of algal and vascular plant organic matter is thus depicted by intermediate values (Meyers 1994; Meyers and Teranes 2001). $\delta^{13}C$ data from bulk sediments can be used to identify sources of organic matter and productivity levels (Talbot 2001). C3 vascular plants and lake algae are characterized by a similar $\delta^{13}C$ under normal conditions (Meyers 1994; Meyers and Teranes 2001). However, $\delta^{13}C$ increase in vascular plants during dry conditions and thus can be read in terms of water use efficiency (Farquhar et al. 1982). Algae preferentially incorporate the lighter C isotopes from the water's dissolved inorganic carbon pool. As a consequence, an algae productivity bloom enriches the carbon isotopic composition of the water, leading to the same isotopic enrichment in the algae from water mass (O'Leary 1988; Wolfe et al. 2001). Low C/N values in the Sierra Nevada records during the Early and Early-Middle Holocene suggested that the source of organic matter was mixed aquatic (algal)/terrestrial and very influenced by algae (Jiménez-Espejo et al. 2014; García-Alix et al. 2017). This points to relatively high lake levels in the LdRS and BdlV wetlands, which agrees with high humid climate conditions interpreted for this time period (Jiménez-Espejo et al. 2014; García-Alix et al. 2017). The highest $\delta^{13}C$ reached during this time interval in BdlV also implies significant algal productivity (García-Alix et al. 2012; Fig. 5). A major shift in the environment between 6000 and 5000 cal yr BP is observed through the significant increase in the C/N records from the Sierra Nevada. In LdRS this increase was interpreted as enhanced sedimentation of terrestrial vegetation with respect to algae related to a lowering of lake level and decrease in lake productivity. In BdlV, this increasing pattern was interpreted as a transition from a lake to a bog in the Middle Holocene, which also agrees with the $\delta^{13}C$ data and pollen and algae records (García-Alix et al. 2017; Fig. 5). Less precipitation and high evaporation rates may have caused the transition from a small lake to a bog in BdlV. Other alpine bogs such as BdlC also developed in the Sierra Nevada during the Middle Holocene at ~4400 cal yr BP. This agrees with the decrease in precipitation deduced by the decrease in arboreal pollen at this time (Fig. 6). C/N > 20 and $\delta^{13}C$ data from BdlV indicate that biomass production was dominated mainly by vascular terrestrial plants between ~5000 and ~4000 cal yr BP. From 4000 cal yr BP on,

C/N values decreased in all the studied records from the Sierra Nevada indicating more aquatic primary production in lakes and the development of occasional aquatic environments in peatlands. In the last 300 cal yr BP, the BdlC and BdlV records show a rapid and significant increase. C/N records in BdlC, LdM and LdRS registered a significant decrease in C/N in the last decades; however, the BdlV record did not register a substantial drop, suggesting that wetlands can display different responses to climate within the same Sierra Nevada region, depending on the landscape and the environmental evolution of the area. In any case, the abrupt environmental responses shown by the C/N records pointed to an amplification of natural trends caused by human pressure (García-Alix et al. 2017; see chapter "Paleolimnological Indicators of Global Change").

TOC content in the bulk sediment records gives insight about how much organic material occurs in the sediments with respect to the detrital fraction, and thus wetland productivity. In addition, the *n*-alkane abundance and distribution has been used as indicators of biological sources of organic matter accumulated in sediment records (e.g., Ficken et al. 2000; Eglinton and Eglinton 2008). *N*-alkanes derived from leaf plant waxes are proven to be extremely resistant to degradation during the transport and to the diagenesis, allowing its study lake records and providing information about vegetation and hydrological changes (e.g., Eglinton and Eglinton 2008). In general, *n*-alkanes with 21, 23 and 25 carbon atoms are associated with submerged and floating aquatic plants, expressed as $P_{aq}$ (Cranwell 1984; Ficken et al. 2000). TOC and $P_{aq}$ (deduced from leaf waxes, *n*-alkanes) values in LdRS are consistent with precipitation changes inferred from the local Sierra Nevada pollen and inorganic geochemistry records throughout the Holocene (Figs. 4, 5 and 6). Maxima in TOC and $P_{aq}$ were reached during the Early and early-Middle Holocene, indicating maxima in algal productivity and deepest lake levels, and since ~6000 cal yr BP values have decreased until present. This was interpreted as related to the climate aridification process previously described in the Middle and Late Holocene (Jiménez-Espejo et al. 2014; Toney et al. 2020).

The fractional abundance of some long chain diols ($C_{28-30}$ 1,13- and 1,15-diols) expressed in the Long chain diol Index (LDI) is proven to be related to the superficial temperature of the water in which the algal sources grew (Rampen et al. 2014). The LDI has been successfully applied in the LdRS in the Sierra Nevada (Fig. 5) (García-Alix et al. 2020; Toney et al. 2020). The very good correlation between mean annual air temperature (MAAT) data of the last ~100 years and the LdRS LDI record permitted the utilization of this index as paleotemperature proxy for the Sierra Nevada (García-Alix et al. 2020). The LDI-inferred temperatures show that MAAT fluctuated around ~2.4 °C during the Holocene. MAAT changed around this mean value during the Early-Middle Holocene but changes occurred during the Middle to Late Holocene transition. Reconstructed temperatures reached a peak between ~5000 and 4200 cal yr BP, recording positive MAAT Anomalies (MAATAs). The especially warm period of the Middle Holocene ended abruptly at around 4150–4100 cal yr BP, reaching a minimum MAAT of −1.5 °C at ~4000 cal yr BP. Reconstructed Late Holocene temperatures oscillated around the average MAATA until ~1020 cal yr BP, when MAATA increased significantly to −0.5 °C during the well-known warm period of the Medieval Climate Anomaly (MCA; Moreno et al. 2012). Later on, temperatures decreased during the Little Ice Age (LIA), reaching the lowest value of the record at ~260 cal yr BP (−2.2 °C MAATA). Estimated temperatures increased later on until present, except for a sharp and short decline at the beginning of the twentieth century (−1.7 °C MAATA). Increasing temperature trends during the twentieth century match the temperature changes measured during the current global warming (García-Alix et al. 2020; IPCC 2013).

## 4.5 Palynological Analysis—Holocene Forest, Landscape and Lake Level Variations Due to Climate Change and Human Impact

Variations in the abundance of arboreal pollen (AP, including Mediterranean tree species) have previously been used in the Sierra Nevada records as a proxy for regional climate changes (Jiménez-Moreno and Anderson 2012; Ramos-Román et al. 2016, 2018a, b; Camuera et al. 2019). Increases in AP in the alpine records indicate enhanced humidity (overall, trees require more soil humidity than most herbs/shrubs) and proximity of forest (i.e., treeline) to the alpine wetlands, thus indicating increases in temperature as treeline moves to higher elevation during climate warming (Camuera et al. 2019).

A transition period from glacial to interglacial climate conditions occurred in the Sierra Nevada area during the earliest Holocene (~11,700 to 10,000 cal yr BP). This is shown in the oldest part of the sedimentary record of LdRS, which indicates steppe-like vegetation (*Artemisia*, Amaranthaceae, *Ephedra*), pointing to very arid and cold conditions. Regionally, similar pollen data have been recorded in Carihuela Cave (Fernández et al. 2007; Carrión et al. 2019), as well as in marine records from the Alboran Sea (Dormoy et al. 2009; Fletcher et al. 2010). In the lower-elevation Padul area, this transition seems to have occurred faster (from 11,700 to 11,000 cal yr BP) and was characterized by pollen assemblages dominated by evergreen *Quercus* and to a lesser extent deciduous *Quercus*. The increase in Mediterranean forest species recorded in the Sierra Nevada pollen records could be interpreted as a regional vegetation response to warmer and more humid climate conditions, generating a displacement of forest species toward higher elevations and more heavily tree-populated forest.

Palynological records indicate that a maximum in temperature and humidity occurred in the Sierra Nevada area between ~10,500 and 7000 cal yr BP (Fig. 6). This is indicated by the highest abundance of tree species and the abundance of algae (*Botryococcus* and *Pediastrum*) in the studied records. The abundance of high-elevation *Pinus* species indicates that the highest elevation of treeline was reached in the Sierra Nevada subalpine area at this time. In the Padul-15–05 record the highest values of Mediterranean forest also occurred at that time, showing an expansion in mesic forest species (e.g., deciduous *Quercus*). Both regional (Burjachs et al. 1997; Fletcher and Sánchez Goñi 2008) and global studies show a humid and warm Early Holocene (Jalut et al. 2009; Brayshaw et al. 2011). A very warm Early Holocene could be explained by maximum summer insolation at this time that would have produced climate warming. The maximum humidity between ~10,500 and 7000 cal yr BP could be explained by an increase in the land/sea temperature contrast in the Mediterranean region during autumn, which would favor an increase in rainfall during autumn/winter seasons (Tuenter et al. 2003; Meijer and Tuenter 2007).

The Sierra Nevada pollen records show a progressive process of forest reduction, increase in xerophytes (i.e., *Artemisia*, Amaranthaceae or Caryophyllaceae), and decrease in aquatic species in the wetlands in the Middle and Late Holocene (Fig. 6). This trend, which begins after ~7000 cal yr BP and intensifies after ~6000–5000 cal yr BP, could be explained by a progressive climatic cooling and, mostly, by an increase in aridity. This aridification process has been observed in other pollen records in this region (e.g., Sierra de Cazorla: Carrión 2002; Alboran Sea: Fletcher et al. 2010; SW Iberian Peninsula: Jiménez-Moreno et al. 2015) and from many other paleoclimatic indicators for the entire Mediterranean region (speleothems, lake levels, river and wind inputs; Jalut et al. 2009). This climate change, decrease in temperature and increase in aridity, is explained by the decrease in summer insolation (Cacho et al. 2002; Renssen et al. 2003). Another evidence of aridification in the Sierra Nevada during the Middle and Late Holocene is the interpretation of decreasing lake levels in some of the longest studied alpine records (Anderson et al. 2011; Jiménez-Moreno and Anderson 2012). This was interpreted after noticing a considerable decline in aquatic algae species.

### 4.5.1 Millennial-Scale Climate Variability

Sierra Nevada pollen records show shorter-scale cyclical changes superimposed on the main climate trend toward increased aridity. Significant forest declines occurred around 7500, 6500, 5000 and 4200, 3000 and 1200 cal yr BP (Fig. 6). Some of these forest declines can be interpreted as generated by severe and persistent drought conditions, which have been recognized regionally and globally at those times (see summary in Jiménez-Moreno and Anderson, 2012; Ramos-Román et al. 2018b). For example, the drought that characterized the MCA (Moreno et al. 2012) is very well documented in the LdlM record (Jiménez-Moreno et al. 2013). Among arid periods there were others that were relatively more humid, translated into enhanced forest development such as that observed in LdlM, BdlC, LH or Padul, coinciding with the maximum humidity of the Iberian Roman Humid Period (IRHP) or the LIA (Jiménez-Moreno et al. 2013; Ramos-Román et al. 2016, 2018a; Mesa-Fernández et al. 2018). These climatic variations were probably related with large-scale cyclical changes in the frequency of the North Atlantic Oscillation (NAO) atmospheric phenomenon and arid periods were due to persistent phases of NAO+ and more humid periods to persistent phases of NAO-. For example, wetter climatic conditions during the LIA period were probably related to persistent NAO- conditions, which at present-day produces a general increase in winter precipitation in the area (Trouet et al. 2009; Mesa-Fernández et al. 2018; Ramos-Román et al. 2016, 2018a; Jiménez-Moreno et al. 2020).

### 4.5.2 Human Impact on Vegetation, Grazing and Cultivation

In the last millennia human impact on the Sierra Nevada vegetation has increased and changes in the vegetation are more difficult to interpret. Climate has been a major driver of vegetation change in this area but human impact cannot be neglected (Anderson et al. 2011; Jiménez-Moreno and Anderson 2012, 2018a; Manzano et al. 2019). Pollen data show that since 3000 cal yr BP humans had an impact in Sierra Nevada with evidence of grazing and cultivation (Anderson et al. 2011; Jiménez-Moreno and Anderson, 2012). However, the anthropogenic impact was relatively small until ~1500 cal yr BP and intensified in the last ~500 years with large-scale olive cultivation at lower altitudes and pine reforestation in the last century (Ramos-Román et al. 2019).

Early evidences of grazing in alpine Sierra Nevada come from the *Sporormiella* spore records. *Sporormiella* is a genus of coprophilous fungi that requires herbivore digestion to complete its life cycle, producing spores in the excrements (Anderson et al. 2011). Their increasing abundance in the last 2700 cal yr BP, becoming very abundant in the last millennium (Anderson et al. 2011; Jiménez-Moreno and Anderson 2012; Ramos-Román et al. 2018a; Manzano et al. 2019) probably indicates intensified grazing in the higher elevations of the Sierra Nevada at this time related with the introduction of livestock on the landscape. Enhanced grazing roughly concurs with the first signs of cultivation with occurrences of Cerealia (cereal grass) pollen, most likely planted at lower elevations (Jiménez-Moreno and Anderson 2012; Manzano et al. 2019), and land-use indicators such as *Rumex*, *Plantago*, *Urtica*, Convolvulaceae or Asteraceae, which are consistent in the pollen records since the last 3000–2000 cal yr BP, and enhanced in the last ~1500 cal yr BP (Ramos-Román et al. 2018a).

*Olea* cultivation boosted in the last 1000 years in the Sierra Nevada area (Ramos-Román et al. 2019). A further increasing trend in *Olea* cultivation happened in the last 150 cal yr BP, during the industrial revolution, which was probably triggered by improvements in agricultural practices. A more recent widespread cultivation and olive oil production increase occurred in Spain in the mid-twentieth century (Ramos-Román et al. 2019). Finally, the twentieth century *Pinus* pollen increase (reflected in the AP; Fig. 6) reproduces the extensive human-induced reforestation with pines, done for forest restoration and to prevent soil erosion previously generated by natural and anthropogenic deforestation.

In summary, pollen data from the Sierra Nevada show that human impact on the Sierra Nevada environments was insignificant until the last ~1500 cal yr BP and was less pronounced than in other lower elevation regions of southern Iberia (Anderson et al. 2011).

## 4.6 Charcoal Analysis—Fire Activity Related to Climate and Fuel Variability

Charcoal analysis is based on the accumulation of charcoal particles in sedimentary records during and after a fire event. Sedimentary layers with abundant charcoal are inferred to result from past fire activity (Whitlock and Anderson 2003). Macroscopic charcoal particles (>100 µm) indicate occurrence of nearby fires, because particles of this size do not travel far from their source (Whitlock and Anderson 2003). The charcoal analysis on the Sierra Nevada sedimentary records, combined with palynological and geochemical proxies, permit us to study how fire regimes were affected by periods of major climate change and vegetation reorganization in the past.

Charcoal particles were generally not abundant in the alpine Sierra Nevada records (Fig. 7). This can be explained by a faraway origin of charcoal particles, probably from fires that occurred at lower elevations in this mountain range, as natural fires are extremely rare above treeline in the Sierra Nevada at present and were probably very rare in the past. Considering this relationship, these charcoal records show that fire activity was relatively high during the Middle Holocene between ~7500 and 6700 cal yr BP, decreasing later on. The decrease in fire activity at ~7000 cal yr BP could have been conditioned by a decrease in forest fuel due to the progressive loss in Mediterranean forest previously discussed and due to enhanced aridification (Fig. 6). However, enhanced fire activity in this region is observed since ~3000 cal yr BP (Fig. 7; Anderson et al. 2011; Jiménez-Moreno et al. 2013; Ramos-Román et al. 2016) and maxima in charcoal concentration occurred between ~2700 to 1600 cal yr BP in LdlM, BdlC and LdRS records, coinciding with the time of the IRHP (2600–1600 cal yr BP, Martín-Puertas et al. 2009), the most humid period detected in the area during the Late Holocene in the Sierra Nevada. Many charcoal records from the western Mediterranean region show maxima in fire activity around that time. For example, the nearby Sierra de Baza record depicts remarkable similarities to the Sierra Nevada records with charcoal maxima at ~2000 cal yr BP (Carrión et al. 2007), but also charcoal maxima around that time was observed in the Sierra de Gádor (Carrión et al. 2003), Alboran Sea (Combourieu Nebout et al. 2009) and Djamila, in northern Morocco (Linstädter and Zielhofer 2010). This regional agreement about enhanced fire activity between 3000 and 2000 cal yr BP may have been related to increased forest, and therefore fuel loads on the landscape due to enhanced humidity. This supports the hypothesis of Mediterranean fire regimes today being controlled by fuel load variations (Daniau et al. 2007; Linstädter and Zielhofer 2010). Charcoal decreased considerably after this IRHP peak, during the Dark Ages (DA) and MCA, when warmer and drier conditions occurred (Moreno et al. 2012; García-Alix et al. 2020; Fig. 7). This was probably due to the decrease in forest fuel, which agrees with the decline in forest species recorded in the Sierra Nevada at that time (Fig. 6; Jiménez-Moreno et al. 2013; Ramos-Román et al. 2016, 2018a; Mesa-Fernández et al. 2018).

Even though we believe that patterns recorded in the charcoal records from the Sierra Nevada mostly responded to natural climate variability, the increase in charcoal in the last millennia could have also been due to increased human activity (i.e., forest burning) due to accelerated mining, cultivation and grazing activities (Anderson et al. 2011; García-Alix et al. 2013; see also discussion above).

# 5 Conclusions

Multi-proxy sedimentary records from the Sierra Nevada wetlands show the warmest and wettest climate conditions during the Early and early-Middle Holocene (from ~11,000 to 7000 cal yr BP), related to Holocene summer insolation maxima. Such optimum climate conditions translated into high detrital sedimentation, maximum Mediterranean forest expansion in density and elevation in the Sierra Nevada, and the highest lake levels in the alpine area.

Multiple evidences show a long-term aridification of the Sierra Nevada environment since ca. 7000 cal yr BP until today, probably related to a decrease in precipitation generated by a decrease in summer insolation during the Middle and Late Holocene. This is mostly deduced by a decrease in runoff, a decrease in forest species, particularly in *Pinus* and *Quercus* and other mesophyte species and lower lake levels in alpine environments. This agrees with enhanced Saharan eolian dust input in the alpine records.

This long-term aridification trend was interrupted by short time-scale millennial-scale variability evident in our records by decreases in forest that could be related to arid and/or cold events at ~7500, 6500, 5200, 4200, 3000 and 1200 cal yr BP. Some of these events, such as the one that affected the area during the MCA, coincide in the timing with droughts that affected the wider Mediterranean area and other distant regions, and could be related to persistent NAO + conditions in the study area.

In addition, humans also influenced environments in the Sierra Nevada in the last millennia. Charcoal increase, lead pollution and cultivar proxies indicate anthropogenic impact since 3000 cal yr BP. Human impact is particularly important in the last 1500 cal yr BP, with a strong increase in the palynological record of cultivars and grazing, significantly stronger in the last millennium with records of massive cultivation of *Olea* at lower elevations and *Pinus* reforestation. Estimated temperatures using the algal-derived lipids in the Sierra Nevada show a very high rate of warming in the last century and that human-derived greenhouse gases seem to be the major temperature driver in these high-elevation sites at present.

Sedimentary archives from the Sierra Nevada wetlands show a fast and very sensitive response of mountain ecosystems to climate change and human impact throughout the Holocene, reflecting climate events on a regional and global scale. Sierra Nevada is, therefore, especially sensitive to global change and important changes will occur in these alpine ecosystems (e.g., eutrophication, aridification, increase in fires) if the current temperature increase is not slowed or reversed. This study shows the importance of past and present monitoring of wetlands in mountain areas for a better understanding of climate variability and future rapid environmental responses.

**Acknowledgements** This study was supported by projects CGL2013-47038-R and CGL2017-85415-R funded by Ministerio de Economía y Competitividad of Spain and Fondo Europeo de Desarrollo Regional FEDER; Séneca Project 20788/PI/18; Junta de Andalucía I+D +i Junta de Andalucía 2020 Retos P-20-00059, FEDER Project B-RNM-144-UGR18, UGR-FEDER B-RNM-144-UGR18 Proyectos I + D + i del Programa Operativo FEDER 2018 and the research group RNM-190 (Junta de Andalucía). M.J.R.R. acknowledges the postdoctoral funding provided by the European Commission/H2020 (ERC-2017-ADG, project number 788616). J.C. acknowledges the postdoctoral funding provided by the Academy of Finland (project number 316702). A.G.-A. acknowledges the Ramón y Cajal fellowship RYC-2015-18966 provided by the Ministerio de Economía y Competitividad of the Spanish Government. M.R.G. acknowledges funding by the Juan de la Cierva-Incorporación program in the University of Granada (IJCI-2017-33755) from Secretaría de Estado de I+D+i, Spain. RSA acknowledges several travel grants from Northern Arizona University to support this work.

## References

Anderson RS, Jiménez-Moreno G, Carrión JS, Pérez-Martínez C (2011) Holocene vegetation history from Laguna de Río Seco, Sierra Nevada, southern Spain. Quat Sci Rev 30:1615–1629

Beug H-J (2004) Leitfaden der Pollenbestimmung für Mitteleuropa und angrenzende Gebiete, Fisch. Stuttg., Leitfaden der Pollenbestimmung für Mitteleuropa und angrenzende Gebiete, Friedrich Pfeil, München

Blaauw M (2010) Methods and code for 'classical' age-modelling of radiocarbon sequences. Quat Geochron 5:512–518

Cacho I, Grimalt JO, Canals M (2002) Response of the Western Mediterranean Sea to rapid climatic variability during the last 50,000 years: a molecular biomarker approach. J Mar Sys 33–34:253–272

Cranwell PA (1984) Lipid geochemistry of sediments from Upton broad, a small productive lake. Org Geochem 7:25–37

Brayshaw DJ, Rambeau CMC, Smith SJ (2011) Changes in Mediterranean climate during the Holocene: insights from global and regional climate modelling. Holocene 21:15–31

Burjachs F, Giralt S, Roca JR et al (1997) Palinología holocénica y desertización en el Mediterráneo occidental. In: Ibáñez JJ, Valero BL, Machado C (eds) El paisaje mediterráneo a través del espacio y del tiempo. Implicaciones en la desertificación, Geoforma Editores, Logroño, pp 379–394

Camuera J, Jiménez-Moreno G, Ramos-Román MJ et al (2018) Orbital-scale environmental and climatic changes recorded in a new ~ 200,000-year-long multiproxy sedimentary record from Padul, southern Iberian Peninsula. Quat Sci Rev 198:91–114

Camuera J, Jiménez-Moreno G, Ramos-Román MJ et al (2019) Vegetation and climate changes during the last two glacial-interglacial cycles in the western Mediterranean: A new long pollen record from Padul (southern Iberian Peninsula). Quat Sci Rev 205:86–105

Camuera J, Jiménez-Moreno G, Ramos-Román MJ et al (2021) Chronological control and centennial-scale climatic subdivisions of the Last Glacial Termination in the western Mediterranean region. Quat Sci Rev 255:106814

Carrión JS (2002) Patterns and processes of Late Quaternary environmental change in a montane region of southwestern Europe. Quat Sci Rev 21:2047–2066

Carrión JS, Sánchez-Gómez P, Mota JF et al (2003) Fire and grazing are contigent on the Holocene vegetation dynamics of Sierra de Gádor, southern Spain. Holocene 13:839–849

Carrión JS, Fernández S, Jiménez J et al (2019) The sequence at Carihuela Cave and its potential for research into Neanderthal ecology and the Mousterian in southern Spain. Quat Sci Rev 217:194–216

Carrión JS, Fuentes N, González-Sampériz P (2007) Holocene environmental change in a montane region of sourthern Europe with a long history of human settlement. Quat Sci Rev 26:1455–1475

Combourieu Nebout N, Peyron O, Dormoy I et al (2009) Rapid climatic variability in the west Mediterranean during the last 25000 years from high resolution pollen data. Clim past 5:503–521

Daniau AL, Sánchez-Goñi MF, Beaufort L et al (2007) Dansgaard-Oeschger climatic variability revealed by fire emissions in southwestern Iberia. Quat Sci Rev 26:1369–1383

Dearing J (1999) Magnetic susceptibility. In: Walden J, Oldfield F, Smith J (eds) Environmental magnetism: a practical guide. Quaternary Research Association, London, pp 35–62

Dormoy I, Peyron O, Combourieu Nebout N et al (2009) Terrestrial climate variability and seasonality changes in the Mediterranean region between 15 000 and 4000 years BP deduced from marine pollen records. Clim past 5:615–632

Eglinton TI, Eglinton G (2008) Molecular proxies for paleoclimatology. Earth Planet Sci Lett 275:1–16

Faegri K, Iversen J (1989) Textbook of Pollen analysis. Caldwell. Blackburn Press, New Jersey, pp 69–90

Farquhar GD, O'Leary MH, Berry JA (1982) On the relationship between carbon isotope discrimination and the intercellular carbon dioxide concentration in leaves. Aust J Plant Physiol 9:121–137

Fernández S, Fuentes N, Carrión JS et al (2007) The Holocene and upper Pleistocene pollen sequence of Carihuela cave, southern Spain. Geobios 40:75–90

Ficken KJ, Li B, Swain D, Eglinton G (2000) An n-alkane proxy for the sedimentary input of submerged/floating freshwater aquatic macrophytes. Org Geochem 31:745–749

Fletcher WJ, Sanchez Goñi MF (2008) Orbital- and sub-orbital-scale climate impacts on vegetation of the western Mediterranean basin over the last 48,000 yr. Quat Res 70:451–464

Fletcher WJ, Sanchez Goñi MF, Peyron O, Dormoy I (2010) Abrupt climate changes of the last deglaciation detected in a Western Mediterranean forest record. Clim past 6:245–264

Fletcher WJ, Debret M, Sanchez Goñi MF (2013) Mid-Holocene emergence of a low frequency millennial oscillation in western Mediterranean climate: implications for past dynamics of the North Atlantic atmospheric westerlies. Holocene 23(2):153–166

García-Alix A, Jiménez-Moreno G, Anderson RS et al (2012) Holocene paleoenvironmental evolution of a high-elevation wetland in Sierra Nevada, southern Spain, deduced from an isotopic record. J Paleolimnol 48:471–484

García-Alix A, Jimenez-Espejo FJ, Lozano JA et al (2013) Anthropogenic impact and lead pollution throughout the Holocene in southern Iberia. Sci Total Environ 449:451–460

García-Alix A, Jiménez-Espejo FJ, Toney JL et al (2017) Alpine bogs of southern Spain show human-induced environmental change superimposed on long-term natural variations. Sci Rep 7:7439

García-Alix A, Jiménez-Espejo FJ, Jiménez-Moreno G et al (2018) Holocene geochemical footprint from Semi-arid alpine wetlands in southern Spain. Sci Data 5:180024

García-Alix A, Toney JL, Jiménez-Moreno G et al (2020) Algal lipids reveal unprecedent warming rates in alpine areas of SW Europe during the Industrial Period. Clim past 16:245–263

Gehrig-Fasel J, Guisan A, Zimmermann NE (2008) Evaluating thermal treeline indicators based on air and soil temperature using an air-to-soil temperature transfer model. Ecol Model 213:345–355

Heegaard E, Birks HJB, Telford RJ (2005) Relationships between calibrated ages and depth in stratigraphical sequences: an estimation procedure by mixed-effect regression. Holocene 15:612–618

IPCC (2013) Climate change 2013: the physical science basis. In: Contribution of working group I to the fifth assessment report of the Intergovernmental Panel on Climate Change. Cambridge University Press, Cambridge, United Kingdom and New York, USA

Jalut G, Dedoubat JJ, Fontugne M, Otto T (2009) Holocene circum-Mediterranean vegetation changes: climate forcing and human impact. Quat Inter 200:4–18

Jiménez-Espejo FJ, García-Alix A, Jiménez-Moreno G et al (2014) Saharan aeolian input and effective humidity variations over Western Europe during the Holocene. Chem Geol 374–375:1–12

Jiménez-Moreno G, Anderson RS (2012) Holocene vegetation and climate change recorded in alpine bog sediments from the Borreguiles de la Virgen, Sierra Nevada, southern Spain. Quat Res 77:44–53

Jiménez-Moreno G, García-Alix A, Hernández-Corbalán MD et al (2013) Vegetation, fire, climate and human disturbance history in the southwestern Mediterranean area during the late Holocene. Quat Res 79:110–122

Jiménez-Moreno G, Rodríguez-Ramírez A, Pérez-Asensio JN et al (2015) Impact of late-Holocene aridification trend, climate variability and geodynamic control on the environment from a coastal area in SW Spain. Holocene 25:607–617

Jiménez-Moreno G, Anderson RS, Ramos-Román MJ et al (2020) The Holocene Cedrus pollen record from Sierra Nevada (S Spain), a proxy for climate change in N Africa. Quat Sci Rev 242:106468

Laskar J, Robutel P, Joutel F, Gastineau M, Correia ACM, Levrard B (2004) A long term numerical solution for the insolation quantities of the Earth. Astron. Astrophys. Nor 428:261–285

Linstädter A, Zielhofer C (2010) Regional fire history shows abrupt responses of Mediterranean ecosystems to centennial-scale climate change (Olea–Pistacia woodlands), NE Morocco. J Arid Environ 74:101–110

Löwemark L, Chen H-F, Yang T-N et al (2011) Normalizing XRF-scanner data: a cautionary note on the interpretation of high-resolution records from organic-rich lakes. J Asian Earth Sci 40:1250–1256

Malanson GP, Resler LM, Butler DR, Fagre DB (2019) Mountain plant communities: uncertain sentinels? Prog Phys Geogr 43:521–543

Manzano S, Carrión JS, López-Merino L et al (2019) A palaeoecological approach to understanding the past and present of Sierra Nevada, a Southwestern European biodiversity hotspot. Global Planet Change 175:238–250

Martín-Puertas C, Valero Garcés BL, Brauer A et al (2009) The Iberian-Roman Humid Period (2600–1600 cal yr BP) in the Zoñar lake varve record (Andalucía, southern Spain). Quat Res 71:108–120

Martínez-Ruiz F, Kastner M, Gallego-Torres D et al (2015) Paleoclimate and paleoceanography over the past 20,000 yr in the Mediterranean Sea Basins as indicated by sediment elemental proxies. Quat Sci Rev 107:25–46

Mesa-Fernández JM, Jiménez-Moreno G, Rodrigo-Gámiz M et al (2018) Vegetation and geochemical response to Holocene rapid climate change in the Sierra Nevada (southeastern Iberia): the Laguna Hondera record. Clim past 14:1687–1706

Meijer PTh, Tuenter E (2007) The effect of precession-induced changes in the Mediterranean freshwater budget on circulation at shallow and intermediate depth. J Mar Sys 68:349–36

Meyers PA (1994) Preservation of elemental and isotopic source identification of sedimentary organic matter. Chem Geol 113:289–302

Meyers PA, Teranes JL (2001) Sediment organic matter. In: Last WM, Smol JP (eds) Tracking environmental changes using lake sediments, vol 2. Kluwer, Dordrecht, pp 239–270

Moreno A, Pérez A, Frigola J et al (2012) The Medieval Climate Anomaly in the Iberian Peninsula reconstructed from marine and lake records. Quat Sci Rev 42:16–32

O'Leary MH (1988) Carbon isotopes in photosynthesis. Bioscience 38:328–336

Oliva M, Gómez-Ortiz A, Palacios D et al (2020) Multiproxy reconstruction of Holocene glaciers in Sierra Nevada (south Spain). Med Geos Rev 2:5–19

Palacios D, Gómez-Ortiz A, Andrés N et al (2016) Timing and new geomorphologic evidence of the last deglaciation stages in Sierra Nevada (southern Spain). Quat Sci Rev 150:110–129

Páscoa P, Gouveia CM, Russo A, Trigo RM (2017) Drought trends in the Iberian Peninsula over the last 112 years. Adv Meteorol 2017:4653126

Pausas JG, Fernández-Muñoz S (2012) Fire regime changes in the western Mediterranean basin: from fuel limited to drought-driven fire regime. Clim Change 110:215–226

Ramos-Román MJ, Jiménez-Moreno G, Anderson RS (2016) Centennial-scale vegetation and North Atlantic Oscillation changes during the late Holocene in the western Mediterranean. Quat Sci Rev 143:84–95

Ramos-Román MJ, Jiménez-Moreno G, Camuera J et al (2018a) Holocene climate aridification trend and human impact interrupted by millennial and centennial-scale climate fluctuations from a new sedimentary record from Padul (Sierra Nevada, southern Iberian Peninsula). Clim past 14:117–137

Ramos-Román MJ, Jiménez-Moreno G, Camuera J (2018b) Millennial-scale cyclical environment and climate variability during the Holocene in the western Mediterranean region deduced from a new multiproxy analysis from the Padul record (Sierra Nevada, Spain). Global Planet Change 168:35–53

Ramos-Román MJ, Jiménez-Moreno G, Anderson RS et al (2019) Climate controlled historic olive tree occurrences and olive oil production in southern Spain. Global Planet Change 182:102996

Reille M (1992) Pollen et Spores d'Europe et Afrique du Nord. Laboratoire de Botanique Historique et Palynologie, Marseille

Reimer PJ, Bard E, Bayliss A et al (2013) IntCal13 and Marine13 radiocarbon age calibration curves 0–50,000 years cal BP. Radiocarbon 55:1869–1887

Renssen H, Brovkin V, Fichefet T, Goosse H (2003) Holocene climate instability during the termination of the African Humid Period. Geophys Res Lett 30:1184

Rodrigo-Gámiz M, Martínez-Ruiz F, Jiménez-Espejo FJ et al (2011) Impact of climate variability in the western Mediterranean during the last 20,000 years: oceanic and atmospheric responses. Quat Sci Rev 30:2018–2034

Snowball I, Sandgren P (2001) Application of mineral magnetic techniques to paleolimnology. In: Last WM, Smol JP (eds) Tracking environmental changes using lake sediments, vol 2. Kluwer Academic Publishers, Dordrecht, pp 217–237

Sousa PM, Trigo RM, Pereira MG et al (2015) Different approaches to model future burnt area in the Iberian Peninsula. Agric for Meteorol 202:11–25

Sousa PM, Barriopedro D, Ramos AM et al (2019) Saharan air intrusions as a relevant mechanism for Iberian heatwaves: the record breaking events of August 2018 and June 2019. Weather Clim Extr 26:100224

Talbot MR (2001) Nitrogen isotopes in palaeolimnology. In: Last WM, Smol JP (eds) Tracking environmental changes using lake sediments: physical and chemical techniques. Kluwer, Dordrecht, pp 401–439

Thuiller W, Lavorel S, Araujo MB et al (2005) Climate change threats to plant diversity in Europe. PNAS 102:8245–8250

Toney JL, García-Alix A, Jiménez-Moreno G et al (2020) New insights into Holocene hydrology and temperature from lipid biomarkers in western Mediterranean alpine wetlands. Quat Sci Rev 240:106395

Trouet V, Esper J, Graham NE et al (2009) Persistent positive north Atlantic oscillation mode dominated the Medieval climate anomaly. Science 324:78–80

Tuenter E, Weber SL, Hilgen FJ, Lourens LJ (2003) The response of the African summer monsoon to remote and local forcing due to precession and obliquity. Glob Planet Change 36:219–235

Webster CE (2018) Late-Pleistocene and Holocene record of fire at Padul peat bog (Granada, Spain). MS Dissertation, Northern Arizona University

Whitlock C, Anderson RS (2003) Fire history reconstructions based on sediment records from lakes and wetlands. In: Veblen TT, Baker WL, Montenegro G, Swetnam TW (eds) Fire and climatic change in temperate ecosystems of the Americas, vol 160. Springer, New York, pp 3–31

Wolfe BB, Edwards TWD, Beuning KRM, Elgood RJ (2001) Carbon and oxygen isotope analysis of lake sediment cellulose: methods and applications. In: Last WM, Smol JP (eds) Tracking environmental changes using lake sediments: physical and chemical techniques. Kluwer, Dordrecht, pp 373–400

Zielhofer C, Köhler A, Mischke S et al (2019) Western Mediterranean hydro-climatic consequences of Holocene ice-rafted debris (Bond) events. Clim past 15:463–475

# Ancient and Present-Day Periglacial Environments in the Sierra Nevada

Marc Oliva, Antonio Gómez-Ortiz, David Palacios, Ferran Salvador Franch, Miguel Ramos, José Juan de Sanjosé-Blasco, Marcelo Fernandes, José M. Fernández-Fernández, Jesús Galindo-Zaldívar, Julia García-Oteyza, Lourdes González, Christian Hauck, Jordi Martín-Díaz, Jordi Nofre, Carlos Sanz de Galdeano, and Luis Miguel Tanarro-García

**Abstract**

The Sierra Nevada constitutes one of the massifs in Europe where periglacial processes have been more extensively and thoroughly examined. Periglacial phenomena in the massif are distributed from the mountaintops at 3300–3400 m a.s.l. down to elevations of 1100–1200 m. Active periglacial dynamics prevail today above 2500 m with a variety of landforms mostly related to frost shattering, cryoturbation and solifluction processes, among others. Besides, inherited periglacial landforms formed during Quaternary glacial phases are also found in the summit plateaus (i.e. patterned ground features) and valley heads (i.e. inactive rock glaciers). Since the early 2000s, a multi-approach research program has been carried out in the Sierra Nevada to monitor present-day periglacial dynamics and frozen ground conditions in the high lands of this massif. Results show evidence of the key role of seasonal frost driving environmental dynamics above 2500 m. Permafrost is spatially limited, and only confined to the areas that were glaciated during the Little Ice Age (LIA), namely in the Veleta and Mulhacén cirques above 3000–3100 m. Buried ice derived from LIA glaciers and isolated permafrost patches developed subsequently are still preserved under the thick debris mantle distributed across the cirque floors as revealed by the monitoring of soil temperature at different depths in the Veleta cirque. Here, geomatic and geophysical surveys of an incipient rock glacier indicate that permafrost conditions are undergoing a process of degradation. Seasonally frozen ground prevails in the rest of the Sierra Nevada, even at the summit surfaces at 3300–3400 m where mean annual soil temperatures are in the order of ~2.5 °C. In other periglacial landforms, positive temperatures were also recorded, ranging between 2 °C (inactive sorted-circles) and 3–4 °C (inactive and weakly active solifluction lobes).

**Keywords**

Sierra Nevada • Periglacial processes • Permafrost • Seasonal frost

M. Oliva (✉) · A. Gómez-Ortiz · F. S. Franch · J. García-Oteyza · J. Martín-Díaz
Department of Geography, Universitat de Barcelona, Barcelona, Catalonia, Spain
e-mail: marcoliva@ub.edu

D. Palacios · L. M. Tanarro-García
Department of Geography, Universidad Complutense de Madrid, Madrid, Spain

M. Ramos
Department of Physics, University of Alcalá, Madrid, Spain

J. J. de Sanjosé-Blasco
Department of Graphical Expression, University of Extremadura, Extremadura, Spain

M. Fernandes · J. M. Fernández-Fernández
Centre for Geographical Studies, IGOT, Universidade de Lisboa, Lisbon, Portugal

J. Galindo-Zaldívar · L. González · C. S. de Galdeano
Instituto Andaluz de Ciencias de La Tierra, CSIC, University of Granada, Granada, Spain

C. Hauck
Department of Geosciences, University of Fribourg, Fribourg, Switzerland

J. Nofre
Interdisciplinary Centre of Social Sciences, NOVA University of Lisbon, Lisbon, Portugal

## 1 The Discovery of Periglacial Landforms

Periglacial studies in the Sierra Nevada have been limited to the glaciated domain during the last glacial cycle, which extended from the Trevélez divide (2882 m) to the Cerro del Caballo (3011 m). Towards the east, the port of La Ragua (2000 m) and Cerro del Chullo (2609 m), the available information is very scarce despite glacionival processes and


R. Zamora and M. Oliva (eds.), *The Landscape of the Sierra Nevada*,
https://doi.org/10.1007/978-3-030-94219-9_8

periglacial morphogenesis were active during Quaternary glaciations (Oliva et al. 2014a).

The interest in understanding the contribution of periglacial dynamics in shaping the landscape of the Sierra is relatively recent. During the first decades of the twentieth century, scholars mostly focused on the spatial distribution of glacial activity and paid little attention to other cold-climate geomorphological processes (Quelle 1908; Obermaier 1916; García-Sainz 1947). Dresch (1937) and Paschinger (1957) were the first highlighting the role of nivation processes in the landscape of the Sierra Nevada high lands. Hempel (1960) was the first scientist using the term *periglacial* in the massif to establish the elevation belts of Pleistocene and present-day geomorphological processes. He noted the existence of relict stratified slopes deposits at altitudes between 800 and 1000 m, with contemporary periglacial processes above 2100 m, and particularly active above 2700 m. Indeed, Ulrich-Brosche (1978) also placed active solifluction at 2650–2700 m where he also observed that incipient patterned ground phenomena developed.

Bruno Messerli's Ph.D. Thesis (1965) constituted a turning point for periglacial research in Spain. This researcher highlighted the importance of both inherited and current periglacial morphogenesis—together with the impact of Quaternary glacial activity—in shaping the landscape of the Sierra Nevada summits. He also referred to the lower limit of the Pleistocene solifluction processes which he established at 700 m. He suggested that this limit was placed at 2100 m in the 1960s, with remarkable activity above 2700 m. Messerli (1965) also described the process of formation of rock glaciers, which he interpreted as cirque moraines affected by solifluction dynamics. Several authors also focused on the development of rock glaciers based on the features located in the Dílar and Poqueira valleys (Lhenaff 1977; Martín-Vivaldi Caballero 1980; Soria and Soria 1987).

In the 1970s, contemporary periglacial dynamics were explained from an integrating perspective between relief-plant-soil. The works of Soutadé and Baudière (1970) in the Guarnón, San Juan, Monachil and Dílar valleys, and Serve (1972) in the western slope of the Veleta Peak, highlighted the connection between cold-climate geomorphological processes, the scarce vegetation cover and the resulting periglacial landforms. Two decades later, Sánchez-Gómez (1990) and Simón et al. (1994) complemented this approach by examining current and past soil processes within the periglacial belt, particularly in inherited patterned ground features at the head of the Lanjarón Valley (3150 m) as well as at the lower periglacial limit at the Mulhacén Valley (2700 m).

Since the 1990s, periglacial research in the Sierra Nevada focusing both on inherited and active landforms incorporated renewed disciplinary approaches and the use of new methodologies and work techniques in parallel to substantial advances on the glacier history of the massif (Gómez Ortiz 2004) (chapter "The Impact of Glacial Development on the Landscape of the Sierra Nevada"). In this sense, over the last three decades research has shown the importance of the periglacial processes reshaping the environment that was glaciated during the last glacial cycle (Oliva et al. 2014c). Among the different recent research lines, we highlight:

- Geomorphological mapping of the glacial and periglacial landforms (Gómez-Ortiz et al. 2002; Palma et al. 2017).
- Chronology of rock glacier formation (Gómez-Ortiz et al. 2012, 2013; Palacios et al. 2016) and Holocene paleoenvironments based on solifluction records (Oliva et al. 2009, 2011, 2019; Oliva and Gómez-Ortiz 2012).
- Surface monitoring of present-day periglacial landforms and nival processes, such as patterned ground features (Gómez-Ortiz et al. 2019), solifluction lobes (Oliva et al. 2009, 2014d), minimal annual snow cover extension and annual pronival landform evolution (Gómez-Ortiz et al. 2003, 2004, 2014, 2019; Tanarro-García et al. 2010).
- Soil thermal regime at different elevations and topographic settings within the periglacial belt (Salvador-Franch et al. 2011; Oliva et al. 2014b, 2016a).
- Recent morphodynamic evolution in the high lands of the Sierra Nevada, particularly in the Mulhacén (Oliva and Gómez-Ortiz 2012; Palacios et al. 2020) and Veleta cirques (Gómez-Ortiz et al. 2004, 2014; Tanarro-García et al. 2010). Here, geomatic and geophysical studies focused on the evolution of the rock glacier formed after the Little Ice Age, which still includes buried ice and permafrost and is undergoing a process of degradation (De Sanjosé et al. 2014; Gómez-Ortiz et al. 2014; Palacios et al. 2019).

## 2 The Geographical Context

The Sierra Nevada, the largest mountainous sector of the Betic Range, constitutes also the highest massif in Southern Europe at latitude 37° N. The highest peaks of the massif are concentrated in its western fringe, where summits exceed 3300–3400 m (Mulhacén, 3479 m; Veleta, 3396 m; and Alcazaba, 3369 m; Fig. 1). The present-day landscape of the high lands of the Sierra Nevada is a result of the regional geological setting composed mainly of Paleozoic micachists (chapter "Geological Setting of Sierra Nevada") reshaped by Quaternary environmental dynamics, namely by glacial activity (chapter "The Impact of Glacial Development on the Landscape of the Sierra Nevada") and periglacial processes, which are examined in detail in this section.

The relief organizes following the distribution of deep valleys, rock ridges and remnants of ancient erosion surfaces that constitute the highest summit lands. The W-E alignment

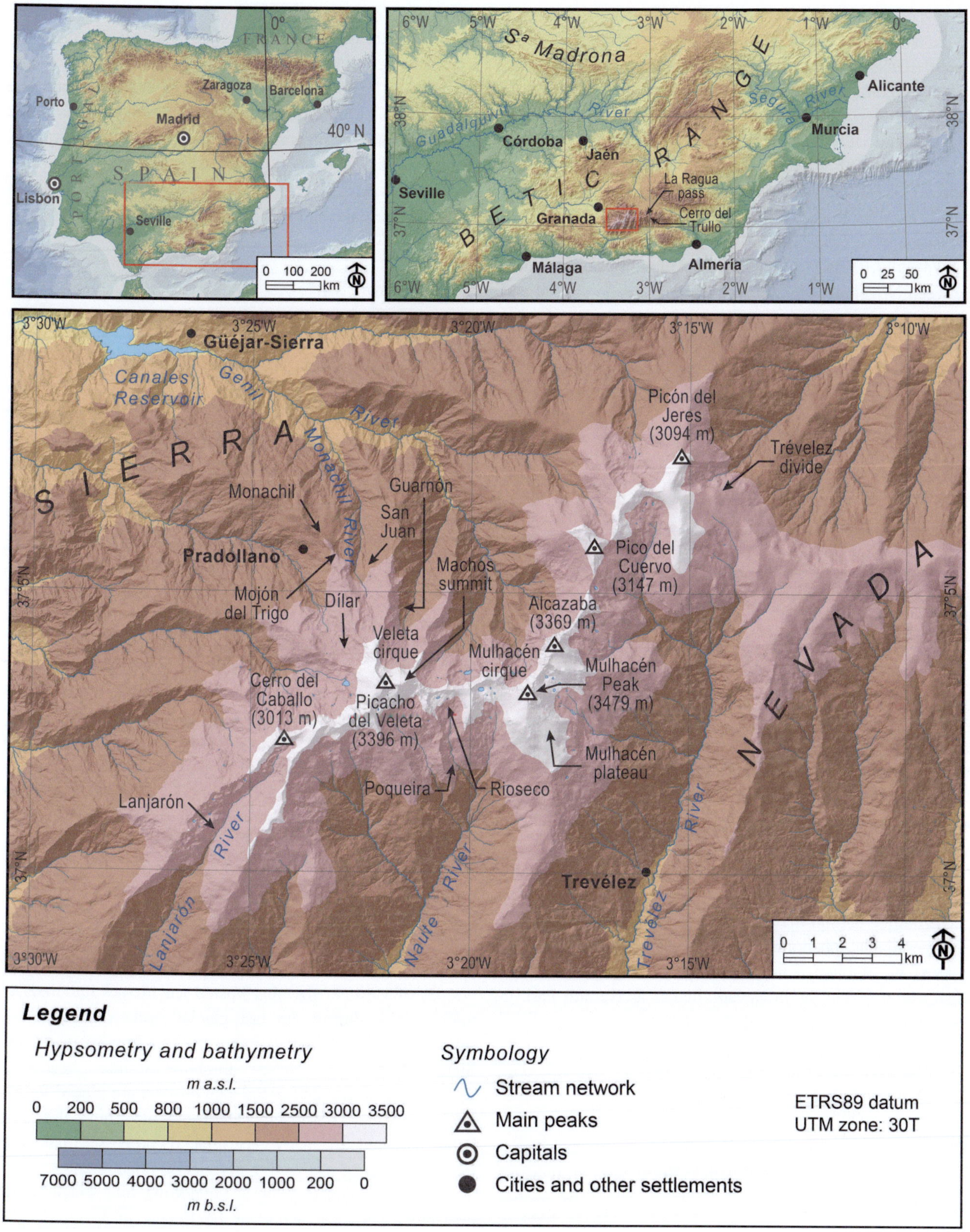


**Fig. 1** Map of Sierra Nevada with the location of the place names mentioned in the text

of Sierra Nevada conditions a radial fluvial system, with rivers descending through steep valleys from the axis of the massif. This rough relief in south Iberia, in a transitional area of climatic influence between the cool Atlantic Ocean and the warm Mediterranean Sea, determines a wide spectrum of topoclimates within the massif as well as in the neighbouring regions. At 2500 m, today, the mean annual air temperature (MAAT) is 4.4 °C, with an annual precipitation of 710 mm, mostly as snow from October to May (Oliva et al. 2014b). At the summit surfaces at 3300–3400 m, the MAAT is 0 °C (Oliva et al. 2016a), with negative monthly temperatures between November and April. The combination of cold and abundant winter precipitation favours the persistence of snow in the ground during 6–8 months per year in the highest lands, and even longer in topographically sheltered areas in the highest northern cirques (Gómez-Ortiz et al. 2019).

The present-day periglacial belt extends from ~2500 m to the highest surfaces (Gómez-Ortiz et al. 2019). The scarce vegetation in this semi-arid massif is subject to the action of cold weather, ice, wind, high solar radiation and insolation, and water availability in the soil. Cryophytes scattered across the landscape comprise the dominant plant species. Altitudes above 2500 m no longer host tree strata, and small-sized juniper trees (*Juniperus communis*, *Cytisus oromediterraneus*, *Arenaria pungens*, etc.) dominate the landscape. From 2900 m and up to the level of the peaks, the soils are stony and poorly-developed, allowing the presence of the psychoxerophilic grassland only, with abundant grasses (*Festuca clementei*, *Festuca pseudoeskia*, *Artemisia granatensis*, etc.). Hygrophilous grasslands (locally named *borreguiles*), which are used by transhumant livestock, are distributed at the headwaters of ravines near late-melting snow patches and on the bottoms of glacial valleys supplied with abundant water, particularly around lakes (chapter "Altitudinal Patterns and Changes in the Composition of High Mountain Plant Communities").

Its unique natural heritage including the variety of glacial and periglacial phenomena existing in the massif in the context of southern Europe was of key importance in the designation of the Sierra Nevada massif as Natural Park in 1989 and later as National Park in 1999.

## 3 The Periglacial Environment

The Sierra Nevada includes periglacial landforms distributed from the mountaintops to the valleys floors until altitudes of 1100–1200 m (Gómez-Ortiz et al. 2002). The broad spatial distribution across an elevation range exceeding >2000 m results from past and present climate variability. During Quaternary cold phases (glacial periods), the periglacial belt decreased in altitude and widespread permafrost conditions prevailed in non-glaciated environments above the glaciated domain and in ice-free slopes above 2500 m, whereas during warm periods (interglacial), such as the current one—the Holocene—periglacial processes shifted in altitude towards the highest lands of the massif (Oliva et al. 2014a). The alternation of cold and warm periods resulted thus in the vertical migration of periglacial dynamics, which explains the co-existence of both inactive and active periglacial landforms under present-day climate regime. These features are distributed across the high lands of the massif (Fig. 2):

### 3.1 Summit Surfaces

The relatively flat surfaces distributed above 2800–3000 m in the western part of the massif have been traditionally considered cryoplanation surfaces that divide glacial valleys (Gómez-Ortiz et al. 2002). They do not preserve geomorphic evidence of having been glaciated during the last glacial cycle, when glaciers filled the surrounding cirques and valley floors. The intense cold prevailing during that time must have been favourable for the development of meter-sized sorted-circles (Fig. 3) composed of heterogeneous blocks (long axis: 30–90) under a permafrost soil thermal regime. These patterned ground features turn into block streams as the slope gradient increases, such as it occurs across the Machos and Mulhacén plateaus at 3300–3400 m (Fig. 1). Both periglacial landforms are inherited from past colder periods; currently, they do not show evidence of activity, as revealed by the abundance of lichens on the boulder surface as well as of sparse cryo-xerophyte herbaceous vegetation in the surface. However, in some relatively flat surfaces, such as in the Chorrillo mountain divide, active smaller features develop, such as stone stripes and sorted circles composed of small gravels (long axis: 1–5 cm) (Fig. 3).

### 3.2 Glacial Cirques

The valley heads in the Sierra Nevada correspond to glacial cirques of variable size that hosted ice masses hundreds of meters thick during the last glacial cycle (chapter "The Impact of Glacial Development on the Landscape of the Sierra Nevada"). These amphitheatre-like headwalls widened by glacial and periglacial erosion are placed within the present-day periglacial environment. Thus, they are being affected by active periglacial processes but also include evidence of inactive periglacial phenomena.

Permafrost-related features—including rock glaciers and protalus lobes—are abundant landforms in formerly glaciated cirques and high valleys of the Sierra Nevada. Rock glaciers are characterized by accumulations of heterometric blocks at the foot of rock walls and display a sequence of transversal ridges and furrows, whereas protalus lobes are

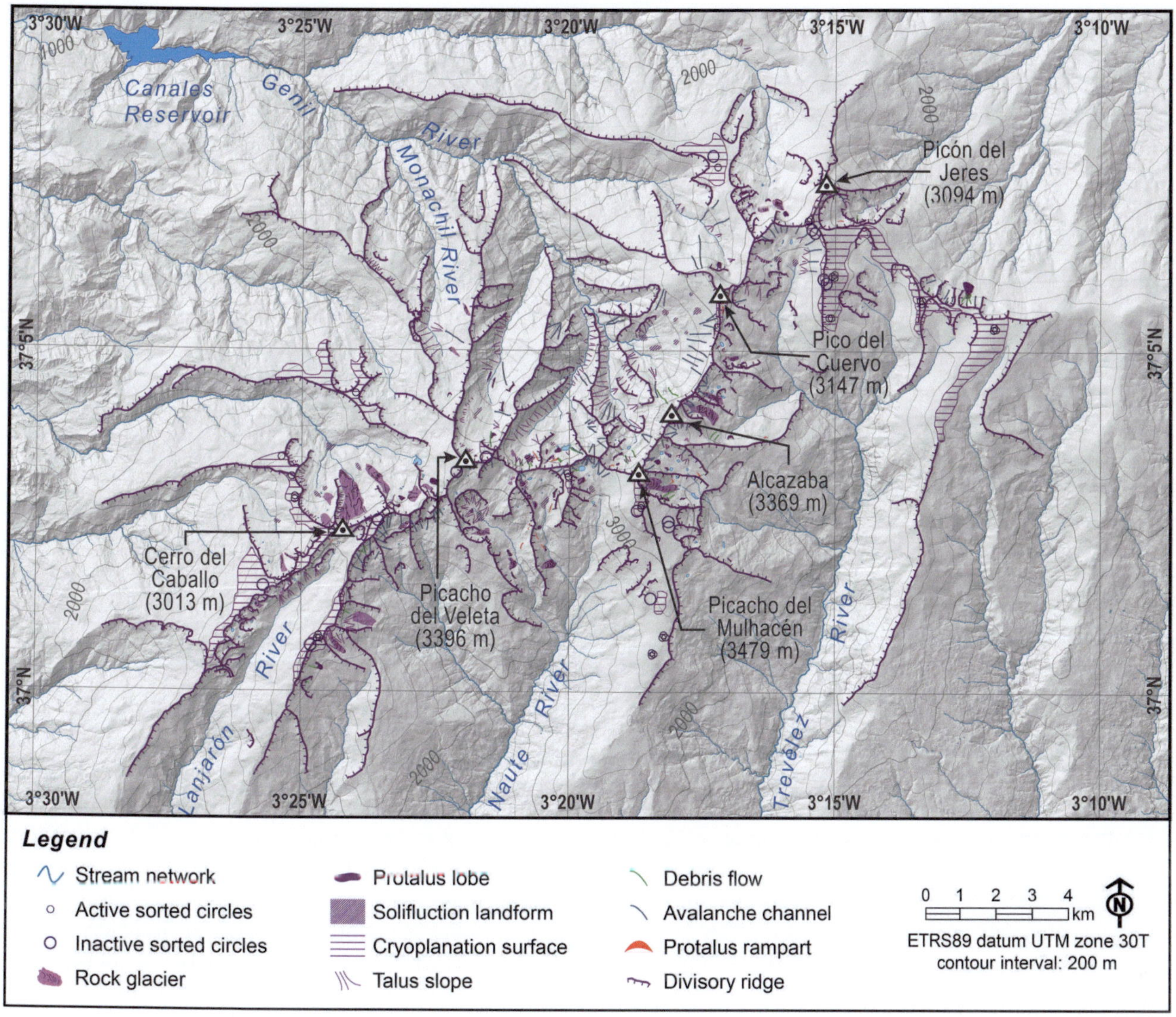


**Fig. 2** Geomorphological map with the distribution of the main periglacial landforms

generally defined by a single ridge composed of coarse material at the foot of steep slopes. A recent study has mapped Up to 50 rock glaciers were mapped in the Sierra Nevada, with only 3 of them being located outside the maximum ice extent of the last glacial cycle; their distribution, thus, suggests that they formed following ice retreat during Termination-1 (Palma et al. 2017). Rock glaciers are the best indicators of the occurrence of permafrost conditions in the Mediterranean mountains (Oliva et al. 2018b). Almost all permafrost landforms are currently inactive and only some features located in the highest northern cirques show activity, such as a rock glacier existing in the Veleta cirque (Gómez-Ortiz et al. 2019) and a protalus lobe located in the Mulhacén cirque (Serrano et al. 2018).

The cirque slopes are being currently reworked by intense geomorphological dynamics: the intensity and frequency of freezing–thawing cycles favour frost shattering processes, and periglacial landforms derived from them are the most widespread active phenomena associated with ground ice in the Sierra Nevada (talus cones, talus slopes, rock fall deposits). These unconsolidated scree slopes are also reshaped by other cryonival processes, such as snow avalanches, debris flows and mud flows. These events are particularly frequent in spring, when soil saturation due to snow melting (and/or rainy events) can mobilize sediment downslope. Protalus ramparts are also observed at the foot of long-lying snow patches in the highest parts of the massif.

## 3.3 Slopes and Valley Floors

Above 2500 m the slopes of the Sierra Nevada are scarcely vegetated, which favours sediment remobilization and frost creep. Steep slopes are generally covered by unconsolidated

**Fig. 3** Example of currently active periglacial landforms in the Sierra Nevada

debris supplied by the rock ridges defining the cirques and *arêtes* dividing the parallel U-shaped valleys. Gentle slopes and valley floors also include a wide range of solifluction landforms that reported a weakly active to inactive pattern (Oliva et al. 2009). In autumn and spring, micromorphological features associated with daily freeze–thaw cycles in the soil develop, such as those produced by pipkrake.

At elevations below 2500 m, as frost weathering becomes less intense, talus screes are gradually less abundant, with a higher proportion of fine-grained particles and less coarse sediments. This favours soil formation and the presence of a denser vegetation cover. At the lowest parts of the valleys, at elevations of only 1100–1200 m in the southern slope of the massif, stratified screes—formed probably during Quaternary glacial phases—have been observed (Gómez-Ortiz et al. 2002).

## 4 Monitoring of Present-Day Periglacial Processes

Research groups from the Universitat de Barcelona, Universidad Complutense de Madrid and Universidad de Extremadura have been monitoring the ground thermal regime and morphodynamics of several periglacial features distributed across the summits of the Sierra Nevada since the early 2000s. Data suggest that periglacial processes prevail above 2500–2600 m, and geomorphological processes are mostly driven by seasonal frost. At present, permafrost conditions are only found in the form of isolated patches in the highest northern cirques (Oliva et al. 2016a, 2018b; Gómez-Ortiz et al. 2019).

As in other mountain regions in Iberia (Palacios et al. 2003) and other mid-latitude mountain environments (Zhang 2005), ground temperatures are strongly controlled by the abundance, duration and thickness of snow in the ground, including also the calendar of the first snowfall. Years with abundant snow fall and long-lying snow cover promote lower soil temperatures, whereas years with less snow favour higher soil temperatures and a longer duration of the thawed soil season. In the semi-arid massif of the Sierra Nevada, the duration of snow cover has major geomorphological, hydrological and edaphic implications but has also a major role on ecological dynamics, as it determines the length of the vegetation growing cycle (Pepin et al. 2015).

Soil temperatures recorded by loggers installed at different depths in several periglacial landforms (Table 1) showed evidence of the occurrence of positive soil temperatures across the massif. In San Juan and Rio Seco valleys, at elevations of 2800–3000 m, mean annual temperatures range between 3 and 4 °C (Table 2), with positive temperatures extending 5–7 months depending on the snow cover (Fig. 4). The monitoring of solifluction landforms in these two valleys yielded annual displacement rates <1 cm $yr^{-1}$, slightly higher in areas with high water supply, particularly near long-lying snow patches (Oliva et al. 2009, 2014d).

**Table 1** Main characteristics of the monitoring sites, geomorphological settings and variables that are being examined in the Sierra Nevada

| Study site | Geomorphological setting | Elevation (m a.s.l.) | Monitoring variables | Dynamic control of periglacial landforms | References |
|---|---|---|---|---|---|
| Veleta Peak (PV-air) | Summit surface | 3396 | Air temperatures | – | Gómez-Ortiz et al. (2019) |
| Veleta Peak (PV) | Summit surface | 3380 | Deep borehole down to 115 m. Control of bedrock temperatures at depths of 0.2, 0.5, 1, 2.5, 4, 7.5, 10, 12.5, 15, 20 and 60 m | – | Oliva et al. (2016) |
| Machos summit (CM) | Summit surface | 3297 | Shallow borehole (0.5 m). Control of soil temperatures at depths of 0.05, 0.2 and 0.5 m | Geomorphic evidence | Gómez-Ortiz et al. (2019) |
| Veleta cirque (CV) | Glacial cirque, rock glacier | 3107 | Shallow borehole (1.5 m). Control of soil temperatures at depths of 0.05, 0.2, 0.5, 1 and 1.5 m | Several geomatics techniques | Salvador-Franch et al. (2011); Gómez-Ortiz et al. (2014, 2019) |
| Río Seco (RS) | Glacial cirque, solifluction lobe | 3005 | Shallow borehole (1 m). Control of soil temperatures at depths of 0.02, 0.1, 0.2, 0.5 and 1 m | Stake measurements | Oliva et al. (2014a, 2014b, 2009) |
| San Juan (SJ) | Glacial valley, solifluction lobe | 2817 | Shallow borehole (1 m). Control of soil temperatures at depths of 0.02, 0.1, 0.2, 0.5 and 1 m | Stake measurements | Oliva et al. (2009, 2014b) |

Average annual soil temperatures in inactive sorted-circles of the Machos mountain divide (3300 m) also recorded positive temperatures of ~2 °C at 0.5 m (Gómez-Ortiz et al. 2019). These values were very similar to those obtained in the 115 m-deep borehole drilled in the bedrock of the Veleta Peak (Table 2), where mean annual soil temperatures are of ~2.5 °C (Oliva et al. 2016a).

Permanently frozen ground conditions have been only observed in the areas that were glaciated during the Little Ice Age (LIA), such as the Veleta and Mulhacén cirques (Oliva et al. 2018b). Geomorphic evidence of permafrost conditions is revealed by the presence of active rock glaciers (Veleta cirque) and protalus lobes (Mulhacén cirque) formed during the paraglacial stage following post-LIA warming. The shrinking glacier favoured intense debris supply from the cirque rock walls that covered the stagnant glacier ice with a ~2-m thick debris layer, and part of it froze in contact with the ice body. This permafrost layer favoured the downslope movement of the surface blocks and the development of these permafrost-related landforms (Gómez-Ortiz et al. 2014; Serrano et al. 2018).

The incipient rock glacier located in the Veleta cirque has been monitored over the last two decades using different geomatic techniques (Table 1). Data loggers revealed mean temperatures of −1.0 °C at 1.5 m depth, with frozen sediments down to the glacial ice layer. The uppermost layer constitutes thus the active layer of the permafrost that freezes and thaws every year. Its variable thickness is largely controlled by the snow cover conditions (Fig. 4). The control of the surface displacement rates of the rock glacier is indicative of the state of the underlying frozen mass (Fig. 5). From 2002 to 2016, the geomatic measurements revealed a continuous subsidence or collapse that has been attributed to the gradual melting of the ice as well as the thawing of the permafrost layer (Gómez et al. 2003; Gómez-Ortiz et al. 2019).

The rate of vertical subsidence has been much greater than the rate of horizontal advance over these years (Fig. 5). The degradation of the subsurficial frozen body was also reflected in the surface of the rock glacier through the formation of multiple collapses. These observations were also validated by geophysical surveys conducted in 1999, 2009 and 2019, which revealed that the frozen layer thinned and fragmented (Figs. 6 and 7). Relict ice and permafrost degradation has been also observed in other high mountains in the Iberian Peninsula, namely in the Picos de Europa and the Pyrenees (Oliva et al. 2016b; Serrano et al. 2018).

The high altitude and topographical configuration of the Veleta Cirque are unique within the Sierra Nevada, as it is the site where currently snow remains longer in the ground. However, over the last decades, snow duration has decreased, and long-lying snow patches have even disappeared some summers (2005–2009, and 2015–2017); only two years it persisted across the cirque floor (2010 and 2013; Fig. 8). Rapid snow melting triggers frequent slumps, mudflows, debris flows, etc., in the talus slopes at the foot of the cirque wall. Recently, the application of the Structure from Motion (SfM) photogrammetry has allowed

**Table 2** Summary of the mean annual temperature values (2001–2016) at different sites and depths, including the number of freezing–thawing cycles (modified from Gómez-Ortiz et al. 2019)

| Study site | Depth (m) | Mean (°C) | FT |
|---|---|---|---|
| Veleta peak (PV-air) | | 0.5 | 108.0 |
| Veleta peak (PV) | 0.2 | 2.8 | 5.5 |
| | 0.5 | 3.2 | 2.2 |
| | 1.0 | 3.1 | 5.4 |
| | 2.5 | 2.8 | – |
| | 4.0 | 2.5 | 0.0 |
| | 7.5 | 2.5 | 0.0 |
| | 10.0 | 2.3 | 0.0 |
| | 12.5 | 2.3 | 0.0 |
| | 15.0 | 2.2 | 0.0 |
| | 20.0 | 2.4 | 0.0 |
| | 60.0 | 2.5 | 0.0 |
| Machos summit (CM) | 0.05 | 2.0 | 65.9 |
| | 0.2 | 1.8 | – |
| | 0.5 | 1.7 | 6.0 |
| Veleta cirque (CV) | 0.05 | 0.5 | 32.2 |
| | 0.2 | 0.4 | 2.8 |
| | 0.5 | 0.3 | 1.4 |
| | 1.0 | 0.1 | 2.0 |
| | 1.5 | −1.0 | 14.8 |
| Río Seco (RS) | 0.02 | 3.4 | 17.9 |
| | 0.1 | 3.3 | 5.9 |
| | 0.2 | 3.5 | 11.6 |
| | 0.5 | 3.4 | 13.0 |
| | 1.0 | 3.8 | 0.0 |
| San Juan (SJ) | 0.02 | 3.0 | 20.3 |
| | 0.1 | 3.3 | 8.1 |
| | 0.2 | 3.7 | 4.4 |
| | 0.5 | 3.4 | 2.4 |
| | 1.0 | 3.9 | 2.8 |

– No data

quantifying sediment redistribution in these areas (Fig. 9), where no permafrost was found as values ranged from 0.232 and 0.01 °C (Tanarro et al. 2018).

The experiments conducted in the high lands of the Sierra Nevada since the early 2000s conclude that present-day geomorphological dynamics above 2500 m are driven by seasonally frozen ground and permafrost is very marginal. Indeed, isolated patches of permanently frozen ground are only present in the environments glaciated during the LIA, and are out of balance with present day climate conditions. Monitoring of ground temperatures and rock glacier kinematics show a progressive volume loss of the buried ice and permafrost patches and an accelerated surface subsidence. Thus, according to the warming anticipated by international reports for the next decades, it is likely that these southernmost patches of permafrost in Europe will disappear in the next decades (Pörtner et al. 2019).

## 5 The Chronology of Inactive Periglacial Landforms

Quaternary cold phases favoured the development of large glaciers in the Sierra Nevada that descended from the high glacial cirques to elevations of ca. 2000 m. The longest alpine glaciers reached a length of 8–9 km in both the northern and southern slopes (chapter "The Impact of Glacial Development on the Landscape of the Sierra Nevada"). Glacier reconstruction models suggest the existence of several ice-free areas above the glaciated domain (Palma et al. 2017), such as summit plateaus and arêtes dividing valleys. The intense wind redistributed the snow and did not favour its accumulation and subsequent transformation into ice (Oliva et al. 2016b). The Equilibrium Line Altitude (ELA) during the maximum ice extent of the last glacial cycle was estimated at 2525 m on the northern slope and 2650 m on the southern slope (Palma et al. 2017). At the highest peaks, thus, MAATs of −4 to −6 °C are estimated for that period, confirming the occurrence of permafrost conditions in the ice-free summits. The present-day inactive meter-sized sorted circles existing in these areas must have probably developed during glacial phases (Fig. 10). Despite annual soil temperatures at 0.5 m are fairly positive (~2 °C), geophysical surveys in 1999 detected relatively high electrical resistivity values (up to 200 kohm m, Fig. 7) that clearly indicate the presence of ice below ~2 m depth. This discrepancy between warm near-surface temperatures and high electrical resistivities (indicative of massive ground ice) point to a permafrost occurrence of glacial origin that is not in equilibrium with current climate conditions. A repeated survey in 2019, albeit with larger electrode spread and therefore larger investigation depth, clearly outlines this permafrost body with a thickness of 10–15 m. Although uncertainties remain as the surveys in 1999 and 2019 were conducted with different geometries, a comparison suggests a decrease of electrical resistivities (from ~200 to 80 kohm m) and a deepening of the active layer from 1.5 m in 1999 to 2–3 m in 2019 (Fig. 7), which is indicative of the degradation of the permafrost conditions.

Some rock glaciers located in slopes that were not glaciated during the last glacial cycle (e.g. rock glacier at the foot of the Mojón del Trigo) point also to the occurrence of permafrost conditions at altitudes down to ~2500 m (Oliva et al. 2014a; Gómez-Ortiz et al. 2019). However, most rock glaciers formed in areas that were heavily glaciated during the last glacial cycle and developed during the paraglacial phase under intense slope

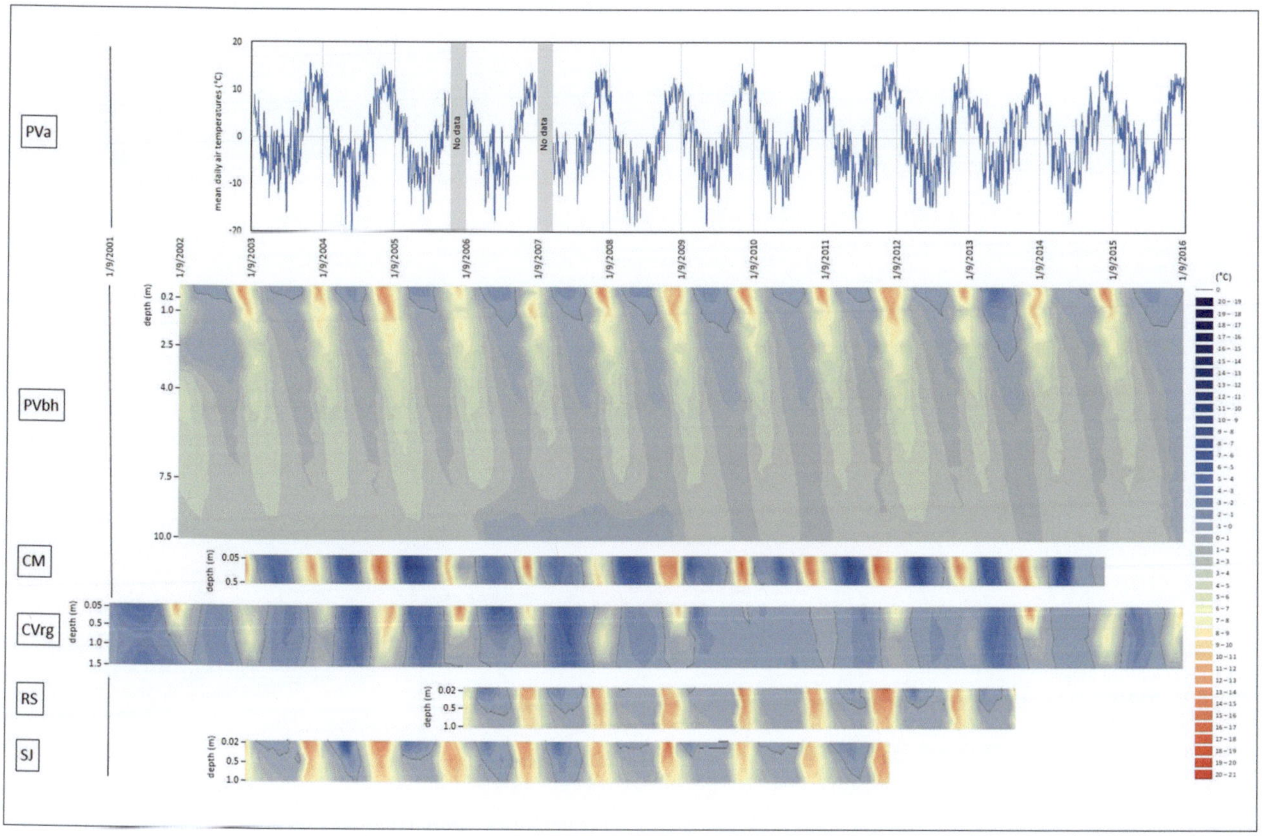


**Fig. 4** Soil temperature evolution in different periglacial landforms

**Fig. 5** Results of the kinematic control of the rock glacier located in the Veleta cirque (generated from Gómez-Ortiz et al., 2019)

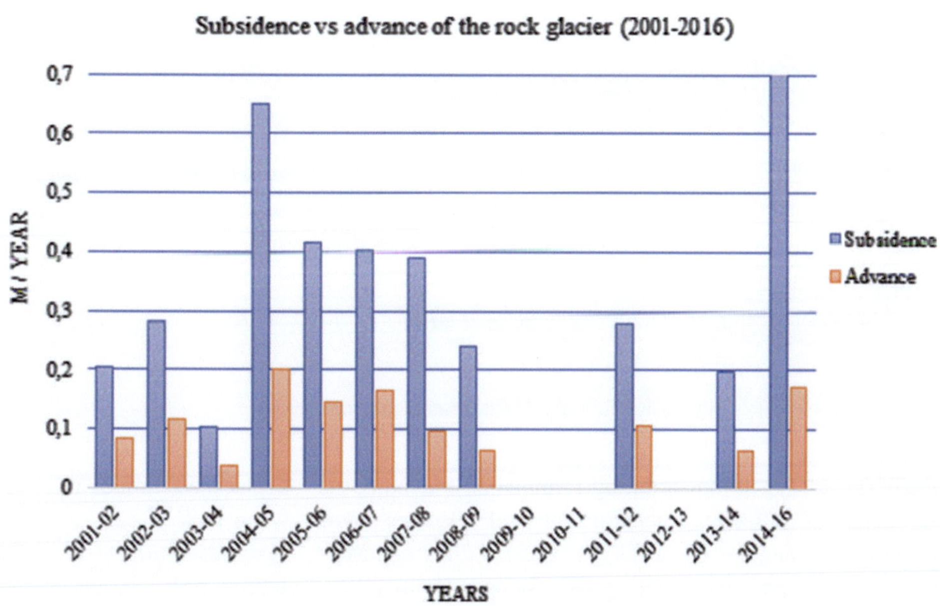

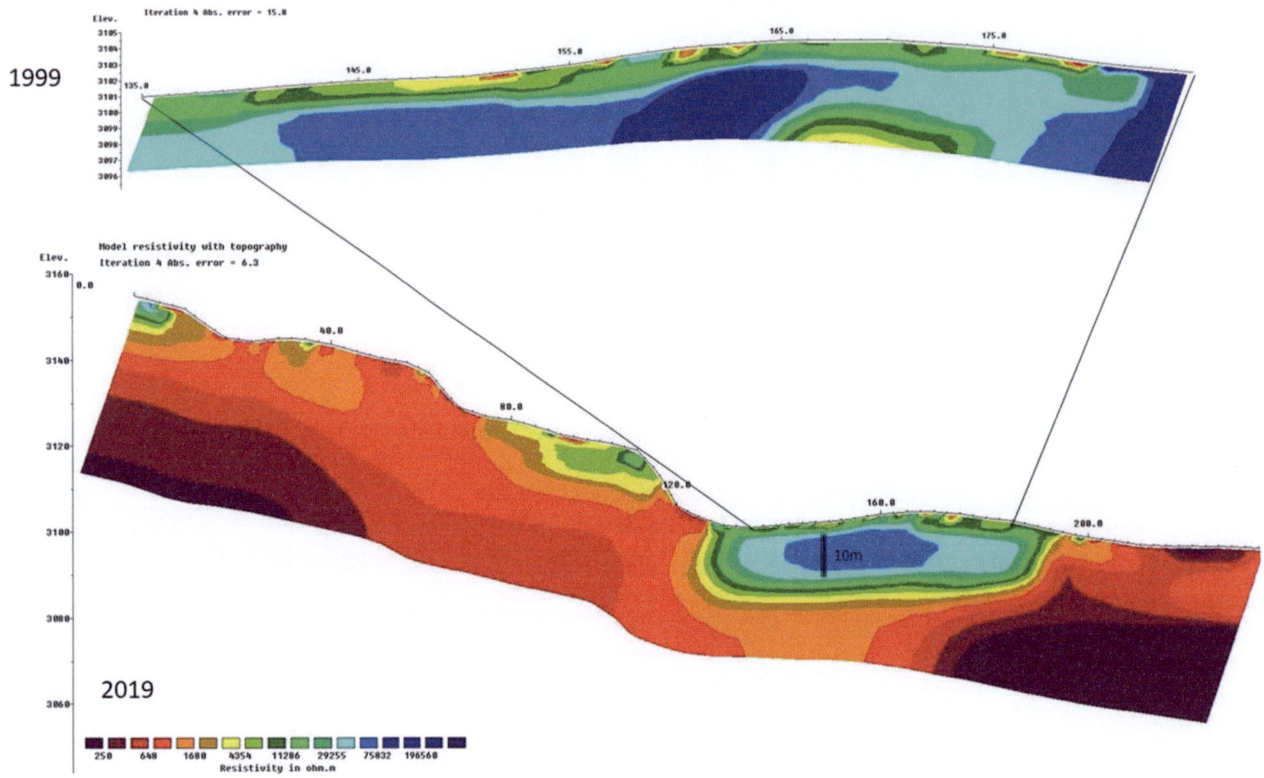


**Fig. 6** Comparison of geophysical surveys along the rock glacier existing in the Veleta cirque (1999 vs 2019)

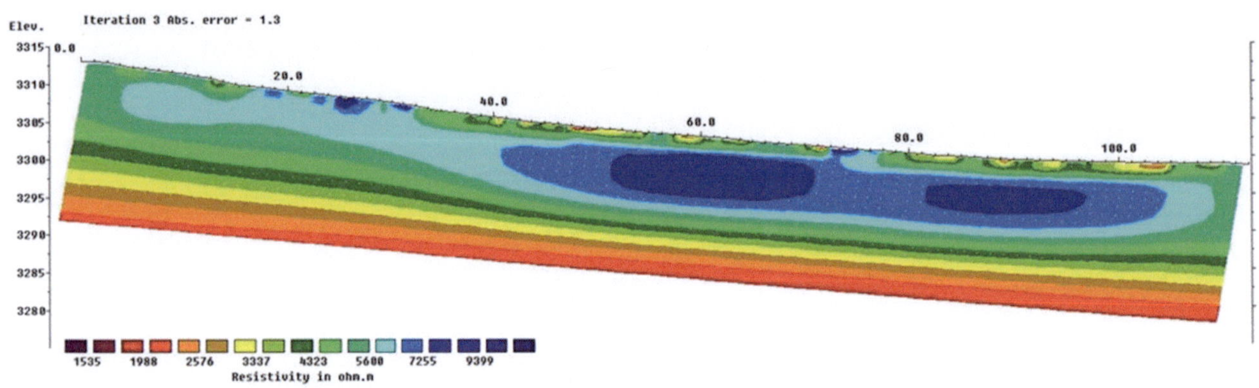


**Fig. 7** Geophysical survey conducted in the Machos summit surface (2019)

readjustment. Glacial retreat during Termination-1 favoured the formation of different generations of rock glaciers: some developed during the Oldest Dryas (17.5–14.7 ka; e.g. Mulhacén valley) and others during the Younger Dryas (12.9–11.7 ka; e.g. Lanjarón, Rio Seco and Dílar valleys) (Gómez-Ortiz et al. 2012, 2013; Palacios et al. 2016). Therefore, permafrost conditions must have occurred in recently deglaciated cirques at ~2800–3000 m during those phases when rock glaciers also formed in the Central and Eastern Pyrenees (Palacios et al. 2015, 2017; Andrés et al. 2018). Indeed, post-LIA glacial disappearance generated also intense paraglacial dynamics that led to the formation also of the rock glacier of the Veleta cirque (Gómez-Ortiz et al. 2019) and the protalus lobe in the Mulhacén cirque (Serrano et al. 2018).

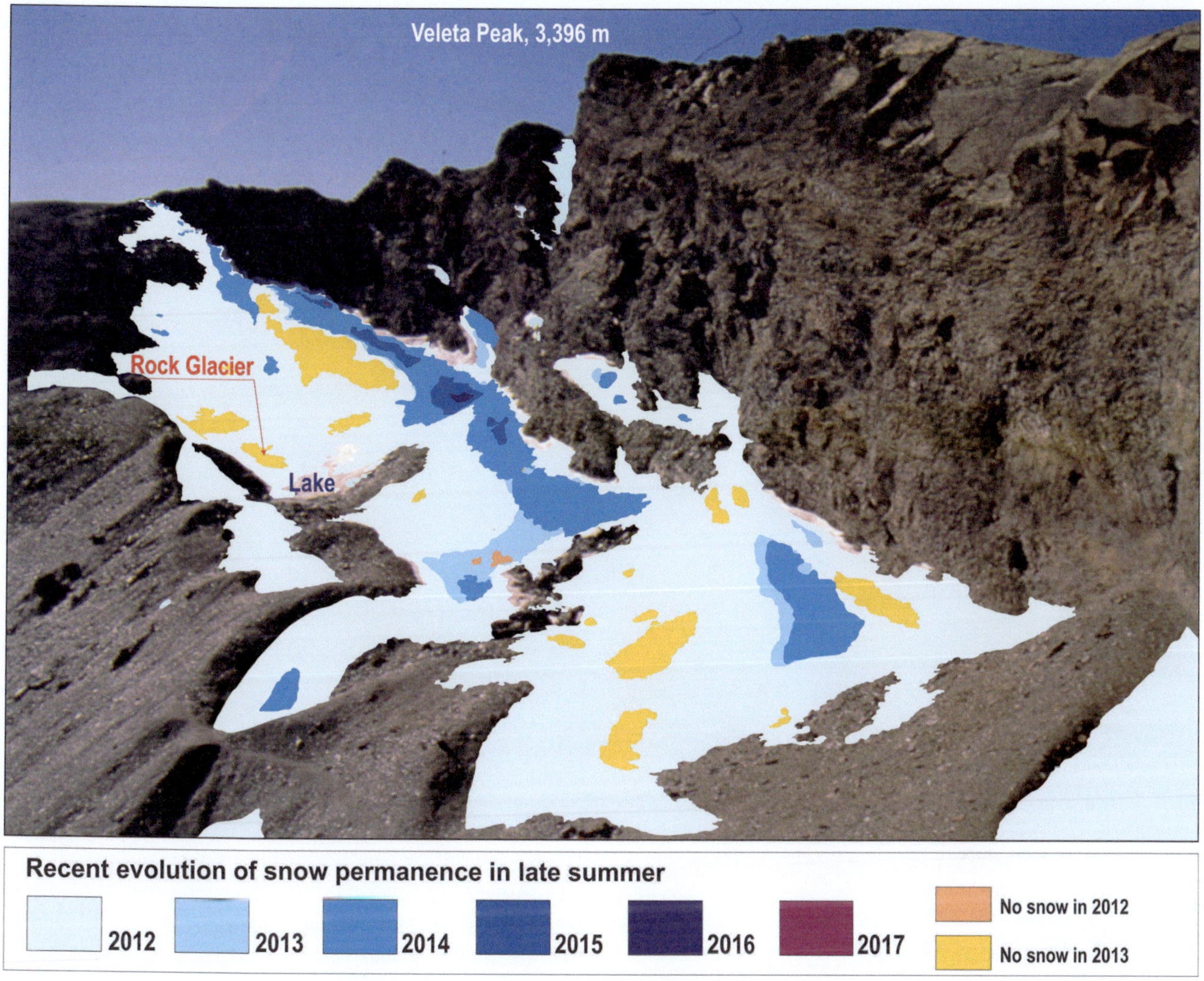


**Fig. 8** Persistence of snow in the Veleta Cirque by the end of summer (2012–2017)

Some rock glaciers remained active during the Early Holocene and stabilized at ca. 7–8 ka when higher temperatures during the Holocene Thermal Maximum favoured the definitive thawing of permafrost in most areas (Palacios et al. 2016). Present-day soil temperatures of 2–2.5 °C in depth at 3300–3400 m suggest that widespread permafrost conditions in the Sierra Nevada did not occur during the Holocene (Gómez-Ortiz et al. 2019). Despite the Late Holocene recorded phases with the formation of small glaciers in the highest northern cirques, such as during the Neoglacial period (2.8–2.7 ka cal BP), the Dark Ages (1.4–1.2 ka cal BP) or the LIA (1300–1850 CE), permafrost conditions must have been marginal and very limited to sheltered areas within the cirques. However, these phases were probably accompanied by an intensification of periglacial processes and their expansion to lower altitudes than nowadays (Oliva et al. 2016b). Post-LIA warming in the Sierra Nevada has been quantified in 1 °C and has brought a migration of colder geomorphological processes to higher elevations (~100–150 m) as well as a reduced activity of periglacial processes (Oliva and Gómez-Ortiz 2012).

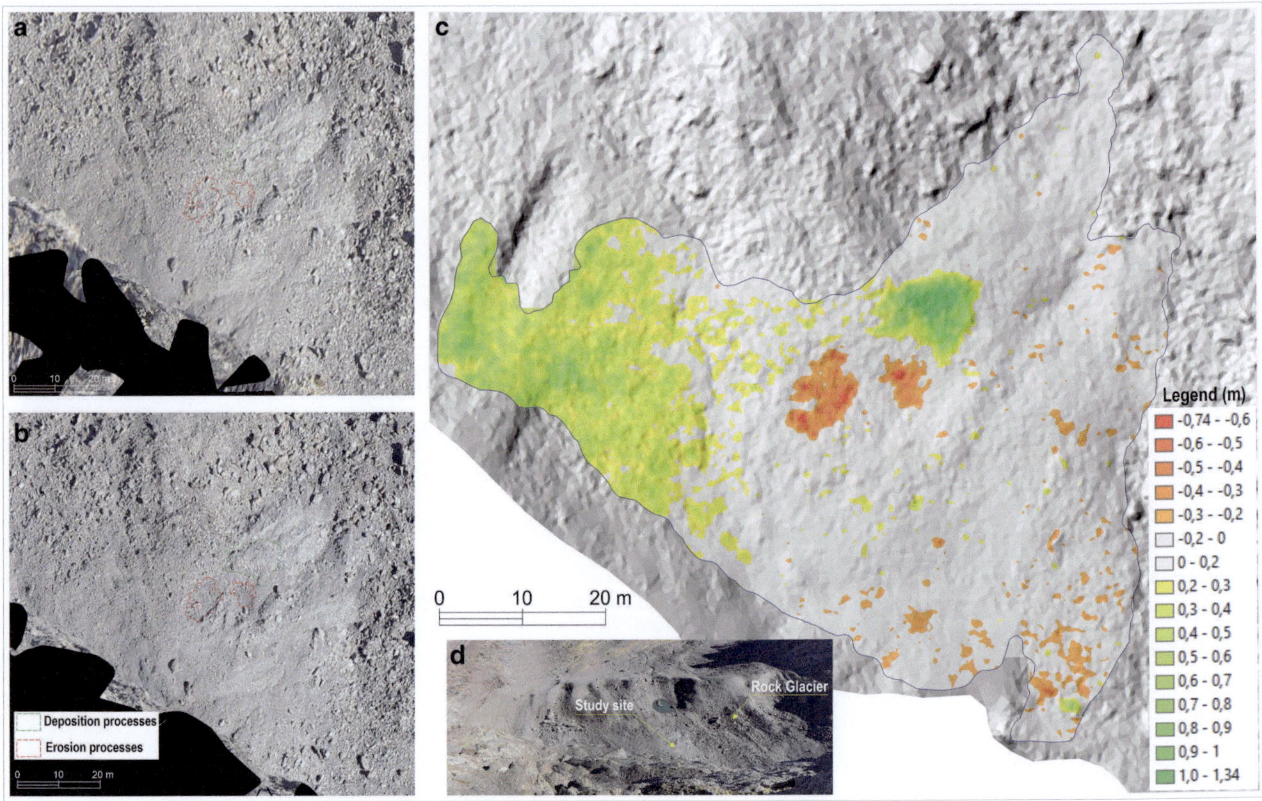


**Fig. 9** Recent geomorphological changes in the talus slope area of the Veleta Cirque based on SfM photogrammetry: **a** orthophoto, 25–09-2015; **b** orthophoto, 17–09-2016; **c** DEM showing the elevation differences between 2016 and 2015; and **d** picture taken from the Veleta Peak showing the location of the talus slope

## 6 Concluding Comments: The Impact of Periglacial Dynamics in the Landscape of the Sierra Nevada

The high lands of the Sierra Nevada show a rich diversity of glacial and periglacial landforms. Present-day climate conditions and the bareness of the ground favour active periglacial dynamics that are reshaping the landscape sculpted by Quaternary glaciers. Intense cryogenic processes above 2500 m determine the geomorphological processes and ecological dynamics prevailing in the summit areas of the massif.

Periglacial processes in the Sierra Nevada are driven by seasonal frost that shows an annual depth range between 0.5 and 2 m depth. Ground temperatures are controlled by topographic conditions as well as by the highly variable calendar, persistence and thickness of the snow cover. The long-term monitoring carried out since the early 2000s shows the absence of widespread permafrost conditions in the Sierra Nevada. MAAT at the summit level at 3300–3400 m yielded 0.5 °C, whereas subsurface bedrock temperatures at this altitude reported 2–2.5 °C and at 2800 m annual values were ~3–4 °C. Negative ground temperatures were only found in areas that were glaciated during the LIA, such as the Veleta cirque. The existence of these isolated permafrost patches is related to paraglacial activity following glacial retreat at the end of the LIA, when the shrinking glacier was covered by a thick debris mantle supplied by the debutressing rock walls. However, nowadays, the positive MAAT favour the gradual melting of the buried ice and the progressive thawing of these permafrost patches.

The singularity of these sites still hosting permafrost environments in the Sierra Nevada justifies the need for the maintenance of the long-term monitoring network of ground temperatures and kinematic control of permafrost-related landforms, which are considered geoindicators of climate change in southern Iberia. Indeed, the uniqueness of these rapidly changing geomorphological settings deserve special protection within the Sierra Nevada National Park as sites of high geomorphological significance of the recent natural history of the massif in order to ensure a proper management of these areas.

**Fig. 10** Example of inactive periglacial landforms in the Sierra Nevada

**Acknowledgements** This work was conducted under the aegis of the research group ANTALP (Antarctic, Arctic, Alpine Environments, 2017-SGR-1102) funded by the Catalan Government. Marc Oliva is supported by the Ramón y Cajal Program (RYC-2015-17597).

## References

Andrés N, Gómez-Ortiz A, Fernández-Fernández JM et al (2018) Timing of deglaciation and rock glacier origin in the southeastern Pyrenees: a review and new data. Boreas 47:1050–1071

De Sanjosé JJ, Berenguer F, Atkinson ADJ et al (2014) Geomatics techniques applied to glaciers, rock glaciers, and ice patches in Spain (1991–2012). Geogr Ann Ser A Phys Geogr 96:307–321

Dresch J (1937) De la Sierra Nevada au Grand Atlas, formes glaciaires et de nivation. Mélanges Géographie d'orientalisme Offer à EF Gautiers 194–212

García-Sainz L (1947) El clima de la España cuaternaria y los factores de su formación. Secretariado de Publicaciones Universidad de Valencia, Valencia

Gómez-Ortiz A, Oliva M, Salvador-Franch F et al (2014) Degradation of buried ice and permafrost in the Veleta cirque (Sierra Nevada, Spain) from 2006 to 2013 as a response to recent climate trends. Solid Earth 5:979–993

Gómez-Ortiz A, Oliva M, Salvador-Franch F et al (2019) Monitoring permafrost and periglacial processes in Sierra Nevada (Spain) from 2001 to 2016. Permafr Periglac Process

Gómez-Ortiz A, Palacios D, Palade B et al (2013) La evolución glaciar de Sierra Nevada y la formación de glaciares rocosos. Bol la Asoc Geogr Esp 139–162

Gómez-Ortiz A, Palacios D, Palade B et al (2012) The deglaciation of the Sierra Nevada (Southern Spain). Geomorphology 159–160:93–105

Gómez-Ortiz A, Palacios D, Ramos M et al (2004) Permafrost, evolución de formas asociadas y comportamiento térmico en el Corral del Veleta (Sierra Nevada, España). Últimos Resultados 99:47–63

Gómez-Ortiz A, Schulte L, Salvador-Franch F et al (2002) Mapa geomorfológico de Sierra Nevada. Morfología glaciar y periglaciar. Consejería de Medio Ambiente, Junta de Andalucía, Granada

Gómez A, Palacios D, Luengo E et al (2003) Talus instability in a recent deglaciation area and its relationship to buried ice and snow cover evolution (Picacho del Veleta, Sierra Nevada, Spain). Geogr Ann Ser A Phys Geogr 85:165–182

Gómez Ortiz A (2004) El conocimiento glaciar de Sierra Nevada. De la descripción ilustrada del siglo XVIII al conocimiento científico actual. Reial Acadèmia de Doctors. Barcelona

Hempel L (1960) Límites geomorfológicos altitudinales en Sierra Nevada. Estudios Geográficos 78:81–93

Lhenaff R (1977) Recherches géomorphologiques sur les Cordillères Bétiques Centro-Occidentales (Espagne). Université de Lille

Martín-Vivaldi Caballero ME (1980) Paisaje físico y la morfología del valle del río Monachil (Sierra Nevada). Cuad. Geográficos La Univ. Granada 10:117–156

Messerli B (1965) Beiträge zur Geomorphologie der Sierra Nevada (Andalusien). Juris Verlag, Zürich

Obermaier H (1916) Los glaciares cuaternarios de Sierra Nevada. Trab Del Mus Nac Ciencias Nat 17:1–68

Oliva M, Gómez-Ortiz A (2012) Late-Holocene environmental dynamics and climate variability in a Mediterranean high mountain environment (Sierra Nevada, Spain) inferred from lake sediments and historical sources. Holocene 22:915–927

Oliva M, Gómez-Ortiz A, Palacios D et al (2019) Multiproxy reconstruction of Holocene glaciers in Sierra Nevada (south Spain). Mediterr Geosci Rev

Oliva M, Gómez-Ortiz A, Salvador-Franch F et al (2016a) Inexistence of permafrost at the top of the Veleta peak (Sierra Nevada, Spain). Sci Total Environ 550:484–494

Oliva M, Gómez Ortiz A, Palacios D et al (2014a) Environmental evolution in Sierra Nevada (South Spain) since the Last Glaciation, based on multi-proxy records. Quat Int 353:195–209

Oliva M, Gómez Ortiz A, Salvador F et al (2014b) Long-term soil temperature dynamics in the Sierra Nevada, Spain. Geoderma 235–236:170–181

Oliva M, Ortiz AG, Estremera DP et al (2014c) El cuaternario en el macizo de Sierra Nevada: Evolución paleoambiental y paisaje a partir de la interpretación de registros naturales y documentos de época. Scr Nov 18:1–16

Oliva M, Ortiz AG, Franch FS, Catarineu MS (2014d) Present-day solifluction processes in the semi-arid range of Sierra Nevada (Spain). Arctic Antarct Alp Res 46:365–370

Oliva M, Ruiz-Fernández J, Barriendos M et al (2018) The Little Ice Age in Iberian mountains. Earth-Science Rev 177:175–208

Oliva M, Schulte L, Gómez-Ortiz A (2009) Morphometry and Late Holocene activity of solifluction landforms in the Sierra Nevada, southern Spain. Permafr Periglac Process 20:369–382

Oliva M, Schulte L, Ortiz AG (2011) The role of aridification in constraining the elevation range of Holocene solifluction processes and associated landforms in the periglacial belt of the Sierra Nevada (southern Spain). Earth Surf Process Landforms 36:1279–1291

Oliva M, Serrano E, Gómez-Ortiz A et al (2016b) Spatial and temporal variability of periglaciation of the Iberian Peninsula. Quat Sci Rev 137:176–199

Oliva M, Žebre M, Guglielmin M et al (2018) Permafrost conditions in the Mediterranean region since the Last Glaciation. Earth Sci Rev 185:397–436

Palacios D, de Andrés N, Luengo E (2003) Distribution and effectiveness of nivation in Mediterranean mountains: Peñalara (Spain). Geomorphology 54:157–178

Palacios D, García-Ruiz JM, Andrés N et al (2017) Deglaciation in the central Pyrenees during the Pleistocene-Holocene transition: timing and geomorphological significance. Quat Sci Rev 162:111–127

Palacios D, Gómez-Ortiz A, Alcalá-Reygosa J et al (2019) The challenging application of cosmogenic dating methods in residual glacial landforms: the case of Sierra Nevada (Spain). Geomorphology 325:103–118

Palacios D, Gómez-Ortiz A, Andrés N et al (2016) Timing and new geomorphologic evidence of the last deglaciation stages in Sierra Nevada (southern Spain). Quat Sci Rev 150:110–129

Palacios D, Gómez-Ortiz A, Andrés N et al (2015) Maximum extent of Late Pleistocene glaciers and last deglaciation of La Cerdanya mountains, Southeastern Pyrenees. Geomorphology 231:116–129

Palacios D, Oliva M, Gómez-Ortiz A et al (2020) Climate sensitivity and geomorphological response of cirque glaciers from the late glacial to the Holocene, Sierra Nevada, Spain. Quat Sci Rev 248

Palma P, Oliva M, García-Hernández C et al (2017) Spatial characterization of glacial and periglacial landforms in the highlands of Sierra Nevada (Spain). Sci Total Environ 584–585:1256–1267

Paschinger H (1957) Las formas glaciares de Sierra Nevada. España. Memorias y Comun Inst Geológico Prov 81–94

Pepin N, Bradley RS, Diaz HF et al (2015) Elevation-dependent warming in mountain regions of the world. Nat Clim Chang 5:424–430

Pörtner HO, Roberts DC, Masson-Delmotte V et al (2019) Summary for Policmakers. Intergov Panel Clim Chang

Quelle O (1908) Beiträge zur Kenntnis der spanischen Sierra Nevada. University of Friedrich-Wilhelm, Germany

Salvador-Franch F, Gómez Ortiz A, Salvà M et al (2011) Caracterización térmica de la capa activa de un glaciar rocoso en medio periglaciar de alta montaña mediterránea : El ejemplo del Corral del Veleta (Sierra Nevada, España). Cuad Investi Geográfica 37:25

Sánchez-Gómez S (1990) Aplicación del estudio de suelos a la dinámica de la cuenca del río Lanjarón: relación suelos-geomorfología. University of Granada

Serrano E, Oliva M, González-García M et al (2018) Post-little ice age paraglacial processes and landforms in the high Iberian mountains: a review. L Degrad Dev 29:4186–4208

Serve L (1972) Recherches comparatives sur quelques groupements végétaux orophiles et leurs relations avec la dinamique periglaciaire dans les Pyrénées et la Sierra Nevada. Centre Universitaire de Perpignan

Simón M, Sánchez S, Cabezas O et al (1994) Configuraciones ordenadas en la alta montaña mediterránea (Sierra Nevada. Granada). In: Gómez-Villar A, García-Ruiz JM, Arnáez-Vadillo J (eds) Geomorfología en España: III Reunión de geomorfología. Sociedad Española de Geomorfología, Logroño, pp 253–264

Soria FJ, Soria JM (1987) Depósitos de glaciares rocosos en Sierra Nevada (Granada). Acta Geológica Hispánica 21–22:123–129

Soutadé G, Baudière A (1970) Végétation et modelés des hauts versants septentrionaux de la Sierra Nevada. Ann Georgr 79:709–736

Tanarro-García LM, Palacios D, Zamorano JJ, Gómez-Ortiz A (2010) Cubierta nival, permafrost y formación de flujos superficiales en un talud detrítico de alta montaña (Corral de Veleta, Sierra Nevada, España). Cuad Investig Geogr 36:39–59

Tanarro LM, Gómez A, Salvador F et al (2018) Twenty years monitoring the degradation of buried ice and permafrost at the Veleta cirque (Sierra Nevada, Spain). In: Deline P., Bodin X, Ravanel L (eds) Book of abstracts of the 5th European conference on permafrost. International Permafrost Association, Chamonix, France

Ulrich-Brosche K (1978) Formas actuales y límites inferiores periglaciares en la Península Ibérica. Estud Geográficos 39:131–162

Zhang T (2005) Influence of the seasonal snow cover on the ground thermal regime: An overview. Rev Geophys 43

# Climate Variability and Trends

María Jesús Esteban-Parra, Matilde García-Valdecasas Ojeda, Eric Peinó-Calero, Emilio Romero-Jiménez, Patricio Yeste, Juan José Rosa-Cánovas, Alicia Rodríguez-Brito, Sonia Raquel Gámiz-Fortis, and Yolanda Castro-Díez

**Abstract**

The climate of Sierra Nevada conditions many relevant aspects for the living systems that inhabit it as well as the water resources of a region with semi-arid characteristics. Climate change in Sierra Nevada can be especially exacerbated by its mountain condition in a Mediterranean area, which makes it a double climate change hotspot. This chapter focuses on describing the main climate characteristics of Sierra Nevada as well as the recent trends and climate change projections for the near future (2020–2050) and the far future (2070–2100) for the principal climate variables. Precipitation in Sierra Nevada is characterized by marked inter and intraannual variability, a typical condition of the Mediterranean climate, and is mainly concentrated between the months from October to April, with a clear gradient west–east. This variability is mainly driven by the North Atlantic Oscillation (NAO) over the western part, meanwhile the eastern part is more dominated by Mediterranean depressions, and particularly by the Western Mediterranean Oscillation (WeMO). On the other hand, the altitude has a limited impact on rainfall distribution. The influence of altitude is clearer for both maximum and minimum temperature, being, in general, lower for minimum temperature. Both temperatures show increasing trends during the last decades, although with a more generalized spatial pattern for minimum ones. According to this increase, significant positive trends are found for extreme event indices associated with warm days as well as a marked increase of potential evapotranspiration (ET0). There is a prevailing decrease in annual and winter precipitation for the whole area, related with significant negative trends over the western part of Sierra Nevada. However, this fact does not translate into an increase in drought episodes characterized by Standardised Precipitation Index (SPI), being the enhanced drought conditions related to a more atmospheric demand. Climate projections from an ensemble of Euro-CORDEX regional models simulations depict clear warming along with drier conditions over Sierra Nevada, more intense for the far future under the RCP8.5 scenario. The results from the ensemble mean reveal reductions in evapotranspiration for most Sierra Nevada, with only moderate increases at higher altitudes in winter and spring probably related by an increase in potential evapotranspiration and increased temperature. The total soil moisture content is projected to decrease under the RCP8.5 scenario for all Sierra Nevada. Drought events are likely to become slightly longer and more frequent in the near future, over the entire Sierra Nevada, with a marked increase in duration and intensity for the far future.

**Keywords**

Sierra Nevada • Climate • Temperature • Precipitation • Potential evapotranspiration • Drought indices • Trends • Climate change projections

M. J. Esteban-Parra (✉) · M. García-Valdecasas Ojeda · E. Peinó-Calero · E. Romero-Jiménez · P. Yeste · J. J. Rosa-Cánovas · A. Rodríguez-Brito · S. R. Gámiz-Fortis · Y. Castro-Díez
Dpto Física Aplicada, Universidad de Granada, Granada, Spain
e-mail: esteban@ugr.es

M. J. Esteban-Parra · P. Yeste · J. J. Rosa-Cánovas · S. R. Gámiz-Fortis · Y. Castro-Díez
Instituto Interuniversitario de Investigación del Sistema Tierra de Andalucía, Granada, Spain

M. García-Valdecasas Ojeda
Istituto Nazionale di Oceanografia e di Geosica Sperimentale (OGS), Sgonico, Italy

Earth System Physics Section, International Centre for Theoretical Physics (ICTP), Trieste, Italy


R. Zamora and M. Oliva (eds.), *The Landscape of the Sierra Nevada*,
https://doi.org/10.1007/978-3-030-94219-9_9

## 1 Introduction

In the current context, mountain regions are particularly vulnerable to climate change, and therefore, constitute natural global observatories where the most obvious signs of global change are shown. According to the latest report of the Intergovernmental Panel on Climate Change (IPCC), global warming is affecting the mountain cryosphere, consequently altering seasonal runoff patterns (IPCC 2021). Changes in the mountain albedo can also modify temperature patterns and land–atmosphere coupling (García-Valdecasas et al. 2020a, b). In this way, hydrological cycles will gradually go from being dominated by snow and ice to being determined by rain (Kohler et al. 2014). This has important implications because mountain areas are valuable sources of water for the adjacent lowlands (Hartmann et al. 2013). Indeed, the percentage of runoff that these contribute to rivers ranges between 40 and 95%, depending on the region (Pittock et al. 2008), which in turn makes them one of the main sources of water supply in the world.

In the case of the high mountain areas of the Mediterranean, where alpine and semi-arid conditions coexist (Giorgi 2006), an increase in the spatio-temporal variability of precipitation events together with the increase in temperatures have affected both the quantity and the snowfall distribution and soil moisture cycle (New et al. 2001; Norrant and Douguédroit 2006). The impact of this climate variability on snow and, therefore, on the hydrological regime can provoke very complex ecological responses (chapter “Snow Dynamics, Hydrology, and Erosion”), which shows the special interest that these regions present to evaluate immediate decision-making processes.

The Sierra Nevada is a clear example of a high mountain area in a semi-arid context, where Mediterranean and Alpine climates coexist only 40 km apart (Polo et al. 2019). This mountainous massif, located in the southeast of the Iberian Peninsula, is the southernmost mountain area on the European continent. Approximately 80 km long in an east–west direction and 27 km wide in a north–south direction, it runs parallel to the coast of the Mediterranean Sea. With a considerable altitude, this constitutes one of the highest reliefs in Europe with a score of peaks higher than 3000 masl, and is recognized as one of the most important reservoirs of biodiversity, with the presence of a large number of endemic species (Heywood and Watson 1995).

Sierra Nevada is located into a transitional region between wet and dry climates, where soil moisture controls the changes in the partitioning of radiative energy into sensible and latent heat fluxes, leading to land–atmosphere feedbacks (García-Valdecasas Ojeda et al. 2021a, b). In these transitional regions, negative anomalies of soil moisture may exacerbate extreme events, such as drought (Quesada et al. 2012; García-Valdecasas Ojeda et al. 2021a, b) and heatwaves (Miralles et al. 2014). Hence, land water storage largely implicates the resulting surface climate, altering the temperature (Vogel et al. 2017), the boundary layer stability (Dirmeyer et al. 2013), and the subsequent precipitation (Guo et al. 2006). In addition, its particular location and the pronounced topographic gradient determine the regime of the climate elements and particularly the rainfall distribution over the mountainous group and its surroundings. The marked variability, both in time and space, of rainfall is associated with the alternative influence of air masses from subtropical and temperate origin, which modify their thermal and hygrometric conditions when they are transported over the Atlantic and Mediterranean waters or over wide territorial extensions of North Africa (Oliva and Moreno 2008).

Some studies of the influence of climate change in this area indicate both an increase in maximum and minimum temperatures, and a decrease in rainfall in practically the entire Sierra Nevada, with these trends being more pronounced at higher elevations (Zamora and Barea-Azcón 2015). These studies also coincide with the more overall estimates from General Circulation Model (GCM) projections for the Mediterranean Basin (Argüeso et al. 2012, García-Valdecasas Ojeda et al. 2021a, b; Pérez-Palazón et al. 2018). However, these trends are often not significant, particularly for precipitation, and show high seasonal variability, indicating important uncertainties in future predictions of rainfall, especially in this region.

Sierra Nevada, as a protected space and observatory of global change, is a privileged region for the observation of global processes, where its ecosystems act as sensors for the early detection of signs of change throughout the planet and mainly in the Iberian Peninsula. Taking these considerations into account, the main objective of this work is to analyse the spatio-temporal climate variability in Sierra Nevada and its surrounding area and to present the main climate projections under different climate change scenarios.

## 2 Observational Data

Although it is important to consider the complex climatic processes in the mountains to understand the climatic trends that affect larger areas, in many mountainous regions the monitoring of climatic variables such as temperature and precipitation remains insufficient (Kohler et al. 2014). On the other hand, most of the meteorological stations in the mountainous regions are located in lower areas with easy access, which means that the information related to the peaks is not adequately represented (Alonso-González et al. 2018).

In recent years, attempts have been made in order to solve this problem in Sierra Nevada, with the inclusion of new sensors and stations at higher altitude areas. However, their records do not cover a period long enough to be considered by themselves in climate variability studies, so they must be complemented by the rest of the observatories that have been used for monitoring the region for a longer time and covering a wider spatial distribution. Another added problem is the dispersion of this information, since it is managed by several networks and projects. Therefore, it is necessary to dedicate efforts to the construction of an integrated database with climatological information of the area that meets the minimum requirements of spatial density and temporal length of the records.

As a response to this problem, recently a database composed of all climatic information has been compiled as an initiative of the Observatory of Global Change of Sierra Nevada, called *ClimaNevada*. It is formed by a total of 295 observatories from 17 observational networks and records of 128 different meteorological variables. Limited to a *buffer* of 15 km around the Sierra Nevada Natural Park, the database is the result of the integration of meteorological information from the region in a PostgreSQL database system. All this information is currently accessible in https://climanevada.obsnev.es/. More details about this can be also found in Pérez-Luque et al. (2021).

The total network has an appreciable number of observatories in the area under study. Shortcomings related to the decrease in the number of stations located at higher altitudes have been alleviated with the inclusion of new automatic sensors in recent years, although the time coverage at high altitudes is still low for performing a climatic evaluation. Precipitation is the variable which presents the highest number of records, with 203 stations from the total containing precipitation data with a variable measurement frequency (minute, hourly, daily and monthly). Much more limited is the number of stations with temperature records, with only 77 locations with thermometric records.

Quality control and homogenization methods were applied in order to avoid non-climatic alterations affecting time series (Aguilar and Llanso 2003) used mainly in studies of climatic trends and climate change. Particularly, for temperature and precipitation data from ClimaNevada, Climatol package (Guijarro 2018) has allowed to detect spurious data, change points as well as the missing data filling and the correction of some inhomogeneities for both variables. Moreover, for temperature, ACMANT software (Domonkos 2014) has been also used to check the data quality. As a result, after this quality assessment, 137 precipitation time series with more than 30 years of data and 39 temperature series with more than 50 years during the last decades have been obtained. The time length of these series varies, reaching more than 80 years in some locations for precipitation and 70 years for temperature.

Beside quality data, the need for accurate and high-resolution descriptions of the spatial distribution of precipitation, with a marked spatial heterogeneity, leads to the construction of a grid for this variable. This is particularly relevant for those mountain regions such as Sierra Nevada, where the topographic complexity and marked variations generate a great diversity of microenvironments, which deserve to be considered (Kumari et al. 2017; Meersmans et al. 2016; Zamora et al. 2021).

In this work, the R package RegRAIN version 0.1.0 (Alzate Velásquez et al. 2017), based on the regionalized rainfall model Regnie (Rauthe et al. 2013), has been used to interpolate the precipitation. It is essentially based on the use of point data of climatic variables such as precipitation or temperature, with their respective geographical location and physiographic information of these locations (elevation, slope and terrain exposure), obtained from a digital elevation model (DEM) of the study region, to calculate quotients with a multiple linear regression. In order to perform an interpolation with better results, the entry of the largest number of stations available is desirable. For this reason, the interpolation was restricted to the years 1990–2019. In addition to the selected stations, 39 more stations with homogeneous records of at least 5 years in this period were used in the interpolation scheme.

The interpolation error evaluation was performed by a cross-validation procedure following a leave one out scheme (Wilks 2011), with good results in terms of correlation and mean square error of the estimation. Finally, the result of implementing the RegRAIN algorithm was a set of rasters in tif format for each month during the period 1990–2019 with the interpolated values in the area under study and the resolution of the input DEM, in this case 400 m $\times$ 400 m. This resolution was considered optimal, taking into account the minimum distance between stations, to avoid that two of them coincide in the same cell.

The main source of these primary climate variables is the National Meteorology Agency (AEMet). These data are included in different climate gridded datasets at different spatial resolutions covering Spain. Considering the size of the area under study, the most reliable gridded data for completing this study are those with a higher resolution, namely SPREAD (Serrano-Notivoli et al. 2017) for precipitation and STEAD (Serrano-Notivoli et al. 2019) for maximum and minimum temperature. Both databases have a 5 km resolution at daily scales. In the frame of CLICES project, Peña-Angulo et al. (2016) have produced a climatology for both maximum and minimum temperature with a resolution of 1 km. Regarding other interesting variables related with water cycle, Tomas-Bruguera et al. (2019) have performed a 1.1 km gridded resolution database for potential evapotranspiration from Penman–Monteith equation. Based on this last data, Vicente-Serrano et al. (2017) has computed several drought indices at the same resolution. Among these, the Standardized Precipitation-Evapotranspiration (SPEI) and the Standardized Precipitation (SPI) indices at 3- and 12-month time scales are also included in this study for Sierra Nevada. These datasets have been used to complete the different analyses performed with ClimaNevada temperature and precipitation database. The time coverage of these datasets differs although they share the period from 1961 to almost present. On the other hand, trends computed with these data are quite similar to those calculated with direct observational series from ClimaNevada database. For this reason, the spatial trend analysis is based on these gridded data meanwhile global and more recent trends are presented using precipitation and temperature data from ClimaNevada.

## 3 Main Climate Characteristics

This section is devoted to describing the climatic conditions on Sierra Nevada. Figure 1 shows the annual climatology of the primary variables, maximum and minimum temperature from Peña-Angulo et al. (2016), precipitation from ClimaNevada interpolated data at 400 m resolution and potential evapotranspiration from Tomás-Bruguera et al. (2019).

The spatial distribution of average annual precipitation in Sierra Nevada and its surroundings, for the period 1990–2019, using the interpolated data from ClimaNevada (Fig. 1) shows a marked east–west gradient that is observed during the whole year, with the highest accumulated precipitation values towards the south and north-west region and minor ones in the eastern area, standing out a minimum to the northeast of the mountainous system. This east–west gradient is observed during the whole year, and only in the summer period a change in the pattern is identified, when this gradient weakens and practically the entire region suffers very low precipitation values (Oliva and Moreno 2008).

Rainfall increases from east to west, and approximately 80% of it occurs between the months from October to April. The higher altitudes of the western half and the orientation of the massif itself in relation to the predominant atmospheric circulation over the west of the study area determine higher accumulated precipitation on its western flank (Pereira and Oliva 2010). Precipitation in this western area is also highly correlated with the North Atlantic Oscillation (NAO) index, while the eastern half of the Sierra has a more Mediterranean influence, with rainfall more correlated with the Western Mediterranean Oscillation (WeMO, Hidalgo-Muñoz et al. 2015). This feature is related to the elongated shape, from west to east, of the massif and its complex orography in the south of the Iberian Peninsula, which impose a redistribution

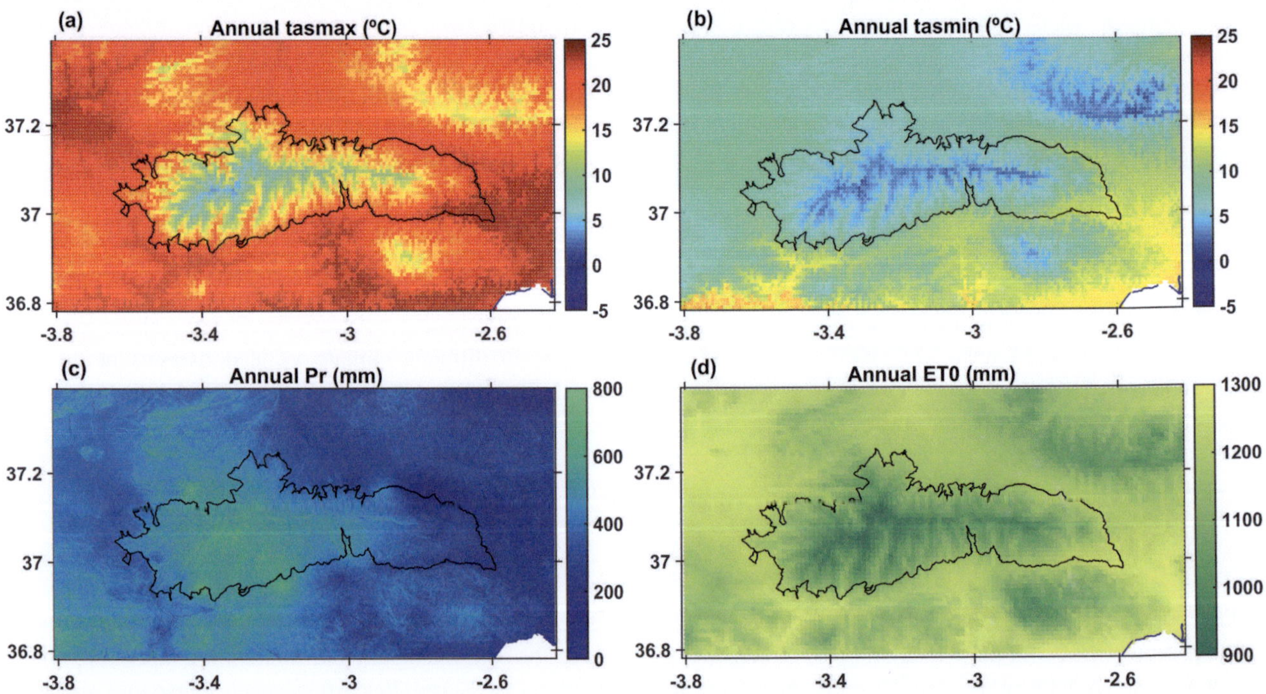


**Fig. 1** Spatial distribution of annual maximum **a** and minimum **b** temperature from Peña-Angulo et al. (2016) climatology between the years 1950–2010, precipitation **c** from interpolated ClimaNevada data from 1990 to 2019 and potential evapotranspiration **d** from Tomás-Burguera et al. (2019) averaged in the period 1961–2014

of the regional wind circulation and consolidate the role played by Sierra Nevada as a climatic frontier. On the other hand, the eastern sector, with less abrupt relief and more moderate altitudes, has a rainfall behaviour that responds more to variations and advections from the Mediterranean Sea.

The maximum annual accumulated averages occur in stations located indistinctly on both sides of the northern and southern slopes of the massif, but towards the western zone. The annual precipitation values in these regions, which range between 400 and 600 mm, greater at higher elevations, contrast with accumulated values that do not exceed 200 mm in areas of the eastern region.

The spatial distribution of rainfall at seasonal and monthly scale (not shown) reveals the usual inter-annual rainfall variability observed in most Andalusia. As a typical characteristic of areas with Mediterranean influence, it highlights the lack of rainfall during the summer months, when precipitation mean values do not exceed 40 mm, being even below 10 mm in many areas.

On the other hand, the winter period constitutes the rainiest season of the year, mainly in the western zone, with the greatest contributions to the annual amount. During this season, in the Alpujarra region, close to the massif, and in higher areas, the largest accumulations are generated, close to 300 mm, which represent almost 50% of the average annual rainfall. The influence of frontal systems from the Atlantic is typical during these months, which together with the orographic conditions constitute the main mechanisms for the generation of rainfall, hence the east–west gradient is further accentuated. The autumn and spring seasons present similarities in terms of the contribution of precipitation. However, there are slight differences between the western and eastern areas. For example, the behaviour to the east of the Natural Park is mainly due to the influence of Mediterranean mechanisms, which is the reason for a secondary autumn rainfall maximum. On the other hand, for areas further west, the secondary maximum occurs towards the spring months, although with not very significant differences. The mean accumulated precipitation values during these seasons range between 40 and 230 mm, highlighting the months of March and November, respectively, as the months with the highest contributions.

Table 1 presents the values of precipitation gradient with altitude, computed as the slope of the linear fitting of precipitation versus altitude. For all the months, these altitudinal gradients are positive and low. Only for November and December, they are significant at the 90% confidence level with values below 2 mm/km, in agreement with the results of Pereira and Oliva (2010), who evaluated the altitudinal gradient of extreme precipitation in Sierra Nevada. Although in mountain areas, precipitation tends to show a positive correlation with altitude (Celleri et al. 2007), the results presented here for Sierra Nevada suggest that the orography has different effects on precipitation and that its distribution depends more on the synoptic systems and general climate conditions.

Compared to other climatologies performed using other databases such as SPREAD, these last databases show higher values of precipitation at higher altitudes. However, the scarce precipitation data above 2500 m included in ClimaNevada are not in agreement with this idea, showing accumulated annual precipitation values more in line with those shown in Fig. 1.

Regarding maximum and minimum temperatures, Fig. 1a and b shows the climatology with a resolution of 1 km

**Table 1** Altitudinal changes in precipitation (mm/km) and temperature (tasmin, tasmax, tasmean, °C/km) computed as the slope of the linear regression fitting precipitation and temperatures with altitude. Bold numbers and * indicate significant values at 95% and 90% confidence levels, respectively

| Month | Temperature (°C/km) | | | Precip (mm/km) |
|---|---|---|---|---|
| | tasmax | tasmin | tasmean | |
| Jan | **−4.97** | **−3.80** | **−4.39** | 1.04 |
| Feb | **−5.31** | **−3.95** | **−4.63** | 1.05 |
| Mar | **−5.59** | **−3.87** | **−4.73** | 1.04 |
| Apr | **−5.75** | **−3.89** | **−4.82** | 0.74 |
| May | **−5.74** | **−3.65** | **−4.70** | 0.44 |
| Jun | **−5.68** | **−3.34** | **−4.51** | 0.16 |
| Jul | **−5.42** | **−2.87** | **−4.14** | 0.09 |
| Aug | **−5.42** | **−2.92** | **−4.17** | 0.14 |
| Sep | **−5.48** | **−3.24** | **−4.36** | 0.59 |
| Oct | **−5.40** | **−3.59** | **−4.49** | 1.43 |
| Nov | **−5.25** | **−3.71** | **−4.48** | 1.59* |
| Dec | **−5.00** | **−3.73** | **−4.37** | 1.91* |

obtained by Peña-Angulo et al. (2016) for the period 1951–2010. As was expected, there is a decrease in temperature with height for both the maximum (tasmax) and minimum (tasmin) temperatures. Thus, the maximum temperature ranges from an average value of around 25 °C in much of the surrounding areas and at the eastern end of Sierra Nevada, to values around 0 °C in the highest areas located to the west, while a large part of it has temperatures close to or above 12 °C. The minimum temperature shows a more uniform pattern throughout the Park, with annual mean values around 0 °C along the west–east transversal axis, reaching again the highest minimum temperatures to the East. The minimum temperature tends to be also higher towards the west and the south of Sierra Nevada, except in the southernmost areas of the eastern Alpujarra region.

At seasonal and monthly scales (results not shown), there is a marked intra-annual variability of tasmax. The lower mean values (around 0 °C) cover an ample area of the Sierra Nevada during February, which lasts until April for altitudes above 2000 m, meanwhile during the summer, most of the area presents values above 20 °C and only the highest elevations at the western part present mean maximum temperatures around 15 °C. Sierra Nevada shows the lower seasonal tasmin during January and February, with values below 0 °C, except the eastern zone and Alpujarras. During summer the coldest temperatures are located along the zonal axis, with values close to 12–14 °C. It is also worth mentioning the longer duration of the winter-summer transition compared to autumn, more markedly for maximum temperatures. Thus, while relatively low temperature values are observed until the first weeks of May, the autumn decline seems faster, with just October as the transition month.

Table 1 also presents the values of the altitudinal monthly temperature gradients estimated from the slope of the regression line of temperature versus height, calculated with the regional average temperature for the different series included in ClimaNevada. In all cases, these gradients are significant at the 95% confidence level and lower than the proposed average value for the vertical temperature gradient, −6.5 °C/km. For mean temperature, the altitudinal gradients computed with ClimaNevada varies from −4.14 to −4.82 °C/km, in agreement with the results from Navarro-Serrano et al. (2018), who find an altitudinal slope of around −4.8 °C/km for the region under study. In general, there are greater altitudinal gradients for maximum temperatures, which intensify or weaken in summer depending on the database used. The gradients also show a seasonal variation, being generally more moderate in summer, especially for minimum temperatures. As was suggested by Navarro-Serrano et al. (2018), these weakest altitudinal changes during summer can be associated with a greater presence of anticyclonic conditions that favour thermal inversions and therefore a weakening of the altitudinal gradients, although with a lesser extent than what occurs in other mountainous areas of Spain, probably due to the greater proximity to the sea.

Taking into account the location of Sierra Nevada in the Mediterranean area and within a semi-arid climatic regime, a particularly interesting variable is the potential evapotranspiration (ET0), crucial for an adequate representation of the hydrological cycle. Figure 1d shows the annual mean values of potential evapotranspiration for the period 1961–2014 obtained from the database of Tomás-Bruguera et al. (2019). As can be seen, the spatial variability pattern of evapotranspiration is closely linked to height, in such a way that the highest areas present lower ET0 values, particularly in the western part of Sierra Nevada where annual accumulations reach 900 mm. The ET0 progressively increases from these high elevations along the hillsides to very high values in the surrounding lowlands, as can be seen in the most north-western part of Sierra Nevada or in the southeastern where the ET0 exceeds 1250 mm per year. This pattern is essentially associated with the temperature spatial variation. Seasonally (results not shown), winter is the season of the year with more moderate potential evapotranspiration values, which do not exceed 70 mm in the highest areas of Sierra Nevada. During the summer, the highest ET0 values are found, although still moderately low in the elevated areas, where ET0 reaches values around 450 mm meanwhile the area to the northwest of Sierra Nevada presents very high values, above 650 mm.

## 4 Atmospheric Circulation Drivers

As was mentioned above, the North Atlantic Oscillation (NAO) has an important role to explain the precipitation regime over the Sierra Nevada area, particularly for the western part. It is well-known how this teleconnection pattern affects precipitation over the Iberian Peninsula, being its main driver, mainly for winter months (Esteban-Parra et al. 1998). During the positive phase of the NAO, prevailing high-pressure conditions over the Iberian Peninsula occur, with blocking situations, and then, low precipitation, meanwhile during the negative phase, the weakness of subtropical Atlantic High favours the presence of Atlantic weather systems and storm tracks over the Iberian Peninsula. This situation determines that the Atlantic flows more directly affect the western margin of the massif, being the high peaks and especially the areas facing the humid synoptic systems coming from the west, those that receive greater amounts of precipitation, fundamentally during the cold semester of the year. On the other hand, Sierra Nevada acts as a natural barrier preventing the passage of cold fronts from the Atlantic Ocean to the south-east, and limiting the influence area of the Mediterranean depressions to the

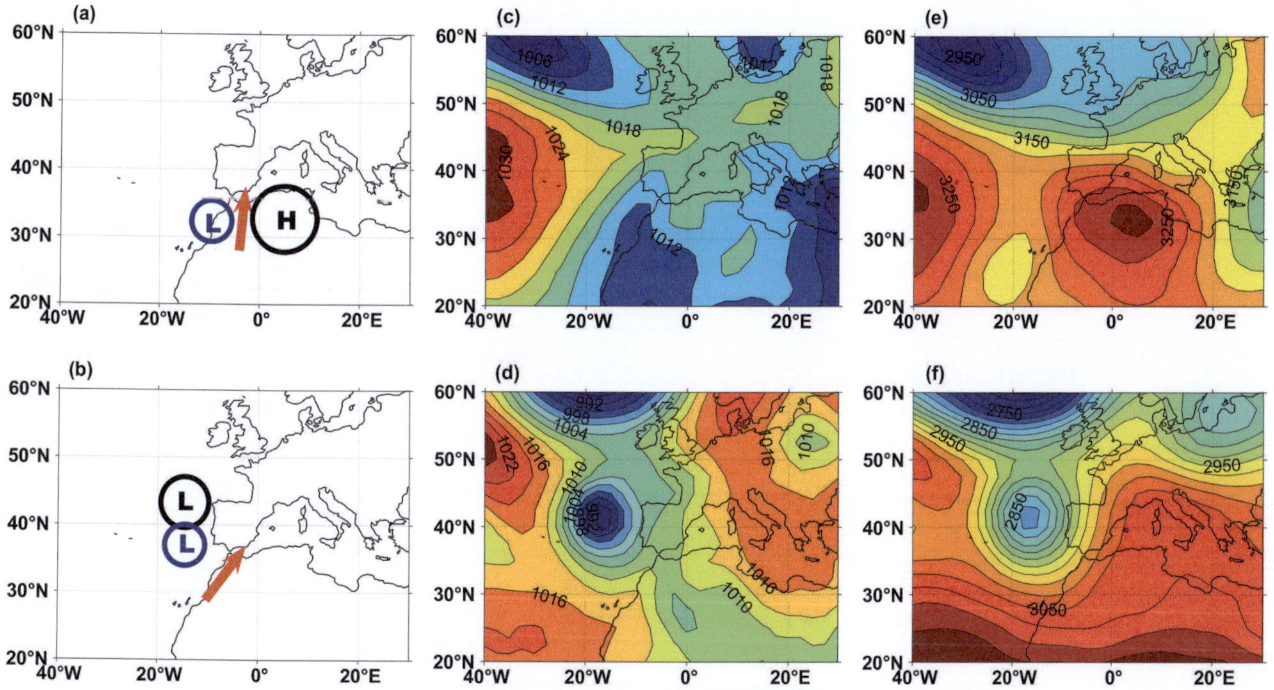


**Fig. 2** Main atmospheric circulation during the African dust incursions over Sierra Nevada: **a** for the North African high in upper-level pattern (NAH-A); **b** for Atlantic depression (AD). Blue circles indicate the approximate location of low pressure centers in the surface and black circles indicate the position of high (H) and low (L) at upper levels

south-east, with a significant impact of the Western Mediterranean Oscillation (WeMO, Martín-Vide and Lopez-Bustins 2006). This pattern describes the pressure gradients between the Cadiz Gulf and North of Italy. During the negative phase of this pattern, strong cyclogenetic activity is presented over the Mediterranean area, with positive precipitation anomalies over the Mediterranean fringe. On the other hand, the positive phase has been shown to trigger air masses from the Atlantic to move into the Iberian Peninsula, while its negative phase is associated with flows from the Mediterranean (Martín-Vide and Lopez-Bustins 2006; Lopez-Bustins et al. 2008). Both patterns, NAO and WeMO are also linked with the main synoptic conditions responsible for the extreme precipitation events in Andalusia (Hidalgo-Muñoz et al. 2011).

Another important characteristic relative with atmospheric circulation is regarding Saharan dust intrusions affecting atmospheric radiation balance as well as several biochemical aspects, being a major climate driver of long-range transported particulate matter over Sierra Nevada, whose deposition adds to that from local sources (Morales-Baquero et al. 2006; Pulido-Villena et al. 2008, chapter "Atmospheric Inputs and Biogeochemical Consequences in High-Mountain Lakes"). These events usually are associated with increased temperature and can occur during all the year with a duration between 1 to 10 days, being particularly more frequent during late spring and summer months on the south of the Iberian Peninsula (Russo et al. 2020). The main synoptic features of these intrusions towards southeastern Iberia are the presence of a thermal surface low pressure center over northwestern Africa with the presence of the high geopotential anomalies at upper levels over North Africa (NAH-H), leading the dust transport in the mid-troposphere (Escudero et al. 2005). This pattern is prevailing during the warm months. Another pattern associated with dust transport and more presents during cold months is characterized by an Atlantic depression (AD), both in surface and at higher levels, with a trough over the Mediterranean with southwesterly flow bringing dust from northwestern Africa to Sierra Nevada. Figure 2 shows a scheme of these patterns along with a synoptic map of an episode.

## 5 Recent Precipitation and Temperature Evolution

In order to evaluate recent changes in annual and seasonal temperature, a trend analysis has been carried out with the Sen slope estimator. The significance of the trends has been evaluated by the Mann–Kendall test. Global average temperature trends for Sierra Nevada and surrounding areas have been computed both using temperature series from Clima-Nevada and STEAD database, obtaining similar results. Therefore, spatial distributions of the trends have been performed from STEAD for the period 1961–2015, meanwhile

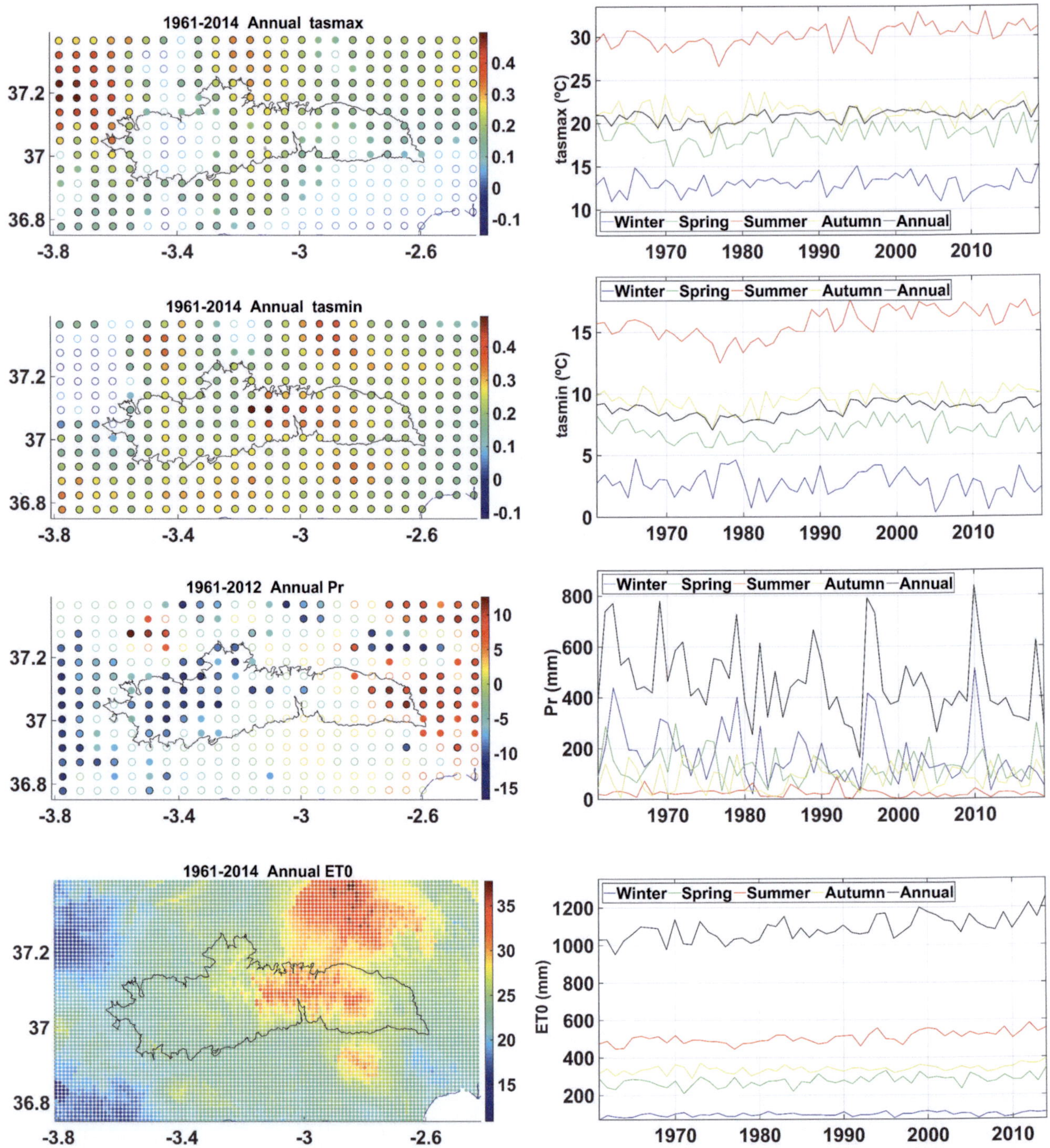


**Fig. 3** Trends in annual and seasonal temperature (tasmax, tasmin), precipitation (pr) and potential evapotranspiration from 1961

regional average trends are presented from ClimaNevada mean series during the period 1961–2019. Figure 3 shows these trends for both tasmax and tasmin, respectively.

As can be seen in Fig. 3, the area is dominated by increasing trends, although the spatial pattern of warming differs between the maximum and minimum temperatures. Warming is more generalized for tasmin, with significant trends at 95% confidence level in almost the entire area, increases between 0.15 and 0.25 °C/decade and with the highest growth rates in the central zone of the Sierra Nevada. On the contrary, the greatest warming of tasmax occurs to the northwest of Sierra Nevada in the area surrounding the city of Granada, with trends being also significant to the southwest, in the western and southwestern area of La Alpujarra, meanwhile a large extension of the western zone and to the east of La Ragua do not present significant tendencies.

However, globally, annual increasing rates are slightly higher for maximum temperatures than for minimum temperatures, the increase in both temperatures being particularly evident from the second half of the 70s (Fig. 3). Seasonally, the greatest increases are observed during the summer with very significant trends above 0.3 °C/decade both for the maximum and minimum temperatures. The growth is also appreciable during the spring for tasmax (0.24 °C/decade) while for tasmin is 0.11 °C/decade. During autumn, growth is also significant around 0.14 °C/decade for both temperatures, while during winter there are no significant trends. These results are partially in agreement with those found by Fernández-Montes and Rodrigo (2015) who reported significant trends in the area but with greater increasing rates for the period 1970–2007 using the different grid database named Spain02.

Regarding annual precipitation, significant negative trends at the 95% confidence level are found to the west side of the Sierra Nevada and in some elevated areas, with values around 10 mm/decade. On the other hand, significant increases appear in the most easterly part, with values above 7 mm/decade. The prevalence of negative trends leads to an overall decrease in the annual precipitation for the area, significant at the 95% confidence level and with a value of −2.7 mm/decade, mainly due to the significant negative trend during winter (−1.3 mm/decade) and to a lesser extent during autumn (−0.9 mm/decade). The significant decrease in Sierra Nevada precipitation has been also reported by Ruiz-Sinoga et al. (2010) who found a significant annual decrease for a station located in Sierra Nevada of −3.3 mm/decade for the period 1962–2007, also associated with a diminution in winter precipitation. Additionally, Fig. 3f also shows the marked rainfall interannual variability. Thus, despite the fact that the period 1990–2019 is clearly less rainy compared to the decades of the 60s and 70s, with the severe drought of 1994 and 1995, there were also years with the highest rainfall values, as 1996 and 2010.

In terms of weather extremes, the widely known extreme indices developed by the Expert Team (ET) on Climate Change Detection and Indices (ETCCDI, Karl et al. 1999; Peterson et al. 2001), have been computed for both temperature and precipitation series and their changes have been also evaluated using the Mann–Kendall test and the Sen's slope. Among the different measures defined by the ETCCDI, a set of nine indices has been selected to analyse extreme precipitation events, meanwhile for temperature the number of computed indices was 7. These indices are summarized in Table 2. The indices have been computed for each time series and were averaged for the complete area. Figure 4 depicts the time evolutions of some indices from ClimaNevada data for the period 1961–2019, considering that their temporal behaviour is quite similar to those obtained from the SPREAD and STEAD gridded data set for the same period.

**Table 2** ETCCDI extreme precipitation and temperature indices selected

| Index | Definition | Units |
|---|---|---|
| RX1day | Maximum 1-day precipitation | mm |
| RX5day | Maximum 5-days consecutive precipitation | mm |
| SDII | Average precipitation on days with accumulated values above 1 mm | mm |
| R10 | Number of days with precipitation above 10 mm | days/year |
| R20 | Number of days with precipitation above 20 mm | days/year |
| CWD | Annual mean maximum number of consecutive days with precipitation $\geq 1$ mm | days/year |
| CDD | Annual mean maximum number of consecutive days with precipitation <1 mm | days/year |
| R95pTOT | Annual sum where daily precipitation exceeds the 95th percentile of daily precipitation in the reference period | % |
| R99pTOT | Annual sum where daily precipitation exceeds the 99th percentile of daily precipitation in the reference period | % |
| ID | Icing days, i.e., tasmax <0 °C | days/year |
| SU | Summer days, i.e., tasmax >25 °C | days/year |
| HD | Hot days, i.e., tasmax >35 °C | days/year |
| FD | Frost days, i.e., tasmin >0 °C | days/year |
| TR | Tropical nights, i.e., tasmin >20 °C | days/year |
| Tmax90p | Count of days with tasmax >90th percentile of the corresponding calendar day in the base period | days/year |
| Tmin10p | Count days with tasmin <10th percentile of the corresponding calendar days in the base period | days/year |

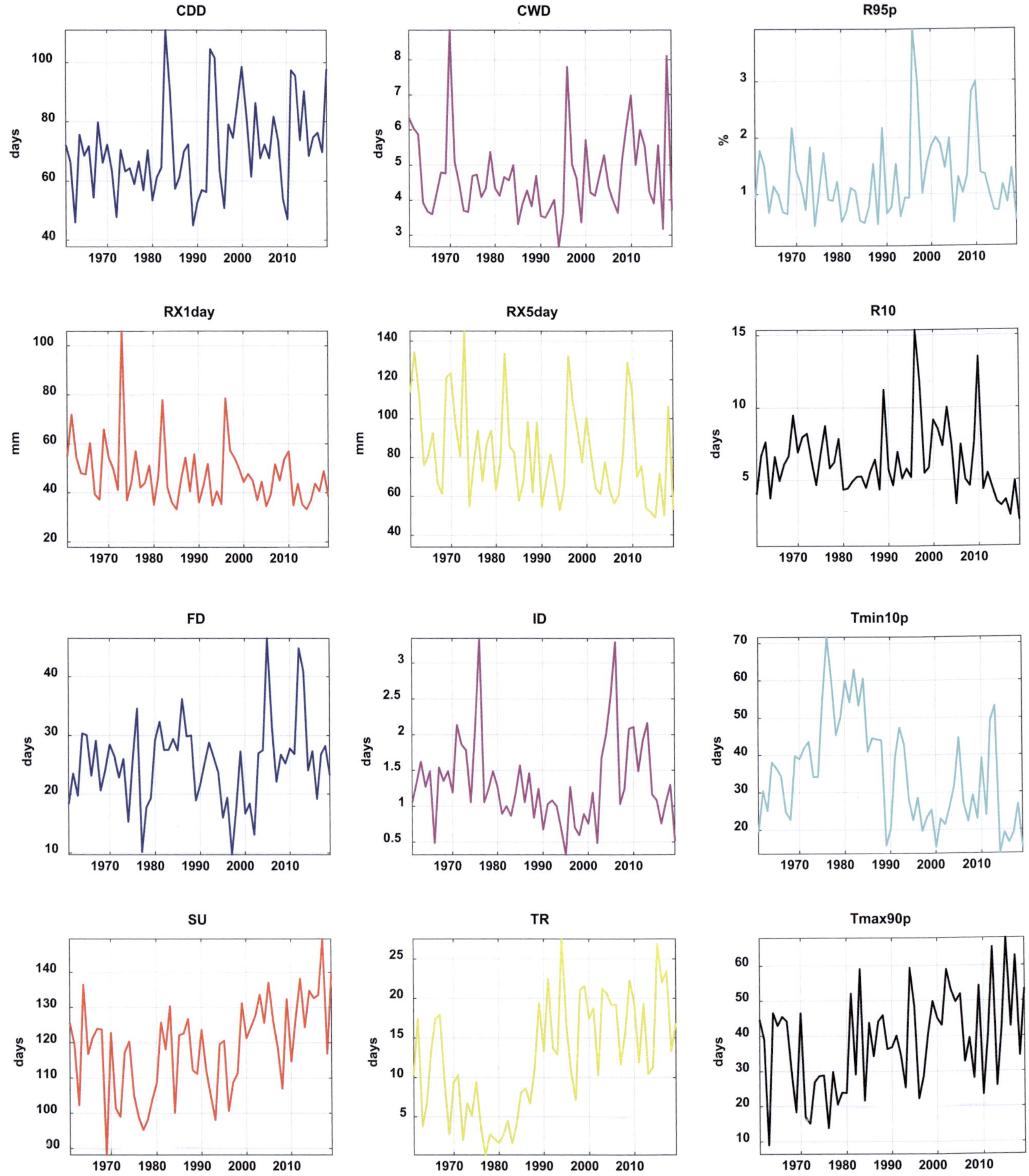


**Fig. 4** Time series for several averaged annual extreme indices: first and second rows for precipitation extreme indices; third and fourth rows for temperature extreme indices

For precipitation extreme indices, there are a significant decrease for RX1day, RX5day and RX3day (not shown) with values from 1.8 mm/decade for RX1day to 5.2 mm/decade for RX5day, revealing a decrease in precipitation amount with a similar evolution for these 3 indices. Thus, for example, these 3 indices present their maximum value in October 1973, when very intense precipitation happened, particularly in the southern part. For the whole area, there are no evidence of an increase of more intense precipitation events, with significant negative trends for R20 and R10 (at the 90% confidence level) and no significant trends for SDII, CWD, R95p and R99p (using 1971–2000 as reference period). These decreasing trends occur along with a significant positive trend for the CDD of 2.4 days/decade, indicating drier conditions over the area. These results are in agreement with those from Hidalgo-Muñoz et al. (2015), who reported a trend towards fewer heavy rainfall episodes during winter and spring, mainly associated to the decrease in the occurrence of the synoptic pattern characterized by a deep low both in surface and in 500 hPa to the west of Ireland with strong zonal flow over Andalusia.

Regarding extreme temperature, the behaviour of the selected indices confirms the warming tendency in the area. Particularly, the major significant trends appear for the summer indices, namely summer days (SU) and tropical nights (TR), with trend values of 3.4 days/decade and 2.5 days/decade, respectively, in concordance with the higher trends for both tasmax and tasmin during summer. Additionally, significant trends are found for Tmin10p (−0.30 days/decade) and Tmax90p (0.34 days/decade). It is worth mentioning the marked change between the 70s and 80s and the period from 1990 onwards in the Tmin10p and TR (Fig. 4), with a predominance of values in the indices indicating lower tasmin in the 70 s and 80 s and a strong increase in this variable during the last 30 years. On the other hand, no significant trends are detected for FD and ID. These indices are characterized by an exacerbated variability during the last 20 years, containing some of the lowest values (i.e., 2002) and the highest ones (i.e., 2005). These conditions seem to be generalized for the Iberian Peninsula, with an increase of the warmer days from 1990 and high presence of cold events during the two previous decades (Rodríguez-Puebla et al. 2010).

Therefore, both temperatures show increasing trends during the last decades, although with a more generalized spatial pattern for minimum ones. According to this increase, significant positive trends are found for extreme events indices associated with warm days. On the other hand, there is a prevailing decrease in annual and winter precipitation for the whole area, related with significant negative trends over the western part of Sierra Nevada, although this trend is not accompanied by significant trends in extreme precipitation extremes.

## 6 Trends in Other Variables Related with the Hydrological Cycle

As for the former variables, Fig. 3 also depicts the spatial distribution of annual ET0 trends and the seasonal and annual ET0 evolution for the average of the entire area of the Sierra Nevada for the period 1961–2014 using the database from Tomás-Burguera et al. (2019). This variable is characterized by generalized significant positive trends above 30 mm/decade in the easterly part of the region and with a reduction in of the trends towards the west, with trends around 20–25 mm/decade in the western of the mountain and with lower increases, 15 mm/decade, over the northwest and southwest. These last trends are more moderate than those reported by Yeste et al. (2018) for stations sited in the Guadalquivir Basin (northwest of the study area) that presented an increased rate above 25 mm/decade for the period 1951–2015. These differences may be due to the use of a different database as well as the use of the Hargraves formula to calculate ET0 instead of the Pemann-Monteith formula. At seasonal scale, significant increases are detected for all the seasons, with the major change rates during spring (7.8 mm/decade) and summer (12.1 mm/decade), meanwhile the changes for autumn and winter are more modest (5.2 and 2.6 mm/decade, respectively). These seasonal trends yield an annual trend of 25 mm/decade for the whole area.

Finally, the temporal behaviour of the drought phenomenon is examined. For this objective, the time evolution of SPI and SPEI indices at 3- and 12-month timescale is evaluated from 1961 to 2019 (Fig. 5). These timescales have been chosen as an adequate way to characterize the meteorological (3-month scale) and hydrological (12-month scale) droughts (Vicente-Serrano et al. 2010). In general, both indices present similar values, detecting the prolonged and intense droughts of the first half of the 1990s and the 2000s, and showing the main differences during the last 20 years, when SPEI shows lower values than SPI, with a prevalence of dry years according the SPEI. Thus, both SPEI at 3-month and 12-month present negative significant trends towards drier conditions, associated with the increase of temperature and ET0, and therefore with a major atmospheric evaporative demand, particularly during the last decade. On the other hand, SPI indices do not present any significant trends, despite the significant trends found for annual and winter precipitation.

Positive trends in ET0 are consistent with temperature increase, being the recent enhanced drought conditions related with a more atmospheric demand than with a decreased annual and winter precipitation as the no significant trends for SPI show.

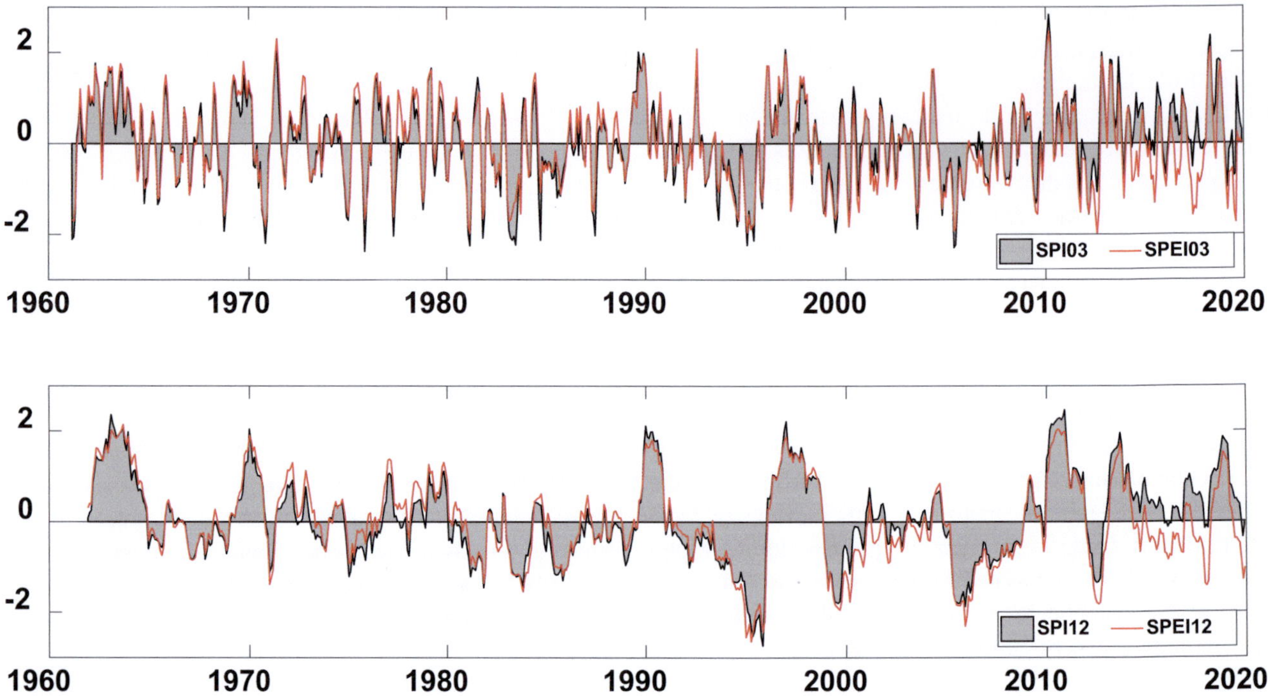


**Fig. 5** Temporal evolution of monthly SPI (grey columns) and SPEI (red line) indices for the period 1961–2019: upper panel for 3-month scale; bottom panel for 12-month scale

## 7 Climate Projections

In this section, climate change projections for Sierra Nevada have been analysed using a multi-model ensemble based on 15 regional climate simulations (see Table 3) within the EURO-CORDEX initiative (Jacob et al. 2014) (http://www.cordex.org). EURO-CORDEX aims to provide regional climate information for impact assessments over the European continent. For that, regional climate models (RCMs) were conducted using different global climate models (GCMs) at various spatial resolutions. In this analysis, regional climate simulations outputs at their finer spatial scale have been selected, corresponding to 0.11° grid resolution

**Table 3** Members of the ensemble (*Not included in the analysis of mrso, evspsbl, and drought indices)

| Experiment member | Institution | RCM | GCM | GCM run |
|---|---|---|---|---|
| ENS01 | SMHI | RCA4 | CNRM-CM5 | r1i1p1 |
| ENS02 | SMHI | RCA4 | EC-EARTH | r12i1p1 |
| ENS03 | SMHI | RCA4 | HadGEM2-ES | r1i1p1 |
| ENS04 | CLMcom | CCLM4-8-17 | CNRM-CM5 | r1i1p1 |
| ENS05 | CLMcom | CCLM4-8-17 | EC-EARTH | r12i1p1 |
| ENS06 | CLMcom | CCLM4-8-17 | HadGEM2-ES | r1i1p1 |
| ENS07 | DMI | HIRHAM5 | NorESM1-M | r1i1p1 |
| ENS08 | DMI | HIRHAM5 | EC-EARTH | r3i1p1 |
| ENS09 | DMI | HIRHAM5 | HadGEM2-ES | r1i1p1 |
| ENS10 | KNMI | RACMO22E | CNRM-CM5 | r1i1p1 |
| ENS11 | KNMI | RACMO22E | EC-EARTH | r12i1p1 |
| ENS12 | KNMI | RACMO22E | HadGEM2-ES | r1i1p1 |
| ENS13* | GERICS | REMO2015 | NorESM1-M | r1i1p1 |
| ENS14* | GERICS | REMO2015 | EC-EARTH | r12i1p1 |
| ENS15* | GERICS | REMO2015 | HadGEM2-ES | r1i1p1 |

(approximately 12 km) to better capture the spatial patterns of the changes in Sierra Nevada. Changes for two periods, the near (2020–2049) and the far (2070–2099) future have been computed related to the recent past period (1980–2009) and under two Representative Concentration Pathways (RCPs), RCP4.5 and RCP8.5.

Figure 6 depicts the multi-model ensemble mean changes in annual means of tasmax, tasmin, and pr for both future periods and RCPs. In this figure, the crossed grid points indicate the areas where less than 75% of the models (12 of the 15 models) agree on the sign of the change represented by the ensemble mean.

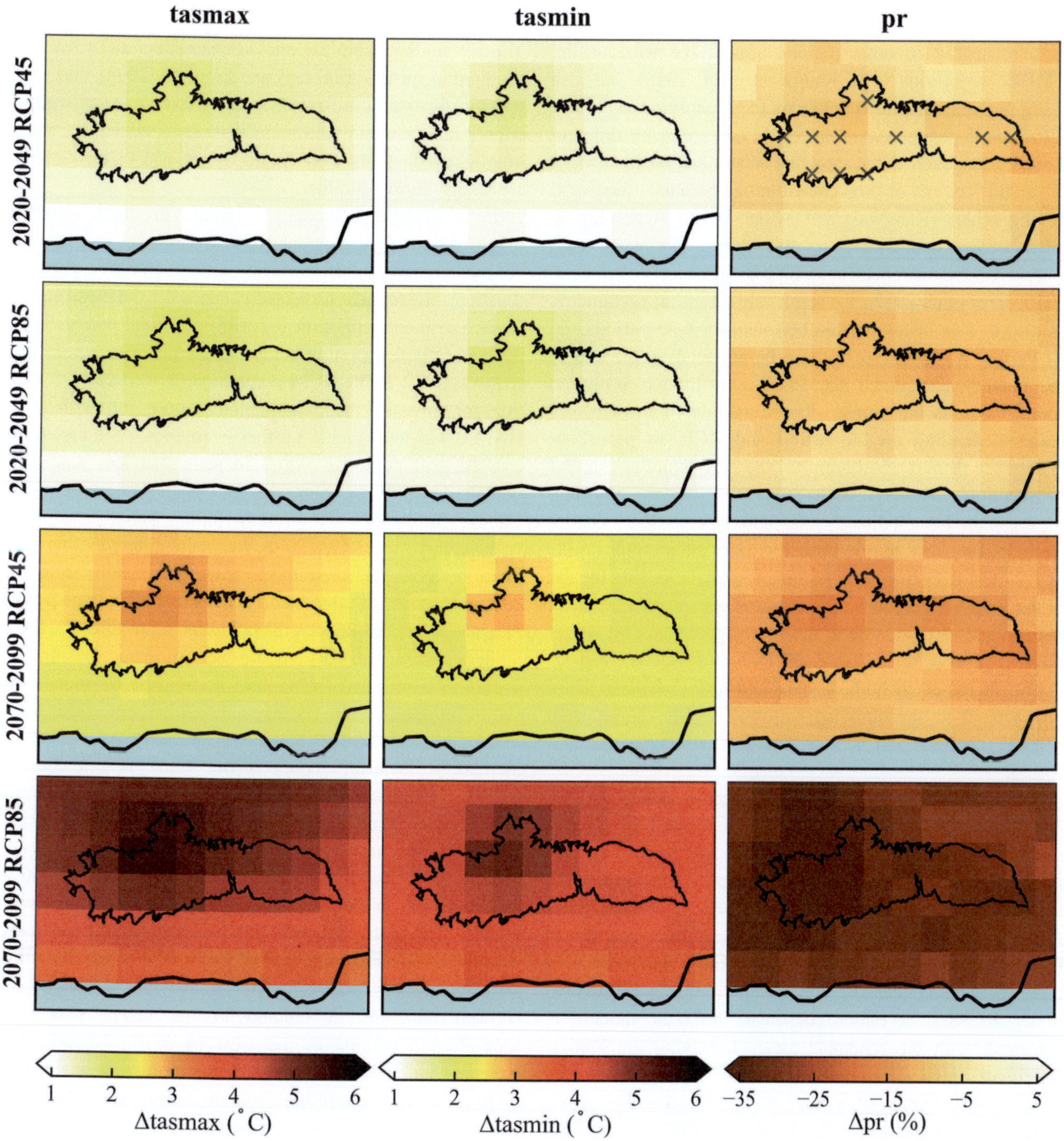


**Fig. 6** Changes in the ensemble mean of mean annual maximum temperature (tasmax, first column), minimum temperature (tasmin, second column), and precipitation (pr, third column). The rows comprise the results for the two periods, the near and the far future compared to the recent past, under both RCPs (2020-2049 RCP4.5, 2020-2049 RCP8.5, 2070-2099 RCP4.5, and 2070-2099 RCP8.5, respectively). The crossed grid-points represent the areas in which less than 12 models (75%) agree in the sign according to the ensemble mean

Sierra Nevada is very likely to undergo an increase in temperature, with all the members of the ensemble showing positive values of change in relation to the recent past in both periods and under both RCPs (first and second column in Fig. 6). According to the ensemble mean, all simulations show changes in tasmax and tasmin with similar spatial patterns, with the most pronounced temperature rises occurring in the high mountain (west of Sierra Nevada). The results suggest an elevation-dependent warming, similar to those observed in other regions around the world using CMIP5 projections (e.g., Rangwala et al. 2013), but also using regional climate simulations (e.g., Giorgi et al. 1997). Elevation-dependent warming could be caused by different mechanisms, such as the snow/ice-albedo feedback, changes in cloud cover, or soil moisture-temperature feedback, among others (Rangwala and Miller 2012). According to other studies conducted over the Iberian Peninsula (e.g., García-Valdecasas Ojeda et al. 2020b; Gómez-Navarro et al. 2010; Jerez et al. 2012), the results also show an asymmetric warming. This behaviour can be found for both periods and RCPs, but it appears to be more pronounced by the end of the century and under the RCP8.5. Although projected spatial patterns are similar, the magnitude of the changes varies depending on the period and RCP. In particular, changes in tasmax and tasmin for the near future are similar in magnitude between RCPs, suggesting that the warming for this period is not directly related to the emission scenario. Both RCPs show temperature increases of around 1.5 °C and 2 °C for tasmax and tasmin, respectively. For the far future, differences between scenarios are already notable, with increases compared to the recent past reaching values up to 3 °C and 5.5 °C under RCP4.5 and RCP8.5, respectively. Increases in both, tasmax and tasmin will be more pronounced in the summer, followed by the autumn. In contrast, changes will be more moderate during winter and spring.

In terms of precipitation (third column in Fig. 6), the results show a slight decrease in the annual mean for the near future compared to the recent past. For this period, the difference between RCPs is slightly more pronounced than it is for temperature. In fact, under RCP4.5, most of Sierra Nevada exhibits a larger inter-model variability, as shows the discrepancy in the sign of the changes for more than three models concerning the ensemble mean. However, under RCP8.5, a major consensus between models is achieved, with the ensemble mean displaying the highest differences related to the recent past over eastern Sierra Nevada with values of around −15%. By the end of the century, the reduction was already highlighted for both RCPs, with almost all the members in agreement with the ensemble mean about the sign of the changes. However, differences between RCPs are clearly visible for this period; RCP4.5 indicates reductions in precipitation, on average, of more than 15%, whereas RCP8.5 shows diminutions of up to 35% over the western Sierra Nevada and some surrounding areas. Differences in this variable also appear throughout the year (results not shown), with the summer being the season when the largest reductions in mean precipitation are shown for the entire Sierra Nevada. In contrast, winter changes appear to be the most moderate, thus the spring and autumn show a transitional behaviour between winter and summer. Similar projected changes for precipitation have been found in other studies using regional climate simulations over the Mediterranean Basin (Giorgi and Lionello, 2008), and more specifically over southern Iberian Peninsula (e.g., Argüeso et al. 2012; García-Valdecasas Ojeda et al. 2020a, b), where highlight reductions were found by the end of the century, especially during summer.

In a changing climate, variations in other variables involved in the hydrological cycle are also expected (chapter "Snow Dynamics, Hydrology, and Erosion"). Figure 7 illustrates the relative projected changes for the mean annual surface evapotranspiration (evspsbl) and total soil moisture content (mrso). For this analysis, the multi-ensemble was reduced to 12 RCM-GCM combinations, because information for three of the combinations (those run with HIRHAM5) was unavailable for these variables. As a result, the crossed grid points in Fig. 7 indicate that less than nine models agree on the sign with that of the ensemble mean. Changes in evspsbl also present similar spatial patterns for both periods and RCPs, with the main difference between them in terms of magnitude. Thus, the results from the ensemble mean reveal reductions in evspsbl in a large part of Sierra Nevada, while several points at high altitudes show moderate increases. This enhancement of evapotranspiration occurs in winter and spring, and it could be caused by the significant temperature rise appeared in the region that leads to an increase in potential evapotranspiration. As a result, an increase in evspsbl occurs if the water is not limited. Evapotranspiration increases at high-altitude, however, are only seen in some models included in the ensemble, indicating that the inter-model variability is large for this variable. In fact, evspsbl changes vary depending on the RCM used, as six of the simulations (those completed with REMO2015 and CCLM4-8-17) indicated a general trend towards the overall reduction of evspsbl, whereas the other six simulations (conducted with RCA4 and RACMO22E) showed increases in relation to the recent past over high-altitude regions in the west of Sierra Nevada. For summer and autumn, however, the reduction is already widespread throughout the mountain, and more marked over the east of Sierra Nevada in summer. Concerning differences between periods, changes in evspsbl for the near future are more moderate (values ranging from around −10 to 5%), when differences in relation to the recent past are very similar between RCPs. In general, differences in magnitude were more pronounced between scenarios for the far future (2070–2099), indicating, in any case, stronger changes than for the near future for both RCPs; under the RCP4.5 most of the region shows reductions not reaching values above 15%, while RCP8.5 shows decrease up to 20% for eastern Sierra Nevada.

The mrso (second column in Fig. 7) is projected to decrease overall, with the agreement in the sign being generalized between models for both periods under the RCP8.5 scenario for all Sierra Nevada. Under the RCP4.5, reductions of up to 6% are found for the near future compared to the recent past, with several points indicating that more than

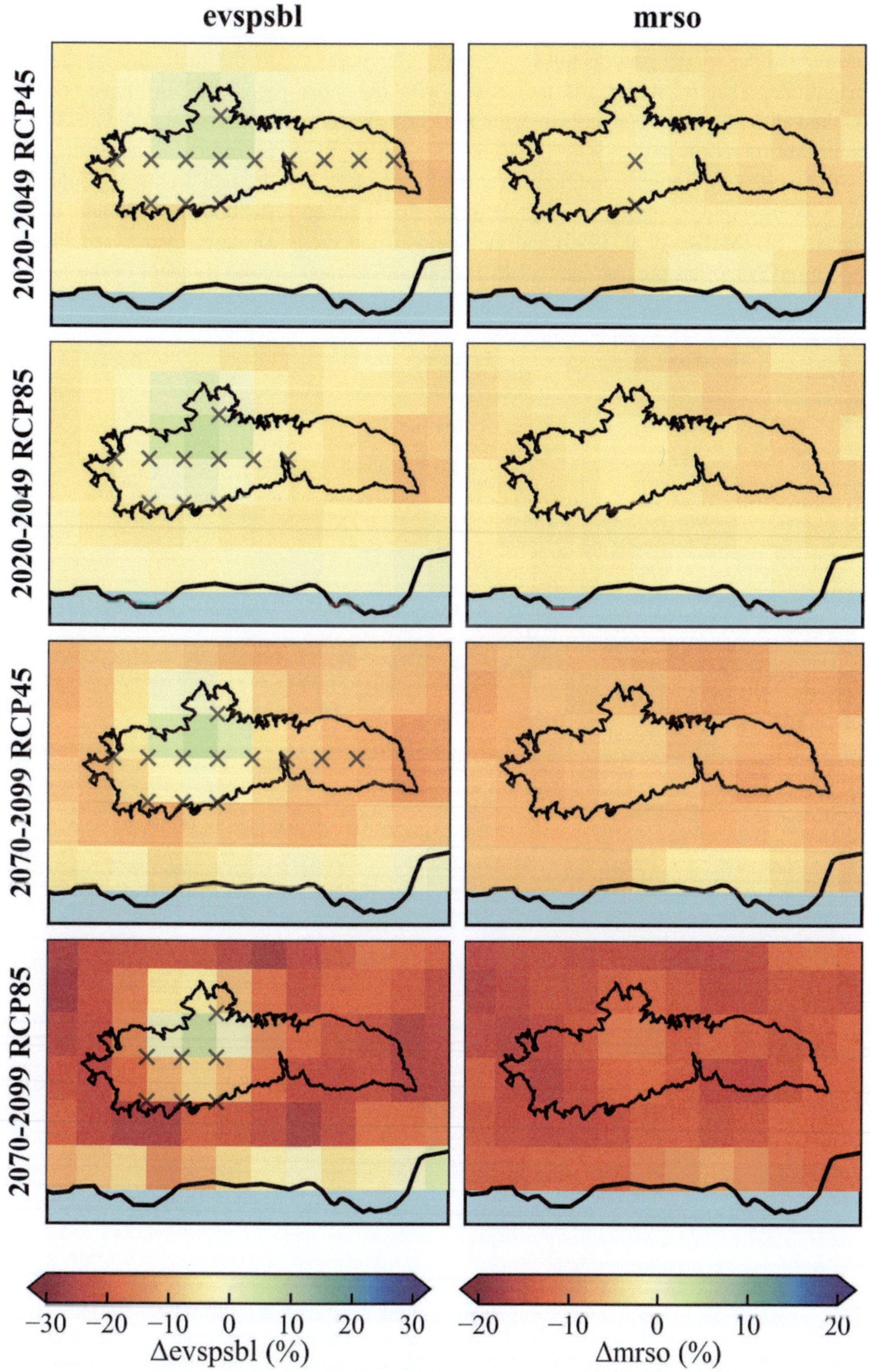


**Fig. 7** Relative projected changes (%) in the ensemble mean of mean annual surface evapotranspiration (evspsbl, first column) and total soil moisture content (mrso, second column). The rows comprise the results for the two periods, the near and the far future compared to the recent past, under both RCPs (2020-2049 RCP4.5, 2020-2049 RCP8.5, 2070-2099 RCP4.5 and 2070-2099 RCP8.5, respectively). The crossed grid points represent the areas in which less than 9 models (75%) agree in the sign according to the ensemble mean

three models presented an increased mean annual soil moisture content. Under the RCP8.5, the mrso reductions are similar with regard to the spatial patterns although slightly larger in magnitude. For this period, both RCPs locate the greatest reduction in the eastern Sierra Nevada, where changes in precipitation are the greatest as well. Stronger decreases were shown for the far future, with RCP4.5 and RCP8.5 indicating reductions, on average, around 5% and 10%, respectively. Changes across seasons (not shown) also present similar spatial patterns but show some differences in magnitude. That is, summer is the season with the most marked differences in comparison with the corresponding values in the recent past.

Projections in drought conditions over the Sierra Nevada have been explored using the Standardized Precipitation Index (SPI) (McKee et al. 1993) and the Standardized Precipitation-Evapotranspiration Index (SPEI) (Vicente-Serrano et al. 2010) at 3- and 12-month timescales. Both, the SPEI and SPI were estimated using the entire period (1980–2099) as baseline. Additionally, a double standardization of the indices was adopted in order to avoid unrealistic values along the period, using for that the mean and standard deviation in the recent past (García-Valdecasas Ojeda et al. 2021b). More details about this methodology can be found in Vicente-Serrano et al. (2020). Then, changes in the occurrence of drought events, which were defined according to the methodology described in García-Valdecasas Ojeda et al. (2021a), have been analysed.

Figure 8 compares the changes in mean drought duration (numbers of consecutive months within a drought event) and drought frequency (number of drought events per decade) for both periods and RCPs using the SPEI drought index. Drought events are likely to become slightly longer and more frequent in the near future, in general, over the entire

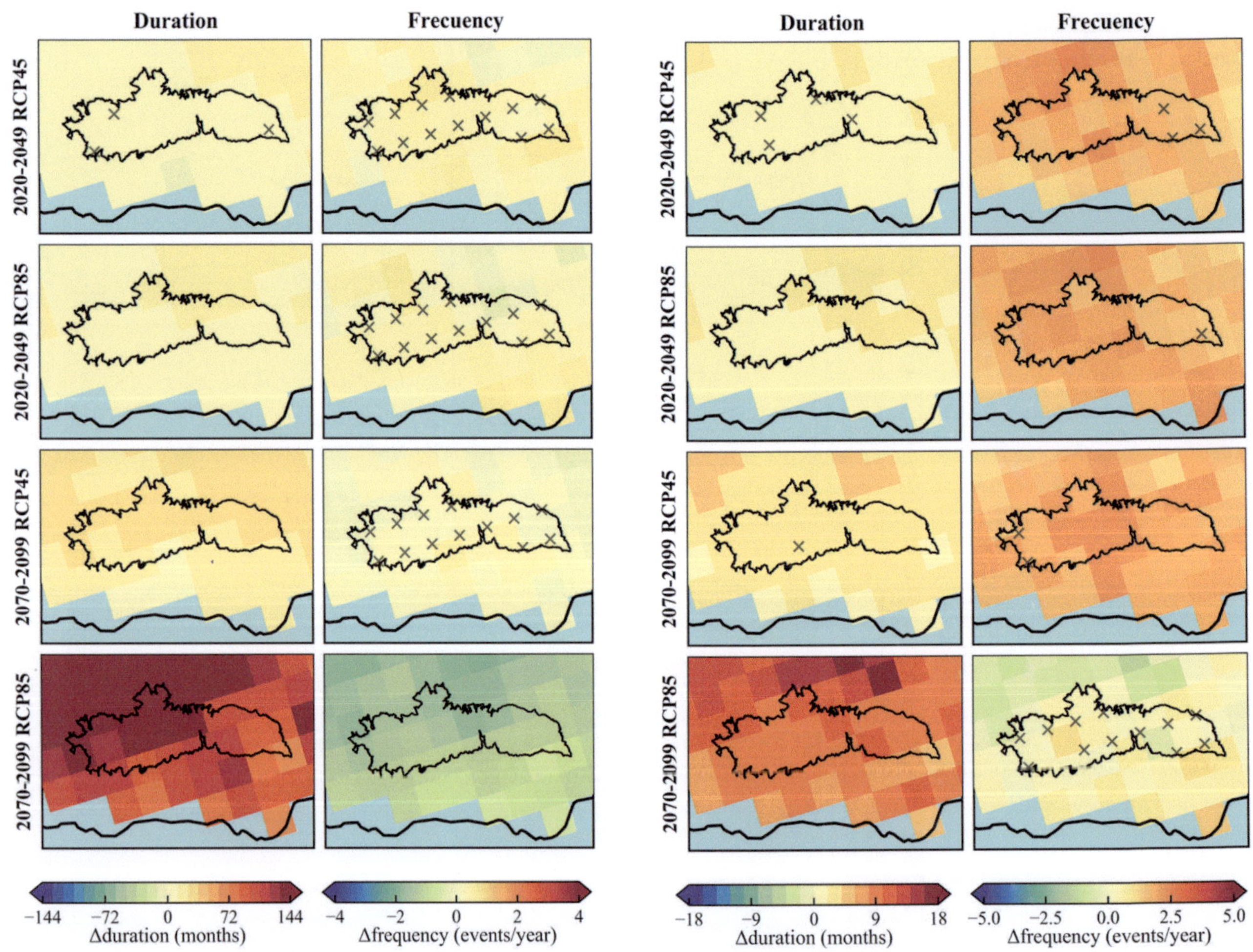


**Fig. 8** Changes in the ensemble mean of mean duration (in months) and frequency (in events per decade) of drought events based on the SPEI computed at 3- and 12-month timescales (3-SPEI duration, 3-SPEI frequency, 12-SPEI duration and 12-SPEI frequency, respectively). The rows comprise the results for the two periods, the near and the far future compared to the recent past (1980-2009), under both RCPs (2020-2049 RCP4.5, 2020-2049 RCP8.5, 2070-2099 RCP4.5, and 2070-2099 RCP8.5, respectively). The crossed grid points represent the areas in which less than nine models (75%) agree in the sign according to the ensemble mean

Sierra Nevada. The RCP scenario does not appear to have a clear effect on drought conditions during this period, as both mean duration and frequency show similar values of change for RCP4.5 and RCP8.5 at both timescales. Increases in mean duration appear to be stronger for indices computed at 12-month timescale, with the length of the events increasing, on average, by around 2 months and 10 months, at 3- and 12-month timescales, respectively. In terms of frequency, however, both timescales show values of change more similar, though indices at 12-months indicate that the drought events will be slightly less frequent and for this timescale, more than 25% of the models in the ensemble show discrepancies in the sign of the changes with respect to the ensemble mean in all Sierra Nevada.

For the far future, differences between the results from the two RCPs and timescales are pronounced. According to RCP4.5, Sierra Nevada is likely to experience an increase in the severity of drought conditions with more frequent and longer-lasting drought events, with increases in mean duration and frequency being up to 3.5 months and two events per decade, respectively, at 3-month timescale. For indices computed at 12-months, highlight increases in the mean duration are found, reaching values around 30 months, on average, whereas the frequency only increases in 0.1 events per decade. Therefore, the results suggest that for this period and RCP, hydrological droughts could cause important impacts on the region mainly due to the increase in lengthening in the events. This effect is even more notorious under the RCP8.5, in which the length of the events, on average, may increase around 125 months, reaching even values up to 144 months over western Sierra Nevada. Such an increase, however, resulted in fewer events for this RCP, with a decrease in the number of events compared to the recent past up to two events per decade. At 3-month timescale, the duration also increases in general (values of increase up to 11 months on average), while the frequency shows both increases and decreases (values of change ranging from −0.3 to 1 events per decade), slight in magnitude. Similar values of change in drought conditions have been observed in other studies. For instance, García-Valdecasas Ojeda et al. (2021a, b), using regional climate projections over the Iberian Peninsula, found strong increases in the mean drought duration for a large part of the IP when the SPEI was used at 12-month timescale. Similarly, strong increases in the severity of drought conditions have been established in other regions across the IP, more marked for the SPEI than for the SPI (Marcos-Garcia et al. 2017). In this regard, the results from SPEI tended to present stronger increases in severity of drought conditions than the SPI, especially at the end of the century and under the RCP8.5 (results not shown). This is because decreases in precipitation projected for this region are accompanied by a significant increase in temperature. Differences between these indices are therefore noteworthy since the SPI is solely based on precipitation data, whereas the SPEI also takes into account the effect of the temperature trend on its calculation. Therefore the results reveal a clear trend towards a dryer and warmer climate in the Sierra Nevada by the end of the century compared to the recent past, particularly under the RCP8.5.

## 8 Conclusions

In this chapter, the main characteristics of the Sierra Nevada climate from different data sources have been shown. Additionally, relevant circulation drivers have been described. Trends in annual and seasonal temperature and precipitation and in several extreme indices have been evaluated along with the trends in other important variables for a semiarid region such as potential evapotranspiration and drought indices. Finally, climate change projections were analysed from an ensemble of regional model simulations within the Euro-CORDEX scheme for the near future (2020–2050) and far future (2070–2100) under the RCP 4.6 and 8.5 scenarios. The main conclusions are briefly summarized as follow:

- Homogenized time series of temperature and precipitation from ClimaNevada have allowed us to evaluate recent trends of these variables. Moreover, for precipitation, a high-resolution database at monthly scale for the period 1990–2019 has been performed. These data complement the information provided by other databases.
- Precipitation in Sierra Nevada is characterized by marked inter and intra-annual variability, a typical condition of the Mediterranean climate and it is mainly concentrated between the months from October to April, with a clear gradient west-east. This variability is mainly driven by the NAO over the western part, meanwhile the eastern part is more dominated by Mediterranean depression, and particularly by the WeMO. On the other hand, the altitude seems to have a limited impact on rainfall distribution.
- Saharan dust intrusions have an important impact on Sierra Nevada, being a main climate driver of long-range transported particulate matter over the area. These intrusions usually are associated with warm events. The main synoptic features of these intrusions are the presence of a thermal surface low pressure center over northwestern Africa with the presence of the high geopotential anomalies at upper levels over North Africa (NAH-H), leading the dust transport in the mid-troposphere. Atlantic depressions (AD), both on the surface and at higher levels, also bring dust from northwestern Africa over Sierra Nevada.
- Altitude controls the spatial temperature patterns, particularly for maximum temperature and in less extension for

the minimum one. This aspect is also revealed by the different altitudinal gradients for both temperatures, with lower values for tasmin. In general, these altitudinal gradients are lower than the accepted mean value for vertical temperature ratio of 6.5 °C/km (for instance, mean temperature ranges from 4.14 to 4.82 °C/km, with smaller values during summer.

- For ET0, its spatial pattern is closely linked to height, with the highest areas presenting lower ET0 values, particularly in the western part of Sierra Nevada, where annual accumulations reach 900 mm, and progressively increase to very high values in the surrounding lowlands, exceeding 1250 mm per year in the southeast. This pattern is essentially associated with the temperature spatial variation.
- There is a prevailing decrease in annual and winter precipitation for the whole area, related with the significant negative trends over the western part of Sierra Nevada. However, this fact does not translate into an increase in drought episodes characterized by SPI, being the enhanced drought conditions related with a more atmospheric demand.
- Significant trends are detected for temperature. Spatially, tasmin shows an increase in almost the entire area, with the highest growth rates in the central zone of Sierra Nevada. On the other hand, globally, annual increased rates are slightly higher for maximum temperatures than for minimum temperatures, Seasonally, the greatest increases are observed during the summer, with very significant trends above 0.3 °C/decade both for the maximum and minimum temperatures. Positive significant trends are also found during spring and autumn.
- Significant trends in several ETCCDI indices suggest an increase of the warm events, with a high variability of the cold events during winter. For precipitation, the results suggest a trend towards drier conditions with less heavy rainfall episodes for the overall region.
- As a result of the temperature rise, significant trends for annual ET0 are found for all the area under study for the period 1961–2014, with higher rates in the eastern part of the region. At a seasonal scale, significant increases are detected for all the seasons, with the major change rates during spring and summer, meanwhile the changes for autumn and winter are more moderate. These seasonal trends yield an annual trend of 25 mm/decade for the ET0 for the whole area.
- Time evolution of SPI and SPEI indices from 1961 to 2019 is similar, detecting the prolonged and intense droughts of the first half of the 1990s and the 2000s, and with the main differences between indices found during the last 20 years, when SPEI shows lower values than SPI. SPEI at 3-month and 12-month present negative significant trends towards drier conditions, associated with the increase of temperature and ET0, particularly during the last decade.
- Climate change projections are analysed from an ensemble of regional model simulations within the Euro-CORDEX scheme for the near future (2020–2050) and far future (2070–2100) under the RCP 4.6 and 8.5 scenarios. These projections depict a clear warming and a precipitation reduction over Sierra Nevada, more intense for the far future under the RCP8.5. The results from the ensemble mean also reveal decreases in evspsbl in most of Sierra Nevada, with only moderate increases in higher altitudes in winter and spring. This fact could be caused by the significant temperature rise and then, increase in potential evapotranspiration with no water limitation. The mrso is projected to decrease overall, with the agreement in the sign being generalized between models for both periods under the RCP8.5 scenario for all Sierra Nevada. Drought events are likely to become slightly longer and more frequent in the near future, with a marked increase in duration and intensity for the far future.

Future research on the climate in Sierra Nevada should be aimed at improving the databases in the area. Furthermore, its topographic complexity requires the use of regional models at very high spatial resolution, which could allow, for example, a more realistic representation of convective processes, particularly relevant in mountain regions. These simulations will be not only relevant to study the Sierra Nevada climate in the future, but also to complete the current climate information.

**Acknowledgements** This research was financed by the Spanish Ministry of Economy, Industry and Competitiveness, with additional support from the European Community Funds (FEDER) [projects CGL2017-89836-R and LifeWatch-2019-10-UGR-01] and by the FEDER/Junta de Andalucía—Ministry of Economy and Knowledge/ Project [B-RNM-336-UGR18].

## References

Aguilar E, Llanso P (2003) Guidelines on climate metadata and homogenization. WMO

Alonso-González E, López-Moreno JI, Gascoin S et al (2018) Daily gridded datasets of snow depth and snow water equivalent for the Iberian Peninsula from 1980 to 2014. Earth Syst Sci Data 10:303–315

Alzate Velásquez DF, Araujo Carrillo GA, Rojas Barbosa EO et al (2017) Interpolacion Regnie para lluvia y temperatura en las regiones andina, caribe y pacífica de Colombia. Colombia for 21:102

Argüeso D, Hidalgo-Muñoz JM, Gámiz-Fortis SR et al (2012) High-resolution projections of mean and extreme precipitation over Spain using the WRF model (2070–2099 versus 1970–1999). J Geophys Res Atmos 117:D12108

Celleri R, Willems P, Buytaert W, Feyen J (2007) Space–time rainfall variability in the Paute basin, Ecuadorian Andes. Hydrol Process 21:3316–3327

Dirmeyer PA, Jin Y, Singh B, Yan X (2013) Trends in land-atmosphere interactions from CMIP5 simulations. J Hydrometeorol 14:829–849

Domonkos P (2014) The ACMANT2 software package. In: Eighth seminar for homogenization and quality control in climatological databases and third conference on spatial interpolation techniques in climatology and meteorology. WMO, WCDMP-84, pp 46–72

Escudero M, Castillo S, Querol X et al (2005) Wet and dry African dust episodes over eastern Spain. J Geophys Res 110:D18S08

Esteban-Parra MJ, Rodrigo FS, Castro-Díez Y (1998) Spatial and temporal patterns of precipitation in Spain for the period 1880–1992. Int J Climatol 18(14):1557–1574

Fernández-Montes S, Rodrigo FS (2015) Trends in surface air temperatures, precipitation and combined indices in the southeastern Iberian Peninsula (1970–2007). Clim Res 63:43–60

García-Valdecasas Ojeda M, Rosa-Cánovas JJ, Romero-Jiménez E et al (2020a) The role of the surface evapotranspiration in regional climate modelling: evaluation and near-term future changes. Atmos Res 237:104867

García-Valdecasas Ojeda M, Yeste P, Gámiz-Fortis SR et al (2020b) Future changes in land and atmospheric variables: an analysis of their couplings in the Iberian Peninsula. Sci Total Environ 722:127902

García-Valdecasas Ojeda M, Gámiz-Fortis SR, Romero-Jiménez E et al (2021a) Projected changes in the Iberian Peninsula drought characteristics. Sci Total Environ 757:143702

García-Valdecasas Ojeda M, Romero-Jiménez E, Rosa-Cánovas JJ et al (2021b) Assessing future drought conditions over the Iberian Peninsula: the impact of using different periods to compute the SPEI. Atmosphere 12:980

Giorgi F (2006) Climate change hot-spots. Geophys Res Lett 33: L08707

Giorgi F, Lionello P (2008) Climate change projections for the Mediterranean region. Glob Planet Change 63:90–104

Giorgi F, Hurrell JW, Marinucci MR, Beniston M (1997) Elevation dependency of the surface climate change signal: a model study. J Clim 10:288–296

Gómez-Navarro JJ, Montávez JP, Jimenez-Guerrero P et al (2010) Warming patterns in regional climate change projections over the Iberian Peninsula. Meteorol Zeitschrift 19:275–285

Guijarro JA (2018) Homogenization of climatic series with Climatol. Version 3.1. 1

Guo Z, Dirmeyer PA, Koster RD et al (2006) GLACE: the global land-atmosphere coupling experiment. Part II: Analysis. J Hydrometeorol 7:611–625

Hartmann DL, Klein Tank AMG, Rusticucci M. et al (2013) Observations: atmosphere and surface. In: Stocker TF, Qin D, Plattner G-K, Tignor M, Allen SK, Boschung J, Nauels A, Xia Y, Bex V, Midgley PM (eds) Climate change 2013: the physical science basis. Contribution of Working Group I to the fifth assessment report of the intergovernmental panel on climate change. Cambridge University Press, Cambridge, United Kingdom and New York, NY, USA

Heywood VH, Watson RT (1995) Global biodiversity assessment. Cambridge University Press, Cambridge

Hidalgo-Muñoz JM, Argüeso D, Gámiz-Fortis SR et al (2011) Trends of extreme precipitation and associated synoptic patterns over the southern Iberian Peninsula. J Hydrol 409(1–2):497–511

Hidalgo-Muñoz JM, Gámiz-Fortis SR, Castro-Díez Y et al (2015) Long-range seasonal streamflow forecasting over the Iberian Peninsula using large-scale atmospheric and oceanic information. Water Resour Res 51:3543–3567

IPCC (2021) Climate change 2021: the physical science basis. In: Masson-Delmotte V, Zhai P, Pirani A, Connors SL, Péan C, Berger S, Caud N, Chen Y, Goldfarb L, Gomis MI, Huang M, Leitzell K, Lonnoy E, Matthews JBR, Maycock TK, Waterfield T, Yelekçi O, Yu R, Zhou B (eds) Contribution of Working Group I to the sixth assessment report of the intergovernmental panel on climate change. Cambridge University Press (In Press)

Jacob D, Petersen J, Eggert B et al (2014) EURO-CORDEX: new high-resolution climate change projections for European impact research. Reg Environ Change 14:563–578

Jerez S, Montavez JP, Gomez-Navarro JJ et al (2012) The role of the land-surface model for climate change projections over the Iberian Peninsula. J Geophys Res Atmos 117:D01109

Karl TR, Nicholls N, Ghazi A (1999) CLIVAR/GCOS/WMO workshop on indices and indicators for climate extremes: Workshop summary. Clim Change 42:3–7

Kohler T, Wehrli A, Jurek M (2014) Las montañas y el cambio climático: una preocupación mundial. Serie sobre el desarrollo sostenible de las regiones de montaña. Berna, Suiza, Centro para el Desarrollo y el Medio Ambiente (CDE), Agencia Suiza para el Desarrollo y la Cooperación (COSUDE), y Geographica Bernensia

Kumari M, Singh CK, Bakimchandra O, Basistha A (2017) DEM-based delineation for improving geostatistical interpolation of rainfall in mountainous region of Central Himalayas, India. Theor App Climatol 130:51–58

Lopez-Bustins JA, Martin-Vide J, Sanchez-Lorenzo A (2008) Iberia winter rainfall trends based upon changes in teleconnection and circulation patterns. Glob Planet Change 63(2–3):171–176

Marcos-Garcia P, Lopez-Nicolas A, Pulido-Velazquez M (2017) Combined use of relative drought indices to analyze climate change impact on meteorological and hydrological droughts in a Mediterranean basin. J Hydrol 554:292–305

Martín-Vide J, Lopez-Bustins JA (2006) The Western Mediterranean Oscillation and rainfall in the Iberian Peninsula. Int J Climatol 26 (11).1455–1475

Meersmans J, Van Weverberg K, De Baets S et al (2016) Mapping mean total annual precipitation in Belgium, by investigating the scale of topographic control at the regional scale. J Hydrol 540:96–105

Miralles DG, Teuling AJ, van Heerwaarden CC et al (2014) Mega-heatwave temperatures due to combined soil desiccation and atmospheric heat accumulation. Nat Geosci 7:345–349

McKee TB, Nolan J, Kleist J (1993) The relationship of drought frequency and duration to time scales. In: Eighth conference on applied climatology. Amer. Meteor. Soc

Morales-Baquero R, Carrillo P, Barea-Arco J et al (2006) Climate-driven changes on phytoplankton–zooplankton coupling and nutrient availability in high mountain lakes of Southern Europe. Freshw Biol 51(6):989–998

Navarro-Serrano F, López-Moreno JI, Azorín-Molina C et al (2018) Estimation of near-surface air temperature lapse rates over continental Spain and its mountain areas. Int J Climatol 38:3233–3249

New M, Todd M, Hulme M, Jones P (2001) Precipitation measurements and trends in the twentieth century. Int J Climatol 21:1889–1922

Norrant C, Douguédroit A (2006) Monthly and daily precipitation trends in the Mediterranean (1950–2000). Theor App Climatol 83:89–106

Oliva M, Moreno I (2008) Sierra Nevada, nexo entre dos patrones de teleconexión: la NAO y la WeMO. In: Sigró J, Brunet M, Aguilar E (eds) Cambio Climático Regional y sus Impactos. Publicaciones de la Asociación Española de Climatología, Tarragona, Serie A, vol 6, pp 199–208

Peña-Angulo D, Brunetti M, Gonzalez-Hidalgo CC, JC (2016) A new climatology of maximum and minimum temperature (1951–2010) in the Spanish mainland: a comparison between three different interpolation methods. Inter J Geograph Inform Sci 30:2109–2132

Pereira P, Oliva M (2010) Análisis comparativo de diferentes métodos de interpolación en Sierra Nevada. Aplicación a la precipitación extrema de otoño e invierno. In: Fernández F, Galán E, Cañada R (eds) Clima, ciudad y ecosistemas, Publicaciones de la Asociación Española de Climatología, Serie A, vol 7, pp 237–248

Pérez-Luque AJ, Peinó Calero E, Rodríguez-Brito A et al (2021) ClimaNevada: Base de datos climática del Observatorio de Cambio Global de Sierra Nevada. Ecosistemas 30(1):2155

Pérez-Palazón MJ, Pimentel R, Polo MJ (2018) Climate trends impact on the snowfall regime in Mediterranean mountain areas: future scenario assessment in Sierra Nevada (Spain). Water 10:720

Peterson TC, Folland C, Gruza G et al (2001) Report on the activities of the working group on climate change detection and related Rapporteurs 1998–2001. WMO, Rep. WCDMP-47, WMO-TD 1071, Geneve, Switzerland, 143 pp

Pittock J, Hansen L, Abell R (2008) Running dry: freshwater biodiversity, protected areas and climate change. Biodiversity 9 (3–4):30–38

Polo MJ, Herrero J, Pimentel R, Pérez-Palazón MJ (2019) The Guadalfeo monitoring network (Sierra Nevada, Spain): 14 years of measurements to understand the complexity of snow dynamics in semiarid regions. Earth Syst Sci Data 11:393–407

Pulido-Villena E, Reche I, Morales-Baquero R (2008) Evidence of an atmospheric forcing on bacterioplankton and phytoplankton dynamics in a high mountain lake. Aquat Sci 70(1–9):1015-1621/08/010001-9

Quesada B, Vautard R, Yiou P, Hirshchi M, Seneviratne SI (2012) Asymmetric European summer heat predictability from wet and dry southern winters and springs. Nat Clim Change 2:736–741

Rangwala I, Miller JR (2012) Climate change in mountains: a review of elevation-dependent warming and its possible causes. Clim Change 114:527–547

Rangwala I, Sinsky E, Miller JR (2013) Amplified warming projections for high altitude regions of the northern hemisphere mid-latitudes from CMIP5 models. Environ Res Lett 8:024040

Rauthe M, Steiner H, Riediger U et al (2013) A Central European precipitation climatology—Part I: Generation and validation of a high-resolution gridded daily data set (HYRAS). Meteorol Zeitschrift 22:235–256

Rodríguez-Puebla C, Encinas AH, García-Casado LA, Nieto S (2010) Trends in warm days and cold nights over the Iberian Peninsula: relationships to large-scale variables. Clim Change 100:667–684

Ruiz Sinoga JD, Garcia Marin R, Martinez Murillo JF et al (2010) Precipitation dynamics in southern Spain: trends and cycles. Int J Climatol 31(15):2281–2289

Russo A, Sousa PM, Durão RM et al (2020) Saharan dust intrusions in the Iberian Peninsula: predominant synoptic conditions. Sci Total Environ 717:137041

Serrano-Notivoli R, Beguería S, Saz MÁ et al (2017) SPREAD: a high-resolution daily gridded precipitation dataset for Spain—an extreme events frequency and intensity overview. Earth Syst Sci Data 9:721–738

Serrano-Notivoli R, Beguería S, de Luis M (2019) STEAD: a high-resolution daily gridded temperature dataset for Spain. Earth Syst Sci Data 11:1171–1188

Tomas-Burguera M, Vicente-Serrano SM, Beguería S et al (2019) Reference crop evapotranspiration database in Spain (1961–2014). Earth Syst Sci Data 11:1917–1930

Vicente-Serrano SM, Beguería S, López-Moreno JI (2010) A multiscalar drought index sensitive to global warming: the standardized precipitation evapotranspiration index. J Clim 23:1696–1718

Vicente-Serrano SM, Tomás-Burguera M, Beguería S et al (2017) A high resolution dataset of drought indices for spain. Data 2(3):22

Vicente-Serrano SM, McVicar TR, Miralles DG et al (2020) Unraveling the influence of atmospheric evaporative demand on drought and its response to climate change. WIRES Clim Chang 11(2):e632

Vogel MM, Orth R, Cheruy F et al (2017) Regional amplification of projected changes in extreme temperatures strongly controlled by soil moisture-temperature feedbacks. Geophys Res Lett 44:1511–1519

Wilks DS (2011) Statistical methods in the atmospheric sciences. Academic Press, 676 pp

Yeste P, Dorador J, Martin-Rosales W et al (2018) Climate-driven trends in the streamflow records of a reference hydrologic network in Southern Spain. J Hydrol 566:55–72

Zamora R, Barea-Azcón JM (2015) Long-term changes in mountain passerine bird communities in the Sierra Nevada (southern Spain): a 30-year case study. Ardeola 62(1):3–18

Zamora R, Pérez-Luque AJ, Guerrero-Alonso PD et al (2021) Linking macro and microclimate in mountain landscapes: a conceptual and instrumental approach. Ecosistemas 30(1):2166

# Snow Dynamics, Hydrology, and Erosion

María J. Polo, Javier Herrero, Agustín Millares, Rafael Pimentel, Antonio Moñino, María-José Pérez-Palazón, Cristina Aguilar, and Miguel A. Losada

**Abstract**

Sierra Nevada is the southernmost snow area in Europe and its singular geographical location, together with its high altitudes, result in strong gradients of the local climate and hydrology, and rich biodiversity. This chapter aims at providing insights into the major components of the energy and water balance in snowpacks in Mediterranean areas and across the different snow domains found in Sierra Nevada. For this, key descriptors of the snow regime are included from both historical observations and physical modelling during the 1961–2015 period, in the framework of different research projects for the last two decades. The observed relevance of water fluxes to the atmosphere as evaposublimation under these conditions, up to 30–35% of the annual water stored as snow, may be enhanced by the generally observed shift towards torrentiality in the precipitation and snowfall regimes in this site. However, the impacts on hydrology are highly non-linear, with in-season timing of snow events, and their duration, being determinant for the hydrological response and the associated pulse-events of sediment production and deposition downstream. The results highlight the singularity of Sierra Nevada in the context of the high mountain areas in Mediterranean-type regions, and the complexity of the snow dynamics in these areas in the current global warming scenario.

**Keywords**

Sierra Nevada • Snow • Mountain hydrology • Mediterranean regions • Erosion

## 1 Sierra Nevada Hydrology in a World Context

High mountain hydrology is dominated by the glaciers and/or snowpack dynamics, whose seasonal, annual, and decadal regime propagates through the water pathways in the territory, and feeds and interacts with the biogeochemical processes on different spatial scales. Despite these common drivers, hydrological processes in high mountain regions in the world substantially differ on a global scale across the latitudinal gradient, the persistence of glaciers, or the proximity to the sea, among others (Viviroli and Weingartner 2004; López-Moreno et al. 2017). Additionally, local factors result in further small-scale spatial variability along the altitudinal, facing, exposure gradients, which is also enhanced by the geology and soil characteristics, the presence and type of vegetation cover, and the expansion and intensity of human activities, including governance schemes (González-Reyes et al. 2017; Lievens et al. 2019).

The role of mountains as "water towers" for ecosystems and society makes them one of the most likely vulnerable areas to the impacts of global warming on hydrology and biodiversity, especially in the Mediterranean-type regions. Sierra Nevada can be generally described as a Mediterranean mountain area and integrates headwaters from different basins representative of the variability of the hydrological processes in the Mediterranean region (Latron et al. 2009; Fayad et al. 2017; Polo et al. 2020). It is, however, a singular high mountain area in the world context due to its geographical location, between two continents and two very different seas. Moreover, it is the southernmost snow area in Europe, and its small area compared to other high mountain

M. J. Polo (✉) · J. Herrero · R. Pimentel · M.-J. Pérez-Palazón · C. Aguilar
Fluvial Dynamics and Hydrology Research Group, Andalusian Institute for Earth System Research, University of Córdoba, Córdoba, Spain
e-mail: mjpolo@uco.es

A. Millares · A. Moñino · M. A. Losada
Environmental Fluid Dynamics Research Group, Andalusian Institute for Earth System Research, University of Granada, Granada, Spain


R. Zamora and M. Oliva (eds.), *The Landscape of the Sierra Nevada*,
https://doi.org/10.1007/978-3-030-94219-9_10

sites makes Sierra Nevada a "snow island" within a semiarid region, where water resources availability downstream has been historically dependent on the provisioning service regulated by the snow presence and its melting. Its strong altitudinal gradients and high peaks above 3000 m a.s.l., its dominant west–east range axis with decreasing height, and its close proximity to the Mediterranean Sea to the South result in a rich variety of physical domains and ecosystems, from the Alpine high areas, to the Mediterranean and semiarid conditions down the valleys.

Snow islands are prone to earlier suffer the impacts of global warming and, thus, constitute strategic observatories of potential change and tipping points. Moreover, the exposure to highly variable climate conditions makes Sierra Nevada a very special example to study the impact of global warming on snowpacks that may represent different currents of future snow regions in the world, and provide an early-alert assessment of their potential future regime, the implications on water resource availability, and strategies for environmental, economical, and social adaptation.

Figure 1 shows the Sierra Nevada massif in the context of major river basins fed by these mountains, where their headwater areas have been delimited in five regions, Adra, Andarax, Fardes, Genil, and Guadalfeo, named under their principal rivers. These local regions are used in this work to illustrate the spatial variability of the snow domains (Pérez-Palazón 2019) and assess key processes in the snow dynamics, and the associated hydrological and erosive processes in this area.

## 2 Snow Dynamics in a Mediterranean Region

Snowfall occurrence is a key descriptor of high mountain areas, but it is the snow persistence that mostly conditions the snow regime and impacts the local hydrology. The governing energy and water balance in the snowpack, driven by the weather and other physical conditions, results in different fractions of evaporation/sublimation and melting fluxes that decrease or even exhaust the accumulated snow between successive snowfall events, and influence the hydrological regime on different time scales.

### 2.1 Snow in a Mediterranean Region: Why a Special Regime?

Snow evolution is not a single hydrological process that develops with a unique or universal structure, but rather each snowy area responds uniquely and individually to the particular combinations of meteorological (precipitation, temperature, solar radiation, relative humidity, cloudiness, wind speed and pressure) and other biophysical drivers (altitude, orientation, soil type, surface roughness, vegetation, stoniness or the presence of dust), with significant effects on different time and spatial scales.

In a global context, Sierra Nevada can be classified as a "snow area in a Mediterranean-type region", which defines mountain areas with shared features of the snow processes

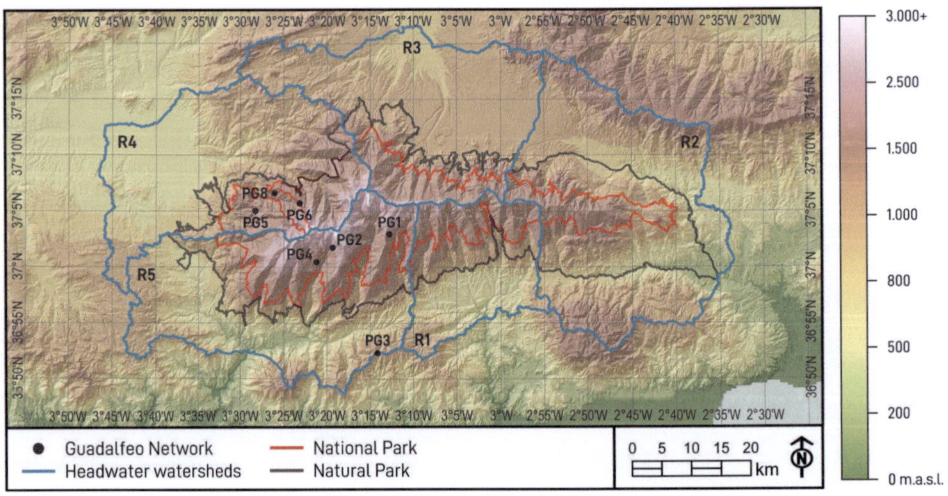


**Fig. 1** The Sierra Nevada and the limits (black line) of the headwaters areas in the five snow-influenced basins in the study area (R1, Adra; R2, Andarax; R3, Fardes; R4, Genil; R5, Guadalfeo), together with the digital elevation model and the limits of the Natural (blue line) and National (violet line) Park areas. The location of the monitoring stations of the Guadalfeo Network (PG) is included

due to their similar climate regime found around the Mediterranean Sea, in the Central Andes of Chile, or in Western-North America (Fayad et al. 2017). These similarities of the snowpacks can be expressed as (1) a highly variable persistence in time and space, (2) the presence of several melting cycles within each annual snow season associated with mild winter weather periods, (3) high snow density and a tendency to create ice layers while evolving rapidly in patches, together with low albedo associated with dust accumulated in the surface (Skiles and Painter 2018), and 4) a not negligible flux of water to the atmosphere as both sublimation and direct evaporation from the snowpack under favourable weather conditions of cold, low air humidity, and strong wind. The latter has a relevant impact on the hydrological regime; in the experimental area of Refugio Poqueira, Herrero and Polo (2016) reported annual "evaposublimation" rates up to a 30–35% of the annual storage of water in the snowpack. Moreover, these areas are particularly exposed to rain-on-snow events (Corripio and López-Moreno 2017) that trigger snowmelt in warm periods, increasing runoff volumes and the possibility of flooding downstream.

When analysing snow properties in detail and comparing with other Mediterranean-type regions of the Earth (López-Moreno et al. 2017), Sierra Nevada appears as a transitional snow domain between the more Alpine character of the Pyrenees and the clear semi-arid character of the Atlas in Morocco or the Chilean Andes. The twentieth century brought the disappearance of the glaciers that shaped the Sierra Nevada valleys above 2000 m during the different glacial periods (Gómez-Ortiz et al. 2012). Only a small rock glacier, formed by buried ice covered by debris, languishes in the Veleta Cirque (Gómez-Ortiz et al. 2014, 2019; see chapter "Ancient and Present-Day Periglacial Environments in the Sierra Nevada"). Moreover, it is extremely unusual that snow persists during the summer season, being 2018 the exception during the past decades. With all, the seasonal variability of weather conditions makes it possible the occurrence of snowfalls in the summits all the year, although they are more likely to produce a sustained snow cover from October to April. However, periods of favourable weather and mild temperatures in the winter months may cause the practically complete ablation of the accumulated snow even in the higher areas of Sierra Nevada as a result of melting and evaposublimation.

Sierra Nevada is a clear paradigm of the usual variability of the snow annual mean and extremal regime in Mediterranean-type regions. There is no story that better exemplifies this than the unavoidable postponement of the celebration of the XXXIII World Alpine Ski Championships in Sierra Nevada, initially planned for 1995, due to the total lack of snow that year, with barely 14 $hm^3$ of snow water equivalent accumulated. Despite the setbacks and uncertain scenario for the next year, the celebration in 1996 was a complete success, with an extraordinary accumulation of more than 20 times the previous year's volume.

This time variability has also a strong spatial component driven by the high altitudinal gradient of Sierra Nevada massif, the different north–south exposure to the sea influence and solar radiation, and its major west-to-east axis parallel to the Mediterranean coastline (Fig. 1). On one side, there is a very strong precipitation gradient from west to east (i.e. average annual precipitation of 900 and 600 mm at 2000 m, respectively) since the Atlantic cyclonic fronts result in much more precipitation, and snowfall, in this area (chapter "Climate Variability and Trends"). On the other side, the greater insolation and exposure to humid winds from the nearby sea impacts the snow duration in the southern face, whereas a generally colder climate favours the persistence of the snowpack in the northern face. As a result, the snowfall north-limit is on average 200–300 m lower than the southern. Another phenomenon very much affected by this north–south difference is the occurrence of hard rime, a striking formation of ice on any surface exposed to the wind during the passage of freezing fog, that is very common on the northern slopes above 2500 m but sporadic to the south.

Finally, it is worth mentioning the exposure of Sierra Nevada to the recurrent Saharan dust events from Africa (see chapter "Atmospheric Inputs and Biogeochemical Consequences in High-Mountain Lakes"). This dust accumulation changes the optical properties of the snowpack surface and enhances melting (Di Mauro et al. 2019). On 23 April 2018, a large-scale Saharan dust event (Fig. 2) lowered the snow albedo from 0.65 to 0.5. The resulting increase of the solar radiation absorption by the dusted snow intensified the melting rates and had an estimated reduction of the snow season of 6 days.

## 2.2 Snow Monitoring in Sierra Nevada: A Historical Perspective

Despite the first precipitation collector gauges dating from the 1940s, the systematic monitoring of the snow in Sierra Nevada was not significant until the twenty-first century, when both ground and remote sensors developed and provided key information for this area. Difficulty of access and the harsh winter conditions are the reasons behind the lack of both direct measurements and reliable sensor records maintained on a long-term basis.

Ground monitoring on an automatic basis is mainly achieved from the different climate stations networks operating in this area. The Meteorological State Agency of Spain (AEMET) network, which recorded daily total precipitation and the standard weather variables, provides the longest locally available data series but mostly from areas below 2000 m, being the snow domain scarcely represented. Since

**Fig. 2** Field monitoring work in Sierra Nevada: left, snow pit at 2500 m in the Monachil Valley after the Saharan dust event on 23 April 2018; right, snow monitoring station PG2-N in Refugio Poqueira, at 2500 m on the southern face of Sierra Nevada

2000, the density of automated stations has been strongly increased by the National and Natural Park network, the Agroclimatic Network of Andalusia (RIA), and others in the Observatory of Global Change in Sierra Nevada. However, these stations are not provided with snowfall collectors, which results in a significant uncertainty of precipitation records during the snow season due to their under catch of snowfall. Since 2004, with the first station at 2500 m in this area, the progressive installation of the automated advanced snow stations in the framework of the Guadalfeo Monitoring Network (PGN; Fig. 1) by the Andalusian Institute for Earth System Research (www.uco.es/dfh/snowmed) has been providing the first continuous monitoring system of snowfall and snowpack properties, and their weather drivers, in Sierra Nevada (Polo et al. 2019; Aguilar et al. 2021). These stations include terrestrial cameras that record different sized scenes (20 to 1000-m cell size) to support the scaling issues for modelling the snowpack evolution in such spatially variable conditions (Pimentel et al. 2015, 2017a) and the validation of data from remote sources (Pimentel et al. 2017b); they provide information on snow cover fraction (SCF), snow deposition or melting patterns, fog, freezing fog, or rain-on-snow events, cloud height or malfunctions in the other sensors.

Among the PGN stations, the Refugio Poqueira station is located in the snow experimental area in the headwaters of the Guadalfeo River basin, which has been operating the only automatic system for monitoring snowpack properties in Sierra Nevada (Fig. 2): thickness, water equivalent, albedo, and surface radiative balance. Additionally, specific field campaigns are achieved on a regular basis to complete the integrated monitoring of snow dynamics in Sierra Nevada, such as in situ measurements of snow density and temperature, surface roughness, grain size and type, layer distribution, among others) by means of snow pits and core samples, or specific essays oriented to modelling calibration (Herrero and Polo 2012, 2016; Algarra and Herrero 2016). These measurements allow for a deeper understanding of local processes, such as the development of ice cover after a snowfall, which may occur after just two–three sunny days of sunshine under moderate daytime temperature, or the efficient compaction of snow due to the sequence of diurnal melting–freezing cycles that quickly increase the initial density of fresh snow from 100–200 to 300–400 g $L^{-1}$, and reaches the spring melting state with values up to 500–600 g $L^{-1}$.

Understanding the processes involved in the energy and water balance that drive the snowpack evolution is key for an appropriate modelling of the snow dynamics in these Mediterranean-type regions. Both model inputs and parameter calibration require high-quality field data sets to reduce the uncertainty of simulations under so many variable conditions. From the initial weather stations in PGN and field campaigns, Herrero et al. (2009) defined the basis of the physical modelling of snow in Sierra Nevada, which has been progressively improved during the last decade (Herrero and Polo 2012, 2016) from the increasingly available information described above, including the field validation of the model estimations of sublimation volumes in this area. The extension of the point calculations to adequately simulate the evolution of the spatial distribution of snow throughout Sierra Nevada was further developed by means of "depletion curves", i.e. experimental relationships between the average snow depth and the snow cover fraction in a given area (Herrero et al. 2011) that can be obtained or improved by the images provided by the time-lapse cameras in the monitoring network (Pimentel et al. 2015). These depletion curves have also shown their capability for describing different types of accumulation–ablation cycles in

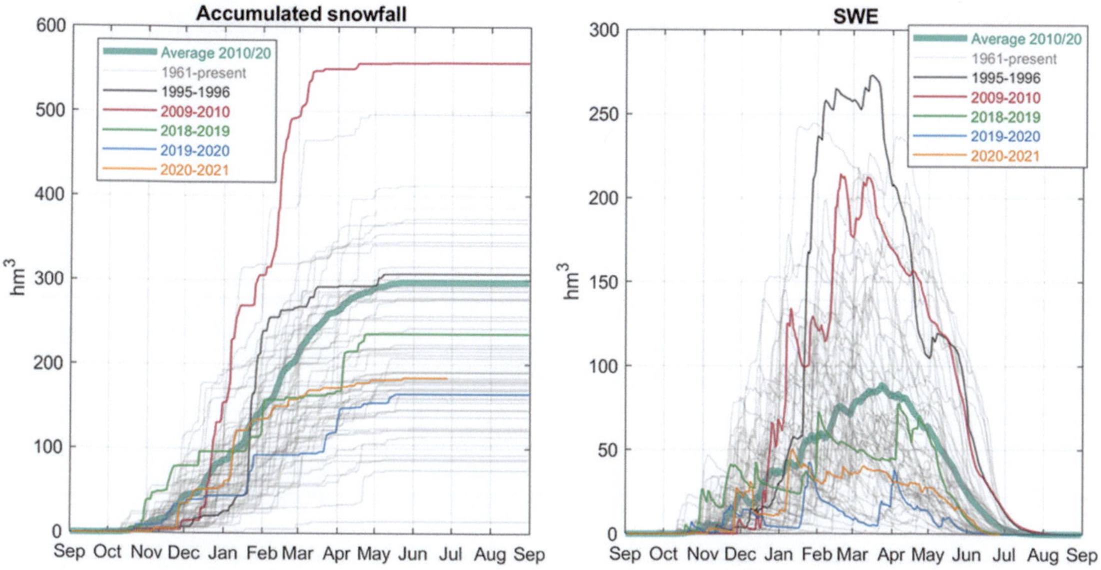


**Fig. 3** Daily evolution of the spatially aggregated (left) annual cumulative snowfall (expressed as liquid water volume) and (right) snow water equivalent (SWE) in Sierra Nevada headwater area (as shown in Fig. 1) for each hydrological year during the 1961–2020 period; selected years are highlighted in the graphs, together with the mean daily values for the last decade. *Note* 100 $hm^3$ is equivalent to a uniform water depth of 21.72 mm throughout the headwater areas

Sierra Nevada, with key applications for modelling the seasonal evolution of even patchy snowpacks during the season (Pimentel et al. 2017a). The distributed modelling of snow in Sierra Nevada was satisfactorily calibrated from SCF values retrieved from Landsat TM 5 and 7 images (Herrero et al. 2011; Pimentel et al. 2017b), with an adequate spatial resolution (30 × 30 m) given the local topographic gradient, and acceptable revisiting period (8–14 days). Currently, different satellite sensors provide high-frequency SCF data (MODIS, daily maps with a 500-m pixel size since the early years of the twenty-first century; Sentinel 2, 5-day maps with 20-m pixel size) for both model validation or time-series analyses of relevant snow-related variables that confirm the complex heterogeneity of the snow regime in this Mediterranean mountain area (Pérez-Luque et al. 2015). This variability is also captured by Fig. 3, which shows the simulated daily evolution of the annual cumulative snowfall and the snow water equivalent (SWE) during each hydrological year in the period 1961–2020 in Sierra Nevada, aggregated over the whole headwater area of its basins (see Fig. 1). Maximum historical values of annual snowfall in 2009–2010 (equivalent to a 558 $hm^3$ of water volume) and SWE in April 1996 (equivalent to 273 $hm^3$) were estimated.

This observation-modelling framework was integrated in 2019 as the Global Monitoring System SNOWMED for the Sierra Nevada area, a freely accessible website (www.uco.es/dfh/snowmed) that provides a daily real-time assessment of the state of the snow cover in Sierra Nevada. For this, the automated collection of the daily weather records from the PGN and other public networks feeds the snow distributed model, corrected with Sentinel 2 SCF maps, and delivers daily 30-m resolution maps of the final SWE and the cumulative fluxes of snowmelt and sublimation throughout the area, together with other variables of interest (e.g. the associated values of aggregated water volumes of precipitation, snowfall, SWE, snowmelt and sublimation for different headwater areas, and their historical annual evolution for comparison).

## 2.3 Recent Evolution and Future Perspectives

Despite point records showing higher precipitation and snowfall values in the northern face, the greatest snow accumulation in Sierra Nevada (and thus, annual snowmelt volume) corresponds to the Guadalfeo River basin (R5 in Fig. 1), in the southern face, due to its largest snow-covered area. However, in this headwater area, the more energy-active snowpack results in lower persistence of snow than on the northern face. This spatial variability of the snow regime is assessed in the long-term by Pérez Palazón (2019), as illustrated in both Fig. 4 and Table 1, that sum up some major annual descriptors of snow occurrence and persistence on a decadal basis throughout the headwater area in Sierra Nevada, as well as averaged per headwater basin (Fig. 1). In this 55-yr analysis, the results show an overall decreasing trend of 2.3 and 0.3 mm $yr^{-1}$ per year of the annual precipitation and snowfall, respectively, in Sierra Nevada, with local differences of values and significance. The detailed trend analysis of recent years made by Pérez-Palazón (2019),

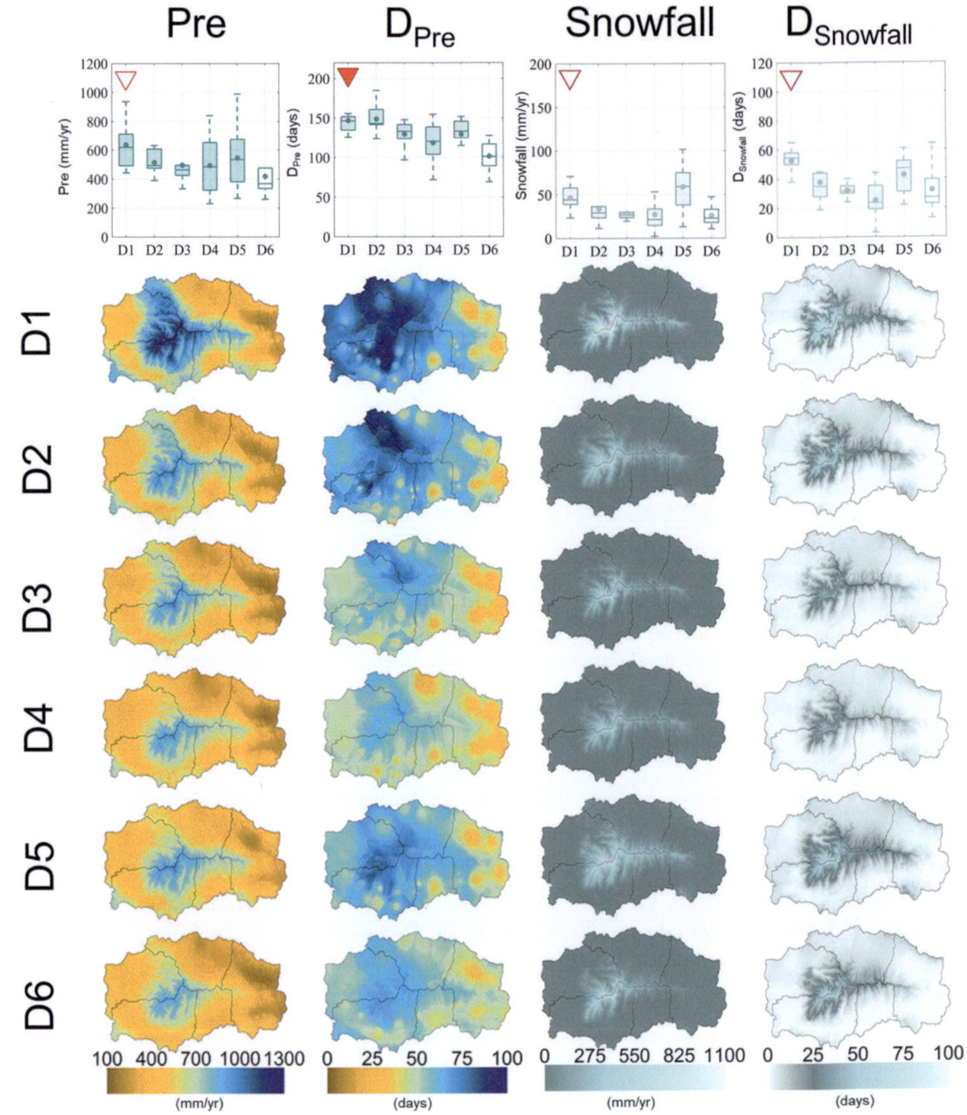


**Fig. 4** Decadal averaged (up) and distributed (down) values of annual precipitation (Pre), number of days with precipitation ($D_{Pre}$), snowfall, and number of days with snowfall ($D_{snowfall}$) during the 1961–2015 period in Sierra Nevada (D1, 1961–70; D2, 1971–80; D3, 1981–1990; D4, 1991–2000; D5, 2001–2010; D6, 2011–2015, uncomplete). Bold arrows indicate statistical trends (Mann–Kendall test with significance level $\alpha = 0.1$). From Pérez-Palazón (2019)

which also offers a glimpse of what future scenarios indicate for the next century (Pérez-Palazón et al. 2018), points to a slight though the steady loss of snow but increasingly torrential snowfalls on the seasonal, annual, and decadal scales. On one side, the northern face, especially Genil (R4 in Fig. 1), shows greater losses of annual snowfall, although in relative values, the eastern area of Andarax (R2), less snowy, is the most affected. On the other hand, the apparent shift towards torrential events may result in negligible changes or even slight local increases of the annual snowfall in the southern areas, Guadalfeo and Adra (R5 and R1). The results highlight, once more, the complex and rich variability of the snow regime in Sierra Nevada, and the strategic opportunities to monitor and track the potential future regime of very different snow domains in this site.

## 3 Hydrology and Snow Signature

High mountain hydrology is determined by the local dynamics of snow, which covers the soil surface and partially or totally the vegetation cover, decreases infiltration rates and direct runoff from precipitation during the cold periods but enhances infiltration volumes and aquifer recharge during melting periods, delays the water delivery to the fluvial network, and increases spring and summer river flows and soil moisture, extending the flow season in ephemeral rivers. This general signature is, however, largely affected by the timing and sequence of wet and dry periods, and their temperature range, during the snow season in Mediterranean areas like Sierra Nevada, due to the specific

**Table 1** Statistics of the annual precipitation and snowfall regime per headwater region (see Fig. 1) in Sierra Nevada during the 1961–2015 period, and annual trend results from Mann–Kendall test (***, α = 0.01; **, α = 0.05; *, α = 0.1; +, no significance). From Pérez-Palazón (2019)

| | Maximum value | Mean value | Minimum value | Standard deviation | Trend ($yr^{-1}$) |
|---|---|---|---|---|---|
| **Annual Precipitation (mm $yr^{-1}$) [$hm^3$]** | | | | | |
| ADRA (R1) | 1242.3 [570.4] | 581.3 [266.9] | 230.8 [106.0] | 206.6 [94.9] | −1.73 (*) |
| ANDARAX (R2) | 894.7 [1045.9] | 388.3 [453.9] | 267.9 [313.2] | 134.6 [157.3] | −1.14 (*) |
| FARDES (R3) | 883.8 [808.2] | 471.2 [430.9] | 171.5 [156.8] | 151.2 [138.3] | −2.18 (**) |
| GENIL (R4) | 1146.0 [1126.7] | 571.05 [561.4] | 215.8 [212.2] | 183.8 [180.7] | −3.88 (**) |
| GUADALFEO(R5) | 1193.3 [1262.5] | 660.43 [698.7] | 217.5 [230.1] | 226.6 [239.7] | −2.66 (**) |
| *TOTAL AREA* | *989.4 [4535.1]* | *526.20 [2412.0]* | *241.5 [1107.0]* | *169.9 [778.8]* | *−2.34 (*)* |
| **Annual number of days with precipitation (days $yr^{-1}$)** | | | | | |
| ADRA (R1) | 147 | 98.0 | 50 | 18.5 | −0.32 (*) |
| ANDARAX (R2) | 146 | 102.1 | 54 | 20.3 | −0.42 (*) |
| FARDES (R3) | 185 | 118.9 | 65 | 22.4 | −0.83 (***) |
| GENIL (R4) | 156 | 109.7 | 56 | 20.6 | −0.65 (***) |
| GUADALFEO(R5) | 164 | 114.9 | 65 | 21.5 | −0.57 (***) |
| *TOTAL AREA* | *185* | *131.8* | *69* | *23.9* | *−0.75 (***)* |
| **Annual number of days with temperature below 0 °C (days $yr^{-1}$)** | | | | | |
| ADRA (R1) | 37 | 9.9 | 0 | 8.9 | 0.09 (+) |
| ANDARAX (R2) | 28 | 9.3 | 0 | 6.4 | 0.02 (+) |
| FARDES (R3) | 104 | 50.4 | 12 | 17.8 | −0.21 (+) |
| GENIL (R4) | 80 | 37.5 | 5 | 16.2 | −0.11 (+) |
| GUADALFEO(R5) | 67 | 19.9 | 1 | 12.4 | 0.15 (+) |
| *TOTAL AREA* | *56* | *22.4* | *1* | *12.2* | *0.05 (+)* |
| **Annual Snowfall (mm $yr^{-1}$) [$hm^3$]** | | | | | |
| ADRA (R1) | 103.1 [47.3] | 29.3 [13.5] | 0.7 [0.3] | 22 [10.1] | −0.25 (+) |
| ANDARAX (R2) | 47.1 [55.1] | 13.5 [15.8] | 0.7 [0.8] | 11.4 [13.3] | −0.21 (+) |
| FARDES (R3) | 81.5 [74.5] | 27.8 [25.4] | 1.6 [1.5] | 18 [16.5] | −0.02 (+) |
| GENIL (R4) | 134.9 [132.6] | 54.9 [54.0] | 3.8 [3.7] | 29.6 [29.1] | −0.36 (+) |
| GUADALFEO(R5) | 150.2 [158.9] | 60.3 [63.8] | 3.2 [3.4] | 34.2 [36.2] | −0.34 (+) |
| *TOTAL AREA* | *101.4 [464.8]* | *37.6 [172.3]* | *2.1 [9.6]* | *21.7 [99.5]* | *−0.30 (+)* |
| **Annual Number of days with snowfall (days $yr^{-1}$)** | | | | | |
| ADRA (R1) | 45 | 23.2 | 1 | 10.3 | −0.22 (**) |
| ANDARAX (R2) | 30 | 16.1 | 1 | 8.5 | −0.36 (**) |
| FARDES (R3) | 55 | 28.6 | 1 | 12.5 | −0.39 (***) |
| GENIL (R4) | 80 | 42.4 | 6 | 15.4 | −0.42 (***) |
| GUADALFEO(R5) | 78 | 42.4 | 4 | 15.3 | −0.36 (**) |
| *TOTAL AREA* | *75* | *38.7* | *4* | *14.8* | *−0.42 (***)* |
| **Annual Mean SWE (mm) [$hm^3$]** | | | | | |
| ADRA (R1) | 3.7 [1.7] | 0.6 [0.3] | 0.00 [0.00] | 0.82 [0.4] | −0.002 (**) |
| ANDARAX (R2) | 2.1 [2.5] | 0.3 [0.3] | 0.00 [0.00] | 0.39 [0.5] | −0.003 (**) |
| FARDES (R3) | 3.1 [2.8] | 0.8 [0.7] | 0.00 [0.00] | 0.78 [0.7] | −0.01 (***) |
| GENIL (R4) | 19.0 [18.7] | 4.6 [4.6] | 0.02 [0.02] | 4.33 [4.3] | −0.11 (***) |
| GUADALFEO(R5) | 19.6 [20.7] | 3.8 [4.0] | 0.02 [0.01] | 4.03 [4.3] | −0.07 (***) |
| *TOTAL AREA* | *8.9 [41.1]* | *2.2 [9.9]* | *0.01 [0.04]* | *2.12 [9.7]* | *−0.04 (***)* |

(continued)

**Table 1** (continued)

| | Maximum value | Mean value | Minimum value | Standard deviation | Trend ($yr^{-1}$) |
|---|---|---|---|---|---|
| **Annual duration of snow (days $yr^{-1}$)** | | | | | |
| ADRA (R1) | 35.9 | 13.6 | 0.6 | 7.2 | −0.05 (*) |
| ANDARAX (R2) | 24.4 | 8.1 | 0.5 | 5.2 | −0.06 (+) |
| FARDES (R3) | 36.2 | 16.4 | 1.1 | 7.81 | −0.08 (+) |
| GENIL (R4) | 51.7 | 28.3 | 3.3 | 10.3 | −0.16 (**) |
| GUADALFEO(R5) | 49.3 | 24.6 | 2.2 | 9.2 | −0.03 (+) |
| *TOTAL AREA* | *37.2* | *18.4* | *1.6* | *7.5* | *−0.08 (+)* |

features of snow dynamics described in Sect. 2. More specifically, the evaposublimation rates (Herrero and Polo 2016) found in these areas constitute a loss to the atmosphere from the point of view of the available seasonal water for rivers and aquifers; this loss is highly variable during the year and among years, with higher volumes usually found when the precipitation is concentrated in less number of events.

## 3.1 The Snow Signature in the Hydrological Regime in Sierra Nevada: A Diverse Scenario

The Sierra Nevada mountain constitutes a massive volume of intrinsically non-permeable and majorly metamorphic materials, with low-depth soils if any in the highest areas, and a dominant rocky surface; the treeline is found approximately at 2300 m, with pastures, shrubs, and other low height covers being the predominant vegetation. Despite the absence of aquifers in a strict sense, restricted to boundary systems in the lowland limits of the massif, the highly and deep fractured nature of these materials results in a highly permeable surface structure that stores precipitation and snowmelt volumes and delays water inflow to the rivers, especially in snowy and cold years. As a result, surface direct runoff is low when compared to other basins with similar pluviometry, being the slower subsurface and base flows significant components of the fluvial regime, that lack a direct connection river-aquifer in this area. This delay is enhanced in the southern face also due to the existence of "acequias de careo", ancient traditional recharge systems dating from the Arab time that still derive snowmelt towards infiltration areas that downstream spring and feed agricultural areas downhill in the valleys (Barberá et al. 2018; Martos-Rosillo et al. 2019; chapter "Singular Cultural Landscapes of the Sierra Nevada"). Moreover, this structure also explains the large amount of springs found in Sierra Nevada (Castillo 2002) which, together with the mountain remaining glacier lagoons (Castillo 2016), constitute indirect sensors of the hydroclimatic seasonal regime every year, as they partially/totally dry or replenish, or get frozen or melt in the case of the lagoons, highly-valued ecosystems directly affected by the climate shifts (González-Olalla et al. 2018). Citizen science-based activities (chapter "Advancing Open Science in Sierra Nevada: Current Citizen Science Campaigns") have largely contributed to a valuable repository of image data sets in an interesting project led by the University of Granada (https://lagunasdesierranevada.es/), from which timing of different states during a wide set of years is feasible to retrieve, providing key information for climate, hydrology and ecology.

With all, the north–south and west–east gradients create a diverse scenario of the hydrological regime in the associated basins of Sierra Nevada, linked to its different snow domains as described in Sect. 2 and summed up in Table 1. In an attempt to simplify, precipitation (and snowfall) decreases on an annual basis along the west–east axis, whereas the seasonal snow duration decreases along the north–south axis. The clear trend towards torrentiality of the precipitation regime (Table 1), with higher significance in the wettest areas, results in longer dry periods between rain-/snowfall events; this affects the dynamics of the energy and water balance in the snowpack and, among others, propitiates the dominance of conditions that favours evaposublimation rates. Table 2 sums up the annual regime of the water fluxes resulting in the energy and water balance in the different regions in Sierra Nevada, and their trends, during the study period 1961–2015. Beyond the interest of the local and global values of snowmelt volumes for assessing water delivery, the results highlight the importance of evaposublimation rates in this mountain area. As a whole, this volume represents approximately 25% of the annual water stored as snow, and it is equivalent to one-third of the melting volume. Interestingly, the variability of the precipitation and temperature patterns in this area on both the annual and, specifically, the seasonal scales, show no clear trend in the annual evaposublimation rates in any region, whereas the corresponding snowmelt values are decreasing in the

**Table 2** Statistics of the annual water fluxes from the snowpack and snow cover extension per headwater region (see Fig. 1) in Sierra Nevada during the 1961–2015 period, and annual trend results from Mann–Kendall test (***, $\alpha = 0.01$; **, $\alpha = 0.05$; *, $\alpha = 0.1$; +, no significance). From Pérez-Palazón (2019)

| | Maximum value | Mean value | Minimum value | Standard deviation | Trend ($yr^{-1}$) |
|---|---|---|---|---|---|
| **Annual Melting (mm $yr^{-1}$) [$hm^3$ $yr^{-1}$]** | | | | | |
| ADRA (R1) | 74.8 [34.3] | 25.4 [11.7] | 0.6 [0.3] | 17.8 [8.2] | −0.07 (+) |
| ANDARAX (R2) | 46.2 [54.1] | 11.7 [13.8] | 0.7 [0.8] | 9.8 [11.5] | −0.05 (+) |
| FARDES (R3) | 87.2 [79.7] | 25.0 [22.9] | 2.1 [1.9] | 16.2 [14.8] | −0.03 (+) |
| GENIL (R4) | 110.6 [108.7] | 46.9 [46.1] | 1.6 [1.6] | 25.7 [25.3] | −0.53 (***) |
| GUADALFEO (R5) | 121.6 [128.7] | 48.4 [51.2] | 1.6 [1.7] | 27.0 [28.6] | −0.27 (**) |
| *TOTAL AREA* | *87.7 [402.2]* | *31.7 [145.7]* | *1.4 [6.7]* | *17.8 [8.2]* | *−0.20 (**)* |
| **Annual Evaposublimation (mm $yr^{-1}$) [$hm^3$ $yr^{-1}$]** | | | | | |
| ADRA (R1) | 39.8 [18.3] | 5.1 [2.3] | 0.04 [0.02] | 7.3 [3.4] | 0.11 (+) |
| ANDARAX (R2) | 16.6 [19.4] | 2.2 [2.6] | 0.00 [0.00] | 2.9 [3.4] | 0.03 (+) |
| FARDES (R3) | 21.8 [19.9] | 3.4 [3.1] | 0.03 [0.03] | 3.9 [3.6] | 0.06 (+) |
| GENIL (R4) | 56.9 [55.9] | 14.2 [14.0] | 0.31 [0.30] | 10.7 [10.5] | 0.19 (+) |
| GUADALFEO (R5) | 71.9 [76.2] | 16.6 [17.6] | 0.27 [0.28] | 14.4 [15.2] | 0.27 (+) |
| *TOTAL AREA* | *41.0 [188.0]* | *8.6 [39.5]* | *0.13 [0.2]* | *7.7 [35.3]* | *0.13 (+)* |
| **Annual Mean Fractional Snow Cover ($m^2m^{-2}$) [$km^2$]** | | | | | |
| ADRA (R1) | 0.04 [18.4] | 0.012 [5.5] | 0.00 [0.0] | 0.009 [4.1] | −2.44·e-5 (+) |
| ANDARAX (R2) | 0.03 [35.1] | 0.006 [7.0] | 0.00 [0.0] | 0.005 [5.8] | −3.47·e-5 (*) |
| FARDES (R3) | 0.04 [36.6] | 0.013 [11.9] | 0.00 [0.0] | 0.008 [7.3] | −6.34·e-5 (+) |
| GENIL (R4) | 0.08 [78.7] | 0.036 [35.4] | 0.00 [0.0] | 0.017 [16.7] | −3.65·e-4 (***) |
| GUADALFEO (R5) | 0.07 [74.1] | 0.032 [33.9] | 0.00 [0.0] | 0.016 [19.6] | −1.67·e-4 (*) |
| *TOTAL AREA* | *0.05 [229.2]* | *0.021 [96.3]* | *0.00 [0.0]* | *0.011 [50.4]* | *−1.42·e-4 (*)* |

snowiest regions, up to a 1 and 0.5%, approximately, of their mean annual rate in Genil (R4, northwest) and Guadalfeo (R5, southwest), respectively. There is not a single underlying process that explains this decrease but rather a concomitant combination of drivers, being snow occurrence and its spatial pattern, the duration of the events, the shift towards torrentiality, and the apparent increase of evaposublimation losses to the atmosphere, the most relevant ones. The observed trends of the associated mean snow cover area (Table 2) support this analysis.

## 3.2 Seasonality of the Fluvial Regime: The Coevolution of a Natural System and Traditional Man-Made Waterways

The seasonal regime of river flow in Sierra Nevada follows the dynamics of snow, with water storage in the snowpack during cold periods preventing runoff generation in the snow-covered areas, water delivery to rivers during melting events by means of both direct runoff and subsurface flow, and aquifer recharge. The occurrence of snowfall laminates peakflows due to direct runoff and delays the flooding periods until warmer conditions, which increases river flow in spring and summer from the expected values under a rainfall-dominated regime, with final recession periods beyond the dates usually found in similar climate catchments. However, this general annual pattern shows a significant variability on a seasonal basis from year to year, depending not only on the timing of snowfall and rainfall events but also on their associated precipitation volumes and spatial distribution. In this context, the increasing importance of evaposublimation annual loss to the atmosphere (see Fig. 5a for the relative partition of the annual water stored in the snowpack into melting and evaposublimation) adds uncertainty to the expected water release to the river system, especially if the recent decade's trend is maintained in the future (Pérez-Palazón et al. 2018).

This annual shift towards enhanced evaposublimation volumes is not uniformly distributed during the year, as Fig. 5b clearly depicts. The general shortening of the snow season can be easily seen from the monthly trends of both melting and evaposublimation being significantly decreasing from April on, and maximum monthly values occurring in winter. Moreover, there is an apparent increase of snowfall events in autumn but with limited persistence and volumes, which causes a very light

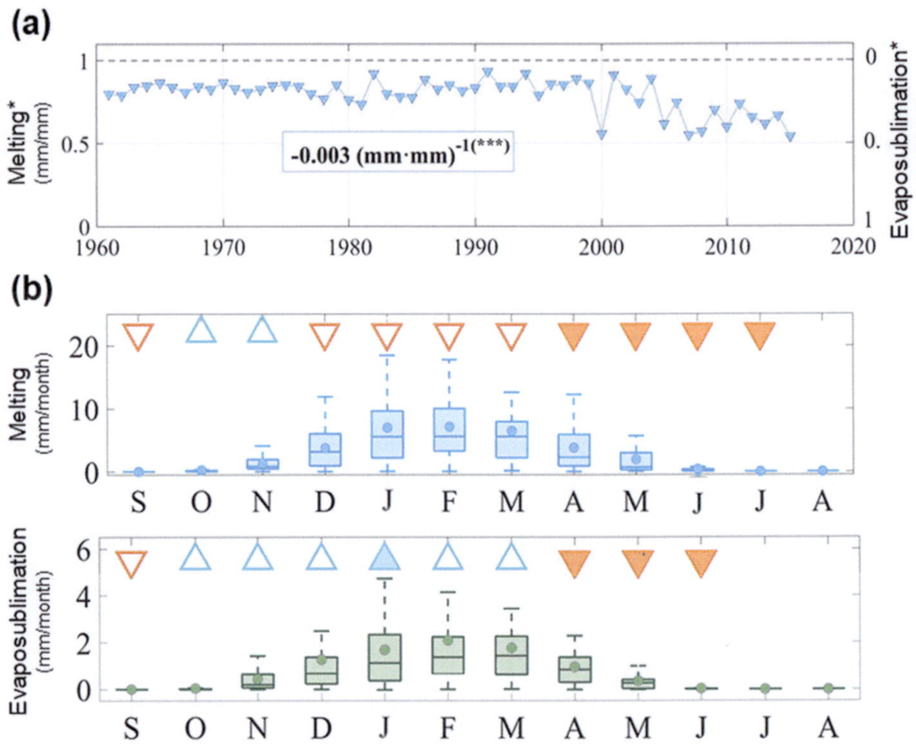


**Fig. 5** Annual and monthly evolution of melting and sublimation rates averaged over the study area in Sierra Nevada (see Fig. 1) during the 1961–2015 period: **a** annual values of the relative melting (melting* = annual melting volume/sum of annual melting and evaposublimation volumes)) and evaposublimation (evaposublimation* = annual evaposublimation volume/sum of annual melting and evaposublimation volumes); **b** mean monthly values of melting and evaposublimation represented by whisker boxes, together with their annual trend (up and down arrows stand for positive or negative trend, respectively, where bold arrows represent statistical significance with $\alpha > 0.1$)

increase of water fluxes during October and November. More interestingly, evaposublimation experiments an apparent increase during all the wintertime in contrast to the decrease shown by melting; the significant trend found for the increased January evaposublimation rate highlights the relevance of this process in Mediterranean mountain areas and its implications for water availability during spring and summer periods. Further work is ongoing to confirm these trends with the observations and modelling of the current and future decades, since the variability of the climate regime and the large spatial gradients in this area condition the statistical significance of the results and pose some key questions for the future of snow in Sierra Nevada, and the future local hydrological regime, without simple cause–effect relationships.

The human modification of the hydrological pathways in Sierra Nevada since ancient times has resulted in a coevolution of the natural environment with man-made derivations of water ("acequias") that mainly recharge the subsurface flow and feed springs and land downstream, and distribute melting water for supply and traditional agriculture system (chapters "Singular Cultural Landscapes of the Sierra Nevada" and "Local Ecological Knowledge and the Sustainable Co-Management of Sierra Nevada's Social-Ecological System"). These traditional uses of water alter the local hydrological regime and transfer water between adjacent catchments, with extreme affections in some cases due to more recent developments of the agricultural land, and make it difficult to model the hydrological response of these head areas due to the lack of gauge records to quantify these abstractions.

This complex hydrological scenario results in an extremely high variability of the annual fluvial discharge in the catchments fed by Sierra Nevada, with non-linear relationships between flow and precipitation on such a time scale. Figure 6 represents the annual discharge recorded at three different rivers in the western basins in Sierra Nevada, which are representative of snow-influenced streams: Órgiva, in the Guadalfeo River basin to the south, and Canales and Quéntar, in the Genil River basin to the north (being Quéntar the less influenced by snow volumes). Unfortunately, the generalized lack of a long time series of river flow makes it not feasible to perform a sound analysis of joint trends of precipitation-discharge, and time series filling clearly dampers the extreme values and masks the intrinsic complexity of these systems. However, two points deserve attention. Firstly, the data do represent the sequence of wet–dry cycles during the study period and highlight the strong variability of the fluvial discharge on a year-to-year basis in these

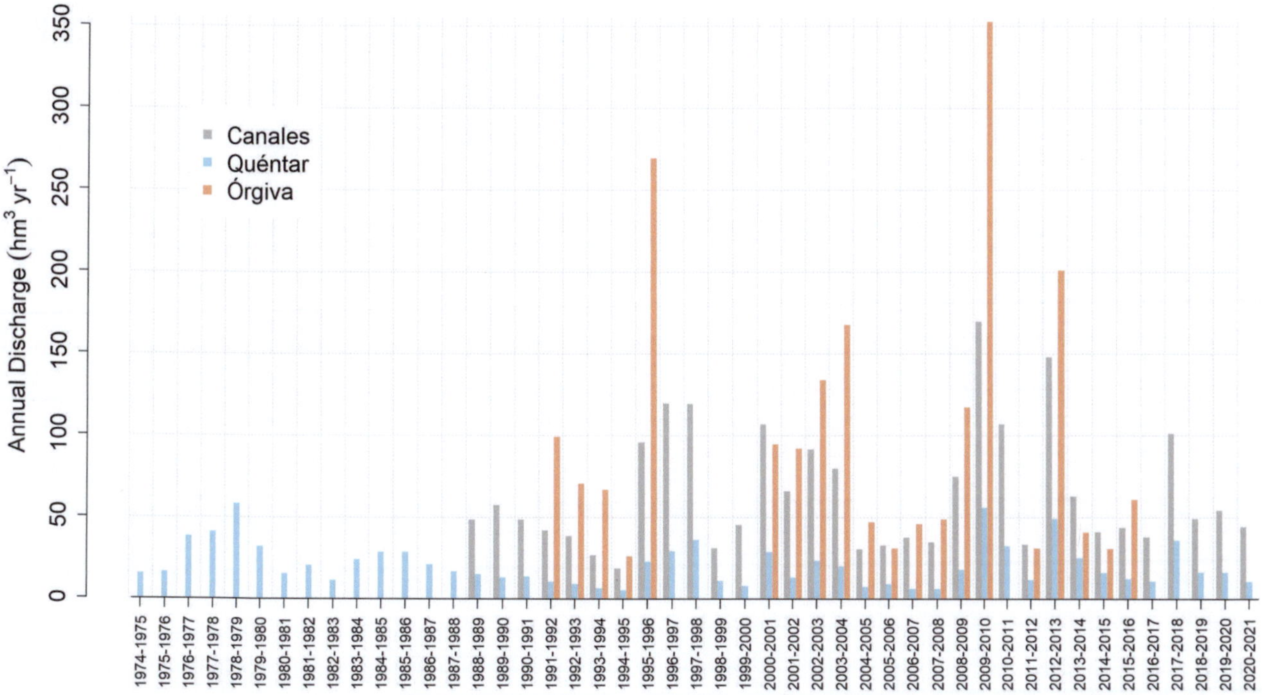


**Fig. 6** Annual discharge in selected river sections in the western regions in the Sierra Nevada area during the period 1974–2020 estimated from observed daily flow series (years with no data correspond to ungauged periods or scarce observations): inflow to the Canales and Quéntar reservoirs in the Genil river basin, and Órgiva gauge station in the Guadalfeo river basin (R4 and R5, respectively, in Fig. 1)

catchments. Secondly, despite the resulting seasonal shifts of the hydrological balance, it cannot be concluded that, on an annual basis, the fluvial discharge is experimenting any clear decrease during the last decades, even though the observations reflect a decrease of precipitation (Table 1), with some very high flows occurring in the last decades. The enhanced runoff due to not only snowfall but also rainfall concentration into less events during the wet season, and the increase of surface soil moisture caused by more frequent rain-on-snow events during the cold period, might be behind this, with overall volumes that likely ameliorate the evapo-sublimation loss impact on the water balance and resource availability. All this merely justify the need for further insight into this singular hydrological system to contrast and validate this and other potentially feasible hypotheses, since the seasonal variations of the fluvial regime pose non-negligible implications from the ecosystem perspective and the water resource management decision-making.

## 4 Soil Erosion and Transport Processes

Mediterranean mountain basins have historically been characterized by significant challenges in the management of erosion and sedimentary processes. The long-term monitoring in these environments, highly sensitive to the effects of global warming, is key to understanding the loss of services associated with soil retention and the impacts they produce downstream. Interest in evaluating soil loss and sediment transport processes of these areas has increased significantly in recent years due to this sensitivity to climatic variations. Changes in the hydrometeorological agents can lead to seasonal variations in water flow as seen in the previous section, from spring snowmelt to direct runoff in winter, with significant variations on erosion and sediment transport (López-Moreno 2005). However, the historical analysis of land-uses reveals that climate drivers are only one part of the erosive and sedimentary dynamics. Large anthropic transformations on the hillslopes of the Sierra Nevada have led to increased soil loss and impacts on fluvial processes with significant costs downstream.

### 4.1 Soil Loss in Semi-arid High Mountainous Environments

From soil disaggregation by raindrop impact, concentrated flow in rills or gullies, to hydrodynamic processes along the channel, sediment flow has to be understood as a concatenation of interrelated processes. In semi-arid high mountain basins, the large spatio-temporal variability of hydrometeorological agents, the torrential nature of rainfall, and the

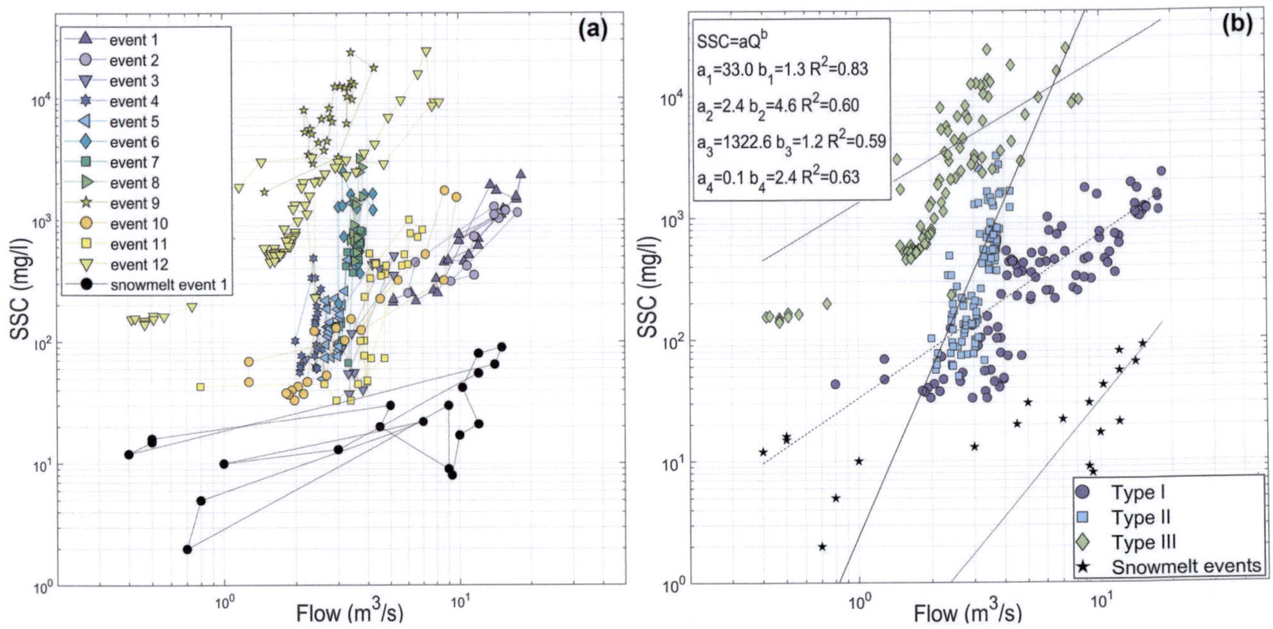


**Fig. 7** Water discharge vs suspended sediment concentration (SSC) for 12 measured events at the Órgiva gauge station, southern side of Sierra Nevada (**a**). Potential model SSC = a$Q^b$, fitted parameters a and b, and determination coefficient are shown for the three identified groups of events and snowmelt periods (**b**) (modified from Millares and Moñino 2020)

dynamics of snowmelt cycles condition very important soil loss and sediment transport rates in hillslopes and river systems (Lana-Renault and Regüés 2009). The attenuation effect of erosive processes related to snowfall instead of rainfall occurrence is subsequently counteracted with rapid snowmelt cycles that can produce very important erosive rates in a timely manner (Ollesch et al. 2006; Tanasienko et al. 2011; Millares and Moñino 2020). At the river network, the torrential nature of the channels of headwater basins conditions transport processes in the form of debris flows that are stored in the form of alluvial fans downstream (Santi et al. 2010). This storage constitutes, in turn, an important source for fluvial contributions from the main rivers.

In Sierra Nevada, the hydrometeorological drivers have been related to up to three different erosion and sediment transport responses by suspended load, comparable with other Mediterranean environments and snowmelt-driven basins (Torres et al. 2013; Millares et al. 2018; Millares and Moñino 2020). Figure 7 shows the relationship between suspended sediment and water discharge in the Guadalfeo river basin, in the southern face of Sierra Nevada. These responses allowed for events typification from their relationship with source areas and drivers. Type I and II events were related to winter events linked to snow dynamics, while type III corresponded to autumn events of high precipitation intensity and modest snowmelt contribution. Although the frequency of type III events is lower in Sierra Nevada, their high erosive rates highlight the importance of snow dynamics in laminating erosive processes. Changes in hydro-meteorological drivers in mountainous areas can lead to seasonal shifts in water flow, from spring thaw to direct runoff in winter, with significant variations in sediment contributions (López-Moreno et al. 2008). These results clearly point out the vulnerability of semi-arid high mountain areas to erosive dynamics and justify their long-term monitoring as hotspots of future challenges for water resource management.

In the fluvial system, the main rivers surrounding Sierra Nevada show a quasi-ephemeral behavior in water and sediment fluxes. The dynamics of snow accumulation-ablation cycles condition the riverbed configuration and, therefore, the processes of equimobility transport of sediment during intense events (Moñino et al. 2011; Millares et al. 2014). These inputs have a relevant impact downstream on river morphodynamics, ecosystems, and existing infrastructures (Arjona et al. 2018).

## 4.2 Sediment Yield in Sierra Nevada and Impacts Downstream

During the past 25 years, different approaches were followed to estimate erosion rates on the hillslopes of Sierra Nevada. From parametric models such as the revised universal soil loss equation (RUSLE) (Renard et al. 1997), the averaged specific sediment yield (SSY) for the southern area of the Sierra Nevada ranges from 5 to 50 t (ha yr)$^{-1}$ (Vázquez 1995; DGMNPF 2007). Plot-scale studies obtained from rainfall simulators provided ranges from 2 to 25 t (ha yr)$^{-1}$ (Francia et al. 2006). These results contrast with the values estimated in lowland areas of Sierra Nevada and surrounding sierras, that exceed 200 t (ha yr)$^{-1}$, and very intense erosive impacts of small-scale landslides and gullying (DGMNPF 2007; Millares et al. 2012, 2019). From 15 years of monitoring of accumulated sediment in the Rules dam, in the Guadalfeo river basin, Millares and Moñino (2018) estimated 19–22 t (ha yr)$^{-1}$ of SSY considering both hillslope and channel contributions. These annual ranges are highly variable over time according to the pulse dynamics of Mediterranean mountain areas. From a historical point of view, the Sierra Nevada basins have evolved under the action of very intense rainfall events. Extreme events such as the 1973 event on the southern slope of Sierra Nevada, with maximum flows estimated at 1000 m$^3$s$^{-1}$, have been capable to generate erosive rates of up to 60–65 t (ha yr)$^{-1}$ (Jabaloy-Sánchez et al. 2014). A certain latitudinal pattern of these ranges can be found when compared with other Mediterranean mountainous areas. From north to south, rates of 1.5–4 t (ha yr)$^{-1}$ have been reported in the Italian Alps (de Vente et al. 2006), 2–5.5 t (ha yr)$^{-1}$ in basins of the Spanish Pyrenees (Batalla and Vericat 2011), 19–24 t (ha yr)$^{-1}$ in Sierra Nevada, and 10–16 t (ha yr)$^{-1}$ in the Moroccan High Atlas (Alahiane et al. 2016). However, local differences in soil erodibility, topography, sediment connectivity, or other anthropic factors make this latitudinal behavior of SSY even more variable.

The impact associated with erosion processes downstream can be analyzed from two points of view. On the one hand, there is a loss of the edaphic profile, in many cases very difficult to recover due to the high topographic gradients and the scarce existing soil thickness. The loss here is very broad and ranges from decreasing subsurface recharge capacity (aquifers recharge, dynamics of subsurface flow, etc.), soil fertility and physical quality, and biodiversity (Orgiazzi and Panagos 2018), to the abandonment of agricultural land and depopulation. On the other hand, transport processes along the river cause severe impacts on aquatic ecosystems and hydraulic structures. In Sierra Nevada, some examples can be related to loss of water supply or changes and destabilization of river morphology. These impacts highlight the importance of headwater basins, in many cases of high natural value, and their influence on the conservation of different services.

### 4.3 Long-Term Assessment of Erosion and Transport Processes

The Sierra Nevada massif has been subjected to major land-use transformations throughout its history which have had a special impact on soil erosion, transport, and sedimentation processes. The intense deforestation developed during the 16th to eighteenth centuries conditioned very high erosive rates that had a very relevant and direct impact on the Mediterranean coastal area downstream the southern face, with historical records of progression rates of the coastline: 0.15 m yr$^{-1}$ during 2000 BC to 1500 AD versus 3.3 m yr$^{-1}$ during 1500 to 1872 AD (Hoffmann 1987). Moreno-Llorca et al. (2020) evaluated the loss of soil retention services in Sierra Nevada compared to other ecosystem services from available land use information for the periods 1956, 1977, 1984, 1999, and 2007. Soil erosion prevention, as t (ha yr)$^{-1}$, was calculated through physically based distributed hydrological modelling by means of the WiMMed model (Polo et al. 2009), with soil loss by rilling and raindrop impacts (Fig. 8) assessed as described in (Millares et al. 2019). It was found that, after a period of deforestation and extensive land use, soil retention services gradually increased as grassland and shrubland expanded. These findings demonstrated that the prevalence of provisioning services is replaced by regulating services after agricultural abandonment as previously reported (Vigl et al. 2016). These results are consistent with the major socio-ecological changes that occurred in Sierra Nevada, mainly strong depopulation and land abandonment of annual and permanent crops (García-Martínez 1995), as well as high elevations.

Future projections based on information from different climate scenarios available in EUROCORDEX highlight the difficulty in estimating impacts on erosion processes. While results from Global Circulation Models (GCMs) point to an increase in erosive processes linked to a drastic decrease in snow presence and higher precipitation intensity, other models point to a decrease in erosive processes due to a significant decrease in precipitation (Millares et al. 2021).

All the above shows the high sensitivity of Sierra Nevada basins to impacts of global change and the interest of these areas as global observatories in order to understand landscape evolution and planning effective management strategies for the near and long-term future.

## 5 Final Remarks

Snow occurrence and persistence are the inner soul of Sierra Nevada environment. The complexity of the snow regime in the Mediterranean and semi-arid regions is here further enhanced by the singular topography of this mountain and its vicinity to the sea, with seasonal, annual, and decadal variability that does not always result in clear trends for the associated weather and hydrological variables. Moreover, the spatial heterogeneity of the snow regime defines different domains across Sierra Nevada, with significant hydrological variations. However, the results gathered in this chapter point

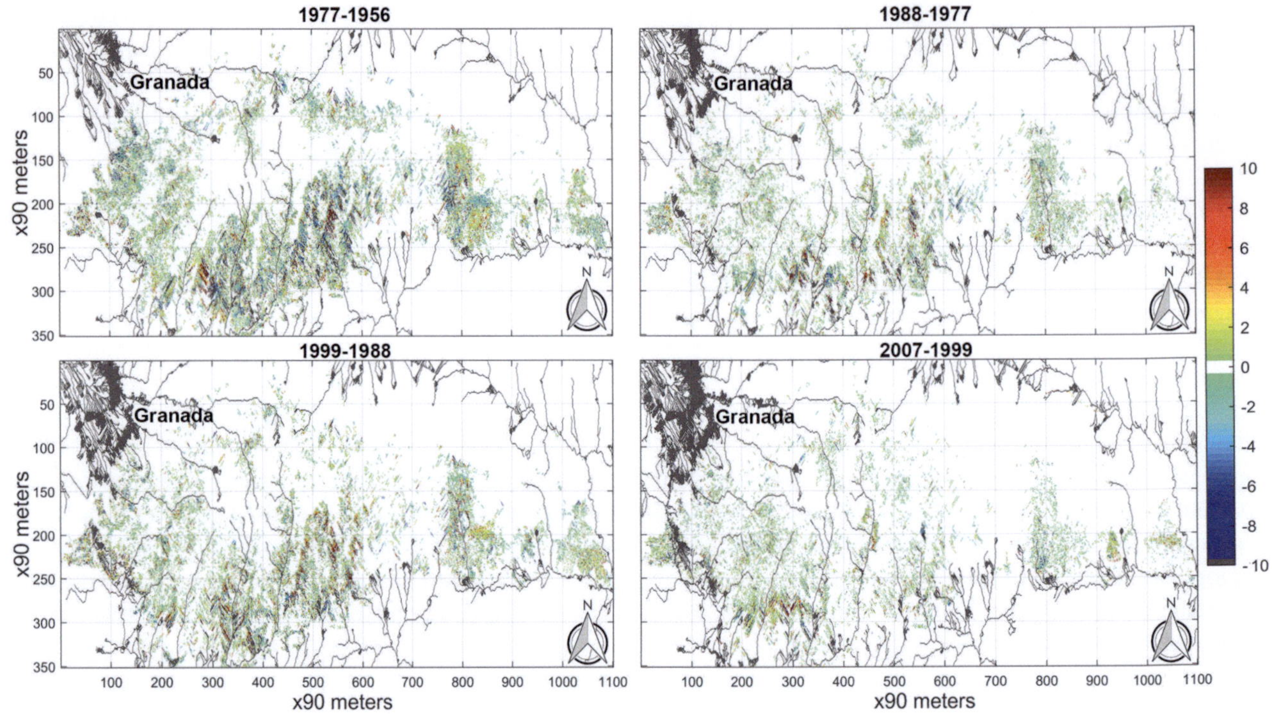


**Fig. 8** Annual soil loss rates (t (ha yr)$^{-1}$) for different historical periods (1956–2007) obtained from physically based and distributed hydrological modeling with WiMMed. In black, urban areas and the main drainage network

out to a shift towards a more torrential regime of precipitation and, specifically, snowfall, which does not necessarily imply a decrease of the annual river discharge but sheds questions on the further impacts of global warming in this site and its valuable ecosystems. Under these conditions, shorter snow seasons, shrinkage of the annual snow cover area, and torrential events bring together potentially higher losses of water to the atmosphere through evaposublimation, but may indirectly favour runoff generation during the wet periods. Torrentiality impacts include, among others, wider ranges of soil moisture and larger spatial variability, enhancement of both hillslope and fluvial erosive processes, and changes in the seasonal river flows, with expected longer low-flow periods. On-going and future research assess whether the coming decades will consolidate the observed trends or will rather confirm the highly variable character of the snow regime in this "snow island" in the southern border of Europe, that may act as an early-alert system of the future of high mountain regions in other parts of the globe.

**Acknowledgements** This work was conducted under the aegis of the research groups DFH (Fluvial Dynamics and Hydrology, TEP248-PAIDI) and DFA (Environmental Fluid Dynamics, TEP209-PAIDI), and funded by the FEDER Program through the Spanish Ministry of Science and Innovation (research project RTI2018-099043-B-I00, OPERA, "Operability in hydrological management under snow torrentiality/drought conditions in the high mountain in semiarid watersheds") and the Spanish Ministry of Economy and Competitiveness (research project CGL 2014-58508R, GMS-SNOWMED "Global monitoring system for snow areas in Mediterranean regions: trends analysis and implications for water resource management in Sierra Nevada"; research project CGL 2011-25632, SNOWMED, "Snow dynamics in Mediterranean regions and its modelling at different scales. Implication for water management"), the European Union's H2020 programme under grant agreement 641762, "ECOPOTENTIAL: improving future ecosystem benefits through earth observations", and by the project AQUACLEW, which is part of ERA4CS, an ERA-NET initiative by JPI Climate "Connecting Climate Knowledge for Europe" Grant 690462, and co-funded by MINECO (Spain). Moreover, the present work was partially developed within the framework of the Panta Rhei Decade of the International Association of Hydrological Sciences (IAHS) (working group Water and energy fluxes in a changing environment). Rafael Pimentel acknowledges funding by the Juan de la Cierva Incorporación Program of the Ministry of Science and Innovation (IJC2018-038093-I). The continuous support of the Natural and National Park of the Sierra Nevada has also been determinant for the development of this line of research since 2002. Rafael Pimentel and María J. Polo are members of DAUCO, Unit of Excellence ref. CEX2019-000968-M, with financial support from the Spanish Ministry of Science and Innovation, the Spanish State Research Agency, through the Severo Ochoa and María de Maeztu Program for Centers and Units of Excellence in R&D.

## References

Aguilar C, Pimentel R, Polo MJ (2021) Two decades of distributed global radiation time series across a mountainous semiarid area (Sierra Nevada, Spain). Earth Syst Sci Data 13:1335–1359

Alahiane N, Elmouden A, Aitlhaj A (2016) Boutaleb S (2016) Small dam reservoir siltation in the Atlas Mountains of Central Morocco: analysis of factors impacting sediment yield. Environ Earth Sci 7512(75):1–11

Algarra JA, Herrero J (2016) Monitoring the physical characteristics of the snow layer. In: Zamora R, Pérez-Luque AJ, Bonet FJ et al (eds) Global change impacts in Sierra Nevada: challenges for conservation. Consejería de Medio Ambiete y Ordenación del Territorio, Junta de Andalucía, Granada, pp 36–38

Arjona S, Millares A, Baquerizo A (2018) Reservoir sedimentation impact downstream in a semi-arid basin with greenhouses cultivation. E3S Web Conf 40:03006

Barberá JA, Jódar J, Custodio E et al (2018) Groundwater dynamics in a hydrologically-modified alpine watershed from an ancient managed recharge system (Sierra Nevada National Park, Southern Spain): insights from hydrogeochemical and isotopic information. Sci Total Environ 640–641:874–893

Batalla RJ, Vericat D (2011) An appraisal of the contemporary sediment yield in the Ebro Basin. J Soils Sediments 11:1070–1081

Castillo A (2002) Manantiales. Granada, Diputación Provincial de Granada

Castillo A (2016) Lagunas de Sierra Nevada, 2nd editio. Universidad de Granada, Granada

Corripio JG, López-Moreno JI (2017) Analysis and Predictability of the Hydrological Response of Mountain Catchments to Heavy Rain on Snow Events: A Case Study in the Spanish Pyrenees. Hydrol (2017) Vol 4. Page 20(4):20

de Vente J, Poesen J, Bazzoffi P et al (2006) Predicting catchment sediment yield in Mediterranean environments: the importance of sediment sources and connectivity in Italian drainage basins. Earth Surf Process Landforms 31:1017–1034

DGMNPF (2007) Inventario Nacional de Erosión de Suelos. Comunidad Autónoma de Andalucía. Granada/Madrid

Di Mauro B, Garzonio R, Rossini M et al (2019) Saharan dust events in the European Alps: role in snowmelt and geochemical characterization. Cryosphere 13:1147–1165

Fayad A, Gascoin S, Faour G et al (2017) Snow hydrology in Mediterranean mountain regions: a review. J Hydrol 551:374–396

Francia JR, Durán VH, Martinez A (2006) Environmental impact from mountainous olive orchards under different soil-management systems (SE Spain). Sci Total Environ 358:46–60

García-Martínez P (1995) La transformación del paisaje y la economía rural en la montaña mediterránea andaluza el caso de la Alpujarra Occidental granadina. Universidad de Granada

Gómez-Ortiz A, Palacios D, Palade B et al (2012) The deglaciation of the Sierra Nevada (Southern Spain). Geomorphology 159–160:93–105

Gómez-Ortiz A, Oliva M, Salvador-Franch F et al (2014) Degradation of buried ice and permafrost in the Veleta cirque (Sierra Nevada, Spain) from 2006-2013. Solid Earth 5:979–993

Gómez-Ortiz A, Oliva M, Salvador-Franch F et al (2019) Monitoring permafrost and periglacial processes in Sierra Nevada (Spain) from 2001 to 2016. Permafrost Periglac Process 30:278–291

González-Olalla JM, Medina-Sánchez JM, Lozano IL et al (2018) Climate-driven shifts in algal-bacterial interaction of high-mountain lakes in two years spanning a decade. Sci Reports 81(8):1–12

González-Reyes Á, McPhee J, Christie DA et al (2017) Spatiotemporal variations in hydroclimate across the Mediterranean Andes (30°–37°S) since the early twentieth century. J Hydrometeorol 18:1929–1942

Herrero J, Polo MJ (2012) Parameterization of atmospheric longwave emissivity in a mountainous site for all sky conditions. Hydrol Earth Syst Sci 16:3139–3147

Herrero J, Polo MJ (2016) Evaposublimation from the snow in the Mediterranean mountains of Sierra Nevada (Spain). Cryosph 10:2981–2998

Herrero J, Polo MJ, Moñino A, Losada MA (2009) An energy balance snowmelt model in a Mediterranean site. J Hydrol 371:98–107

Herrero J, Polo MJ, Losada MA (2011) Snow evolution in Sierra Nevada (Spain) from an energy balance model validated with Landsat TM data. In: Neale CMU, Maltese A (eds) SPIE remote sensing, p 817403

Hoffmann G (1987) Holozänstratigraphie und Küstenlinienverlagerung an der andalusischen Mittelmeerküste. Bremen

Jabaloy-Sánchez A, Lobo FJ, Azor A et al (2014) Six thousand years of coastline evolution in the Guadalfeo deltaic system (southern Iberian Peninsula). Geomorphology 206:374–391

Lana-Renault N, Regüés D (2009) Seasonal patterns of suspended sediment transport in an abandoned farmland catchment in the Central Spanish Pyrenees. Earth Surf Process Landforms 34:1291–1301

Latron J, Llorens P, Gallart F (2009) The hydrology of Mediterranean mountain areas. Geogr Compass 3:2045–2064

Lievens H, Demuzere M, Marshall H-P et al (2019) Snow depth variability in the Northern Hemisphere mountains observed from space. Nat Commun 101(10):1–12

López-Moreno JI (2005) Recent variations of snowpack depth in the Central Spanish Pyrenees. Arctic Antarct Alp Res 37:253–260

López-Moreno JI, Beniston M, García-Ruiz JM (2008) Environmental change and water management in the Pyrenees: facts and future perspectives for Mediterranean mountains. Glob Planet Change 61:300–312

López-Moreno JI, Gascoin S, Herrero J et al (2017) Different sensitivities of snowpacks to warming in Mediterranean climate mountain areas. Environ Res Lett 12:74006

Martos-Rosillo S, Ruiz-Constán A, González-Ramón A et al (2019) The oldest managed aquifer recharge system in Europe: new insights from the Espino recharge channel (Sierra Nevada, southern Spain). J Hydrol 578:124047

Millares A, Moñino A (2018) Sediment yield and transport process assessment from reservoir monitoring in a semi-arid mountainous river. Hydrol Process 32:2990–3005

Millares A, Moñino A (2020) Hydro-meteorological drivers influencing suspended sediment transport and yield in a semi-arid mountainous basin. Earth Surf Process Landforms 45:3791–3807

Millares A, Gulliver Z, Polo MJ (2012) Scale effects on the estimation of erosion thresholds through a distributed and physically-based hydrological model. Geomorphology 153–154:115–126

Millares A, Polo MJ, Moñino A et al (2014) Bedload dynamics and associated snowmelt influence in mountainous and semiarid alluvial rivers. Geomorphology 206:330–342

Millares A, Moñino A, Arjona S, Baquerizo A (2018) Suspended sediment dynamics by event typology and its siltation effects in a semi-arid snowmelt-driven basin. E3S Web Conf 40:04008

Millares A, Díez-Minguito M, Moñino A (2019) Evaluating gullying effects on modeling erosive responses at basin scale. Environ Model Softw 111:61–71

Millares A, Cantalejo M, Galve JP (2021) Respuesta erosiva y sedimentaria al cambio climático y global en cuencas mediterráneas de montaña. In: CONAMA, Congreso Nacional de Medio Ambiente. Madrid

Moñino A, Millares A, Herrero J et al (2011) Measurement of infilling rate in a small reservoir in a Mediterranean semi-arid area. Geophys Res Abstr 13:2011–10756

Moreno-Llorca R, Vaz AS, Herrero J et al (2020) Multi-scale evolution of ecosystem services' supply in Sierra Nevada (Spain): an assessment over the last half-century. Ecosyst Serv 46:101204

Ollesch G, Kistner I, Meissner R, Lindenschmidt K-E (2006) Modelling of snowmelt erosion and sediment yield in a small low-mountain catchment in Germany. CATENA 2–3:161–176

Orgiazzi A, Panagos P (2018) Soil biodiversity and soil erosion: it is time to get married. Glob Ecol Biogeogr 27:1155–1167

Pérez-Luque AJ, Pérez-Pérez R, Bonet-García FJ, Magaña PJ (2015) An ontological system based on MODIS images to assess ecosystem functioning of Natura 2000 habitats: a case study for Quercus pyrenaica forests. Int J Appl Earth Obs Geoinf 37:142–151

Pérez-Palazón MJ (2019) Análisis de tendencias en los flujos de agua y energía de la capa de nieve a diversas escalas en Sierra Nevada. Universidad de Córdoba, UCOPress

Pimentel R, Herrero J, Zeng Y et al (2015) Study of snow dynamics at subgrid scale in semiarid environments combining terrestrial photography and data assimilation techniques. J Hydrometeorol 16:563–578

Pérez-Palazón MJ, Pimentel R, Polo MJ et al (2018) Climate trends impact on the snowfall regime in Mediterranean mountain areas: future scenario assessment in Sierra Nevada (Spain). Water 10:720

Pimentel R, Herrero J, Polo MJ (2017a) Subgrid parameterization of snow distribution at a Mediterranean site using terrestrial photography. Hydrol Earth Syst Sci 21:805–820

Pimentel R, Herrero J, Polo MJ (2017b) Quantifying snow cover distribution in semiarid regions combining satellite and terrestrial imagery. Remote Sens 9:995

Polo M, Herrero J, Aguilar C et al (2009) WiMMed, a distributed physically-based watershed model (I). Environ Hydraul Theor Exp Comput Solut 225–228

Polo MJ, Herrero J, Pimentel R, Pérez-Palazón MJ (2019) The Guadalfeo Monitoring Network (Sierra Nevada, Spain): 14 years of measurements to understand the complexity of snow dynamics in semiarid regions. Earth Syst Sci Data

Polo MJ, Pimentel R, Gascoin S, Notarnicola C (2020) Mountain hydrology in the Mediterranean region. Water Resour Mediterr Reg 51–75

Renard KG, Foster GR, Weesies GA et al (1997) Predicting soil erosion by water: a guide to conservation planning with the Revised Universal Soil Loss Equation (RUSLE). Agriculture Handbook, 703

Santi PM, Hewitt K, VanDine DF, Barillas Cruz E (2010) Debris-flow impact, vulnerability, and response. Nat Hazards 561(56):371–402

Skiles SM, Painter TH (2018) Assessment of radiative forcing by light-absorbing particles in snow from in situ observations with radiative transfer modeling. J Hydrometeorol 19:1397–1409

Tanasienko AA, Yakutina OP, Chumbaev AS (2011) Effect of snow amount on runoff, soil loss and suspended sediment during periods of snowmelt in southern West Siberia. CATENA 87:45–51

Torres R, Millares A, Aguilar C et al (2013) Zonal characterization of hillslope erosion processes in a semi-arid high mountain catchment. Geophys Res Abstr 15:2013–13894

Vázquez F (1995) Restauración hidrológico-forestal de la cuenca alimentadora del embalse de Rules en el río Guadalfeo. Granada, Confederación Hidrográfica del Sur de España, Málaga

Vigl LE, Schirpke U, Tasser E, Tappeiner U (2016) Linking long-term landscape dynamics to the multiple interactions among ecosystem services in the European Alps. Landsc Ecol 319(31):1903–1918

Viviroli D, Weingartner R (2004) The hydrological significance of mountains: from regional to global scale. Hydrol Earth Syst Sci 8:1017–1030

# Concluding Notes: The Physical Setting: From Natural Geomorphologic and Climatic Dynamics to Anthropogenic Changes

**M. Oliva, José M. Fernández-Fernández, M. J. Esteban Parra, J. Galindo-Zaldívar, G. Jiménez-Moreno, and M. J. Polo**

Research on environmental and climatic processes in the Sierra Nevada has experienced substantial advances over the last decades of the 20th and early twenty-first century. The chapters of *Section 2—The physical setting* show evidence of the current broad scientific comprehension of past and present-day Earth system dynamics that has shaped the landscape of the massif until present.

The uniqueness of the natural heritage of the Sierra Nevada is a combination of the abiotic and biotic aspects, which is crucially reflected on its singular geographical characteristics. The high altitudes at latitude 37° N in the SW extreme of the European continent provide a singular physical setting that constitutes a key site of hotspot biodiversity in Europe. Indeed, in this unique mountain system, a large number of endemic species have developed, favouring the declaration of this massif as a Natural Park in 1989 and National Park in 1999 in order to protect its unique and sensitive ecosystems, characteristic of the high semiarid Mediterranean mountain environments.

Six chapters of Section 2 summarized the key elements that characterize the physical setting of the Sierra Nevada. The first chapter focuses on the role of tectonics in mountain building dynamics and how geomorphological processes shaped this geological setting (chapter "Geological Setting of Sierra Nevada"). Indeed, the second chapter is centred on past glacial dynamics, one of the exogenous geomorphological processes that played a more prominent role on the current landscape of the massif (chapter "The Impact of Glacial Development on the Landscape of the Sierra Nevada"). Colder climate conditions during the Quaternary favoured glacier inception and the formation of the large glacial cirques and valleys of the high lands of the Sierra Nevada. As glaciers retreated, some overdeepened basins within the glacial cirques were occupied by tens of lakes and wetlands, whose sediments have been used to accurately reconstruct past climate conditions since the last deglaciation, occurred at $\sim$12 ka cal BP. Albeit, these lacustrine records can tell us not only about past climates, but also about how the environment in the Sierra Nevada responded to changes in climate, land use or human pressure (chapter "Reconstruction of Past Environment and Climate Using Wetland Sediment Records from the Sierra Nevada"). In parallel to the formation of these lakes at the transition between the Late Pleistocene and the Early Holocene, the high lands of the massif were affected by periglacial processes, which replaced glaciers as the main agent shaping the summit areas and highest slopes of the massif. Over the last millennia, periglacial dynamics driven by seasonal frost have remodelled the terrain glaciated during the last glacial cycle (chapter "Ancient and Present-Day Periglacial Environments in the Sierra Nevada"). Together with ground ice, the thickness and duration of the snow cover in the massif has, today, a

M. Oliva (✉)
Department of Geography, Universitat de Barcelona, Barcelona, Catalonia, Spain
e-mail: marcoliva@ub.edu

J. M. Fernández-Fernández
Centre for Geographical Studies, IGOT, Universidade de Lisboa, Lisbon, Portugal

M. J. Esteban Parra
Department of Physics, Universitad de Granada, Granada, Andalusia

J. Galindo-Zaldívar
Department of Geodynamics, Universidad de Granada, Granada, Spain
Instituto Andaluz de Ciencias de la Tierra, CSIC-Universidad de Granada, Granada, Spain

G. Jiménez-Moreno
Department of Stratigraphy and Paleontology, Universidad de Granada, Granada, Andalusia, Spain

M. J. Polo
Fluvial Dynamics and Hydrology Research Group – Andalusian Institute for Earth System Research (IISTA), University of Córdoba, Córdoba, Andalusia, Spain

prominent role on geoecological dynamics (chapter "Snow Dynamics, Hydrology, and Erosion"). Indeed, within the regional context of Andalusia, where water is a scarce resource, snow has had a major importance as a water supply to large areas surrounding the Sierra Nevada, increasingly demanded for human activities. And this may be even more important in the present context of accelerated climate change in the Iberian Peninsula as well as in the entire Mediterranean region, which may be larger than the global average. In this sense, an accurate comprehension of the past and recent climate trends is needed to better frame the anticipated warmer and drier climate scenarios for the end of the twenty-first century (chapter "Climate Variability and Trends").

Following a pedagogic and integrative perspective, these chapters provide a recent and comprehensive overview of the current scientific knowledge of the formation of the Sierra Nevada as well as its subsequent environmental transformation. In turn, the chapters also point to some pending scientific gaps in our comprehension of these processes, at both long- and short-term scales:

- Traditional geological observations combined with recent gravity approaches and seismic refraction profiles have shed light on the timing, phases and factors controlling the formation of the Betic Range at regional to local scale, including the Sierra Nevada. Evidence reveals that this massif including the highest peaks in the Iberia Peninsula keeps rising, at maximum elevation rates up to 0.5 mm $yr^{-1}$. However, there are still major discrepancies to explain the mountain building processes, such as models and relative timing proposed for the subduction-exhumation of the Nevado-Filabride and Alpujarride complexes.
- An enhanced coupling between the endogenous versus exogenous processes can provide a better comprehension of landscape response in the Sierra Nevada. Long-term geomorphic dynamics is constrained by Quaternary tectonics, but the chronological control of endogenous versus exogenous events is, in some cases, still missing.
- Quaternary dynamics in the Sierra Nevada has been crucial to understand why the massif is a hotspot of biodiversity and a refuge for a large number of endemic species. However, the relationship between glacial processes and flora dynamics in the massif (e.g. time of plant colonization, species diversification, biodiversity changes over time, etc.) is yet to be explored, as has been done in other European mountain areas.
- Glacial dynamics is, probably, the best-known Quaternary environmental process in the Sierra Nevada. This knowledge is particularly well-known for the western fringe of the massif, where the impact of glaciations in the landscape was larger. However, little is known about glacial processes in the central and eastern parts of the Sierra Nevada.
- The last deglaciation brought major environmental shifts in the high lands of the massif and the landscape readjusted to the new setting. The transition from glacial to periglacial landsystems seems to have been a fast process, as in other mountain environments, but further research needs to confirm how deglaciated mountain slopes react to the newly exposed ice-free areas.
- Indeed, glacial, periglacial, and lacustrine records in the Sierra Nevada have been usually studied separately, which has favoured the reconstruction of palaeoenvironments based on specific landsystems. Future research should favour the integration of knowledge from different environmental sources to generate a more integrative picture of the transition from the "glacial" Sierra Nevada to the present-day landscape.
- The Quaternary environmental response of the connection between the highlands and the lowlands is not yet well-understood. Glacial phases must have significantly altered hydrological dynamics, particularly, fluvial and alluvial processes in the flatlands surrounding the massif, though this has been little investigated to date.
- Lakes and peat bogs formed after the deglaciation preserve the record of postglacial palaeoclimatic and environmental changes. An accurate reconstruction of past climate conditions can provide background for recent climate trends. The linkage between palaeoclimatic reconstructions from natural archives in the Sierra Nevada and instrumental records from the longest climatic series in the region should be further explored to better frame the magnitude of the recent climate variability.
- Recent studies focusing on climate change are constrained by the existing climatic series in the region, which are scarcely distributed and short above 2000 m. Most stations are located in the surrounding lowlands and the climatic signal is thus referred to changes occurred in these regions. Future research should include data from stations installed over the last two decades in order to explore if the climate signals observed at high altitude show the same temporal pattern as those of the lowlands. Furthermore, the physical complexity of the Sierra Nevada requires an increase in spatial resolution in climate modeling schemes to improve the knowledge of the current climate, as well as climate change projections.

– As it had in the past, the snow still presents a major role for socio-economic activities in the regional setting. Its recent evolution, framed within the large interannual and intraanual variability characteristic of the Mediterranean region, will have to be assessed within the context of global change. Particularly, the impacts of the partition between snowmelt and evaposublimation on the water storage by snow in a warmer and more torrential climatic regime will be relevant for the short- and medium-term evolution of the ecohydrological conditions in this area.

**Author**: Jose Ignacio Cuenca

# Biotic Responses to Recent Changes: Terrestrial Ecosystems as Laboratories of Global Change

**Author**: Andrés Ureña

# Altitudinal Patterns and Changes in the Composition of High Mountain Plant Communities

Juan Lorite, Andrea Lamprecht, Julio Peñas, Katy Rondinel-Mendoza, Rosa Fernandez-Calzado, Blas Benito, and Eva Cañadas

**Abstract**

Sierra Nevada, comprising 2348 vascular flora taxa (including 95 endemic taxa) is considered one of the most important plant hotspots within the Mediterranean region. Sierra Nevada presents 362 taxa inhabiting the alpine area (ca. 242 km$^2$), 75 endemic species (62 endemic plus 13 sub-endemic) among them, constituting ca. 79% of the endemism of the entire area. This high-mountain has preserved many species, allowing the current presence of many artic-alpine species, including twelve cold-adapted species with their southernmost limit here. There are 23 nano-hotspots, most of them occurring at the highest altitude, at the coldest parts. Altogether, they host 30% of the Baetic endemic flora in just 0.07% of the area. Plant communities are also original, and they are composed of a mixture of Alpine and Mediterranean species. Climate change is strongly impacting alpine biota leading to an adaptation to the new conditions. When this adaptation capacity is overcome species are forced to migrate to avoid extinction. Some responses are already noticeable in alpine areas, such as: phenological changes, altitudinal movements, increasing competition and hybridization, and changes in plant assemblages. Direct impact related to human activities such as livestock grazing, use of fire to manage alpine pasturelands, mountain agriculture, outdoor activities, and infrastructure construction have additive effects to climate change, and altogether they can exacerbate negative changes. Monitoring, evaluating, and understanding the effect of global change in the Mediterranean mountains is a top priority. We offer guidelines to orient the conservation agenda at Sierra Nevada: To (i) establish an early warning indicators system, (ii) preserve plant species and habitats, (iii) preserve threatened plant species ex situ, (iv) promote adaptive management measures, (v) evaluate outdoor recreation activities, and (vi) control and regulate activities.

**Keywords**

Sierra Nevada • Alpine flora • Climate change • Phenology • Migration • Competition • Hybridization • Conservation • Adaptation

J. Lorite (✉) · J. Peñas · K. Rondinel-Mendoza · R. Fernandez-Calzado · E. Cañadas
Department of Botany, University of Granada, Granada, Spain
e-mail: jlorite@ugr.es

J. Lorite
Interuniversity Institute for Earth System Research, University of Granada, Granada, Spain

A. Lamprecht
GLORIA Coordination, Department of Integrative Biology and Biodiversity Research, University of Natural Resources and Life Sciences, Vienna, Austria

A. Lamprecht
GLORIA coordination, Department of Integrative Biology and Biodiversity Research, University of Natural Resources and Life Sciences, Vienna & Institute for Interdisciplinary Mountain Research, Austrian Academy of Sciences, Vienna, Austria

B. Benito
Department of Ecology and Multidisciplinary, Institute for Environment Studies Ramon Margalef, University of Alicante, Alicante, Spain

## 1 Characteristics of the Alpine Belt

Sierra Nevada takes part of an heterogeneous group of Mediterranean mountains (Nagy and Grabherr 2009). The typical high seasonality of these temperate mountains is accentuated by the Mediterranean climate type (Nogués Bravo et al. 2008), characterized by a severe summer drought and precipitation occurring mainly in the winter. Thus, a considerable amount of snow may accumulate and remain in early-, and even mid-summer in sheltered areas (Nagy and Grabherr 2009). While in alpine areas the length of the growing season at treeline generally ranges from 90 days in the Arctic to ca. 365 days at Equatorial mountains, it is only ca. 90 days at


R. Zamora and M. Oliva (eds.), *The Landscape of the Sierra Nevada*,
https://doi.org/10.1007/978-3-030-94219-9_11

Mediterranean high mountains, where snow disappears in July and then growing is water-limited (Körner 2012).

Mediterranean mountains comprise not only mountain areas around the Mediterranean sea such as: Sierra Nevada, High and Middle Atlas, Mount Etna, Apennines, Dinarides, Mt. Olympus and Pindos, mountains of Crete, Southern Taurus, and Lebanese mountains, but also, other mountainous systems with Mediterranean-type climate around the world, including: The Sierra Nevada of California, Central Chilean Andes, Maloti-Drakensberg of Southern Africa, and Australian Snowy mountains (Nagy and Grabherr 2009). Since the earliest botanical explorations in Sierra Nevada, some authors (i.e. Boissier 1839; Willkomm and Lange 1861) expressed the idea that Mediterranean mountains (Sierra Nevada in particular) have certain singularities that "*do not fit with the general patterns observed in Alps or other European mountains*". In fact, the most typical communities of the Mediterranean-alpine area are cushion-like communities (represented in Sierra Nevada by the genera *Hormatophylla*, *Genista*, *Arenaria*, etc.) and short grasses tussock-forming (mainly belonging to *Festuca* genus), as a response to the characteristic yearly two stress periods (winter and summer).

One of the most controversial issues in Sierra Nevada, as in other Mediterranean mountains, is the delimitation of alpine area, which is widely acknowledged as the area above the climatic, high elevation treeline, including treeline ecotone (Körner 2021). Treeline is defined as the upper limit of the biological potential of tree species growing into tree size, regardless of whether trees are present or not (Körner 2021). This potential treeline is dominated by conifers in temperate mountains together with scattered broadleaves (Nagy and Grabherr 2009). A similar pattern might be observed in Mediterranean mountains, with different conifer species taking this role, such as: *Abies alba* in Corsican mountains (Gamisans 2003), *Cedrus libani* at Lebanese mountains (Hajar et al. 2010), *Cedrus atlantica, Juniperus thurifera, Pinus nigra* at Moroccan Atlas (Charco 1999), *Pinus nigra, P. heldrechii, Abies borisii-regis* and *A. cephalonica* in southern Mediterranean Balkans (Nagy 2006; Körner 2012), or *Cupresssus sempervirens* at Crete (Brullo et al. 2001). However, the natural treeline has been largely disturbed and artificially lowered down several hundred meters in many European mountains (Nagy et al. 2003; Körner 2012). To this, for several Mediterranean mountains, a lack of treeline species since the last glaciation is assumed, resulting in tree limits at exceptionally high temperatures and thus at lower elevations than expected (Körner 2012).

This pattern is similar in Sierra Nevada where treeline and treeline ecotone are completely gone and the only remaining patches of *Pinus sylvestris* var. *nevadensis* on the Western slope reach only 1900 m a.s.l. (Lorite 2002). For geographical and historical reasons, the high human pressure in the Sierra Nevada, long inhabited by nomadic (transhumant) herders at least during the last 5000–4000 years (Redondo and Martínez-Rodríguez 2021), have caused the disappearance of the typical treeline. These nomadic herders felled the trees as firewood and managed the dwarf and creeping shrubs with fire to obtain pasturelands. Fire occurrences have proved to be particularly high since ca. 3700 years ago, in response to regional human population expansion (Anderson et al. 2011). The local and regional impact of humans increased substantially after ca. 2700 years ago, and presumably in combination with aridification, almost causing the loss of *Pinus* forest within the mountain range (Anderson et al. 2011). Spanish botanists who found this absence of conifers tree belt considered the area above the timberline (i.e. the line formed by the close forest, composed by *Quercus ilex* and *Q. pyrenaica* in Sierra Nevada) as a treeless area (e.g. Rivas Martínez 1961; Losa Quintana et al. 1986), grouping this potential gymnosperm belt with the alpine cushion-like shrubs, naming this belt as Oromediterranean and, using the term together with Cryoromediterranean (equivalent to the sub-nival belt) to designate alpine-Mediterranean vegetation zone (Rivas-Martínez 1987).

There are two candidates in Sierra Nevada, *Pinus sylvestris* and *P. nigra,* which could have formed this treeline. Together with *Sorbus aria*, *S. hybrida* and *Acer opalus* subsp. *granatense*, all of them having nowadays scant and scattered stands close to the treeline (Lorite 2002). Temperature and length of the growing season, together with the presence of these isolated tree individuals, allow us to establish the beginning of alpine area at ca. 2400 m (2300–2600 m) for the area (Lorite et al. 2020) in line with interpretations from other authors (see Piper et al. 2016).

## 2 Alpine Flora of Sierra Nevada

Sierra Nevada in southern Spain (from 36° 50′ 24″ to 37° 15′ 0″ N in latitude and 3° 44′ 24″ to 2° 35′ 24″ W in longitude) is a paradigmatic example of a small-scaled isolated Mediterranean high massif, constituting the only true alpine region between the North-African mountains (High and Middle Atlas), the Central Range, and the Pyrenees, all being several hundreds of kilometers away (Lamprecht et al. 2021). Within its limited surface area of 2100 km$^2$, it shows a complex orography with an ample altitudinal range (from 200 m to 3479 m). The climate is Mediterranean, characterized by cold and wet winters, and hot and dry summers (with strong summer drought in July–August; Gómez-Ortiz 2002, see chapter "Climate Variability and Trends"). Average temperatures are usually below 0 °C during winter with a snow cover that can persist up to 8 months in the highest areas (occasionally up to 10 months in small patches of snowbeds; Lamprecht et al. 2021). This mountain marks the southernmost limit of the influence of the Quaternary glaciations in Europe, when it was covered with glaciers only in areas above 2500 m., while large areas of it remained free of glacial ice (Gómez-Ortiz et al. 2013, see chapters "The Impact of Glacial Development on the Landscape of the Sierra Nevada" and "Ancient and Present-Day Periglacial Environments in the Sierra Nevada"). Sierra Nevada acted as a refuge for many plant species during glacial ages as well as for isolated populations that have evolved under particular conditions, not only climatic, but also specific soil types or isolated summit habitats (Blanca et al. 1998; Médail and Diadema 2009). This has encouraged speciation, resulting in 2348 taxa from a total of 756 genera and 146 families, with 95 endemic or subendemic taxa (Lorite et al. 2007, 2020; Lorite 2016; see Table 1). For this reason, it is considered one of the most important plant hotspots within the Mediterranean region (Blanca et al. 1998; Medail and Quezel 1999; Médail and Diadema 2009; Cañadas et al. 2014; Peñas and Lorite 2019), while the total Mediterranean basin already is regarded as 'hyper-hot candidates for conservation support in light of their exceptional totals of endemic plants' (Myers et al. 2000).

As alpine species, we considered all the species appearing in the alpine belt of Sierra Nevada, including: (1) Species restricted to alpine zone, (2) species centered in the alpine zone that can be found scattered at lower altitudes (specially at the treeline ecotone), (3) Species mainly distributed in montane areas but spreading both to lowland areas and to alpine areas as well, and finally (4) Lowland species spread into montane areas and above (usually only reaching the lower alpine belt).

Overall, there are 362 taxa inhabiting the alpine area (Table 1), despite alpine area covering only 242 km$^2$. This number of species is in line with those observed for other alpine areas in other mountains around the world (Fig. 1). Vascular plant species are mostly flowering plants (ca. 94%),

**Table 1** Composition of the flora of Sierra Nevada (whole area vs. alpine area). Extracted from Lorite et al. (2020)

| | Whole area | Alpine area (>2400 m) |
|---|---|---|
| Taxa (taxa rank) | 2348 (species: 1951; subspecies: 363; hybrid 362) | 362 (species: 275; subspecies: 82; hybrid 5) |
| Major groups (Phylum) | Magnoliophyta: 2269; Pteridophyta: 50; Pinophyta: 29 | Magnoliophyta: 339; Pteridophyta: 19; Pinophyta: 4 |
| Origin | Native: 2214; Alien: 134 | Native: 360; Alien: 2 |
| Main families | Asteraceae: 274; Fabaceae: 208; Poaceae: 207; Brassicaceae: 126; Caryophyllaceae: 114; Lamiaceae: 112; Apiaceae: 92; Rosaceae: 73; Veronicaceae: 53; Orchidaceae: 51; Ranunculaceae: 48; Cistaceae: 46; Boraginaceae: 40; Rubiaceae: 39; Cyperaceae: 38; Orobanchaceae: 34; Chenopodiaceae: 30; Geraniaceae: 28; Crassulaceae: 26; Juncaceae: 25; Euphorbiaceae: 24; Salicaceae: 24; Polygonaceae: 23; Fumariaceae: 19; Campanulaceae: 17; Primulaceae: 17; Alliaceae: 15; Aspleniaceae: 15; Convolvulaceae: 15; Malvaceae: 15; Pinaceae: 15; Plantaginaceae: 15 | Asteraceae: 52; Poaceae: 28; Brassicaceae: 27; Caryophyllaceae: 27; Fabaceae: 13; Veronicaceae: 13; Crassulaceae: 12; Ranunculaceae: 12; Cyperaceae: 11; Lamiaceae: 11; Rosaceae: 11; Apiaceae: 10; Rubiaceae: 10 |
| Main genera | *Trifolium*: 28; *Silene*: 27; *Helianthemum*: 26; *Carex*: 25; *Galium*: 25; *Ranunculus*: 25; *Centaurea*: 24; *Vicia*: 24; *Astragalus*: 23; *Festuca*: 21; *Orobanche*: 21; *Juncus*: 20, *Sedum*: 19; *Salix*: 18; *Thymus*: 18; *Erodium*: 16; *Ononis*: 16; *Senecio*: 16; *Teucrium*: 16; *Allium*: 15; *Euphorbia*: 15; *Linaria*: 15; *Plantago*: 15 | *Carex*: 10; *Sedum*: 10; *Galium*: 9; *Festuca*: 8; *Ranunculus*: 8; *Arenaria*: 6; *Asplenium*: 6; *Cerastium*: 6; *Senecio*: 6; *Allium*: 5; *Artemisia*: 5; *Cuscuta*: 5; *Epilobium*: 5; *Juncus*: 5; *Saxifraga*: 5; *Veronica*: 5 |
| Distribution | Wide: 1119; Mediterranean: 348; Iberian-NAfrican: 243; W Mediterranean: 198; Iberian: 174; Baetic: 112; Sierra Nevada endemic: 78; Baetic-NAfrican: 35; SE Iberian: 24; Sierra Nevada subendemic: 17 | Wide: 149: Sierra Nevada endemic: 62; Iberian-NAfrican: 40; Iberian: 34; Baetic: 21; W Mediterranean: 16; Mediterranean: 5; Sierra Nevada subendemic: 13; Baetic-NAfrican: 12 |
| Biotypes | Therophyte: 783; hemicryptophyte: 732; chamaephyte: 337; phanerophyte: 255; geophyte: 211; hydrophyte: 18; helophyte: 12 | Hemicryptophyte: 218; chamaephyte: 64; therophyte: 39; geophyte:29; phanerophyte: 10; helophyte: 1; hydrophyte: 1 |
| Major plant community types | • Synanthropic, fringe and megaphorbic: 714<br>• Grasslands and meadows: 667<br>• Heathland and scrub vegetation: 324<br>• Forest and woodlands: 211<br>• Rock crevices and screes 186<br>• Amphibious vegetation of fresh-waters:102<br>• Supratimberline veg. (geliturbated soils): 48<br>• Continental halophilous vegetation: 19<br>• Aquatic vegetation (floating or rooted): 17 | • Grasslands and meadows: 103<br>• Rock crevices and screes: 84<br>• Synanthropic, fringe and megaphorbic: 46<br>• Supratimberline veg. (geliturbated soils): 46<br>• Heathland and scrub vegetation: 38<br>• Amphibious vegetation of fresh-waters: 34<br>• Forest and woodlands: 10 |

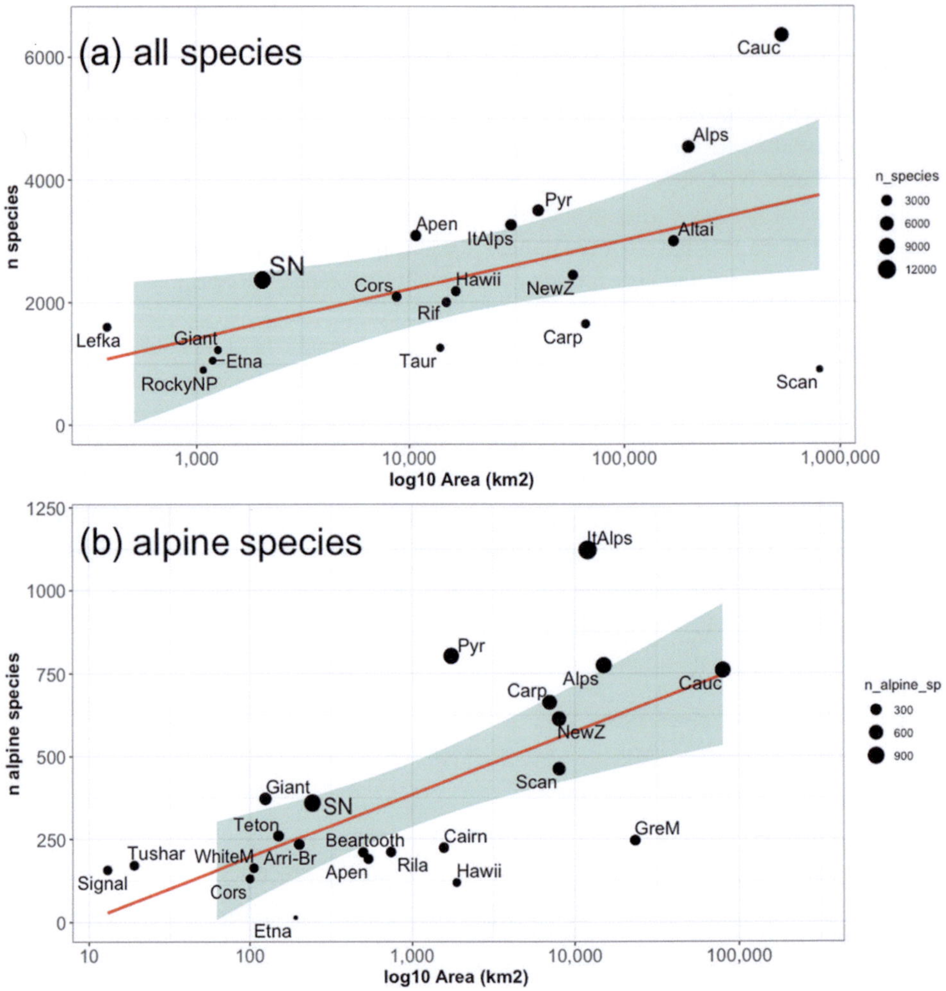


**Fig. 1** Relationship between area and plant species richness for different mountain ranges, taking into account **a** all the species and **b** only species inhabiting alpine area. (Red line shows a linear fit and and shaded area the standard error of the mean). Alps = Alps, Altai = Altai mountains, Apen = Central Apennines, Arri-Br = Arrigeth, Brooks range (Alaska), Balkan = Balkan Peninsula mountains, Cauc = Caucasus, Cors = Corsica high mountains, Etna = Etna (Sicily), Giant = Giant Mountains, GreM = Greece high mountains, Hawii = Hawaii mountains, ItAlps = Italian Alps, Lefka = Lefka Ori (Crete), NewZ = New Zealand Alps, Olym = Olympic Range (USA), RockyNP = Rocky Mountains National Park (USA), Pyr = Pyrenees, Rila = Rila Mountains, Scan = Scandes (Scandinavian mountains), Cairn = Cairngorns (Scottish Highlands), SN = Sierra Nevada (Spain), Signal = Signal Mountain (Canadian Rockies), Carp = South-Eastern Carpatians, Taur = Taurus, Tushar = Tushar Mountains (SW Utah, USA), Ural = Urals, Ruby = Ruby Range (Colorado, USA), Teton = Teton Range (Wyoming, USA), Beartooth = Beartooth Plateau (Montana, USA), WhiteM = White Mountains (California, USA)

together with 19 fern species (ca. 5%), and only four gymnosperms. Main families are similar to the whole mountain flora, except for the typical Mediterranean Fabaceae family with lower relative weight in the alpine zone. Regarding the main genera, the differences among summit areas and the whole massif are radical. Genera such as *Carex*, *Sedum*, *Galium*, *Festuca* or *Ranunculus* being the most frequent in alpine area, highlighting the role as areas containing original and isolated floras that encourage speciation. In fact, there are 75 endemic species (62 endemic plus 13 subendemic), which constitutes ca. 79% of the endemism of the entire area in just 11.5% of the whole area. To this, the alpine area acted as conservative allowing the current presence of many artic-alpine species, including twelve cold-adapted species with their southernmost limit in the Sierra Nevada (Abeli et al. 2018), such as *Ranunculus glacialis* or *Saxifraga oppositifolia*. Regarding biotypes within alpine species (Table 1), they are mainly hemicryptophytes (i.e. plants having perennating buds at or near the soil surface, including biennial species) and chamaephytes (i.e. plants having buds on persistent shoots under 50 cm), the high relative abundance of annual plants (i.e. therophytes) is remarkable compared to other European mountains (Table 2).

The richest major plant communities in alpine areas are grasslands and meadows associated with lagoons, springs and streambeds (ca. 83% of the total), followed by rock and

**Table 2** Raunkiaer life forms spectra (in percentage) in Sierra Nevada compared with other European mountains

| | Lat. (°) | Phanerophyte | Chamaephyte | Hemicryptophyte | Criptophyte | Therophyte |
|---|---|---|---|---|---|---|
| Sierra Nevada[1] | 37 | 2.8 | 17.7 | 60.2 | 8.6 | 10.8 |
| Corsica[5] | 42 | 1.5 | 18.3 | 71.8 | 7.6 | 0.8 |
| Apennines[3] | 42 | 3.7 | 14.3 | 66.8 | 11.1 | 4.1 |
| Pyrenees[3] | 42 | 1.9 | 21.5 | 64.6 | 9.0 | 3.0 |
| Caucausus (Elbrus)[4] | 43 | – | 9.0 | 72.0 | 4.0 | – |
| Italian Alps[3] | 45 | 2.8 | 14.7 | 68.0 | 10.9 | 3.6 |
| South-east Carpathians[3] | 47 | 2.5 | 13.6 | 70.0 | 7.5 | 6.5 |
| Scottish higlands[3] | 57 | 1.0 | 16.0 | 67.0 | 12.0 | 4.0 |
| Scandes[2] | 67 | 5.0 | 18.0 | 59.0 | 16.0 | 2.0 |
| **Average** | | **2.7** | **15.9** | **66.6** | **9.6** | **4.4** |

[1] From Lorite et al. (2020)
[2] Virtanen (in Nagy et al. 2003)
[3] Nagy and Grabherr (2009)
[4] Subnival area, Nakhutsrishvili (in Nagy et al. 2003)
[5] Gamisans (in Nagy et al. 2003)

scree communities (ca. 23%). Plants associated with human activities (i.e. synanthropic species) are noteworthy (ca. 10% of the total) despite being less frequent than in lowlands (ca. 30% of the total).

In a previous study (Cañadas et al. 2014), we identified that the endemic-vascular-plant richness showed a marked spatial pattern in Sierra Nevada, with higher numbers of taxa concentrated in the core summit areas of the massif and decreasing towards the boundaries. This pattern is closely linked to altitude (Fig. 2b). Thus, in general terms, we found that endemic-vascular-plant richness (EVPR) increased with altitude, annual precipitation, precipitation of driest period, and low temperature, these variables explaining almost 50% of the variation in species richness. In line with the pattern observed for mountain systems throughout the globe, plant diversity often declines from mid to high altitudes (Sanders and Rahbek 2012), but the percentage of endemic species usually increases with altitude (Essl et al. 2009; Fernández Calzado et al. 2012; Steinbauer et al. 2012).

Nevertheless, EVPR variability was not totally explained on the basis of altitudinal and/or climatic factors in Sierra Nevada, where we identified 23 nano-hotspots (i.e. 1-km$^2$ grid cell accounting for more than 5% of the total regional endemic vascular taxa; Cañadas et al. 2014). Most of the nano-hotspots identified occurred at the highest altitude specially at the western part of Sierra Nevada (e.g. Mulhacén, Cerro de los Machos, Veleta, Tozal del Cartujo, Lavaderos de la Reina, Tajos Negros), and in the coldest and

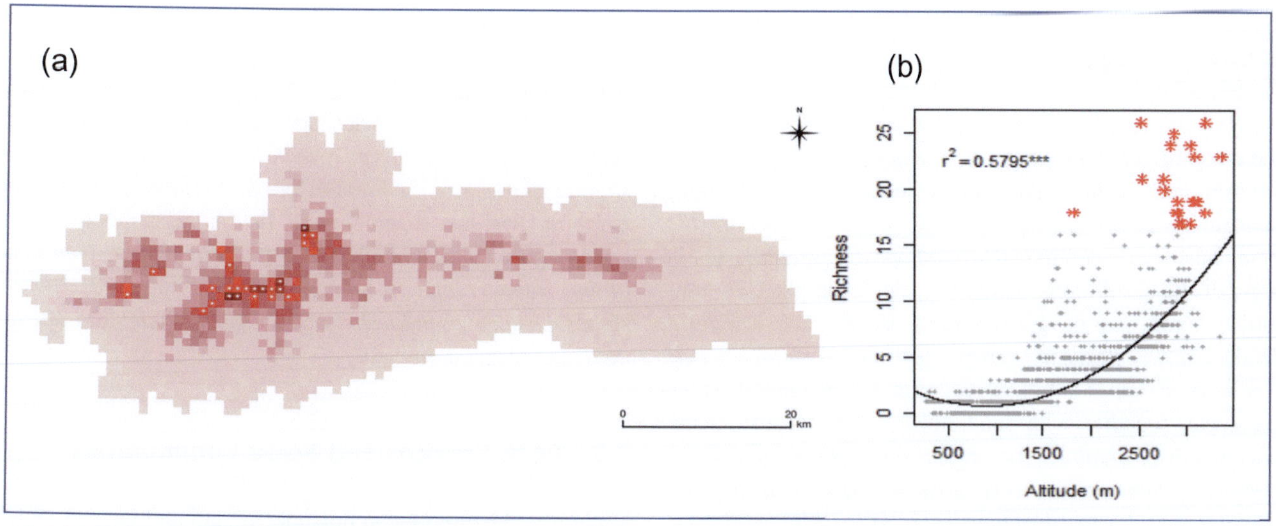


**Fig. 2** **a** Distribution of endemic richness in Sierra Nevada, white crosses symbolize nano-hotspots. **b** Relationships between endemic-vascular-plant richness (EVPR) and altitude. Stars symbolize the nano-hotspots (from Cañadas et al. 2014)

most rainy cells, but some of the nano-hotspots corresponded to lower calcareous outcrops within the siliceous bedrock that predominates, outstanding among them some dolomitic outcrops (e.g. Trevenque) harbouring also a high number of endemic taxa, despite the much lower altitude (Fig. 2). This case agrees with other studies reporting higher endemicity rate or general richness, on calcareous bedrock than on other types of bedrock (Wohlgemuth 1998; Essl et al. 2009). Actually, climate and geological history greatly contribute to the distribution patterns of endemics (Jansson 2003; Ohlemüller et al. 2008).

The fact that most of the nano-hotspots identified in Sierra Nevada are located in high peaks, makes them particularly vulnerable to climatic change, as explained below. Altogether, these nano-hotspots hosted 30% of the Betic endemic flora, despite that they represented only 0.07% of the Betic Range surface area.

## 3 Alpine Plant Communities

Hereinafter, we offer a summary with the major plant community types described for alpine area in Sierra Nevada (according to Lorite 2002; Lorite et al. 2003; Fernández--Calzado and Molero 2011).

### 3.1 Low Shrubland Cushion-Like Communities

This community type should cover treeline transition and lower alpine area from 1900–2000. to 2800–3000 m, being replaced by grasses (psicroxerophylous pasturelands) in the summit areas. As the aciculifolious belt is absent, as above referred, this community type occupies this wide belt appearing as a treeless area. Small shrubs and thyme scrub species exhibit the typical cushion-like form, characteristic of these cold-adapted and xerophyte communities (Nagy and Grabherr 2009).

Well-preserved communities are typically composed by the creeping juniper (*Juniperus communis*), savin (*Juniperus sabina*) and cushion-like and frequently thorny legumes *Genista versicolor, Cytisus galianoi* or *Erinacea anthyllis*, being also very common *Hormathophylla spinosa*. In the gaps of the shrub communities appear some plant communities, composed of grasses such as: *Festuca indigesta*, *Deschampsia flexuosa* subsp. *iberica*, *Avenula laevis*, *Koeleria crassipes* subsp. *nevadensis*, *Corinephorus canescens*, etc. together with small thickets of *Arenaria tetraquetra* subsp. *amabilis, Leontodon boryi, Leucanthemopsis pectinata, Senecio boissieri, Jasione crispa* subsp. *amethystina,* and *Jurinea humilis*.

When the soils are scarce, thyme-scrub communities become dominant with: *Sideritis glacialis, Arenaria pungens* subsp. *pungens, Thymus serpylloides, Astragalus sempervirens* subsp. *nevadensis, Scabiosa turolensis, Erysimum baeticum, E. nevadense, Anthyllis vulneraria* subsp. *pseudoarundana, Hormathophylla spinosa,* and *Acinos alpinus* subsp. *meridionalis.* Ultramafic rocks outcrops favor dominance of certain species such as *Astragalus granatensis, Poa ligulata, Erodium cheilanthifolium,* or *Androsace vitaliana* subsp. *nevadensis*.

Traditional management by transhumant shepherds consisted of burning the *Juniperus-Genista* patches to obtain *Festuca* pasturelands, that after 8–10 years were reburned again. The low recruitment capacity of juniper and savin under Mediterranean conditions (García et al. 1999) led to a growing scarcity of these species, replaced by the better fire-adapted Mediterranean plants, mostly being seeders and/or resprouters following fires (Lorite 2002). Since this management technique was banned in the early 80 s, nowadays it is still possible to observe juniper and savin sheltered in rocky places. This management, associated with soil loss, may have caused the current prevalence of *Festuca* and thyme scrub communities.

### 3.2 Psicroxerophylous Pasturelands

Appearing usually over 2800 m (above 2900–3000 m on southern exposures), they constitute one of the most original vegetation types of the area. Despite the richness of these pasturelands is relatively low, plant species are mostly endemics or boreo-alpine disjunctions (Lorite et al. 2020).

As explained above, plants in Mediterranean mountains have to cope not only with the typical alpine habitat constraints, but also with typical Mediterranean constraints (lower water availability, most irregular snowfall pattern and derived frozen damages; Lorite 2002). This community appears under the most extreme environmental conditions within this mountain. It is mostly composed of hemicryptophytes (i.e. plants having perennating buds at, or near the soil surface) with small cushion-like or little tussock shape, such as: *Festuca clementei, Hormathophylla purpurea, Artemisia granatensis, Chaenorrhinum glareosum, Leontodon boryi, Viola crassiuscula* or *Erigeron frigidus*, together with Iberian-high mountain, or boreo-alpine species, such as: *Scutellaria alpina, Gentiana alpina, Papaver lapeyrousianum*, etc. Over stony and exposed slopes *Festuca pseudoeskia* and *Festuca paniculata* subsp. *moleroi* became dominant.

### 3.3 Rock Crevices and Scree Communities

Within the Mediterranean hotspot (Medail and Quezel 1999) cliffs have acted as anthropogenic and environmental refuges, resulting in unique habitats with a high number of rare and endemic species (Thompson 2005). Rock crevices harbour specialised plant species, the so-called rupicolous or chasmophytes (Thompson 2005; Lorite et al. 2017).

Despite their extremely harsh conditions, the low herbivory pressure and low competition favour that cliffs acted as preservative habitats (Larson et al. 2000). The species composition is very diverse and strongly depends on exposure, slope inclination, and resulting insolation and moisture. Moreover, cliffs at summit areas are free of snow for a wider period than hollows, where snowbeds may remain until mid-summer. This habitat acted in Sierra Nevada as refuge, being rich in endemic species, such as: *Saxifraga nevadensis, Chaenorrhinum glareosum*, and *Holcus caespitosus*. Also, microclimatic conditions in shaded cliffs help make these shelters rich in interesting artic–alpine species, such as: *Draba dubia* subsp. *laevipes*, *Saxifraga oppositifolia, Valeriana apula, Androsace vandellii, Ranunculus glacialis*, etc. together with ferns such as: *Asplenium viride, Polystichum lonchitis, Dryopteris abbreviata, D. thyrrena, D. filix-mas*, or *Cryptogramma crispa*. Moreover, at treeline transitional area and lower alpine belt appear some endemic species such as *Centranthus nevadensis, Moehringia fontqueri* or *Sarcocapnos speciosa*.

Screes (i.e., stony places, formed by broken rock fragments at the base of crags and mountain cliffs, resulting from frost shattering) constitute a very harsh habitat. Unstable rocks usually move downslope, so plants can be easily buried. Under these conditions, plant cover is especially low and plants are specialist with fragile shoots, particularly at root neck, in order to avoid the uprooting of the entire plant. Once buried, they quickly regrow and wind around the stones reaching the surface. Most characteristic species are: *Linaria glacialis* and *Viola crassiuscula*, together with the non-exclusive of screes *Silene boryi, Chaenorrhinum glareosum, Holcus caespitosus, Crepis oporinoides, Reseda complicata*, and *Senecio nevadensis*, among others. Scree fields are the habitat of one of the most threatened plant species in the area, *Arenaria nevadensis*.

## 3.4 Synanthropic Communities

Due to the growing impacts in the mountain areas, some species that in natural conditions may have been restricted to heavily grazed areas or landslides, become more frequent. In fact, there is an ongoing increasing pressure in the area linked to ski-slopes construction and management, hiking and overgrazing by domestic ungulates (Blanca et al. 1998; Lorite et al. 2007, 2010; Lamprecht et al. 2019). These activities promote fast-growing colonizers that are often equipped with physical defences (e.g., thistles) or chemical defenses. Some of them are endemics, such as *Carduus carlinoides* subsp. *hispanicus, Eryngium glaciale, Verbascum nevadense, Dactylis juncinella, Reseda complicata*, while others are wide distributed mountain taxa, such as *Cirsium acaule* subsp. *gregarium, Cirsium odontolepis, Eryngium bourgatii, Marrubium supinum* or *Artemisia absinthium*.

## 3.5 Alpine Hygrophilous Pasturelands "Borreguiles"

In the alpine belt, close to streams and lagoons appear hygrophilous pasturelands locally known as "borreguiles" (from Spanish word "borrego" = lamb). These pasturelands are a mosaic of different plant micro-communities placed along the moisture gradient. Different communities appear from the lagoon shore or stream to the border (in contact with climate-dependent communities; Fig. 3). Directly in contact with streams or lagoons appear cushion-like bright-green coloured communities of bryophytes (*Philonotis seriata, Polytrichum juniperinum, Brachythecium rivulare, Bryum schleicheri, Pellia epiphylla*), with *Saxifraga stellaris* subsp. *robusta, Veronica nevadensis, Cerastium cerastoides, Ranunculus angustifolius* subsp. *alismoides, Sedum melanantherum* and *Festuca rivularis*.

Damp areas, constructed by incipient peat bogs are rich mostly in Cyperaceae species: *Carex nigra, C. echinata, C. nevadensis, Eleocharis quinqueflora, Festuca frigida, Triglochin palustris*, and *Juncus alpino-articulatus*. Places with permanent moisture, but not damped, show a dense pastureland with *Nardus stricta, Festuca trichophylla* subsp. *scabrescens, Scorzoneroides microcephala, Luzula hispanica, Ranunculus demissus, Gentiana boryi, G. sierrae, G. pneumonanthe* subsp. *depressa* and *Campanula herminii*. In stony, slightly elevated micro-habitats appear *Vaccinium uliginosum* subsp. *nanum* and *Ranunculus acetosellifolius*, together with *Plantago nivalis*. In transitional areas to the psicro-xerophilous pasturelands, over places with moisture at the beginning of the summer, yet becoming dry by mid- to late-summer, there is a pastureland with medium cover (see Table 3) including species such as *Agrostis nevadensis, Armeria splendens, Potentilla nevadensis, Plantago nivalis* and *Arenaria tetraquetra* subsp. *amabilis*.

**Fig. 3** Sequence of micro-communities composing high-mountain hygrophilous pasturelands, from the damped areas (left) to the temporary moisture ones (right). (1) Bryophyte communities with *Sedum melanantherum* and *Saxifraga stellaris* subsp. *robusta*, (2) Incipient peat bogs with *Cyperaceae*, (3) Pasturelands with *Nardus stricta* and *Festuca frigida*, (4) Pasturelands with *Agrostis nevadensis* and *Armeria splendens* (from Salazar and Valle 2004). **a** *Saxifraga stellaris* subsp. *robusta*, **b** *Armeria splendens*, **c** *Plantago nivalis*

At scattered thick soils with medium to high herbivore pressure, this community is enriched with megaforbic herbs such as *Aconitum burnatti* or *A. vulparia* subsp. *neapolitanum*.

## 4 Current and Projected Changes in Alpine Vegetation of Sierra Nevada

Mountain ecosystems are experiencing strong responses to climate change (Lenoir et al. 2008; Steinbauer et al. 2018). In fact, they are projected to suffer more severe climatic warming consequences than other ecosystems (Rixen and Wipf 2017), such that high-mountain environments are experiencing more rapid changes in temperature than lowland ones (Pepin et al. 2015). According to the literature this trend is exacerbated in the Mediterranean (Nogués-Bravo et al. 2008), especially in the transitional areas and mountains (Thuiller et al. 2005; Nogués-Bravo et al. 2008). Together with the high biological diversity of Mediterranean-type ecosystems (Médail and Quézel 1999), make those Mediterranean high mountains are among the most vulnerable to global warming (Grabherr et al. 1994; Nogués-Bravo et al. 2008).

Climate change may impact biota at four levels: metabolism, phenology, evolution, and spatial distribution (Hughes 2000). The first three may lead to an adaptation (both via individual adaptation and/or natural selection) to the new conditions, the last one could pose local extinctions and/or migration. As mountain plant species usually are cold-adapted, migration leads mainly to an upward movement as thermal and bioclimatic suitable conditions migrate upward as well (Steinbauer et al. 2018). Also, the upward migration of the suitable conditions for many mountain species encompasses a reduction of the suitable habitats, and could lead to an increasing competition. This fact could be very detrimental for alpine species that could be outnumbered, particularly in small massifs with a limited vertical extent. Especially endemic ones could be pushed to the brink of extinction.

**Table 3** Main features (vegetation type, subtype, main species, altitudinal range, mean cover and total number of endemics) of the alpine plant communities in Sierra Nevada. (N = 10 relevès in all cases. *Source* Sistema de Información de la Vegetación Ibero-Macaronésica SIVIM- http://www.sivim.info/sivi/)

| Type | Subtype | Main species | Alt. range (m) | Mean cover (%) | n total endemic |
|---|---|---|---|---|---|
| Low shrubland Cushion-like communities | Juniper-Genista | *Juniperus communis*<br>*Juniperus sabina*<br>*Genista versicolor*<br>*Cytitus galianoi* | 2000–2900 | 55.9 | 15 |
| | *Festuca* pasturelands | *Festuca indigesta*<br>*Arenaria tretraquetra* subsp. *amabilis*<br>*Koeleria crassipes* | 2000–2900 | 33.2 | 18 |
| | Thyme scrub matorral | *Thymus serpylloides*<br>*Sideritis glacialis*<br>*Arenaria pungens* | 2000–2900 | 45.1 | 11 |
| Psicroxerophylous pasturelands | | *Festuca clementei*<br>*Hormathophylla purpurea* | 2800–3478 | 16.2 | 29 |
| Rock crevices and screes | Rock crevices | *Saxifraga nevadensis*<br>*Draba dubia* subsp. *laevipes*<br>*Centranthus nevadensis*<br>*Moehringia fontqueri*<br>*Sarcocapnos speciosa* | 2000–3478 | 5.5 | 4 |
| | Screes | *Linaria glacialis*<br>*Viola crassiuscula* | 2300–3400 | 7.6 | 12 |
| Synanthropic communities | | *Carduus calinoides* subsp. *hispanicus*<br>*Eryngium glaciale* | 2000–3478 | 20.5 | 9 |
| Alpine higrophylous pasturelands | Bryophyte cushion-like | *Philonotis seriata*<br>*Polytrichum juniperinum*<br>*Saxifraga stellaris* subsp. *robusta*<br>*Veronica nevadensis* | 2300–3000 | 76.0 | 7 |
| | Cyperaceae communities | *Carex nigra*<br>*C. echinata*<br>*C. nevadensis,*<br>*Eleocharis quinqueflora*<br>*Festuca frigida* | 2500–3000 | 90.6 | 9 |
| | *Narduus stricta* communities | *Nardus stricta*<br>*Festuca trichophylla* subsp. *scabrescens*<br>*Scorzoneroides microcephala* | 2400–3000 | 99.0 | 11 |
| | *Agrostis-Plantago* pasturelands | *Agrostis nevadensis*<br>*Armeria splendens*<br>*Plantago nivalis* | 2400–3100 | 68.8 | 15 |

Diverse ecological changes could already be recorded in Sierra Nevada, such as: upward migration of treelines and alpine plants (e.g. Benito et al. 2011, 2014; Gottfried et al. 2012; Pauli et al. 2012; Steinbauer et al. 2018), shrub expansion (Fernández-Calzado and Molero 2013; Lamprecht et al. 2021; see chapters "Forest Dynamics Under Land-Use and Climate Change Scenarios" and "Managing the Uniqueness of Sierra Nevada Ecosystems Under Global Change: The Value of in situ Scientific Research"), or phenological events occurring earlier (e.g. germination, leaf unfolding, or flowering). They are consequences and clear indicators of these profound changes in climate (Walther et al. 2005). Following, we explore phenological changes, altitudinal movements, increasing competition and hybridization, as well as changes in plant assemblages at summit areas.

## 4.1 Phenological Changes

Plant phenology (i.e. the seasonal timing of life-history events such as flowering and leaf-out) is a key determinant of plant success and ecosystem productivity (Willis et al. 2017). Changes in the timing of phenological events are among the most important indicators of global warming (Parmesan and Yohe 2003). Climate warming is expected to alter seasonal biological phenomena such as plant growth and flowering or animal migration, which depend on accumulated temperature (i.e. the total heat required for a given organism to develop one or more stages of its life cycle; Parmesan 2007). These so-called phenological changes are likely to have important consequences for ecological processes, having a wide range of outcomes influencing agriculture, forestry or human health (Peñuelas and Filella 2001).

For plants, altered phenology may result in temporal mismatch and reduced complementarity between species, modifying interspecific competition for resources or plant pollination success (Smith et al. 2012). Failure in pollination reduces fruit set and causes great alterations in trophic chains (Amano et al. 2010) causing a decay at the community level, and the lost, in the worst scenario, of ecological niche for many species (Willis et al. 2017).

The majority of plant phenology research, specially large-scale surveys, has been conducted in lowland areas (Hülber et al. 2010). However, alpine plants that are limited by a short growing season and extreme environmental conditions (Körner 2003), can be highly vulnerable to modification of limiting factors (Thuiller et al. 2005). High alpine plants endure a cold climate with short growing seasons entailing severe consequences of an improper timing of development. Hence, their flowering phenology is rigorously controlled by climatic factors, being temperature modulated by photoperiod and snow cover duration (closely related to temperature and precipitation) the main controlling factors (Hülber et al. 2010). An earlier onset of growth and longer growing seasons may initially increase plant productivity, but most likely may result in longer periods with drought stress (Nogués Bravo et al. 2008), resulting in nutrient reserve depletion, higher reproductive effort and eventual failures in reproduction (Wookey et al. 1993).

Despite the importance of understanding the changes in phenology, and the effort in collecting comprehensive data in the past years, long-term records are still unavailable for most regions and species (Primack et al. 2004). In line with this issue, data on phenological responses for Mediterranean mountain plants are almost negligible. Even though changes in phenology of Mediterranean alpine species are expected to be even stronger than in temperate regions. For Sierra Nevada, we have only two studies focused on plant species phenology based on the same dataset (Pérez-Luque et al. 2015a, b). They included 20 species in three plots along an altitudinal gradient and offered phenological data for these species. Though, they present a comparison between 1990 and 2010, so no clear tendencies can be established beyond these punctual data.

Taking advantage of the existence of herbarium samples covering a timespan of 100 years for Sierra Nevada, we are carrying out a study (Rondinel et al. Unpubl. Data) reviewing phenological changes. We have collected data for 2650 herbarium samples from different regional herbaria belonging to 120 endemic species, registering flowering and fruiting both quantitatively (number of flowers and fruits), and qualitatively. As preliminary results for all species, we found a clear tendency to earlier flowering and fruiting events for the whole dataset (Fig. 4). For instance, in the past 100 years, the mean date of flowering peak has moved 45 days earlier. This response is stronger than those obtained with similar approaches for lowland areas (e.g. Amano et al. 2010). The relative sensitivity of the different plant species and the relationship with the changes in temperature are still unanalysed.

## 4.2 Altitudinal Movements

Mountain plant species are already experiencing important responses to climate change, being one of the most remarkable the upwards shifts in distribution (Lenoir et al. 2008; Pauli et al. 2012; Steinbauer et al. 2018; Lamprecht et al. 2021). Species distribution models predict that this will lead to a contraction or even a total loss of many alpine species' ecological niche (Engler et al. 2011), while they might suffer an increasing competence, and an eventual replacement by lowland species also migrating upwards (Engler et al. 2011). For example, species distribution models predict losses of over one-third of the total alpine plant species for some regions of the Alps, and even higher extinction rates in other European mountains (Engler et al. 2011).

For the Sierra Nevada, a mean-temperature increase of 4.8 °C at the end of the twenty-first century was simulated, indicating a vertical shift of the suitable habitats for Sierra Nevada key species at a rate of 11.57 m/year (Benito et al. 2011). This rate is much higher than the observed patterns of actual movement for alpine species, for instance, 1–4 m per decade in Alps (Grabherr et al. 1994), or 2.9 m for European mountains (Lenoir et al. 2008). It is foreseeable that this pace will result in the extinction or local extinction of an important part of the upward migrating species (Benito et al. 2014). Specifically, model projections suggest a complete loss of current suitable conditions under all warming scenarios for *Juniperus communis* (see also chapter "Forest

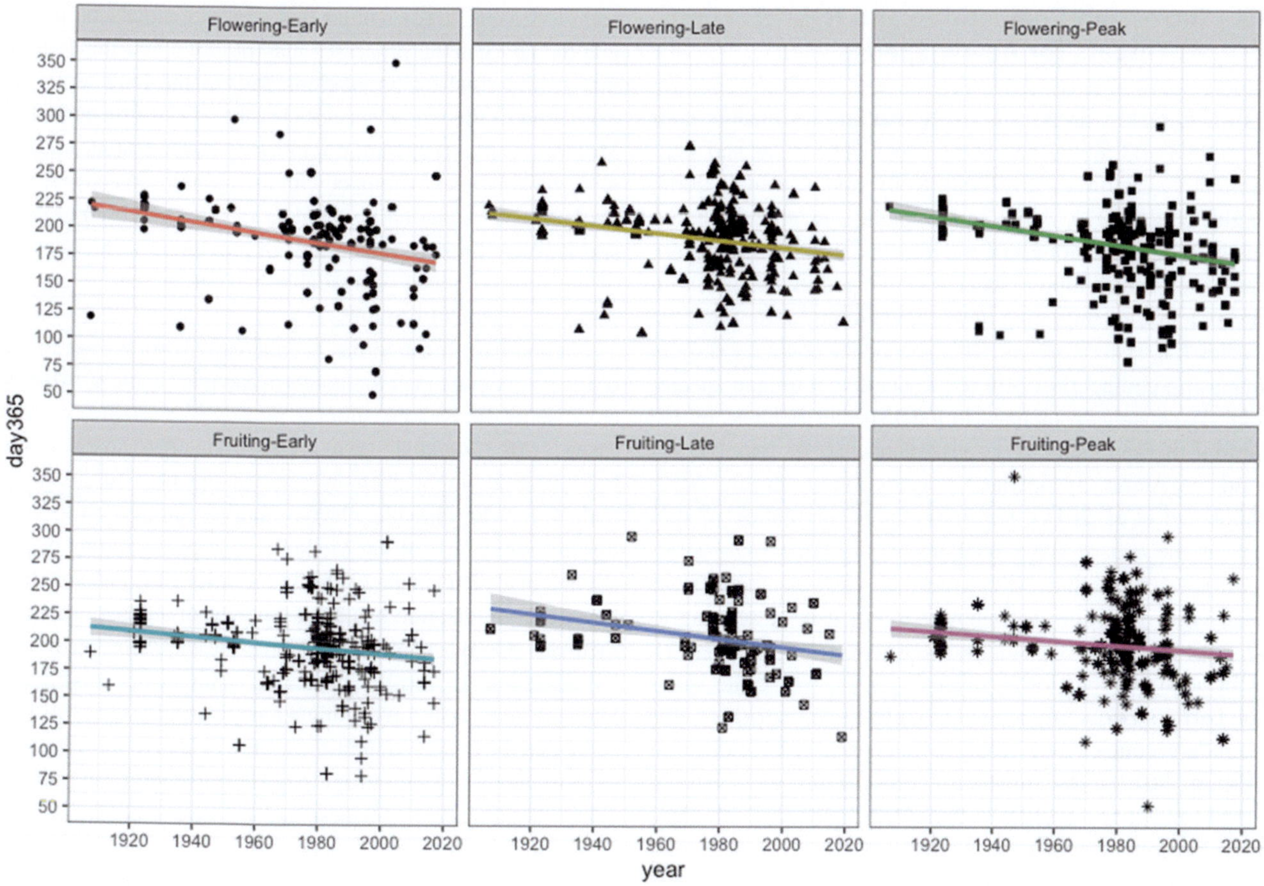


**Fig. 4** Scatterplot panel showing the date (1–365 days) for different phenophases (from early flowering to late-fruiting) for the period 1906–2020, derived from 2650 herbarium samples (from GDA, MGC, ALME and JAEN) of 120 endemic or subendemic species to Sierra Nevada. Line shows a linear trend for the period and shaded area the standard error of the mean (derived from generalized linear mixed models)

Dynamics Under Land-Use and Climate Change Scenarios"), *Genista versicolor* by 2100 and *Festuca clementei* by 2050 (Fig. 5). Results do not necessarily mean the extinction of these species, but the thermal niche they currently inhabit. In this way, it is foreseeable a population's decline as the current niches are dwindling.

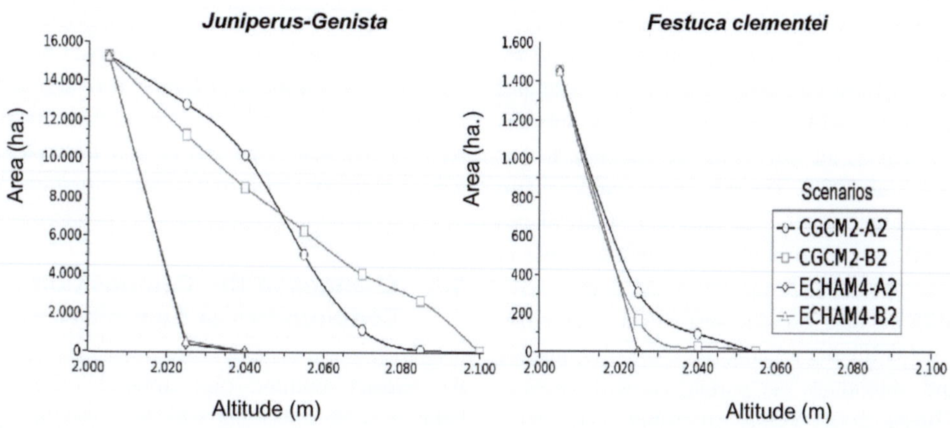


**Fig. 5** Projected loss of the suitable habitat area (in ha.) for three key species (*Juniperus communis, Genista versicolor* and *Festuca clementei*) from alpine and treeline ecotone in Sierra Nevada (from Benito et al. 2011). Shown are changes in the area under four scenarios (see Benito et al. 2011 for further information on the scenarios) of climatic change for 2000–2100

## 4.3 Increasing Competition and Hybridization as Threats for Alpine Species

As the climate warms, it is likely that species change their latitudinal and altitudinal distributions (Theurillat and Guisan 2001; Lenoir et al. 2008; Kelly and Goulden 2008; Engler et al. 2011). Altitudinal migration is expected to be much faster than latitudinal movements (Jump et al. 2009). Therefore, many alpine species could get in contact with lowland species migrating upward, including closely related species (i.e. sister species). This secondary contact led to an increasing competition between lowland and upland species (Fig. 6), causing an increase in local species richness on the summit of mountains as a consequence of the massive upward migration by lowland species (Pauli et al. 2012; Steinbauer et al. 2018). However, the accumulation of species at the top of the mountains would exacerbate interspecific competition (Roux et al. 2012), being the foreseeable outcome that alpine species, usually tending to be weak competitors would be the losers (Roux et al. 2012).

In addition, for closely related species, where the isolation barriers frequently are not complete (e.g. extrinsic), this secondary contact may result in interspecific hybridization (Abbott 2017). As mountain ecosystems are rich in narrow-endemic species, the consequences of this hybridization would have both positive (i.e. increasing genetic diversity and leading new adaptations) and negative effects (i.e. genetic swamping) for the alpine species (Gómez et al. 2015).

Copious empirical evidence shows that climate change modifies the altitudinal ranges of plant species (Grabherr et al. 1994; Lenoir et al. 2008; Benito et al. 2011, 2014; Engler et al. 2011; Rumpf et al. 2018; see chapter "Forest Dynamics Under Land-Use and Climate Change Scenarios"). Since altitudinal ecological gradients are narrower than latitudinal gradients, the distance necessary to migrate in searching suitable habitats is much smaller along the altitudinal than the latitudinal axis. Moreover, migrating among different mountains is unlikely in isolated mountains, as mountain habitats are surrounded by unsuitable habitats for most mountain species (La Sorte and Jetz 2010; Benito et al. 2014). For most montane species, upward movement will be the only possible response to climate change (Jump et al. 2009). Though, some authors claim that horizontal migration in suitable habitats (e.g. shady places under rock crevices) may have played an important role in persistence of some species (Körner 2021). As plants move uphill, and suitable habitats become lost at the rear edge, species distributions will contract and, therefore, the number of populations and individuals per population will decrease (Theurillat and Guisan 2001; Vegas-Vilarrúbia et al. 2012; Rumpf et al. 2018). In many cases, a shrinking suitable habitat will be also associated with a genetic diversity loss (Franks and Hoffmann 2012). Under this scenario, stochastic demographic and genetic processes may provoke first the extirpation of the smallest populations and, in the long term, the extinction of the whole species (Franks and Hoffmann 2012). Moreover, a rapid upward migration of lower-altitude species will probably lead to contact with high-altitude species before the latter go extinct (Jump et al. 2009). In fact, high-altitude species have limited possibilities of moving upward, because of the shrinking area of available habitats, given the conical shape of mountain summits.

Natural hybridization has been already reported for 26 out of the 95 endemic taxa to the Sierra Nevada (Table 4). That is, ca. 25% of the endemic flora is already hybridizing, in most cases, with lowland taxa (Table 4). The intensity of hybridization might even increase in the near future given that: (1) most climatic change models predict an overall upward movement of the Mediterranean plants in this massif (Benito et al. 2011); and (2) for two-thirds of the endemic taxa lineages hybridization has been reported in at least one of its members (Gómez et al. 2015). Under these circumstances (i.e. hybridization is mostly occurring with wider distributed lowland taxa, also having higher population sizes), we presume that the observed hybridization will lead to introgression in most cases. In fact, introgressive hybridization can produce complete admixture or widespread introgression (Fig. 6).

Complete admixture has proved negative outcomes triggering the extinction of rare, endemic taxa (Levin et al. 1996). Hybridization, besides the aforementioned negative effects, may also increase the diversity of the Sierra Nevada flora by giving rise to new species. For example, *Armeria filicaulis* subsp. *nevadensis* has been suggested as a new hybrid taxon that originated as a consequence of the hybridization of *A. splendens* and *A. filicaulis* subsp. *filicaulis,* due to altitudinal migrations within a contraction–expansion model (Gutierrez Larena et al. 2002). Similarly, *Nepeta boissieri* would have originated from the hybridization between *N. granatensis* and *N. amethystina* subsp. *laciniata* (Blanca et al. 2001). It seems that hybridization, among other ecological consequences of global warming, will have more impact on the Sierra Nevada endemic flora in the near future. It is urgent to determine the role of hybridization in the loss of plant diversity in the Sierra Nevada (Gómez et al. 2015), and to explore potential mitigation strategies, such as assisted migration (Vitt et al. 2010).

## 4.4 Changes in the Composition of Plant Communities at Summit Areas

As already pointed out, upward migration for suitable habitats is an important way for mountain plants to persist when phenological and ecological plasticity are overcome (Walther 2003). The expected changes at the community

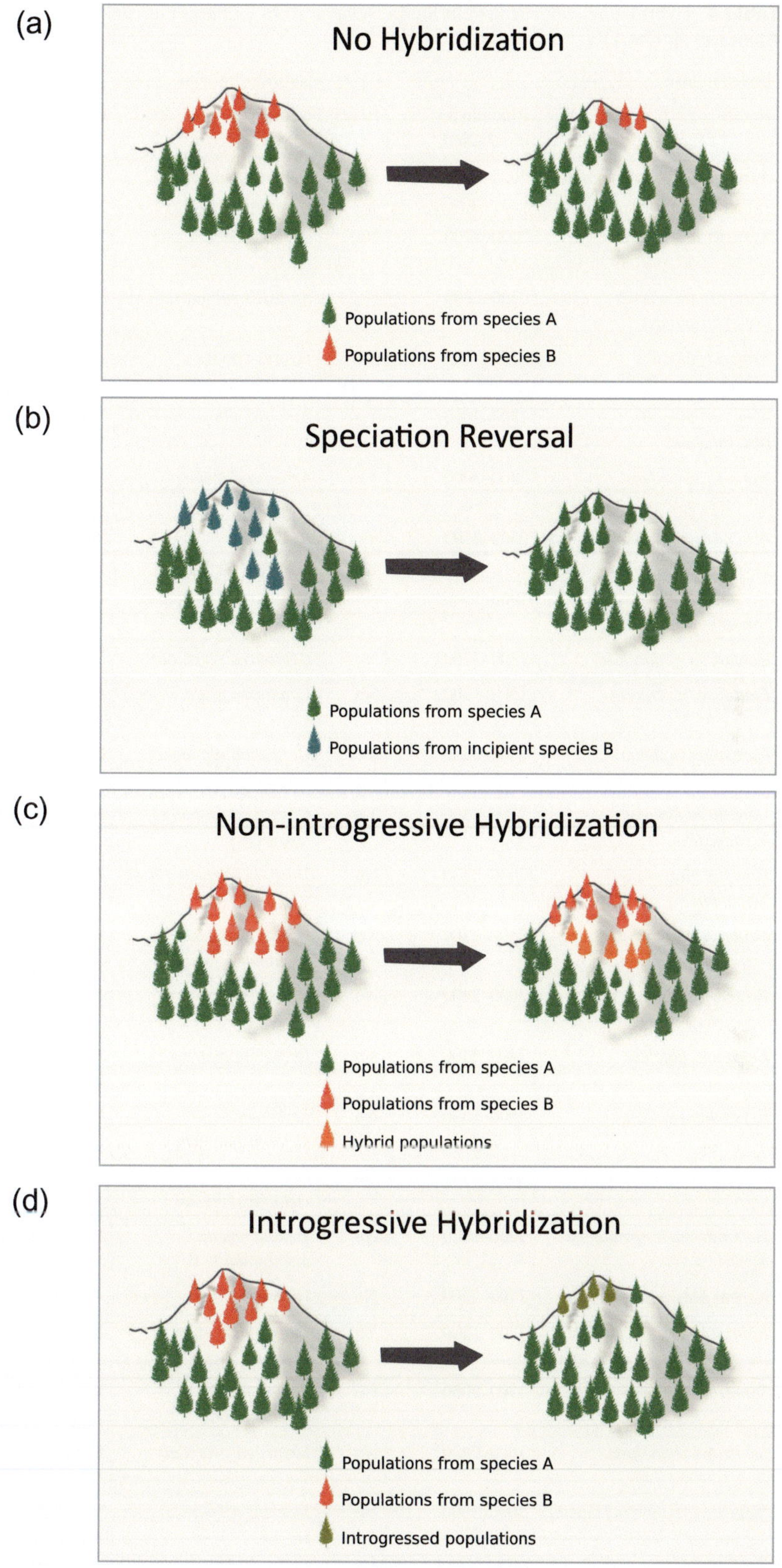


**Fig. 6** Potential consequences of the contact between lowland (species A) and high-mountain (species B) plant species. **a** *No hybridization,* the outcome is the demographic decline of the less competitive species. **b** *Speciation reversal* occurs after the primary contact between an incipient species and the widespread species. The result is a single, genetically homogeneous, species. **c** *Non-introgressive hybridization* occurs when two species contact and hybridize without between-species transfer of genes. The outcome is a hybrid zone in the contact area. **d** *Introgressive hybridization* occurs when two species contact and there is a net transfer of genes from one species to the other. When the two species in contact differ markedly in their abundance, the rarer may be completely replaced by hybrids (from Gómez et al. 2015)

**Table 4** Hybridizing endemic taxa in Sierra Nevada (from Gómez et al. 2015). Information on altitudinal and geographical distribution is from Blanca et al. (2001)

| Endemic taxa | | | Hybridizing with | | | |
|---|---|---|---|---|---|---|
| Name | Altitudinal distribution | Relict | Name | Geographical distribution | Altitudinal distribution in Sierra Nevada | Relative altitudinal location |
| *Agrostis canina granatensis* | 2000–2900 | No | *Agrostis canina* s. sl | Eurasia | 600–2200 | Lower altitudinal belt |
| *Agrostis nevadensis* | 2000–3300 | No | *Agrostis canina* s. sl | Eurasia | 600–2200 | Lower altitudinal belt |
| *Anthyllis vulneraria pseudoarundana* | 2200–3200 | No | *Anthyllis vulneraria* s.sl | Eurasia | 50–2000 | Lower altitudinal belt |
| *Armeria filicaulis nevadensis* | 2300–3000 | Yes | Hybrid species | | | |
| *Armeria filicaulis trevenqueana* | 1700–2000 | Yes | *Armeria villosa bernisii* | Baetic mountains | 900–2200 | Same altitudinal belt |
| *Armeria splendens* | 2400–3000 | Yes | *Armeria filicaulis* | Baetic mountains | 900–1900 | Lower altitudinal belt |
| *Artemisia granatensis* | 2500–2700 | Yes | *Artemisia umbelliformis* | Alpine mountains | 2800–3000 | Same altitudinal belt |
| *Centaurea bombycina xeranthemoides* | 1000–1500 | No | *Centaurea bombycina* s. sl | Baetic mountains | 200–1600 | Lower altitudinal belt |
| *Centaurea gadorensis* | 1300–1700 | No | *Centaurea pulvinata* | Sierra Nevada | 1200–2000 | Same altitudinal belt |
| *Centaurea pulvinata* | 1500–1900 | No | *Centaurea gadorensis* | Penibaetic mountains | 1100–1900 | Lower altitudinal belt |
| *Cerastium alpinum aquaticum* | 2500–3300 | Yes | *Cerastium alpinum nevadense* | Sierra Nevada | 2700–3300 | Same altitudinal belt |
| *Cerastium alpinum nevadense* | 2700–3300 | Yes | *Cerastium alpinum aquaticum* | Sierra Nevada | 2500–3300 | Same altitudinal belt |
| *Dactylis glomerata juncinella* | 2000–3300 | No | *Dactylis glomerata* s. sl | Eurasia | 600–1500 | Lower altitudinal belt |
| *Draba hispanica laderoi* | 2500–3200 | Yes | *Draba hispanica hispanica* | Western Mediterranean | 1000–2200 | Lower altitudinal belt |
| *Erigeron frigidus* | 3000–3400 | No | *Erigeron major* | Baetic mountains | 1800–3000 | Lower altitudinal belt |
| *Eryngium glaciale* | 2400–3400 | No | *Eryngium bourgatii* | Western Mediterranean | 1500–3000 | Lower altitudinal belt |
| *Erysimum baeticum* | 1600–2600 | No | *Erysimum mediohispanicum* | Iberian Peninsula | 700–1900 | Lower altitudinal belt |
| *Erysimum nevadense* | 2000–2800 | No | *Erysimum mediohispanicum* | Iberian Peninsula | 700–1900 | Lower altitudinal belt |
| *Helianthemum apenninum estevei* | 1300–1800 | No | *Helianthemum apenninum* s. sl | Baetic mountains | 0–1800 | Lower altitudinal belt |
| *Nepeta boissieri* | 1700–2200 | No | Hybrid species | | | |
| *Pinus sylvestris nevadensis* | 1700–2200 | Yes | *Pinus sylvestris* s. sl | Eurasia | 700–1900 | Lower altitudinal belt |
| *Santolina elegans* | 1700–2000 | No | *Santolina rosmarinifolia* | Western Mediterranean | 300–2100 | Lower altitudinal belt |
| *Saxifraga trabutiana* | 1600–2400 | No | *Saxifraga granulata* | Eurasia | 200–2800 | Lower altitudinal belt |
| *Sideritis arborescens luteola* | 1000–1600 | No | *Sideritis arborescens* s. sl | Western Mediterranean | 500–1000 | Lower altitudinal belt |
| *Sideritis glacialis* | 2000–3000 | No | *Sideritis hirsuta* | Western Mediterranean | 50–2000 | Lower altitudinal belt |
| *Thymus serpylloides serpylloides* | 2000–3000 | No | *Thymus serpylloides gadorensis* | Iberian Peninsula | 1500–2200 | Lower altitudinal belt |

level are initially in the relative abundance of plant species, followed by changes in composition and species richness (Rixen and Wipf 2017; Steinbauer et al. 2018, 2020).

Model projections of massive extinction rates of alpine species at one hand (e.g. Engler et al. 2011; Benito et al. 2014), and the observational findings of a strong increase in summits species numbers (e.g. Pauli et al. 2012; Steinbauer et al. 2018) on the other hand, seem to be contradictory. However, usually, long-lived alpine species may persist with increasing competition through lowland species and in climatically unsuitable habitats, resulting in an initial enrichment of species composition, accumulating an extinction of the formerly local species (Engler et al. 2009; Dullinger et al. 2012).

To study climate change induced effects on alpine plant diversity, the international long-term monitoring network GLORIA (Global Observation Research Initiative in Alpine Environments, see https://www.gloria.ac.at) was established in 2001. It started with 18 European mountain regions, including the Sierra Nevada, and has been extended to a global network with about 130 regions involved so far (Pauli et al. 2015).

For now, there are two GLORIA study regions at the Sierra Nevada (SNE: Sierra Nevada—West, SNN: Sierra Nevada—Northeast), including eight summits, arranged along an elevation gradient between 2668 and 3327 m (Fig. 7a, Table 5; Lamprecht et al. 2021). On each summit, eight summit area sections and four 3 m × 3 m quadrat clusters (in each cardinal direction) were established and have been investigated three times so far by using a globally applicable and comparable standardized protocol (Pauli et al. 2015).

Results obtained after two decades showed in parts a warring trend between the periods (Lamprecht et al. 2021).

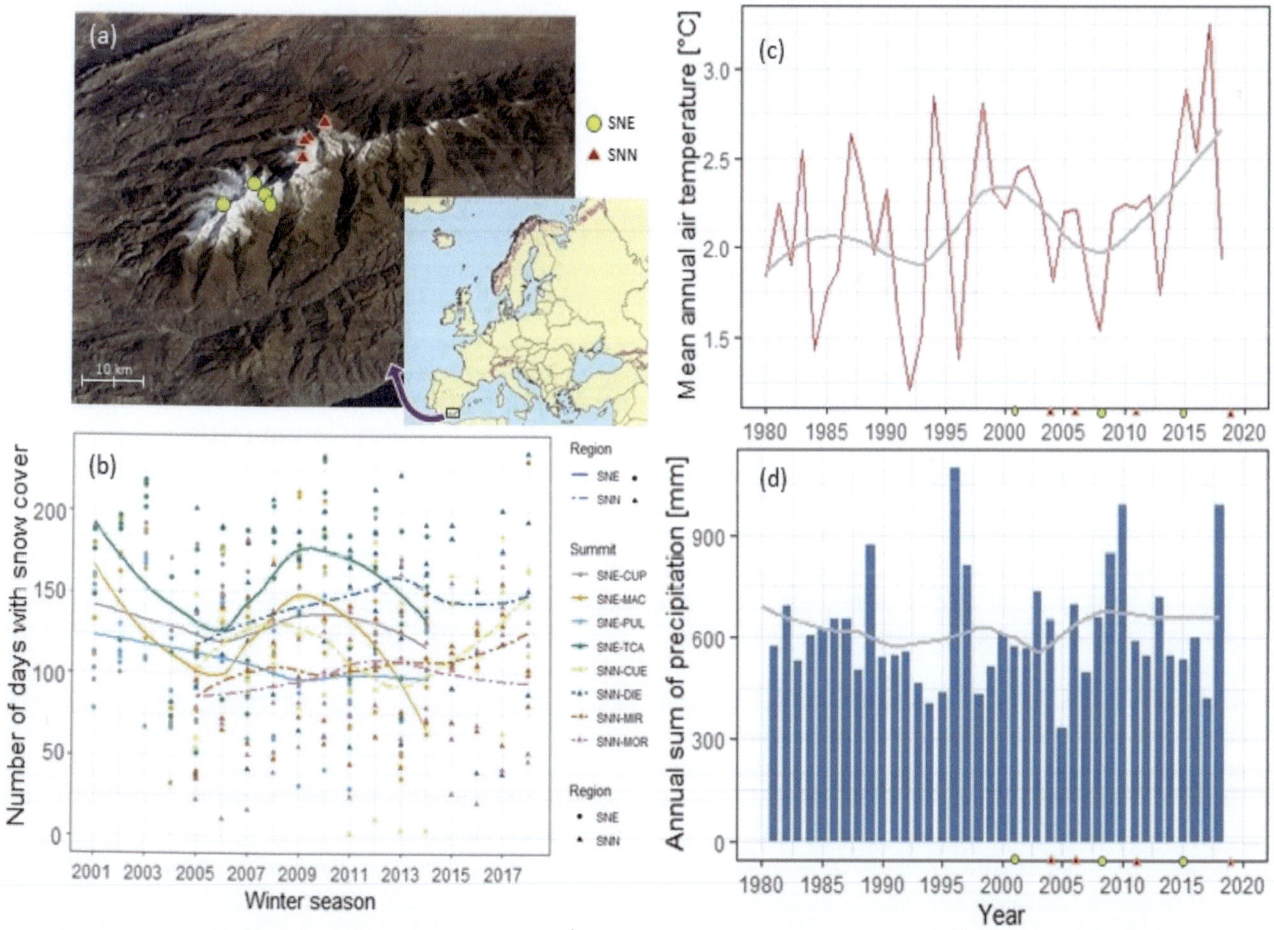


**Fig. 7** Location of the GLORIA summits in Sierra Nevada, Spain, and the climatic conditions during the recent decades (from Lamprecht et al. 2021). **a** Geographic location of eight GLORIA summits of the two study regions Sierra Nevada—West (SNE) and Sierra NevadaNortheast (SNN). **b** Days with snow cover, i.e. days with a maximum of 0.5 °C and a minimum of −5 °C and a maximum daily difference of 2 °C (after Teubner et al. 2015) derived from soil temperature measurements in each cardinal direction on each summit (dots and triangles), as well as the loess local weighted regression (lines). **c** Mean annual air temperature and **d** annual precipitation of the Sierra Nevada, derived from ERA5 data (https://www.ecmwf.int/en/forecasts/datasets/reanalysis-datasets/era5). The grey line represents the 20-year Gaussian low-pass filter smooth trend. On the x-axis, vegetation survey years in the two study regions are shown (dots and triangles)

**Table 5** Number of vascular plant species per summit and survey in the Sierra Nevada (from Lamprecht et al. 2021). Shown are the number of perennial species (annuals in brackets) recorded per summit in the two study regions Sierra NevadaWest (SNE) and Sierra NevadaNortheast (SNN) for each of three surveys (SNE: 2001, 2008, 2015; SNN: 2004 (2006 on MIR), 2011, 2019)

| Region | Summit name | Summit code | Elevation (m) | Number of species in survey | | |
|---|---|---|---|---|---|---|
| | | | | 1 | 2 | 3 |
| SNE | Pulpito | PUL | 2778 | 45 (2) | 41 (4) | 44 (4) |
| SNE | Cúpula | CUP | 2968 | 51 (1) | 47 (2) | 51 (1) |
| SNE | Pico del Tosal Cartujo | TCA | 3150 | 39 (1) | 37 (1) | 43 (2) |
| SNE | Cerro de los Machos | MAC | 3327 | 18 (0) | 16 (0) | 18 (1) |
| SNN | Monte Rosa | MOR | 2668 | 36 (4) | 40 (4) | 40 (4) |
| SNN | Miron | MIR | 2717 | 55 (8) | 61 (7) | 62 (5) |
| SNN | Diegisa | DIE | 2800 | 41 (5) | 49 (7) | 55 (5) |
| SNN | Cuervo | CUE | 3144 | 20 (5) | 23 (5) | 27 (5) |

Plant species richness per plot and per summit showed a decline between 2001 and 2008, and increased thereafter. Species cover followed the same patterns, and increased slightly but significantly over the whole period. 39% of the species occurring on the GLORIA summits were local endemic species of the Sierra Nevada. The total cover of the endemics underwent losses proportional to non-endemics, and there were significantly more disappearance events among endemic species than among non-endemic species. This is of great concern, considering the special situation in this small-scaled and isolated mountain range, and in some cases the extremely low numbers of individuals for some species ($\sim$2000 of *Artemisia granatensis*; Lorite et al. 2007). At the same time, not only a continued increase in cover of some shrub species could be observed (see chapter "Managing the Uniqueness of Sierra Nevada Ecosystems Under Global Change: The Value of in situ Scientific Research"), but also an increase of their proportional cover sum. The findings indicate a consistent expansion of the upper distribution ranges of shrubs to high-elevation habitats (cf. Fernández-Calzado and Molero 2013), commencing a "shrubification" of the higher vegetation zones in the Sierra Nevada.

All observed changes in plant species composition and diversity showed strong correlations to water supply components. Rising temperatures showed only a relationship to the increase in species richness. Rapid warming was highlighted as the main driver for accelerating increases in species richness on summits all over Europe (Steinbauer et al. 2018). However, precipitation and snow as limited resources during the vegetative period in the Mediterranean mountains seemed to be the more important drivers of vegetation changes. After a downward trend since 1960 (Ruiz Sinoga et al. 2011), annual precipitation tended to increase in the Sierra Nevada during the last decade (Polo et al. 2019; Lamprecht et al. 2021). Higher amounts of precipitation coincided not only with species richness increase, but also with an increase in colonization events and cover, and a decrease in disappearance events. This tight synchronization of rapid responses of alpine species with climate change, notably with water availability, gives reason to apprehend detrimental effects on this water-limited high-mountain flora of the Sierra Nevada. Future climate projections predict not only continuing warming, but also a decrease in mean precipitation in Mediterranean regions (Pérez-Palazón et al. 2018). Considering the special situation of the Sierra Nevada with an outstanding set of unique plants, consequences may be of global significance.

## 5 Direct Anthropogenic Drivers of Changing Patterns in Plant Communities

Mediterranean mountains have a long history of human intervention, which has modified land cover and resulted in numerous land use changes over time, with increasing or decreasing pressure depending on socio-economic framework (Davis 1976). The Sierra Nevada, as other Mediterranean mountain regions, has been subjected to a high anthropic pressure in the last millennia (Jiménez-Olivencia et al. 2010). The combining result of anthropogenic land-use with the formerly described patterns can modulate, and often magnify, the changing patterns formerly described (Lamprecht et al. 2019).

In the Sierra Nevada, a series of disturbing factors stands out for their important impact on ecosystems, that can be divided into: (i) traditional and operating nowadays (i.e. livestock grazing), (ii) traditional but no longer operating, despite that imprinting the current ecosystems, (i.e. management of vegetation with fire or high-mountain agriculture), or (iii) novel impacts (mainly derived from outdoor activities and infrastructure construction).

## 5.1 Livestock Grazing

Pastoralism has been one of the main economic activities in the Mediterranean Basin since humans domesticated sheep, goats, and cattle in the Neolithic period, using them as a source of food and fibres (Vogiatzakis 2012). At the same time, Mediterranean mountains have been used for millennia by seasonal herders in order to make use of their forage resources during the summer, when the lowlands could not provide feed to the animals due to the seasonality of the Mediterranean climate (Vogiatzakis 2012). The complementary exploitation of highlands and lowlands pasturelands alternatively for ages, known as transhumance, became a common practice across all Mediterranean mountains and resulted in a complex and widespread network of sheep trail routes called "cañadas" in Spain (Ruiz and Ruiz 1986).

In the Sierra Nevada, as in Spain, transhumance reached its peak during the 5 to fifteenth centuries but after the Industrial Revolution, it started to decline, accelerating dramatically during the last three decades of twentieth century, due to a profound change in socio-cultural and economic factors (Ruiz et al. 2020; chapter "Singular Cultural Landscapes of the Sierra Nevada"). These seasonal herders started to introduce important changes in the vegetation. In fact, there are signals of pastoral burning ca. 3000 years ago, afterwards intensified in Roman times and beyond (Jiménez-Moreno et al. 2013; chapter "Reconstruction of Past Environment and Climate Using Wetland Sediment Records from the Sierra Nevada"). Fire was used not only in open grazing lands but also to suppress undesirable woody vegetation in rangelands. This burning had to play a determinant role in the disappearance of the treeline in the Sierra Nevada. They felled and harvested the scarce wood as firewood and they used fire to promote pasturelands, causing the disappearance of the spotted trees, and a decline of the creeping juniper communities (Lorite 2002).

Particularly important is the role played by "Borreguiles" in the traditional livestock management. These hygrophilous pasturelands, highly productive in summer, have been intensively used in the transhumance system at the Sierra Nevada highlands as a trophic reserve for livestock in this season (Robles et al. 2016). This traditional management is rapidly changing in the past years, leading to a grazing intensification in some accessible spots, while others lack domestic ungulates at all for the first time in the last millennia (Fernández-Fernández 2018; Peñas and Lorite 2019).

Wild goats have largely increased in number in the Sierra Nevada since the 1970s as a result of the lack of predators, a decreasing human pressure, and a hunting regulation dating back from the 1960s (Fandos et al. 2010). This species poses nowadays an important threat for many endemic and/or endangered plant species in the area (Peñas and Lorite 2019), as in other Betic mountains (Lorite et al. 2007).

Another issue concerns "acequias de careo", an intricated irrigation system that nowadays consists of more than 550 km of irrigation channels (Cano-Manuel and Ortiz-Moreno 2010; chapter "Singular Cultural Landscapes of the Sierra Nevada") established by the moors around the ninth century (Martos-Rosillo et al. 2019b). This system extended and artificially maintained summer pasturelands by irrigating many mountain slopes (Martos-Rosillo et al. 2019a). The abandonment of this practice and their progressive transformation into conventional water channels have tended to reduce the surface area of these irrigated pasturelands, consequently reducing their availability in summer and causing overgrazing in hygrophilous natural pasturelands (Peñas and Lorite 2019).

## 5.2 Mountain Agriculture

Mediterranean mountains have experienced both expansions and contractions of their human population, being this fact connected with agricultural changes in these areas (Vogiatzakis 2012). Hence, mountain populations began to grow in the late eighteenth century after the introduction of maize and potato, which together with the traditional pastoralism made the survival of people much easier (McNeill 1992).

High mountain agriculture is nowadays almost absent in the Sierra Nevada. However, potatoes and ray croplands at ca. 2500 m (Castellon 2008). They remained until the 1950s or 1960s when rural depopulation of almost 50% resulted in the abandonment of alpine crops that have been converted to grassland, or they have suffered matorral encroachment (Douglas et al. 1996).

## 5.3 Outdoor Activities and Infrastructure Construction

Since the 1950s, recreation and tourism have expanded exponentially in natural areas worldwide, both in the number of practitioners as well as in the number of activities practiced (Cordell et al. 2012). Mountain outdoor activities (including hiking, mountain climbing, rock climbing, skiing, cross-country skiing, camping, mountain biking, orienteering, cross-country racing, extreme trails, etc.) have exponentially risen in the last 20 years in Spain and the Sierra Nevada following the general trend. In this line, recreation activities in the Sierra Nevada are strongly increasing since the designation as National Park (Lamprecht et al. 2019).

Thus, the number of visitors has risen from 302,520 in 2002 to 732,657 in 2017, reaching the peak with 790,520 visitors in 2020 (VV.AA. 2020). These activities cause many direct and indirect impacts on vegetation such as trampling, pulling (i.e. pulling due to the climbing activity, Lorite et al. 2017), construction of infrastructure, illegal collection of plants, waste deposition, or water pollution. Being the effects for the vegetation an important cover loss and subsequent erosion (Lorite et al. 2010). Moreover, due to the harsh Mediterranean-alpine environment recovery is very slow, this being particularly harmful to rare and unique species, such as narrow endemic and arctic-alpine species (Peñas and Lorite 2019).

Especially harmful is building and maintaining ski resorts that imply the use of heavy machinery during summer for the levelling and drainage of the slopes. Ski pistes are machine-graded to smooth surface, whereby rocks, obstacles, the natural vegetation, and most of the organic topsoil, are removed. This results in major disturbance (often complete) to plant and soil (Roux-Fouillet et al. 2011; Casagrande Bacchiocchi et al. 2019). The Sierra Nevada ski station has 97 km of ski slopes. Its construction has caused the destruction of very valuable vegetation patches in the core area of the mountain. Meanwhile, restoration efforts have failed in restoring the original composition and structure of the vegetation so far (Lorite et al. 2010, see also chapter "Managing the Uniqueness of Sierra Nevada Ecosystems Under Global Change: The Value of in situ Scientific Research").

## 6 Concluding Remarks and Guidelines for Conservation and Adaptive Management

The Sierra Nevada contains a unique alpine flora. However, the conservation of this outstanding natural heritage is facing important challenges, some of them coming from the increasing effect of climate change, which is being especially severe in the Mediterranean mountains. However, different impacts caused by human activities may act synergistically and exacerbate these changes.

Monitoring, evaluating, and understanding the effect of global change in Mediterranean mountains have to be a priority in conservation agendas. Only the adoption of measures with a solid scientific basis can help to mitigate the combined effects of climatic and land use changes.

We want to point the main guidelines resulting from the revision made in this chapter, that can act as aims to orient the conservation agenda at the Sierra Nevada and other Mediterranean mountains:

(i) To establish an early warning indicators system: Following key processes and species can help to predict and anticipate changes, especially negative ones.
(ii) To preserve plant species and habitats: Preserving species and habitats at their best will make them more robust to face the climatic changes.
(iii) To preservee ex situ threatened plant species: Species preservation outside the habitats is an extreme but sometimes necessary measure when preservation in the habitat is not feasible. This measure can help to take other adaptation measures (i.e. assisted colonization).
(iv) To promote adaptive management measures: They can be very varied, going from measures intended to restore the habitats that have been damaged, altered or destroyed, to assisting the colonization of suitable habitat by facilitating natural colonization, or even seeding or planting the species in the new suitable areas.
(v) To evaluate outdoor recreation activities: The aim is to determine carrying capacity and limits of acceptable change (i.e. the change acceptable as a compromise between resource/visitor experience protection and recreation use goals; Hammitt et al. 2015).
(vi) To establish measures for controlling and regulating outdoor activities: Establishing necessary measures to minimize damages. Including severe limitation of access to sensitive habitats, or to forbid some activities that are not compatible with conservation.

**Acknowledgements** To all the researchers and naturalists that have been studying the Sierra Nevada during the last 150 years. Project B1-RNM-163-UGR18-Programa Operativo FEDER, 2018 partially funded this research. The Direction and the staff of the National Park of Sierra Nevada provided permissions and greatly assisted the fieldwork. The staff of the Agencia de Medio Ambiente y Agua de Andalucía collaborated in some ongoing and past projects that supported this chapter. GLORIA coordination team at the University of Vienna established the permanent plots to follow the changes in plant communities, and it still supports and gives technical advice.

## References

Abbott RJ (2017) Plant speciation across environmental gradients and the occurrence and nature of hybrid zones. J Syst Evol 55:238–258
Abeli T, Vamosi JC, Orsenigo S (2018) The importance of marginal population hotspots of cold-adapted species for research on climate change and conservation. J Biogeogr 45:977–985
Amano T, Smithers RJ, Sparks TH, Sutherland WJ (2010) A 250-year index of first flowering dates and its response to temperature changes. Proc R Soc Ser B Biol Sci 277:2451–2457

Anderson RS, Jiménez-Moreno G, Carrión JS, Pérez-Martínez C (2011) Postglacial history of alpine vegetation, fire, and climate from Laguna de Río Seco, Sierra Nevada, southern Spain. Quat Sci Rev 30:1615–1629

Benito B, Lorite J, Peñas J (2011) Simulating potential effects of climatic warming on altitudinal patterns of key species in Mediterranean-alpine ecosystems. Clim Change 108:471–483

Benito BM, Lorite J, Pérez-Pérez R et al (2014) Forecasting plant range collapse in a mediterranean hotspot: when dispersal uncertainties matter. Divers Distrib 20:72–83

Blanca G, Cueto M, Martínez-Lirola MJ, Molero-Mesa J (1998) Threatened vascular flora of Sierra Nevada (southern Spain). Biol Conserv 85:269–285

Blanca G, López-Onieva M, Lorite J et al (2001) Flora amenazada y endémica de Sierra Nevada. Universidad de Granada, Granada

Boissier E (1839) Voyage botanique dans le midi de l'Espagne pendant l'année 1837. Gide et Cie Librairies-Éditeurs, Paris

Brullo S, Giusso Del Galdo G, Guarino R (2001) The orophilous communities of the Pino-Juniperetea class in the Central and Eastern Mediterranean area. Feddes Repert 112:261–308

Cañadas EM, Fenu G, Peñas J et al (2014) Hotspots within hotspots: endemic plant richness, environmental drivers, and implications for conservation. Biol Conserv 170:282–291

Cano-Manuel J, Ortiz-Moreno E (2010) Inventario de acequias de Sierra Nevada. In: El agua domesticada, los paisajes de los regadíos de montaña en Andalucía. Consejería de Medio Ambiente, Junta de Andalucía, Sevilla, pp 520–525

Casagrande Bacchiocchi S, Zerbe S, Cavieres LA, Wellstein C (2019) Impact of ski piste management on mountain grassland ecosystems in the Southern Alps. Sci Total Environ 665:959–967

Castellon F (2008) Sierra Nevada: Un escenario natural y humano. La Alpujarra Oriental: la gran desconocida. Editorial Universidad de Almería, Almería, pp 25–38

Charco J (1999) El bosque mediterráneo en el norte de África: biodiversidad y lucha contra la desertificación

Cordell HK, Betz CJ, Mou SH, Gormanson DD (2012) Outdoor recreation in the Northern United States, Gen. Tech. Department of Agriculture, Forest Service, Northern Research Station, Newtown Square

Davis P (1976) Land use. McGraw-Hill, New York

Douglas T, Critchley D, Park G (1996) The deintensification of terraced agricultural land near Trevelez, Sierra Nevada, Spain. Glob. Ecol. Biogeogr Lett 5:258–270

Dullinger S, Gattringer A, Thuiller W et al (2012) Extinction debt of high-mountain plants under twenty-first-century climate change. Nat Clim Change 2:619–622

Engler R, Randin CF, Vittoz P et al (2009) Predicting future distributions of mountain plants under climate change: does dispersal capacity matter? Ecography (cop) 32:34–45

Engler R, Randin CF, Thuiller W et al (2011) 21st century climate change threatens mountain flora unequally across Europe. Glob Chang Biol 17:2330–2341

Essl F, Staudinger M, Stöhr O et al (2009) Distribution patterns, range size and niche breadth of Austrian endemic plants. Biol Conserv 142:2547–2558

Fandos P, Arcinegui P, Lora MA et al (2010) Evolución demográfica de la cabra montés en Andalucía en los últimos 100 años. Galemys 22:347–358

Fernández-Fernández E (2018) La trashumancia en Sierra Nevada y su patrimonio cultural inmaterial. In: Titos-Martínez M, Luque T, Navarro-Llena JM (eds) Cimas. Montañas; fuentes de vida y de futuo. Editorial Universidad de Granada, Granada, pp 321–335

Fernández-Calzado MR, Molero J (2011) The cartography of vegetation in the cryoromediterranean belt of Sierra Nevada: a tool for biodiversity conservation. Lazaroa 32:101–115

Fernández Calzado MR, Molero J (2013) Changes in the summit flora of a Mediterranean mountain (Sierra Nevada, Spain) as a possible effect of climate change. Lazaroa 34:65–75

Fernández Calzado MR, Molero Mesa J, Merzouki A, Casares Porcel M (2012) Vascular plant diversity and climate change in the upper zone of Sierra Nevada, Spain. Plant Biosyst 146:1044–1053

Franks SJ, Hoffmann AA (2012) Genetics of climate change adaptation. Annu Rev Genet 46:185–208

Gamisans J (2003) The vegetation of the Corsican high mountains. In: Nagy L, Grabherr G, Körner C, Thompson DBA (eds) Alpine biodiversity in Europe. Springer, Berlin Heidelberg, Berlin, Heidelberg, pp 105–111

García D, Zamora R, Hódar JA, Gómez JM (1999) Age structure of Juniperus communis L. in the Iberian peninsula: conservation of remnant populations in Mediterranean mountains. Biol Conserv 87:215–220

Gómez-Ortiz A (2002) Geomorphological map of Sierra Nevada; glacial a periglacial geo-morphology. Consejería de Medio Ambiente. Junta de Andalucía y Universidad de Barcelona, Sevilla

Gómez-Ortiz A, Oliva M, Salvà-Catarineu M, Salvador-Franch F (2013) The environmental protection of landscapes in the high semiarid Mediterranean mountain of Sierra Nevada National Park (Spain): historical evolution and future perspectives. Appl Geogr 42:227–239

Gómez JM, González-Megías A, Lorite J et al (2015) The silent extinction: climate change and the potential hybridization-mediated extinction of endemic high-mountain plants. Biodivers Conserv 24:1843–1857

Gottfried M, Pauli H, Futschik A et al (2012) Continent-wide response of mountain vegetation to climate change. Nat Clim Change 2:111–115

Grabherr G, Gottfried M, Pauli H (1994) Climate effects on mountain plants. Nature 369:448–448

Gutierrez Larena B, Fuertes Aguilar J, Nieto Feliner G (2002) Glacial-induced altitudinal migrations in Armeria (Plumbaginaceae) inferred from patterns of chloroplast DNA haplotype sharing. Mol Ecol 11:1965–1974

Hajar L, François L, Khater C et al (2010) Cedrus libani (A. Rich) distribution in Lebanon: past, present and future. Comptes Rendus Biol 333:622–630

Hammitt WE, Cole DN, Monz CA (2015) Wildland recreation: ecology and management, 3rd edn. Wiley, Ltd, Chichester

Hughes L (2000) Biological consequences of global warming: is the signal already apparent? Trends Ecol Evol 15:56–61

Hülber K, Winkler M, Grabherr G (2010) Intraseasonal climate and habitat-specific variability controls the flowering phenology of high alpine plant species. Funct Ecol 24:245–252

Jansson R (2003) Global patterns in endemism explained by past climatic change. Proc R Soc London Ser B Biol Sci 270:583–590

Jiménez-Moreno G, García-Alix A, Hernández-Corbalán MD et al (2013) Vegetation, fire, climate and human disturbance history in the southwestern Mediterranean area during the late Holocene. Quat Res 79:110–122

Jiménez-Olivencia Y, Porcel-Rodríguez L, Píñar-Álvarez A (2010) Evolución Histórica de los paisajes del Parque Nacional Sierra Nevada y su entorno. In: Proyectos de Investigación en Parques Nacionales 2006–2009. Serie Investigación en la Red: Naturaleza y Parques Nacionales. Organismo Autónomo de Parques Nacionales, Madrid, pp 2006–2009

Jump AS, Mátyás C, Peñuelas J (2009) The altitude-for-latitude disparity in the range retractions of woody species. Trends Ecol Evol 24:694–701

Kelly AE, Goulden ML (2008) Rapid shifts in plant distribution with recent climate change. Proc Natl Acad Sci 105:11823–11826

Körner C (2003) Alpine plant life, 2a edn. Springer, Berlin Heidelberg, Berlin, Heidelberg

Körner C (2012) Alpine treelines. Springer Basel, Basel

Körner C (2021) Alpine plant life. Springer International Publishing, Cham

La Sorte FA, Jetz W (2010) Projected range contractions of montane biodiversity under global warming. Proc R Soc B Biol Sci 277:3401–3410

Lamprecht A, Rutzinger M, Pauli H et al (2019) Disentangling anthropogenic drivers of climate change impacts on alpine plant species: Alps vs Mediterranean mountains. Vienna

Lamprecht A, Pauli H, Fernández Calzado MR et al (2021) Changes in plant diversity in a water-limited and isolated high-mountain range (Sierra Nevada, Spain). Alp Bot 131:27–39

Larson DW, Matthes U, Kelly PE (2000) Cliff ecology. Cambridge University Press, Cambridge

le Roux PC, Virtanen R, Heikkinen RK, Luoto M (2012a) Biotic interactions affect the elevational ranges of high-latitude plant species. Ecography (cop) 35:1048–1056

Lenoir J, Gégout JC, Marquet P, et al (2008) A significant upward shift in plant species optimum elevation during the 20th century. Science 320:1768–1771

Levin DA, Francisco-Ortega J, Jansen RK (1996) Hybridization and the extinction of rare plant species. Conserv Biol 10:10–16

Lorite J (2002) La vegetación de Sierra Nevada. In: Blanca G (ed) Flora amenazada y endémica de Sierra Nevada. Editorial Universidad de Granada, Granada, pp 23–45

Lorite J (2016) An updated checklist of the vascular flora of Sierra Nevada (SE Spain). Phytotaxa 261:1–57

Lorite J, Valle F, Salazar C (2003) Síntesis de la vegetación edafohigrófila del Parque Natural y Nacional de Sierra Nevada. Monogr Flora y Veg Béticas 13:47–110

Lorite J, Navarro FB, Valle F et al (2007) Estimation of threatened orophytic flora and priority of its conservation in the Baetic range (S. Spain). Plant Biosyst 141:1–14

Lorite J, Molina-Morales M, Cañadas EM et al (2010) Evaluating a vegetation-recovery plan in Mediterranean alpine ski slopes: a chronosequence-based study in Sierra Nevada (SE Spain). Landsc Urban Plan 97:92–97

Lorite J, Serrano F, Cañadas EM et al (2017) Rock climbing alters plant species composition, cover, and richness in Mediterranean limestone cliffs. PLoS ONE 12:1–14

Lorite J, Ros-Candeira A, Alcaraz-Segura D, Salazar-Mendías C (2020) FloraSNevada: a trait database of the vascular flora of Sierra Nevada, southeast Spain. Ecology 101:ecy.03091

Losa Quintana JM, Casares MMJ, M., Pérez-Raya F (1986) El paisaje vegetal de Sierra Nevada. La cuenca alta del Río Genil. Servicio de Publicaciones de la Universidad de Granada, Granada

Martos-Rosillo S, Antonio G-R, Ruiz-Constán A et al (2019a) El manejo del agua en las cuencas de alta montaña del Parque Nacional de Sierra Nevada (Sur de España). Un ejemplo ancestral de Gestión Integral del Agua. BOLETÍN GEOLÓGICO Y Min 130:729–742

Martos-Rosillo S, Ruiz-Constán A, González-Ramón A et al (2019) The oldest managed aquifer recharge system in Europe: new insights from the Espino recharge channel (Sierra Nevada, southern Spain). J Hydrol 578:124047

McNeill J (1992) The mountains of the Mediterranean world. Cambridge University Press, Cambridge

Médail F, Diadema K (2009) Glacial refugia influence plant diversity patterns in the Mediterranean Basin. J Biogeogr 36:1333–1345

Medail F, Quezel P (1999) Biodiversity hotspots in the Mediterranean Basin: setting global conservation priorities. Conserv Biol 13:1510–1513

Myers N, Mittermeier RA, Mittermeier CG et al (2000) Biodiversity hotspots for conservation priorities. Nature 403:853–858

Nagy L (2006) European high mountain (alpine) vegetation and its suitability for indicating climate change impacts. Biol Environ Proc R Irish Acad 106:335–341

Nagy L, Grabherr G (2009) The biology of alpine habitats. Oxford University Press, Oxford

Nagy L, Grabherr G, Körner C, Thompson DBA (2003) Alpine biodiversity in Europe. Springer, Berlin Heidelberg, Berlin, Heidelberg

Nogués-Bravo D, Araújo MB, Lasanta T et al (2008) Climate change in Mediterranean mountains during the 21st century. Ambio 37:280–285

Ohlemüller R, Anderson BJ, Araújo MB et al (2008) The coincidence of climatic and species rarity: high risk to small-range species from climate change. Biol Lett 4:568–572

Parmesan C (2007) Influences of species, latitudes and methodologies on estimates of phenological response to global warming. Glob Chang Biol 13:1860–1872

Parmesan C, Yohe G (2003) A globally coherent fingerprint of climate change impacts across natural systems. Nature 421:37–42

Pauli H, Gottfried M, Dullinger S et al (2012) Recent plant diversity changes on Europe's mountain summits. Science (80–)336:353–355

Pauli H, Gottfried M, Lamprecht A, et al (2015) The GLORIA field manual—standard Multi-Summit approach, supplementary methods and extra approaches, 5th edn. GLORIA-Coordination, Austrian Academy of Sciences & University of Natural Resources and Life Sciences, Vienna

Peñas J, Lorite J (2019) Biología de la conservación de plantas en Sierra Nevada. Principios y retos para su preservación. Editorial Universidad de Granada, Granada

Peñuelas J, Filella I (2001) Phenology: responses to a warming world. Science (80–)294:793–795

Pepin N, Bradley RS, Diaz HF et al (2015) Elevation-dependent warming in mountain regions of the world. Nat Clim Change 5:424–430

Pérez-Luque AJ, Sánchez-Rojas CP, Zamora R et al (2015a) Dataset of phenology of Mediterranean high-mountain meadows flora (Sierra Nevada, Spain). PhytoKeys 46:89–107

Pérez-Luque AJ, Sánchez-Rojas CP, Zamora R, Bonet FJ (2015b) Cambios temporales en la diversidad, abundancia y fenología de las comunidades vegetales: un estudio de 25 años en los borreguiles. In: Zamora R, Pérez-Luque AJ, Bonet-García FJ et al (eds) La huella del cambio global en Sierra Nevada: Retos para la conservación. Consejería de Medio Ambiente y ordenación del Territorio, Junta de Andalucía, Sevilla, pp 126–127

Pérez-Palazón M, Pimentel R, Polo M (2018) Climate trends impact on the snowfall regime in Mediterranean Mountain areas: future scenario assessment in Sierra Nevada (Spain). Water 10:720

Piper FI, Viñegla B, Linares JC et al (2016) Mediterranean and temperate treelines are controlled by different environmental drivers. J Ecol 104:691–702

Polo MJ, Herrero J, Pimentel R, Pérez-Palazón MJ (2019) The Guadalfeo Monitoring Network (Sierra Nevada, Spain): 14 years of measurements to understand the complexity of snow dynamics in semiarid regions. Earth Syst Sci Data 11:393–407

Primack D, Imbres C, Primack RB et al (2004) Herbarium specimens demonstrate earlier flowering times in response to warming in Boston. Am J Bot 91:1260–1264

Redondo JJ, Martínez-Rodríguez F (2021) Primeros datos sobre un petroglifo prehistórico de estilo atlántico hallado en Peña Madura, cuenca alta del Río Dílar, Sierra Nevada (Granada). Cuad Arte Prehistórico 11:124–160

Rivas Martínez S (1961) Los pisos de la vegetación de Sierra Nevada. Boletín Real Soc Española Hist Nat Secc Biol 59:55–64
Rivas-Martínez S (1987) Memoria del mapa de series de vegetación de España, Serie Técn. ICONA, Ministerio de Agricultura, Pesca y Alimentación, Madrid
Rixen C, Wipf S (2017) Non-equilibrium in Alpine plant assemblages: shifts in Europe's Summit Floras. In: Catalan J, Ninot J, Aniz M (eds) High mountain conservation in a changing world. Springer, Cham., pp 285–303
Robles AB, Ramos ME, Salazar C, González Rebollar JL (2016) Response of vegetation to exclusion and grazing in Mediterranean high-mountain wet pastures (Sierra Nevada, Granada, Spain). Options Méditerranéennes Ser A Mediterr Semin
Roux-Fouillet P, Wipf S, Rixen C (2011) Long-term impacts of ski piste management on alpine vegetation and soils. J Appl Ecol 48:906–915
Ruiz M, Ruiz JP (1986) Ecological history of transhumance in Spain. Biol Conserv 37:73–86
Ruiz FA, Vázquez M, Camuñez JA et al (2020) Characterization and challenges of livestock farming in Mediterranean protected mountain areas (Sierra Nevada, Spain). Spanish J Agric Res 18:e0601
Ruiz Sinoga JD, Garcia Marin R, Martinez Murillo JF, Gabarron Galeote MA (2011) Precipitation dynamics in southern Spain: trends and cycles. Int J Climatol 31:2281–2289
Rumpf SB, Hülber K, Klonner G et al (2018) Range dynamics of mountain plants decrease with elevation. Proc Natl Acad Sci 115:1848–1853
Salazar C, Valle F (eds) (2004) Series de vegetación edafohigrófila de Andalucía. Consejería de Medio Ambiente, Junta de Andalucía, Sevilla
Sanders NJ, Rahbek C (2012) The patterns and causes of elevational diversity gradients. Ecography (cop) 35:1–3
Smith J, Sconiers W, Spasojevic M et al (2012) Phenological changes in alpine plants in response to increased snowpack, temperature, and nitrogen. Arctic Antarct Alp Res 44:135–142
Steinbauer MJ, Otto R, Naranjo-Cigala A et al (2012) Increase of island endemism with altitude—speciation processes on oceanic islands. Ecography (cop) 35:23–32
Steinbauer MJ, Grytnes J-A, Jurasinski G et al (2018) Accelerated increase in plant species richness on mountain summits is linked to warming. Nature 556:231–234
Steinbauer K, Lamprecht A, Semenchuk P et al (2020) Dieback and expansions: species-specific responses during 20 years of amplified warming in the high Alps. Alp Bot 130:1–11
Teubner IE, Haimberger L, Hantel M (2015) Estimating snow cover duration from ground temperature. J Appl Meteorol Climatol 54:959–965
Theurillat J, Guisan A (2001) Potential impact of climate change on vegetation in the European Alps: a review. Clim Chang 77–109
Thompson JD (2005) Plant evolution in the Mediterranean. Oxford University Press
Thuiller W, Lavorel S, Araújo MB et al (2005) Climate change threats to plant diversity in Europe. Proc Natl Acad Sci USA 102:8245–8250
Vegas-Vilarrúbia T, Nogué S, Rull V (2012) Global warming, habitat shifts and potential refugia for biodiversity conservation in the neotropical Guayana Highlands. Biol Conserv 152:159–168
Vitt P, Havens K, Kramer AT et al (2010) Assisted migration of plants: changes in latitudes, changes in attitudes. Biol Conserv 143:18–27
Vogiatzakis IN (2012) Mediterranean mountain environments. Wiley, Chichester, UK
VV.AA. (2020) Memoria de actividades y resultados 2020 del Parque Nacional y Parque Natural de Sierra Nevada. Sevilla
Walther G-R (2003) Plants in a warmer world. Perspect Plant Ecol Evol Syst 6:169–185
Walther G-R, Beißner S, Burga C (2005) Trends in the upward shift of alpine plants. J Veg Sci 16:541–548
Willis CG, Ellwood ER, Primack RB et al (2017) Old Plants, new tricks: phenological research using Herbarium specimens. Trends Ecol Evol 32:531–546
Willkomm M, Lange J (1861) Prodromus florae Hispanicae: seu Synopsis methodica omnium plantarum in Hispania, sponte nascentium vel frequentius cultarum quae innotuerunt/auctoribus Mauritio Willkomm et Joanni Lange. E. Schweizerbart, Stuttgartiae
Wohlgemuth T (1998) Modelling floristic species richness on a regional scale: a case study in Switzerland. Biodivers Conserv 7:159–177
Wookey PA, Parsons AN, Welker JM et al (1993) Comparative responses of phenology and reproductive development to simulated environmental change in sub-Arctic and high Arctic plants. Oikos 67:490–502

# Responses of Animal Populations and Communities to Climate Change and Land-Use Shifts

Regino Zamora, José Antonio Hódar, Antonio J. Pérez-Luque, and José Miguel Barea-Azcón

**Abstract**

Changes in the distribution and abundance of animal populations and communities signal a clear response to environmental alterations. A number of changes in climate and land use are taking place in the Sierra Nevada, such as higher temperatures and greater forest cover. In this chapter, we analyse the responses of animal populations and communities to these changes, for which long-term data are available in Sierra Nevada. In the first part of the chapter, several examples illustrate spatio-temporal changes in the distribution and abundance of bird communities (passerines) in 3 types of habitats for which information has been available since the 1980s: (1) Oak groves, (2) high-mountain juniper, (3) high-mountain summits. The results indicate a continuous turnover within the bird community, in an ecological setting that has changed little, especially in the high-mountain scrubland and summit areas. Moreover, the results also show a sharp decrease in bird density during the 40-year study period, chiefly affecting the dominant species of the 1980s in the Pyrenean oak woodland and in the high-mountain juniper scrubland. The outcome of these processes is a community in continuous flux, both in composition and abundance. The second part uses the pine processionary moth (*Thaumetopoea pityocampa*) as a biological sensor of the changes occurring in the climate and land use, considering not only spatial (elevational as well as regional) but also temporal. The pine processionary moth (*T. pityocampa*) exemplifies the way in which some defoliators of Spanish forests benefit from global warming and land-use change. As its larvae develop during the winter, higher temperatures may benefit this insect, accelerating its development. The analyses confirm such benefits due to rising temperatures in the medium–high elevations of Sierra Nevada, since the climate there has been the most limiting factor for this pest until now, as natural predators are scarce at these altitudes and it is also the area where most of the pine plantations are found.

**Keywords**

Animal population and community responses • Climatic change • Land use change • Altitudinal migration • Temporal changes • Phenological shifts • Mountain birds • *Thaumetopoea pityocampa*

## 1 Introduction

Mountain ranges harbour exceptionally high biodiversity, which is now under threat from rapid environmental change (Quintero and Jezt 2018; Rahbek et al. 2019). Numerous hypotheses have been implicated as underlying causes of elevational diversity patterns, such as land area, climate and productivity (Colwell et al. 2004; McCain 2009; Koh et al. 2006; Nogués-Bravo et al. 2008; Sanders and Rahbek 2012; Price et al. 2014). On mountains, abiotic conditions and biological communities change rapidly over short altitudinal distances, and for this reason, mountain areas are excellent scenarios to detect such change because they encompass both upper and lower limits of species distribution. Moreover, biological communities change also rapidly over time in mountain areas as a result of population fluctuations as well as local colonizations and extinctions of species (species turnover, Legendre and Gauthier 2014). In this heterogeneous ecological context, climate change introduces an

R. Zamora (✉) · J. A. Hódar
Department of Ecology, University of Granada, Granada, Spain
e-mail: rzamora@ugr.es

R. Zamora · J. A. Hódar · A. J. Pérez-Luque
Andalusian Institute for Earth System Research (IISTA), University of Granada, Granada, Spain

J. M. Barea-Azcón
Agencia de Medio Ambiente y Agua, Consejería de Agricultura, Ganadería, Pesca y Desarrollo Sostenible, Seville, Spain


R. Zamora and M. Oliva (eds.), *The Landscape of the Sierra Nevada*,
https://doi.org/10.1007/978-3-030-94219-9_12

additional level of complexity, causing massive disruptions in biological systems due to shifts in distribution ranges of the species, phenology and ecological interactions. Therefore, the spatio-temporal monitoring of communities and populations along altitudinal gradients is an essential tool to evaluate biotic responses within a context of global change (Zamora et al. 2016, 2017).

A number of changes in climate and land use are taking place in the Sierra Nevada (Spain), such as higher temperatures and greater forest cover (see chapters "Climate Variability and Trends" and "Forest Dynamics Under Land-Use and Climate Change Scenarios"). In this chapter, we analyse the responses of animal populations and communities to these changes, for which long-term data are available in Sierra Nevada. In the first part of the chapter, examples from Sierra Nevada illustrate spatio-temporal changes in the distribution and abundance of bird communities (passerines) in 3 types of habitats for which information has been available since the 1980s: (1) Oak groves, (2) high-mountain juniper, (3) high-mountain summits. The second part uses the pine processionary moth (*Thaumetopoea pityocampa*) as a case study of a biological sensor of the changes occurring in the climate and land use, considering not only spatial (elevational as well as regional) but also temporal changes.

Among mountain vertebrates, birds represent the most taxonomically diverse group, and show high species turnover (Terborgh 1971; Terborgh and Weske 1975; Prodon et al. 2002; Ruggiero and Hawkins 2006). Due to their colourfulness and easy identification, birds constitute a classic study group in community ecology (Wiens 1989). A large part of community theory is based on studies examining the ecological structure of avian assemblages (from the seminal studies of MacArthur 1972 and Diamond and Cody 1975 to the present day). Undoubtedly, birds constitute an important component of biological diversity, and their ecological, cultural, recreational and economic benefits are universally recognized (Sekercioglu et al. 2016). Furthermore, as vital links in many food webs, they often serve as highly visible biological indicators of ecosystem health.

Elevation is typically the main agent driving species assemblies in montane bird communities, and, consequently, a plethora of studies have addressed how the composition and structure of ecological communities vary with elevation (e.g. Graham et al. 2009; Dehling et al. 2014; García-Navas et al. 2020). To explain these elevational patterns, numerous hypotheses concerning the complex interactions among environmental factors and biotic responses have been proposed, but these remain controversial (Lomolino 2001; Rahbek 2005; Rahbek et al. 2007; McCain 2009; Pan et al. 2016). Climatic context is considered the main driver of bird diversity in mountains (McCain 2009), and temperature follows a distinct pattern that decreases with increasing elevation, which directly affects the physiological tolerance of birds (Currie et al. 2004; Pan et al. 2016) and indirectly affects birds by influencing vegetation and food resources. Climate can directly set limits on species distributions given a species' physiological tolerances, while species distributions can be indirectly influenced by temperature and precipitation, which affect photosynthetic activity and rates of biological processes in the environment. On the other hand, habitat heterogeneity providing a greater variety of habitat types per unit area and a greater complexity of vegetation structure lead to strong diversity changes along altitudinal gradients (MacArthur and MacArthur 1961; Pan et al. 2016).

Recent studies concerning the impact of climate change in montane regions have variously documented bird species richness as declining or increasing with elevation (Popy et al. 2010). All these changes may be reversed, masked or exacerbated by land-use changes (Clavero and Brotons 2010), causing non-random species distributional shifts in mountain areas. These questions are examined here within a global-change framework, relating the population and community dynamics of birds to changes in climate change and land use over the past 40 years in Sierra Nevada.

In Sect. 3 of this chapter, as a case study, the biotic responses of pine processionary moth (hereafter PPM) to global change in Sierra Nevada was chosen for three basic reasons: (1) the PPM is an organism that has had detailed population records over a long time period (Ros-Candeira et al. 2019). (2) This lepidopteran develops its larval phase in winter, so it is expected to be strongly limited by temperature (Netherer and Schopf 2010). (3) As an herbivore, it feeds on pine trees, which have massively expanded their cover in recent decades as a result of reforestation programmes.

In the winter larval development of the processionary, winter temperatures have commonly been considered the main limiting factor for the subsequent development of the moth and its geographical expansion, mainly in the northern part of its geographical distribution in Europe (Huchon and Démolin 1971; Battisti et al. 2005). As a consequence of this temperature rise, a higher survival rate and stronger larval growth have been reported (Battisti et al. 2005) and, therefore, greater winter survival (Hoch et al. 2009).

The changes in the bird populations and communities and the population changes of the PPM are analysed below within the same conceptual framework, seeking to help elucidate the biotic responses of the communities and animal populations to climate change and land-use changes in a Mediterranean high mountain.

## 2 Avian Communities Case Study

Here, we compare bird censuses conducted within the same locations of Sierra Nevada over a 40-year period. This comparison focuses on species composition, abundance and distribution on a spatial (elevational gradient) scale, considering representative mountain habitats, from relict Pyrenean oak *Quercus pyrenaica* woodland and high-mountain scrubland, to the high-mountain summits.

A large majority (88%) of the species that inhabit the oak woodland are sedentary (Zamora and Camacho 1984), spending more than 9 months in this woody habitat. On the contrary, all the species of the community of Passeriformes inhabiting the juniper scrubland and the high mountain summits have a temporary presence. These constitute three groups: the species that nest in Sierra Nevada and that are long-distance migrants; those that also reproduce in the mountains and make elevational displacements of a certain magnitude; and those that undertake long-distance wintering (Zamora 1987, 1988a). Therefore, the species only temporarily present in Sierra Nevada are subjected to selective pressures of different kinds regarding wintering, migration, and reproduction sites. In this sense, we hypothesise that birds making long-distance migrations must be doing worse at any elevation than residents and short-distance migrants (Vickery et al. 2014). On the other hand, since high-elevation communities have a temporary presence and are typically composed of mountain specialists as well as of generalists expanding their distribution range upwards (Zamora 1987; Zamora and Barea-Azcón 2015; García-Navas et al. 2020), we predict a higher temporal turnover in alpine environments. On the contrary, we predict stronger temporal stability in forest bird communities compared to those from high-mountain scrubland and alpine habitats, since forests are more predictable environments for bird communities in Sierra Nevada in comparison with alpine habitats (Zamora and Camacho 1984; Zamora 1988a).

In a previous study, Zamora and Barea-Azcón (2015) analysed shifts in composition and abundance of bird communities along an elevational gradient in the Sierra Nevada, comparing censuses made at the same locations in the 1980s and during 2008–2012. Bird censuses were made during the 1981 breeding season (May, June, and July; Zamora and Camacho 1984) in the Pyrenean oak woodland, during the 1984 and 1985 breeding seasons in the high-mountain scrubland (Zamora 1987, 1988a), and during the 1982 breeding season on the high-mountain summits (Zamora 1987). Historical transects were marked on maps. Recent data for bird communities of the same three habitats were collected in 2008–2019 as well as during the breeding seasons by the Global-Change Observatory of Sierra Nevada (https://obsnev.es/en/), this entailing resurveys of the same historical sites using the same methodology as Zamora and Camacho (1984) and Zamora (1987, 1988a, b, 1990). The sampling procedure involved the line-transect method (Verner 1985, see Zamora and Barea-Azcón 2015 for further details). Below, a description of the three study sites is provided (see Fig. 1):

- Site 1 (1700 m a.s.l). Mediterranean oak woodland composed mainly of Pyrenean oak forest. *Quercus* woodlands were strongly managed for charcoal, pastureland creation, and wood production until the 1950s, so that the current trees are mostly resprouts 60–70 years old. Dense and cleared woodland patches alternate with pasturelands and scrublands, mainly at the upper limit of the wooded area.
- Site 2 (2200 m a.s.l). The vegetation is composed mainly of common juniper *Juniperus communis* and *Genista versicolor*. The scrub rarely exceeds 60 cm in height and is intermingled with pastureland, rocks, and stony ground.
- Site 3 (3100 m a.s.l). A typical alpine landscape is characterised by rocky outcrops (that originated from glacial activity), pastureland, small snowfields, and small glacial lakes.

With respect to land-use change, in Pyrenean oak woodland (site 1), Zamora and Barea-Azcón (2015) calculated a 9% increase in tree density (trees/ha) and a 5% increase in cover (proportion of ground occupied by trees) from 1984 to 2012. In high-mountain scrublands (site 2), the area covered by common juniper underwent little change (<5% of juniper coverage increase), being the increase of *G. versicolor* coverage the most outstanding change (see chapter "Managing the Uniqueness of Sierra Nevada Ecosystems Under Global Change: The Value of in situ Scientific Research"). On high-mountain summits (site 3) no changes were detected. Overall, land use has not significantly changed since the early 1980s, except for the growth and densification of the oak woodlands, resulting in a wooded area with a more closed canopy. On the other hand, the climate over the study periods (1981–1986 vs. 2008–2019) registered a moderate increase in temperature and a decrease in precipitation (see chapter "Climate Variability and Trends").

The monitoring (from the 1980s to 2012) considered in Zamora and Barea-Azcón (2015), was extended in this chapter to 2019, which represents a sampling period of almost 40 years. In general, the patterns detected in Zamora and Barea-Azcón (2015) in and climatic and land-use change, as well as the trends detected in the biotic responses of the birds, persisted throughout the extension period, both at the community and at the population level (Fig. 2).

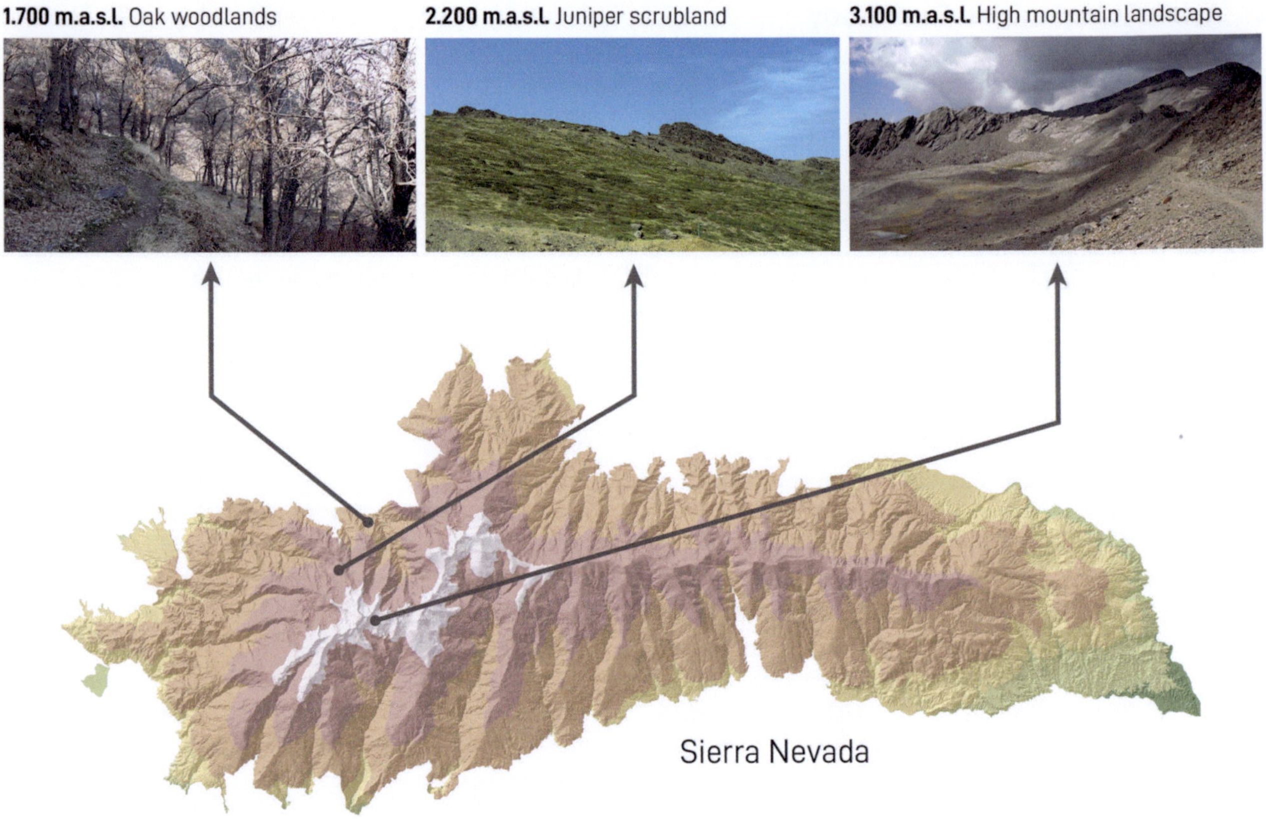


**Fig. 1** Location in Sierra Nevada of the three study sites (*Quercus pyrenaica* forest, *Juniperus communis* scrubland and High mountain summits) for the bird community study

In the Pyrenean oak woodland (site 1), the surveys in the early 1980s recorded 21 passerine species whereas the 2008–2019 surveys registered a total of 31 species, of which 18 species were common to both periods. Some species that were absent in the 1980s are frequently detected today: the Long-tailed Tit *Aegithalos caudatus*, Subalpine Warbler *Sylvia cantillans*, Crested Tit *Lophophanes cristatus*, Firecrest *Regulus ignicapilla*, and Eurasian Nuthatch *Sitta europaea*. Conversely, the Golden Oriole *oriolus*, and European Goldfinch *C. carduelis* were common in the early 1980s but absent during 2008–2019. In addition, a conspicuous decline in population densities was also detected. The most striking examples of this include Bonelli's Warbler *Phylloscopus bonelli*, Blue Tit *Cyanistes caeruleus*, Eurasian Jay *Gurrulus glandarius*, and Common Blackbird *Turdus merula*. Such a notable decline in formerly dominant species has led to a thinning of the overall community density over the past three decades in the Pyrenean oak woodland (from 108.1 birds/10 ha in the 1980s to nearly 30 birds/10 ha in 2009 and 2018, see Fig. 2). However, in the period 2009–2018, stability was observed in the bird community of the Pyrenean oak woodland, both in composition and abundance (Fig. 2).

In the juniper scrub habitat (site 2), we recorded 12 species in the 1980s and 14 species in 2008–2019, 9 of these being common to both periods. Thekla's Lark *Galerida theklae*, Common Stonechat *Saxicola rubicola*, and Spectacled Warbler *Sylvia conspicillata* were detected in 2008–2012 but not in the 1980s, whereas three species were detected three decades ago but not in 2008–2012: the Common Whitethroat *Sylvia communis*, Ortolan Bunting *Emberiza hortulana*, and Common Blackbird *T. merula*. In this ecosystem, 7 of the species show a clear negative trend in abundance. Consequently, the overall community density in the juniper scrubland has decreased from 28 to 17 birds/10 ha (Fig. 2). Particularly striking is the reduction in the density of the major dominant passerine species since the 1980s: the Northern Wheatear *O. oenanthe*. The negative population trend for this species was statistically significant (Mann–Kendall tau = −0.901). In the case of the Northern Wheatear, a long-distance migrant, the population change is consistent with a widespread decline (−71%) throughout Europe (Pan

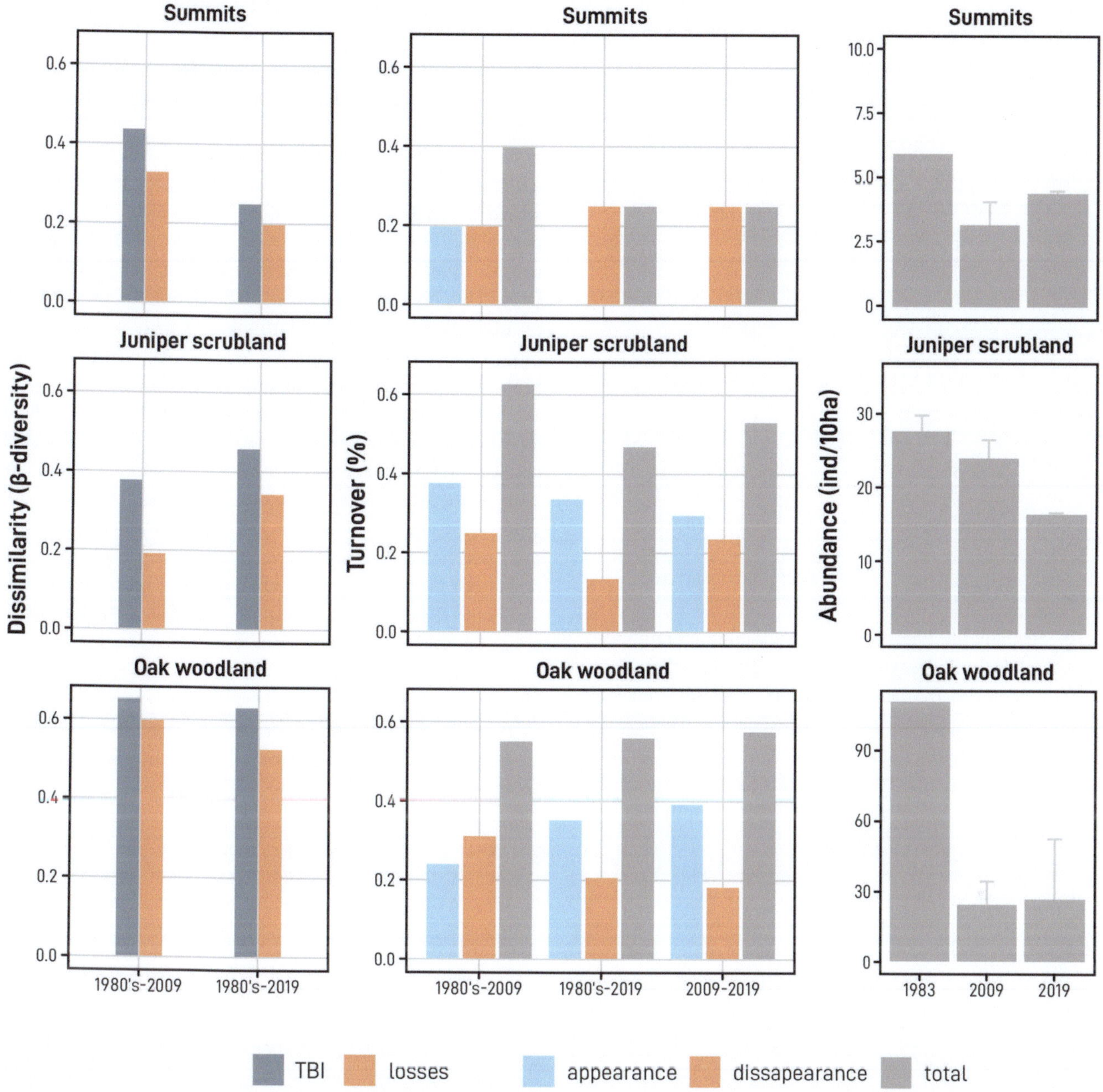


**Fig. 2** Change in composition (β diversity), turnover and abundance of bird communities along the altitudinal gradient in Sierra Nevada, comparing 3 periods: 1980s, 2008–2009 season, and the 2017–2018 season. TBI (Temporal betadiversity index, sensu Legrende 2019)

European common bird monitoring scheme 1980–2017) and may be related to ecological constraints outside the Sierra Nevada, in the wintering areas (Vickery et al. 2014).

At the high-mountain summits (site 3), we counted 5 species during the breeding season in the 1980s and 3 to 5 species in 2008–2019; 3 of these were common to both periods. The Alpine Accentor *Prunella collaris*, Northern Wheatear *O. oenanthe*, Black Redstart *Phoenicurus ochruros* and Common Linnet *Carduelis cannabina* were present both in the 1980s and during 2008–2019, whereas the Dunnock *Prunella modularis* is a newcomer occasionally present on the high-mountain summits. Overall, no major changes in density were detected, *P. ochuros* populations expanded while those of *O. oenanthe* contracted.

**Box 1: Bird-Community Changes**

**Community changes**: We have analysed the change in composition (β-diversity), turnover and abundance for the three bird communities (oak woodland, juniper scrubland, high-mountain summits) over time, comparing 3 periods: 1980s, 2008–2009 season, and the 2017–2019 season (see Fig. 2).

β-diversity: The temporal β-diversity index (sensu Legendre 2019) was calculated for each habitat for the initial period (1980s), and for the other two periods combined (2008–2009 and 2017–2019). The results (scaled from 0 to 1) indicate that the greatest dissimilarity in β-diversity was registered in the oak woodland bird community both between 1980 and 2008 as well as between the 1980s and 2019. In the three habitats (oak woodland, juniper scrubland, and summits), the greatest contribution of differences in the β-diversity index between time periods was due to the loss of species.

Turnover: The turnover analysis (Hillebrand et al. 2018) for the three bird communities over time showed the greatest variation for the oak woodland bird community. The comparison between the three time points reveals high turnover values in the oak area between 1980 and 2008 (0.55%) and between 1980 and 2019 (0.57%), due to losses and gains of species. In the case of the juniper bird community, turnover values proved high between 1980 and 2008 (0.62%), with the same losses and gains of species, registering values of 0.46% between the 1980s and 2019, with similar contributions of exits of old species and entries of new ones. Finally, between the 1980s and 2019 the turnover rate in the bird community of the high-mountain summits was 0.25, and 0.40% between 1980 and 2008, changes due primarily to the loss of species.

Abundance: An abrupt decline in abundance was detected in the oak woodland between the 1980s and 2009, although the temporal pattern of abundance subsequently became more stable. Also, the abundance of the bird community progressively declined in the juniper scrubland. By contrast, the high mountain bird community remained relatively stable.

**Population trends**: We analysed the temporal trend (τ) for each species in each of the habitats using non-parametric Mann-Kendall analysis. The results showed the following:

Oak woodland: both negative and positive trends appeared among the 20 species analysed, with a notably significant decrease of *C. caeruleus* and *G. glandarius* ($\tau = -0.535$).

Juniper scrubland: many species (7) followed negative trends, especially in the case of *O. oenanthe* ($\tau = -0.901$), the latter trend being significant in the study period.

Summits: no significant trends were detected for the 4 species analysed in this habitat.

## 2.1 Shifts in Elevational Distribution

We compared bird censuses conducted within the same locations of Sierra Nevada considering three representative mountain habitats: a Pyrenean oak woodland, a high-mountain juniper scrubland, and the high-mountain summits (Fig. 1), so we had limited spatial resolution with which to analyse spatial shifts. Nevertheless, when comparing the 2008–2019 data with the historical distribution records (1980s), we found certain evident changes, particularly in the high-mountain zone, where some species showed an upward shift in their distribution. Thekla's Lark, Common Stonechat, and Spectacled Warbler were sighted in the high-mountain juniper scrubland in 2008–2019 but not in the 1980s, presumably signifying an upward shift in their distribution. On the other hand, both the Black Redstart and Common Linnet increased their breeding populations both in the high-mountain summits, and in the oak woodland, thus showing an upward and downward shift in their elevational distribution, respectively, whereas the density of the breeding population of these species in juniper scrubland has thinned (Zamora and Barea-Azcón 2015) (Fig. 3).

**Box 2. Population Trends of Seed-Dispersing Birds in Oak Woodlands and Juniper Scrublands**

Seed-dispersing birds are of paramount importance in the maintenance and regeneration of the forests and scrublands of Sierra Nevada, which are made up of many woody species that produce fleshy fruits (Zamora and Matias 2014; García et al. 2010, 2011). Acorns and seed dispersal are essential for the regeneration of established oak and juniper formations, and for the colonization of degraded mountain habitats where oaks and junipers were formerly present. For this reason, it is relevant to know the availability and temporal changes of seed-dispersing bird populations, due to the ecosystem service they provide by dispersing the seeds. In Sierra Nevada, the main disperser of oak acorns is the Eurasian Jay (*G. glandarius*) a sedentary species, while the main disperser of juniper seeds is the Ring Ouzel (*T. torquatus*), a migratory one. We determined the temporal changes in the abundance of these two seed-dispersing bird species by conducting censuses during the autumn (September–November) in the oak woodland and juniper scrubland from 1983 to the present, in the same places and using the same methodology described for the censuses of the bird breeding community.

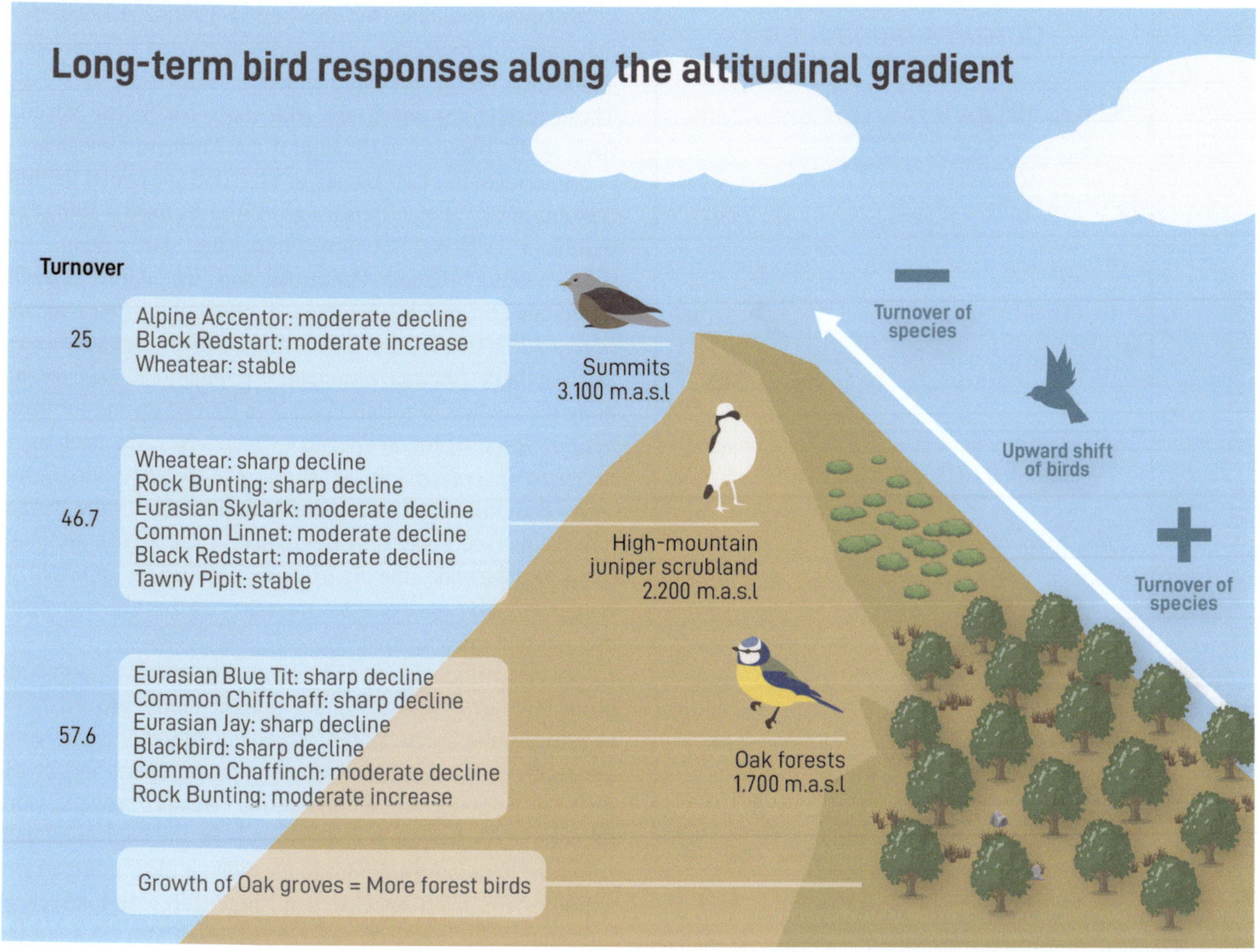


**Fig. 3** Temporal and elevational changes in breeding bird populations of Sierra Nevada. Species turnover and population trends are indicated

With respect to *J. communis*, cone production varied considerably between years (Jordano 1993; García et al. 2000). This strong variation in reproductive output appears to be typical of several *Juniperus* species and may be related to the long cone-growth period as well as their masting behaviour, which appears to be influenced by climate, resource levels and trade-offs involving previous reproductive events (Jordano 1993; García 1998; Chambers et al. 1999). Avian dispersers (mostly *T. torquatus*) are able to respond to the spatio-temporal variation in cone abundance at a regional scale, discriminating more between juniper populations than between individual plants within a population (García et al. 2001). *Q. pyrenaica*, like other *Quercus* species, also markedly varies in annual acorn production (Gea-Izquierdo et al. 2006). In fact, in Sierra Nevada, several studies have found that Pyrenean oak has a highly variable spatio-temporal acorn-production cycle (Leal 2013), both between and within years and individual trees. Most of the acorns dispersed by *G. glandarius* are quickly consumed by wild boar (*Sus scrofa*) and woodmouse (*Apodemus sylvaticus*, Gómez et al. 2003).

Therefore, both juniper and oak present a spatio-temporal mosaic in the production of acorns and cones, respectively, which their consumers have to track. In the case of juniper, this spatio-temporal match is increasingly hampered by the progressive disappearance of the wintering population of the main disperser, *T. torquatus*. Figure 4 reflects a clear temporal decline in the abundance of the latter species, whose populations shrink much more than those of *G. glandarius*, a sedentary species with few interannual variations in its abundance. In fact, in the last two decades, *T. torquatus* has hardly been seen in the Sierra Nevada (Fig. 4), while 30–40 years ago, large numbers of these birds were seen every year wintering in the Sierra Nevada juniper scrublands, as in other Mediterranean high-mountain ranges (Jordano 1993; Zamora 1990; García et al. 2001).

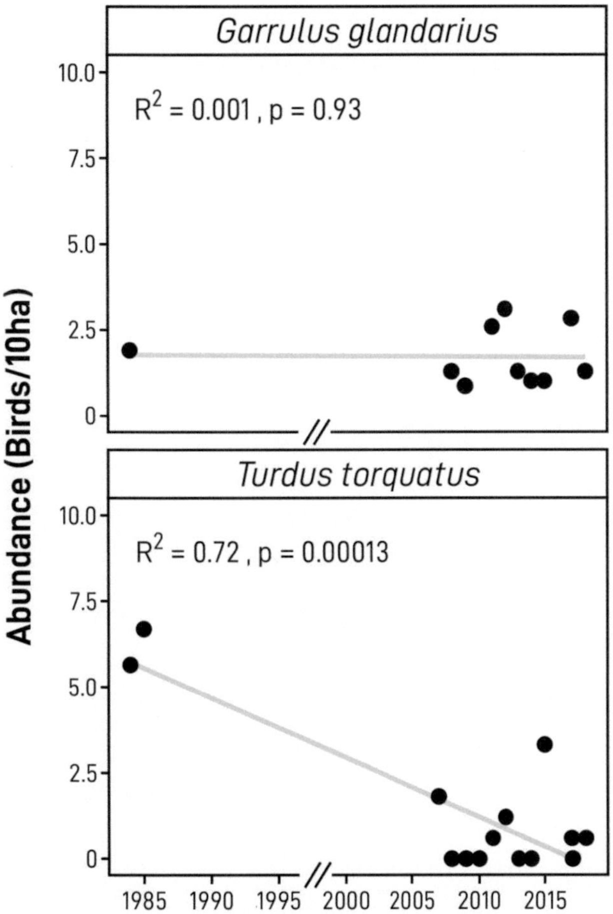


**Fig. 4** Temporal evolution of the density of Eurasian Jay (*G. glandarius*) in the oak woodland (upper panel), and of Ring Ouzel (*T. torquatus*) in the juniper scrubland (lower panel) during autumn. Grey line indicates the temporal trend

## 2.2 Temporal Shifts Observed in Sierra Nevada in a Regional and Continental Context

In a regional context, the information available (since 1998) indicates that many bird species significantly fluctuate in population abundance, both in Andalusia (SEO/BirdLife 2012) as well as in the rest of the Iberian Peninsula and other peninsular mountain ranges (SEO/ BirdLife 2013). Recent studies report a general decline in passerine birds in Europe, a trend that appears to be more pronounced for common species than for scarce and rare ones. Inger et al. (2015) put this decline at 421 million individuals for the period 1980–2009, indicating a more pronounced population decline during the first half of the study period (1980–1994), followed by a final period of stability. This pattern is consistent with the one detected in Sierra Nevada, with a sharp decline from the 1980s to 2008 followed by greater stability in the communities detected from 2008 to the present. Overall, the information available for the period 1996–2013 indicates that many bird species significantly fluctuated in population abundance in the Iberian Peninsula (SEO/BirdLife 2013). These results are consistent with those for Sierra Nevada, such as the population decline of the Northern Wheatear.

More recently, Lehikoinen et al. (2018) provided the first account of population trends for 44 species from 4 European mountain regions: Fennoscandia, the UK plain, the south-western Iberian Peninsula and the Alps. Overall, mountain bird species declined significantly (-7%) over the period 2002–2014, especially in the northern (Fennoscandia) and southern (Iberian mountains) populations, this resembling the decline in common birds in Europe during the same period. Some common species in Sierra Nevada, such as *A. arvensis*, *P. collaris*, *P. ochruros*, *O. oenanthe* and *E. cia* showed significant negative population trends for the Iberian mountains (see Table 2 of Lehikoinen et al. 2018), although for Sierra Nevada, the *A. arvensis* population appeared to remain steady or even to increase. Only *C. cannabina* populations grew in size.

As shown in Fig. 5, the population trend of passerine birds in Sierra Nevada in the present study proved more negative in the Pyrenean oak woodland than that observed in the south of the Mediterranean region of Spain and in the rest of Europe, while population trends proved more favourable in the juniper scrubland. In the high-mountain summits, data is insufficient to establish a clear comparison, although the main species that show gains and losses of individuals (*P. ochruros* and *O. oenanthe*) also do so in the rest of the geographical regions studied.

Many studies show a remarkable decline in migrant species, this being the group of passerine birds that has undergone the highest temporal shifts during recent decades (Sanderson et al. 2006; Vickery et al. 2014). By contrast, sedentary species should be beneficiaries of these declines and in fact are already colonizing ecological niches left by the loss of reproductive territories of long-distance migrants (Pearce-Higgins et al. 2015). However, our data do not indicate a differential trend between migratory and sedentary species, with population increases or losses being evenly distributed between the two groups regardless of the migratory behaviour of the species (U Mann Whitney = 4.36, $p = 0.907$). Thus, our results in Sierra Nevada fail to match this pattern except for some specific migrant species having clear negative trends during recent decades. The most striking example is the Northern Wheatear, which shows a significant decline in the juniper scrubland. In this habitat, this species was dominant three decades ago. Currently, its territories have been occupied by the Common Stonechat, a sedentary species that appears to have benefited from the progressive disappearance of the Northern Wheatear.

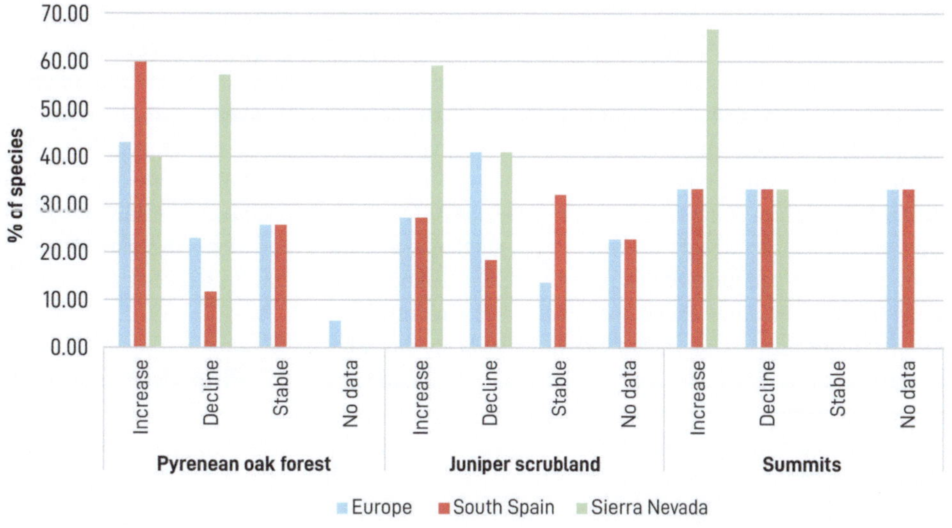


**Fig. 5** Classification of bird population trends at the European level (1980–2017, https://pecbms.info/trends-of-common-birds-in-europe-2019-update/), the southern Mediterranean region of Spain (1996–2013, SEO/BirdLife 2013) and the Sierra Nevada (80s-2017, own data)

## 2.3 Long-Term Bird Responses to Climatic and Land-Use Changes

Many factors can alter avian community composition and structure, land-use change and vegetation shifts being a major cause of avian species turnover and richness shifts (Yahner 1997). In Sierra Nevada, a trend towards vegetation spread and denser cover in the Pyrenean oak woodland is evident. On the other hand, juniper scrub cover has only slightly increased, and no changes are evident at the summits (Zamora and Barea-Azcón 2015). The growth and densification of Pyrenean oaks during the past 40 years must have significantly altered the landscape structure from the standpoint of its suitability as a breeding habitat for many species, leading to declines in some species more associated with open forest, e.g. the Eurasian Jay and Common Blackbird, Blue Tit, Common Chaffinch, and Bonelli's Warbler, while increases in certain other species are associated more with mature forests, as in the case of the Eurasian Nuthatch or even the Long-tailed Tit, which are newcomers (Zamora and Barea-Azcón 2015). In fact, in the 1980s the breeding-bird density was far higher on the forest edges in comparison with the denser and more homogeneous Pyrenean oak stands (Zamora and Camacho 1984). Thus, forest maturation has been fundamental in shaping bird-community changes over the past 40 years in the Sierra Nevada, driving significant interannual species turnover, with 10 more species now than in the 1980s. These results go counter to our initial hypothesis, which assumed a higher temporary turnover in alpine environments, and a higher stability in forest bird communities. However, what we observe is just the opposite of our initial hypothesis: the highest turnover and higher β-diversity values appear in the oak woodland bird community (Fig. 2). Thus, the process of land-use change, due to the growth and densification of the oak woodland, is so important in S Nevada that it strongly determines the composition and structure of its bird community, to the point that this is the community that changes the most over time.

Our results also show a significant turnover of the bird community in the high-mountain juniper scrubland and in the summits. This uncoupling between minor environmental changes and shifts in the communities may be related to the unsaturated character of high-mountain communities, exposed to constant entries and exits of species, causing a high level of taxonomic diversity. Strong temporal changes also occur in the structure of the community, which is characterized at any point in time by a low number of species with high dominance (Zamora 1988a). Overall, an average turnover of 24–58% in the species composition of the breeding communities during a 40-year period is noteworthy (Fig. 2), being clearly greater than reported by other authors in the Alps (Popy et al. 2010; Archaux 2004) and even higher than to the overall 35% species turnover found in the Sierra Nevada mountains of California over the last 100 years (Tingley and Beissinger 2013).

The climate over the study periods (1981–1986 vs. 2008–2019) registered steady increases in temperature and decreases in precipitation (see chapter "Climate Variability and Trends"). By contrast, bird responses to temperature changes were not straightforward. Recent studies have indicated that birds are tracking climate warming, but not fast enough (Devictor et al.2008a, 2012). Bird species may exhibit an accumulated delay in responding to temperature changes attributable to an inability to track climate change at its current rate (Devictor et al. 2012). The resulting climatic debt may still be determining the structure of the montane bird communities of Sierra Nevada after the most recent natural climatic perturbation, the Little Ice Age, a period of intense cooling that occurred from the medieval period until the mid-nineteenth century and which has left traces of periglacial activity in Sierra Nevada (Gómez-Ortiz et al. 2014, see chapter "Ancient and Present-Day Periglacial Environments in the Sierra Nevada"). Since the Little Ice Age, the high-mountain bird community of Sierra Nevada appears to have lost its alpine character, blending with the Mediterranean conditions of its surroundings. The clearest example at present is the Alpine Accentor as the only current survivor of the bird community adapted to the high-mountain, but hints of this process are also found in the disappearance of other alpine species that were recorded by the first naturalists who visited Sierra Nevada in the nineteenth century, such as the Wallcreeper (*Trychodroma muraria*), Alpine Chough (*Pyrrhocorax graculus*), the White-winged Snow Finch (*Montifringilla nivalis*), the Snow Bunting (*Plectrophenax nivalis*) or the Water Pipit (*Anthus spinoletta*) (Vaughan 1955). These latter two species are still seen with some frequency as winter visitors in Sierra Nevada (Mellone and Mellado 2020). Moreover, a Water Pipit chick has recently been identified in 2021, which represents very recent evidence of reproduction of this species in Sierra Nevada (Hódar 2021), demonstrating that there may be sporadic recolonization events of alpine species even in the current conditions of interglacial aridity. This long-term dynamic registered at Sierra Nevada is consistent with habitat models for four alpine bird species (*M. nivalis, P. collaris, P. graculus* and *T. muraria*) at the southwestern edge of the Palearctic ecozone (Gabriel Hernando et al. 2021). This study shows that climate is the main limiting factor to explain habitat suitability. Furthermore, these models conclude that potential breeding sites are currently limited by their size, landscape connectivity and habitat quality, predicting a substantial reduction of suitable habitats for the period 2041–2060. Thus, long-term climate warming, together with the isolation of Sierra Nevada respecting other Iberian Mountain ranges are explaining that alpine accentor is currently the unique surviving species.

We conclude that the Sierra Nevada is acting as an ecological island for bird communities (MacArthur 1972), displaying strong spatial and temporal dynamics that may be accelerating because of land-use change and global warming. As a result, threats to high-mountain species offer opportunities to species from lower elevations. In fact, the ecomorphological characteristics of the breeding passerines in the juniper scrubland and high-mountain summits do not differ greatly from those non-forest passerines inhabiting the Iberian Peninsula. Therefore, most of the species in the regional peninsular pool might, in terms of morphological traits, occupy Sierra Nevada high-mountain habitats (Zamora 1988b). For instance, current conditions also offer opportunities for generalist species, such as the Black Redstart or the Common Linnet to colonize the high-mountain summits. These spatio-temporal community dynamics do not necessarily lead to significant diversity loss, because the maximum diversity of Mediterranean bird communities frequently occurs in agro-silvo pastoral mountain landscapes, which are often occupied by open-habitat birds (Clavero and Brotons 2010). This would explain the observed surge in Mediterranean and steppe bird species in parallel to the decline in the mountain specialist species, resulting in a functional homogenization of bird communities along the elevational gradient (see also Devictor et al. 2008b; Clavero and Brotons 2010; Barnagaud et al. 2011).

## 3 Pine Processionary Moth Case Study

The second part of this chapter centres on the pine processionary moth (PPM, *Thaumetopoea pityocampa*) as a case study, because of its role as a biological sensor (Netherer and Schopf 2010). Here, we analyse the demographic and phenological changes in land use and climatic change in Sierra Nevada using a broad time series on infestation data in pine forests of this mountain range and link this information with climatic data. We also investigate the changes in intensity and phenology of pine forest infestation depending on altitude and pine host. These questions are examined within a global-change framework, relating PPM population dynamics to changes in the following two drivers: climatic change and land-use change occurring over the past 30 years in the Sierra Nevada. We predicted an upward movement of PPM in this mountain and some phenological changes as consequences of recent temperature increases, these elevational changes having been fuelled by the changes in landscape due to human activity.

### 3.1 Temporal Population Shifts (Outbreaks) in Relation to Climatic Change

Sierra Nevada is a highly appropriate environment to analyse biotic responses related to global change, due to its extraordinary elevational gradient and wide coverage of natural and reforested pine forests. The incidence of PPM was monitored in Andalusia by the Andalusian Environmental Council, which evaluates the intensity of annual PPM defoliation in a set of pine forest stands distributed throughout the region (Ros-Candeira et al. 2019). Altogether, some 4300 pine forest stands, covering an area of ca. 770,000 ha, are monitored every year throughout Andalusia. Defoliation intensity is scored, for each plot, in a semi-quantitative scale from 0 (no defoliation) to 5 (complete defoliation, Montoya and Hernández 1991). For this analysis, we used the stands located in Sierra Nevada (we considered all stands within a buffer of 20 km of Sierra Nevada), for a total of 989 stands, representing 152,000 ha of pine forest, reforested mainly in the second half of the

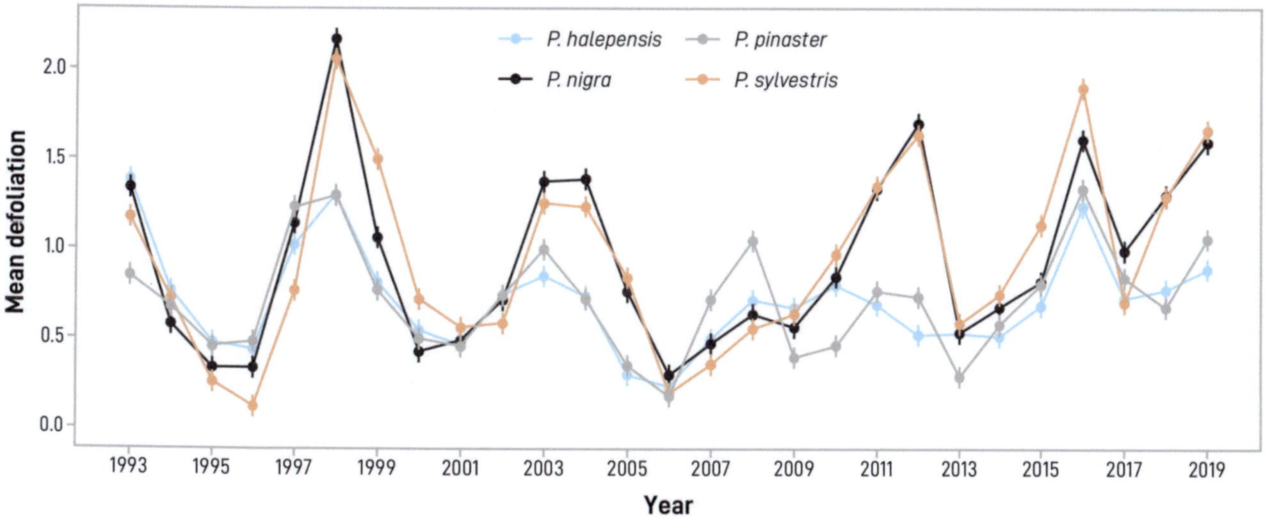


**Fig. 6** Population trend (average defoliation index per winter) of PPM in Sierra Nevada. Trends are provided separately for each pine species

twentieth century. Four species fill the gradient: the Aleppo pine (*Pinus halepensis*), which ranges from sea level to 1300 m; the Maritime pine (*Pinus pinaster*), which is situated at intermediate levels of 1000–1500 m; and the Black (*Pinus nigra*) and Scots (*Pinus sylvestris*) pines, which grow in the highest part of the forested gradient, 1400–2300 m.

Defoliation by PPM in Sierra Nevada follows notably cyclic dynamics (Hódar et al. 2012b). Approximately every 5–6 years, defoliation intensifies, alternating with periods of very low infestation. The intensity of the infestation peaks varies substantially, although the four pine species show synchrony in their defoliation peaks (Fig. 6). Two peaks from the temporal series are noteworthy: 1998–1999, for its intensity among all 4 species; and 2011–2012, for contrasting intensity, i.e. the attack was severe for the 2 high-elevation pines (*P. nigra* and *P. sylvestris*), but mild for the two species inhabiting the low and intermediate elevations (*P. pinaster* and *P. halepensis*).

Given that the pine forests sampled have remained almost the same since 1992, any trend of change in the incidence of PPM could be attributed primarily to the effect of climate change over the last 30 years. We analysed the population trend with the Mann–Kendall tau (Fig. 7), which presents values between −1 (negative) and +1 (positive). A large majority of the stands show a positive trend (79.2%, tau > 0) of which between 14 and 19% show strong and significant positive trends (tau > 0.55, see Fig. 7). Thus, we identified a clear net change towards stronger defoliations of pine forests by PPM over the years.

The stands located at high elevations (*P. nigra* and *P. sylvestris*) consistently showed greater differences between the years of maximum and minimum infestation than did pine stands located at lower elevations (*P. halepensis* and *P. pinaster*). Scots and black pine were the preferred hosts by PPM instead of maritime pine (Hódar et al. 2003). This behaviour suggests that pine woodlands located at a lower elevation show a more constant biological control (Hódar et al. 2012b, 2015, 2021), although this control is not by itself able to avoid some population fluctuations. On the contrary, defoliation of pine forests located at higher altitudes depended more on climate than on biological drivers (Figs. 6 and 7; see also Hódar et al. 2012b, 2021).

**Box 3: PPM Population Outbreaks and the North Atlantic Oscillation (NAO)**

Many forest pest species strongly depend on the temperature in their population dynamics, so that rising temperatures worldwide as a consequence of climatic change are leading to increased frequencies and intensities of insect-pest outbreaks. In the Mediterranean area, the climatic conditions are strongly linked to the effects of the North Atlantic Oscillation (NAO). The synchrony between the outbreak events along the altitudinal gradient suggests an exogenous climatic factor as a determinant of the population dynamics of PPM. Defoliation dynamics by PPM in southern Spain is determined by a global-scale climate index, the NAO (Hódar et al. (2012b), but the global effects are strongly modulated by the local conditions, mainly the pine species and the elevation.

With the current 30-year database, this trend persisted (Fig. 8), especially for the pines in the upper part of the gradient (*P. sylvestris* $R^2 = 0.25$, $p = 0.0076$, *P. nigra* $R^2 = 0.24$, $p = 0.01$), while no clear relationship was found for pine woodlands located at a

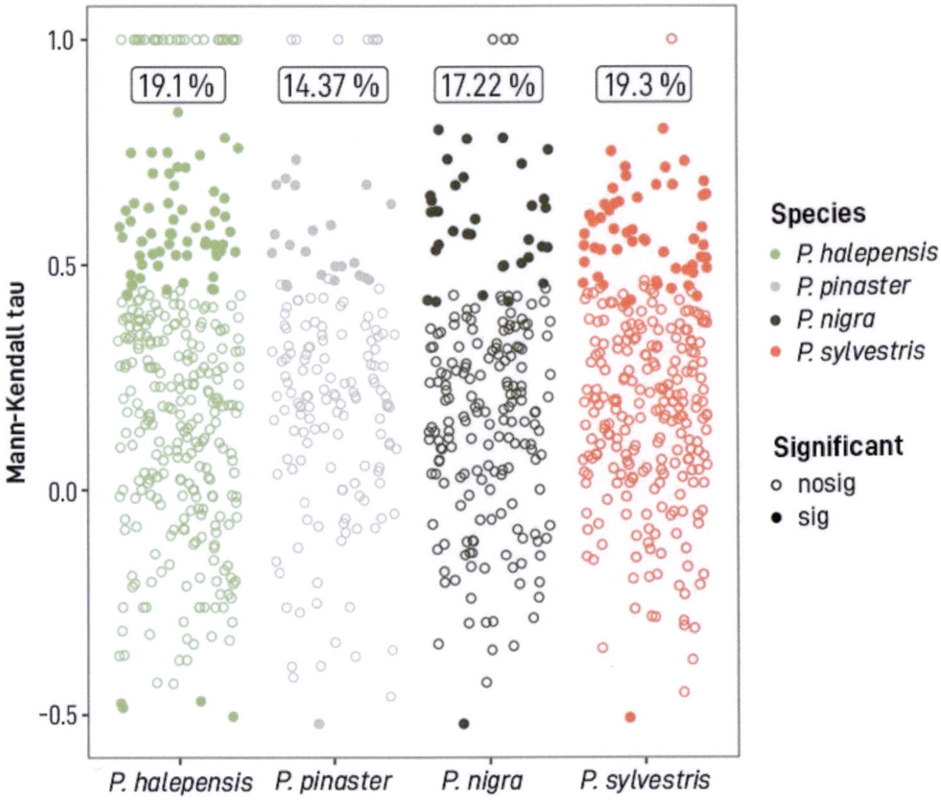


**Fig. 7** Population trend by stand (tau index) with PPM in Sierra Nevada. Analyses are provided separately for each pine species. Filled circles indicated significant trends. The percentage of pixels with significant trends are indicated

lower altitude (*P. pinaster* $R^2$ = 5.4e-05, p = 0.97, *P. halepensis* $R^2$ = 7e-07, p = 1). This result implies that any future change in NAO trends due to climatic change will predictably affect PPM population dynamics. These results highlight the necessity of using global drivers in order to understand the overall consequences of global change on population dynamics at the regional scale, helping to reduce the uncertainty in regional predictions, and opens the window to the development of NAO-based predictive models on a regional scale while incorporating local drivers (Fig. 8).

## 3.2 Shifts in Elevational and Host Pine Distribution

An elevational migration has been confirmed for the processionary moth, which showed a clear displacement in Sierra Nevada together with a phenological advance as a consequence of global warming (Hódar et al. 2003; Hódar and Zamora 2004). Other insects, such as dung beetles (Menéndez et al. 2014) also have recently changed his altitudinal distribution, with a 400 m average elevational ascent in Sierra Nevada over 25 years, coinciding with a rise of the same magnitude in the butterfly Apollo (*Parnasius apollo nevadensis*, Barea-Azcón 2016). Meanwhile, ants (*Formicidae, Hymenoptera*) have also changed altitudinal distribution, at least in the case of two species (*Proformica longiseta* and *Formica fusca/lemani*), which have expanded their upper distribution limits by about 200 m on the southern slopes of Sierra Nevada (González-Megías et al. 2016).

In reference to the winter larval development of PPM, it was suggested long ago that winter temperatures would determine PPM performance and distribution (Buffo et al. 2007; Hoch et al. 2009). As a consequence of the climatic warming over the last few decades, the natural distribution of PPM has expanded both in elevation in the mountains and in latitude in Europe (Hódar and Zamora 2004; Battisti et al. 2005; Robinet et al. 2007, 2013). In this process, PPM found new host plants with which previous interaction has been limited or null. In Sierra Nevada and Sierra de Baza (Granada), one of the plant species forming the treeline is a relict subspecies of the widely distributed Scots pine, *Pinus sylvestris nevadensis*. These southernmost populations are isolated in high mountains, which until recently had offered good protection against defoliation by PPM. However, as the PPM caterpillar moved upwards in mountains during recent decades (the insect has recently been recorded at 2300 m in altitude), the frequency and intensity of PPM defoliations on *P. sylvestris nevadensis* has increased. This promotes a new interaction of a non-specialist Mediterranean caterpillar pest,

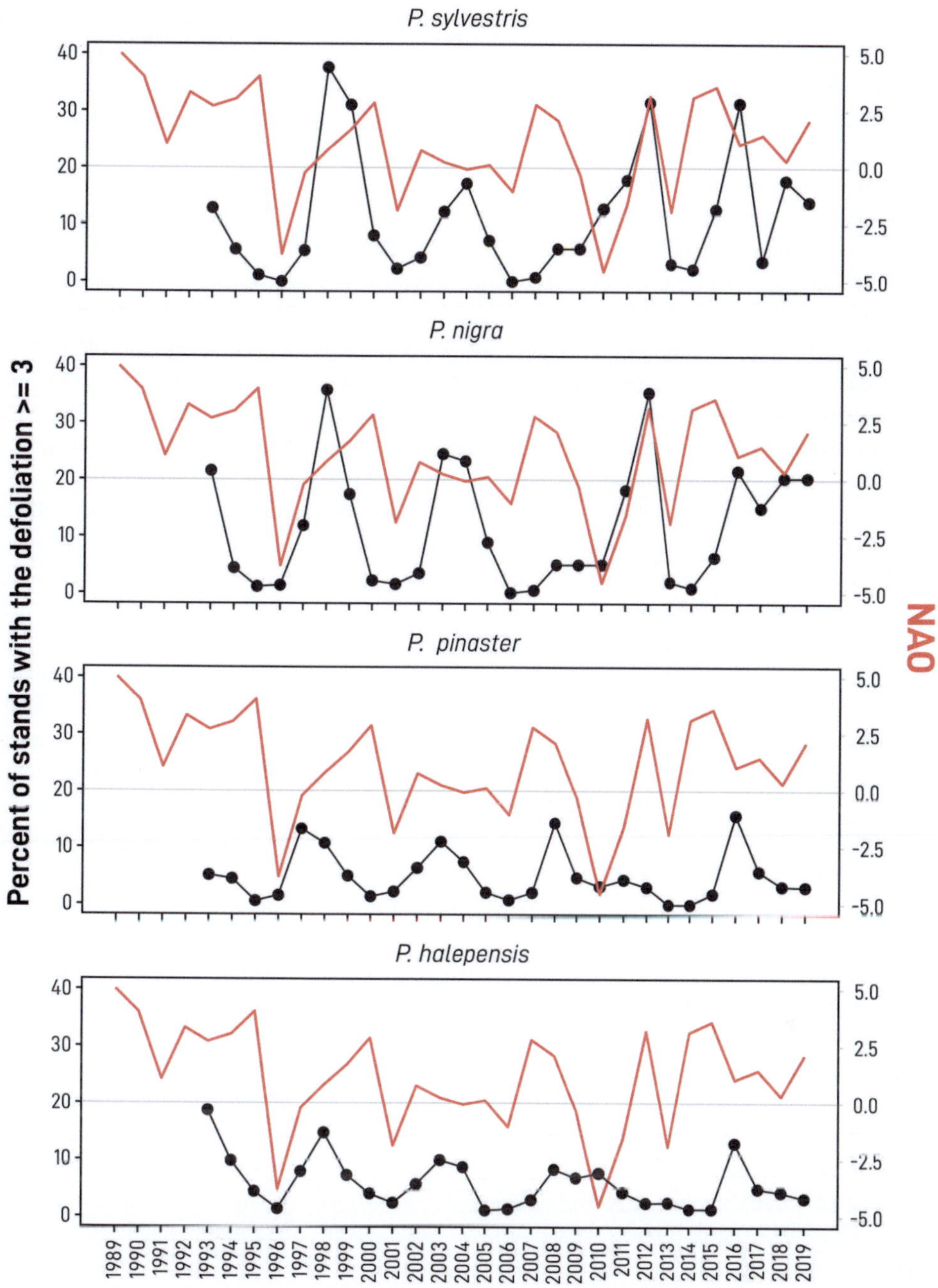


**Fig. 8** Relationship between the Hurrell's winter NAO (red line) and the incidence of the PPM (black line) in Sierra Nevada, estimated as a percentage of stands with an infestation score $\geq 3$

PPM, with a suitable boreal host occupying the Mediterranean high-mountain, the Scots pine (Hódar et al. 2003; Hódar and Zamora 2004). Defoliation thus constitutes a new source of stress to add to the usual summer drought in Mediterranean summers (Rivas-Ubach et al. 2016a, b, 2017, 2019), and it has been demonstrated that PPM defoliation drastically reduces the growth and the reproductive capacity of *P. sylvestris nevadensis*. Pines react to defoliation by reducing PPM larval survival, although the exact mechanism behind this response is not clear (Hódar et al. 2015; Achotegui-Castells et al. 2013). Defoliated juvenile trees grow, on average, less than half the amount that undefoliated ones grow, whereas defoliated adult trees produce 50% fewer seeds and have seeds almost 40% lighter than those of undefoliated trees (Hódar et al. 2003; Hódar and Zamora 2004). Defoliation, therefore, alters the pattern of growth and reproduction of trees and can hamper tree survival, due either directly to the lack of photosynthetic tissues or indirectly to greater susceptibility to other negative factors such as drought or pathogen attack (Zamora et al. 2001; Rivas-Ubach et al. 2016a, 2017; Linares et al. 2014). Although pines seem for now able to tolerate the present rate of defoliation (Linares et al. 2014), a steady increase of winter temperatures may promote a dangerous impact of PPM in these relict populations by further reducing the weak pine regeneration capacity (Castro et al. 2004).

The upward movement of PPM in mountains has a climatic background but has been fuelled by the changes in landscape due to human activity. In Mediterranean mountains, the remnant natural pine woodlands are small fragments of low-density forest, usually mixed with other species (e.g. *Quercus* spp.; Mesón and Montoya 1993). A forest structure with suitable hosts in a density low enough and scattered among different, unsuitable tree species hosts, would dilute the effect of the PPM population explosion. By contrast, the massive pine plantations in the Mediterranean mountains during the last century (Hódar et al. 2012a, 2015) have created a new scenario prone to pest spread. An extensive stand of uniformly planted pines, characterized by monospecific stands of similar age with high density, regular spacing, and low variability among individuals, has created a habitat in which PPM can easily find suitable hosts. Therefore, the potential negative consequences of this new, emerging interaction are the combined result of climatic warming, which encourages uphill displacements, and homogeneous pine plantations, which represent a culture medium for outbreaks. Clearly, an ecologically healthy way to mitigate PPM outbreaks in its new high-mountain woodland habitat is to break up the homogeneity of these stands, drastically reducing pine density in afforested woodlands and promoting spatial heterogeneity at the landscape level (Hódar et al. 2012a, 2015; Hódar 2015). Additionally, fostering biodiversity in pine plantations by, for instance, increasing the proportion of broadleaved trees and shrubs in forest stands (particularly at the stand borders), would bolster the resilience of these systems to pest outbreaks (Jactel and Brockerhoff 2007; Dulaurent et al. 2012a; Hódar 2015). This is particularly applicable in our study region, where the recovery of woody species diversity is being considered as part of forest management plans for the next 50 years, given that forest logging is no longer profitable, and forest values other than timber are gaining importance (Palahi et al. 2008; and references therein). Reinforcement of parasitoid insects and predators that attack PPM also offers a highly recommendable option (Jactel and Brockerhoff 2007; Dulaurent et al. 2012b; Hódar 2013; Hódar et al. 2021). In fact, parasitoids are also favoured by high forest diversity and heterogeneity (Jactel and Brockerhoff 2007).

## 3.3 Response of PPM to Changing Scenarios

Predictions of future climate-change scenarios suggest that a good part of the Mediterranean basin will undergo rising temperatures and diminishing rainfall (Doblas-Miranda et al. 2017, chapter "Climate Variability and Trends"), which would lead to a transformation of many of the native forests and reforestation into open forest formations with a predominance of scrublands (Christensen et al. 2007; Matías et al. 2012, see chapter "Forest Dynamics Under Land-Use and Climate Change Scenarios"). This could be advantageous for one of the critical points in the PPM life cycle, which is the selection of the burial site to pupate. According to Markalas (1989) and Torres-Muros et al. (2017), cleared land or areas with little plant cover and with lower levels of soil moisture, aid the survival of processionary pupae in the soil. However, in many areas of the Mediterranean, this trend of change in the landscape, impelled by the climatic effector, is being counterbalanced by a human tendency to abandon traditional uses. As a result, these areas are being colonized by scrubs and forests (Peña et al. 2007), a trend that could be disadvantageous for PPM. This leads to two possible and opposed scenarios: (1) where aridification exerts a greater influence than changes in land use, PPM will have more places to successfully bury itself, since it will find a greater number of optimal sites for pupation; and (2) where land-use shifts are stronger than climate-driven changes, denser shrub and tree cover will make it more difficult for PPM to survive and emerge (Fig. 9). We should note, however, the effect that such changes can promote in the prolonged diapause of PPM buried pupae. More severe aridification would foreseeably boost the number of pupae entering a prolonged diapause (Salman et al. 2019).

PPM has shown great demographic plasticity in contrasting situations, from the biennial strategy of Corsica to the preferential prolonged diapause in very cold and very hot sites (Salman et al. 2019), leaving the univoltine for intermediate situations, and all this modulated by local characteristics of the habitat (open, shrubland, forest). In Sierra Nevada, clearings in the forest with dry soil imply better conditions for pupation and survival during the diapause, whether univoltine or prolonged. All these options represent a complex spatio-temporal scenario for PPM interactions in the future (Arnaldo et al. 2011; Robinet et al. 2015), highlighting the need to consider all phases within the life cycle of an organism in order to adequately forecast the effects of global change (Suárez-Muñoz et al. 2019).

With populations on both sides of the Mediterranean Sea, PPM has its southern limit bordering the Sahara and the northern one in Central Europe. At both limits, the effect of climatic change is already becoming evident, with retractions along the southern edge (Bourougaaoui et al. 2021) and advances along the northern one (Battisti et al. 2005, 2015). Sierra Nevada is a natural laboratory of the first order, since it allows an altitudinal analysis of the phenomena that occur along a much broader latitudinal gradient, minimizing possible local differences between PPM populations at different altitudes (Fig. 10). The highest pine plantations in Sierra

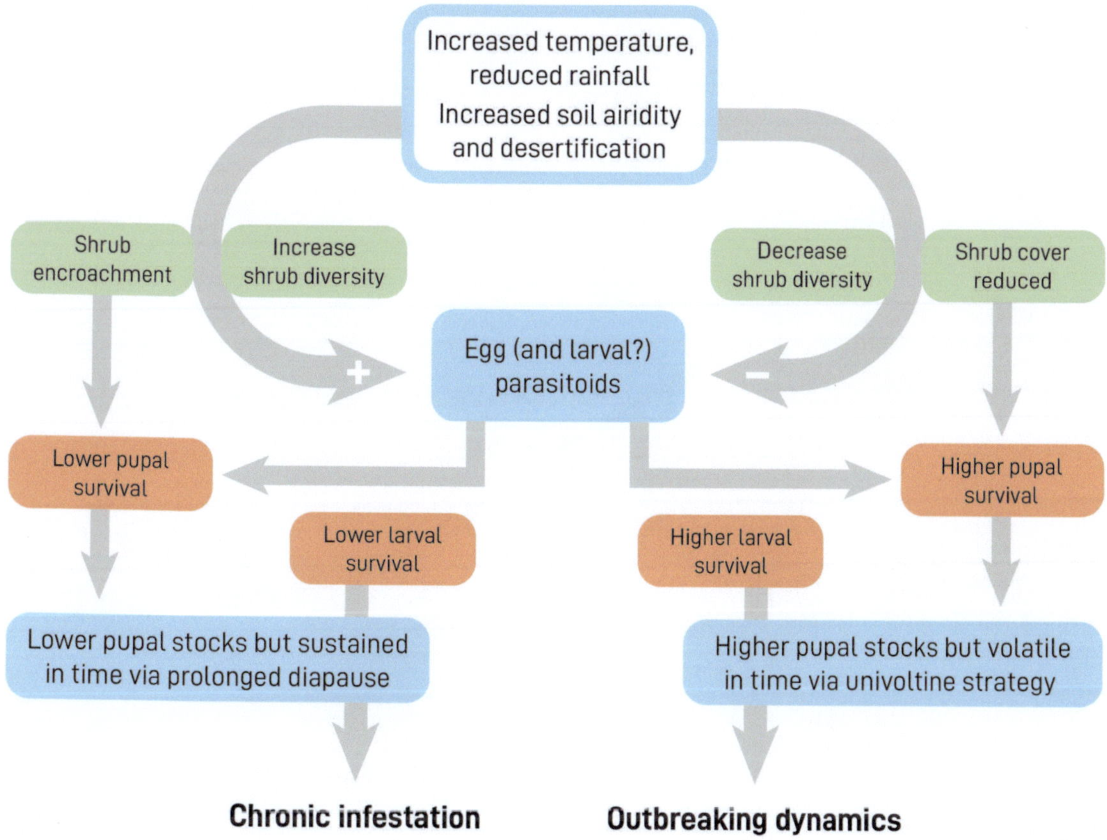


**Fig. 9** Conceptual flowchart of the possible response of PPM prolonged diapause and population dynamics to variations in tree and shrub cover as a consequence of climatic change

Nevada, at 2500 m, are separated from the coastal Aleppo pine forests by barely 40 km in a straight line, and barely 100 km if we consider the plantations in the semi-desert of Almería, already close, due to its climatic conditions, to what would be the pest's retraction limit. Taking advantage of the man-made pine plantations, PPM manages to thrive along the gradient (Fig. 10) and fit its phenology and life cycle to all these contrasting situations, providing the opportunity for researchers to learn how this insect adapts to the ongoing changes.

## 4 Concluding Comments

In this chapter, we have analysed responses of animal populations and communities to land use and climatic changes in Sierra Nevada. The results showed strong spatio-temporal changes in the distribution and abundance of bird communities in 3 types of habitats for which information has been available since the 1980s: (1) Oak groves, (2) high-mountain juniper, (3) high-mountain summits. We also detected relevant biotic responses of pine processionary moth (*Thaumetopoea pityocampa*) as a biological sensor of the changes occurring in the climate and land use, considering not only spatial (elevational as well as regional) but also temporal and phenological population changes.

**Responses to climatic change**: In the case of birds, the shifts in the composition and abundance of the high-mountain communities seem to respond mainly to the natural climatic variability since the last glacial period, and particularly since the last cooling episode (Little Ice Age, fifteenth–nineteenth centuries in Sierra Nevada) until the current period of interglacial aridity. Some species, such as *O. oenanthe* and the *T. torquatus* show the most evident population declines. The bird community also seems to respond to rising temperatures and current aridity, provoking an elevational climb in some species of Mediterranean passerines from steppe open habitats to the high-mountains, such as *S. rubicola*. In the case of PPM, population fluctuations show a strong link with cyclical global climatic phenomena such as the NAO, in addition to a clear elevational displacement in Sierra Nevada and a latitudinal one in Europe, together with a phenological advance as a consequence of global warming.

**Responses to land-use change**: A strong determining factor in the distribution and abundance of forest birds and

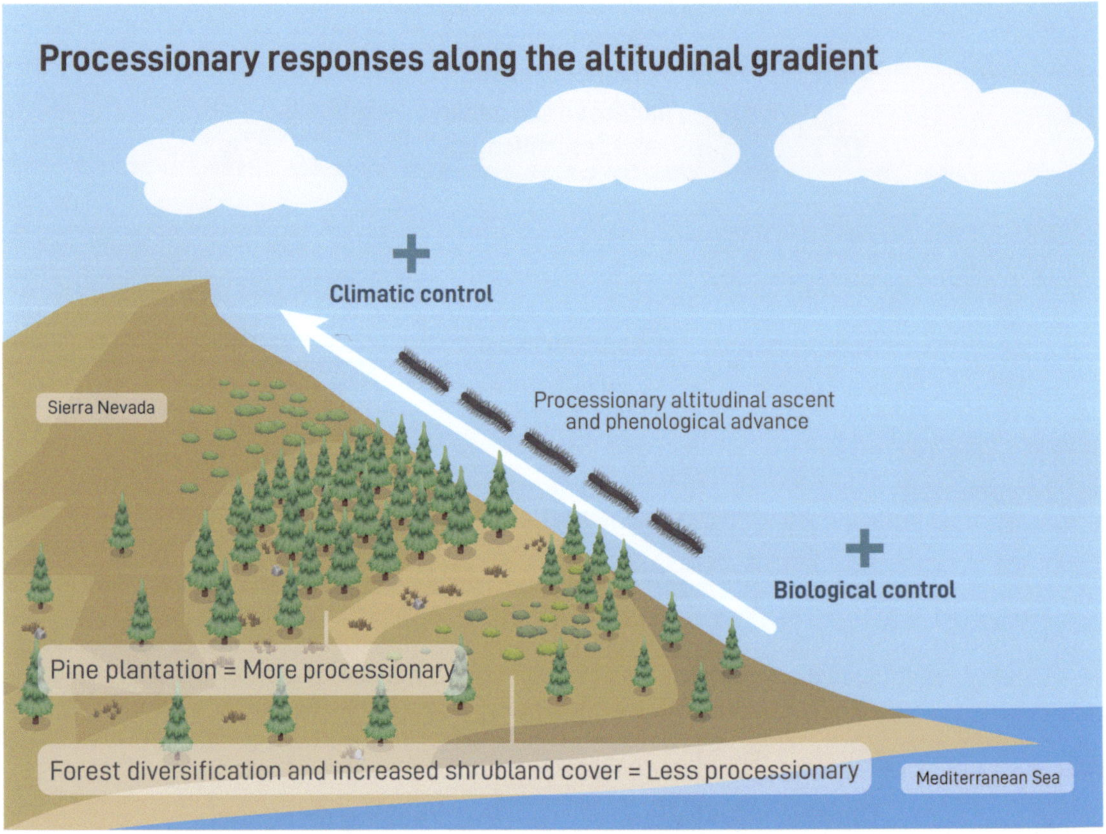


**Fig. 10** Elevational changes in the PPM population in relation to major drivers (land use change and climatic change) in Sierra Nevada

PPM in Sierra Nevada is the land-use change. In the case of birds, the increase in coverage and densification of oak woodlands has favoured forest species, which register a higher temporal turnover than in birds of the juniper scrublands and high summits. In the case of PPM, the extensive and monospecific plantations of conifers have helped its survival, magnifying its defoliating impact, especially at high elevations due to warming. PPM displays a very wide spectrum of possible responses to climatic change in interaction with land-use change, highlighting the need to consider all spatio-temporal scales within the life cycle of an organism in order to adequately forecast the effects of climate change.

Overall, we conclude that both passerine birds and pine processionary moths are displaying a strong spatial and temporal "natural" dynamics that is now accelerating because of land-use change and global warming in Sierra Nevada under the Anthropocene.

**Acknowledgements** This research work was conducted in the collaborative framework of the "Sierra Nevada Global Change Observatory" monitoring program http://obsnev.es promoted by the Council of Agriculture, Livestock, Fishing and Sustainable Development of the Junta de Andalucía and the University of Granada. We thank the LIFE-ADAPTAMED (LIFE14 CCA/ES/000612): Protection of key ecosystem services by adaptive management of Climate Change endangered Mediterranean socio-ecosystems for the funding support, the H2020 project European Long-Term Ecosystem and socio-ecological Research Infrastructure (eLTER), the Thematic Center on Mountain Ecosystem & Remote sensing, Deep learning-AI e-Services University of Granada-Sierra Nevada (LifeWatch-2019-10-UGR-01), and PROPIFEN (PGC2018-101773-B-I00) from MCIN/AEI/10.13039/501100011033, for partial funding. The Andalusian Network for Monitoring Damage to Forest Ecosystems (RED SEDA) provided the databases of PPM defoliation. We thank the management and technical team of the Sierra Nevada National Park (especially Pablo Galdo, Francisco C. Parody, Fanny Archila, Carmen Pulido and many others, and to the Agency of Environment and Water of Andalucia team, Mariano Guerrero, Rogelio López and Javier M. Jaramillo) for their support of our work in Sierra Nevada for many years and their continued collaboration within the framework of the joint activities that we develop in the Global Change Observatory.

## References

Achotegui-Castells A, Llusià J, Hódar JA, Peñuelas J (2013) Needle terpene concentrations and emissions of two coexisting subspecies of Scots pine attacked by the pine processionary moth (*Thaumetopoea pityocampa*). Acta Physiol Plantarum 35:3047–3058

Archaux F (2004) Breeding upwards when climate is becoming warmer: no bird response in the French Alps. Ibis 146:138–144

Arnaldo PS, Oliveira I, Santos J, Leite S (2011) Climate change and forest plagues: the case of the pine processionary moth in Northeastern Portugal. For Syst 20:508–515

Barea-Azcón JM (2016) The phenology of butterflies in Sierra Nevada. In: Zamora et al (eds) Global change impacts in Sierra Nevada: challenges for conservation. Consejería de Medio Ambiente y Ordenación del Territorio. Junta de Andalucía

Barnagaud JY, Devictor V, Jiguet F, Archaux F (2011) When species become generalists: on-going large-scale changes in bird habitat specialization. Global Change Biol 20:630–640

Battisti A, Stastny M, Netherer S et al (2005) Expansion of geographic range in the pine processionary moth caused by increased winter temperatures. Ecol Appl 15:2084–2096

Battisti A, Avcı M, Avtzis DN et al (2015) Natural history of the processionary moths (*Thaumetopoea* spp.): new insights in relation to climate change. *Thaumetopoea pityocampa*. In: Roques A (ed) Processionary moths and climate change: an update. Springer Netherlands and Éditions Quæ, Versailles, France, pp 15–80

Bourougaaoui A, Ben Jamâa ML, Robinet C (2021) Has North Africa turned too warm for a Mediterranean forest pest because of climate change? Clim Change 165:46

Buffo E, Battisti A, Stastny M, Larsson S (2007) Temperature as a predictor of survival of the pine processionary moth in the Italian Alps. Agric For Entomo l9:65–72

Castro J, Zamora R, Hódar JA, Gómez JM (2004) Seedling establishment of a boreal tree species (*Pinus sylvestris*) at its southernmost distribution limit: consequences of being in a marginal, Mediterranean habitat. J Ecol 92:266–277

Chambers JC, Vander Wall SB, Schupp EW (1999) Seed and seedling ecology of piñon and juniper species in the Pigmy Woodlands of western North America. Bot Rev 65:1–38

Christensen JH, Hewitson B, Busuioc A et al (2007) Regional climate projections. Climate change 2007: the physical science basis. Contribution of Working Group I to the Fourth Assessment Report of the Intergovernmental Panel on Climate Change, Chapter 11 University Press, UK

Clavero M, Brotons L (2010) Functional homogenization of bird communities along habitat gradients: accounting for niche multidimensionality. Global Ecol Biogeogr 19:684–696

Colwell RK, Rahbek C, Gotelli NJ (2004) The mid-domain effect and species richness patterns: what have we learned so far? Am Nat 163:1–23

Currie DJ, Mittelbach GG, Cornell HV et al (2004) Predictions and tests of climate-based hypotheses of broad-scale variation in taxonomic richness. Ecol Lett 7:1121–1134

Dehling MD, Fritz SA, Töpfer T et al (2014) Functional and phylogenetic diversity and assemblage structure of frugivorous birds along an elevational gradient in the tropical Andes. Ecography 37:1047–1055

Devictor V, Julliard R, Couvet D, Jiguet F (2008a) French birds lag behind climate warming. Proc R Soc B 275:2743–2748

Devictor V, Julliard R, Clavel J et al (2008b) Functional biotic homogenization of bird communities in disturbed landscapes. Global Ecol Biogeogr 17:252–261

Devictor V, van Swaay C, Brereton T et al (2012) Differences in the climatic debts of birds and butterflies at a continental scale. Nat Clim Chang 2:121–124

Diamond JM, Cody ML (1975) Assembly of species communities. In: Diamond JM, Cody ML (eds) Ecology and evolution of communities. Harvard University Press, London, pp 342–444

Doblas-Miranda E, Alonso R, Arnan X et al (2017) A review of the combination among global change factors in forests, shrublands and pastures of the Mediterranean Region: beyond drought effects. Glob Planet Change 148:42–54

Dulaurent AM, Porté AJ, van Halder I et al (2012a) Hide and seek in forests: colonization by the pine processionary moth is impeded by the presence of nonhost trees. Agric for Entomol 14:19–27

Dulaurent AM, Rossi JP, Deborde C et al (2012b) Honeydew feeding increased the longevity of two egg parasitoids of the pine processionary moth. J Appl Entomol 135:184–194

Gabriel Hernando M, Fernández-Gil J, Roa IE et al (2021) Warming threatens habitat suitability and breeding occupancy of rear-edge alpine bird specialists. Ecography 44:1191–1204

García D (1998) Interaction between juniper Juniperus communis L. and its fruit pest insects: pest abundance, fruit characteristics and seed viability. Acta Oecologica 19:517–525

García D, Zamora R, Gómez JM et al (2000) Geographical variation in seed production, predation and abortion in *Juniperus communis* throughout its range in Europe. J Ecol 88:436–446

García D, Zamora R, Gómez JM, Hódar JA (2001) Frugivory at Juniperus communis depends more on population characteristics than on individual attributes. J Ecol 89:639–647

García D, Zamora R, Amico GC (2010) Birds as suppliers of seed dispersal in temperate ecosystems: conservation guidelines from real-world landscapes. Conserv Biol

García D, Zamora R, Amico GC (2011) The spatial scale of plant-animal interactions: effects of resource availability and habitat structure. Ecol Monogr 81(1):103–121

García-Navas V, Sattler T, Schmid H, Ozgul A (2020) Temporal homogenization of functional and beta diversity in bird communities of the Swiss Alps. Divers Distrib 26:900–911

Gea-Izquierdo GI, Cañellas I, Montero G (2006) Acorn production in Spanish holm oak woodlands. Invest Agrar: Sist Recur for 15 (3):339–354

Gómez J, García D, Zamora R (2003) Impact of vertebrate acorn- and seedling-predators on a Mediterranean Quercus pyrenaica forest. For Ecol Manag 180:125–134

Gómez-Ortiz A, Oliva M, Salvador-Franch F et al (2014) Degradation of buried ice and permafrost in the Veleta cirque (Sierra Nevada Spain), from 2006 to 2013 as a response to recent climate trends. Solid Earth 5:979–993

González-Megías A, Menéndez R, Tinaut A (2016) Shifts in the elevational ranges of insects in Sierra Nevada: evidence of climate change. In Zamora et al (eds) Global change impacts in Sierra Nevada: challenges for conservation. Consejería de Medio Ambiente y Ordenación del Territorio. Junta de Andalucía

Graham CH, Parra JL, Rahbek C, McGuire JA (2009) Phylogenetic structure in tropical hummingbird communities. PNAS 106:19673–19678

Hillebrand H, Blasius B, Borer ET et al (2018) Biodiversity change is uncoupled from species richness trends: consequences for conservation and monitoring. J Appl Ecol 55:169–184

Hoch G, Toffolo E, Netherer S, Battisti A, Schopf A (2009) Survival at low temperature of larvae of the pine processionary moth *Thaumetopoea pityocampa* from an area of range expansion. Agric for Entomol 11:313–320

Hódar JA (2013) Expansión de patógenos por cambios en el clima y el uso del suelo: efectos de la procesionaria del pino en el paisaje. In: Doblas-Miranda E (ed) Conservar Aprovechando. Cómo integrar el cambio global en la gestión de los montes españoles. CREAF, Barcelona, pp 124–127

Hódar JA (2015) Incidencia de la procesionaria del pino como consecuencia del cambio climático: previsiones y posibles soluciones. In: Herrero A, Zavala MA (eds) Los Bosques y la Biodiversidad frente al Cambio Climático: Impactos, Vulnerabilidad y Adaptación en España. Ministerio de Agricultura, Alimentación y Medio Ambiente, Madrid, pp 295–302

Hódar JA (2021) Bisbita alpino *Anthus spinoletta*. Ardeola 68:297

Hódar JA, Zamora R (2004) Herbivory and climatic warming: a Mediterranean outbreaking caterpillar attacks a relict, boreal pine species. Biodivers Conserv 13:493–500

Hódar JA, Castro J, Zamora R (2003) Pine processionary caterpillar *Thaumetopoea pityocampa* as a new threat for relict Mediterranean Scots pine forests under climatic warming. Biol Conserv 110:123–129

Hódar JA, Zamora R, Cayuela L (2012a) Cambio climático y plagas: algo más que el clima. Ecosistemas 21:73–78

Hódar JA, Zamora R, Cayuela L (2012b) Climate change and the incidence of a forest pest in Mediterranean ecosystems: can the North Atlantic Oscillation be used as a predictor? Clim Change 113:699–711

Hódar JA, Aspizua R, Bollullos C, Zamora R (2015) Dinámica poblacional de la procesionaria del pino: respuestas a los cambios en el clima y en el manejo forestal. In: Zamora R, Pérez Luque AJ, Bonet FJ, Barea-Azcón JM, Aspizua R (eds) La huella del Cambio Global en Sierra Nevada: retos para la conservación. Consejería de Medio Ambiente y Ordenación del Territorio. Junta de Andalucía, pp 159–161

Hódar JA, Cayuela L, Heras D et al (2021) Expansion of elevational range in a forest pest: Can parasitoids track their hosts? Ecosphere 12(4):e03476

Huchon H, Démolin G (1971) La bioécologie de la processionnaire du pin. Dispersion Potentielle. Dispersion Actuelle. Phytoma 23:11–20

Inger R, Gregory R, Duffy JP et al (2015) Common European birds are declining rapidly while less abundant species's numbers are rising. Ecol Lett 18:28–36

Jactel H, Brockerhoff EG (2007) Tree diversity reduces herbivory by forest insects. Ecol Lett 10:835–848

Jordano P (1993) Geographical ecology and variation of plant seed disperser interactions: southern Spanish junipers and frugivorous thrushes. Vegetatio 107(108):85–104

Leal J (2013) Análisis del crecimiento vegetativo e inversión reproductiva de Quercus pyrenaica Willd. en tres localidades de Sierra Nevada. Master's thesis. Universidad de Granada

Legendre P (2019) A temporal beta-diversity index to identify sites that have changed in exceptional ways in space–time surveys. Ecol Evol 9:3500–3514

Legendre P, Gauthier O (2014) Statistical methods for temporal and space–time analysis of community composition data. Proc R Soc B 281:20132728

Lehikoinen A, Brotons L, Calladine J et al (2018) Declining population trends of European mountain birds. Glob Change Biol 25:577–588

Linares JC, Senhadji K, Herrero A, Hódar JA (2014) Growth patterns at the southern range edge of Scots pine: disentangling the effects of drought and defoliation by the pine processionary caterpillar. For Ecol Manag 315:129–137

Lomolino MV (2001) Elevation gradients of species-density: historical and prospective views. Global Ecol Biogeogr 10:3–13

Koh CN, Lee PF, Lin RS (2006) Bird species richness patterns of northern Taiwan: primary productivity, human population density, and habitat heterogeneity. Divers Distrib 12:546–554

MacArthur RH, MacArthur JW (1961) On bird species diversity. Ecology 42:594–598

MacArthur RH (1972) Geographical ecology: patterns in the distribution of species. Harper and Row, New York

Markalas S (1989) Influence of soil moisture on the mortality, fecundity and diapause of the pine processionary moth *Thaumetopoea pityocampa* Schiff. J Appl Entomol 107:211–215

Matías L, Castro J, Zamora R (2012) Effect of simulated climate change on soil respiration in a mediterranean-type ecosystem: rainfall and habitat type are more important than temperature or the soil carbon pool. Ecosystems 15:299–310

McCain CM (2009) Global analysis of bird elevational diversity. Global Ecol Biogeogr 18:346–360

Menéndez R, González-Megías A, Robert P, Marquez-Ferrando R (2014) Climate change and elevational range shifts: evidence from dung beetles in two European mountain ranges. Glob Ecol Biogeogr 23(6):646–657

Mesón M, Montoya M (1993) Selvicultura mediterránea. Mundi-Prensa, Madrid

Mellone U, Mellado A (2020) Escribano nival en Sierra Nevada

Montoya R, Hernández R (1991) La procesionaria del pino. In: Romanyk N, Cadahia D (eds) Plagas de insectos en las masas forestales españolas. MAPA, Madrid, pp 59–73

Netherer S, Schopf A (2010) Potential effects of climate change on insect herbivores in European forests. General aspects and the pine processionary moth as specific example. For Ecol Manag 259:831–838

Nogués-Bravo D, Araújo MB, Romdal TS, Rahbek C (2008) Scale effects and human impact on the elevational species richness gradients. Nature 453:216–219

Palahi M, Mavsar R, Gracia C, Birot Y (2008) Mediterranean forests under focus. Int for Rev 10:676–688

Pan X, Ding Z, Hu Y (2016) Elevational pattern of bird species richness and its causes along a central Himalaya gradient, China. PeerJ 4: e2636

Pearce-Higgins JW, Eglington SM, Martay B, Chamberlain DE (2015) Drivers of climate change impacts on bird communities. J Anim Ecol 84:943–954

Peña J, Bonet A, Bellot J et al (2007) Driving forces of land-use change in a cultural landscape of Spain. A preliminary assessment of the human-mediated influences. In: Koomen E, Stillwell J, Bakema A, Scholten HJ (eds) Modelling land-use change, progress and applications. Springer, Netherlands, pp 97–115

Popy S, Bordignon L, Prodon R (2010) A weak upward elevational shift in the distributions of breeding birds in the Italian Alps. J Biogeogr 37:57–67

Price TD, Hooper DM, Buchanan CD et al (2014) Niche filling slows the diversification of Himalayan songbirds. Nature 509:222–225

Prodon R, Thibault JC, Dejaifve PA (2002) Expansion vs. compression of bird altitudinal ranges on a Mediterranean island. Ecology 83:1294–1306

Quintero I, Jetz W (2018) Global elevational diversity and diversification of birds. Nature 555:246–250

Rahbek C (2005) The role of spatial scale and the perception of large-scale species-richness patterns. Ecol Lett 8:224–239

Rahbek C, Gotelli NJ, Colwell RK et al (2007) Predicting scale patterns of bird species richness with spatially explicit models. Proc R Soc B 274:165–174

Rahbek C, Borregaard MK, Colwell RK et al (2019) Humboldt's enigma: what causes global patterns of mountain biodiversity? Science 365:1108–1113

Rivas-Ubach A, Sardans J, Hódar JA et al (2016a) Similar local but different systemic metabolomic responses of close related pine subspecies to folivory by caterpillars of the processionary moth. Plant Biol 18:484–494

Rivas-Ubach A, Hódar JA, Sardans J et al (2016b) Are the metabolomic responses to folivory of closely related plant species linked to macroevolutionary and plant-folivore coevolutionary processes? Ecol Evol 6:4372–4386

Rivas-Ubach A, Sardans J, Hódar JA et al (2017) Close and distant: contrasting the metabolomes of two subspecies of Scots pine under the effects of summer drought and folivory. Ecol Evol 7:8976–8988

Rivas-Ubach A, Peñuelas J, Hódar JA et al (2019) We are what we eat: a stoichiometric and ecometabolomic study of caterpillars feeding on two pine subspecies of *Pinus sylvestris*. Int J Mol Sci 20:59

Robinet C, Baier P, Pennerstorfer J et al (2007) Modelling the effects of climate change on the potential feeding activity of *Thaumetopoea pityocampa* (Den. and Schiff.) (Lep., Notodontidae) in France. Global Ecol Biogeogr 16:460–471

Robinet C, Rousselet J, Roques A (2013) Potential spread of the pine processionary moth in France: preliminary results from a simulation model and future challenges. Ann for Sci 71:149–160

Robinet C, Laparie M, Rousselet J (2015) Looking beyond the large scale effects of global change: local phenologies can result in critical heterogeneity in the pine processionary moth. Front Physiol 6:334

Ros-Candeira A, Pérez-Luque AJ, Suárez-Muñoz M et al (2019) Dataset of occurrence and incidence of processionary moth in Andalusia (South Spain). ZooKeys 852:125–136

Ruggiero A, Hawkins BA (2006) Mapping macroecology. Global Ecol Biogeogr 15:433–437

Salman MHR, Bonsignore P, El Fels MAEA et al (2019) Winter temperature predicts prolonged diapause rate in pine processionary moth across its geographic range. PeerJ 7:e6530

Sanders NJ, Rahbek C (2012) The patterns and causes of elevational diversity gradients. Ecography 35:1–3

Sanderson FJ, Donald PF, Pain DJ, Burfield IJ, van Bommel FPJ (2006) Long-term population declines in Afro-Palearctic migrant birds. Biol Cons 131:93–105

Sekercioglu CH, Wenny DG, Whelan CJ (2016) Why Birds matters. Avian ecological function and ecosystem services. University of Chicago Press

SEO/BirdLife (2012) Seguimiento de las poblaciones de aves comunes en Andalucía, con especial interés de las Fringílidas de interés canoro. Seguimiento de las poblaciones de aves comunes reproductoras en Andalucía. Informe 2012. SEO/BirdLife, Madrid

SEO/BirdLife (2013) Tendencias de las aves en primavera. Resultados del Programa Sacre 1966–2013. SEO/BirdLife, Madrid

Suárez-Muñoz M, Bonet-García FJ, Hódar JA et al (2019) INSTAR: an agent-based model that integrates existing knowledge to simulate the population dynamics of a forest pest. Ecol Model 411:108764

Torres-Muros L, Hódar JA, Zamora R (2017) Effect of habitat type and soil moisture on pupal stage of a Mediterranean forest pest (*Thaumetopoea pityocampa*). Agric for Entomol 19:130–138

Terborgh J (1971) Distribution on environmental gradients: theory and a preliminary interpretation of distributional patterns in the avifauna of the Cordillera Vilcabamba, Peru. Ecology 52:23–40

Terborgh J, Weske JS (1975) The role of competition in the distribution of Andean birds. Ecology 56:562–576

Tingley MW, Beissinger SR (2013) Cryptic loss of montane avian richness and high community turnover over 100 years. Ecology 94:598–609

Vaughan R (1955) A note on the birds of the Sierra Nevada, South Spain. Ibis 97:159

Verner J (1985) Assessment of counting techniques. In: Ornithology C (ed) Johnston RF. Plenum Press, New York, pp 247–302

Vickery JA, Ewing SR, Smith KW et al (2014) The decline of Afro-Palaearctic migrants and an assessment of potential causes. Ibis 156:1–22

Wiens J (1989) The ecology of bird communities. Cambridge University Press, Cambridge

Yahner RH (1997) Responses of bird communities to early successional habitat in a managed landscape. Wilson Bull 115:292–298

Zamora R (1987) Variaciones altitudinales en la composición de las comunidades nidificantes de aves de Sierra Nevada. Doñana Acta Vertebrata 14:83–106

Zamora R (1988a) Composición y estructura de la comunidad de Passeriformes de la alta montaña de Sierra Nevada. Ardeola 35:197–220

Zamora R (1988b) Estructura morfológica de una comunidad de Passeriformes de alta montaña (Sierra Nevada, SE de España). Ardeola 35:71–95

Zamora R (1990) Seasonal variations of a passerine community in a mediterranean high-mountain. Ardeola 37:219–228

Zamora R, Barea-Azcón JM (2015) Long term changes in mountain passerine bird communities in the Sierra Nevada (Southern Spain): a 30-year case study. Ardeola 62:3–18

Zamora R, Camacho I (1984) Evolución estacional de la comunidad de aves en un robledal de Sierra Nevada. Doñana Acta Vertebrata 11:129–150

Zamora R, Matías L (2014) Seed dispersers, seed predators, and browsers act synergistically as biotic filters in a mosaic landscape. PLoS ONE 9:e107385

Zamora R, Gómez JM, Hódar JA et al (2001) Effect of browsing by ungulates on sapling growth of Scots pine in a Mediterranean environment: consequences for forest regeneration. For Ecol Manag 144:33–42

Zamora R, Pérez-Luque AJ, Bonet et al FJ (eds) (2016) Global change impacts in Sierra Nevada: challenges for conservation. Consejería de Medio Ambiente y Ordenación del Territorio. Junta de Andalucía

Zamora R, Pérez-Luque AJ, Bonet FJ (2017) Monitoring global change in high mountains. In: Catalan J, Ninot JM, Aniz MM (eds) High mountain conservation in a changing world. Springer International Publishing, pp 385–413

# Forest Dynamics Under Land-Use and Climate Change Scenarios

Luis Matías, Antonio J. Pérez-Luque, and Regino Zamora

**Abstract**

Mediterranean forests provide a diverse set of ecosystem services and harbour a large amount of biodiversity. The Sierra Nevada massif hosts extensive areas of forest with high conservation interest, although they have undergone an intense historical transformation due to land management and are vulnerable to further alterations in response to ongoing changes in climate. The main objective of this chapter is to analyse the state of the art concerning forest dynamics in response to past shifts in land use and changes in climate, both those that have been forecasted and those already registered. Starting with the scientific studies conducted in Sierra Nevada during recent decades, we review the main implications of pest management strategies on current forest structure and the consequences for forest functioning. In addition, we analyse the impact of climate aridification on the regeneration capacity and growth of tree species, and the relevance of biotic interactions for community dynamics. We end this chapter with management suggestions aimed at improving the resilience of Sierra Nevada forests to global change.

L. Matías (✉)
Department of Plant Biology and Ecology, University of Seville, Seville, Spain
e-mail: lmatias@us.es

A. J. Pérez-Luque · R. Zamora
Andalusian Institute for Earth System Research (IISTA), University of Granada, Granada, Spain

R. Zamora
Department of Ecology, University of Granada, Granada, Spain

A. J. Pérez-Luque
Service of Assesment, Restoration and Protection of Mediterranean Agrosystems, Estación Experimental del Zaidín (EEZ-CSIC), Spanish Council for Scientific Research, Granada, Spain

**Keywords**

Sierra Nevada • Forest ecology • Recruitment • Growth • Climate and land-use change • Altitudinal migration • Biotic interactions • Land management

## 1 The Forests of Sierra Nevada

Forests not only provide essential ecosystem services including climate regulation, air and water depuration, and erosion prevention (Iverson et al. 2018; Noce and Santini 2018), but also harbour a large amount of the biodiversity of Mediterranean ecosystems. In Sierra Nevada, forests represent one of the most abundant habitats, covering about 63,000 ha within the protected area (37% of the surface area; Bonet et al. 2009). For its geographical position and elevation, Sierra Nevada has acted as a refuge for many tree species that migrated during the successive arid and cold periods of the Quaternary (Carrión 2002). As a result, this massif currently hosts a broad array of tree species with diverse biogeographical origins, varying from typical Mediterranean species to temperate or boreal ones that left behind their core distribution ranges during the past glacial and interglacial periods (Blanca et al. 1998; Médail and Diadema 2009; Manzano et al. 2019).

The main native forests of Sierra Nevada are dominated by the evergreen holm oak *Quercus ilex* subsp. *ballota* (Desf.) Samp., occupying low and medium mountain areas (8800 ha), and deciduous broadleaf forests containing Pyrenean oak *Quercus pyrenaica* Willd., Granada maple *Acer opalus* subsp. *granatense* (Boiss.) Font Quer, and Rothm., and whitebeam *Sorbus aria* (L.) Crantz. The forests containing these species reach their southernmost European limit in Sierra Nevada, with populations ranging between 1100 and 2000 m a.s.l. and occupying about 2400 ha. These forests, distributed on siliceous soils both in the north-western and southern slopes of Sierra Nevada, are


R. Zamora and M. Oliva (eds.), *The Landscape of the Sierra Nevada*,
https://doi.org/10.1007/978-3-030-94219-9_13

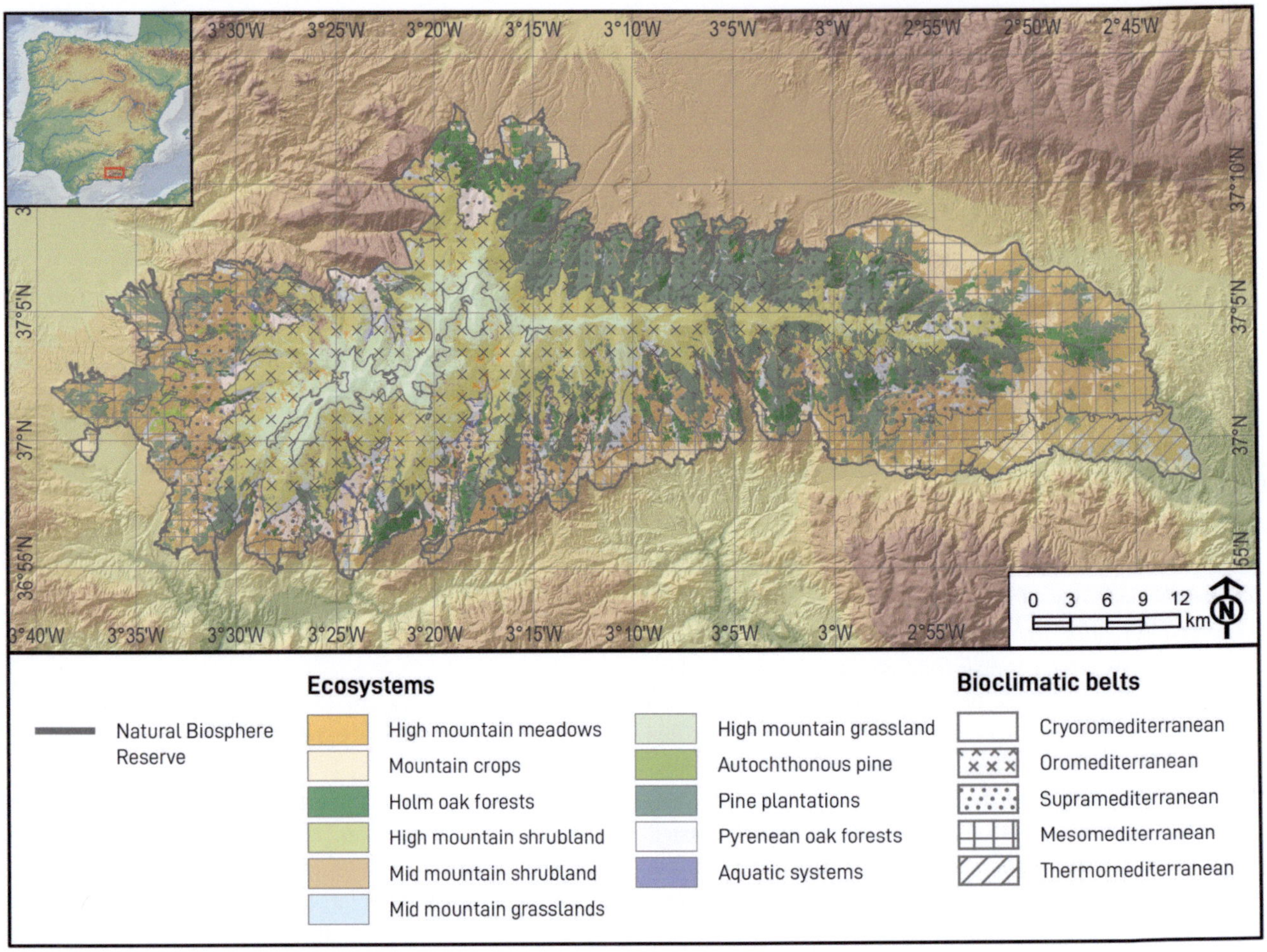


**Fig. 1** Distribution of forest ecosystems in Sierra Nevada

often associated with major river valleys. The Pyrenean oak woodlands in this mountain represent the rear edge of its habitat distribution (Hampe and Petit 2005), with high levels of intraspecific genetic diversity (Valbuena-Carabaña and Gil 2014). The conservation status of this species for southern Spain is considered "Vulnerable" and it is expected to suffer from climate change, potentially reducing its suitable habitats in the near future (see below). This oak is highly sensitive to moisture availability but both primary and secondary growth expresses high resilience to drought over the short and the long term (Pérez-Luque et al. 2021a).

Forests of autochthonous Scots pine *Pinus sylvestris* var. *nevadensis* can also be found in small patches at high altitudes with a characteristically low tree cover. Sierra Nevada harbours the southernmost populations of Scots pine (Boratynski 1991), where populations are ascribed to the subspecies *nevadensis* Christ (Molotkov and Patlaj 1991; Blanca et al. 1998). Native Scots pine populations occupy a small area, each roughly 5–10 km$^2$ (Catalán 1991), making them particularly vulnerable to the risk of extinction (Rivas-Martínez et al. 1991; Blanca et al. 1998). Today, these Scots pine forests are still threatened by several factors that hamper regeneration, such as high rates of seed predation (Castro et al. 1999; Matías et al. 2009), extremely high rates of seedling mortality due basically to summer drought (Castro et al. 2004), and high herbivory pressure from ungulates (Gómez et al. 2001; Zamora et al. 2001). Despite the relatively small sizes of these native oak and pine forests, they have key ecological relevance for the high biodiversity they support (Pérez-Luque et al. 2014). However, as a consequence of the intense land use that this mountain has undergone for centuries (see Sect. 2), the surface area of these ecosystems is at present critically restricted (Fig. 1).

Most of the forest cover in Sierra Nevada is currently dominated by pine plantations of Aleppo pine *Pinus halepensis* Mill., maritime pine *Pinus pinaster* Ait., black pine *Pinus nigra* Arnold subsp. *salzmannii* (Dunal) Franco and Scots pine *Pinus sylvestris* L. now covering approximately 42,000 ha (84% of the total forest area; Mesa-Garrido 2019). These plantations were mostly established in the period 1940–1980 (Ortuño 1990; Mesa-Garrido 2019) as a means to halt soil erosion in recently abandoned agricultural areas. Therefore, pines were planted in high densities, but favourable climate conditions and lack of appropriate post-planting management have resulted in highly dense monospecific even-aged stands. Consequently, these forest plantations are currently at extreme risk from climate change and forest pests, resulting in processes of decline and massive mortality (Sánchez-Salguero et al. 2012).

The geographical location, representing the rear distribution edge for several important tree species, and the high altitudinal gradient found in forests at Sierra Nevada make them especially sensitive to changes in climate and/or in land use. These characteristics, together with the long and intense management history of this area (see below) make Sierra Nevada a unique area to monitor the ecological consequences that these alterations pose for forest dynamics. Thus, a great number of studies have been conducted to monitor or predict the changes in forest structure and functioning in response to environmental alterations, producing a great amount of scientific output. These include ca. 12 Ph.D. theses, more than 200 papers in scientific journals (some of them are highly cited papers), several outreach publications, many management recommendations, and national and international networks of research, monitoring and conservation. In the sections below, we summarise the main findings of these studies.

## 2 Changes in Forest Structure and Management Over the Last Century

For millennia, human activity has strongly influenced the structure of the landscape in the Sierra Nevada (Anderson et al. 2011). From ca. 3000 cal yr BP, this human activity intensified in this mountain region with an increase in the frequency of fires, grazing, and mining (Jiménez-Moreno and Anderson 2012). Several palaeorecords also registered evidence of olive cultivation on a large scale at the lowest elevations (Jiménez-Moreno 2015; see also chapter "Reconstruction of Past Environment and Climate Using Wetland Sediment Records from the Sierra Nevada" of this book). The long-term human pressure in the Sierra Nevada, at least during the past 5000 years (Redondo and Martínez-Rodríguez 2021), has caused the disappearance of the typical treeline, as occurs in other Mediterranean mountains (Körner 2012). Fire occurrences have proved to be particularly high since ca. 3700 years ago, in response to regional human population expansion (Connor et al. 2019). The local and regional impact of humans increased substantially after ca. 2700 cal yr BP, and presumably in combination with aridification, almost causing the loss of *Pinus* forest within the mountain range (Anderson et al. 2011, see chapter "Altitudinal Patterns and Changes in the Composition of High Mountain Plant Communities").

The changes of the mountain landscapes of Sierra Nevada during the twentieth century are closely related to socioeconomic, demographic and cultural developments that intensified after the 1950s, e.g. land abandonment, depopulation, and the impoverishment of rural lifestyles (Jiménez-Olivencia et al. 2015a; Valbuena-Carabaña et al. 2010; see chapter "Singular Cultural Landscapes of the Sierra Nevada"). We highlight two important socioecological processes that appear to have shaped the current landscapes in this mountainous area: pine afforestation and the abandonment of traditional activities (Jiménez-Olivencia et al. 2015a; Navarro-González et al. 2013; Pérez-Luque et al. 2021a).

By the beginning of the twentieth century, landscapes of the Iberian Peninsula had been drastically deforested (Valbuena-Carabaña et al. 2010), and the remnant forests presented deplorable conditions (Douglas et al. 1996; Navarro and Medina-Achirica 2009). In Sierra Nevada, forest cover regeneration was an urgent social need for Spanish authorities, and several economic and political efforts were made to reforest the main degraded areas (Pemán-García et al. 2017). In fact, during the twentieth century, many large reforestation projects were conducted in the Mediterranean region, where different pine species were planted (Pausas et al. 2004). For instance, in Spain, 3.8 million ha were converted to pine plantations between 1945 and 1986 (Ortuño 1990). These activities were particularly intense in mountainous areas, where plantations were intended to boost forest productivity as a way not only to control soil erosion but also to increase employment in rural areas (Pausas et al. 2004). Pine stands in Sierra Nevada were planted mainly to palliate the extreme erosion of their watersheds, some of them having the highest erosion rates in the Iberian Peninsula (Mesa-Garrido 2019). Currently, pine plantations in Sierra Nevada account for more than 80% of its forest coverage, and are composed essentially of a single pine species (Bonet et al. 2016; Mesa-Garrido 2019). These pine plantations drastically reduced soil erosion but generated large homogeneous afforested areas that are highly vulnerable to pests and droughts, while other perturbations such as fire are being exacerbated by climate warming (Hódar and Zamora 2004; Choat et al. 2012; Sánchez-Salguero et al. 2013). In addition, the afforestation of arid and semi-arid areas was focused on promoting only a few ecosystem services (i.e. greater soil-carbon sequestration and control of soil erosion). However, forest ecosystems provide a wide range of ecosystem services, including timber production, carbon storage, local climate regulation, and many cultural services associated with nature experience and recreation (Gamfeldt et al. 2013; Felipe-Lucía et al. 2018; Iverson et al. 2018). Furthermore, planted forests can be managed to be resilient ecosystems (Puettmann et al. 2013) that maximize the multifunctionality of the ecosystem services they provide (Cruz-Alonso et al. 2019; Freer-Smith et al. 2019; Pérez-Luque and Zamora 2021).

The ecological relevance of these pine plantations is controversial in the scientific community. While some studies consider tree plantations as "biological deserts" with impoverished biodiversity and lack of natural dynamics, others find them useful for restoring degraded landscapes (Kanowski et al. 2005; Stephens and Wagner 2007). Gómez-Aparicio et al. (2009), analysing the biodiversity and regeneration capacity of the pine plantations in Sierra Nevada, found notable heterogeneity across different sites depending on stand density, local climate, and distance to natural seed sources. Plantations located at low to mid-elevations present a more diverse and abundant sapling bank, dominated by holm oak *Q. ilex* and accompanied by several pine species. These areas, therefore, show a high potential for natural replacement with the mixed pine-oak forest over the mid-term, while monospecific Scots pine plantations are expected to self-perpetuate by natural succession at high elevations as a consequence of the cold limitations for broadleaf species (Gómez-Aparicio et al. 2009). Although plantations can drastically reduce resource availability in the understory (i.e. light, soil water, and nutrients), negatively affecting the diversity and performance of native plant species (Bellot et al. 2004), adequate management can transform these monocultures into naturalized and diverse forests. To accelerate the transition from planted stands to natural forests with higher biodiversity and resistance to perturbations, suggestions include a thinning of dense stands to between 500 and 1000 trees/ha, allowing the entrance of light, diminishing competition, and maximizing regeneration (Gómez-Aparicio et al. 2009). Additionally, management plans meant to favour heterogeneous structures in contact with natural vegetation patches encourage the entrance of propagules mediated by zoochorous bird dispersers (Zamora et al. 2010), fortifying the naturalization capacity of these plantations.

As in other mountain areas, Sierra Nevada has been subject to extensive human use over the centuries. The pressures on the land were often extreme, and every usable piece of land was submitted to agriculture in the Alpujarra region (southern slopes of Sierra Nevada) (Douglas et al. 1996). Agriculture was the main way of life in the Sierra Nevada (Bosque-Maurel 1999), and therefore deforestation and land clearance for agriculture were the main factors shaping landscapes of this mountain ever since Moorish settlers entered the region (Douglas et al. 1996). The intense livestock use of this territory gradually depleted forests until the near disappearance of this activity towards the end of the nineteenth century (Laguna 1870). All these anthropic uses have weakened the structure, functioning, and the resilience of forests (Pérez-Luque et al. 2021a; Alba-Sánchez et al. 2021, Jimenez-Olivencia et al. 2015a; Padilla et al. 2010). In some cases, this pressure has resulted in up to 90% loss of broadleaf *Quercus* tree cover at medium and low elevations, i.e. in those areas close to the villages (Jimenez-Olivencia et al. 2015a). For instance, Pérez-Luque et al. (2021a) have revealed, using a combination of forest monitoring and review of historical documents, that forest clearings, firewood removal, charcoal production and mining have strongly affected the Pyrenean oak forest structure and dynamics in Sierra Nevada.

However, a de-intensification and abandonment of farming, forest activities, and animal grazing have been recorded in this region since the 1950s due to the rural exodus (Douglas et al. 1996; Jiménez-Olivencia et al. 2015b; Calatrava and Sayadi 2019). Calatrava and Sayadi (2019), in a detailed review of farming systems in Sierra Nevada, highlighted that the main features of this period were the abandonment of agriculture, a decline of the mountain livestock with a decrease of summer grazing in high-elevation areas, and a disappearance of the traditional activities. The decline in agricultural practices followed a spatial pattern, from the highlands to the areas closest to the villages (Jiménez-Olivencia et al. 2015b), extending even to the driest areas of the Sierra Nevada massif (Padilla et al. 2010). The abandonment of the traditional practices has caused some alterations in the structure of the landscape. Abandoned crop fields are being colonized by pioneer vegetation and shrublands (Jiménez-Olivencia et al. 2015b). In some cases, native tree species such as Pyrenean oak are also colonizing certain areas, although this process depends on the presence of livestock (Pérez-Luque et al. 2021b; 2021c). In addition, the densification of the forested area has been observed. For instance, several studies have indicated an increase in Pyrenean oak density since the 1950s (Navarro-González et al. 2012; Jiménez-Olivencia et al. 2015b; Zamora and Barea-Azcón 2015; Zamora et al. 2017). Paradoxically, some of these oak woodlands present a state of advanced degradation (growth stagnation, poor regeneration), having stands with high densities and high biomass accumulation, these conditions exacerbating the fire risk (Valbuena-Carabaña and Gil 2014). In view of this situation, the identification of the ecosystem services provided by these oak woodlands are pointed out as crucial to develop management strategies for Pyrenean oak forests in Sierra Nevada (Pérez-Luque et al. 2021b2021b).

## 3 Forest Responses to Recent Changes in Climate

The Mediterranean region is expected to be one of the most sensitive areas to the future climate changes projected (Giorgi 2006; Doblas-Miranda et al. 2017; IPCC 2021). Air temperatures in the region are predicted to rise by the end of this century more than 4 °C, especially during summer months. In addition, a progressive decline in precipitation is expected in all seasons during the twenty-first century, but especially during summer, where a reduction of about 28% has been predicted by 2100 (Giorgi and Lionello 2008; see chapter "Climate Variability and Trends"). These fast changes in climate, together with the above-mentioned changes in land use during the past century, can disrupt the conditions needed for the maintenance of numerous woody species, inducing changes in phenology, growth, demography, and ecological interactions (Parmesan 2006). These transformations are especially evident at species range edges, where ecological conditions are typically already at the species' limit of tolerance and where the most intense effects are predicted to occur (Hampe and Petit 2005; Parmesan 2006). Since several major tree species reach their range limit (either in elevation or in latitude) in the Sierra Nevada, this mountain represents a valuable study system to evaluate the consequences that the ongoing changes in climate and land use exert on forest dynamics. During recent decades, extraordinary research effort has been devoted to quantifying and analysing the magnitude of these impacts on forest dynamics in the Sierra Nevada protected area, joining observational, experimental, and modelling approaches. In this section, we summarise the most noteworthy responses of the Mediterranean forests of Sierra Nevada to the climate changes that have been both forecasted and registered.

## 3.1 Recruitment Alterations in Response to Climate Changes

The early life stages of plant species are vital for the maintenance of current populations. In this sense, the study of plant regeneration dynamics provides invaluable information concerning the demographic dynamics of plant species and communities. From a demographic perspective, the seedling stage is critical for recruitment (Lloret et al. 2004), and the analysis of its response to different global-change scenarios is crucial for an accurate forecast of future plant-community dynamics for two main reasons: first, because of the high sensitivity to the environmental conditions of seedlings due to their limited root system compared to adults, and second, the quick response to climate alterations compared to the adult phase (Lloret et al. 2009). Given that adult mortality events often occur in short time periods (Allen et al. 2010), the seedling and sapling population of the forest understory (i.e. the recruitment bank) can be used as a reliable indicator of future community composition and abundance. Thus, for a proper assessment of the dynamic response of forest ecosystems to rapid changes related to global climate change, we should pay special attention to the recruitment bank, considering the diverse representation of the different functional groups that inhabit Mediterranean forests (Matías et al. 2011).

Forest-regeneration studies have a long history in the Sierra Nevada, ranging from species to community approaches, and most of them have identified summer drought as the most critical factor currently controlling plant-regeneration dynamics (Castro et al. 2004; Mendoza et al. 2009a; Matías et al. 2012). During the seedling stage, precipitation deficit plays a major role by diminishing seedling growth and accelerating mortality. Therefore, the predicted reduction of precipitation for this area, especially threatening during the summer period, can induce decisive changes in the regeneration dynamics of woody species. An experimental approach simulating different climate scenarios (drier climate, exceptional rainy summers, and current conditions) has demonstrated that heavier precipitation enhances natural seedling emergence in terms of abundance and diversity (Matías et al. 2011), while lighter precipitation does not significantly reduce emergence (Fig. 2). This result highlights the relevance of the sporadic wet summers in the Mediterranean mountains such as Sierra Nevada for promoting regeneration and maintaining biodiversity. In addition, it points to the lack of sensitivity of woody species to reduced precipitation for seedling emergence, presumably because emergence starts at the end of the spring when soil moisture is still not limiting. However, it is through seedling survival that the precipitation pattern determines the outcome of the regeneration process. After the first summer, the expected increase in the frequency of dry periods in the Sierra Nevada range can drastically reduce the abundance and diversity of established seedlings, whereas the full regeneration potential of these forests is reached in exceptionally wetter summers (Matías et al. 2011). This pattern might have as an ultimate consequence a progressive species loss in successive years, with the less abundant species being the more prone to disappear (Lloret et al. 2004), consequently reinforcing the dominance of some drought-tolerant species in the community.

From a species-specific perspective, not all the species comprising Sierra Nevada forests respond equally to changes in precipitation or temperature. Mild or wetter summers in Sierra Nevada are essential for the establishment of some major late-successional tree species with a boreal-alpine origin such as Scots pine, yew *Taxus baccata* or Granada maple, their regeneration capacity being almost completely hampered under drier conditions (Mendoza et al. 2009a; Matías et al. 2012). By contrast, tree species with a typical Mediterranean origin (e.g. holm oak or Pyrenean oak) or mid-successional species (e.g. hawthorn *Crataegus monogyna* or Spanish barberry *Berberis hispanica*) and pioneer shrubs (e.g. broom *Cytisus scoparius* or Spanish sage *Salvia lavandulifolia*) are able to maintain a high regeneration capacity even under the drier conditions (Matías et al. 2012; Fig. 3). These studies indicate that the predicted lower frequency of wet summers in the Mediterranean region has far-reaching consequences for forest dynamics in Sierra Nevada. An unusually wet summer has a quantitative impact on Mediterranean forests, as the number of recruits surges in most species, ensuring a high maintenance capacity for the forest community. In addition, a

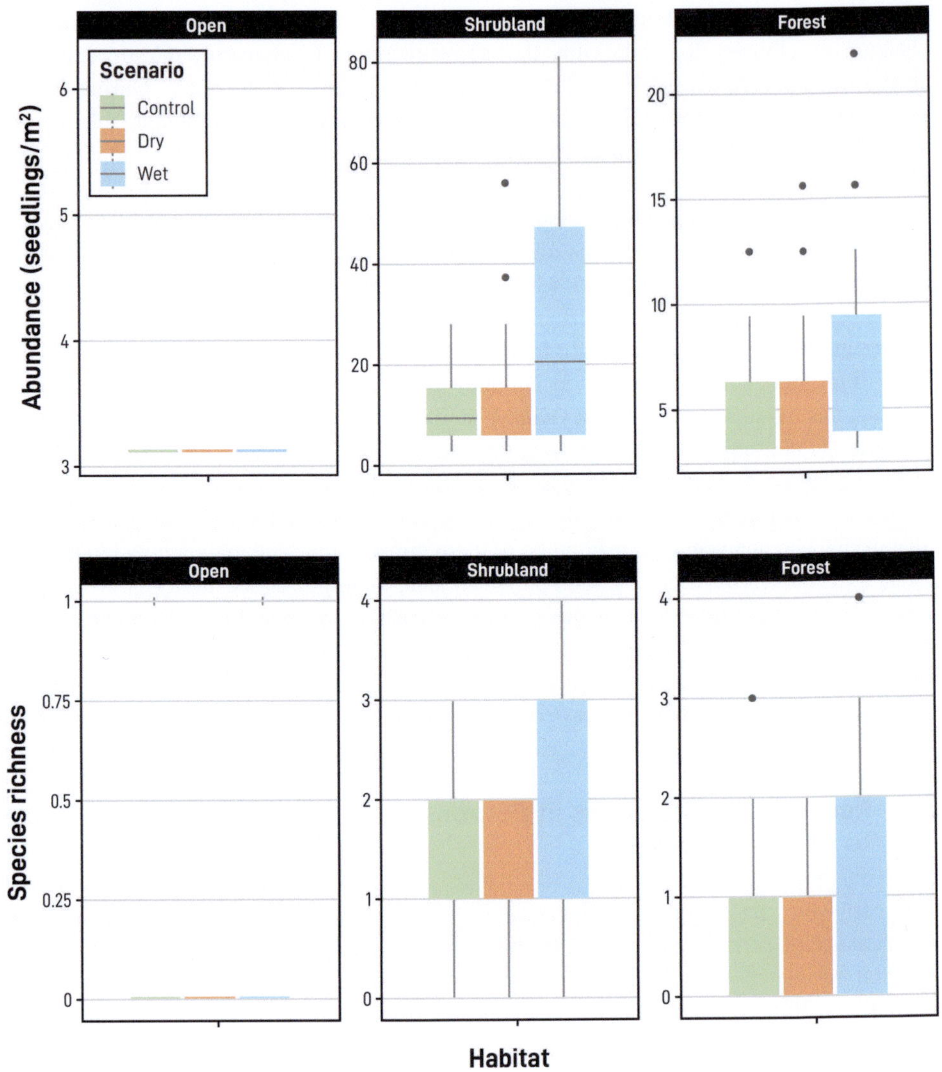


**Fig. 2** Box plots of seedling abundance and species richness under different climate scenarios (current climate, control; 30% precipitation reduction, dry; rainy summer, wet) across the main habitat types found in Sierra Nevada forests (continuous tree canopy, forest; under shrub cover, shrubland; open areas outside tree or shrub canopies, open). Modified from Matías et al. (2011)

wet summer also offers a qualitative effect, being the only window for recruitment in species that have higher water requirements, such as species with a boreal-alpine origin at the southern limit of their distribution area.

Although seedling responses to projected changes in temperature have been less explored, experimental studies under controlled conditions have determined that the regeneration capacity of the tree species in Sierra Nevada can be deeply altered in response to warmer conditions. In the case of Scots pine, seedling survival was reduced to approximately half when subjected to a 5 °C increase, and the expected reductions in precipitation further reduced survival probability to values close to zero when the forecasted temperature increase and rainfall reduction were combined (Matías and Jump 2014). All this evidence implies that, if mild summers become less frequent in coming decades and warmer and drier summers become the norm, as expected from the forecasting models (IPCC 2021, see chapter "Climate Variability and Trends"), we can expect a shift in the community structure involving a diversity loss. With warmer and drier summers, a healthy and diverse recruitment bank able to compensate for tree mortality would be less likely to persist, and thus the woody community may be prone to drastic changes in composition and dominance pattern. As a consequence, the impact of climate change on the diversity and dynamics of forest communities will depend not only on the expected aridity increase but also on the lower frequency of wet growing seasons. For the coming decades, both constraints may converge in an overall loss of forest diversity and greater dominance of the most drought-tolerant species.

## 3.2 Observed and Forecasted Changes in Tree Growth

In addition to the aforementioned consequences for the forest regeneration capacity, the already recorded changes in

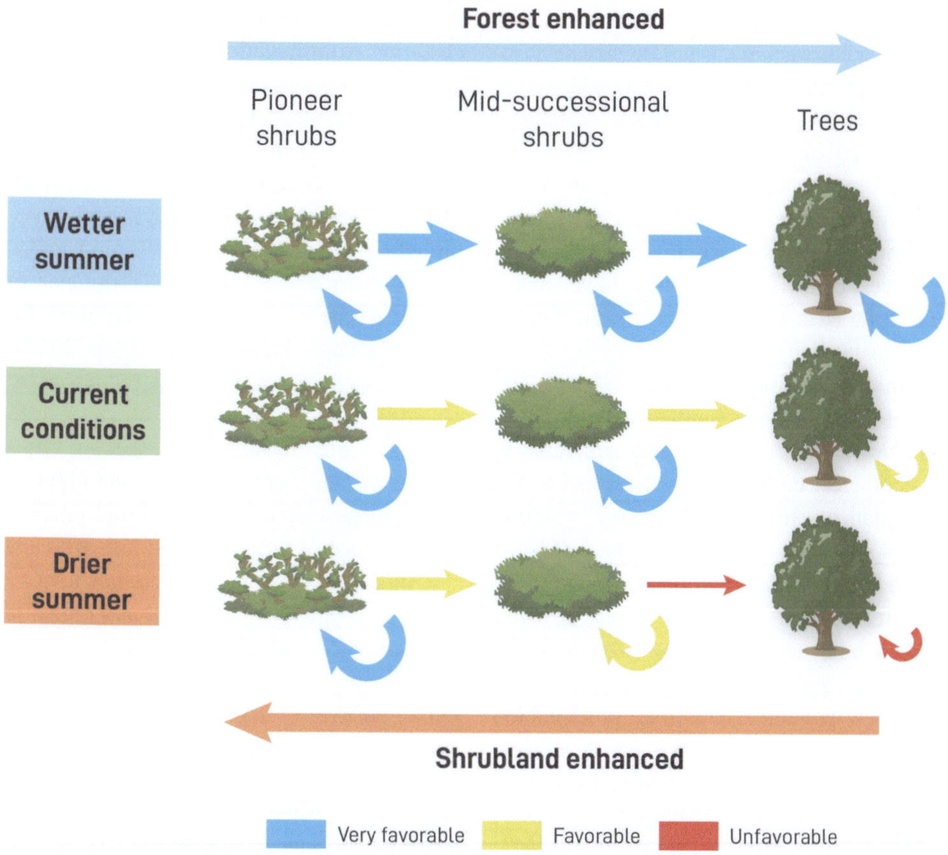


**Fig. 3** Schematic diagram representing the community dynamics within the different climate scenarios predicted for the coming decades in Mediterranean areas. Arrows represent the hypothesised dynamics: within functional groups, represents the self-regeneration capacity; between groups, represents the probability of transition between stages. Arrow width is proportional to success probability, from unfavourable to very favourable. Studies suggest that: (1) wet summer conditions would represent a good opportunity for tree species regeneration, allowing forest maintenance and colonization or expansion to nearby degraded habitats; (2) a drier summer scenario, however, would limit forest regeneration and probably hinder the colonization of nearby open and shrubland habitats; (3) a drier scenario would in turn be very favourable for shrubland expansion, enhancing both shrubland maintenance and colonization and expansion, even to nearby forest environments. Modified from Matías et al. (2012)

climate and the projections for further changes can deeply alter the growth trends of established adult trees at Sierra Nevada forests. Although seedlings of the main tree species have shown marked sensitivity to warming and more intense drought, slowing in growth (Matías et al. 2012; Herrero et al. 2013b; Matías and Jump 2014), dendrochronological approaches enable the study of the response of adult trees to the climate conditions during the past decades or even centuries. Two of the most abundant tree species in the massif, the widespread Scots pine and the deciduous Pyrenean oak, have attracted most of the research aimed at analysing growth responses to climate. These studies provide evidence for contrasted intraspecific responses and sensitivity to climate variability across elevations or slopes (Gea-Izquierdo and Cañellas 2014; Matías et al. 2017; Rubio-Cuadrado et al. 2018; Pérez-Luque et al. 2021a).

In the case of the Pyrenean oak, secondary growth is limited mainly by water availability and, to a lesser extent, by high temperatures during summer months (Gea-Izquierdo and Cañellas 2014; Rubio-Cuadrado et al. 2018; Pérez-Luque et al. 2021a). Populations with a southern exposure on Sierra Nevada, subjected to milder conditions, have followed positive growth trends since the late 1970s, while growth has remained constant on the northern exposure (Pérez-Luque et al. 2021a). This rising growth trend in Pyrenean oaks under warming climate conditions would be expected as characteristic of elevated locations where wet, cool conditions prevail and mitigate drought stress (Rubio-Cuadrado et al. 2018). However, aside from this general trend, Pyrenean oak populations from different locations throughout the Sierra Nevada massif showed differential sensitivity to extreme drought events due to differences in ecological conditions and/or past management legacies. While the northern (and drier) oak populations showed lower resistance to drought in terms of growth than did the southern populations under milder conditions, they also registered a higher recovery capacity (Pérez-Luque et al. 2021a). In addition, other factors such as tree size or population structure can alter post-drought growth, with larger trees or pure oak stands showing higher resistance (Rubio-Cuadrado et al. 2018). This overall high resistance and resilience to the drought of Pyrenean oaks in Sierra Nevada points to high-stress tolerance of these populations, probably related to local adaptation (Rubio-Cuadrado et al. 2018; Pérez-Luque et al. 2021a) and may have major implications for the long-term persistence of these forests, especially under a climate-change scenario.

For the case of the Scots pine, the Sierra Nevada and Sierra de Baza massifs represent the southernmost distribution edge of the species, its populations being subjected to highly contrasting ecological conditions with respect to the core-distribution populations. In these mountains, seedling growth is determined primarily by soil moisture (Matías et al. 2012; Matías and Jump 2014), while a combination of temperature and precipitation controls adult growth. Although both factors influence Scots pine growth at the species' distribution limit, their effects strongly vary across temporal and altitudinal gradients. Summer precipitation consistently promotes tree growth, but the effect of temperature on radial growth changes from negative during the previous autumn to ring formation to positive during winter and early spring (Herrero et al. 2013a; Linares et al. 2014; Matías et al. 2017; Rubio-Cuadrado et al. 2018). Overall, Scots pine populations displayed relatively stable growth trends during the first half of the twentieth century followed by a steep growth increase since the 1950s, although with notable changes across elevations (Matías et al. 2017). It bears mentioning that natural and planted Scots pine populations showed differential trends across elevations, changing from negative to positive as the elevation rose in planted stands and from positive to negative in natural stands (Rubio-Cuadrado et al. 2018). These contrasting trends suggest that planted pines respond better to colder conditions, while natural Scots pine populations had a higher tolerance for warmer conditions and were less affected by drought than planted pines (Rubio-Cuadrado et al. 2018). This appears to be a direct consequence of the local adaptations to the climate already reported for this species in Sierra Nevada (Matías and Jump 2014), and have two noteworthy implications. First, the negative growth trends in planted stands could lead to warming-induced decline episodes at lower elevations and, second, this different response of planted and natural populations highlights the relevance of the conservation of these natural pines in order to bolster the resilience of Sierra Nevada forests to a more arid climate.

Climate-forecast models predict lower summer precipitation (and thus lower growth) and higher temperatures over the year (Giorgi and Lionello 2008; Guiot and Cramer 2016) with varying impacts on tree growth across the seasons (either positive or negative). Under this changing scenario, it is difficult to evaluate the performance of tree species over the coming decades. A possible approach to evaluate future responses of Sierra Nevada forests to climate is through species-distribution modelling (SDM), a useful tool to evaluate future habitat suitability for key species in order to predict possible changes in distribution. Benito et al. (2011) reported a consistent decrease of habitat suitability for the two dominant oak species in Sierra Nevada, Pyrenean and holm oak until the end of the present century, even under the most optimistic scenarios (Fig. 4). This suitability decline is

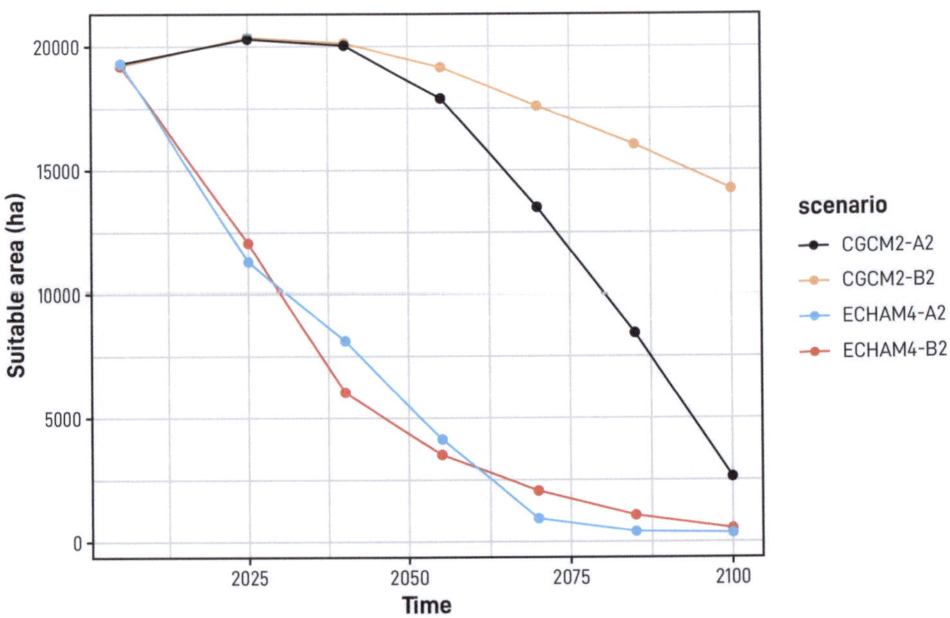


**Fig. 4** Time course of habitat suitability for *Quercus pyrenaica* in Sierra Nevada according to the predictions of different climate-change scenarios (A2 and B2 scenarios from models CGCM2 and ECHAM4). Modified from Benito et al. (2011)

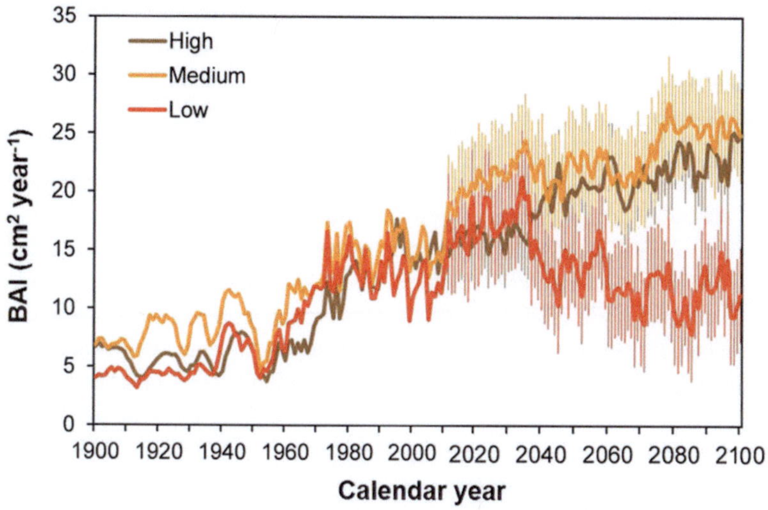


**Fig. 5** Mean basal area increment (BAI) for *P. sylvestris* populations across the elevational gradient in Sierra Nevada (lowermost elevation limit, Low; centre of the elevational distribution, Medium; treeline, High). Values for the 1900–2011 series are the site means of observed individual BAI (N = 60), whereas for the 2012–2100 period are the predictions using climate data predicted by the General Circulation Models, scenario ECHAM A2 (IPCC 2013). Vertical lines represent ± SD for the model predictions based on the simulation of 1000 trees. Modified from Matías et al. (2017)

expected to be especially severe from the year 2060 onward, possibly leading to fast adult mortality events or recruitment collapse, resulting in relict populations if adult mortality thresholds are not surpassed. An alternative way to forecast future forest dynamics is the use of climate–growth relationships to feed predictive models aimed at projecting the likely consequences of climate changes for tree growth. In a study predicting Scots pine growth trends at different elevations, Matías et al. (2017) found that treeline and mid-elevation populations are expected to follow a similar growth trend, with basal-area increases during the first part of the twenty-first century and a stabilized period during the second half (Fig. 5). This model suggests the persistence of natural Scots pine populations at least at medium and high elevations. Nevertheless, a completely different pattern appeared for low elevations, with a short phase of growth increase up to approximately the year 2040, followed by a consistent decline until the end of the century. A negative growth trend is a strongly negative indicator of population persistence, suggesting a strong population decline at lowland populations over the coming decades.

### 3.3 Implications for Elevational Migrations of Forests

All these changes in recruitment and growth trends imply far-reaching consequences for the elevational distribution of forests in Sierra Nevada. As an example of one of the most representative tree species in this massif, Scots pine has been demonstrated to differ in population structure across its elevational distribution, with a significantly reduced proportion of young individuals at the lowermost populations, where older age classes predominate with very low regeneration (Rabasa et al. 2013; Benavides et al. 2013, 2015; Matías and Jump 2015). This demographic structure clearly contrasts with the pattern found at the treeline, where the higher proportion of saplings is a clear indicator of population expansion. However, we must take into account that the treeline in Sierra Nevada is the result of climate and historical management and, in consequence, thermal limits might be over the current tree limit (Körner 2021). In addition, a higher mortality in the lowland populations than at the other altitudes has been reported, together with a lower growth rate and reduced reproductive investment (Matías and Jump 2015). This pattern agrees with the above-mentioned growth-forecasting models, which also predict a growth surge at medium and high elevations but declining growth trends lower on the massif (Matías et al. 2017). All this evidence indicates that a rapid depopulation of low-elevation Scots pine is in process, which, together with the signals found for treeline expansion, suggests an ongoing range migration upslope where physical conditions allow this species to prosper, or a range contraction from low-elevation areas over the coming decades.

Complementarily, the SDM approach also indicated a high potential for upland migration for the holm and Pyrenean oaks (Benito et al. 2011). However, these models predict that elevational migrations need to be rapid to maintain an optimal habitat-suitability range, this implying additional challenges, such as efficient seed dispersal, overcoming establishment limitations (Mendoza et al. 2009b), and coping with biotic interactions (see below). In addition, the area available diminishes with the rise in elevation, and the topography becomes more hostile (steeper slopes), while the soil loses the power to sustain shrub and tree species, therefore limiting uphill forest expansion. All these findings also imply that forest stands at Sierra Nevada can persist at medium or high elevations, although a substantial decline in lowland populations is likely to occur. Moreover, this lowland decline might be even more severe than described under extreme climatic events that are projected, if tipping points are surpassed or if the reduced tree vitality makes these populations more susceptible to pests and pathogens.

## 4 Consequences of Climate and Land-Use Changes for Biotic Interactions

Even minor changes in temperature or in the precipitation pattern could imply sweeping alterations in plant phenology, elevational distribution, development or defence capacity. Consequently, the above-mentioned changes in forest structure and dynamics can also alter the currently established biotic interactions, and such changes in mutualistic (e.g. pollination, seed dispersal, facilitation) and antagonistic interactions (e.g. herbivory, seed predation) could trigger dramatic alterations of the current ecosystem functioning. Biotic interactions play a key role in the functioning and maintenance of forest communities, with vital implications for the maintenance of biodiversity (Bascompte et al. 2006) and the shaping of species' responses to environmental alterations (Tylianakis et al. 2008). Plant-plant and plant-animal interactions have been extensively studied in Sierra Nevada forests, and here we sum up some of the main findings.

The ongoing changes in environmental conditions are leading to the establishment of new ecological interactions as a consequence of the changes in species' phenology or distribution. The best-known example in Sierra Nevada is the case of the pine processionary moth *Thaumetopoea pityocampa* (PPM). The altitudinal distribution of PPM has been limited in Sierra Nevada to about 1600 m, coinciding with the transition between black pine (a natural host of the herbivore with natural chemical defences) and Scots pine. However, the milder winter temperatures recorded since the final decades of the twentieth century allow PPM to develop at higher elevations (up to 2000–2100 m), reaching Scots pine populations, which have never undergone herbivory from this insect before and therefore have not developed defences against it (Hódar and Zamora 2004, see chapter "Responses of Animal Populations and Communities to Climate Change and Land-Use Shifts"). This new interaction has resulted in sharp reductions in seed production and growth for Scots pine (Hódar et al. 2003; Linares et al. 2014), which, together with the problems discussed above regarding growth and recruitment under a drier climate, threaten the persistence of this endangered relict tree species in Sierra Nevada.

Mediterranean forests have historically been decimated by cutting, fires and overgrazing, being reduced at present to a mosaic of small fragmented forests. The effect of herbivorous mammals on forest regeneration is expected to be especially dramatic in Mediterranean environments, where the typical slow sapling growth rate of most species worsens the impact of ungulates (Zamora et al. 2001; Baraza et al. 2007). Where these animals are mostly browsers (e.g. goats), overgrazing severely depresses regeneration due to the selective consumption of saplings and resprouts. Ungulates, therefore, exert a severe impact on sapling performance by stunting the growth of pre-reproductive individuals, a problem that is magnified when herbivory consumption becomes chronic (Zamora et al. 2001; Matías and Jump 2015). In addition, seed predators might also induce marked changes in the plant community by differentially consuming certain groups of species, such as trees with respect to shrubs (Matías et al. 2009). These two biotic interactions act in a similar direction, limiting the natural regeneration of Sierra Nevada forests. Vertebrate post-dispersal predators impose many filters on tree recruitment, by consuming most seeds and seedlings and by repeatedly browsing juveniles (Gómez et al. 2003; Matías et al. 2009; Baraza et al. 2007). Consequently, regeneration via sexual reproduction is highly constrained in these Mediterranean mountains, with resprouting being the main way that some species such as Pyrenean oak can regenerate at present. In addition, seed and seedling predators can also constrain forest expansion into adjacent shrublands. Dispersed seeds are quickly consumed by several species of predators, particularly wild boar *Sus scrofa* and wood mouse *Apodemus sylvaticus* (Castro et al. 1999; Gómez et al. 2003; Matías et al. 2009), suggesting that the forest regeneration in Mediterranean mountains is strongly limited by herbivores acting at several life-history stages (Zamora and Matías 2014).

In Mediterranean environments, the regeneration of a great number of woody species, both trees and shrubs, present a spatial pattern associated with established plants, suggesting a net positive balance in plant-plant interactions (Castro et al. 2002, 2006; Gómez Aparicio et al. 2004; Zamora et al. 2004). Given their ecological relevance, we could highlight the importance of shrub species for the maintenance of biotic interactions in these forests, especially under a context of global change. The successive forest clearings and later abandonment of cultivated areas during the past two centuries have led to the expansion of ecosystems dominated by shrubby species (ca. 40% of the Sierra Nevada protected area; Bonet et al. 2009), establishing influential interactions with tree species. Firstly, shrubs are commonly used as perches by frugivorous bird species which disperse under their canopies the seeds of fleshy fruited trees such as yew, whitebeam or tree honeysuckle *Lonicera arborea*, while rodents and jays actively accumulate oak acorns under shrubs (Gómez 2004). Although seed predation is usually higher under shrubs (Matías et al. 2009), a great portion of these seeds survive predation, generating aggregated seed dispersal kernels associated with shrubs and promoting the colonization of these areas by tree species (García et al. 2000; Mendoza et al. 2009b). Secondly, shrub canopies modify microclimatic conditions, resulting in higher soil moisture and nutrient availability with lower temperatures during summer, while allowing the entrance of a moderate light intensity. These changes result in greater abundance and diversity of seedlings germinating under shrubs with respect to less covered habitats (Gómez-Aparicio et al. 2004, 2005a, b, 2006, 2008; Mendoza et al. 2009a), improving their survival probabilities even under drier conditions expected for the coming decades (Matías et al. 2011, 2012). Finally, shrubs also play a fundamental role in the protection of tree saplings from ungulate herbivory, a factor seriously limiting tree recruitment in Sierra Nevada forests (Zamora et al. 2001; Herrero et al. 2012; Matías and Jump 2015) and able to alter the community structure by differential pressure over certain species (Zamora and Matías 2014). Ungulate herbivores avoid feeding upon unpalatable shrubs (those with chemical defences or low nutritional value) or spiny ones (Baraza et al. 2006, 2007) and, consequently, tree saplings growing under this type of shrubby species suffer significantly lower herbivory damage. This association is especially critical for species such as yew, whitebeam or Granada maple, whose regeneration niche in Sierra Nevada is currently limited to sites protected from herbivores, such as under shrubs with thorns or low palatability such as Spanish barberry, hawthorn or common juniper *Juniperus communis* (García et al. 2000; Gómez-Aparicio et al. 2004). All this evidence implies that facilitation by shrubs is an essential process for the regeneration of some tree species, pointing to the use of nurse plants as a promising restoration tool of degraded areas (Gómez-Aparicio et al. 2004; chapter "Managing the Uniqueness of Sierra Nevada Ecosystems Under Global Change: The Value of in situ Scientific Research"). This highlights the vital role of the colonization of drought-tolerant shrubs in degraded or abandoned areas for the later establishment of tree species.

### 4.1 Biotic Interactions as Plant-Community Filters

Several studies (Mendoza et al. 2009a, b; Zamora et al. 2010; Matías et al. 2012) have analysed the magnitude and the relative importance of abiotic and biotic factors on recruitment limitation (seed vs. establishment limitation) for the woody community. Overall, recruitment is severely limited in terms of both seed dispersal and establishment. However, tree species are more establishment-limited than shrubs. Trees and shrubs are the two dominant types of woody-plant architecture in a given terrestrial forest landscape, their proportions depending on the impact of biotic and abiotic filters on vegetation dynamics. In this respect, Zamora and Matías (2014) analysed the combined effects of mutualistic and antagonistic animals in a mosaic landscape composed of patches with native forest and different types of degraded areas representative of Mediterranean mountains for the various kinds of plant-animal interactions found in the same area at the same time. Observational and experimental results suggest that plant–animal interactions have a consistent influence on the functional components of the vegetation by the array of animals with which the plants interact. Browsers constrain the successional trajectory in a native forest, changing the probabilities of transition from the recruitment pattern of the sapling bank to the adults of the canopy (see also Gómez et al. 2001; Zamora et al. 2001; Gómez-Aparicio et al. 2008). Furthermore, irrespective of the habitat type considered (whether native forest, pine plantations or shrubland), seed dispersal, seed predation, and browsing act synergistically to foster shrub recruitment and discourage tree recruitment (Zamora and Matías 2014).

In conclusion, mutualistic and antagonistic animals in fact act jointly to induce a higher abundance of zoochorous shrubs in Mediterranean mountains. This main plant–animal interaction pattern is constant among the different landscape units, irrespective of the previous management history. In addition, this type of novel community is better adapted to cope with the expected shifts in climate for the coming decades in this area under a global-change scenario (Matías et al. 2011, 2012; Fig. 3), because shrubs are less vulnerable to drought and browsing than are tree species. Over the long term, biotic and abiotic factors may converge to promote increased dominance of the most drought-tolerant species, encouraging shrubland expansion beyond their current limits in the overall mountain landscape.

## 5 Concluding Remarks

Mediterranean forest ecosystems of Sierra Nevada are privileged settings for studying the causes and consequences of global change. These ecosystems have a high diversity of woody species in a mosaic of ecological conditions. Historical and biogeographic factors have powerfully contributed to the coexistence of boreal species that arrived during the glaciations, together with paleotropical relicts and species that originated under the current Mediterranean climate. The resulting environmental heterogeneity is both natural and anthropogenic, as a consequence of the ancient human impact. These forest ecosystems of the Sierra Nevada are therefore a privileged framework for investigating ecological processes in control scenarios, little altered, compared to other nearby forest scenarios highly anthropized. In the lines below, we list some of the main risks associated with global change that Mediterranean forests of Sierra Nevada are suffering, with solid scientific information collected during the last decades:

- **Drought**: The Mediterranean basin is a hot spot for potential changes due to climatic changes (chapter "Climate Variability and Trends"). The forecasts include significant increases in temperature coupled with reduced rainfall, which will affect the growth and regeneration of forest stands.
- **Fires**: It is very likely that fires will become more frequent if temperatures continue increasing and rainfall decreases. To this fact must be added the abandonment of forest use, and the increase in recreational activities, which will contribute to increasing the fires (chapter "Restoration of Mediterranean Forest Ecosystems After Major Disturbances: The Lanjarón Post-fire Experiment Over 15 Years of Succession").
- **Phenological changes**: Changes in temperature will cause alterations in the ecological interactions and the phenology of the species (flowering and fruiting dates, etc.). Many relationships between species can be altered and cause changes in the ecosystem with drastic consequences.
- **Increase in emerging pests and diseases**: Rising temperatures will facilitate the proliferation of forest pests and diseases, causing forest decay. Some of the pest species will lengthen their period of activity and others may extend their distribution area to new territories.
- **Changes in the composition of communities**: Many forest species differ in their ability to withstand drought. Those that have a shorter life cycle will adapt better to changing conditions than those with a longer life. For this reason, it is expected that shrub species will be favoured to detriment of trees. On the other hand, the future climate may be ideal for foreign species that now inhabit lower elevation sites.

Only through an adequate management plan aimed at fomenting the heterogeneity of habitat structure and ecological interactions of forest ecosystems, will we be able to increase the resistance and resilience of Sierra Nevada forests to global change. This management plan is urgently needed for the massive pine plantations, where a drastic reduction of pine density combined with the active conservation of adjacent natural forest and shrubland patches would simultaneously achieve the following: (1) encourage the entrance of propagules through zoochorous seed vectors, (2) favour the establishment of a diverse array of woody species, (3) reinforce tree growth and drought resistance, and (4) reduce the spread of pests such as PPM. In forestry science, recent works suggest the need to consider the structure of the forest (density, size classes and spatial pattern) as an attribute able to modulate the response of forest ecosystems to climate change events. In this sense, the speed with which environmental conditions are expected to change during the twenty-first century requires the establishment of mitigation and adaptation measures.

One of the most evident biotic responses derived from global warming is the altitudinal movements of species and communities in order to continue in the range of suitable climatic conditions. However, today's rapid rates of climate change may prevent many woody species from migrating fast enough to continue in their band of optimal conditions. Furthermore, the impacts of climate change can be exacerbated in a landscape that is fragmented and deeply altered by human activities as occurs in Sierra Nevada.

Despite the uncertainties about climate projections and the possible vegetation responses, there is already a sufficient ecological base to manage the gradual transition from our mountain forests towards a warmer and drier climate (chapter "Managing the Uniqueness of Sierra Nevada Ecosystems Under Global Change: The Value of in situ Scientific Research"). If we continue carrying out conventional management (or no management) under a climate-change scenario, it can lead to a permanent loss of biodiversity, an increase in the risk of fires and diseases and a collapse of many of the ecosystem services that forests provide. If we intend to ensure the future of the forests in Sierra Nevada, we not only must actively manage them, but we will have to do so under sustainability criteria that facilitate the adaptation of forest ecosystems to the new climatic conditions. For example, under the assumption of a higher frequency of drought conditions, a less dense forest (with fewer individuals per hectare) will be more resistant, since trees and plants will compete less with each other for a resource such as water. On the other hand, to increase the resilience (capacity of recovery) of a forest ecosystem it will be necessary to do a management that favours tree and shrub biodiversity. The approach to adaptation measures is mandatory, but without losing sight of the fact that the poor situation of our forest ecosystems has been produced by our inadequate land use during the past decades, not by climatic issues. With the foreseeable changes in climatic conditions, synergistic effects will occur, which might cause profound changes in the forest cover of Sierra Nevada.

In a climate change scenario, it is a priority to establish proactive action plans that anticipate the problems that forests will face in Sierra Nevada, with the aim of preserving their natural diversity, ecological interactions and the ecosystem services they provide. In the design of this type of management plan, predictive models are fundamental tools. In fact, predictive models of spatial distribution based on ecological ranges and their evolution according to IPCC projections under different climate change scenarios have opened a way to assess the consequences that climate change can have on forest distribution and diversity. The development of information technologies, remote sensors and geographic information systems have given rise to a new generation of modelling tools capable of offering plausible answers on the potential behaviour of forest communities in the face of different climate change scenarios.

**Acknowledgements** This research work was conducted in the collaborative framework of the "Sierra Nevada Global Change Observatory" monitoring program http://obsnev.es promoted by the Environmental and Regional Planning Council of the Regional Government of Andalusia and the University of Granada. There are many projects of the Spanish National Research Programme, the OAPN, the Junta de Andalucía and the European Union that have financed our research in the last decades. We can highlight the following projects MIGRAME "Global Change, Migration and colonization of hábitats altitudinal gradients on Mediterranean mountains" from the Excellence Research Group Programme of the Andalusian Government, the LIFE-ADAPTAMED (LIFE14 CCA/ES/000612): Protection of key ecosystem services by adaptive management of Climate Change endangered Mediterranean socio-ecosystems, the H2020 project European Long-Term Ecosystem and socio-ecological Research Infrastructure (eLTER), the MEADCLIM (PID2019-108288RA-I00): "Mechanisms implied on the adaptation of Mediterranean forests to climate change", and the project DECAFOR (US-1380871): "Analysis of the causes and ecological consequences of forest decline in response to global change" (FEDER-US). LM was supported by the fellowship II.5B from VI-PPIT-U. Sevilla.The research summarized in this chapter derives from the field works applied to conservation and restoration carried out by many researchers belonging to the Terrestrial Ecology Research Group RNM-220, including José Antonio Hódar, José María Gómez, Daniel García, Lorena Gómez Aparicio, Elena Baraza, Jorge Castro, Irene Mendoza, José Luis Quero and Asier Herrero among others. They have provided very valuable information, and have contributed to maintaining an atmosphere of permanent collaboration with the managers of the National Park. We thank the managers of the Sierra Nevada Natural Space for their determined support of our work in Sierra Nevada and their continued collaboration within the framework of the joint activities that we develop at the Global Change Observatory.

## References

Alba-Sánchez F, Abel-Schaad D, López-Sáez JA, Sabariego-Ruiz S, Pérez-Díaz S, Luelmo-Lautenschlaeger R, Garrido-García JA (2021) Early anthropogenic change in western Mediterranean mountains (Sierra Nevada, SE Spain). Anthropocene 33:100278

Allen CD, Macalady AK, Chenchouni H et al (2010) A global overview of drought and heat-induced tree mortality reveals emerging climate change risks for forests. For Ecol Manage 259:660–684

Anderson SR, Jiménez-Moreno G, Carrión JS, Pérez-Martínez C (2011) Postglacial history of alpine vegetation, fire, and climate from Laguna de Río Seco, Sierra Nevada, southern Spain. Quat Sci Rev 30:1615–1629

Baraza E, Zamora R, Hódar JA (2006) Conditional outcomes in plant–herbivore interactions: neighbours matter. Oikos 113:148–156

Baraza E, Zamora R, Hódar JA, Gómez JM (2007) Plant–herbivore interaction: beyond a binary vision In: Pugnaire FI, Valladares F (eds) Handbook of functional plant ecology. Marcel Dekker Inc, pp 677–718

Bascompte J, Jordano P, Olesen JM (2006) Asymmetric coevolutionary networks facilitate biodiversity maintenance. Science 312:431–433

Bellot J, Maestre FT, Chirino E, Hernández N, de Urbina JO (2004) Afforestation with Pinus halepensis reduces native shrub performance in a Mediterranean semiarid area. Acta Oecologica 25:7–15

Benavides R, Rabasa SG, Granda E, Escudero A, Hódar JA, Martinez-Vilalta J et al (2013) Direct and indirect effects of climate on demography and early growth of *Pinus sylvestris* at the rear edge: changing roles of biotic and abiotic factors. PLoS ONE 8: e59824

Benavides R, Escudero A, Coll L, Ferrandis P, Gouriveau F, Hódar JA et al (2015) Survival vs. growth trade-off in early recruitment challenges global warming impacts on Mediterranean mountain trees. Perspect Plant Ecol Evol Syst 17:369–378

Benito B, Lorite J, Peñas J (2011) Simulating potential effects of climatic warming on altitudinal patterns of key species in Mediterranean-alpine ecosystems. Clim Change 108:471–483

Blanca G, Cueto M, Martínez-Lirola MJ, Molero-Mesa J (1998) Threatened vascular flora of Sierra Nevada (southern Spain). Biol Cons 85:269–285

Bonet FJ, Aspizua R, Navarro J (2016) History of Sierra Nevada forest management: implications for adaptation to global change. In: Zamora R, Pérez-Luque AJ, Bonet FJ, Barea-Azcón JM, Aspizua R (eds) Global change impacts in Sierra Nevada: challenges for conservation. Consejería de Medio Ambiente y Ordenación del Territorio. Junta de Andalucía, pp 153–156

Bonet FJ, Villegas I, Navarro J, Zamora R (2009) Breve historia de la gestión de los pinares de repoblación en Sierra Nevada. Una aproximación desde la ecología de la regeneración. In: Actas del 5º Congreso Forestal Español-Montes y Sociedad: Saber qué hacer. Ref: *5CFE01–184*, pp 1–13

Boratynski A (1991) Range of natural distribution. In: Giertich M, Mátyás C (eds) Genetics of Scots Pine, Akadémiai Kiadó, Budapest, pp 19–30

Bosque-Maurel J (1999) El hombre y el medio en Sierra Nevada (Andalucía) crisis y cambio. Estud. Geogr. 60:207–248

Calatrava J, Sayadi S (2019) Evolution of farming systems in the Mediterranean high mountain: the case of the Alpujarra alta (Spain). Sustainability 11:704

Carrión JS (2002) Patterns and processes of Late Quaternary environmental change in a montane region of southwestern Europe. Quat Sci Rev 21:2047–2066

Castro J, Gómez JM, García D, Zamora R, Hódar JA (1999) Seed predation and dispersal in relict Scots pine forests in southern Spain. Plant Ecol 145:115–123

Castro J, Zamora R, Hódar JA et al (2002) Use of shrubs as nurse plants: a new technique for reforestation in Mediterranean mountains. Restor Ecol 10:297–305

Castro J, Zamora R, Hódar JA, Gómez JM (2004) Seedling establishment of a boreal tree species (Pinus sylvestris) at its southernmost distribution limit: consequences of being in a marginal Mediterranean habitat. J Ecol 92:266–277

Castro J, Zamora R, Hódar JA (2006) Restoring the forests of Quercus pyrenaica Willd. using pioneer shrubs as nurse plants. Appl Veg Sci 9:137–142

Catalán G (1991) Regiones de procedencia de Pinus sylvestris L., P*inus nigra* Arn *subsp. salzmannii* (Dunal) Franco. MAPA, 31
Choat B, Jansen S, Brodribb TJ et al (2012) Global convergence in the vulnerability of forests to drought. Nature 491:752–755
Connor SE, Vannière B, Colombaroli D et al (2019) Humans take control of fire-driven diversity changes in Mediterranean Iberia's vegetation during the mid–late Holocene. Holocene 29:886–901
Cruz-Alonso V, Ruiz-Benito P, Villar-Salvador P, Rey-Benayas JM (2019) Long-term recovery of multifunctionality in Mediterranean forests depends on restoration strategy and forest type. J Appl Ecol 56:745–757
Doblas-Miranda E, Alonso R, Arnan X et al (2017) A review of the combination among global change factors in forests, shrublands and pastures of the Mediterranean Region: beyond drought effects. Global Planet Change 148:42–54
Douglas T, Critchley D, Park G (1996) The Deintensification of terraced agricultural land near Trevelez, Sierra Nevada, Spain. Glob Ecol Biogeogr Lett 5:258
Felipe-Lucía MR, Soliveres S, Penone C et al (2018) Multiple forest attributes underpin the supply of multiple ecosystem services. Nat Comm 9:4839
Freer-Smith P, Muys B, Bozzano M, Drössler L, Farrelly N, Jactel H, Korhonen J, Minotta G, Nijnik M, Orazio C (2019) Plantation forests in Europe: challenges and opportunities. European Forest Institute
Gamfeldt L, Snäll T, Bagchi R et al (2013) Higher levels of multiple ecosystem services are found in forests with more tree species. Nat Comm 4:1340
García D, Zamora R, Hódar JA, Gómez JM, Castro J (2000) Yew (*Taxus baccata* L.) regeneration is facilitated by fleshy-fruited shrubs in Mediterranean environments. Biol Cons 95:31–38
Gea-Izquierdo G, Canellas I (2014) Local climate forces instability in long-term productivity of a Mediterranean oak along climatic gradients. Ecosystems 17:228–241
Giorgi F (2006) Climate change hot-spots. Geophys Res Lett 33(8)
Giorgi F, Lionello P (2008) Climate change projections for the Mediterranean region. Glob Planet Change 63:90–104
Gómez JM (2004) Importance of microhabitat and acorn burial on Quercus ilex early recruitment: non-additive effects on multiple demographic processes. Plant Ecol 172:287–297
Gómez JM, García D, Zamora R (2003) Impact of vertebrate acorn-and seedling-predators on a Mediterranean *Quercus pyrenaica* forest. For Ecol Manage 180:125–134
Gómez JM, Hódar JA, Zamora R, Castro J, García D (2001) Ungulate damage on Scots pines in Mediterranean environments: effects of association with shrubs. Can J Bot 79:739–746
Gómez-Aparicio L, Zamora R, Gómez JM, Hódar JA, Castro J, Baraza E (2004) Applying plant facilitation to forest restoration: a meta-analysis of the use of shrubs as nurse plants. Ecol Appl 14:1128–1138
Gómez-Aparicio L, Gómez JM, Zamora R, Boettinger J (2005a) Canopy vs soil effects of shrubs facilitating tree seedlings in Mediterranean montane ecosystems. J Veg Sci 16:191–198
Gómez-Aparicio L, Gómez JM, Zamora R (2005b) Microhabitats shift rank in suitability for seedling establishment depending on habitat type and climate. J Ecol 93:1194–1202
Gómez-Aparicio L, Valladares F, Zamora R (2006) Differential light responses of Mediterranean tree saplings: linking ecophysiology with regeneration niche in four co-occurring species. Tree Physiol 26:947–958
Gómez-Aparicio L, Zamora R, Castro J, Hódar JA (2008) Facilitation of tree saplings by nurse plants: microhabitat amelioration or protection against herbivores. J Veg Sci 19:161–172
Gómez-Aparicio L, Zavala MA, Bonet FJ, Zamora R (2009) Are pine plantations valid tools for restoring Mediterranean forests? An assessment along abiotic and biotic gradients. Ecol Appl 19:2124–2141
Guiot J, Cramer W (2016) Climate change: The 2015 Paris Agreement thresholds and Mediterranean basin ecosystems. Science 354:465–468
Hampe A, Petit RJ (2005) Conserving biodiversity under climate change: the rear edge matters. Ecol Lett 8:461–467
Herrero A, Zamora R, Castro J, Hódar JA (2012) Limits of pine forest distribution at the treeline: herbivory matters. Plant Ecol 213:459–469
Herrero A, Rigling A, Zamora R (2013a) Varying climate sensitivity at the dry distribution edge of *Pinus sylvestris* and *P. nigra*. For Ecol Manage 308:50–61
Herrero A, Castro J, Zamora R, Delgado-Huertas A, Querejeta JI (2013b) Growth and stable isotope signals associated with drought-related mortality in saplings of two coexisting pine species. Oecologia 173:1613–1624
Hódar JA, Castro J, Zamora R (2003) Pine processionary caterpillar *Thaumetopoea pityocampa* as a new threat for relict Mediterranean Scots pine forests under climatic warming. Biol Cons 110:123–129
Hódar JA, Zamora R (2004) Herbivory and climatic warming: a Mediterranean outbreaking caterpillar attacks a relict, boreal pine species. Biodiv Cons 13:493–500
IPCC (2013) Climate change 2013: The physical science basis. Contribution of Working Group I to the Fifth Assessment Report of the Intergovernmental Panel on Climate Change [Stocker TF, Qin D, Plattner G-K, Tignor M, Allen SK, Boschung J, Nauels A, Xia Y, Bex V, and Midgley PM (eds)]. Cambridge University Press, Cambridge and New York, NY
IPCC (2021) Climate Change 2021: The Physical Science Basis. Contribution of Working Group I to the Sixth Assessment Report of the Intergovernmental Panel on Climate Change [Masson-Delmotte V, P Zhai, A Pirani, SL Connors, C Péan, S Berger, N Caud, Y Chen, L Goldfarb, MI Gomis, M Huang, K Leitzell, ELonnoy, JBR Matthews, TK Maycock, T Waterfield, O Yelekçi, R Yu, and B Zhou (eds)]. Cambridge University Press. In Press
Iverson LR, Perera AH, Pastur GM, Peterson U (2018) Ecosystem services from forest landscapes: where we are and where we go. In: Perera AH, Peterson U, Pastur GM, Iverson LR (eds) Ecosystem services from forest landscapes: broadscale considerations. Springer International Publishing, Cham, pp 249–258
Jiménez-Moreno G (2015) Reconstrucción de la vegetación a partir del análisis palinológico. In: Zamora R, Pérez-Luque A, Bonet F, Barea-Azcón J, Aspizua R (eds) La Huella Del Cambio Global En Sierra Nevada: Retos Para La Conservación. Consejería de Medio Ambiente y Ordenación del Territorio. Junta de Andalucía, pp 48–50
Jiménez-Moreno G, Anderson RS (2012) Holocene vegetation and climate change recorded in alpine bog sediments from the Borreguiles de la Virgen, Sierra Nevada, southern Spain. Quat Res 77:44–53
Jiménez-Olivencia Y, Porcel L, Caballero A (2015a) Medio siglo en la evolución de los paisajes naturales y agrarios de Sierra Nevada (España). Boletín de la Asociación de Geógrafos Españoles 68:205–232
Jiménez-Olivencia Y, Porcel-Rodríguez L, Caballero-Calvo A, Bonet-García FJ (2015b) Evolución de los usos del suelo en Sierra Nevada en los últimos 50 años y cambios en el paisaje. In: Zamora R, Pérez-Luque A, Bonet F, Barea-Azcón J, Aspizua R (eds) La Huella Del Cambio Global En Sierra Nevada: Retos Para La Conservación. Consejería de Medio Ambiente y Ordenación del Territorio. Junta de Andalucía, pp 54–56

Kanowski J, Catterall CP, Wardell-Johnson GW (2005) Consequences of broadscale timber plantations for biodiversity in cleared rainforest landscapes of tropical and subtropical Australia. For Ecol Manage 208:359–372

Körner C (2012) Alpine treelines. Springer Basel, Basel

Körner C (2021) Alpine plant life. Springer International Publishing, Cham

Laguna M (1870) Comisión de la Flora Forestal Española. Resumen de los trabajos verificados por la misma los años de 1867 y 1868. Tomo I. Imprenta del Colegio Nacional de Sordo-Mudos y de Ciegos. Madrid. España

Linares JC, Senhadji K, Herrero A, Hódar JA (2014) Growth patterns at the southern range edge of Scots pine: Disentangling the effects of drought and defoliation by the pine processionary caterpillar. For Ecol Manage 315:129–137

Lloret F, Peñuelas J, Estiarte M (2004) Experimental evidence of reduced diversity of seedlings due to climate modification in a Mediterranean-type community. Glob Change Biol 10:248–258

Lloret F, Peñuelas J, Prieto P, Llorens L, Estiarte M (2009) Plant community changes induced by experimental climate change: seedling and adult species composition. Persp Plant Ecol Evol Syst 11:53–63

Manzano S, Carrión JS, López-Merino L, Jiménez-Moreno G, Toney JL, Amstrong H, García-Alix A, Guerrero-Pérez JL, Sánchez-Mata D (2019) A palaeoecological approach to understanding the past and present of Sierra Nevada, a Southwestern European biodiversity hotspot. Glob Planet Change 175:238–250

Matías L, Jump AS (2014) Impacts of predicted climate change on recruitment at the geographical limits of Scots pine. J Exp Bot 65:299–310

Matías L, Jump AS (2015) Asymmetric changes of growth and reproductive investment herald altitudinal and latitudinal range shifts of two woody species. Glob Change Biol 21:882–896

Matías L, Mendoza I, Zamora R (2009) Consistent pattern of habitat and species selection by post-dispersal seed predators in a Mediterranean mosaic landscape. Plant Ecol 203:137–147

Matías L, Zamora R, Mendoza I, Hódar JA (2010) Seed dispersal patterns by large frugivorous mammals in a degraded mosaic landscape. Rest Ecol 18:619–627

Matías L, Zamora R, Castro J (2011) Repercussions of simulated climate change on the diversity of woody-recruit bank in a Mediterranean-type ecosystem. Ecosystems 14:672–682

Matías L, Zamora R, Castro J (2012) Sporadic rainy events are more critical than increasing drought intensity for woody species recruitment in a Mediterranean community. Oecologia 169:833–844

Matías L, Linares JC, Sánchez-Miranda Á, Jump AS (2017) Contrasting growth forecasts across the geographical range of Scots pine due to altitudinal and latitudinal differences in climatic sensitivity. Glob Change Biol 23:4106–4116

Médail F, Diadema K (2009) Glacial refugia influence plant diversity patterns in the Mediterranean basin. J Biogeogr 36:1333–1345

Mendoza I, Gómez-Aparicio L, Zamora R, Matías L (2009a) Recruitment limitation of forest communities in a degraded Mediterranean landscape. J Veg Sci 20:367–376

Mendoza I, Zamora R, Castro J (2009b) A seeding experiment for testing tree-community recruitment under variable environments: implications for forest regeneration and conservation in Mediterranean habitats. Biol Cons 142:1491–1499

Mesa-Garrido MÁ (2019) Reforestación, silvicultura e incendios forestales en la dinámica del paisaje del Espacio Natural de Sierra Nevada (1881–2018). Investigaciones Geográficas 71:209–234

Molotkov PI, Patlaj IN (1991) Systematic position within the genus Pinus and intraspecific taxonomy. In: Developments in plant genetics and breeding, vol 3. Elsevier, pp 31–40

Noce S, Santini M (2018) Mediterranean forest ecosystem services and their vulnerability (ver 2.0). Foundation Euro-Mediterranean Center on Climate Change

Navarro J, Medina-Achirica J (2009) La restauración hidrológico forestal del Marquesado del Zenete (Sierra Nevada). Crónica histórica. Cuadernos De La Sociedad Española De Ciencias Forestales 30:297–304

Navarro-González I, Bonet-García FJ, Pérez-Luque AJ (2012) Cartografía histórica de la vegetación mediante ortofotos, in: Aspizua R, Barea-Azcón J, Bonet F, Pérez-Luque A, Zamora R (eds) Observatorio de Cambio Global de Sierra Nevada. Metodologías de Seguimiento. Consejería de Medio Ambiente. Junta de Andalucía, pp 16–17

Navarro-González I, Pérez-Luque AJ, Bonet FJ, Zamora R (2013) The weight of the past: land-use legacies and recolonization of pine plantations by oak trees. Ecol Appl 23:1267–1276

Ortuño F (1990) El plan para la repoblación forestal de España del año 1939. Análisis y comentarios. Ecología (Extra 1):373–392

Parmesan C (2006) Ecological and evolutionary responses to recent climate change. Annu Rev Ecol Evol Syst 37:637–669

Padilla FM, Vidal B, Sánchez J, Pugnaire FI (2010) Land-use changes and carbon sequestration through the twentieth century in a Mediterranean mountain ecosystem: Implications for land management. J Env Manage 91:2688–2695

Pausas JG, Bladé C, Valdecantos A, Seva JP, Fuentes D, Alloza JA, Vilagrosa A, Bautista S, Cortina J, Vallejo R (2004) Pines and oaks in the restoration of Mediterranean landscapes of Spain: new perspectives for an old practice: a review. Plant Ecol 171:209–220

Pérez-Luque AJ, Zamora R (2021) diveRpine: Diversification of pine plantations in Mediterranean mountains. An interactive R tool to help decision-makers. Ecography (in review)

Pérez-Luque AJ, Bonet FJ, Pérez-Pérez R, Aspizua R, Lorite J, Zamora R (2014) Sinfonevada: Dataset of floristic diversity in Sierra Nevada forests (SE Spain). PhytoKeys 35:1

Pérez-Luque AJ, Gea-Izquierdo G, Zamora R (2021a) Land-use legacies and climate change as a double challenge to oak forest resilience: mismatches of geographical and ecological rear edges. Ecosystems 1–19

Pérez-Luque AJ, Barea-Azcón JM, Aspizua R, Cano-Manuel FJ, Zamora R (2021b) Los robledales de Sierra Nevada: Valoración de servicios ecosistémicos y recomendaciones de gestión. (SE España): diversidad, funcionamiento y gestión. Consejería de Agricultura, Ganadería, Pesca y Desarrollo Sostenible. Junta de Andalucía, 112 pp

Pérez-Luque AJ, Bonet-García FJ, Zamora R (2021c) Colonization pattern of abandoned croplands by Quercus pyrenaica in a Mediterranean mountain region. Forests 12:1584. https://doi.org/10.3390/f12111584

Pemán-García J, Iriarte-Goñi I, Lario-Leza FJ (2017) La restauración forestal de España: 75 años de una ilusión. Ministerio de Agricultura y Pesca, Alimentación y Medio Ambiente. Sociedad Española de Ciencias Forestales

Puettmann K, Messier C, Coates K (2013) Managing forests as complex adaptive systems. In: Messier C, Puettmann K, Coates K (eds) Managing forests as complex adaptive systems: building resilience to the challenge of global change. The Earthscan Forest Library, New York

Rabasa SG, Granda E, Benavides R, Kunstler G, Espelta JM, Ogaya R et al (2013) Disparity in elevational shifts of European trees in response to recent climate warming. Global Change Biol 19:2490–2499

Redondo JJ, Martínez-Rodríguez F (2021) Primeros datos sobre un petroglifo prehistórico de estilo atlántico hallado en Peña Madura, cuenca alta del Río Dílar, Sierra Nevada (Granada). Cuad Arte Prehistórico 11:124–160

Rivas-Martínez S, Asensi A, Molero-Mesa J, Valle F (1991) Endemismos Vasculares De Andalucía. Rivasgodaya 6:77–100

Rubio-Cuadrado Á, Camarero JJ, Aspizua R, Sánchez-González M, Gil L, Montes F (2018) Abiotic factors modulate post-drought growth resilience of Scots pine plantations and rear-edge Scots pine and oak forests. Dendrochronologia 51:54–65

Sánchez-Salguero R, Navarro-Cerrillo RM, Camarero JJ, Fernández--Cancio Á (2012) Selective drought-induced decline of pine species in southeastern Spain. Clim Change 113:767–785

Sánchez-Salguero R, Camarero JJ, Dobbertin M, Fernández-Cancio A, Vilà-Cabrera A, Manzanedo RD, Zavala MA, Navarro-Cerrillo RM (2013) Contrasting vulnerability and resilience to drought-induced decline of densely planted vs. natural rear-edge *Pinus nigra* forests. For Ecol Manage 310:956–967

Stephens SS, Wagner MR (2007) Forest plantations and biodiversity: a fresh perspective. J Forestry 105:307–313

Tylianakis JM, Didham RK, Bascompte J, Wardle DA (2008) Global change and species interactions in terrestrial ecosystems. Ecol Lett 11:1351–1363

Valbuena-Carabaña M, Gil L (2014) Efectos de la gestión selvícola pasada y presente sobre la diversidad genética actual y futura de *Quercus pyrenaica* Willd. en Sierra Nevada. Ecosistemas 23:90–98

Valbuena-Carabaña M, Heredia UL, Fuentes-Utrilla P, González-Doncel I, Gil L (2010) Historical and recent changes in the Spanish forests: a socio-economic process. Rev Palaeobot Palynol 162:492–506

Zamora R, Matías L (2014) Seed dispersers, seed predators, and browsers act synergistically as biotic filters in a mosaic landscape. PLoS One 9(9):e107385

Zamora R, Barea-Azcón JM (2015) Long-term changes in mountain passerine bird communities in the Sierra Nevada (Southern Spain): a 30-year case study. Ardeola 62:3–18

Zamora R, Gómez JM, Hódar JA, Castro J, García D (2001) Effect of browsing by ungulates on sapling growth of Scots pine in a Mediterranean environment: consequences for forest regeneration. For Ecol Manage 144:33–42

Zamora R, García-Fayos P, Gómez-Aparicio L (2004) Las interacciones planta-planta y planta-animal en el contexto de la sucesión ecológica. In: Valladares F (ed) Ecología del bosque Mediterráneo en un mundo cambiante. Editorial Parques Nacionales, Madrid, ES, pp 371–393

Zamora R, Hódar JA, Matías L, Mendoza I (2010) Positive adjacency effects mediated by seed disperser birds in pine plantations. Ecol Appl 20:1053–1060

Zamora R, Pérez-Luque AJ, Bonet FJ (2017) Monitoring global change in high mountains. In: Catalan J, Ninot JM, Aniz MM (eds) High mountain conservation in a changing world. Springer.

# Restoration of Mediterranean Forest Ecosystems After Major Disturbances: The Lanjarón Post-fire Experiment Over 15 Years of Succession

Alexandro B. Leverkus and Jorge Castro

**Abstract**

Sierra Nevada faces the stark challenge of preserving and restoring biodiversity and the provision of ecosystem services in the face of ongoing shifts in natural disturbance regimes. As wildfires become more frequent, severe and widespread under human land-use changes and climate change, there is a need to understand the mechanisms that promote ecological resilience after the fire and the effects of management on such mechanisms. After the 2005 Lanjarón fire in Sierra Nevada, we established an experiment in a burnt pine stand to assess how the ecosystem responded to three management schemes: (a) salvage logging, a common strategy consisting in the felling and extraction of burnt trees; (b) partial cut, where 90% of trees were felled and cut in pieces but left scattered on the ground; and (c) non-intervention, where no action was taken. For 15 years, we monitored how different components of biodiversity and ecosystem processes and services responded to the post-fire treatments, and assessed how management affected the mechanisms that drive natural regeneration. In this chapter, we describe the Lanjarón experiment and its key scientific contributions. We first describe the insights related to the role of dead wood in promoting regeneration. This includes processes such as seed dispersal and predation, fertilization through the gradual decomposition of dead wood, microclimatic amelioration, and herbivory. Second, we portray how the community of birds, vascular plants and soil insects responded to the post-fire treatments. Whereas diversity itself was affected by management, the key responses were related to community composition. And third, we address the functions of the ecosystem related to its capacity to provide benefits to human society. We mainly address regulating ecosystem services but also analyze some provisioning services, including their economic value. In each of these three sections, we end by providing a broader, global view on the effects of salvage logging as derived from reviews that have subsequently been made under international collaborations. Altogether, the Lanjarón experiment in Sierra Nevada constitutes a unique research infrastructure that has broadened our understanding of the role of dead wood in promoting ecological resilience and whose findings have contributed to the integration of knowledge about post-fire dynamics in an international context. Additionally, ongoing monitoring aims to fill the research gap of addressing the long-term effects of a critical post-disturbance management strategy in a world facing novel disturbance regimes.

**Keywords**

Wildfire • Salvage harvest • Compounded disturbances • Dead wood • Resilience • Ecosystem service

## 1 Introduction

Research in Sierra Nevada has helped reveal global challenges for ecosystems resulting from novel climatic conditions (as explained in previous chapters). Shifting temperature and rainfall regimes are producing an unprecedented need for local species to migrate or adapt (Zamora et al. 2015). Climate change also has consequences for environmental management programs, which must increasingly adapt to new conditions for instance by selecting species adapted to projected future conditions for revegetation (Hoegh-Guldberg et al. 2008; Leverkus et al. 2015). But while climate change alone is already expected to produce profound changes in biotic communities worldwide (Trisos et al. 2020), the coupled alterations in the frequency,

A. B. Leverkus (✉) · J. Castro
Department of Ecology, University of Granada, Granada, Spain
e-mail: leverkus@ugr.es

A. B. Leverkus · J. Castro
Andalusian Institute for Earth System Research (IISTA), University of Granada, Granada, Spain


R. Zamora and M. Oliva (eds.), *The Landscape of the Sierra Nevada*,
https://doi.org/10.1007/978-3-030-94219-9_14

severity, extent, and seasonality of disturbances such as wildfires and pests (Seidl et al. 2017) demand further measures for adaptation (Leverkus et al. 2021c). To succeed in managing ecosystems under changing climate and disturbance regimes, we need a better understanding of the mechanisms that favor the recovery of ecosystem functioning and biodiversity after disturbances.

Ecosystems are characterized by the occurrence of natural disturbances of particular types, frequencies, and severities —also termed disturbance regimes. In Mediterranean ecosystems, the disturbance regime is generally characterized by wildfires of moderate frequency and severity (Pausas et al. 2008). Many Mediterranean species have evolved adaptations that allow their regeneration after fire, such as the ability to survive under thick protecting bark, to resprout from surviving underground tissues, or to germinate following stimulation from the heat (Pausas and Keeley 2014). However, the capacity of ecosystems to recover after wildfire (which we here term *resilience*; Nimmo et al. 2015) may be compromised under novel disturbance conditions. Resilience is promoted by the legacies from the affected ecosystem that remain after the fire (Johnstone et al. 2016). These legacies include seed banks, patches with surviving vegetation and animals, resprouting plants, and soil fungi (Swanson et al. 2011; Johnstone et al. 2016). Other organic structures and biologically generated spatial patterns also constitute legacies that set the baseline for regeneration, including remaining organic soil, animal burrows, and—in the case of forests—large amounts of deadwood. However, this deadwood is generally removed from disturbed ecosystems through silvicultural management (Lindenmayer et al. 2008; Müller et al. 2019). The question thus arises whether such additional alteration of the disturbed ecosystem through management that removes and modifies disturbance legacies may compromise ecological resilience.

The practice of removing dead trees after disturbances is named salvage logging. Worldwide, it is conducted primarily for economic motivations (recovering some of the affected forest capital or paying for restoration activities; Müller et al. 2019; Castro 2021). Other frequent justifications include the perception that it helps restore the ecosystem, reduce fuel loads, minimize safety hazards from falling trees, mitigate pest outbreaks, and remove the ugliness of the deadwood (Müller et al. 2019; Castro 2021). Such motivations often lead to unusually large areas being subjected to tree harvesting after large disturbances (Radeloff et al. 2000; Gustafsson et al. 2019) and even protected areas being logged (as happened after the Lanjarón fire here described but also elsewhere, e.g. Schiermeier 2016; Leverkus et al. 2017. Until the early 2000s, there was scant scientific knowledge about the ecological implications of this already common practice (McIver and Starr 2000). A subsequent call for a thorough evaluation of the ecological impacts of salvage logging (Lindenmayer et al. 2004) was followed by intense controversy around the topic (e.g. Karr et al. 2004; DellaSala et al. 2006; Donato et al. 2006; Lindenmayer et al. 2017) and instigated an array of research projects worldwide. Recent advances have highlighted the potential for unforeseen, interacting effects among the natural disturbance and the subsequent disturbance by logging (Buma and Wessman 2011; Leverkus et al. 2018a; Lindenmayer and Sato 2018), and subsequent reviews have aimed to provide broader views on the topic (e.g. Royo et al. 2016; Leverkus et al. 2018b; Thorn et al. 2018).

Here, we explain how the post-fire experiment after the 2005 Lanjarón fire, which burned more than 3000 ha of pine stands and shrublands in Sierra Nevada, has contributed to our understanding of the role of deadwood in promoting ecosystem resilience and of the ecological impacts of salvage logging. The experimental setup, with three post-fire management treatments replicated within and across four blocks at different elevations, constitutes one of the most statistically sound experimental designs on post-fire management (Leverkus et al. 2018b) and has produced a great amount of scientific outputs. These include 3 Ph.D. theses, more than 20 papers in scientific journals, several outreach publications and management recommendations, and strong international networks. We also explain how the data and insights from this experiment have subsequently been combined with those from other experiments to produce broader insights into the drivers of the effects of salvage logging on biodiversity and ecosystem services.

## 2 The Lanjarón Post-fire Experiment

From the 22nd to the 24th of September 2005, the Lanjarón fire burned 3425 ha of shrublands, grasslands and pine afforestations in the Sierra Nevada Natural and National Park. The burnt pines were 35–45 years old at the time of the fire. The climate in the area is Mediterranean, with hot, dry summers and wet, mild winters (Table 1). The area surrounding the burnt site was dominated by shrublands.

The local forest service planned the removal of the burnt wood across all the burnt area, and in cooperation with this administration, we implemented an experimental design to study the effect of post-fire salvage logging in pine forests on the ecosystem. The experiment was implemented in four blocks of approximately 25 ha each, located across an elevational gradient (Fig. 1, Table 1). The pine species in each block differed according to their ecological requirements along this elevation and moisture gradient. The cluster pine (*Pinus pinaster*) and black pine (*P. nigra*) dominated in Block 1, black pine in Block 2, and Scots pine (*P. sylvestris*) in Blocks 3 and 4. All three species are native in the region, although they were extensively planted in the area.

**Table 1** Key features of the experimental blocks

| | Block[f] | | |
|---|---|---|---|
| | 1 | 2 | 3 |
| UTM coordinates (x; y)[a] | 456070E−4089811N | 455449E −4091728 N | 457244E–4091551N |
| Block area (ha) | 17.7 | 23.9 | 31.7 |
| Plot area (ha)[b] | 2.0 ± 0.15 | 2.7 ± 0.18 | 3.5 + 0.30 |
| Elevation[*] | 1477 | 1698 | 2053 |
| Slope (%)[c] | 30.3 | 28.7 | 31.4 |
| Mean daily min. temp. (ºC)[d] | 6.8 ± 0.2 | 5.6 ± 0.2 | 3.4 ± 0.2 |
| Mean daily max. temp. (ºC)[d] | 17.1 ± 0.2 | 16.2 ± 0.2 | 13.4 ± 0.2 |
| Mean ann. precip. (mm)[d] | 501 ± 49 | 550 ± 40 | 630 ± 42 |
| Dominant species | *Pinus pinaster/P. nigra* | *Pinus nigra* | *Pinus sylvestris* |
| Tree density (trees/ha)[e] | 1477 ± 46 | 1064 ± 67 | 1051 ± 42 |
| Tree basal diameter (cm)[e] | 17.7 ± 0.2 | 18.3 ± 0.1 | 15.7 ± 0.1 |
| Tree diameter at 1.30 m (cm)[e] | 13.3 ± 0.2 | 14.5 ± 0.2 | 10.7 ± 0.2 |
| Tree height (m)[e] | 6.3 ± 0.1 | 6.6 ± 0.1 | 6.2 ± 0.1 |

[a] Coordinates and elevation measured at the centroid of each block (UTM zone 30 N, Datum: ED-50)
[b]There was no significant difference in plot area among treatments (Kruskal Wallis test; P > 0.05)
[c]Mean slope of the nine plots within each block
[d]Data obtained from interpolated maps of Sierra Nevada (1981–2010) created at Centro Andaluz de Medio Ambiente except precipitation in Block 1, which is an empirical value obtained at Block 1 (1988–2011)
[e]Measured after the fire. Density sampled in each plot by counting the trees in four randomly placed 25 × 25 m quadrats. Basal tree diameter was measured on 30 randomly chosen trees in these quadrats, thus 120 trees per plot
[f]Block 4 was at an elevation of 2200 m a.s.l. and had a different configuration: its surface was larger and it contained a single plot per treatment to have surfaces large enough to monitor $CO_2$ fluxes with eddy covariance towers (see Serrano-Ortiz et al. (2011) for details)

Within each block, we implemented three replicates (plots) of the following burnt-wood management procedures (*treatments* hereafter; Fig. 2) in a random spatial distribution: (1) nonintervention (NI), where all burnt trees were left standing; (2) partial cut plus lopping (PC), where 90% of burnt trees were felled and their main branches lopped off, but all the cut biomass was left on the ground; and (3) experimental salvage logging (SL), where trees were cut and the trunks cleaned of branches with chainsaws, trunks were manually piled (groups of 10–15) and the woody debris was chopped with a mechanical chopper. In the salvage logging plots, we planned to extract the trunks with a log forwarder, but this step was canceled by the forest service because of the difficulties of operating machinery within the spatial arrangement of the experiment. Given that the salvage logging treatment could not be carried out with the same intensity as in the area surrounding the experimental blocks, for some of the studies we also included three replicates per block (of about 3 ha each) in the extraction matrix surrounding each of the experimental blocks (matrix salvage logging, MSL; Castro et al. 2010). The treatments were established between 21 April and 28 June 2006. The four blocks were located on siliceous soils, coming from mica schists of the Nevado-Filábride complex (see Table 1 for details, and chapter "Geological Setting of Sierra Nevada"). Within this experimental setup, we measured the effect of post-fire management on different aspects of the regeneration and functioning of the ecosystem, including: woody species regeneration, plant-animal interactions (seed dispersal, seed predation and herbivory), diversity of plants and birds, wood decomposition and its effect on soil properties, soil respiration, carbon exchange with the atmosphere, associated ecosystem services, and the economic and restoration implications of post-fire logging.

The Lanjarón experiment was among the first where the effect of post-fire salvage logging was assessed in a true experimental design that included random allocation of treatments to spatial units and a size large enough to study processes at ecosystem scale (see chapter "Managing the Uniqueness of Sierra Nevada Ecosystems Under Global Change: The Value of in situ Scientific Research"). The absence of experimental studies has in fact been a major constraint to obtaining robust conclusions on the impacts of post-fire salvage logging, so the design established in the Sierra Nevada was a milestone in this regard.

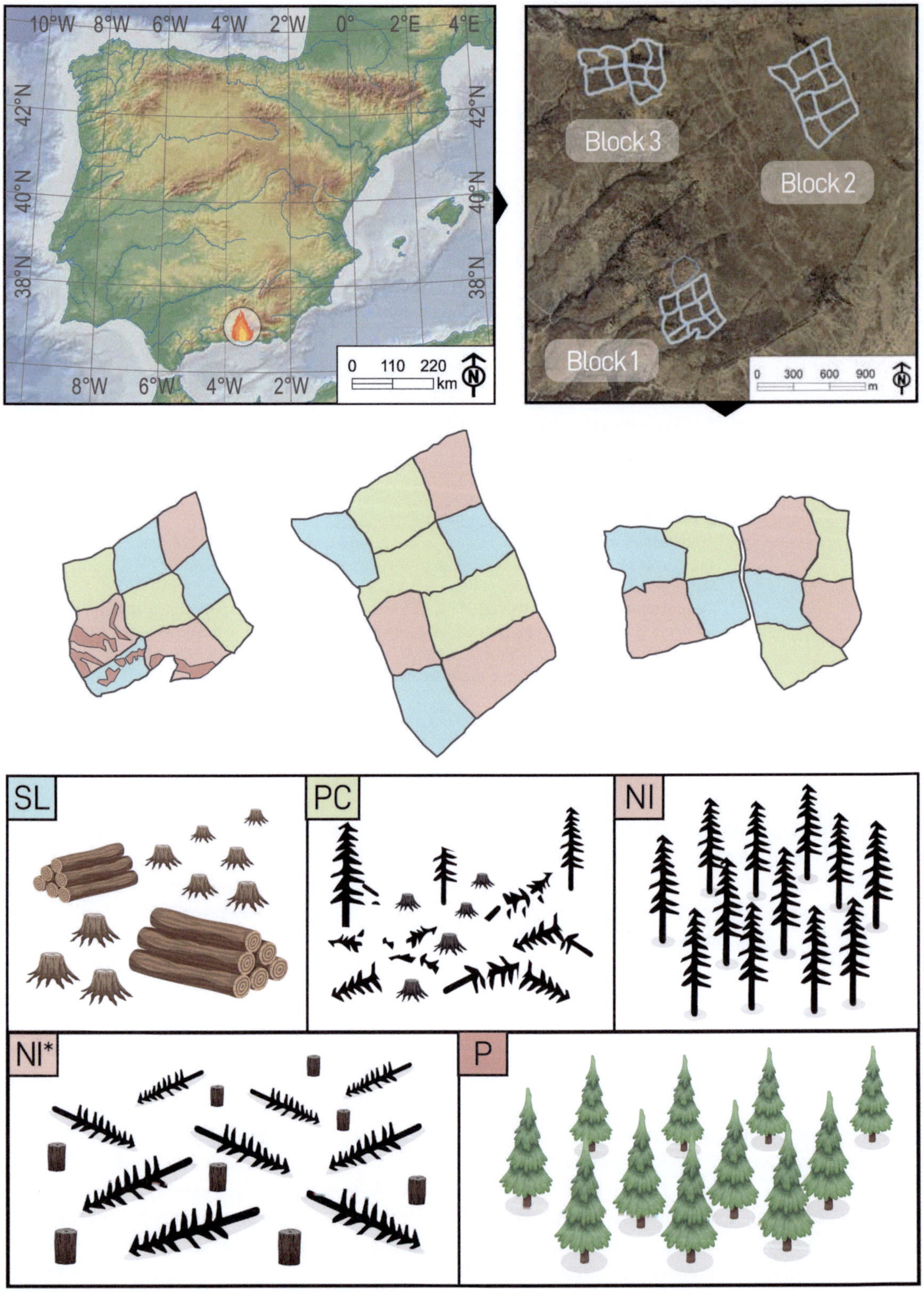


**Fig. 1** Location and set-up of the Lanjarón experiment. Top-left, location of the Lanjarón fire in the Iberian Peninsula; top-right, ortho-rectified photo showing the three experimental blocks at different elevations; middle, experimental design with the allocation of post-fire treatments to three plots per block; bottom, diagram of the experimental treatments. In the top-right map, the blue polygon above Block 1 surrounds a population of unburnt holm oaks from which seed dispersal was monitored. PC = Partial Cut, SL = Salvage Logging, NI = Non--Intervention, NI* = Non-Intervention treatment after 6 years, by when all the standing dead trees had naturally collapsed; P = patches of surviving pines in Block 1 in which demographic dynamics of oaks after dispersal were monitored. Block 4, which had one replicate per treatment and was located at the greatest elevation, is not represented here

**Fig. 2** Photo of post-fire treatments shortly after their implementation in early 2006. Foreground: Partial Cut treatment, where 90% of burnt trees were felled, cut in pieces, and spread on the ground. Middle: Salvage Logging, where all trees were felled, branches were masticated, and the trunks were piled. Back: Non-Intervention treatment, where burnt trees were left standing

## 3 The Role of Burnt Wood as a Disturbance Legacy: Mechanisms that Promote Resilience

From an ecological standpoint, extracting burnt wood implies the removal of large amounts of disturbance legacies—i.e. the organisms, organic materials and biologically derived patterns that persist after disturbance and set the baseline for regeneration (Franklin et al. 2000; Johnstone et al. 2016). As a first logical consideration, extracting burnt wood implies the removal of an important pool of nutrients accumulated over the lifetime of trees. At the Lanjarón site, we found that the macro- and micronutrient content in burnt pine trunks was similar to that in live specimens (Marañón-Jiménez et al. 2013a). Leaves and fine twigs are usually consumed during wildfires and their nutrients are released to the atmosphere or to the ground as ashes. However, coarse branches and trunks are isolated from the heat, and a considerable pool of nutrients may remain in the wood. These nutrients are protected from sudden losses from runoff or wind after the fire and are gradually released into the soil during decomposition, thereby enhancing soil fertility during succession. For instance, the nutrient pool remaining in the burnt wood four years after the fire was 2–9 times higher than the pool in the first 10 cm of the soil for elements such as Na, Mn, Fe, Zn or Cu (Marañón-Jiménez and Castro 2013), and microbiological parameters such as dissolved organic carbon, dissolved organic nitrogen, microbial biomass and microbial N and P content also benefited from the presence of deadwood. At a greater time scale, the content of K and P—which are particularly limiting at the study site—were still higher where the burnt wood was left in situ 10.5 years after the fire (Juan-Ovejero et al. 2021).

Higher soil nutrient pools were transferred to pine (*Pinus pinaster* Ait.) saplings regenerating from seed (Fig. 3a). These grew more in the treatments with abundant decomposing wood while keeping a similar nutrient concentration as in salvaged areas (Marañón-Jiménez et al. 2013b), suggesting that a higher nutrient availability led to a higher total nutrient pool in the plants. The reproductive status of the pines after six years was also improved by the presence of deadwood, and the analysis of stable isotopes pointed to greater water-use efficiency in pines growing in the presence of burnt wood due to enhanced nutritional status (Marañón-Jiménez et al. 2013b). Long-lasting effects reported in recent studies (Bowd et al. 2019) reinforce the role of burnt wood as a biological legacy that improves soil fertility.

The burnt wood may also act as a nurse object. Mediterranean ecosystems are characterized by an abundance of facilitative interactions among plants (Gómez-Aparicio et al. 2004; Siles et al. 2010), which are driven by the mitigation of abiotic stress as seedlings grow sheltered under other plants, especially shrubs. Facilitation in Mediterranean-type ecosystems arises primarily from the reduction of high radiation intensity by the canopy of the nurse plant, which improves water status in the beneficiary plants by reducing temperature and wind speed and increasing soil moisture and air–water vapour content. However, these positive aspects are at least partly counterbalanced by the impact of competition for resources between the nurse plant and the facilitated seedling (Maestre et al. 2003; Gómez-Aparicio et al. 2005;

**Fig. 3** **a** *Pinus pinaster* seedlings regenerating under the shelter of fallen burnt trees; **b** naturally regenerating plant community in the non-intervention treatment; contrast this with **c** plant community in the salvage logging treatment; **d** extraction of soil cores for macroarthropod sampling

Rey Benayas et al. 2005). The burnt wood can act as a nurse object in the sense that it reduces water stress but without producing competition (Castro et al. 2011; Ginzburg and Steinberger 2012; Marcolin et al. 2019; Fig. 2b). This constitutes a mechanism through which burnt wood may improve post-fire regeneration (Castro et al. 2011; Marzano et al. 2013; Taboada et al. 2018; Urretavizcaya and Defosse 2019). Moreover, as mentioned above, the wood provides nutrients during decomposition and it can retain moisture, organic matter, and fine sediments by acting as a barrier against runoff. Burnt trees also reduced wind speed at the study site (Serrano-Ortiz et al. 2011), and acted as a physical barrier against herbivores (Castro 2013). Thus, the burnt wood acted as a nurse object that provided better conditions for two critical factors that limit the regeneration of woody species in Mediterranean-type environments: summer drought and herbivore pressure.

The presence of burnt wood can also favour mutualistic interactions for the natural regeneration of the forest after fire. Many frugivorous birds use standing burnt trees, felled trees or piles of branches as perches (e.g. Castro et al. 2010; Rost et al. 2010). This may increase the entry of tree and shrub seeds of intermediate stages of succession—many of them producing fleshy fruits—with respect to areas where burnt wood is removed (e.g. Rost et al. 2010; Cavallero et al. 2013). A relevant case of mutualism for post-fire regeneration is the dispersal of acorns by the Eurasian jay (*Garrulus glandarius*). Jays disperse acorns in temperate forests of the Holarctic, as they scatterer-hoard seeds across the landscape for later consumption. A fraction of the acorns remains unrecovered, providing the opportunity for seed germination and tree recruitment (Pesendorfer et al. 2016). The Eurasian jay dwells in relatively dense forests and avoids open landscapes; pine forests, including relatively dense plantations, are among the habitats most used by jays to hide acorns (Mosandl and Kleinert 1998; Gómez 2003). In the Lanjarón area we documented that jays still used the burnt pine forest to cache acorns as long as the trees remained standing (Castro et al. 2012). Thus, standing dead trees constituted a crucial structural element for the colonisation of the burnt area by oaks (Leverkus et al. 2016), which in turn represents an ecosystem service with quantifiable economic value (Leverkus and Castro 2017).

The plethora of mechanisms through which dead wood may improve regeneration suggest that a major ecological impact of salvage logging may be its potential to affect forest regeneration and thereby compromise resilience (Van Nieuwstadt et al. 2001; Noss and Lindenmayer 2006; Donato et al. 2006). To assess this, we conducted a meta-analysis on regeneration data from Lanjarón and other European and North American studies after wildfires, windthrows, and insect outbreaks. This study found that there is no generalized response of regeneration to salvage logging across all tree species, and that initial impacts of salvage logging on tree regeneration density tend to decrease over time (Leverkus et al. 2021b). However, the lack of overall impacts was accompanied by the finding that more related tree species tend to respond more similarly to salvage logging than more distantly related species, and that the measured effects are quite specific to study sites (Leverkus et al. 2021b). This highlights the importance of understanding the mechanisms through which salvage logging may influence tree natural regeneration under local set of circumstances and species, adding value to detailed studies such as those described in this section for the development of local mitigation strategies.

## 4 Biodiversity Response to Post-fire Management

Salvage logging can affect biodiversity through a range of mechanisms. These include the direct destruction of organisms during working operations, the removal of larvae and propagules in the dead trunks, the elimination of substrate and food in the form of dead wood, the change in structural properties of the habitat, and a reduction in the variety of niches (Lindenmayer et al. 2008; Thorn et al. 2015, 2020b). Still, it can be expected that some species also benefit from an open habitat structure (e.g. Leverkus et al. 2013; Hagge et al. 2019), so the overall impact on biodiversity ultimately results from the balance between winners and losers.

In Lanjarón, biodiversity studies have addressed plant, bird, and soil macroarthropod communities. Vascular plants, sampled two years after the fire, showed a reduction in species richness, Shannon diversity and ground cover after salvage logging (Fig. 3b, c; Leverkus et al. 2014). But the composition of communities also changed. The abundance of species that regenerate through seeds was most negatively affected, likely because by the time salvage logging was conducted those species were already emerging; trampling by people and machinery therefore depleted the seedling bank and compromised regeneration. Tree and shrub species were also negatively affected, which could suggest a slowing down of succession (Leverkus et al. 2014). Re-sampling of vegetation is planned ~17 years after the fire to assess whether these impacts are long-lasting.

Bird communities (Castro et al. 2010), sampled in two years after the fire, showed a higher diversity and abundance in the treatments where the wood was left than in the salvage logging treatment. Moreover, the composition of the community changed depending on the treatment. Species that typically inhabit forests were more abundant in the non-intervention treatment, species from open landscapes and shrublands were most abundant in the salvage-logging treatment, and a mixed species composition was observed in the treatment where 90% of trees were cut and felled (Castro et al. 2010). Of particular relevance was that the non-intervention treatment had the highest abundance of species that disperse mid- and late-successional shrubs and trees that produce fleshy fruits, such as thrushes and Eurasian jays.

Finally, soils were sampled 10 years after the fire to assess the effect of remaining wood on the diversity and composition of soil macroarthropods (Fig. 3d). The salvage logging treatment resulted in an overall reduction in the abundance and richness of macroarthropod communities compared to the partial-cut treatment, in which decomposing dead trunks were still abundant (Molinas-González et al. 2019). This suggests that the legacy of dead wood also affects this hidden—albeit conspicuous—community in the soil, and potentially the ecological functions that it performs.

Besides the Lanjarón biodiversity data serving as an individual case study, international collaborations have allowed learning from this and other experiments worldwide about the extent to which the biodiversity responses to salvage logging are generalized, stochastic, or dependent on factors such as taxonomic group and local conditions. A meta-analysis involving 134 original species abundance datasets found that salvage logging after natural disturbances (including wildfire, insect outbreaks and storms) decreased the species richness of 8 out of 24 taxonomic groups that were assessed, and that significant alterations in community composition occurred in 40% of the groups (Thorn et al. 2018). This impact was mostly observed on saproxylic species; on the contrary, species groups typically associated with open habitats profited from salvage logging (Thorn et al. 2018). More detailed analysis on bird species assemblages found that dissimilarities in the occurrence of species between salvaged and unsalvaged areas are strongest, and most persistent, for rare species (Georgiev et al. 2020). This suggests that common and dominant species are comparatively less affected by salvage logging, and that assessments focusing only on these groups are likely to underestimate impacts on species of high conservation value. Finally, the global biodiversity dataset was used to ask what proportion of an area affected by a natural disturbance needs to be protected from salvage logging to preserve a certain proportion of the biodiversity specific to disturbed yet unsalvaged forest (Thorn et al. 2020a). Newly developed statistical tools helped estimate that 75% of disturbed areas are required to maintain 90% of their unique species assemblages, whereas retaining 50% of a disturbed area would protect 73% of the species unique to disturbed forests. The analyses allow estimating the proportion of disturbed areas that needs to remain unlogged to conserve different proportions of particular taxonomic groups, according to the needs of local managers (Thorn et al. 2020a).

## 5 Effect of Post-fire Management on Ecosystem Services

Ecosystems produce benefits to society, including the provision of materials, the regulation of biogeochemical cycles, and scientific and spiritual values—i.e. ecosystem services (MA 2005). Natural disturbances disrupt the functioning of ecosystems, including a temporary reduction in primary production, a change from subsurface water flow to surface runoff, and an alteration of aesthetic values, among others. Contrary to overall increases in biodiversity, disturbances may thus result in a reduction in the provision of ecosystem services (Thom and Seidl 2016). In fact, a global assessment —which included the case of Sierra Nevada—revealed that the reasons that motivate salvage logging in the first place include mitigating the expected reduction in key ecosystem services such as fire and pest control, timber production, and scenic value (Müller et al. 2019). However, key ecosystem services are not always reduced by natural disturbances (Beudert et al. 2015), and the capacity of post-disturbance management to mitigate potential ecosystem service losses is unclear (e.g. Donato et al. 2006). As explained in this section, the Lanjarón experiment, along with subsequent international reviews, has provided insights into how several classes of ecosystem services (Haines-Young and Potschin 2018) may be affected by salvage logging.

Provisioning ecosystem services are at the core of the salvage logging question. This is greatly because wildfires and other disturbances produce a sudden shift from a forest gradually increasing its economic value (as trees increase their volume through growth) to the value rapidly decreasing (as the remaining wood decays after tree death). Although disturbance-affected wood is usually of lesser quality and value, and wood prices commonly drop after large disturbances due to pulses in offer that saturate local markets (Peter and Bogdanski 2010), the economic incentive for salvaging the deadwood is still usually high—hence the term *salvage* logging (Lindenmayer et al. 2008). Another economic incentive involves the cost of subsequent restoration activities such as revegetation, which are assumed to be easier, and hence less expensive, to conduct. However, an assessment after the Lanjarón fire revealed that the high management costs associated with salvage logging operations were neither compensated by the benefits from selling the wood nor by the reduction in the cost of revegetation (Leverkus et al. 2012). A subsequent assessment showed that salvage logging produced additional losses through the reduction of natural regeneration, which increased the potential need for reforestation (Leverkus and Castro 2017). Of course, several stakeholders such as forestry operators do obtain economic gains from salvage logging, yet the economic assessment in Lanjarón challenged the overall view that salvage logging is generally profitable, particularly in unproductive Mediterranean forests or if considering the value of non-wood ecosystem services such as natural regeneration. This situation may change with ongoing developments such as the growing market for biomass pellets, which can make salvaged wood more profitable (Pons and Rost 2016). But also new forest growth models in more productive forests support that refraining from salvage logging does not necessarily produce the feared detrimental economic impacts (Knoke et al. 2021).

Many of the undesired impacts of salvage logging fall in the realm of regulating ecosystem services. As indicated in Sect. 4 above, a major effect of extracting deadwood is the elimination of food, substrate, and habitat for a plethora of deadwood-dependent species (Thorn et al. 2020b), which fall in the realm of lifecycle maintenance services (Haines-Young and Potschin 2018). But also the structural attributes of salvaged and non-salvaged ecosystems define habitat suitability; the Lanjarón experiment has demonstrated opposing outcomes of such changes for different guilds of mammals (Puerta-Piñero et al. 2010; Leverkus et al. 2013). Effects on habitat provision can carry over to other ecosystem services produced by animals such as seed dispersal by birds (Castro et al. 2012; Leverkus et al. 2016), and to disservices such as seed predation (Puerta-Piñero et al. 2010; Leverkus et al. 2013). Additionally, services of soil quality and formation are improved by dead wood, as described above (Marañón-Jiménez and Castro 2013). Ultimately, the effects of salvage logging on vegetation, microclimate and soil functioning may influence the ecosystem service of climate regulation at larger spatial scales. The measurement of soil respiration and carbon fluxes with eddy-covariance towers (Marañón-Jiménez et al. 2011; Serrano-Ortiz et al. 2011) helped reveal that the areas subjected to salvage logging acted as net carbon sources, whereas areas devoid of salvage logging acted as carbon sinks. The mechanisms underlying this counterintuitive result are likely related to the increase in primary production in nonintervention areas for the reasons described in Sect. 3, which overcompensated the larger carbon emissions from deadwood decomposition. These results, combined with those from several other studies in a global meta-analysis (Leverkus et al. 2020), showed that salvage logging produces a negative overall impact of moderate magnitude on regulating ecosystem services. Further, the meta-analysis produced the key insight that waiting before logging after natural disturbances can mitigate this negative effect—a conclusion that is unfortunately difficult to reconcile with the desire to quickly recover the deadwood (Leverkus et al. 2020).

One key concern about disturbed forests is the risk of further disturbances that may produce additional impacts on either the ecosystem or human populations and infrastructures. This is because natural disturbances can modify the likelihood of occurrence, extent and severity of further disturbances, thereby producing what is termed a linked disturbance (Buma 2015). As salvage logging produces additional alterations to the disturbed ecosystem, it can both mitigate some additional disturbances and trigger others (Leverkus et al. 2018a). In Lanjarón, an evident example of the latter is drought stress. The removal of the deadwood, which otherwise produced shade and retained moisture, increased the physiological stress suffered by plant seedlings, as measured by the ratios of stable isotopes in leaves (Marañón-Jiménez et al. 2013b). Also browsing disturbance was greater in salvaged areas, as these were devoid of the scattered trunks and branches that otherwise impeded access by large herbivores (Puerta-Piñero et al. 2010; Castro 2013). On the contrary, salvage logging removed the hazard from falling dead trees. However, we point out that the collapse of dead trees occurred mostly during stormy events and along a relatively narrow period of time (Molinas-González et al. 2017), so that the risk related to falling trees was generally small except for those events. In terms of larger-scale disturbances, salvage logging frequently aims to reduce ecosystem fuels as well as breeding grounds for pest insects (Müller et al. 2019). However, an additional meta-analysis combining data from multiple studies around the world revealed that the effectiveness of salvage logging in reducing fuel loads depends on how particular types of fuels respond to salvage logging over time (Leverkus et al. 2020). Salvage logging was found to be most effective in reducing the amount of large downed logs at intermediate to long time-frames, while it produced strong increases in small—and thus more flammable—fuels in the first years after logging (Leverkus et al. 2020). A broader international review highlighted that predicting the effects of post-disturbance management on subsequent disturbance risk requires evaluating the effects of management on multiple, not always evident ecosystem properties. For instance, predicting the temporal trajectories of fuel loads also requires understanding the response of vegetation to management (Leverkus et al. 2021a). The review highlights that managing disturbed ecosystems to avoid additional disturbances already implies imposing an additional disturbance (namely logging) and may also result in other, unforeseen disturbances (such as hydrologic disturbances and mass flows); it proposed a decision-making framework to avoid hasty decision-making after natural disturbances for the sake of reducing hazards (Leverkus et al. 2021a).

## 6 Conclusions

The Lanjarón experiment has played a fundamental role in understanding the implications of post-fire management for ecosystem resilience. Extracting the deadwood from a recently disturbed ecosystem produces an array of impacts that ultimately affect the capacity for natural regeneration, biodiversity, and the ecosystem services produced by burnt forests. As found in subsequent global reviews, some of these effects are generalized, yet others require detailed assessment of local species and conditions to fully understand—and properly manage—the drivers of management effects. New studies ought to transcend the plot scale to address the landscape-scale implications of different types of management, and integrate that information to obtain optimal landscape configurations that maximize the delivery of different ecosystem service and biodiversity elements. Also, long-term monitoring is needed to assess successional trajectories across succession under multiple combinations of natural and anthropogenic disturbances. The broad conclusion of the Lanjarón experiment is that post-disturbance management strategies that avoid superimposing additional disturbances may help promote resilience and thereby reduce the risk of ecosystem collapse in a world facing shifting disturbance regimes.

**Acknowledgements** We are grateful to numerous colleagues and students who helped conceptualize and conduct this experiment. The Direction of the Natural Park of Sierra Nevada and the Consejería de Medio Ambiente from the Andalusian Regional Government provided invaluable support. Key funding was provided by grants 10/2005 from Organismo Autónomo de Parques Nacionales, SUM2006-00010-00-00 from INIA, CGL2008-01671 from Ministerio de Ciencia e Innovación, P12-RNM-2705 from Junta de Andalucía, and an ongoing ERIC LifeWatch Mountain Thematic Center, supplemented by several other grants and personal fellowships that are acknowledged in individual publications.

## References


Beudert B, Bässler C, Thorn S et al (2015) Bark beetles increase biodiversity while maintaining drinking water quality. Conserv Lett 8:272–281

Bowd EJ, Banks SC, Strong CL, Lindenmayer DB (2019) Long-term impacts of wildfire and logging on forest soils. Nat Geosci 12:113–118

Buma B (2015) Disturbance interactions: characterization, prediction, and the potential for cascading effects. Ecosphere 6:Art70

Buma B, Wessman CA (2011) Disturbance interactions can impact resilience mechanisms of forests. Ecosphere 2:art64

Castro J (2021) Post-fire restoration of Mediterranean pine forests. In: Ne'eman G OY (ed) Pines and their mixed forest ecosystems in the Mediterranean Basin. Springer, p In press

Castro J (2013) Postfire burnt-wood management affects plant damage by ungulate herbivores. Int J for Res 2013:965461

Castro J, Allen CD, Molina-Morales M et al (2011) Salvage logging versus the use of burnt wood as a nurse object to promote post-fire tree seedling establishment. Restor Ecol 19:537–544

Castro J, Moreno-Rueda G, Hódar JA (2010) Experimental test of postfire management in pine forests: Impact of salvage logging versus partial cutting and nonintervention on bird-species assemblages. Conserv Biol 24:810–819

Castro J, Puerta-Piñero C, Leverkus AB et al (2012) Post-fire salvage logging alters a key plant-animal interaction for forest regeneration. Ecosphere 3:art90. 1890/ES12-00089.1

Cavallero L, Raffaele E, Aizen MA (2013) Birds as mediators of passive restoration during early post-fire recovery. Biol Conserv 158:342–350

DellaSala DA, Karr JR, Schoennagel T et al (2006) Post-fire logging debate ignores many issues. Science (80-) 314:51–52

Donato DC, Fontaine JB, Campbell JL et al (2006) Post-wildfire logging hinders regeneration and increases fire risk. Science (80-) 311:352

Franklin JF, Lindenmayer D, MacMahon JA et al (2000) Threads of continuity. Conserv Pract 1:8–17

Georgiev KB, Chao A, Castro J et al (2020) Salvage logging changes the taxonomic, phylogenetic and functional successional trajectories of forest bird communities. J Appl Ecol 57:1103–1112

Ginzburg O, Steinberger Y (2012) Salvage logging versus natural regeneration post-fire practices in a forest: soil chemical and microbial aspects. Open J Ecol 2:29–37

Gómez-Aparicio L, Gómez JM, Zamora R, Boettinger JL (2005) Canopy vs soil effects of shrubs facilitating tree seedlings in Mediterranean montane ecosystems. J Veg Sci 16:191–198

Gómez-Aparicio L, Zamora R, Gómez JM et al (2004) Applying plant facilitation to forest restoration: a meta-analysis of the use of shrubs as nurse plants. Ecol Appl 14:1128–1138

Gómez JM (2003) Spatial patterns in long-distance dispersal of Quercus ilex acorns by jays in a hetereneous landscape. Ecography (Cop) 5:573–584

Gustafsson L, Berglind M, Granström A et al (2019) Rapid ecological response and intensified knowledge accumulation following a north European mega-fire. Scand J for Res 34:234–253

Hagge J, Müller J, Bässler C et al (2019) Deadwood retention in forests lowers short-term browsing pressure on silver fir saplings by overabundant deer. For Ecol Manage 451:117531

Haines-Young R, Potschin MB (2018) Common international classification of ecosystem services (CICES) V5.1 and guidance on the application of the revised structure. www.cices.eu

Hoegh-Guldberg O, Hughes L, McIntyre S et al (2008) Assisted colonization and rapid climate change. Science (80-) 321:345–346

Johnstone JF, Allen CD, Franklin JF et al (2016) Changing disturbance regimes, ecological memory, and forest resilience. Front Ecol Environ 14:369–378

Juan-Ovejero R, Molinas-González CR, Leverkus AB et al (2021) Decadal effect of salvage logging on the soil nutrient stocks of a wildfire-affected Mediterranean forest. For Ecol Manage 498:119570

Karr JR, Rhodes JJ, Minshall GW et al (2004) The effects of postfire salvage logging on aquatic ecosystems in the American West. Bioscience 54:1029

Knoke T, Gosling E, Thom D et al (2021) Economic losses from natural disturbances in Norway spruce forests—a quantification using Monte-Carlo simulations. Ecol Econ 185:107046

Leverkus AB, Buma B, Wagenbrenner J et al (2021a) Tamm review: does salvage logging mitigate subsequent forest disturbances? For Ecol Manage 481:118721

Leverkus AB, Gustafsson L, Lindenmayer DB et al (2020) Salvage logging effects on regulating ecosystem services and fuel loads. Front Ecol Environ 18:391–400

Leverkus AB, Lindenmayer DB, Thorn S, Gustafsson L (2018a) Salvage logging in the world's forests: interactions between natural disturbance and logging need recognition. Glob Ecol Biogeogr 27:1140–1154

Leverkus AB, Polo I, Baudoux C et al (2021b) Resilience impacts of a secondary disturbance: meta-analysis of salvage logging effects on tree regeneration. J Ecol in Press

Leverkus AB, Rey Benayas JM, Castro J et al (2016) Shifting demographic conflicts across recruitment cohorts in a dynamic post-disturbance landscape. Ecology 97:2628–2639

Leverkus AB, Rey Benayas JM, Castro J et al (2018b) Salvage logging effects on regulating and supporting ecosystem services—a systematic map. Can J for Res 48:983–1000

Leverkus AB, Thorn S, Gustafsson L et al (2021c) Environmental policies to cope with novel disturbance regimes—steps to address a world scientists' warning to humanity. Environ Res Lett 16:021003

Leverkus AB, Castro J (2017) An ecosystem services approach to the ecological effects of salvage logging: valuation of seed dispersal. Ecol Appl 27:1057–1063

Leverkus AB, Jaramillo-López PF, Brower LP et al (2017) Mexico's logging threatens butterflies. Science 358:1008

Leverkus AB, Lorite J, Navarro FB et al (2014) Post-fire salvage logging alters species composition and reduces cover, richness, and diversity in Mediterranean plant communities. J Environ Manage 133:323–331

Leverkus AB, Puerta-Piñero C, Guzmán-Álvarez JR et al (2012) Post-fire salvage logging increases restoration costs in a Mediterranean mountain ecosystem. New for 43:601–613

Leverkus AB, Castro J, Delgado-Capel MJ et al (2015) Restoring for the present or restoring for the future: enhanced performance of two sympatric oaks (Quercus ilex and Quercus pyrenaica) above the current forest limit. Restor Ecol 23:936–946

Leverkus AB, Castro J, Puerta-Piñero C, Rey Benayas JM (2013) Suitability of the management of habitat complexity, acorn burial depth, and a chemical repellent for post-fire reforestation of oaks. Ecol Eng 53:15–22

Lindenmayer D, Thorn S, Banks S (2017) Please do not disturb ecosystems further. Nat Ecol Evol 1:art31

Lindenmayer DB, Burton PJ, Franklin JF (2008) Salvage logging and its ecological consequences. Island Press, Washington, D.C.

Lindenmayer DB, Foster DR, Franklin JF et al (2004) Salvage harvesting policies after natural disturbance. Science 303:1303

Lindenmayer DB, Sato C (2018) Hidden collapse is driven by fire and logging in a socioecological forest ecosystem. Proc Natl Acad Sci 115:5181–5186

MA [Millennium Ecosystem Assessment] (2005) Ecosystems and human well-being: synthesis. Island Press, Washington, DC

Maestre FT, Bautista S, Cortina J (2003) Positive, negative, and net effects in grass-shrub interactions in Mediterranean semiarid grasslands. Ecology 84:3186–3197

Marañón-Jiménez S, Castro J (2013) Effect of decomposing post-fire coarse woody debris on soil fertility and nutrient availability in a Mediterranean ecosystem. Biogeochemistry 112:519–535

Marañón-Jiménez S, Castro J, Fernández-Ondoño E, Zamora R (2013a) Charred wood remaining after a wildfire as a reservoir of macro- and micronutrients in a Mediterranean pine forest. Int J Wildl Fire 22:681–695

Marañón-Jiménez S, Castro J, Kowalski AS et al (2011) Post-fire soil respiration in relation to burnt wood management in a Mediterranean mountain ecosystem. For Ecol Manage 261:1436–1447

Marañón-Jiménez S, Castro J, Querejeta JI et al (2013b) Post-fire wood management alters water stress, growth, and performance of pine regeneration in a Mediterranean ecosystem. For Ecol Manage 308:231–239

Marcolin E, Marzano R, Vitali A et al (2019) Post-fire management impact on natural forest regeneration through altered microsite conditions. Forests 10:1014

Marzano R, Garbarino M, Marcolin E et al (2013) Deadwood anisotropic facilitation on seedling establishment after a stand-replacing wildfire in Aosta Valley (NW Italy). Ecol Eng 51:117–122

McIver JD, Starr L (2000) Environmental effects of postfire logging: literature review and annotated bibliography. USDA For Serv Gen Tech Rpt PNW-GTR 72

Molinas-González CR, Castro JJ, González-Megías A, Leverkus AB (2019) Effects of post-fire deadwood management on soil macroarthropod communities. Forests 10:art1046

Molinas-González CR, Leverkus AB, Marañón-Jiménez S, Castro J (2017) Fall rate of burnt pines across an elevational gradient in a Mediterranean mountain. Eur J For Res 136

Mosandl R, Kleinert A (1998) Development of oaks (Quercus petraea (Matt.) Liebl.) emerged from bird-dispersed seeds under old-growth pine (Pinus silvestris L.) stands. For Ecol Manage 106:35–44

Müller J, Noss RF, Thorn S et al (2019) Increasing disturbance demands new policies to conserve intact forest. Conserv Lett 12: e12449

Nimmo DG, Mac Nally R, Cunningham SC et al (2015) Vive la résistance: reviving resistance for 21st century conservation. Trends Ecol Evol 30:516–523

Noss RF, Lindenmayer DB (2006) The ecological effects of salvage logging after natural disturbance. Conserv Biol 20:946–948

Pausas JG, Keeley JE (2014) Evolutionary ecology of resprouting and seeding in fire-prone ecosystems. New Phytol 204:55–65

Pausas JG, Llovet J, Anselm R, Vallejo R (2008) Are wildfires a disaster in the Mediterranean basin?—a review. Int J Wildl Fire 17:713–723

Pesendorfer MB, Sillett TS, Koenig WD, Morrison SA (2016) Scatter-hoarding corvids as seed dispersers for oaks and pines: a review of a widely distributed mutualism and its utility to habitat restoration. Condor 118:215–237

Peter B, Bogdanski B (2010) The economics of salvage harvesting and reforestation in British Columbia's mountain pine beetle-affected forests. Nat Res Canada, Canada

Pons P, Rost J (2016) The challenge of conserving biodiversity in harvested burned forests. Conserv Biol 31:226–228

Puerta-Piñero C, Sánchez-Miranda A, Leverkus A, Castro J (2010) Management of burnt wood after fire affects post-dispersal acorn predation. For Ecol Manage 260:345–352

Radeloff VC, Mladenoff DJ, Boyce MS (2000) Effects of interacting disturbances on landscape patterns: Budworm defoliation and salvage logging. Ecol Appl 10:233–247

Rey Benayas JM, Navarro J, Espigares T et al (2005) Effects of artificial shading and weed mowing in reforestation of Mediterranean abandoned cropland with contrasting Quercus species. For Ecol Manage 212:302–314

Rost J, Clavero M, Bas JM, Pons P (2010) Building wood debris piles benefits avian seed dispersers in burned and logged Mediterranean pine forests. For Ecol Manage 260:79–86

Royo AA, Peterson CJ, Stanovick JS, Carson WP (2016) Evaluating the ecological impacts of salvage logging: can natural and anthropogenic disturbances promote coexistence? Ecology 97:1566–1582

Schiermeier Q (2016) Pristine forest at risk. Nature 530:393

Seidl R, Thom D, Kautz M et al (2017) Forest disturbances under climate change. Nat Clim Chang 7:395–402

Serrano-Ortiz P, Marañón-Jiménez S, Reverter BR et al (2011) Post-fire salvage logging reduces carbon sequestration in Mediterranean coniferous forest. For Ecol Manage 262:2287–2296

Siles G, Rey PJ, Alcántara JM (2010) Post-fire restoration of Mediterranean forests: testing assembly rules mediated by facilitation. Basic Appl Ecol 11:422–431

Swanson ME, Franklin JF, Beschta RL et al (2011) The forgotten stage of forest succession: early-successional ecosystems on forest sites. Front Ecol Environ 9:117–125

Taboada A, Fernández-García V, Marcos E, Calvo L (2018) Interactions between large high-severity fires and salvage logging on a short return interval reduce the regrowth of fire-prone serotinous forests. For Ecol Manage 414:54–63

Thom D, Seidl R (2016) Natural disturbance impacts on ecosystem services and biodiversity in temperate and boreal forests. Biol Rev 91:760–781

Thorn S, Bässler C, Bernhardt-Römermann M et al (2015) Changes in the dominant assembly mechanism drives species loss caused by declining resources. Ecol Lett 19:109–215

Thorn S, Bässler C, Brandl R et al (2018) Impacts of salvage logging on biodiversity—a meta-analysis. J Appl Ecol 55:279–289

Thorn S, Chao A, Georgiev KB et al (2020a) Estimating retention benchmarks for salvage logging to protect biodiversity. Nat Commun 11:4762

Thorn S, Seibold S, Leverkus AB et al (2020b) The living dead–acknowledging life after tree death to stop forest degradation. Front Ecol Environ 18:505–512

Trisos CH, Merow C, Pigot AL (2020) The projected timing of abrupt ecological disruption from climate change. Nature 580:496–501

Urretavizcaya MF, Defosse GE (2019) Restoration of burned and post-fire logged Austrocedrus chilensis stands in Patagonia: effects

of competition and environmental conditions on seedling survival and growth. Int J Wildl Fire 28:365–376
Van Nieuwstadt MGL, Sheil D, Kartawinata K (2001) The ecological consequences of logging in the burned forests of East Kalimantan, Indonesia. Conserv Biol 15:1183–1186
Zamora RJ, Pérez Luque AJ, Bonet García FJ et al (eds) (2015) La huella del Cambio Global en Sierra Nevada: Retos para la conservación. Consejería de Medio Ambiente y Ordenación del Territorio, Junta de Andalucía

**Author**: Luis Ordoñez

# Biotic Responses to Recent Changes: Aquatic Ecosystems as Laboratories of Global Change

**Author**: Andrés Ureña

# Aquatic Animal Communities of Watercourses from Sierra Nevada

M. J. López-Rodríguez, J. Alba-Tercedor, M. Galiana-García, J. E. Larios-López, C. E. Sainz-Cantero Caparrós, J. M. Tierno de Figueroa, M. Villar-Argaiz, and C. Zamora-Muñoz

**Abstract**

The geographical situation of Sierra Nevada and its great altitudinal gradient generate particular environmental conditions in the watercourses that flow through the massif, which determines the composition and structure of the animal community inhabiting them. Regarding invertebrates, macroinvertebrates have been the organisms more widely studied in this massif and, within them, four orders of insects: Ephemeroptera, Plecoptera, Trichoptera and Coleoptera (EPTC). In aquatic vertebrates, most studies have focused on brown trout, the most characteristic high mountain fish species of this biogeographic area. A total of 189 taxa of EPTC have been recorded up to now in the massif: 36 taxa of Ephemeroptera, 24 of Plecoptera, 41 of Trichoptera and 88 of lotic aquatic Coleoptera, showing a great diversity from the biogeographical point of view, but with only a few of them endemic to Sierra Nevada. All these animals are subject to several threats in the massif, many of them related not only to climate change, but also to human-induced pollution and alterations, such as dams, pollution from the ski resort, water diversion, or even diffuse pollution due to high stocking densities. Some species will be able to cope with changing conditions throughout particular adaptations, while others without those strategies will be more vulnerable and the first to disappear. These disturbances, together with the introduction of exotic species such as rainbow trout, also affect brown trout populations. At the community scale, few studies have accomplished the analysis of whole communities of Sierra Nevada watercourses. Most data come from the application of biological indexes to assess the ecological status of streams and rivers, though some investigations have focused on particular biocoenosis, such as those of Plecoptera or Trichoptera. All of them concluded that macroinvertebrate communities under particular climate change scenarios will probably reduce their taxa richness in comparison to the present, that generalist taxa will move upstream to higher altitude reaches, if possible, and that vulnerable taxa will reduce their distribution area. Despite all this knowledge, many gaps still remain to be fulfilled, some of them discussed in this chapter. In this sense, data coming from new research at different organization levels, from managers, and even from citizen science initiatives, will contribute to improving the knowledge and conservation measures to be developed in Sierra Nevada.

**Keywords**

Aquatic insects • Brown trout • Climate change • Macroinvertebrates • River • Stream

## 1 The Lotic Ecosystems of Sierra Nevada

The particularities of the mountain system of Sierra Nevada have an important impact on the watercourses that run along its slopes. Geologically, it is part of the Betic System and the second European mountain system in altitude, after the Alps. Due to its meridional situation within Europe, it has a representation of five out of six of the bioclimatic zones present in the Mediterranean region (from the thermo-mediterranean to the crioro-mediterranean). This massif has an East–West arrangement, with a north slope draining to the Guadalquivir

M. J. López-Rodríguez (✉) · M. Villar-Argaiz
Department of Ecology, University of Granada, Granada, Spain
e-mail: manujlr@ugr.es

J. Alba-Tercedor · C. E. Sainz-Cantero Caparrós · J. M. Tierno de Figueroa · C. Zamora-Muñoz
Department of Zoology, University of Granada, Granada, Spain

M. Galiana-García
Agencia de Medio Ambiente y Agua de Andalucía, Granada, Spain

J. E. Larios-López
Dnota Medio Ambiente S.L., Granada, Spain


R. Zamora and M. Oliva (eds.), *The Landscape of the Sierra Nevada*,
https://doi.org/10.1007/978-3-030-94219-9_15

basin (which drains into the Atlantic Ocean) and a south face draining to several of the Mediterranean river basins of southern Andalusia (which are small and steep independent basins draining into the Mediterranean Sea). Nonetheless, in the westernmost part of the mountain range, watercourses have a more radial arrangement.

The core of the mountain is siliceous, while the outermost parts are calcareous (see chapter "Geological Setting of Sierra Nevada"). Due to this geologic composition, there is more superficial water than subterranean. This has important implications, not only in the hydrological and physico-chemical characteristics of the water but also in the community of organisms that inhabits each kind of stream, as we will see below. Administratively, the water courses are managed by two administrative river basin management agents: the *Demarcación Hidrográfica del Guadalquivir* and the *Demarcación Hidrográfica de las Cuencas Mediterráneas Andaluzas* (Fig. 1).

Because of the high altitudinal gradient of the river basins that originate in Sierra Nevada, many of them with altitudes ranging from above 2500 m a.s.l. to the sea level, drastic changes occur, in some cases, in less than 50 km of the river course. This is the case of the Adra River basin, for instance, where a clear strong zonation from an epirhithron to an epipotamon has been described (Alba-Tercedor et al. 1986b). Thus, the aquatic ecosystems of Sierra Nevada have characteristics more frequently found in streams and rivers from higher latitudes. Such a great altitudinal gradient of the massif and its particular situation in the Iberian Peninsula have contributed to the special characteristics these watercourses have in a region, the Mediterranean, where other kinds of streams should be more frequent.

Sierra Nevada watercourses change their characteristics rapidly with altitude. In the core of the massif, at high altitude, where the main waterways arise, the forest is absent and the vegetation is scarce, mainly in the form of hydrophilic communities such as high-mountain grasslands and peat bogs (Salazar Mendías and Valle Tendero 2019). Disturbances in these environments are locally concentrated, mainly related to cattle, tourism, both aestival and winter (the latter due to the ski resort), and water diversion (*careos*, ditches traditionally used in high mountain streams to slow down runoff and recharge the aquifers). The Sierra Nevada streams feed on snow and precipitation, so their regime is mostly nivo-pluvial (mainly in the upper reaches, above 2000 m). On the other side, the mid and low reaches of these streams flow throughout partially disturbed riparian forests. In the siliceous part of the massif, riparian vegetation is mainly composed of alder, willow and ash forests, while in the calcareous ring of the massif, elm, poplar, and willow groves predominate (Molero Mesa et al. 1992). The fluvial regime of mid to low altitude reaches changes to pluvio-nival (Pulido 1980; Alba-Tercedor and Jiménez-Millán 1987). Overall, the effect of the snow on the streams and rivers of Sierra Nevada is more important in the westernmost part of the mountain (Castillo Martín 2001). This fact conditions the period in which the maximum discharge of water is registered on each side. Thus, western streams have discharge peaks from April to July, and minimums mainly in September and August, while eastern

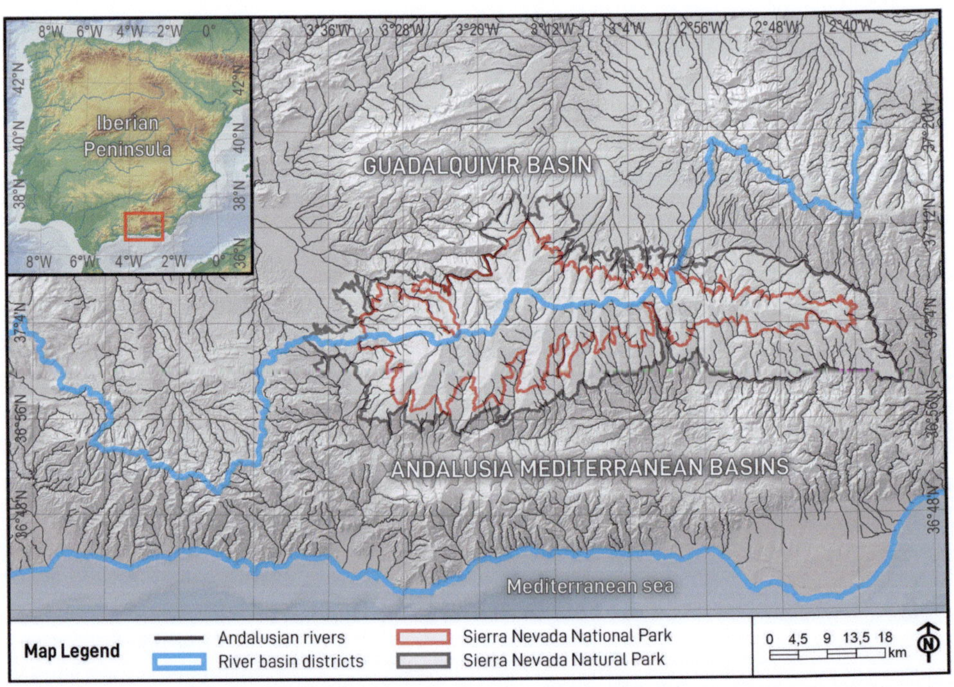


**Fig. 1** Map of Sierra Nevada showing the main watercourses that drain through it and the river basins they belong to

streams, where thaw occurs before, have peaks in February and minimums in September (Castillo Martín 2001). In general, western streams have higher discharges than eastern ones, and in the latter losses due to evapotranspiration are higher (Castillo Martín 2001).

All these environmental characteristics determine the communities of fluvial organisms found in Sierra Nevada, as well as the set of interactions that take place among the organisms that compose them. The following sections of this chapter will be devoted to them.

## 2 Who Lives There?

### 2.1 State of the Art: Current Knowledge of the Best-Studied Fluvial Macroinvertebrate Groups

Macroinvertebrates have been the focus of most studies on the invertebrate fluvial fauna of Sierra Nevada. Several M. Sc. or Ph.D. theses have been developed, totally or partially, during the past decades in the massif, which have opened different lines of research at different levels of biological organization. Within macroinvertebrates, those better studied and with a longer history of research in Sierra Nevada are mayflies (Ephemeroptera), stoneflies (Plecoptera), caddisflies (Trichoptera) and beetles (Coleoptera) (henceforth, EPTC; Fig. 2). In this sense, different authors carried out an in-deep review on the knowledge of these aquatic insect groups (in addition to other aquatic and terrestrial taxa) from Sierra Nevada (in Ruano et al. 2013). Within this review, Romero Martín and Alba-Tercedor (2013) compiled and assessed the state of knowledge (until 2010, not included) of the invertebrates from streams and rivers of Sierra Nevada after an intensive analysis of more than 300 references, including "grey literature". As pointed out by these authors, one-off records of lotic invertebrates date from the mid of the nineteenth century, but it was at the end of the 70s of the twentieth century when surveys and studies started to rise significantly. These coincide with the development of some of the M.Sc. and Ph.D. theses previously mentioned and the scientific papers derived from them, mainly on the taxonomy and biology of mayflies, stoneflies, beetles and caddisflies. Nonetheless, some interesting data come also from more general studies concerning the ecological status of water bodies and biomonitoring (e.g., Zamora-Muñoz and Alba-Tercedor 1992; Alba-Tercedor 1996; Jáimez-Cuéllar 2004; Bello and Alba-Tercedor 2005). After the intensive revision by Romero Martín and Alba-Tercedor (2013), it was concluded that despite a total of 617 species of aquatic macroinvertebrates having been listed, many taxa still remain to be recorded, because many records are based on identifications to family or genus level. Especially, there is an important lack of knowledge on the macroinvertebrate fauna at high mountain reaches (above 1700 m), and the easternmost basins are the poorest studied (the Guadiana Menor River and the Andarax River).

From then to now, some other studies have been carried out in Sierra Nevada, even expanding in some cases the study of macroinvertebrates to new fields never analyzed in these mountains before. Particularly regarding the best studied groups (i.e., EPTC), these studies have focused on different aspects of the organisms, such as behaviour (Tierno de Figueroa et al. 2009b; Tierno de Figueroa et al. 2014), life cycles and nymphal trophic ecology (López-Rodríguez et al. 2012; Sáinz-Bariáin and Zamora-Muñoz 2012; Tierno de Figueroa et al. 2019), taxonomy (Múrria et al. 2010; Olah et al. 2014; Sáinz-Bariáin and Zamora-Muñoz 2015), molecular and physiological aspects (Sanz et al. 2010, 2017; Finn et al. 2014; Boumans and Tierno de Figueroa 2016) and several other fields (Tierno de Figueroa et al. 2009a; Sáinz-Bariáin et al. 2016b; Garcia-Raventós et al. 2017; Múrria et al. 2020; Villar-Argaiz et al. 2020). Moreover, organisms collected in Sierra Nevada for other purposes have been also employed in more general studies to respond to scientific questions in a wider geographical area (Bonada et al. 2009; Hering et al. 2009; Tierno de Figueroa et al. 2010; Luzón-Ortega et al. 2013; Zhou et al. 2016; Ferreira et al. 2020), and to develop the MEDPACS (MEDiterranean Prediction And Classification System). MEDPACS is a predictive approach to the analysis of the macroinvertebrate communities of the Mediterranean watercourses (Poquet et al. 2009), a model used as a basis to predict the distributional shifts of river macroinvertebrate communities under future global change scenarios in the Spanish Mediterranean area (Alba-Tercedor et al. 2017).

### 2.2 Biodiversity of EPTC

A total of 189 species and subspecies of EPTC inhabit the fluvial environments of Sierra Nevada. For mayflies, 36 taxa have been recorded in this massif, representing one-third of the mayflies recorded in the Iberian Peninsula, though it is known that still remains new species to be described, especially Heptageniidae (Alba-Tercedor 2013). Three mayflies species present in the massif are endemic to the Iberian Peninsula, and two of them are endemic to Sierra Nevada. Regarding stoneflies, there are 24 species recorded from the 146 species cited in the Iberian Peninsula (Tierno de Figueroa et al. 2018), two of them recently found and cited in the massif (Tierno de Figueroa et al. 2013b; Fajardo Merlo 2021). Seven stoneflies species present in the massif are endemic to the Iberian Peninsula, none of them endemic to Sierra Nevada. The caddisfly diversity in Sierra Nevada is poorer than in other mountain systems probably due to the

**Fig. 2** Pictures from Sierra Nevada showing: **a** a mayfly (author: J. Alba-Tercedor); **b** two stoneflies mating (author: J. R. Fernández Cardenete); **c** a caddisfly (author: J. J. Soler); **d** an aquatic beetle (author: I. Flores Arcas); and **e** brown trout (author: E. Sofos Naveros)

harsh climate and the dominance of siliceous composition in headwater streams, which limit the distribution of caddisfly species specialized in non-siliceous ecotype. There are 41 species of caddisflies reliably cited; nine species are Iberian endemism, two of them endemic from Sierra Nevada (Sáinz-Bariáin et al. 2013). Finally, of the total number of aquatic Coleoptera species catalogued in Sierra Nevada (Sáinz-Cantero 2013), 88 of them inhabit the running waters of the massif, which constitute an important component of the macroinvertebrate communities present in this type of epicontinental aquatic environment. Of them, 11 are Iberian endemisms and one is endemic to Sierra Nevada. Most aquatic Coleoptera associated with flowing waters are typically rheophilic, among which predominates representatives of the families Hydraenidae (20 spp.), followed by Elmidae and Dytiscidae (with 10 and 8 spp, respectively). However, slightly more than 40% of these species can also be found in other freshwater environments of lentic nature (Millán et al. 2014) such as ponds, small lagoons, springs or fountains. Among this latter group of species, representatives of the families Dytiscidae (17 spp.) and Hydrophilidae (12 spp.) clearly predominate. The presence of species both in lentic and lotic environments in Sierra Nevada also occurs in Ephemeroptera (11 species) and Trichoptera (17 species).

## 2.3 Biogeographical Overview of EPTC

From a biogeographical point of view, in Fig. 3 are represented the six main elements detected in the fauna of these four aquatic insect orders in Sierra Nevada, which are based on the biogeographical categories proposed by Ribera et al. (1998) and Vigna Taglianti et al. (1999). Species with a wide geographical distribution are slightly predominant in Sierra Nevada (36.5%). These are mainly elements widely distributed in the Palaearctic region together with some representatives of Coleoptera and Ephemeroptera known from other biogeographic regions such as the Nearctic or Afrotropical. Most of them fit the Trans-Iberian chorotype defined by Ribera et al. (1998) and in general, their dispersal ability (among other biological characteristics) seems to be the explanatory factor on the basis of the available data. However, a similar proportion (31.8%), corresponds to species of European distribution according to the criteria of Vigna Taglianti et al. (1999) (Northern chorotype *sensu* Ribera et al. 1998), most of which find their geographical distribution limit in the south of the Iberian Peninsula, which clearly exceeds the elements with a Mediterranean distribution (13.8%) that are generally widespread in the southern sector of the Iberian Peninsula, despite the latitudinal location of Sierra Nevada. The orographic characteristics of the area considered, with a wide altitudinal and consequently thermal gradient, could explain this discordance in terms of the ecological preferences of the species with respect to the latter factor. The representativeness values of Iberian endemisms account for 18% of the river species catalogued, a value which is higher than that recorded for the rest of the aforementioned southern species. Of these, 41.2% are restricted to the south-eastern peninsular area (Ribera et al. 1998), including the five taxa exclusive to Sierra Nevada: *Limnebius monfortei*, *Ephemerella ikonomovi nevadensis* (formerly *Serratella spinosa nevadensis*, though this is under discussion), *Ecdyonurus* (*Ecdyonurus*) *baeticus*, *Annitella iglesiasi* and *Limnephilus obsoletus*.

## 2.4 Threats and Conservation Status of EPTC

One of the main threats for these animals is the altered environmental conditions due to climate change. These organisms will face climate change and some of them will cope with it through different strategies, avoiding or mitigating the effect of adverse conditions. The temporal evolution of precipitation in Sierra Nevada over the past 50 years has shown a general decline (Pérez-Luque et al. 2016, see chapter "Climate Variability and Trends"), the river water flow has decreased, and the average water temperature has risen by almost 2 °C (1.6) over the past 20 years. These changes have apparently had a drastic effect on the aquatic macroinvertebrate communities (Sáinz-Bariáin et al. 2016a, b). Due to this, many species of Trichoptera have been seen to move up in altitude in search of colder

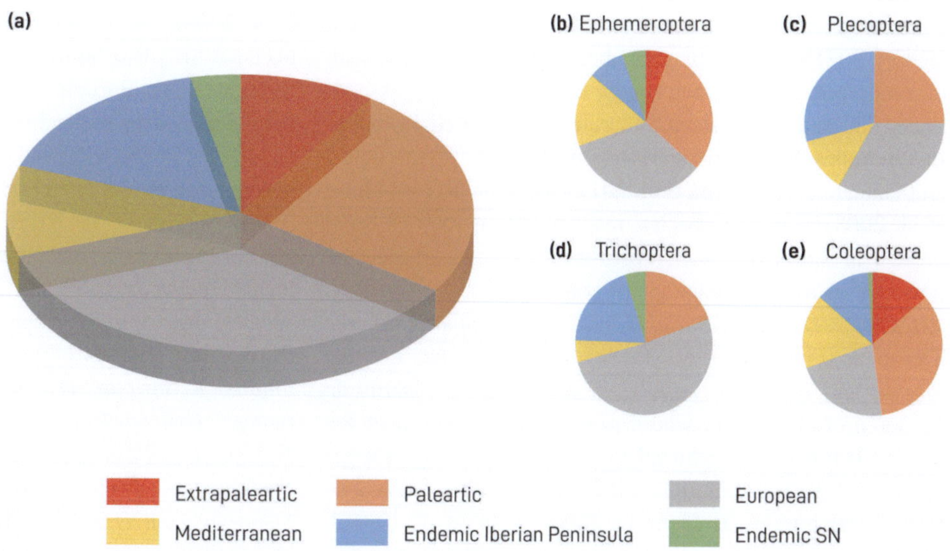


**Fig. 3** Pie charts showing the proportion of species of: **a** the four aquatic insect orders analyzed; **b** mayflies (Ephemeroptera); **c** stoneflies (Plecoptera); **d** caddisflies (Trichoptera); and **e** beetles (Coleoptera) from Sierra Nevada in each biogeographical category considered. SN: Sierra Nevada

waters. This has led to an increase in the number of species inhabiting the protected area of the Sierra Nevada National Park as a result of movements up to an altitude of 800 m in search of more suitable temperature conditions (Sáinz--Bariáin et al. 2016a, b). In relation to this, a strong change in the macroinvertebrate communities has been predicted at different global warming scenarios (Alba-Tercedor et al. 2017). This would imply that the set of interactions in each community could change, and this could have an effect on ecosystem processes such as the cycle of matter and the flux of energy. Other strategies would be those related to the alteration of the life cycle of certain species, either through the displacement of the growth period, modifications in their phenology and/or modifications of their voltinism (e.g., Sáinz-Bariáin and Zamora-Muñoz 2012). Some examples of changes of these kinds have already been related to temperature and water regime of the streams from Sierra Nevada (e.g., López-Rodríguez et al. 2008; Sáinz-Bariáin et al. 2016b), some of them triggered by the temperature regime changes induced by dams (e.g., Alba-Tercedor 1990a, b). Thus, those without strategies such as some of the ones pointed out here would be those more vulnerable in the near future (Múrria et al. 2020).

Despite the difficulties in categorizing a species as threatened, because in some cases it is not possible to determine whether what is observed is the actual status of the species or an effect of limited sampling, some Sierra Nevada mayflies species with restricted global distribution are in the red list (Verdú and Galante 2006).

Stoneflies are, due to their general high stenoecity and low dispersal capacity (favoring that many species are isolated as small populations), one of the most vulnerable and threatened animal groups in the frame of the current Global Change (Fochetti and Tierno de Figueroa 2006, 2008; Tierno de Figueroa et al. 2010). Nonetheless, only four of the approximately 3800 extant species (DeWalt et al. 2020) have been currently included in the IUCN Red data list (IUCN 2021), neither of them belonging to the European fauna despite the high percentage of species from this continent that can be considered threatened. The Red Books of Invertebrates from Spain and Andalusia include eight (Verdú and Galante 2006; Verdú et al. 2011) and four (Barea-Azcón et al. 2008) stoneflies species, respectively, neither of them present in Sierra Nevada. Nevertheless, it is remarkable the presence of relict populations of some species such as *Perlodes microcephalus*, *Perla grandis* (or *P. bipunctata*, see Tierno de Figueroa et al. 2013b), *Capnia nigra*, *Leuctra inermis* and *Taeniopteryx hubaulti* in this massif (Tierno de Figueroa et al. 2013b), whose presence in the south of the Iberian Peninsula is restricted to Sierra Nevada or, in some cases as *C. nigra*, *L. inermis* and *P. grandis/bipunctata*, also to close mountains as Sierra de Baza or Sierra de Castril. Those five species could be considered threatened at a regional scale in Andalusia or, in the case of *T. hubaulti*, in all the Iberian Peninsula (Tierno de Figueroa et al. 2013b). This threat situation is particularly true in the case of *T. hubaulti* and *P. microcephalus*, whose scarce populations have been detected in high altitude (at 2000 m or higher) in Sierra Nevada (although punctually a few nymphs have been collected at lower altitudes, what could be a consequence of the drift) and that makes those populations particularly vulnerable to the effects of climate change (see the vulnerability criteria reported in Tierno de Figueroa et al. 2010).

Caddisflies present a high taxonomic, ecological and functional diversity, being able to be found in most of the freshwater ecosystems of the world (Wiggins 2004; Morse et al. 2019). They are considered organisms sensitive to climate change mainly because: (1) they have narrow ecological niches and low dispersal capacity; (2) many species inhabit upper reaches of rivers; and (3) are cold-stenothermic species (Hering et al. 2009; Kernan et al. 2010). Around 68% of caddisfly populations are declining, more than any other group of aquatic insects (Sánchez-Bayo and Wyckhuys 2019). Of the 956 species found in the Mediterranean area, 425 are considered Mediterranean endemisms (Tierno de Figueroa et al. 2013a; Morse et al. 2019). In the Iberian Peninsula and the Balearic Islands, 368 species have been recorded, one-third endemic (González and Martínez-Menéndez 2011; Martín 2017). Only five of the approximately 16,300 extant species (Morse 2021) have been currently included in the IUCN red list (IUCN 2021), neither of them belonging to the Iberian or Mediterranean fauna. There is no caddisfly species incorporated in the Red Books of Invertebrates from Spain (Verdú and Galante 2006; Verdú et al. 2011) and, although 11 species were included in the Red Book of Invertebrates from Andalusia, 10 were catalogued in the "data deficient" category. The other one, *Annitella esparraguera*, a species present in Sierra Nevada and nearby mountains, was classified as "endangered" (Barea-Azcón et al. 2008), although this classification should be reviewed because its distribution range is wider than previously thought (Sáinz-Bariáin et al. 2013). Long-term changes have been detected in the richness and altitudinal distribution of caddisflies in Sierra Nevada, probably related to the increase in temperature and decrease in the water regime of the rivers (Sáinz-Bariáin et al. 2016a, b). It has been observed that species richness is increasing in altitude (specially at sites located between 1800 and 2000 m), as a consequence that species from middle reaches have enlarged their range of distribution towards higher elevations (*Rhyacophila meridionalis, R. nevada, Hydroptila vectis, Philopotamus montanus, Hydropsyche infernalis, Micrasema moestum, Halesus tessellatus* and *Sericostoma vittatum*), and the colonization of species with good dispersal abilities (e.g., *Stenophylax nycterobius, Allogamus mortoni*) from nearby mountain ranges (Sáinz-Bariáin et al. 2016b). Species inhabiting headwaters in high-altitude ecosystems do not have the opportunity to accommodate ecological requirements in scenarios of climate change, and could be threatened by more generalist species that are able to migrate in altitude and compete for resources. Hence, endemic species may be more vulnerable than more general species due to their lower dispersive capacity and narrow ecological requirements (Múrria et al. 2020). It has been estimated that about 50% of the Iberian species of caddisfly could be severely affected by the potential impact of climate change (Hering et al. 2009). In a study on the vulnerability to climate change for two endemic species, high-elevation and low-dispersive species of *Annitella* (*A. esparraguera* and *A. iglesiasi*) in Sierra Nevada, it was found that both species showed low genetic diversity but only *A. esparraguera* exhibit locally unique haplotypes, indicating limited gene flow. For *A. esparraguera*, modelled future habitat suitability showed 88.4% range contraction by 2050 (RCP scenario 8.5) and a displacement of 41.5% of the current potential distribution to higher elevations. Populations of *A. esparraguera* are predicted to be lost because of the reduction of optimal habitat and limited propensity for tracking future suitable conditions (Múrria et al. 2020).

Regarding the conservation status of aquatic beetles, with the exception of *Hydroporus decipiens*, which is widely distributed in the Iberian Peninsula, the rest of the endemic Coleoptera present in watercourses of Sierra Nevada have moderate to high vulnerability values according to the number of localized populations. Because of demographic data available for each of them, as well as the particular characteristics of their habitat (Sánchez-Fernández et al. 2008), a total of five endemic species present in this mountain massif are considered worthy of priority attention in terms of conservation. Among them, *Limnebius monfortei,* which is only known from its type locality in Sierra Nevada (Barranco de Las Víboras, 1500 m), is particularly noteworthy.

Apart from climate change, other important threat factors for Sierra Nevada EPTC are: (1) construction of dams, reservoirs and water diversion (*careos*), which interrupt or interfere with natural water flow, as it is the case of Canales dam in the Genil River Basin (Bello and Alba-Tercedor 2005) or the Benínar dam in the Adra River Basin (Alba-Tercedor et al. 1986a, 1986b); (2) construction of small electric plants, as in the Poqueira River; (3) punctual pollution from tourist complexes (camping sites, ski resort, etc.) or villages which alter natural water quality (Sánchez-Ortega and Tierno 1996; Zamora-Muñoz and Alba-Tercedor 1992); and (4) diffuse pollution in some high mountain streams due to, mainly, high stocking densities.

## 2.5 The Brown Trout: Threats and Conservation Status

Regarding freshwater vertebrates, fish, and particularly the brown trout, have also a relatively long history of studies. The low temperatures of the streams and rivers that flow through the Sierra Nevada massif allow the maintenance of populations of brown trout (*Salmo trutta*) as the only high mountain species characteristic of this biogeographic area (Fig. 2). In mid and low altitude reaches, outside the protected area, the increase of water temperature allows the appearance of cyprinic species such as andalusian barbel (*Luciobarbus sclateri*), southern straight-mouth nase (*Pseudochondrostoma willkommii*) and southern Iberian chub (*Squalius pyrenaicus*). However, due to the introduction of exotic species by fishermen, rainbow trout (*Oncorhynchus mykiss*) appeared inside the National Park in the mid-twentieth century and, outside it, exotic cyprinids such as north American largemouth bass (*Micropterus salmoides*), Northern pike (*Esox lucius*) or common carp (*Cyprinus carpio*), among others. At present, all these invasive alien species are naturalized (they have their "own reproductive capacity") and compete with the endemic ones with which they coexist, causing their displacement (Larios-López et al. 2015a).

After cataloguing the Andalusian populations of brown trout as threatened (Franco Ruíz and Rodríguez de los Santos 2001), the regional government started in 2005 the "Recovery Program for brown trout populations in Andalusia". In addition, this species was included in 2007 as a bioindicator of climate change within the project Global Change Observatory of Sierra Nevada (Larios-López et al. 2018). All the results obtained for this salmonid in both projects are recorded in "gray literature", but it is necessary to highlight those referring to its historical distribution (Sáez et al. 2010), the detection of high genetic diversity, with new haplotypes discovered (Almodóvar et al. 2010), the carrying capacity of rivers (Barquín et al. 2010) and estimates of ecological flow and trophic preferences (Barquín et al. 2015). Later, Larios-López et al. (2015a) identified that the upper limits of distribution of these populations are of natural origin (populations reaching the headwaters of the rivers, the highest possible altitude, or impassable waterfalls), while the lower ones are due to anthropogenic causes (impassable dams, water abstraction for irrigation and human consumption, or synergistic effects of agriculture, water diversion, water pollution and habitat fragmentation). Also, Larios-López et al. (2015b) reported the most extensive spawning period described for this species to date (from early October through late April or early May) as a result of lack of anadromy, the characteristic unpredictability of the

Mediterranean climate and the temperature range of the rivers inhabited by the species, including those of Sierra Nevada. Moreover, Larios-López et al. (2021) found that the recruitment of trout populations in the study region was a density-independent process and synchronized by means of a Moran effect. Larios-López (2017) also proposed particular adaptive management measures for these southernmost populations in Europe, probably some of the most threatened in the entire natural range of the species.

Larios-López et al. (2015a) described the main threat factors for brown trout in Sierra Nevada. In the high reaches inhabited by this species (between 1500 and 2000 m), they detected the presence of numerous specific water diversions (the aforementioned *careos*), which cause the desiccation of some stretches of the river, sometimes for more than half a kilometer away, within the Sierra Nevada Natural Park (e.g., Trevélez or Bérchules rivers). In addition, in the Monachil River, due to both the poor water quality after passing through the ski resort and the intense extraction of water that it suffers in its upper and middle section, the upper limit of this species is located at 1300 m. However, brown trout has an upper distribution limit of around 2000 m in all the other rivers of this sub-basin (Genil).

It is in the middle reaches of Sierra Nevada (700–1500 m), where most of the factors that threaten the populations of brown trout are concentrated. Uncontrolled water withdrawals greatly increase near the villages and, in summer, they cause the complete drying of some rivers. This fact limits ten populations in the lowest part of their distribution range (e.g., Bérchules, Bayárcal or Alhama de Lugros). The effect of water extraction, coupled with the decrease in quality due to the presence of numerous untreated discharges, generates synergistic effects (Gasith and Resh 1999) which lead to loss of the species in six other rivers (e.g., Monachil, Lanjarón or Dílar). In addition, poaching in these middle reaches (Larios-López et al. 2019) and the unsafe effects of the "catch and release" fishing modality (Arlinghaus et al. 2007; Cooke et al. 2013), affect the behavior, growth and reproduction of fish (Policansky 2002). Moreover, the presence of four large dams in Sierra Nevada is notable, fragmenting the populations of the Genil (Canales reservoir), Aguas Blancas (Quéntar reservoir), Guadalfeo (Rules reservoir) and Adra (Benínar reservoir) river basins. The populations that inhabit downstream of these reservoirs are significantly unstructured and disappear after a few kilometers. Associated with reservoirs, the introduction of invasive exotic fish species by fishermen is a fact (American bass, pike and common carp, among others) and, in the case of rivers, rainbow trout, that was introduced during the second half of the twentieth century in this region. Their reproductive and competitive capacity against brown trout has been proven in Sierra Nevada (Larios-López 2017). This exotic salmonid has displaced the native species of several river stretches (e.g., Genil, Alcázar), as well as in entire rivers (e.g., Arroyo del Pueblo, Ohanes). In 2009, the Regional Council for the Environment of Junta de Andalucía began an eradication program for naturalized populations of rainbow trout in Sierra Nevada with great success, by unifying this program with the introduction of populations of brown trout with genetic background typical of the region.

However, the current context of climate change anticipates extremely drastic shifts in the dynamics of these regulatory external drivers (mainly precipitation) in the study region (IPCC 2013). Longer, wetter and stormier winters are expected. These storms will also be more intense in Mediterranean systems (Ulbrich et al. 2006; Giorgi and Lionello 2008), so that the damage to fish populations will be greater since fish species are more sensitive to floods if they occur out of the natural flood regime. In this sense, Jensen and Johnsen (1999) verified the huge impact of floods reducing entire age classes of brown trout. Considering this fact, as well as that these southernmost brown trout populations of the Iberian Peninsula have population synchrony (Larios-López et al. 2021), negative environmental disturbances would have the capacity to endanger their resilience. This phenomenon would be even greater in Sierra Nevada, both due to the devastating predicted effects of climate change and the high degree of isolation that the species presents on this distribution edge.

## 3 Changes at the Population and Community Level

### 3.1 Key Ecological Factors Acting on Macroinvertebrate Communities

Most macroinvertebrate community studies have been carried out at the family level due to the complexity of reaching a lower taxonomic resolution with many aquatic stages of several groups (Romero Martín and Alba-Tercedor 2013). Furthermore, most of this information comes from analyses conducted to determine the ecological status of the water bodies through biotic indexes such as the IBMWP, which only needs identification of macroinvertebrates at the family level (Alba-Tercedor and Sánchez-Ortega 1988; Alba-Tercedor 1996, 2000; Alba-Tercedor et al. 2004). Despite the lack of more specific information on these approaches, they can provide a good general overview of the status of the macroinvertebrate communities of streams and rivers from Sierra Nevada, mainly when analyzed together with the major physicochemical parameters of the water.

Key abiotic features of the environment that usually drive the abundance and distribution of macroinvertebrates are those related to current, substrate, temperature, and sometimes water chemistry variables such as alkalinity and dissolved oxygen (Allan et al. 2021). In the context of climate change, water temperature and flow are probably two of the most important, as they will likely be affected in future scenarios due to global warming and changes in the precipitation patterns (IPCC 2013; Alba-Tercedor et al. 2017). Both temperature and spatial variation in hydraulic parameters have been shown to affect the local distribution and abundance of stream macroinvertebrates (Vannote and Sweeney 1980; Allan et al. 2021). This may be especially noteworthy in high mountain streams such as those of Sierra Nevada, where increased temperatures and decreased river water flow over the last 40 years have been already detected (Sáinz-Bariáin et al. 2016b).

Few investigations have accomplished the study of whole communities of Sierra Nevada watercourses, and those that have done it have focused mainly on the application of biotic indexes to assess the ecological status of streams and rivers, as mentioned before. For instance, Zamora-Muñoz and Alba-Tercedor (1996) surveyed the macroinvertebrate communities of several streams of the Genil River basin, measured water physicochemical parameters and assessed the ecological status of the waters using biotic indices, as the BMWP' (posteriorly called IBMWP). The studies carried out in the Monachil River revealed that sewages from the ski resort located at the head of the river drastically reduced the diversity of macroinvertebrates, collecting only taxa highly tolerant to organic pollution (Zamora-Muñoz and Alba-Tercedor 1992; Zamora-Muñoz et al. 1993). In the Genil River Basin, nutrient content and water hardness were the main factors influencing macroinvertebrate distribution. Most of the sites in the study area were polluted, and consequently the most frequently collected taxa were tolerant species. Moreover, the worsening of the quality downstream was accompanied by a substitution of species, sometimes within the same genus, from less to more pollution tolerant (Zamora-Muñoz and Alba-Tercedor 1996). On the other hand, Alba-Tercedor et al. (1986a, b) and Jáimez-Cuéllar (2004) surveyed the ecological status of two basins, Adra and Guadalfeo river basins, using the IBMWP index and, additionally, carried out an assessment of the β-diversity of the communities in both rivers, as well as a study on the relationship between the biocoenosis of the main aquatic insect orders and the physicochemical characteristics of the studied reaches. These authors found that β-diversity decreased downstream (with one occasional exception in the Adra River) due to progressive deterioration of streams and rivers. In a more recent study performed as part of a scientific project in four basins of Sierra Nevada (Dílar, Genil, Trevélez and Poqueira Rivers) during two seasons (autumn and spring), some unpublished preliminary data showed that, though the overall ecological status and α-diversity were high in most reaches (Fig. 4), those situated under particular sources of disturbances suffered a drastic reduction in these parameters (see, for instance, the T3 sampling station in Fig. 4, situated downstream of the Trevélez village). In this study, communities were affected by anthropogenic impacts, but some studies have also projected the effect on macroinvertebrate communities of climate change related impacts, such as temperature and flow reduction. This is the case of a research using the MEDPACS predictive approach simulating three different climatic scenarios of water temperature increase and water flow reduction, in which Alba-Tercedor et al. (2017) concluded that the lotic macroinvertebrate communities of the Iberian Peninsula will probably reduce their taxa richness in comparison to present. Moreover, generalist taxa were predicted to move upstream to lower temperature reaches and vulnerable taxa would reduce their distribution area. A similar conclusion was reached previously by Tierno de Figueroa et al. (2010) and by Sáinz-Bariáin et al. (2016b) studying stoneflies and caddisflies biocoenosis, respectively.

## 3.2 Key Ecological Factors Acting on the Macroinvertebrate Populations

At the population level, again water temperature and flow are two of the most important extrinsic factors modulating the biology of macroinvertebrates in general, and of aquatic insects in particular (Sweeney 1984; Allan et al. 2021). To study this, some species of Ephemeroptera, Plecoptera and Trichoptera have been monitored in the massif, several of them during a whole year. In particular, López-Rodríguez et al. (2008) determined that differences in temperature regime (measured as day-degrees accumulated by the species) affected population dynamics of several species of mayflies, either due to changes in the growth rate (as occurred with *Serratella ignita* and *Ephemerella ikonomovi nevadensis*, reported as *S. spinosa nevadensis*), or due to changes in voltinism (as in *Baetis muticus*, reported as *Alainites muticus*, and *Baetis alpinus*). These authors also concluded that the effect of temperature on the studied stonefly species was less drastic, just advancing, delaying or displacing the life cycle some months.

Other important factors that act on the populations of several aquatic insects are those related to pollution and habitat degradation (Fig. 5). Though Sierra Nevada is a protected area, some anthropogenic impacts can still be found in the mountains, the most important of them related to the ski resort within the Natural Park. In this sense, is the Monachil River the fluvial axis in which the effect of the pollution generated in the ski resort of Sierra Nevada becomes more apparent (Zamora-Muñoz and Alba-Tercedor 1992). Alba-Tercedor et al. (1991) analyzed the impact of this area on the mayfly and stonefly biocoenoses of several reaches in the Monachil River and they found that the

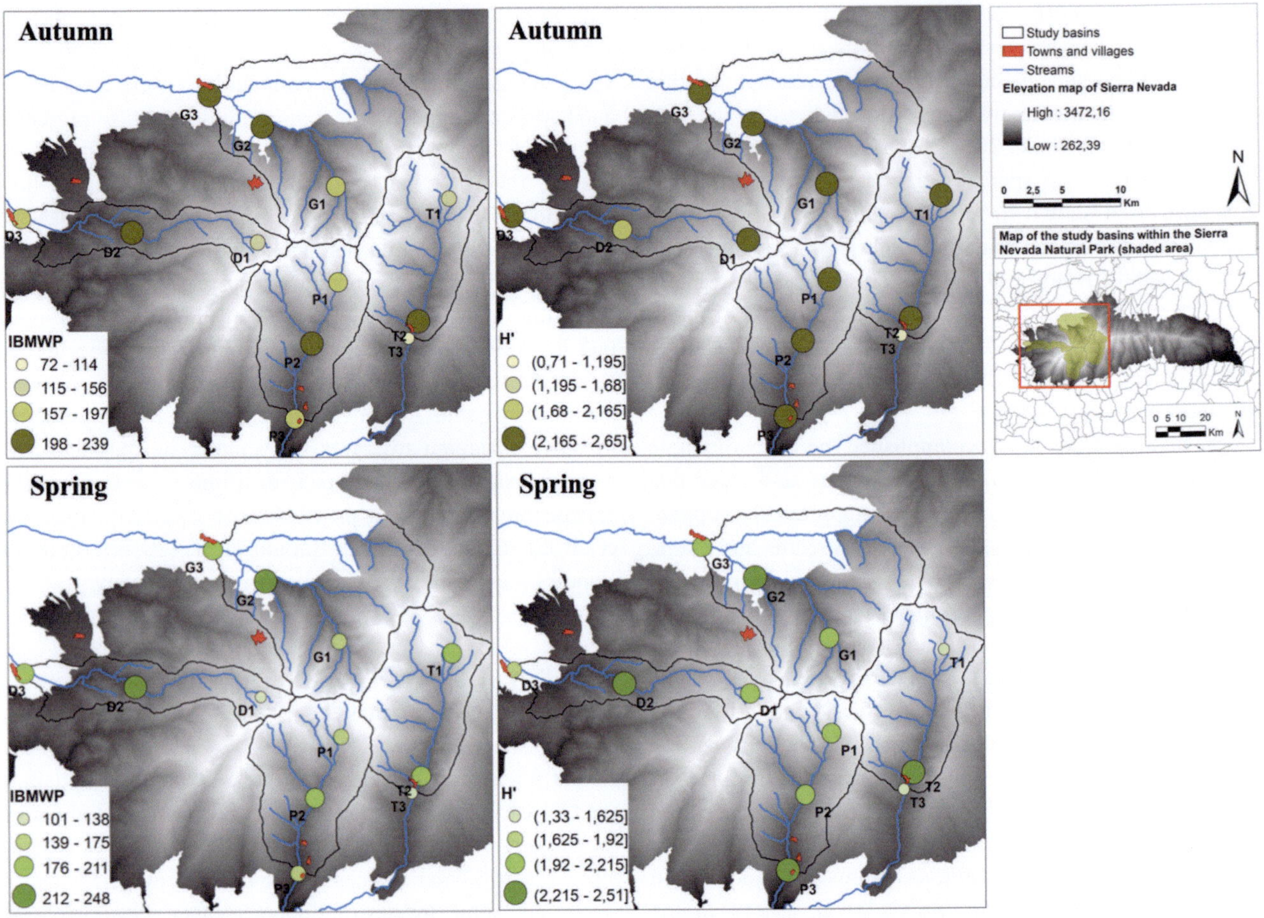


**Fig. 4** Map showing reaches from the Dílar (D), Genil (G), Trevélez (T), and Poqueira (P) Rivers in which both IBMWP and Shannon–Wiener diversity (H′) were assessed (figure courtesy of Ander Congil Ross). Variations in both indexes occur within watercourses and between seasons (spring and autumn) due to changes in environmental conditions, and in some cases due to inefficient wastewater purification (e.g., T3)

populations of most species declined with altitude and in relation to pollution of lower altitude sites. Afterward, Zamora-Muñoz et al. (1993) analyzed the effect of several physicochemical parameters on the distribution of mayflies and stoneflies in the same river. These authors found that the distribution of ten of the 15 species of mayflies and two of the 12 species of stoneflies studied by them was significantly influenced by one of the environmental parameters assessed. These parameters were, in order of importance, water temperature, content of calcium, nitrates, dissolved oxygen, pH and phosphates, some of them related to organic pollution, and the most affected species by the organic pollution were *Baetis rhodani*, *Isoperla nevada* and *Leuctra inermis*.

Regarding habitat degradation, one of the main and more obvious impacts on stream connectivity are dams (Allan et al. 2021). Depending on their size, they can isolate stream reaches almost completely, so they have an important effect on the structure of communities downstream (e.g., Bello 1997; Bello and Alba-Tercedor 2005) and even on ecosystem-scale processes such as leaf decomposition (Casas et al. 2000).

## 3.3 Key Ecological Factors Acting on the Brown Trout Populations

The effect of ecological parameters on aquatic vertebrates, particularly brown trout, has also been accomplished in streams and rivers of Sierra Nevada. With this goal, Larios-López et al. (2021) investigated the drivers of their population dynamics in the region between 2006 and 2014. They proved that recruitment is not regulated by density-dependent mechanisms and that there is not even a positive effect on the reproductive stock. Specifically, rainfall

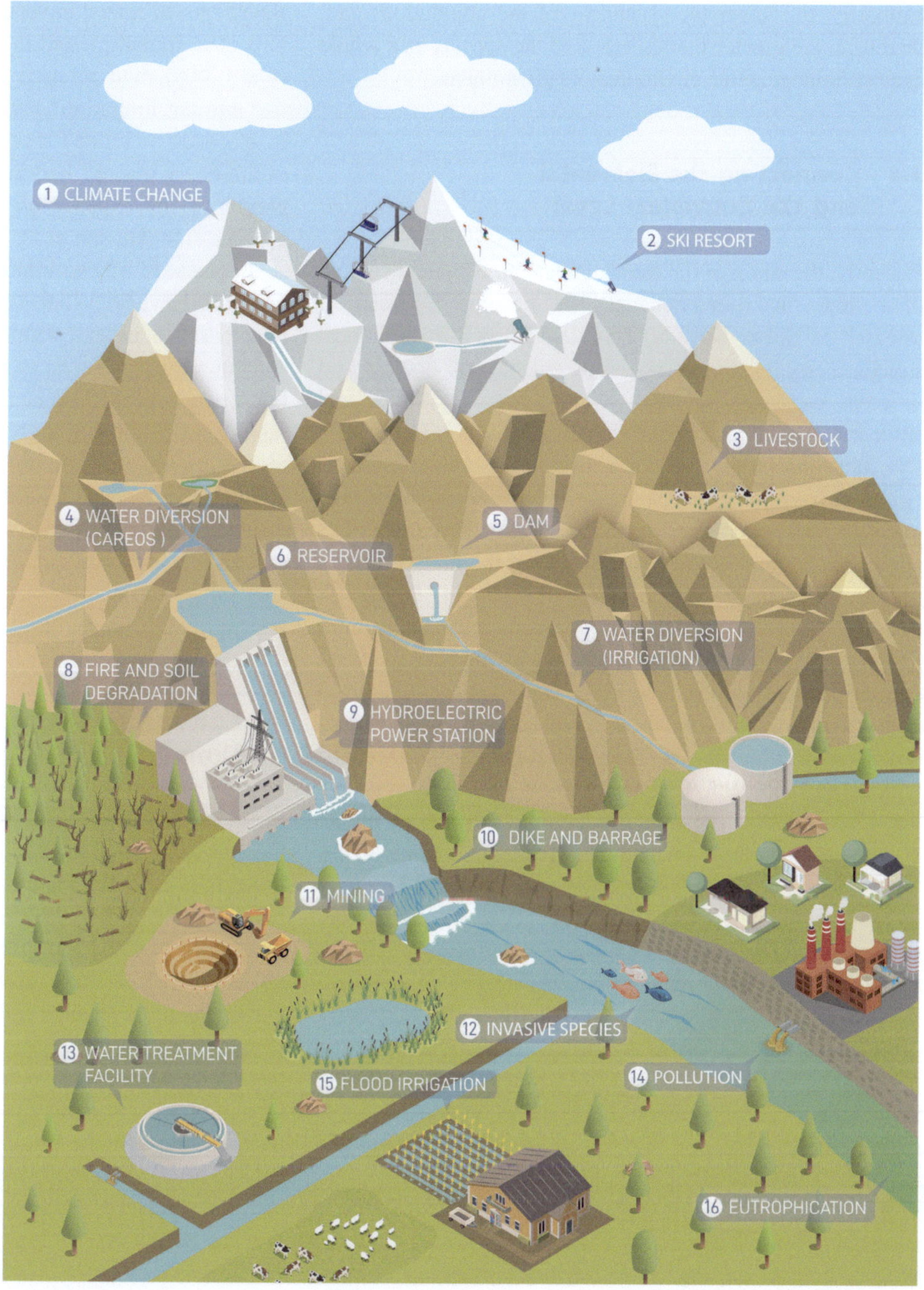


**Fig. 5** Schematic diagram of the main threats to aquatic fauna of Sierra Nevada. The key risks to aquatic fauna include climate change (1), water regulation and extraction processes (2, 4, 7, 9, 15), interception activities including reservoirs and dikes (5, 6, 10), land use degradation (8), impacts of mining, industry and water pollution (11, 14), invasive species (12), and eutrophication due to livestock and inefficient sewage water purification (3, 13, 16)

is the only external driver that explains the recruitment dynamics in this area (negatively affected by winter rainfall and positively affected by spring rainfall). Moreover, juvenile density is determined by the density of young-of-the-year during the previous year, while there is no significant effect of the density of juveniles during the previous year on the density of adults in a given year. All these results could be expected in populations inhabiting their rear edges of distribution (*sensu* Hampe and Petit 2005), where the extreme conditions prevent populations reaching an equilibrium density. This is the case of brown trout in Sierra Nevada, where densities are below the carrying capacity and relatively free of intraspecific competition, unlike the central European core areas where the density-dependent factors are involved in regulating populations. Moreover, Larios-López et al. (2021) showed that brown trout populations in this European southern region are synchronized as a result of a Moran effect (as previously mentioned) directed mainly by winter precipitation and conditioned by the habitat similarity between populations (highlighting altitude, distance to the upper limit, geographic distance and result of IHF index; see Pardo et al. 2002). Therefore, the greater the differences in these factors of habitat similarity, the lower the synchrony of the populations and, therefore, the greater the resilience of the species as a whole against homogeneous environmental phenomena.

## 3.4 Connecting the Elemental and the Ecosystem Level

Recently, the study of the elemental and biochemical content of benthic macroinvertebrates has become a focus of research in Sierra Nevada. This field known as biological stoichiometry examines how the balance of energy and different elements influence living systems (Sterner and Elser 2002). The analysis of elemental carbon (C), nitrogen (N) and phosphorus (P) composition of 436 specimens collected from four high mountain streams of Sierra Nevada revealed that macroinvertebrate C:N:P stoichiometry differs among taxonomic groups but not among functional feeding groups, and is largely due to changes in P content (Villar-Argaiz et al. 2020). Although numerous studies have documented similar results at the species level (e.g., Liess and Hillebrand 2005), very few studies have examined ontogenetic variation within a given taxon and still a consensus is lacking in the literature regarding ontogenetic patterns in P content. Thus, while most studies have shown that for a given taxa small individuals have invariably higher P content than large individuals (Back and King 2013), Villar-Argaiz et al. (2020) reported over threefold increases in the P content from small to large *Dinocras cephalotes* (Plecoptera). The findings of positive relationships between P content and body mass in Sierra Nevada are undocumented in the literature and, most intriguingly, they were exclusively found for hemimetabolous insects. Because elemental composition reflects body demands, the study of stoichiometry can therefore be used to understand how organisms respond to changes in resource quality and quantity facing current global change.

Knowledge of how P content varies across ontogeny is essential because P plays a key role in the growth rate, a most crucial life trait driving the evolutionary fitness of animals (Sterner and Elser 2002). The observation that rapidly growing organisms have a high P content compared to slower growing organisms gave rise to the "growth rate hypothesis", which states that elevated growth rates demand large amounts of P linked to the RNA needed to sustain rapid protein synthesis (Elser et al. 2003). The success of the growth-rate hypothesis in improving our understanding of the mechanistic variation between P content, C:P stoichiometry and growth rate has paved the road to the extrapolation of biological stoichiometry to the domain of evolutionary ecology. Collectively, these studies suggest that variations in C:N:P ratios in macroinvertebrates reflect underlying allocations to major molecules such as RNA versus DNA. This idea, initially tested in crustacean zooplankton (Hessen et al. 2008), and referred as to the "growth rate-genome size-nutrient limitation" hypothesis, claims that P limitation in chronically limited environments could be behind the evolutionary tradeoff between P allocation to RNA for rapid growth at the expense of low DNA content and reduced genome size (Hessen et al. 2009). In a study on the nucleic acid content of 639 specimens of benthic macroinvertebrates in Sierra Nevada, Villar-Argaiz et al. (2021) extended this hypothesis to insects with different metamorphosis modes, and found that differences in allocation between RNA and DNA may reflect fundamental evolutionary tradeoff between rapid growth rate and high RNA content in holometabolans at the expense of diminished genome sizes relative to hemimetabolans (Villar-Argaiz et al. 2021). By connecting elements, macromolecules, and key life-history traits such as growth rate and organisms evolutionary fitness, these studies contribute to bridge the gap between different layers of biological organization from genes to ecosystems.

# 4 Concluding Remarks: What Else Can Be Done?

Sierra Nevada streams and rivers are subject to a wide range of pressures and impacts (a schematic summary of them can be found in Fig. 5), and there is still much to be studied about the macroinvertebrates of the massif watercourses and the effect of these pressures on them. Romero Martín and Alba-Tercedor (2013) underlined that many groups are still very poorly known, and that detailed and systematic studies of large areas above 2000 m are needed. In fact, though in the last decade there have been intensive surveys of, for instance, particular groups of aquatic insects in the massif, new species are still being recorded for the first time in the mountains (e.g., *Leuctra geniculata*, *L. cazorlana*, *Stenophylax nycterobius*, *Allogamus mortoni*, *Adicella reducta*, *Agapetus fuscipes*, *Glossosoma boltoni*, *Melademа coriacea*, *Limnebius* (*Limnebius*) *bacchus*, *L.* (*Limnebius*) *ignarus*, *Hydrochus grandicollis*). This could be a mix of changes in the distribution ranges of these species and of an intensification of the samplings during the past years. In this sense, the Global Change Observatory of Sierra Nevada, with the support of the Environment and Water Agency of Andalusia, is performing a monitoring program that is generating interesting data (both faunistic and in relation with the ecological status of several water stream reaches) and that should be maintained in the future to reach a long-term record of macroinvertebrate and fish communities. This kind of data will allow us to analyze the effect of climate change on these communities and to propose the most realistic mitigation measures.

Some important gaps in the assessment of lotic macroinvertebrate communities from Sierra Nevada are related to the analysis of the role of these organisms in the food webs they belong to. Studies of these aspects still lack in Sierra Nevada, and are fundamental to actually understand the flux of energy and matter in these particular environments. Inferences from other sites or from different taxonomic levels have proved not to work properly (e.g., Tierno de Figueroa et al. 2019), so specific analyses of gut contents or other techniques (such as stable isotopes) should be applied to these organisms in each studied reach.

In the case of the brown trout, the generation of climate models for the Sierra Nevada massif under different climate change scenarios would be an interesting tool. In this way, it would be possible to predict its effect on river courses and, therefore, foresee changes in the current distribution of the species, as in studies carried out in other peninsular regions (Almodóvar et al. 2012; Ayllón et al. 2017; Clavero et al. 2017; Santiago et al. 2017). This research would allow the detection of specific river stretches in Sierra Nevada that could behave, in the future, as thermal shelters (Elliott 2000) against climate change (Daigle et al. 2014). Furthermore, it would be very interesting to expand the investigations initiated by Larios-López et al. (2021), with the intention of verifying the effect of precipitation and summer temperature on the population dynamics of the species.

Finally, the implication of the whole society is essential for maintaining the ecological status of watercourses from Sierra Nevada and, mainly, for avoiding their deterioration. In this sense, the "citizen science" initiatives are an important tool to reach these objectives. The collaboration among society, scientists and managers is the only possible way to face future Global Change scenarios and to properly handle and carry out the mitigation measures needed to stop the inertia of deterioration of these wonderful and interesting environments.

**Acknowledgements** The authors would like to thank Julio Luzón-Ortega for his critical comments in a draft version of the manuscript, to Ander Congil Ross for the realization of Fig. 4 and to Manuel Merino for the realization of Figs. 1 and 5.

## References

Alba-Tercedor J (1990a) Life cycles and ecology of some species of Ephemeroptera from Spain. In: Campbell IC (ed) Mayflies and stoneflies: life histories and biology. Kluwer Academic Publishers, Dordrecht, Boston & London, pp 13–16

Alba-Tercedor J (1990b) Life cycles and ecology of mayflies from Sierra Nevada (Spain). IV. Limnetica 6:23–34

Alba-Tercedor J (1996) Macroinvertebrados acuáticos y calidad de las aguas de los ríos. In: IV Simposio del Agua en Andalucía (SIAGA), Almería, vol II. Instituto Tecnológico Geominero de España, Madrid, pp 203–213

Alba-Tercedor J (2000) BMWP' un adattamento spagnolo del British Biological Monitoring Working Party (BMWP) score system. Biologia Ambientale 14:65–67

Alba-Tercedor J (2013) Los efemerópteros (Ephemeroptera). In: Ruano F, Tierno de Figueroa M, Tinaut A (eds) Los Insectos de Sierra Nevada: 200 años de historia, vol 1. Asociación Española de Entomología, Granada, pp 100–113

Alba-Tercedor J, Guisasola I, Sánchez-Ortega A (1986a) Variaciones estacionales de las características físico-químicas y de la calidad biológica de las aguas del Río Guadalfeo (Granada). In: Pulido et al (eds) II Simposio del Agua en Andalucía (SIAGA), vol I, Granada, pp 235–247

Alba-Tercedor J, Jáimez-Cuéllar P, Álvarez M et al (2004) Caracterización del estado ecológico de ríos mediterráneos ibéricos mediante el índice IBMWP (antes BMWP'). Limnetica 21 (2002):175–185

Alba-Tercedor J, Jiménez-Millán F (1987) Evaluación de las variaciones estacionales de la calidad de las aguas del Río Guadalfeo, basada en el estudio de las comunidades de macroinvertebrados acuáticos y de los factores físico-químicos. LUCDEME III. Monografía 48 del ICONA. Ministerio de Agricultura Pesca y Alimentación-Instituto Nacional para la Conservación de la Naturaleza, Madrid

Alba-Tercedor J, Sáinz-Bariáin M, Poquet JM, Rodríguez-López R (2017) Predicting river macroinvertebrate communities distributional shifts under future global change scenarios in the Spanish Mediterranean area. PLoS ONE 12:e0167904

Alba-Tercedor J, Sánchez-Ortega A (1988) Un método simple para evaluar la calidad biológica de las aguas corrientes basado en el de Hellawell (1978). Limnetica 4:51–56

Alba-Tercedor J, Sánchez-Ortega A, Guisasola I (1986b) Caracterización de los cursos permanentes de agua de la Cuenca del Río Adra: factores físico-químicos, macroinvertebrados acuáticos y calidad de las aguas. Proyecto LUCDEME, Estudio integrado del medio físico de la Cuenca del Río Adra. ICONA, unpublished report

Alba-Tercedor J, Zamora-Muñoz C, Sánchez-Ortega A, Guisasola I (1991) Mayflies and stoneflies from the Río Monachil (Sierra Nevada, Spain) (Ephemeroptera and Plecoptera). In: Alba-Tercedor J, Sánchez-Ortega A (eds) Overview and strategies of ephemeroptera and plecoptera. The Sandhill Crane Press, Gainesville, Florida, pp 529–538

Allan JD, Castillo MM, Capps KA (2021) Stream ecology: structure and function of running waters, 3rd edn. Springer, Cham

Almodóvar A, Nicola GG, Ayllón D, Elvira B (2012) Global warming threatens the persistence of Mediterranean brown trout. Glob Change Biol 18:1549–1560

Almodóvar A, Nicola, GG, Leal S, Elvira B (2010) Análisis genético de las poblaciones de Trucha común *Salmo trutta* en la Comunidad Autónoma de Andalucía. Memoria Final Proyecto Egmasa. Junta de Andalucía & Universidad Complutense de Madrid, Sevilla

Arlinghaus R, Cooke SJ, Lyman J et al (2007) Understanding the complexity of catch and release in recreational fishing: an integrative synthesis of global knowledge from historical, ethical, social, and biological perspectives. Rev Fish Sci 15:75–167

Ayllón D, Railsback SF, Vincenzi S et al (2017) InSTREAM-Gen: modelling eco-evolutionary dynamics of trout populations under anthropogenic environmental change. Ecol Model 326:36–53

Back JA, King RS (2013) Sex and size matter: ontogenetic patterns of nutrient content of aquatic insects. Freshw Sci 32:837–848

Barea-Azcón JM, Ballesteros-Duperón E, Moreno D (coords) (2008) Libro Rojo de los Invertebrados de Andalucía. Consejería de Medio Ambiente, Junta de Andalucía, Sevilla, 4 vols

Barquín J, Álvarez-Cabria M, Peñas F (2015) Estimación de caudales ecológicos y estudio de las preferencias tróficas para los distintos grupos de edad de la trucha común en los ríos de Andalucía. Egmasa-Junta de Andalucía y IH-Cantabria. Proyecto Restauración de las poblaciones de trucha común en Andalucía, Memoria Final 2010–2015. Consejería de Medio Ambiente de la Junta de Andalucía, Sevilla

Barquín J, Álvarez-Cabria M, Peñas F, Revilla JA, Fernández F, Álvarez C (2010) Servicio para la estima de la capacidad de carga de la trucha común *Salmo trutta* en los ríos de Andalucía. Egmasa-Junta de Andalucía y IH-Cantabria. Proyecto Restauración de las poblaciones de trucha común en Andalucía, Memoria Final 2005–2009. Consejería de Medio Ambiente de la Junta de Andalucía, Sevilla

Bello CL (1997) Alteraciones de procesos ecológicos en un río de montaña como consecuencia de su regulación. Incidencia sobre los macroinvertebrados bentónicos. PhD thesis. Universidad de Granada, Granada

Bello CL, Alba-Tercedor J (2005) Efecto de la regulación de la cabecera del Río Genil (Sierra Nevada, España) sobre la comunidad de macroinvertebrados acuáticos, y la dieta larvaria de *Rhyacophila nevada* (Insecta: Trichoptera). Limnetica 23(2004):361–370

Bonada N, Múrria C, Zamora-Muñoz C et al (2009) Using community and population approaches to understand how contemporary and historical factors have shaped species distribution in river ecosystems. Glob Ecol Biogeogr 18:202–213

Boumans L, Tierno de Figueroa JM (2016) Introgression and species demarcation in western European *Leuctra fusca* and *L. digitata* (Plecoptera, Leuctridae). Aquat Insects 37:115–126

Casas J, Zamora-Muñoz C, Archila F, Alba-Tercedor J (2000) The effect of a headwater dam on the use of leaf bags by invertebrate communities. Regul Rivers-Res Manage 16:577–591

Castillo Martín A (2001) Clima e hidrología. In: García-Canseco V (coord) Parque Nacional de Sierra Nevada. Canseco Editores, Talavera de la Reina, pp 57–72

Clavero M, Ninyerola M, Hermoso V et al (2017) Historical citizen science to understand and predict climate-driven trout decline Historical citizen science to understand and predict climate-driven trout decline. Proc R Soc B-Biol Sci 284:20161979

Cooke SJ, Raby GD, Donaldson MR et al (2013) The physiological consequences of catch-and-release angling: perspectives on experimental design, interpretation, extrapolation and relevance to stakeholders. Fisheries Manag Ecol 20:268–287

Daigle A, Jeong DI, Lapointe MF (2014) Climate change and resilience of tributary thermal refugia for salmonids in eastern Canadian rivers. Hydrol Sci J 60:1044–1063

DeWalt RE, Maehr MD, Hopkins H, Neu-Becker U, Stueber G (2020) Plecoptera Species File Online. Version 5.0/5.0. http://Plecoptera.SpeciesFile.org. Accessed 20 Mar 2020

Elliott JM (2000) Pools as refugia for brown trout during two summer droughts: trout responses to thermal and oxygen stress. J Fish Biol 56:938–948

Elser JJ, Acharya K, Kyle M et al (2003) Growth rate–stoichiometry couplings in diverse biota. Ecol Lett 6:936–994

Fajardo Merlo MC (2021) Primeras citas de *Leuctra cazorlana* (Aubert, 1962) y *Leuctra geniculata* (Stephens, 1836) (Plecoptera, Leuctridae) en el macizo de Sierra Nevada (España). Bol Asoc Esp Ent 45:119–121

Ferreira S, Tierno de Figueroa JM, Martins FMS et al (2020) The InBIO barcoding initiative database: contribution to the knowledge on DNA barcodes of Iberian Plecoptera. Biodiver Data J 8:e55137

Finn DS, Zamora-Muñoz C, Múrria C et al (2014) Evidence from recently deglaciated mountain ranges that *Baetis alpinus* (Ephemeroptera) could lose significant genetic diversity as alpine glaciers disappear. Freshw Sci 33:207–216

Fochetti R, Tierno de Figueroa JM (2006) Notes on diversity and conservation of the European fauna of Plecoptera (Insecta). J Nat Hist 40:2361–2369

Fochetti R, Tierno de Figueroa JM (2008) Global diversity of stoneflies (Plecoptera; Insecta) in freshwater. Hydrobiologia 595:365–377

Franco Ruíz A, Rodríguez de los Santos M (2001) Libro rojo de los vertebrados amenazados de Andalucía. Consejería de Medio Ambiente Junta de Andalucía, Sevilla

Garcia-Raventós A, Viza A, Tierno de Figueroa JM et al (2017) Seasonality, species richness and poor dispersion mediate intraspecific trait variability in stonefly community responses along an elevational gradient. Freshw Biol 62:916–928

Gasith A, Resh VH (1999) Streams in Mediterranean climate regions: abiotic influences and biotic responses to predictable seasonal events. Annu Rev Ecol Syst 30:51–81

Giorgi F, Lionello P (2008) Climate change projections for the Mediterranean region. Glob Planet Change 63:90–104

González M, Martínez-Menéndez J (2011) Checklist of the caddisflies of the Iberian Peninsula and Balearic Islands. Zoosymposia 5:115–135

Hampe A, Petit RJ (2005) Conserving biodiversity under climate change: the rear edge matters. Ecol Lett 8:461–467

Hering D, Schmidt-kloiber A, Murphy J et al (2009) Potential impact of climate change on aquatic insects: A sensitivity analysis for European caddisflies (Trichoptera) based on distribution patterns and ecological preferences. Aquat Sci 71:3–14

Hessen DO, Ventura M, Elser JJ (2008) Do phosphorus requirements for RNA limit genome size in crustacean zooplankton? Genome 51:685–691

Hessen DO, Punidan D, Jeyasingh PD et al (2009) Genome streamlining and the elemental cost of growth. Trends Ecol Evol 25:75–80

IPCC (2013) Climate Change 2013: The Physical Science Basis. Contribution of Working Group I to the Fifth Assessment Report of the Intergovernmental Panel on Climate Change. Cambridge University Press, Cambridge & New York

IUCN (2021) The IUCN red list of threatened species. Version 2021-1. https://www.iucnredlist.org. Accessed 7 Apr 2021

Jáimez-Cuéllar P (2004) Caracterización físico-química, macroinvertebrados acuáticos y valoración del estado ecológico de dos cuencas mediterráneas de influencia nival (ríos Guadalfeo y Adra), según los criterios de la Directiva Marco del Agua. PhD thesis. Universidad de Granada, Granada

Jensen AJ, Johnsen B (1999) The functional relationship between peak spring floods and survival and growth of juvenile Atlantic Salmon (*Salmo salar*) and Brown Trout (*Salmo trutta*). Funct Ecol 13:778–785

Kernan M, Battarbee RW, Moss B (2010) Climate change impacts on freshwater ecosystems. Wiley-Blackwell, Oxford

Larios-López JE (2017) La trucha común [*Salmo trutta* (Linnaeus, 1758)] en Andalucía. Distribución, fenología reproductiva, procesos reguladores y propuestas de gestión de sus poblaciones. Ph.D. thesis. Universidad de Granada, Granada

Larios-López JE, Alonso González C, Galiana-García M, Tierno de Figueroa JM (2021) Driving factors of synchronous dynamics in brown trout populations at the rear edge of their native distribution. Ecol Freshw Fish 30:4–17

Larios-López JE, Galiana-García M, Alonso C, Tierno de Figueroa JM (2018) La trucha común como indicador del cambio global en Sierra Nevada. Quercus 390:24–30

Larios-López JE, Galiana-García M, Alonso C, Tierno de Figueroa JM (2019) Consideraciones sobre la trucha común y la pesca sin muerte. Quercus 395:80–81

Larios-López JE, Tierno de Figueroa JM, Alonso C, Nebot Sanz B (2015a) Distribution of brown trout (*Salmo trutta* Linnaeus, 1758) (Teleostei: Salmonidae) in its southwesternmost European limit: possible causes. Ital J Zool 82:404–415

Larios-López JE, Tierno de Figueroa JM, Galiana-García M et al (2015b) Extended spawning in brown trout (*Salmo trutta*) populations from the Southern Iberian Peninsula: the role of climate variability. J Limnol 74:394–402

Liess A, Hillebrand H (2005) Stoichiometric variation in C:N, C:P, and N:P ratios of littoral benthic invertebrates. J N Am Benthol Soc 24:256–269

López-Rodríguez MJ, Luzón-Ortega JM, Tierno de Figueroa JM (2012) On the biology of two high mountain populations of stoneflies (Plecoptera, Perlodidae) in Southern Iberian Peninsula. Limnetica 31:205–212

López-Rodríguez MJ, Tierno de Figueroa JM, Alba-Tercedor J (2008) Life history and larval feeding of some species of Ephemeroptera and Plecoptera (Insecta) in the Sierra Nevada (Southern Iberian Peninsula). Hydrobiologia 610:277–295

Luzón-Ortega JM, López-Rodríguez MJ, Tierno de Figueroa JM (2013) Contribution to the knowledge of the stoneflies from Spain (Insecta, Plecoptera). Bol Asoc Esp Ent 37:225–275

Martín L (2017) Biodiversidad y conservación de los tricópteros (Insecta: Trichoptera) de la península ibérica y la Macaronesia. Ph. D. thesis. Universidade de Santiago de Compostela, Santiago de Compostela

Millán A, Sánchez-Fernández D, Abellán P, et al (2014) Atlas de los coleópteros acuáticos de España peninsular. Ministerio de Agricultura, Alimentación y Medio Ambiente, Madrid

Molero Mesa J, Pérez Raya F, Valle Tendero C (coord) (1992) Parque Natural de Sierra Nevada: Paisaje, Fauna, Flora, Itinerarios. Editorial Rueda, S.L. Madrid

Morse JC (ed) (2021) Trichoptera world checklist. http://entweb.clemson.edu/database/trichopt/index.htm. Accessed 5 May 2021

Morse JC, Frandsen PB, Graf W, Thomas JA (2019) Diversity and ecosystem services of trichoptera. Insects 10:125

Múrria C, Sáinz-Bariáin M, Vogler AP et al (2020) Vulnerability to climate change for two endemic high-elevation, low-dispersive *Annitella* species (Trichoptera) in Sierra Nevada, the southernmost high mountain in Europe. Insect Conserv Divers 13:283–295

Múrria C, Zamora-Muñoz C, Bonada N et al (2010) Genetic and morphological approaches to the problematic presence of three *Hydropsyche* species of the *pellucidula* group (Trichoptera: Hydropsychidae) in the westernmost Mediterranean Basin. Aquat Insects 32:85–98

Olah J, Chvojka P, Coppa G et al (2014) The genus *Allogamus* Schmid, 1955 (Trichoptera, Limnephilidae): revised by sexual selection-driven adaptive, non-neutral traits of the phallic organ. Opusc Zool Budapest 45:33–82

Pardo I, Álvarez M, Casas J et al (2002) El hábitat de los ríos mediterráneos. Diseño de un índice de diversidad de hábitat. Limnetica 21(2002):115–133

Pérez-Luque AJ, Pérez-Pérez R, Bonet FJ (2016) Climate change over the last 50 years in Sierra Nevada. In: Zamora R, Pérez-Luque AJ, Bonet FJ et al (eds) Global change impacts in Sierra Nevada: Challenges for conservation. Consejería de Medio Ambiente y Ordenación del Territorio, Junta de Andalucía, Granada, pp 24–26

Policansky D (2002) Catch–and–release recreational fishing: a historical perspective. In: Pitcher TJ, Hollingwort CE (eds) Recreational fisheries: ecological, economic and social evaluation. Blackwell Science, Oxford, pp 74–94

Poquet JM, Alba-Tercedor J, Punt T et al (2009) The MEDiterranean Prediction And Classification System (MEDPACS): an implementation of the RIVPACS/AUSRIVAS predictive approach for assessing Mediterranean aquatic macroinvertebrate communities. Hydrobiologia 623:153–171

Pulido A (1980) Datos hidrogeológicos sobre el borde occidental de Sierra Nevada. Serie Universitaria, Fundación Juan March, Madrid

Ribera I, Hernando C, Aguilera P (1998) An annotated checklist of the Iberian water beetles (Coleoptera). Zapateri 8:43–111

Romero Martín A, Alba-Tercedor J (2013) Estatus y evolución histórica del conocimiento de los invertebrados acuáticos de Sierra Nevada. In: Ruano F, Tierno de Figueroa M, Tinaut A (eds) Los Insectos de Sierra Nevada: 200 años de historia, vol 1. Asociación Española de Entomología, Granada, pp 26–64

Ruano F, Tierno de Figueroa M, Tinaut A (eds) (2013) Los Insectos de Sierra Nevada: 200 años de historia, vols 1 and 2. Asociación Española de Entomología, Granada

Sáez P, Menor A, Galindo J et al (2010) El estudio de las distribuciones históricas aplicado a la recuperación de especies: el caso de la trucha común (*Salmo trutta* Linnaeus, 1758). XV Congresso da Associação Ibérica de Limnologia, Azores

Sáinz-Bariáin M, Fajardo-Merlo M, Zamora-Muñoz C (2016a) Changes in composition and abundance of benthic invertebrate communities. In: Zamora R, Pérez-Luque AJ, Bonet FJ et al (eds) Global change impacts in Sierra Nevada: challenges for conservation. Consejería de Medio Ambiente y Ordenación del Territorio, Junta de Andalucía, Granada, pp 75–78

Sáinz-Bariáin M, Zamora-Muñoz C (2012) The larva and life history of *Stenophylax nycterobius* (McLachlan, 1875) (Trichoptera: Limnephilidae) in high mountain streams (Sierra Nevada, Spain) and key to the Iberian larvae of the genus. Zootaxa 81:71–81

Sáinz-Bariáin M, Zamora-Muñoz C (2015) Larval descriptions of *Annitella esparraguera* (Schmid 1952) and *Annitella iglesiasi* González & Malicky 1988 (Trichoptera: Limnephilidae), two endemic species from Southern Europe. Zootaxa 4006:347–360

Sáinz-Bariáin M, Zamora-Muñoz C, González M (2013) Los Tricópteros (Trichoptera). In: Ruano F, Tierno de Figueroa JM, Tinaut A (eds) Los Insectos de Sierra Nevada. 200 años de Historia, vol 1. Asociación española de Entomología, Granada, pp 203–230

Sáinz-Bariáin M, Zamora-Muñoz C, Soler JJ et al (2016b) Changes in Mediterranean high mountain Trichoptera communities after a 20-year period. Aquat Sci 78:669–682

Sáinz-Cantero CE (2013) The aquatic Coleoptera (Coleoptera, Adephaga and Polyphaga). In: Ruano F, Tierno de Figueroa M, Tinaut A (eds) Los Insectos de Sierra Nevada: 200 años de historia, vol 1. Asociación Española de Entomología, Granada, pp 324–349

Salazar Mendías C, Valle Tendero C (2019) Las formaciones vegetales de Sierra Nevada y su conservación. In: Peñas J, Lorite J (eds) Biología de la conservación de plantas en Sierra Nevada. Editorial Universidad de Granada, Granada, Principios y retos para su preservación, pp 211–232

Sánchez-Bayo F, Wyckhuys KAG (2019) Worldwide decline of the entomofauna: a review of its drivers. Biol Conserv 232:8–27

Sánchez-Fernández D, Bilton DT, Abellán P et al (2008) Are the endemic water beetles of the Iberian Peninsula and the Balearic Islands effectively protected? Biol Conserv 141:1612–1627

Sánchez-Ortega A, Tierno JM (1996) Current situation of stonefly fauna (Insecta: Plecoptera) in the Iberian Peninsula and Balearic Islands. Mitt Schweiz Ent Ges 69:77–94

Santiago JM, Muñoz-Mas R, Solana J et al (2017) Waning habitats due to climate change: effects of streamflow and temperature changes at the rear edge of the distribution of a cold-water fish. Hydrol Earth Syst Sci 21:4073–4101

Sanz A, López-Rodríguez MJ, García-Mesa S et al (2017) Are antioxidant capacity and oxidative damage related to biological and autoecological characteristics in aquatic insects? J Limnol 76:170–181

Sanz A, Trenzado CE, López-Rodríguez MJ et al (2010) Study of the antioxidant defence in four species of Perloidea (Insecta, Plecoptera). Zool Sci 27:952–958

Sterner RW, Elser JJ (2002) Ecological stoichiometry. Princeton University Press, Princeton

Sweeney BW (1984) Factors influencing life-history patterns of aquatic insects. In: Resh VH, Rosemberg DM (eds) The ecology of aquatic insects. Praeger, New York, pp 56–100

Tierno de Figueroa JM, López-Rodríguez MJ, Bo T, Fenoglio S (2009a) Allometric versus isometric growth in European stoneflies (Plecoptera). J Freshw Ecol 24:581–585

Tierno de Figueroa JM, López-Rodríguez MJ, Fenoglio S et al (2013a) Freshwater biodiversity in the rivers of the Mediterranean Basin. Hydrobiologia 719:137–186

Tierno de Figueroa JM, López-Rodríguez MJ, Lorenz AW et al (2010) Vulnerable taxa of European Plecoptera in the context of climate change. Biodivers Conserv 19:1269–1277

Tierno de Figueroa JM, López-Rodríguez MJ, Luzón-Ortega JM (2013b) Los Plecópteros (Plecoptera). In: Ruano F, Tierno de Figueroa M, Tinaut A (eds) Los Insectos de Sierra Nevada: 200 años de historia, vol 1. Asociación Española de Entomología, Granada, pp 126–138

Tierno de Figueroa JM, López-Rodríguez MJ, Villar-Argaiz M (2019) Spatial and seasonal variability in the trophic role of aquatic insects: an assessment of functional feeding group applicability. Freshw Biol 64:954–966

Tierno de Figueroa JM, Luzón-Ortega JM, López-Rodríguez MJ (2009b) First record of male drumming call of the genus *Capnioneura* Ris, 1905 (Plecoptera, Capniidae). Entomol Sci 12:359–362

Tierno de Figueroa JM, Luzón-Ortega JM, López-Rodríguez MJ (2014) First record of the drumming signals of stoneflies *Capnopsis* Morton, 1896 and *Protonemura* Kempny, 1898 genera (Plecoptera, Capniidae and Nemouridae). Entomol Sci 17:302–308

Tierno de Figueroa JM, Luzón-Ortega JM, López-Rodríguez MJ (2018) Checklist de Fauna Ibérica. Orden Plecoptera (Arthropoda: Insecta) en la península ibérica e islas Baleares (2018 edition). In: Ramos MA, Sánchez Ruiz M (eds) Documentos Fauna Ibérica, 5. Museo Nacional de Ciencias Naturales, CSIC, Madrid, pp 2 (sn) + 15

Ulbrich U, May W, Li L, Lionello P, Pinto JG, Somot S (2006) The Mediterranean climate change under global warming. In: Lionello P, Malanotte-Rizzoli P, Boscolo R (eds) Developments in earth and environmental sciences. Elsevier, Amsterdam, pp 399–415

Vannote RL, Sweeney BW (1980) Geographic analysis of thermal equilibria: a conceptual model for evaluating the effect of natural and modified thermal regimes on aquatic insect communities. Am Nat 115:667–695

Verdú JR, Galante E (eds) (2006) Libro Rojo de los Invertebrados de España. Dirección General para la Biodiversidad, Ministerio de Medio Ambiente, Madrid

Verdú JR, Numa C, Galante E (eds) (2011) Atlas y Libro Rojo de los Invertebrados amenazados de España (Especies Vulnerables). Dirección General de Medio Natural y Política Forestal, Ministerio de Medio Ambiente, Medio Rural y Marino, Madrid

Vigna Taglianti A, Audisio PA, Biondi M et al (1999) A proposal for a chorotype classification of the near East fauna, in the framework of the Western Paleartic region. Biogeographia 20:31–59

Villar-Argaiz M, López-Rodríguez MJ, Tierno de Figueroa JM (2020) Body P content increases over ontogeny in hemimetabolous macroinvertebrates in a Mediterranean high mountain stream. Aquat Ecol 54:1185–1200

Villar-Argaiz M, López-Rodríguez MJ, Tierno de Figueroa JM (2021) Divergent nucleic acid allocation in juvenile insects of different metamorphosis modes. Sci Rep 11:10313

Wiggins GB (2004) Caddisflies: the underwater architects. University of Toronto Press, Toronto, Buffalo & London

Zamora-Muñoz C, Alba-Tercedor J (1992) Caracterización y calidad de las aguas del Río Monachil (Sierra Nevada, Granada). Factores físico-químicos y comunidades de macroinvertebrados acuáticos. Agencia del Medio Ambiente, Granada

Zamora-Muñoz C, Alba-Tercedor J (1996) Bioassessment of organically polluted Spanish rivers, using a biotic index and multivariate methods. J N Am Benthol Soc 15:332–352

Zamora-Muñoz C, Sánchez-Ortega A, Alba-Tercedor J (1993) Physico-chemical factors that determine the distribution of mayflies and stoneflies in a high-mountain stream in Southern Europe (Sierra Nevada, Southern Spain). Aquat Insects 15:11–20

Zhou X, Frandsen PB, Holzenthal RW et al (2016) The Trichoptera barcode initiative: a strategy for generating a species-level Tree of Life. Philos Trans R Soc B-Biol Sci 371:20160025

# High Mountain Lakes as Remote Sensors of Global Change

Juan Manuel Medina-Sánchez, Marco J. Cabrerizo, Juan Manuel González-Olalla, Manuel Villar-Argaiz, and Presentación Carrillo

**Abstract**

This chapter addresses the question of why the high mountain lakes of Sierra Nevada (Spain) are unique remote sensors of global change. The answers arise from a set of peculiar physico-chemical and biological features shared by most of these water bodies. Their physical characteristics stem from their altitude, geographic location, remoteness, small catchment areas, high ultraviolet radiation (UVR) fluxes and great diversity of water bodies within a small-scale mountain range, making these lakes to be highly connected with atmospheric processes in a climate change hot-spot. Their biota is simple and highly sensitive and responsive to the interaction among global-change stressors, as revealed by the variations observed in the structural and functional biotic variables in response to environmental stressors. This allows us to understand the underlying mechanisms of changes in their communities and anticipate induced state changes in these and other ecosystems over the world. The feasibility of studying the physical and biological features of these lakes through observational and experimental approaches at short, mid- and long-term scales provides empirical evidence that supports the vision that these ecosystems are excellent and unique sentinel ecosystems of global change.

**Keywords**

High mountain lakes • Remote sensors • Phytoplankton • Zooplankton • Bacterioplankton • Mixotrophy • Stressors • Aerosols • Nutrient inputs • Carbon flux • Functional biodiversity

## 1 Introduction

An exhaustive review of the research on alpine lakes of Sierra Nevada from 1975 to the present has identified the role of these ecosystems as sentinels of change (Villar-Argaiz and Bullejos 2016; Medina-Sánchez et al. 2016). Unlike other European mountain lakes, the geographic location at a climate change hot spot (Mediterranean region, South of Spain) and geological history of Sierra Nevada cause their high mountain lakes to be exposed simultaneously to several environmental stressors at the largest recorded range: climatic anomalies (temperature and precipitation), ultraviolet radiation, aerosol dust deposition due to atmospheric transport of Saharan aerosols and allochthonous nutrient (phosphorus, calcium, etc.) input linked to these aerosols. The high water transparency, low nutrient content, and narrow but highly variable temperature ranges found in the high mountain lakes of Sierra Nevada qualifies them as integrated biophysical sensors of global change (Medina-Sánchez et al. 2016; Villar-Argaiz and Bullejos 2016). This quality stems from a peculiar biota, characterized by a great simplicity of their biological communities but, paradoxically, establishing complex ecological interactions (Carrillo et al. 2006). These traits help us to assess the impact of these environmental stressors at seasonal and interannual scales on the dynamics of the organisms and communities that inhabit these water bodies, as well as the ecosystem functioning (biodiversity patterns, carbon flux, ecosystem metabolism, etc.).

J. M. Medina-Sánchez (✉) · M. Villar-Argaiz
Department of Ecology, Faculty of Sciences, University of Granada, Granada, Spain
e-mail: jmmedina@ugr.es

J. M. Medina-Sánchez · M. J. Cabrerizo · J. M. González-Olalla · M. Villar-Argaiz · P. Carrillo
University Institute of Water Research, University of Granada, Granada, Spain

M. J. Cabrerizo
Departamento de Ecología y Biología Animal, Facultad de Ciencias del Mar, Universidad de Vigo, Vigo, Spain

Centro de Investigación Mariña da, Universidade de Vigo (CIM-UVigo), Vigo, Spain


R. Zamora and M. Oliva (eds.), *The Landscape of the Sierra Nevada*,
https://doi.org/10.1007/978-3-030-94219-9_16

As an example of the above, long-term monitoring of the physico-chemical parameters and pelagic plankton in La Caldera Lake, the largest high mountain lake of Sierra Nevada, indicates that phytoplankton, but not their herbivorous zooplankton consumers, increased in parallel with the increase in the intensity and frequency of atmospheric aerosols, and experimental approaches corroborated an increase in the greenness and less quality of water lake under large inputs of phosphorus (Villar-Argaiz and Bullejos 2016). These changes were coupled with an interannual increase of the seasonal presence and persistence of microalgae (phytoplankton) with a strict autotrophic metabolism, and the experimental approaches corroborated the development of a simplified and very low diverse algal community dominated by a strict autotrophic algal species under moderate and large inputs of phosphorus, reaching densities of up to three orders of magnitude greater than in the unaltered ecosystem. This response was concomitant with a decrease and even complete loss of the microalgae with a mixotrophic metabolism (a combination of autotrophy and heterotrophy in the same organism) observed in the experiments (Medina-Sanchez et al. 2016). This functional algal group (mixotrophs) is characteristic and even dominant in the phytoplankton inhabiting these lakes in their pristine status, so that their potential loss in favor to blooms of green algae due to a higher nutrient input forcing (e.g., those linked to an increase in frequency and intensity of atmospheric aerosol events) would imply a loss in functional biodiversity and alteration in ecosystem functioning (Medina-Sánchez et al. 2016).

The above results underscore the importance of continuing with long-term monitoring programmes in high mountain lakes of Sierra Nevada and, importantly, underline their role as unique ecosystem sensors of global change. In trying to arrive at a satisfactory answer to the question of why the high mountain lakes of Sierra Nevada are unique sentinels of global change, the guiding thread of this chapter will review and update the mounting observational and experimental scientific evidence derived from the special physico-chemical and biological features of these ecosystems.

## 2 Physico-chemical Features

From a geographic and physical perspective, high mountain lakes of Sierra Nevada are witnesses of past climate and sentinels of future global change, due to several reasons (Fig. 1).

### 2.1 Elevation and Catchment Areas

The high elevation of Sierra Nevada lakes (between 2800 and 3100 m) implies that they are subjected to prolonged ice and snow cover during the annual cycle. Although all these water bodies are fed by snowmelt, snow cover shows pronounced interannual variability. This is due to their location in the south of the Mediterranean region, characterized by a climate of more moderate winter temperatures than in higher latitudes, scarce total annual precipitation, and a high frequency of severe droughts. The meteorological station located in Sierra Nevada at 2507 m reports a mean annual air temperature of 3.9 °C and annual precipitation of 693 mm, with 80% falling as snow between October and April, and ca. 95% as snow above 2500 m altitude (Pérez-Martínez et al. 2020). The ice-cover season in the Sierra Nevada lakes typically lasts from November to June, shorter than in other northern latitudes, although there have been great inter-annual differences (Villar-Argaiz et al. 2001; Pérez-Martínez et al. 2020). The lakes are exposed to generally low, but highly variable temperatures during the ice-free period. As an example, La Caldera Lake, one of the biggest permanent lakes of Sierra Nevada (2.10 ha of maximum lake area, Morales-Baquero et al. 1999; up to 42,500 $m^3$ of water volume, Díaz-Hernández and Herrera-Martínez 2019) has recorded a seasonal variation of mean water-column temperature of ca. 10 °C in the last decades (Medina-Sánchez et al. 1999; Delgado-Molina et al. 2009) and this range of variation is increasing in the past years, with maximum water-column temperature breaking records (>18 °C, unpublished data). The lakes undergo high incident solar visible and ultraviolet radiation (UVR), which increases about 8% (total irradiance), 9% (UV-A, 320–400 nm), and 18% (UVB-erythemal, 280–320 nm) per 1000 m of altitude under clear-sky conditions in summer (Blumthaler et al. 1997; Blumthaler 2005). UVR usually reaches the bottom of these lakes due to the high transparency of their water columns, with low dissolved organic carbon concentration (DOC <200 μmol $L^{-1}$; Reche et al. 2001, 2005; Durán et al. 2016) and, hence, with low diffuse attenuation coefficients for UVA and UVB (<0.5 $m^{-1}$; Carrillo et al. 2008a; Medina-Sánchez et al. 2013). Therefore, these clear-water lakes can forecast what would happen if the incidence of UVR augments in a hypothetical return to lowering stratospheric ozone scenario or due to a higher thermal stratification of their water column due to global warming (Helbling et al. 2013).

Along with their altitude, the Sierra Nevada lakes are located on small catchment areas above the treeline, with

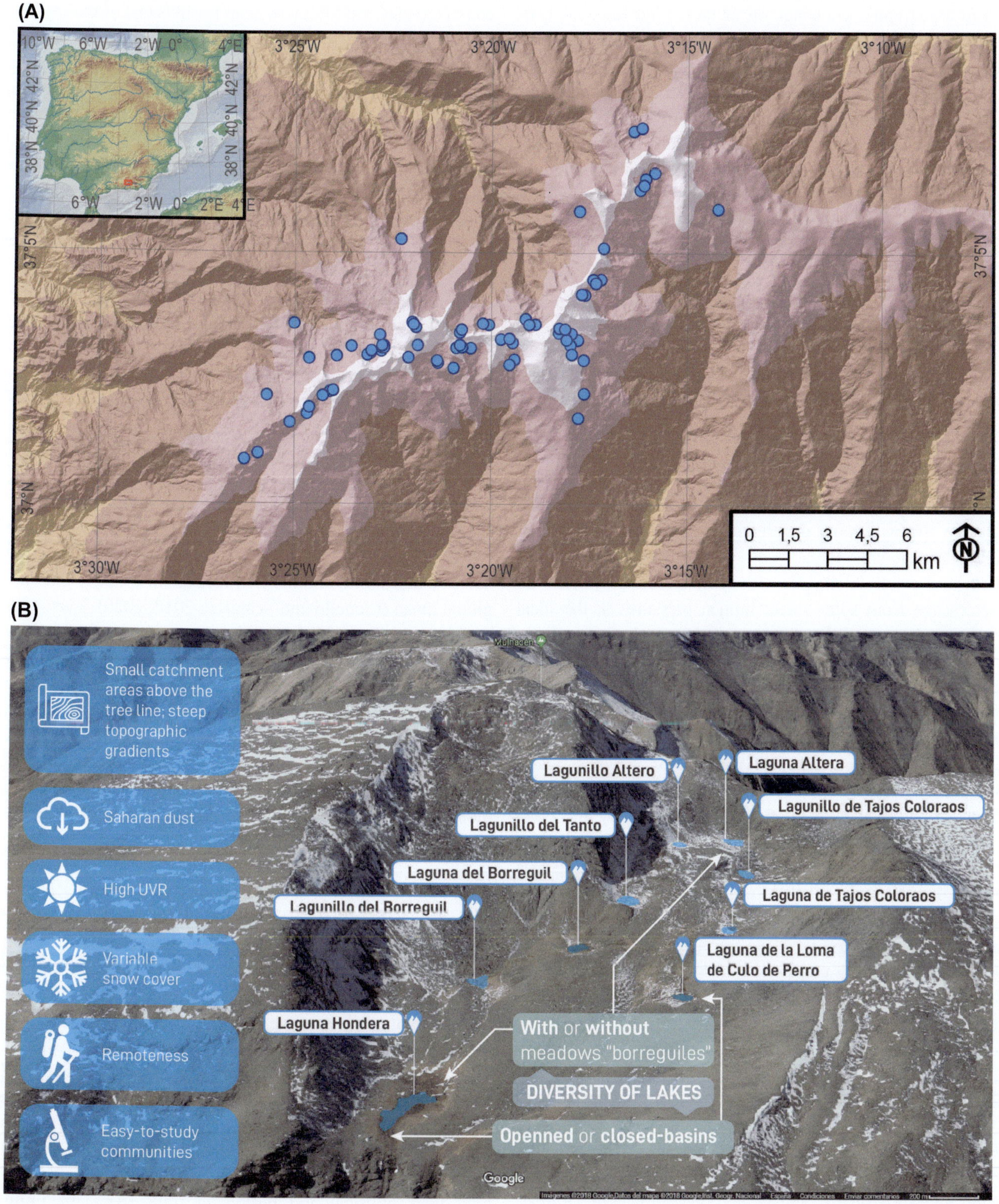


**Fig. 1** **a** Geographical location of the high mountain lakes of Sierra Nevada. Maps taken from the website https://lagunasdesierranevada.es/ and Wikipedia (inset). **b** Infographics showing some features and external factors shaping the high mountain lakes of Sierra Nevada (see text). Modified image taken from the website https://lagunasdesierranevada.es/

steep topographic gradients, on metamorphic siliceous bedrock with soils poorly developed. These features favor the oligotrophy, low salinity and high transparency of their waters, because inputs of organic carbon (C) and nutrients from terrestrial surroundings are small in comparison with those on inland waters placed in lowlands. These physical characteristics, some of them shared with most high mountain lakes in the world, shape their ecosystem structure, diversity and productivity (Carrillo et al. 2006; Catalan et al. 2006; Moser et al. 2019).

## 2.2 Remoteness

In addition to their high elevation, Sierra Nevada lakes are remote because they are far from populated and agricultural areas. Such remoteness has preserved them from intense direct human impact in such way that they can be still considered as pristine water bodies (e.g., Afonina et al. 2020), and consequently, excellent sites to study climate change effects because the climate signal is not as obscured by other human impacts as in more populated areas (Jiménez et al. 2019). Nevertheless, these ecosystems do not escape from some direct human pressures at the local scale, such as traditional mountain livestock, currently in declining trend, and tourism and visitors, currently in augmenting trend. These trends are linked to profound cultural and socioeconomical shifts that favor the public interest and access to mountain ecosystems (Catalan et al. 2017), even though transport with motor vehicles within the National Park of Sierra Nevada is currently forbidden. Both pressures negatively affect the water quality and biota of the high mountain lakes (Derlet et al. 2010, 2012), due to increased load of nutrients and also of artificial materials (e.g., plastics) transported from the watersheds subjected to pasturage (livestock) or trekking (visitors). These pressures are accentuated during the ice-free period, as both agents tend to behaviorally concentrate on the border of the water bodies to perform their activities (e.g., eating, excreting, resting).

## 2.3 Location in a Climate Change Hot Spot

Sierra Nevada is located in the Mediterranean region, which climatologically constitutes a transition zone between the subtropical (African) climate and the temperate (Eurosiberian) climate. This region is affected by the interaction of climate processes of mid and tropical latitudes favoring extreme events such as heat and cold waves, storms and droughts (Giorgi and Lionello 2008; chapter "Climate Variability and Trends"). Further, due to its climatological characteristics, the Mediterranean Basin is one of the regions of the world particularly sensitive to the effects of climate change (Stocker et al. 2013; Masson-Delmotte et al. 2021), in such way that small variations in the general circulation can result in major changes in its regional climate (Garcia-Herrera et al. 2014). Therefore, the Mediterranean region is considered a climate change hot spot, although the causes are not yet fully understood (Tuel and Eltahir 2020). Observed rates of climate change in this region exceed global trends for most variables. Thus, mean annual air temperatures are now 1.4 °C above those of the late nineteenth century. Heatwaves are more frequent, and the intensity and frequency of droughts have increased since 1950 (Cramer et al. 2018). Further, marine storms wave intensity is increasing (Amarouche and Akpinar 2021) as well as the likelihood of heat-induced large wildfires (Ruffault et al. 2020). There are consistent predictions of a generalized warming by the end of the century (between 1 and 5 °C with respect to the 1986–2005 reference period), which is expected to be strongest during summer (up to 7 °C). Thus, future warming in the Mediterranean region is expected to exceed global rates by 25%, particularly accentuated during summer, with rates 40% larger than the global mean (Lionello and Scarascia 2018, 2020). Even with a global warming of 1.5 °C, it is projected a 2.2 °C increase in regional daytime maxima temperature (Seneviratne et al. 2016). A general drying (between 10 and 40%) is also inferred for the Mediterranean region (Zittis et al. 2019), with a causal dependence on the season (Brogli et al. 2019). Notably, warming rate is amplified in high-elevation areas compared to the global average, making these areas to be more sensitive to climate changes (Wang et al. 2014; Pepin et al. 2015). Sierra Nevada has shown rapid changes due to recent climate warming, with the disappearance of glacial ice from the highest north-facing cirques (Oliva et al. 2016), the degradation of the last permafrost patches (Gómez-Ortiz et al. 2019), a trend in declining mean annual rainfall (Ruiz-Sinoga et al. 2011), and a reduction of snow and ice cover since the 1960s (Pérez-Palazón et al. 2015), which is being accentuated since the twenty-first century (Bonet et al. 2016, chapter "Snow Dynamics, Hydrology, and Erosion"). The high mountain lakes of Sierra Nevada may reflect and even magnify the effects of climatic change at a regional/local scale, reinforcing the role of these ecosystems as sensors and forecasters of global change. This is supported by a great body of studies on these lakes showing how their physical and biological traits, e.g., maximum depth, nutrient concentration, composition and biomass of planktonic and benthic communities, etc., vary as a result of the high inter- and intra-annual climatic variability (e.g., Villar-Argaiz et al. 2001; Sánchez-Castillo et al. 2008; García-Jurado et al. 2011; González-Olalla et al. 2018).

## 2.4 Proximity to Major Sources of Limiting Nutrients

The Mediterranean region is exposed to an increasing input of nutrient-rich aerosols from the Sahara Desert (De la Paz et al. 2013). Sierra Nevada is particularly receptive to these intrusions (Morales-Baquero et al. 2006; chapter "Atmospheric Inputs and Biogeochemical Consequences in High-Mountain Lakes") because of its proximity to the original source and its altitude (>3000 m), being the first geographic barrier to intercept Saharan dust transported by atmosphere towards Western Europe. The mainstream of Saharan dust transport is in the free troposphere, between 1500 and 4000 m (Talbot et al. 1986), and because the Sierra Nevada high mountain lakes are located at an elevation above the planetary boundary layer (PBL) on this region (annual mean of $1700 \pm 500$ m, Granados-Muñoz et al. (2012), the influence of regional land-use in dust deposition is low compared to that of atmospheric transport. Dust aerosols are the predominant source of phosphorus (P) in the atmosphere (Mahowald et al. 2008) and also transport nitrogen (N) and micronutrients (e.g., calcium [Ca], iron [Fe]) adsorbed on the particles (Dentener et al. 1996; Morales-Baquero et al. 2013; Rodriguez-Navarro et al. 2018). Therefore, wet and dry deposition of these aerosols provokes increases of nutrient inputs on Sierra Nevada, mainly P and Ca (Morales-Baquero et al. 2013; Morales-Baquero and Pérez-Martínez 2016) that can be particularly critical in the severely nutrient-limited Sierra Nevada lakes (Carrillo et al. 2015; Cabrerizo et al. 2017). In fact, dust deposition or its equivalent amount of nutrients has been shown to increase chlorophyll-*a* and the pool of dissolved P, Ca and organic matter (Villar-Argaiz et al. 2001; Morales-Baquero et al. 2006; Pulido-Villena et al. 2006; Mladenov et al. 2011), bacterial growth (Reche et al. 2009), as well as to alter algae-bacteria and algae-zooplankton trophic interactions (Villar-Argaiz et al. 2002, 2012; Medina-Sánchez et al. 2006, 2013; Cabrerizo et al. 2017; González-Olalla et al. 2018; see below) in these lakes.

## 2.5 Diversity of Lakes and Surrounding Habitats

The high mountain lakes of Sierra Nevada constitute a diverse set of water body types that embrace a wide gradient of ecosystem sensitivity and responsiveness to global change on a small spatial scale. This is because the local features of the lakes determine how their biota is affected by the environmental drivers (Kernan et al. 2009; Moser et al. 2019). The set includes numerous (75–80) permanent and temporary water bodies located within in a small mountain range (80 km length, 15–30 km width, >2000 km$^2$ area) on two main hydrological catchments, the Genil River (Atlantic watershed) and Guadalfeo River (Mediterranean watershed). The lakes have a glacier origin, either by glacial overdeepening or damming by glacial eroded rocky thresholds or moraine deposits (Díaz-Hernández and Herrera-Martínez 2019; chapter "The Impact of Glacial Development on the Landscape of the Sierra Nevada").

The character permanent or temporal of the lakes and the evolution of their water volume depends on the snow cover during winter and spring. Also, it is critical the seepage rate through the rocks and the evaporation rate during the usual strong summer aridity that characterizes the cryo–oromediterranean bioclimate, in contrast to the wetter summer in temperate alpine mountains of the Euro-Siberian region (Díaz-Hernandez and Herrera-Martínez 2019). As a consequence, although most lakes are generally small water bodies (e.g., lake surface area <0.5 Ha, maximum depth <3 m, Morales-Baquero et al. 1999; Durán et al. 2016), it is also true that have a great variability of size, either among lakes (some reaching >8 m depth, Durán et al. 2016) or within a same lake over an inter-annual scale (e.g., La Caldera Lake with a maximum depth of 2–14 m, Medina-Sanchez et al. 1999). To this variability of lake size, it is added a high geomorphological diversity. Many lakes are surrounded by a peripheral fringe of meadows, locally called "borreguiles", that contrast to the extreme summer aridity of the surrounding landscape of cryo-oromediterranean mountain, whereas other water-bodies lack of 'borreguiles', including some of the biggest lakes (e.g., La Caldera, Las Yeguas, Vacares, Altera). Díaz-Hernandez and Herrera-Martínez (2019) distinguish three types of water bodies: (i) without *green fringes*; (ii) with *small green fringes*, i.e., of less than 1000 m$^2$, restricted to small, gently sloping steps around the water's edge suitable for narrow bands of vegetation; and (iii) with *extensive green fringes*, i.e., extending to over 1000 m$^2$ and almost entirely surrounding the water-bodies with green vegetation over summer. Besides, the lakes can differ according to their hydraulic washout, and have been classified as: (i) *open-basin lakes*, i.e., those with persistent superficial diffuse or visible inlets/outlets; (ii) *open-closed basin lakes*, i.e., those with temporary inlets/outlets which disappear with the progression of summer); and (iii) *closed-basin lakes*, i.e., those without visible surface inlets or outlets (Pérez-Martinez et al. 2020). The hydraulic washout, linked to the type of basin, has been shown to be an important regulating factor of current and past (subfossil) phyto- and zooplankton populations in Sierra Nevada lakes (Morales-Baquero et al. 2019; Pérez-Martinez et al. 2020).

### 2.6 Easy Applicability of Monitoring Approaches

The physico-chemical features of Sierra Nevada high mountain lakes, either particular or shared with other high mountain lakes worldwide, allow their study by means of monitoring methodologies. For example, the material preserved in lake sediments can be monitored following the paleolimnological approach that overcomes the lack of continuous records of historical data on the lakes. This material is an excellent archive of long-term environmental changes and allows reconstruction of past environments from limnological, ecological and geochemical lake sediment proxies (Catalan et al. 2013; see chapters "Reconstruction of Past Environment and Climate Using Wetland Sediment Records from the Sierra Nevada" and "Paleolimnological Indicators of Global Change").

The lake morphology, area, water-color, etc., can be monitored by remote sensing (Fig. 1b) through satellite images of high spatial and temporal resolution, which allows us to study the intra- and interannual evolution of the lakes, as already it is being done for the vegetation of Sierra Nevada (Alcaraz-Segura et al. 2016; Zamora et al. 2017) and aquatic ecosystems of Iberian Peninsula (Doña et al. 2016; Franch-Gras et al. 2017). This approach can overcome the lack of continuous and simultaneous monitoring data on all lakes (spatial scale). The satellite's images can be particularly easy to acquire on the Sierra Nevada high mountain lakes, due to the relative proximity among them and the high annual frequency of cloudless days on Sierra Nevada massif.

## 3 Biological Features

Along with their physico-chemical traits, most of the high mountain lakes of Sierra Nevada also share a peculiar biotic structure, being a common trait its simplicity, with a low number of trophic compartments (Carrillo et al. 2006) and the absence of a major size overlap between the organisms of each trophic compartment (Echevarría et al. 1990; Medina-Sánchez et al. 1999, 2002; Villar-Argaiz et al. 2002). This allows their easy segregation, e.g., by filtration, which eases the study of the structural (e.g., elemental composition, lipids content) and functional (e.g., primary and secondary production, respiration, ecoenzyme activities) features for each biological fraction separately (Fig. 1b), as many observational and experimental studies have shown (Medina-Sánchez et al. 2002, 2017; Villar-Argaiz et al. 2009, 2012; Carrillo et al. 2015; Velasco Ayuso et al. 2017).

The peculiar biotic structure is the result of the adaptation of their communities to live in these severe environmental conditions. This is the case of nanoflagellates algae with mixotrophic feeding modes that are integrant of phytoplankton and establish a paradoxical relationship with bacterioplankton. For example, La Caldera Lake, a water body studied over the past decades, shows a consistent successional pattern in which phytoplankton typically is dominated by mixotrophic algae (e.g., *Chromulina*, *Ochromonas*, *Rhodomonas*, *Amphidinium*, etc.) during most of the ice-free period, excepting during a variable period at mid-summer, when strict autotrophic non-flagellated algae (e.g., *Dictyosphaerium*, *Monoraphidium*, *Cyclotella*, etc.) can become dominant (Medina-Sánchez et al. 2004; Delgado-Molina 2009).

Mixotrophic algae combine photosynthesis and phagotrophy within the same cell, a nutrition mode phylogenetically widespread among many phytoplankton groups. Recently, it has been found that these organisms can grow as photoheterotrophs by simultaneously using phototrophy to produce energy (ATP) and reducing power (NADPH) from sunlight without C incorporation, and phagotrophy to obtain C or mineral nutrients (N, P) through consumption of bacteria (Wilken et al. 2014; Cabrerizo et al. 2019a). Based on this metabolic capability, the ecological relationship established between mixotrophic algae and bacterioplankton in the high mountain lakes of Sierra Nevada was poetically defined for the first time as "neither with nor without you" (Medina-Sánchez et al. 2004). In this relationship, bacteria feed on dissolved organic C released by mixotrophic algae ("without you I cannot live", commensalism) and, simultaneously, bacteria are eaten by their own feeders ("with you I die", predation). When sunlight is plentiful but mineral nutrients are scarce—the common conditions in these ecosystems—mixotrophs farm bacteria by feeding them with "cheap" organic C and harvest from them the "expensive" mineral nutrients packed in the bacteria that are ingested by mixotrophs through bacterivory. In the basis of this relationship is the fact that bacterial growth really depends on this freshly released C, rather than on either 'old' autochthonous or the scarce allochthonous C, as repeatedly found in observational and experimental approaches in Sierra Nevada lakes (Carrillo et al. 2002; Medina-Sánchez et al. 2002; Durán et al. 2016). This mutualism has been recently recognized to be of widespread importance, e.g., for oligotrophic ocean, and implies an important shift in the paradigm of aquatic primary producers-decomposers interaction (Mitra et al. 2014). In the traditional paradigm, the organic C release from phytoplankton supports the growth of bacteria that, because of their high surface: volume ratio, advantageously uptake the dissolved mineral nutrients present at low availability in oligotrophy; then, bacterivory by microzooplankton (i.e., heterotrophic flagellates and ciliates) is the way to regenerate the nutrients immobilized in the bacterial cells given the inability of bacterivorous to assimilate all nutrients from bacterial biomass consumed. In the new paradigm, the organic C release by mixotrophic algae also supports the growth of bacteria, but algal bacterivory provides the mineral nutrients contained in bacteria needed to support the primary production (PP), nutrients that would otherwise be unavailable in oligotrophy (Mitra et al. 2014).

The mixed feeding mode of mixotrophs and the mutualistic interaction established with bacterioplankton have profound consequences on the biotic structure of the Sierra Nevada lakes. First, mixotrophic algae have adaptive advantage upon other planktonic components such as (i) strict autotrophic algae, because in oligotrophy the latter hardly can directly acquire the nutrients, mostly immobilized by bacterioplankton, (ii) heterotrophic nanoflagellates and ciliates, because mixotrophic algae, by combining phago- and phototrophy, can thrive with lower bacterial prey densities than their heterotrophic counterparts, and (iii) the autotrophic picoplankton, because they are more sensitive to the high UVR fluxes on high mountain lakes (Callieri et al. 2001). Mixotrophic algae, by feeding on bacteria as a source of limiting mineral nutrients and organic C, can overcome the UVR stress that impairs C acquisition by photosynthesis and mineral nutrient uptake (Medina-Sánchez et al. 2004). This makes mixotrophy an energy-saving and a compensatory mechanism to gain C and nutrients, particularly relevant under UVR and/or warming conditions, as found in experiments on cultured mixotrophic species (Wilken et al. 2013; Cabrerizo et al. 2019a).

Second, the above-described microbial mutualism provided the best explanation of an intriguing question posed (Medina-Sanchez et al. 2004) from the findings on the biotic structure of many European high mountain lakes in the framework of MOLAR (Mountain Lake Research) Project: why the composition of the microbial communities deviates from the general pattern of a predominance of heterotrophic microbial biomass and processes over autotrophic ones in oligotrophic ecosystems (Biddanda et al. 2001; Cotner and Biddanda 2002) in many high mountain lakes in Sierra Nevada, Pyrenees, Alps and even Sierra Nevada of California (e.g., Felip et al. 1999; Camarero et al. 1999; Straškrabová et al. 1999; Thomas et al. 1991). Thus, bacterivory by mixotrophic algae acts as a "by-pass" of C flux towards the herbivorous zooplankton, preventing the development of the heterotrophic microbial food web—the "microbial loop"—and, therefore, explaining that deviation of the general pattern (Fig. 2a).

Third, this by-pass also improves the efficiency of energy transfer towards high trophic levels given the reduction in the number of trophic steps (Medina-Sanchez et al. 2004), favoring the development of large zooplankton consumers and, hence, the mean organism size (Ward and Follows 2016). In fact, mixotrophs are excellent food quality for zooplankton (e.g., Weithoff and Wacker 2007). Therefore, zooplankton consumers co-dominate along with phytoplankton the biomass of the entire biotic community of these lakes exhibiting a well-developed grazing chain (Fig. 2a).

The fate of the mixotrophic algae in Sierra Nevada lakes under global-change scenarios may anticipate potential shifts in the structure and functioning of other oligotrophic ecosystems worldwide given that algal mixotrophy is widespread in these ecosystems (González-Olalla et al. 2021). In fact, mixotrophic algae and their dual control on bacteria proved to be quite sensitive to global-change drivers, as found in observational and experimental approaches at different time scales in Sierra Nevada lakes (see below).

## 4 Observational Approaches

Recent studies performed in the lakes of Sierra Nevada have shown an interannual decline in the percentage of mixotrophic algae coupled with a proportional increase of strict autotrophs (Fig. 3a, Carrillo et al. 2017). From a functional point of view, the studies have found a weakening of algal bacterivory on bacteria and a predominance of autotrophic metabolism (higher particulate PP) along with the reinforcement of commensalistic algae-bacteria relationship through algal excretion of dissolved organic C, which becomes the main C source regulating heterotrophic bacterial production (González-Olalla et al. 2018). These structural and functional changes are associated with an increase of mean air temperature in summer and a rising in the intensity and frequency of Saharan aerosol-dust depositions on Sierra Nevada (Fig. 3b; Carrillo et al. 2017; González-Olalla et al. 2018). In fact, during the last four decades, particularly since the 1990s, the magnitude of aerosol intrusions has quintupled and their frequency tripled; further, strong dust intrusions in winter have occurred in recent years (Fig. 3c), events that until recently were very rare in winter on this region (Villar-Argaiz and Bullejos 2016; Carrillo et al. 2017; González-Olalla et al. 2018).

Interestingly, over an inter-annual scale (i) the intensity and frequency of aerosol intrusions positively correlated to phytoplankton biomass, indicating that the aerosols can provide fertilizing nutrients to which phytoplankton, mainly the strict autotrophs, responded (Bullejos et al. 2010; Villar-Argaiz and Bullejos 2016); (ii) the algal bacterivory positively correlated to UVB/total dissolved P ratio (González-Olalla et al. 2018); (iii) the proportion of mixotrophic algae correlated negatively to total phosphorus (TP) and positively to maximum depth, which in turn is inversely related to TP (Delgado-Molina 2009); (iv) mixotrophs persisted under low but frequent nutrient inputs, whereas strict autotrophs dominated in response to high P-inputs (Cabrerizo et al. 2017). All these findings support the conception

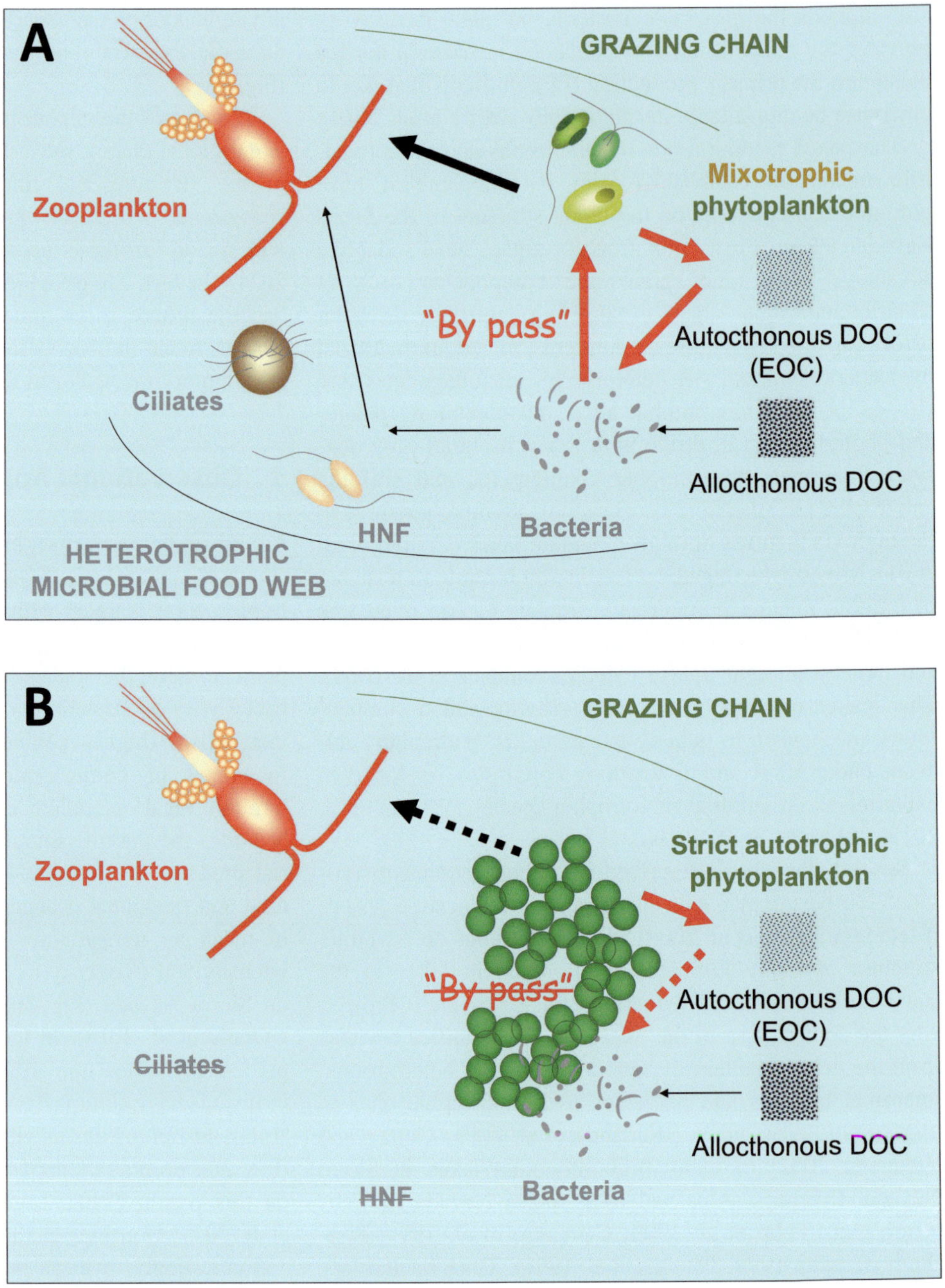


**Fig. 2** **a** Diagram of pelagic food web in La Caldera lake. All arrows represent carbon flux, and their thickness indicates its relative importance. Red arrows also indicate the algae dual control on bacteria (algae-bacteria mutualism) which can be intensified after P-enrichment (see text). DOC: dissolved organic carbon. EOC: excretion of organic carbon by algae. HNF: heterotrophic nanoflagellates. Modified from Medina-Sánchez et al. (2004). **b** Changes in the pelagic food web in La Caldera lake after the experiment of P-enrichment gradient. The algal dual control on bacteria (algae-bacteria mutualism) is broken due to the bloom of strict autotrophic algae that displace mixotrophs, suppress the heterotrophic microbial food web and impair the grazing chain (see text)

that mixotrophy is an adaptive strategy of nanoflagellates algae to thrive under low P and high UVR levels (Medina-Sánchez et al. 2004, Carrillo et al. 2017; Cabrerizo et al. 2017; González-Olalla et al. 2018).

The main implication of the sharp decline of this metabolic trait (algal mixotrophy) is the impoverishment of the functional biodiversity in the high mountain lakes, which alters ecosystem functioning through a weakening of the C by-pass between the microbial loop and the grazing chain supported by the mixotrophs. Then, how does this shift affect higher trophic levels?

This shift in the primary producers diminishes the energy-transfer efficiency towards upper trophic levels, driving to a lesser prevalence of the grazing chain in these remote lakes. Comparative analyses of the community dynamics over the past decades in La Caldera Lake indicate a break in the phytoplankton–zooplankton coupling. Zooplankton biomass remained greater than phytoplankton biomass until the mid-1990s. From this period, a lower accrual of zooplankton matched with an increase in phytoplankton biomass, mainly of strict autotrophs, which became dominant thereafter (Fig. 4a), coinciding with the increase in aerosol

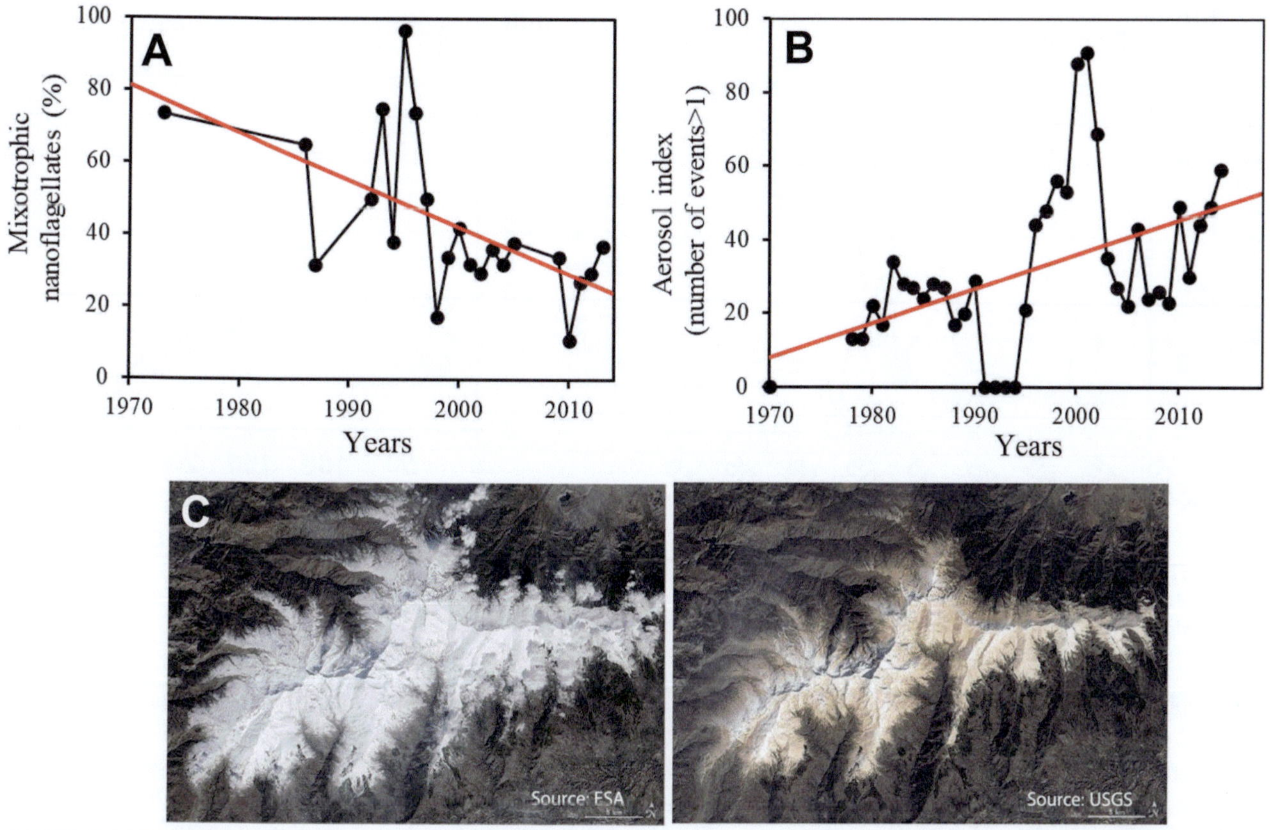


**Fig. 3** Interannual trends in the percentage of mixotrophic nanoflagellates (**a**) and Aerosol index (AI), number events >1 (measured as a sum of daily events of AI >1) (**b**) during ice-free seasons in the last decades in Lake La Caldera. Red lines are fitted linear trends. Modified from Carrillo et al. (2017). **c** Aerial photography captured by the Sentinel 2-A (Left; courtesy of European Space Agency, ESA) and Landsat 8 (Right; courtesy of the U.S. Geological Survey) satellites corresponding to February 18 and February 27, 2017, after an intrusion of atmospheric dust over the Sierra Nevada National Park. Published in González-Olalla et al. (2018)

intrusions. Consequently, the inter-annual relation between the phytoplankton and the zooplankton biomass was not linear but rather unimodal (Fig. 4b), indicating a phytoplankton threshold (100–150 μg fresh weight $L^{-1}$) above which zooplankton decreases (Bullejos et al. 2010; Villar-Argaiz and Bullejos 2016). This uncoupling implies a weakening in the top-down control of zooplankton on primary producers, favoring the proliferation of phytoplankton —mainly of strict autotrophs—that decreases the transparency of the waters, turning them greener, indicative of eutrophication and less water quality of these remote ecosystems (Villar-Argaiz and Bullejos 2016). Besides, these changes also affect the composition of zooplankton. Long-term dynamic analyses (1975–2007) of zooplankton in La Caldera Lake (Villar-Argaiz et al. 2012) showed that precipitation, aerosol index and UVR explained the variance of zooplankton biomass (62%, 26% and 9%, respectively), which was dominated by calanoid copepods (i.e., *Mixodiaptomus laciniatus*), as in other oligotrophic freshwater (Carney and Elser 1990) and oceanic (Nuwer et al. 2008) ecosystems. The negative relationship found between zooplankton biomass and precipitation could be explained by the fact that, during wet years (snowfall in winter), copepods development may be delayed by late thaw and colder water temperatures, while during dry years, an early ice-melting, longer growth windows and warmer water temperatures favor an advanced development of zooplankton. The negative influence of aerosol index on copepod biomass is consistent with interannual findings for this lake, where years of exceptionally high atmospheric loads caused reduced copepod populations but enhanced populations of cladocerans and rotifers (Villar-Argaiz et al. 2001). Therefore, changes in the intensity and frequency of aerosols and a drier precipitation regime could substantially affect the most genuine zooplankton community in these lakes, impairing calanoid copepods populations and favoring other consumers like rotifers or cladocerans (Villar-Argaiz et al. 2012).

While the observational approaches have allowed us evaluating how communities respond to the natural environmental variability, it is challenging to discern what are the main abiotic drivers (e.g., nutrients linked to dust-aerosols, UVR, temperature) explaining the decline of

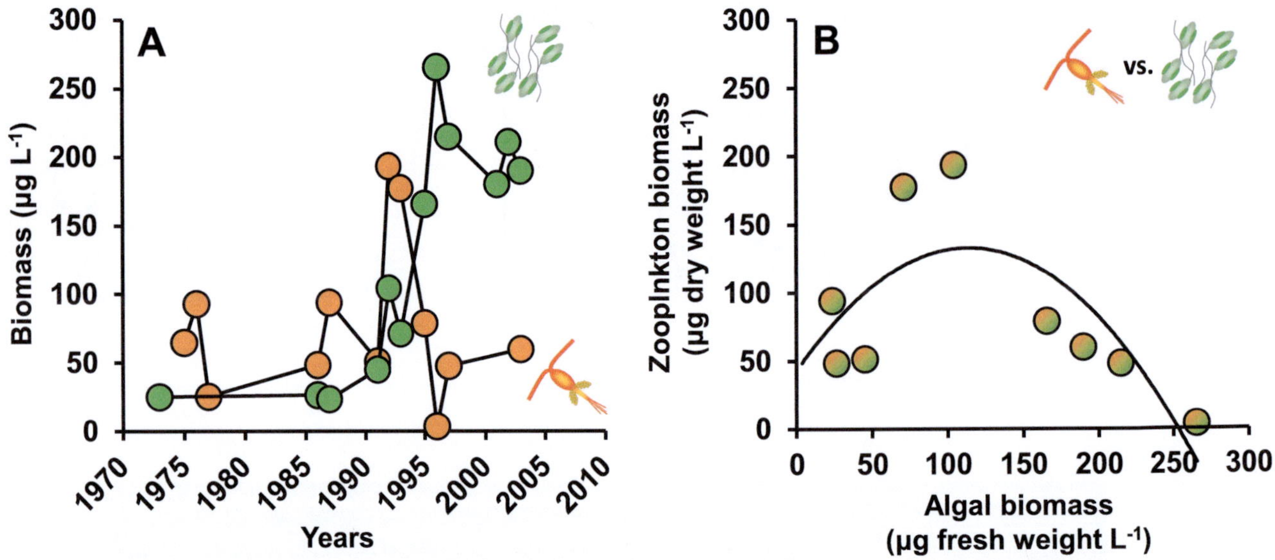


**Fig. 4** Interannual dynamics (**a**) and the relationship between phytoplankton and zooplankton biomass (**b**) in La Caldera Lake. Modified from Bullejos et al. (2010) and Villar-Argaiz and Bullejos (2016)

mixotrophs and their dual control on bacteria (mutualism), and the consequences upon zooplankton consumers. To disentangle all these uncertainties, different experimental approaches, manipulating these abiotic factors (P-addition, UVR), have been performed at short- and mid-term scales.

## 5 Experimental Approaches

### 5.1 P-addition and Role of Stoichiometry

Short-term (hours) experiments were performed in situ when phytoplankton was dominated by mixotrophs, with the addition of inorganic P simulating balanced and unbalanced Redfield's N:P ratio (i.e., molar N:P of 16 vs. 5 or > 180) under different qualities of UVR. Our results showed a reinforcement of the dual algal control on bacteria after P-addition when the initial elemental composition of algae was relatively poor in P ($N{:}P_{algae} > 30$; Medina-Sánchez et al. 2006). This enhanced dual control lies in an increased photosynthetic C release due to transitory metabolic adjustments of algae to grow, and an increased algal bacterivory under light conditions. By contrast, when the elemental composition of algae was P-sufficient ($N{:}P_{algae} < 10$), the dual control was not enhanced, although it persisted. The consequences of these results are multiple: (i) the intensification of the dual control when algae are P-deficient constitutes an adaptive response of mixotrophs that allows their growth, persistence and dominance in these (and other) clear-water oligotrophic ecosystems facing sporadic inputs of P, as those linked to Saharan dust atmospheric transport (*but see* paragraphs below); (ii) the higher excretion of photosynthetic C and nutrient transfer from bacteria to mixotrophs through bacterivory are the mechanisms that explain the surprising low sestonic C:P ratio values found in the Sierra Nevada lakes (Villar-Argaiz et al. 2002; Medina-Sánchez et al. 2006; Carrillo et al. 2008a, b); (iii) the low algal C:P ratio implies a nutrient-rich food quality for zooplankton. This helps to explain the high development of the grazing chain found in Sierra Nevada lakes, dominated by copepods (Medina-Sánchez et al. 1999; Villar-Argaiz et al. 2001, 2012).

Mid-term microcosm experiments performed in La Caldera Lake in three periods of the seasonal succession of the pelagic community (thaw, mid and late ice-free periods) showed marked changes in the dynamic of microbial food web during 9–15 days after the manipulation of inorganic nutrients as in the short-term experiments. The most notable finding was that the ciliates, usually very scarce in the lake, experienced an acute development, mainly at thaw, when bacteria were P-rich (bacterial N:P of 20–24) and algal stoichiometry was balanced, near to Redfield's ratio (algal $N{:}P \approx 16$). In fact, the bacterial and algal N:P ratios proved to be unimodal (no linear) predictors of heterotrophic microbial food web development (Carrillo et al. 2008b). The results also showed that phytoplankton, when dominated by strict autotrophs (at mid-ice-free period), experienced a sharp increase of abundance in the P-enriched treatments; by contrast, when was dominated by mixotrophs, experienced either a modest (at thaw) or no increase of abundance coinciding with a higher cell volume and P cell quota (at late ice-free period), indicative of either moderate algal growth or P-storage strategies. The algal community composition and the abundance of bacterioplankton remained quite invariant among the treatments (control, N:$P_{16}$, N:$P_5$) despite the pronounced variations in their stoichiometry (e.g., cell N:P ratio, P-cell quote, P-incorporation rate) in response to P-inputs. The strict autotrophic algae were the dominant group only at mid-ice-free period, and they controlled bacteria by competition for P, impairing the bacterial development also in this period.

Overall, these results from both short- and mid-term experiments underline the importance of cell stoichiometry in determining the microbial plankton responses to P-inputs. Further, the results support the interpretation that mixotrophic algae exhibit a high metabolic plasticity, following different strategies (e.g., growth, P-storage) that favor their persistence in the ecosystem, even when are facing P-inputs that may benefit bacteria and strict autotrophic algae against them.

## 5.2 P-addition Gradient and UVR

A long-term (>2 months) in situ mesocosm experiment using a P-addition gradient in the presence versus absence of solar UVR (Fig. 5a) evidenced pronounced shifts in the dynamic of communities (Carrillo et al. 2008a; Delgado-Molina et al. 2009; Medina-Sánchez et al. 2013, 2016). Thus, during the first 20 days of the experiment, when added-P was available (P-availability period), mixotrophic algae (mainly *Chromulina nevadensis*, Chrysophyceae) dominated phytoplankton biomass in all experimental conditions, while the heterotrophic microbial food web (bacteria, ciliates and viruses) transitorily developed only in the P-enriched mesocosms. Heterotrophic nanoflagellates (HNF) were absent, probably because mixotrophic algae occupy the ecological niche of HNF in these lakes. From 20 days after the P pulses and up to the end (70 days), matching with the total consumption of P by the community (P-depleted period), we found in the P-enriched mesocosms a bloom of *Dictyosphaerium chlorelloides,* one green algae species of rapid growth and relatively high tolerance to UVR. This resulted in a sharp decrease of phytoplankton diversity (Delgado-Molina et al. 2009; Medina-Sánchez et al. 2016) and the collapse of the heterotrophic microbial food web that transitorily had developed during the P-availability period (Figs. 2b and 5b).

The heterotrophic microbial food web, during its development, followed a consistent unimodal response to the P-enrichment gradient, peaking at intermediate P-enrichment levels and, unexpectedly, more accentuated under UVR (interaction UVR × P-enrichment; Medina-Sánchez et al. 2013; Fig. 5b). This positive response to UVR can be explained by the good adaptation to the high UVR fluxes of bacteria and ciliates in these lakes (Carrillo et al. 2002, 2008b, 2013; Medina-Sánchez et al. 1999, 2002). The positive UVR effect on viruses reflects the activation of lytic cycles, prompted by the stimulus of bacterial growth with P-enrichment under UVR (Medina-Sánchez et al. 2013). The unimodal response of the heterotrophic microbial food web in the P-availability period contrasts with the positive linear response shown by the strict autotrophic algae to the P enrichment gradient during its acute development in the P-depleted period (Fig. 5b). This response evidenced a moderate negative effect of UVR on strict autotroph abundance and biomass, an effect that augmented as the P-enrichment accentuated (Carrillo et al. 2008a; Medina-Sánchez et al. 2013). These findings from Sierra Nevada lakes were the first available report of unimodal responses of heterotrophic microbes (and not only of bacteria, but also of their predators and even non-metabolic parasites—viruses—) to a P-gradient, comparable to those observed for macroscopic heterotrophic organisms such as fish, mollusks or insects (Boersma and Elser 2006). In all these findings, heterotrophs' growth diminishes when faced with a high P content in their food, indicating that "too much of a good thing" is detrimental due to increased metabolic costs derived from unbalanced nutrient content, even though the excess is of a limiting nutrient (Boersma and Elser 2006). This interpretation agrees with the "stoichiometric knife-edge" hypothesis (Elser et al. 2005), which predicts a decline in consumer performance from excessively P-rich food.

What consequences for zooplankton had the displacement of mixotrophs by the bloom of a strict-autotrophic algae? This question is relevant, as zooplankton, mainly the calanoid copepods, are the top-down controllers of phytoplankton, and the long-term experiment allowed us to know the fate of zooplankton due to their longer generation time. In the P-depleted period, when the strict autotroph *D. chlorelloides* dominated, zooplankton followed a consistent unimodal response to the P-enrichment gradient (translated into food terms as a linear gradient of algal biomass and sestonic P content). Thus, *M. laciniatus* biomass peaked at intermediate levels of P-enrichment, similarly to the response of heterotrophic microbial food web in the P-availability period; however, the zooplankton response was milder under UVR (Fig. 5b). This negative UVR effect was rather direct because the effect of UVR on food quantity and quality was not sufficient to restrict zooplankton development (Bullejos et al. 2010; Villar-Argaiz et al. 2009). The direct UVR-effect was surprising, given the defense strategies of copepods in Sierra Nevada, as their ability to migrate at deepwater layers, i.e., refuges of UVR (Medina-Sánchez et al. 1999), and the high antioxidant enzymatic capability counteracting their relatively low content of micosporine-like compounds (Souza et al. 2010). The unimodal pattern proved that zooplankton was mainly constrained by food at both ends of the nutrient gradient. If we

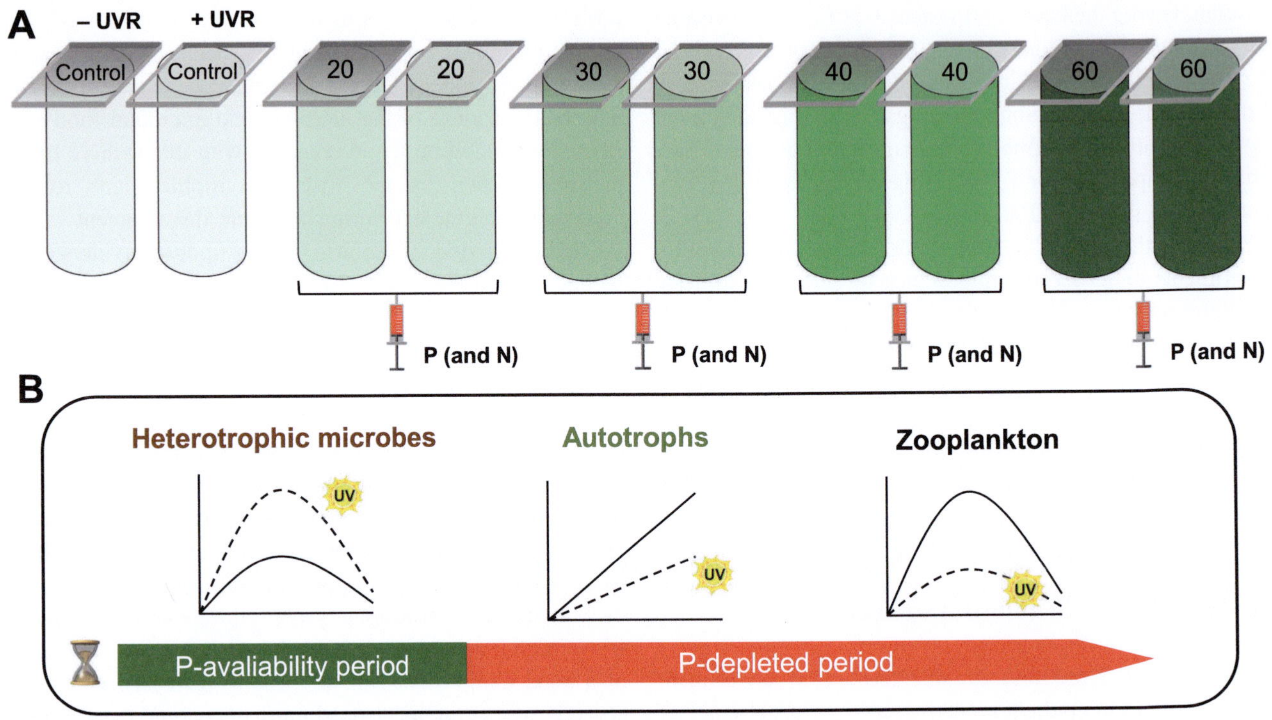


**Fig. 5** Mesocosm experiment of P-enrichment gradient. **a** Factorial design [5 × 2]: 5 levels of P-enrichment (no addition [control], 20, 30, 40, and 60 μg P $L^{-1}$) simulating a gradient of current and future scenarios of allochthonous nutrient inputs associated to Saharan aerosols and two levels of solar-radiation quality (absence and presence of UVR). The P was added to each mesocosm (polyethylene bags of 1 m in diameter × 7 m length; volume of 2.7 $m^3$) as one-time pulse at the beginning of the experiment, accompanied of addition of $NH_4NO_3$ to reach a N:P molar ratio of 30 to avoid the limitation of the organisms by N. After the pulses of nutrients, the mesocosms were incubated in situ for 70 days, and the periodicity of the sampling of the mesocosms was adjusted to the generation times of the different organisms studied (heterotrophic microbes, phytoplankton, zooplankton). Modified from Villar-Argaiz et al. (2012). **b** Graphical scheme of the response of each biotic compartment (Y-axes: biomass) to P-enrichment gradient (X-axes) under +UVR (dashed lines) and –UVR (solid lines) in each period (see text)

look at the low range of P-enrichment gradient, the increase of zooplankton biomass up to intermediate P-enriched levels indicated a positive effect mainly of food quantity and secondarily of food quality (through omega3-polyunsaturated fatty acids, Villar-Argaiz et al. 2012). By contrast, if we look at the high range of P-enrichment gradient, the strong increase in food quantity failed to yield higher zooplankton biomass, evidencing a strong detrimental effect of food "in excess", even despite being composed of the edible green algae *D. chlorelloides*. This was probably due to the high secretions of polysaccharides by the blooming green algae (Delgado-Molina et al. 2009) that may have a clogging effect on copepods by saturating their filtering capacity, resulting in a lesser growth rate as shown in parallel bioassays (Villar-Argaiz et al. 2012). Also, food quality could exert a modulating role in the zooplankton response. Thus, although the concentration of essential biochemicals in food, such as total fatty acids and omega3-polyunsaturated fatty acids, was high in all the P-enriched treatments (Villar-Argaiz et al. 2009, 2012), an excessively P-rich food, i.e., low algal C:P and N:P ratio (Carrillo et al. 2008a; Delgado-Molina et al. 2009), could also explain the detrimental effect, in agreement with the mentioned "stoichiometric knife-edge" (Elser et al. 2005) hypothesis. This was corroborated in a more recent study (Bullejos et al. 2014) on *M. laciniatus*, whose growth (measured as RNA and RNA:DNA ratio) responded to the food quality following a unimodal pattern, i.e., intermediate sestonic C:nutrient ratios yielded the maximum stage-specific growth, which also increased during the copepod ontogeny.

## 5.3 P-addition Mode and UVR

All experiments above described show the organisms' responses to a single pulse of nutrients. Hence, a question arises: Can a greater frequency of nutrient inputs alter the responses of the trophic web? To answer it, mesocosm experiments were performed in La Caldera Lake to evaluate the effect of repeated (press: 5 μg P $L^{-1}$ × 6 times) compared to a one-time (pulse: 30 μg P $L^{-1}$) P-inputs with the same final quantity under two sunlight qualities (3 × 2 factorial design: no addition [control], *press*, and *pulse*; presence and absence of UVR). The results showed that *press* compared to *pulse* under UVR prompted higher increases in primary (PP) than in bacterial production (BP) coupled with a replacement of the initially dominant strict autotrophs (mainly *Monoraphidium* sp.) by mixotrophic nanoflagellates (mainly *Chromulina nevadensis*). Measuring diverse functional (e.g., PP and BP, excretion of organic C, bacterial growth efficiency, bacterial respiration, bacterial assimilation of excreted organic C) and stoichiometric (e.g., sestonic P and C:P ratio) variables allowed to reveal the mechanism underlying these algal responses: it was precisely the dual control exerted by mixotrophic nanoflagellates on bacteria, which were kept scarcely developed due to this control, particularly relevant in absence of ciliates and HNF (Cabrerizo et al. 2017). However, under *pulse* conditions, the strict autotrophs dominated the community and exhibited a characteristic domed-curve dynamic in contrast to the early and sustained development of mixotrophs under the *press* conditions, particularly under UVR. These results suggest that mixotrophs are most successful when solar radiation is plentiful and nutrients are scarce, i.e., below a threshold eliciting the bloom of strict autotrophs. Therefore, a high frequency of low nutrient inputs under UVR, as occurred up to recently in Sierra Nevada lakes, benefit the growth of mixotrophs and reinforce their dual control on bacteria, favoring the by-pass of C flux from bacteria to zooplankton through mixotrophic algae (Medina-Sánchez et al. 2004). This explains the inter-annual persistence of mixotrophs in these clear-water ecosystems despite their decreasing trend against the strict autotrophs, who are benefited by an increase of intense nutrient pulses linked to the increased frequency and intensity of aerosol-dust deposition events registered over the study area (Carrillo et al. 2017; Cabrerizo et al. 2017).

## 6 Integrating the Observational and Experimental Approaches

The integration of observational and experimental results reveals three main outcomes. First, biotic interactions, such as predation (bacterivory), parasitism (lithic viral cycles) and competition with algae for P, play a key role in restricting the heterotrophic microbial food web development. These biotic controls, beyond bacterivory by mixotrophic algae, explain the weakly developed heterotrophic microbial food web in these clear-water high mountain lakes, deviating from the general pattern in oligotrophy (Biddanda et al. 2001; Cotner and Biddanda 2002). By contrast, the strict autotrophs are regulated mainly by abiotic factors (nutrients, UVR), as proved by the linear response pattern of their structural (chlorophyll *a*, abundance, biomass…) and functional variables (PP, EOC) to a P-enrichment gradient (Carrillo et al. 2008a; Delgado-Molina et al. 2009) or by their acute growth response to a pulse of P when they were dominant (Carrillo et al. 2008b; Cabrerizo et al. 2017).

Second, the transitory development of a heterotrophic microbial food web, triggered by P-enrichment when mixotrophic algae are dominant, may be considered as a self-organization mechanism of ecosystem resilience, favoring the persistence of mixotrophs (in transitory coexistence with ciliates) as the main mobilizers of energy (via PP) and nutrients (via bacterivory) in these clear-water ecosystems. In absence of heterotrophic microbes (ciliates, HNF), the persistence of mixotrophs would be assured providing that P-inputs are repeated but of low intensity (Cabrerizo et al. 2017). However, under scenarios of high external P loads, a transitory development of a heterotrophic microbial food web could not avoid a strong growth of strict autotrophs. These can displace mixotrophs, resulting in a loss of functional biodiversity, whose effects propagate toward higher trophic levels.

Third, zooplankton growth is impaired by an excess of food quantity and even quality, after P-inputs. The unimodal zooplankton response to excessive food quantity and quality under experimental P-inputs concords with the unimodal relation between phytoplankton and zooplankton biomass observed over interannual scales in La Caldera Lake. This underlines that the external nutrient loads linked to increasing aerosol-dust inputs on Sierra Nevada lakes can break the phytoplankton–zooplankton coupling in these water bodies (Bullejos et al. 2010; Villar-Argaiz and Bullejos 2016).

*Further insights from more complex experiments and novel approaches*

The impacts of global change drivers on organisms and ecosystems have been often studied independently but mounting evidence—also provided by the experimental approaches above reviewed—points towards the generality of non-additive effects (Villar-Argaiz et al. 2018a). For the reasons presented in this chapter, high mountain lakes are excellent ecosystems to test complex interactive effects. Recent mesocosm experiments performed in Sierra Nevada lakes followed more complex experimental designs, i.e., full simultaneous and split-plot factorial ones, with manipulation of up to three drivers of global change (e.g., UVR × P × C; UVR × P × temperature) to study their net and interactive effects.

For example, depending on lake features, the interaction among UVR, P and temperature had a different effect on phytoplankton-bacteria coupling, i.e., the commensalistic interaction (Durán et al. 2016; González-Olalla et al. 2018). In a cold and highly UVR-transparent lake (La Caldera

Lake), UVR × P × temperature increased BP and decreased PP and EOC, which was not sufficient to satisfy the bacterial carbon demand, leading to a weakening of commensalistic (and predatory) algae–bacteria relationship (Durán et al. 2016, 2020). Contrarily, in a warmer and less UVR-transparent lake (Las Yeguas Lake), the phytoplankton-bacteria coupling was accentuated by the interaction UVR × P × temperature (Durán et al. 2016). These contrasting responses of the phytoplankton–bacteria coupling may have consequences on the microbial loop development, becoming reinforced in warmer and less UVR-transparent high mountain lakes, but weakened in colder and more UVR-transparent high mountain lakes, with implications in the C flux through the microbial food web of these sentinel ecosystems in a scenario of global change.

In another recent study performed in La Caldera Lake (Cabrerizo et al. 2019b), the stability of a great array of structural and functional variables of the microbial plankton was quantified in a complex experimental design through well-established resistance and resilience indexes (Fig. 6a). Under the most complex treatment (UVR × C × P), representative of a global-change scenario, the metabolic balance of the ecosystem, measured as the ratio between total PP and respiration (P/R ratio), proved strongly resistant due to the high resistance of both mixotrophs and the commensalistic interaction (Fig. 6b). By contrast, the high resilience of both phototrophs and algae-bacteria commensalistic interaction can drive (and explain) the long-term maintenance of the autotrophic nature (high P:R ratio) and ultimately, the sink-C capacity of these ecosystems. This sink capacity can be aided by the bacterial co-limitation by C and P, whose intensity is subsequently governed by ecological interactions such as bacterivory (Dorado-García et al. 2014).

The application of "omic" technologies in connection with ecosystem processes has revealed new insights into ecosystem functioning. For example, a recent study (Castellano-Hinojosa et al. 2017) shows that La Caldera Lake harbors a high microbial diversity that drives unexpected denitrification rates despite being a mostly aerobic ecosystem. As a consequence, the lake functions as a "dissimilative biological N-pump", an auto-depurative mechanism that allows the removal of the N excess that arrives at the lake from the atmosphere.

All these outcomes underline the extreme usefulness and versatility of Sierra Nevada high mountain lakes as excellent venues of advanced ecological research, providing us with valuable information of ecosystem responses to global change drivers. Furthermore, the findings reinforce the growing view that the still little-altered ecosystems in

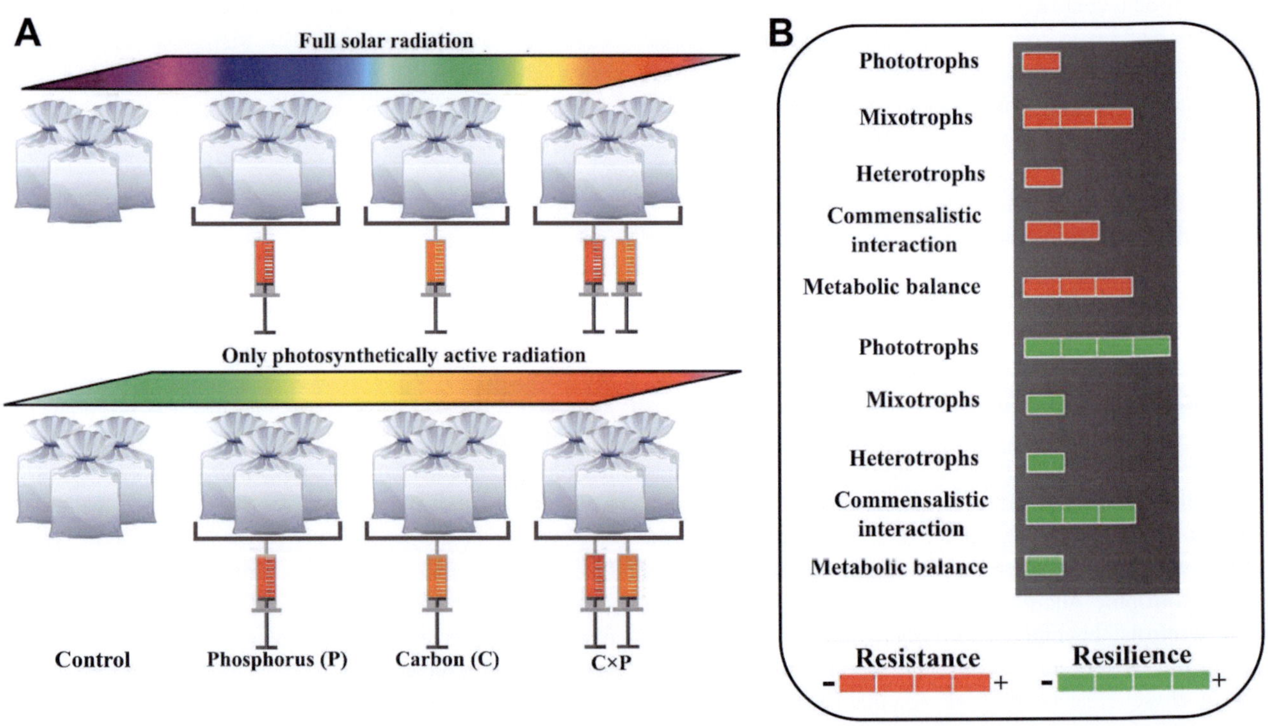


**Fig. 6** Experiment on stability. **a** Graphical scheme of the experimental design: plankton exposed to radiation (full solar radiation [>280 nm] or only photosynthetically active radiation [>400 nm]) and nutrient additions (control, P, C, and C × P). **b** Graphical scheme showing the relative magnitude of the resistance (red bars) and resilience (green bars) indexes for functional groups (photo-, mixo- and heterotrophs), commensalistic interaction and metabolic balance of ecosystem under the most complex scenario (UVR × C × P). Figure modified from Cabrerizo et al. (2019b)

protected areas also play a protective role for humans, in a changing world subjected to abrupt alterations of global biogeochemical cycles and biodiversity loss up to levels exceeding the safe planetary boundaries for life (Steffen et al. 2015).

## 7 Conclusions

The information exposed throughout this chapter allows us to answer the main question posed at the beginning: Why are high mountain lakes of Sierra Nevada unique remote sensors of global change?

(a) Because of their physical characteristics—highly connected with atmospheric processes—stemmed from their altitude, geographic location, remoteness, small catchment areas, high UVR fluxes, great diversity of water bodies within a small-scale mountain range, and processes operating at short- and long-term scales.
(b) Because of the sensitivity and responsiveness of their biota to the interaction among global-change stressors, revealed from the structural and functional biotic variables, whose study is facilitated by the features of their biota. This allows us to understand the underlying mechanisms of changes in their communities and anticipate induced state changes in other ecosystems over world.

As most of the high mountain lakes in the world, considered the "eyes" of the global change (Moser et al. 2019), the high mountain lakes of Sierra Nevada are excellent sentinels of global change even with more emphasis than other alpine lakes due to their southern location, serving as "crystal ball" (Villar-Argaiz et al. 2018b) for forecasting what is to come in lower altitudes, particularly of Mediterranean Region, in the near future.

**Acknowledgements** We thank the Sierra Nevada National Park staff for field assistance, Regino Zamora and Marc Oliva for reviewing a draft of the manuscript, Manuel Merino for improving Figure 1, and Ritu Chandwani for editing the final version of the chapter. This study was funded by PID2020-118872RB-I00 (MICIN/AEI/10.13039/501100011033) and FEDER/Junta de Andalucía-Consejería de Transformación Económica, Industria, Conocimiento y Universidades/Proyecto (B-RNM-310-UGR20) projects.

## References

Afonina EY, Tashlykova NA, Kuklin AP, Tsybekmitova GTs (2020) Environmental features and dynamics of plankton communities in a mountain glacial moraine lake (Baikal lake basin, Russia). Nature Conserv Res 5(3):23–36

Alcaraz-Segura D, Reyes A, Cabello J (2016) Changes in vegetation productivity according to teledetection In: Zamora Rodríguez RJ, Pérez Luque AJ, Bonet FJ, Barea-Azcón JM, Aspizúa R (eds) Impacts of global change in Sierra Nevada: challenges for conservation, pp 142–145

Amarouche K, Akpınar A (2021) Increasing trend on storm wave intensity in the Western Mediterranean. Climate 9:11

Biddanda B, Ogdahl M, Cotner J (2001) Dominance of bacterial metabolism in oligotrophic relative to eutrophic waters. Limnol Oceanogr 46:730–739

Blumthaler M (2005) Ambient levels of UV radiation. BMC Plant Biol 5(1):S5

Blumthaler M, Ambach W, Ellinger R (1997) Increase in solar UV radiation with altitude. J Photochem Photobiol B Biol 39:130–134

Boersma M, Elser JJ (2006) Too much of a good thing: on stoichiometrically balanced diets and maximal growth. Ecology 87:1325–1330

Bonet FJ, Pérez-Luque AJ, Pérez-Pérez R (2016) Trend analysis (2000–2014) of the snow cover by satellite (MODIS sensor). In: Zamora Rodríguez RJ, Pérez Luque AJ, Bonet FJ, Barea-Azcón JM, Aspizúa R (eds) Impacts of global change in Sierra Nevada: challenges for conservation, p 43

Brogli R, Sørland SL, Kröner N, Schär C (2019) Causes of future Mediterranean precipitation decline depend on the season. Environ Res Lett 14:114017

Bullejos FJ, Carrillo P, Villar-Argaiz M, Medina-Sánchez JM (2010) Roles of phosphorus and ultraviolet radiation in the strength of phytoplankton-zooplankton coupling in a Mediterranean high mountain lake. Limnol Oceanogr 55:2549–2562

Bullejos FJ, Carrillo P, Gorokhova E, Medina-Sánchez JM, Balseiro E, Villar-Argaiz M (2014) Shifts in food quality for herbivorous consumer growth: multiple golden means in the life history. Ecology 95:1272–1284

Camarero L, Felip M, Ventura M, Bartumeus F, Catalan J (1999) The relative importance of the planktonic food web in the C cycle of an oligotrophic mountain lake in a poorly vegetated catchment (Redó, Pyrenees). J Limnol 58:203–212

Callieri C, Morabito G, Huot Y, Neale PJ, Litchman E (2001) Photosynthetic response of pico- and nanoplanktonic algae to UVB, UVA and PAR in a high mountain lake. Aquat Sci 63:286–293

Cabrerizo MJ, Medina-Sánchez JM, Dorado I, Villar-Argaiz M, Carrillo P (2017) Rising frequency of resource pulses under solar UVR strengthen microbial interactions. Sci Rep 7:43615

Cabrerizo MJ, González-Olalla JM, Hinojosa-López VJ, Peralta-Cornejo FJ, Carrillo P (2019a) A shifting balance: responses of mixotrophic marine algae to cooling and warming under UVR. New Phytol 221:1317–1327

Cabrerizo MJ, Medina-Sánchez JM, Dorado I, Villar-Argaiz M, Carrillo P (2019b) Interplay between resistance and resilience governs the stability of a freshwater microbial food web under multiple stressors. Sci Total Environ 691:908–918

Carney HJ, Elser JJ (1990) Strength of zooplankton-phytoplankton coupling in relation to trophic state. In: Tilzer M, Serruya C (eds) Large lakes: ecological structure and function. Springer, pp 615–631

Carrillo P, Medina-Sánchez JM, Villar-Argaiz M (2002) The interaction of phytoplankton and bacteria in a high mountain lake: importance of the spectral composition of solar radiation. Limnol Oceanogr 47:1294–1306

Carrillo P, Medina-Sánchez JM, Villar-Argaiz M, Delgado-Molina JA, Bullejos FJ (2006) Complex interactions in microbial food webs: stoichiometric and functional approaches. In: Armengol J, Muñoz I (eds) The ecology of the Iberian inland waters: homage to Ramón Margalef. Asociación Española de Limnología, Madrid. ISSN 0213-8409, pp 189–204

Carrillo P, Delgado-Molina JA, Medina-Sánchez JM, Bullejos FJ, Villar-Argaiz M (2008a) Phosphorus inputs unmask negative effects of UV Radiation on algae in a high mountain lake. Global Change Biol 14:423–439

Carrillo P, Villar-Argaiz M, Medina-Sánchez JM (2008b) Does microorganism stoichiometry predict microbial food web interactions after a phosphorus pulse? Microb Ecol 56:350–363

Carrillo P, Medina-Sánchez JM, Duran C, Herrera G, Villafañe VE, Helbling WE (2015) Synergistic effects of UVR and simulated stratification on commensalistic algal-bacterial relationship in two optically contrasting oligotrophic Mediterranean lakes. Biogeosciences 12:697–712

Carrillo P, Medina-Sanchez JM, Villar-Argaiz M, Bullejos FJ, Durán C, Bastidas-Navarro M, Souza MS, Balseiro EG, Modenutti BE (2017) Vulnerability of mixotrophic algae to nutrient pulses and UVR in an oligotrophic Southern and Northern Hemisphere lake. Sci Rep 7:6333

Castellano-Hinojosa A, Correa-Galeote D, Carrillo P, Bedmar EJ, Medina-Sánchez JM (2017) Denitrification and biodiversity of denitrifiers in a high-mountain Mediterranean lake. Front Microbiol 8:1911

Catalan J, Camarero L, Felip M, Pla S, Ventura M et al (2006) High mountain lakes: extreme habitats and witnesses of environmental changes. In: Armengol J, Muñoz I (eds) The ecology of the Iberian inland waters: Homage to Ramón Margalef. Asociación Española de Limnología, Madrid. ISSN 0213–8409, pp 123–145

Catalan J, Pla-Rabes S, Wolfe AP, Smol JP, Ruhland KM et al (2013) Global change revealed by paleolimnological records from remote lakes: a review. J Paleolimnol 49:513–535

Catalan J, Ninot JM, Aniz MM (2017) The high mountain conservation in a changing world In: Catalan J, Ninot JM, Aniz MM (eds) High mountain conservation in a changing world. Springer. ISSN 2215-1621

Cotner JB, Biddanda BA (2002) Small players, large role: microbial influence on biogeochemical processes in pelagic aquatic ecosystems. Ecosystems 5:105–121

Cramer W, Guiot J, Fader M, Garrabou J, Gattuso J-P et al (2018) Climate change and interconnected risks to sustainable development in the Mediterranean. Nat Clim Change 7:972–980

De la Paz D et al (2013) Modelling Saharan dust transport into the Mediterranean basin with CMAQ. Atmos Environ 70:337–350

Delgado-Molina JA, Carrillo P, Medina-Sánchez JM, Villar-Argaiz M, Bullejos FJ (2009) Interactive effects of phosphorus loads and ambient ultraviolet radiation on the algal community in a high-mountain lake. J Plankton Res 31:619–634

Dentener FJ, Carmichel GR, Zhang Y, Lelieveld J, Crutzen PJ (1996) Role of mineral aerosol as a reactive surface in the global troposphere, J Geophys Res 101:22869–22889

Delgado-Molina JA (2009) Efectos interactivos de radiación UV y pulsos de nutrientessobre la red trófica microbiana: aproximación estructural,estequiométrica y funcional. PhD thesis. University of Granada

Derlet RW, Goldman CR, Connor M (2010) Reducing the impact of summer cattle grazing on water quality in the Sierra Nevada mountains of California: a proposal. J Water Health 8:326–333

Derlet RW, Richards JR, Tanaka LL, Hayden C, Ger KA, Goldman CR (2012) Impact of summer cattle grazing on the Sierra Nevada watershed: aquatic algae and bacteria. J Environ Public Health ID760108

Díaz-Hernandez JL, Herrera-Martínez AJ (2019) Hydrological characteristics and paradoxes of Mediterranean high-mountain water-Bodies of the Sierra Nevada, SE Spain. Hydrology 6:59

Doña C, Chang NB, Caselles V, Sánchez JM, Pérez-Planells L, Bisquert MDM, García-Santos V, Imen S, Camacho A (2016) Monitoring hydrological patterns of temporary lakes using remote sensing and machine learning models: case study of La Mancha Húmeda Biosphere Reserve in Central Spain. Remote Sensing 8:618

Dorado-García I, Medina-Sánchez JM, Herrera G, Cabrerizo MJ, Carrillo P (2014) Quantification of C and phosphorous co-limitation in bacterioplankton: new insights on an old topic. PLoS ONE 9(6): e99288

Durán C, Medina-Sánchez JM, Herrera G, Carrillo P (2016) Changes in the phytoplankton-bacteria coupling triggered by joint action of UVR, nutrients, and warming in Mediterranean high-mountain lakes. Limnol Oceanogr 61:413–429

Durán C, Medina-Sánchez JM, Carrillo P (2020) Uncoupled phytoplankton-bacterioplankton relationship by multiple drivers interacting at different temporal scales in a high-mountain Mediterranean lake. Sci Rep 8:350

Echevarría F, Carrillo P, Jiménez F, Sánchez-Castillo P, Cruz-Pizarro L, Rodríguez J (1990) The size abundance distribution and taxonomic composition of plankton in an oligotrophic, high mountain lake (La Caldera, Sierra Nevada, Spain). J Plankton Res 12:415–422

Elser JJ, Schampel JH, Kyle M, Watts J, Carson EW, Dowling TA, Tang C, Roopnarine PD (2005) Response of grazing snails to phosphorus enrichment of modern stromatolitic microbial communities. Fresh Biol 50:1826–1835

Felip M, Bartumeus F, Halac S, Jordi Catalan J (1999) Microbial plankton assemblages, composition and biomass, during two ice-free periods in a deep high mountain lake (Estany Redó, Pyrenees). J Limnol 58:193–202

Franch-Gras L, García-Roger EM, Franch B, Carmona MJ, Serra M (2017) Quantifying unpredictability: a multiple-model approach based on satellite imagery data from Mediterranean ponds. PlosOne 12(11):e0187958

García-Herrera RF, Lionello P, Ulbrich U (2014) Preface: understanding dynamics and current developments of climate extremes in the Mediterranean region. Nat Hazards Earth Syst Sci 14:309–316

García-Jurado F, Jiménez-Gómez F, Guerrero F (2011) Effects of a dry period on the limnological characteristics of a Mediterranean high mountain lake. Limnetica 30:0005–0016

Giorgi F, Lionello P (2008) Climate change projections for the Mediterranean region. Glob Planet Change 63:90–104

Gómez-Ortiz A, Oliva M, Salvador-Franch F, Palacios D, Tanarro LM, Sanjosé-Blasco JJ, Salvà-Catarineu M (2019) Monitoring permafrost and periglacial processes in Sierra Nevada (Spain) from 2001 to 2016. Permafr Periglac Process 30(4):278–291

González-Olalla JM, Medina-Sánchez JM, Lozano IL, Villar-Argaiz M, Carrillo P (2018) Climate-driven shifts in algal-bacterial interaction of high-mountain lakes in two years spanning a decade. Sci Rep 8:10278

González-Olalla JM, Medina-Sánchez JM, Carrillo P (2021) Fluctuation at high temperature combined with nutrients alters the thermal dependence of phytoplankton. Microb Ecol

Granados-Muñoz MJ, Navas-Guzmán F, Bravo-Aranda JA, Guerrero-Rascado JL, Lyamani H, Fernández-Gálvez J, Alados-Arboledas L (2012) Automatic determination of the planetary boundary layer height using lidar: one-year analysis over southeastern Spain J Geophys Res: Atmospheres 117(D18) n/a-n/a

Helbling EW, Carrillo P, Medina-Sánchez JM, Durán C, Herrera G, Villar-Argaiz M, Villafañe VE (2013) Interactive effects of vertical mixing, nutrients and ultraviolet radiation: in situ photosynthetic responses of phytoplankton from high mountain lakes of Southern Europe. Biogeosciences 10:1037–1050

Jiménez L, Conde-Porcuna JM, García-Alix A, Toney JL, Anderson RS, Heiri O, Pérez-Martínez C (2019) Ecosystem responses to climate-related changes in a Mediterranean alpine environment over the last ~180 years. Ecosystems 22:563–577

Kernan M, Ventura M, Bitušík P, Brancelj A, Clarke G, Velle G et al (2009) Regionalisation of remote European mountain lake ecosystems according to their biota: Environmental versus geographical patterns. Fresh Biol 54:2470–2493

Lionello P, Scarascia L (2018) The relation between climate change in the Mediterranean region and global warming. Reg Environ Change 18:1481–1493

Lionello P, Scarascia L (2020) The relation of climate extremes with global warming in the Mediterranean region and its north versus south contrast. Reg Environ Change 20:31

Mahowald N et al (2008) The global distribution of atmospheric phosphorus sources, concentrations and deposition rates and anthropogenic impacts, Global Biogeochem Cy 22, GB4026

Masson-Delmotte V, Zhai P, Pirani A, Connors SL, Péan C, Berger S, Caud N, Chen Y, Goldfarb L, Gomis MI, Huang M, Leitzell K, Lonnoy E, Matthews JBR, Maycock TK, Waterfield T, Yelekçi O, Yu R, Zhou B (eds) (2021) Climate Change 2021: The Physical Science Basis. Contribution of Working Group I to the Sixth Assessment Report of the Intergovernmental Panel on Climate Change. Cambridge University Press

Medina-Sánchez JM, Villar-Argaiz M, Cruz-Pizarro L, Sánchez-Castillo P, Carrillo P (1999) Structure changes in a planktonic food web: biotic and abiotic controls. J Limnol 58:213–222

Medina-Sánchez JM, Villar-Argaiz M, Carrillo P (2002) Modulation of the bacterial response to spectral solar radiation by algae and limiting nutrients. Fresh Biol 47:2191–2204

Medina-Sánchez JM, Villar-Argaiz M, Carrillo P (2004) Neither with nor without you: a complex algal control on bacterioplankton in a high mountain lake. Limnol Oceanogr 49:1722–1733

Medina-Sánchez JM, Villar-Argaiz M, Carrillo P (2006) Solar radiation nutrient interaction enhances the resource and predation algal control on bacterioplankton: a short-term experimental study. Limnol Oceanogr 51:913–924

Medina-Sánchez JM, Delgado-Molina JA, Bratbak G, Bullejos FJ, Villar-Argaiz M, Carrillo P (2013) Maximum in the middle: nonlinear response of microbial plankton to ultraviolet radiation and phosphorus. PLoS ONE 8(4):e60223

Medina-Sánchez JM, Delgado-Molina JA, Carrillo P (2016) Sentinels of global change (I): mixotrophic algae in La Caldera Lake. In: Zamora R, Pérez-Luque AJ, Bonet-García FJ, Barea-Azcón JM, Aspizua-Cantón R (eds) Observatorio Cambio Global Sierra Nevada. Global change impacts in Sierra Nevada: challenges for conservation. Department of the Environment and Urban Planning. Junta de Andalucía. ISBN: 978-84-338-5960-0, pp 83–85

Medina-Sánchez JM, Herrera G, Durán C, Villar-Argaiz M, Carrillo P (2017) Optode use to evaluate microbial planktonic respiration in oligotrophic ecosystems as an indicator of environmental stress. Aquat Sci 79:529–541

Mitra A, Flynn KJ, Burkholder JM, Berge T, Calbet A, Raven JA, Granéli E, Glibert PM, Hansen PJ, Stoecker DK, Thingstad F, Tillmann U, Våge S, Wilken S, Zubkov MV (2014) The role of mixotrophic protists in the biological C pump. Biogeosciences 11:995–1005

Mladenov N, Sommaruga R, Morales-Baquero R, Laurion I, Camarero L et al (2011) Dust inputs and bacteria influence dissolved organic matter in clear alpine lakes. Nat Comm 2:405

Morales-Baquero R, Carrillo P, Reche I, Sánchez-Castillo P (1999) Nitrogen-phosphorus relationship in high mountain lakes: effects of the size of catchment basins. Can J Fish Aquat Sci 56(10):1809–1817

Morales-Baquero R, Pulido-Villena E, Reche I (2006) Atmospheric inputs of phosphorus and nitrogen to the southwest Mediterranean region: biogeochemical responses of high mountain lakes. Limnol Oceanogr 51:830–837

Morales-Baquero R, Pulido-Villena E, Reche I (2013) Chemical signature of Saharan dust on dry and wet atmospheric deposition in the south-western Mediterranean region. Tellus B 65:18720

Morales-Baquero R, Pérez-Martínez C (2016) Saharan versus local influence on atmospheric aerosol deposition in the southern Iberian Peninsula: significance for N and P inputs. Glob Biogeochem Cy 30:501–513

Morales-Baquero R, Pérez-Martínez C, Ramos-Rodríguez E, Sánchez-Castillo P, Villar-Argaiz M, Conde-Porcuna JM (2019) Zooplankton advective losses may affect chlorophyll-a concentrations in fishless high-mountain lakes. Limnetica 38:55–65

Moser KA, Baron JS, Brahney J, Oleksy IA, Saros JE et al (2019) Mountain lakes: eyes on global environmental change. Glob Planet Change 178:77–95

Nuwer ML, Frost BW, Armbrust EV (2008) Population structure of the planktonic copepod *Calanus pacificus* in the North Pacific Ocean. Mar Biol 156:107–115

Oliva M, Gómez-Ortiz A, Salvador-Franch F et al (2016) Inexistence of permafrost at the top of the Veleta peak (Sierra Nevada, Spain). Sci Total Environ 550:484–494

Pepin N et al (2015) Elevation-dependent warming in mountain regions of the world. Nat Clim Change 5:424e430

Pérez-Palazón MJ, Pimentel R, Herrero J, Aguilar C, Perales JM, Polo MJ (2015) Extreme values of snow-related variables in Mediterranean regions: trends and long term forecasting in Sierra Nevada (Spain). Proc Int Assoc Hydrol Sci 369:157–162

Pérez-Martínez C, Conde-Porcuna JM, Moreno E, Ramos-Rodríguez E, Jiménez L (2020) Cladoceran assemblage distribution in shallow alpine lakes of Sierra Nevada (Spain) and its relationship with environmental variables. Aquat Sci 82

Pulido-Villena E, Reche I, Morales-Baquero R (2006) Significance of atmospheric inputs of calcium over the southwestern Mediterranean region: high mountain lakes as tools for detection. Glob Biogeochem Cy 20, GB2012

Reche I, Pulido-Villena E, Conde-Porcuna JM, Carrillo P (2001) Photoreactivity of dissolved organic matter from high-mountain lakes of Sierra Nevada, Spain. Arctic Antarct Alp Res 33:426–434

Reche I, Pulido-Villena E, Morales-Baquero R, Casamayor EO (2005) Does ecosystem size determine aquatic bacterial richness? Ecology 86:1715–1722

Reche I, Ortega-Retuerta E, Romera O, Villena EP, Baquero RM, Casamayor EO (2009) Effect of Saharan dust inputs on bacterial activity and community composition in Mediterranean lakes and reservoirs. Limnol Oceanogr 54:869–879

Rodriguez-Navarro C, di Lorenzo F, Elert K (2018) Mineralogy and physicochemical features of Saharan dust wet deposited in the Iberian Peninsula during an extreme red rain event. Atmos Chem Phys 18:10089–10122

Ruffault J, Curt T, Moron V, Trigo RM, Mouillot F et al (2020) Increased likelihood of heat-induced large wildfires in the Mediterranean Basin. Sci Rep 10:13790

Ruiz-Sinoga JD, Garcia-Marin R, Martinez-Murillo JF, Gabarron-Galeote MA (2011) Precipitation dynamics in southern Spain: trends and cycles. Int J Climatol 31:2281–2289

Sánchez-Castillo P, Linares-Cuesta E, Fernández Moreno D (2008) Changes in epilithic diatom assemblages in a Mediterranean high mountain lake (Laguna de La Caldera Sierra Nevada, Spain) after a period of drought. J Limnol 67:49–55

Seneviratne SI, Donat MG, Pitman AJ, Knutti R, Wilby RL (2016) Allowable $CO_2$ emissions based on regional and impact-related climate targets. Nature 529:477–483

Souza MS, Modenutti BE, Carrillo P, Villar-Argaiz M, Medina-Sánchez JM, Bullejos FJ, Balseiro EG (2010) Stoichiometric dietary constraints influence the response of copepods to

ultraviolet radiation-induced oxidative stress. Limnol Oceanogr 55:1024–1032
Steffen W, Richardson K, Rockström J, Cornell SE et al (2015) Planetary boundaries: guiding human development on a changing planet. Science 347:1259855
Stocker TF, Qin D, Plattner GK, Tignor M, Allen SK, Boschung J, Nauels A, Xia Y, Bex V, Midgley PM (eds) (2013) Climate change 2013: the physical science basis. In: Intergovernmental Panel on Climate Change, Working Group I Contribution to the IPCC Fifth Assessment Report (AR5). Cambridge Univ Press
Straskrabová V, Callieri C, Carrillo C, Cruz-Pizarro L, Fott J, Hartman P, Maceck M, Medina-Sánchez JM, Nedoma J, Simek K (1999) Investigations on pelagic food webs in mountain lakes-aims and methods. J Limnol 58:77–87
Talbot RW, Harriss RC, Browell EV, Gregory GL, Sebacher DI, Beck SM (1986) Distribution and geochemistry of aerosols in the tropical north Atlantic troposphere: relationship to Saharan dust. J Geophys Res 91:5173–5182
Thomas WH, Cho BC, Azam F (1991) Phytoplankton and bacterial production and biomass in sub-alpine Eastern Brook Lake, Sierra-Nevada, California. 2. Comparison with other high-elevation lakes. Arctic Alp Res 23:296–302
Tuel A, Eltahir EAB (2020) Why Is the Mediterranean a Climate Change Hot Spot? J Clim 33:5829–5843
Velasco Ayuso S, Medina-Sánchez JM, Guénon R, Carrillo P (2017) Ecoenzyme activity ratios reveal interactive effects of nutrient inputs and UVR in a Mediterranean high-mountain lake. Biogeochem 132:71–85
Villar-Argaiz M, Medina-Sánchez JM, Cruz-Pizarro L, Carrillo P (2001) Inter- and intra-annual variability in the phytoplankton community of a high mountain lake: the influence of external (atmospheric) and internal (recycled) sources of phosphorus. Freshw Biol 46:1017–1034
Villar-Argaiz M, Medina-Sánchez JM, Carrillo P (2002) Microbial plankton response to contrasting climatic conditions: insights from community structure, productivity and fraction stoichiometry. Aquat Microb Ecol 29:253–266
Villar-Argaiz M, Medina-Sánchez JM, Bullejos FJ, Delgado-Molina JA, Ruíz-Pérez O, Navarro JC, Carrillo P (2009) UV radiation and phosphorus interact to influence the biochemical composition of phytoplankton. Freshw Biol 54:1233–1245
Villar-Argaiz M, Bullejos FJ, Medina-Sánchez JM, Ramos-Rodríguez E, Delgado-Molina JA, Carrillo P (2012) Disentangling food quantity and quality effects in zooplankton response to P-enrichment and UV radiation. Limnol Oceanogr 57:235–250
Villar-Argaiz M, Bullejos FJ (2016) Sentinels of global change (II): herbivorous consumers in La Caldera Lake. In: Zamora R, Pérez-Luque AJ, Bonet-García FJ, Barea-Azcón JM, Aspizua-Cantón R (eds) Observatorio Cambio Global Sierra Nevada. Global change impacts in Sierra Nevada: challenges for conservation. Department of the Environment and Urban Planning. Junta de Andalucía. ISBN: 978-84-338-5960-0, pp 86–91
Villar-Argaiz M, Medina-Sánchez JM, Biddanda BA, Carrillo P (2018a) Predominant non-additive effects of multiple stressors on autotroph C:N: P ratios propagate in freshwater and marine food webs. Front Microbiol 9:69
Villar-Argaiz M, Medina-Sánchez JM, Corral-Arredondo E, Carrillo P, Biddanda B (2018b) Glacial lakes of Sierra Nevada as a Crystal Ball. EOS, series #Postcards from the field
Wang Q, Fan X, Wang M (2014) Recent warming amplification over high elevation regions across the globe. Clim Dynam 43:87–101
Ward BA, Follows MJ (2016) Marine mixotrophy increases ecological efficiency, mean organism size, and vertical carbon flux. Proc Natl Acad Sci USA 113:2958–2963
Weithoff G, Wacker A (2007) The mode of nutrition of mixotrophic flagellates determines the food quality for their consumers. Funct Ecol 21:1092–1098
Wilken S, Schuurmans JM, Matthijs HCP (2014) Do mixotrophs grow as photoheterotrophs? Photophysiological acclimation of the chrysophyte *Ochromonas danica* after feeding. New Phytol 204:882–889
Wilken S, Huisman J, Naus-Wiezer S, Van Donk E (2013) Mixotrophic organisms become more heterotrophic with rising temperature. Ecol Lett 16:225–233
Zamora R, Pérez-Luque AJ, Bonet FJ, Barea-Azcón JM, Aspizua R et al (2017) Networks, centers, observatories, and field stations. Bull Ecol Soc Am 98:157–164
Zittis G, Hadjinicolaou P, Klangidou M, Proestos Y, Lelieveld J (2019) A multi-model, multi-scenario, and multi-domain analysis of regional climate projections for the Mediterranean. Reg Environ Change 19:2621–2635

# Paleolimnological Indicators of Global Change

Carmen Pérez-Martínez, José María Conde-Porcuna, Eloísa Ramos-Rodríguez, Emilio Moreno, Kathleen M. Rühland, Adam Jeziorski, John P. Smol, Antonio García-Alix, Oliver Heiri, Eulogio Corral-Arredondo, and Laura Jiménez

**Abstract**

Anthropogenic climate change and the recent increase of Saharan dust deposition are potentially affecting Sierra Nevada alpine lakes. In this chapter, we summarize the results of paleolimnological research conducted to track recent environmental and ecological changes in the lakes and their catchments during the last two centuries. We analyzed several independent paleolimnological indicators preserved in highly resolved dated lake sediment cores including spectrally inferred chlorophyll-*a* concentration, leaf waxes (n-alkanes), and the subfossil remains of diatoms, cladocerans, and chironomids. Our results are indicative of significant changes in the lakes and their catchments with subtly starting over a century ago that accelerated in the 1960s–1970s, concurrent with trends in rising regional air temperature, declining precipitation, and increased Saharan dust deposition in the region. Our biological indicators registered pronounced changes in the composition of aquatic communities and a recent increase in algal biomass (inferred from chlorophyll-*a*). Temperature was identified as the main predictor of the observed changes, whereas Saharan dust deposition drivers were secondary explanatory variables. The synchronous change among the paleolimnological proxies and climatic variables analyzed in the Sierra Nevada study lakes indicated that this is a regional shift. The nature of the change in these independent proxies is interpreted as a response to a lengthening of the lake ice-free period, an increase in lake water temperature, as well as a reduction in water availability in the catchments, which affected the volumes and water residence times. All these processes reflect the intensification of summer drought in the Sierra Nevada summit area over the last 50–60 years. In addition, distinct changes in species composition indicate an alkalinization of lake waters. Projected increases in global temperature, decreasing precipitation, and possible increases in Saharan dust inputs will further exacerbate the changes observed so far in these valuable aquatic ecosystems.

**Keywords**

Drought • Alpine lakes • Saharan dust deposition • Climate change • Alkalinization • Sediments

## 1 Suitability of Sierra Nevada Lakes to Study Global Change

The Mediterranean region is considered one of the world's most sensitive areas to climate change. Meteorological data indicate a warming and drying trend in the climate of the Mediterranean region in the last 50–60 years (Drobinski et al. 2020). In fact, the Mediterranean has experienced significant drought over the last century (Hoerling et al. 2012; Kelley et al. 2012). Global climate models indicate that the impact of climate change may be particularly severe in this region (Giorgi and Lionello 2008; Ciscar et al. 2018, see chapter "Climate Variability and Trends"), noting that warming and drying are likely to continue. Thus, studies on the effects of climate change in the Mediterranean region

C. Pérez-Martínez (✉) · J. M. Conde-Porcuna · E. Ramos-Rodríguez · E. Moreno · E. Corral-Arredondo · L. Jiménez
Department of Ecology and Institute of Water Research, University of Granada, Granada, Spain
e-mail: cperezm@ugr.es

K. M. Rühland · A. Jeziorski · J. P. Smol
Paleoecological Environmental Assessment and Research Lab (PEARL), Department of Biology, Queen's University, Kingston, ON, Canada

A. García-Alix
Department of Stratigraphy and Paleontology, University of Granada, Granada, Spain

O. Heiri
Department of Environmental Sciences, University of Basel, 4056 Basel, Switzerland


R. Zamora and M. Oliva (eds.), *The Landscape of the Sierra Nevada*,
https://doi.org/10.1007/978-3-030-94219-9_17

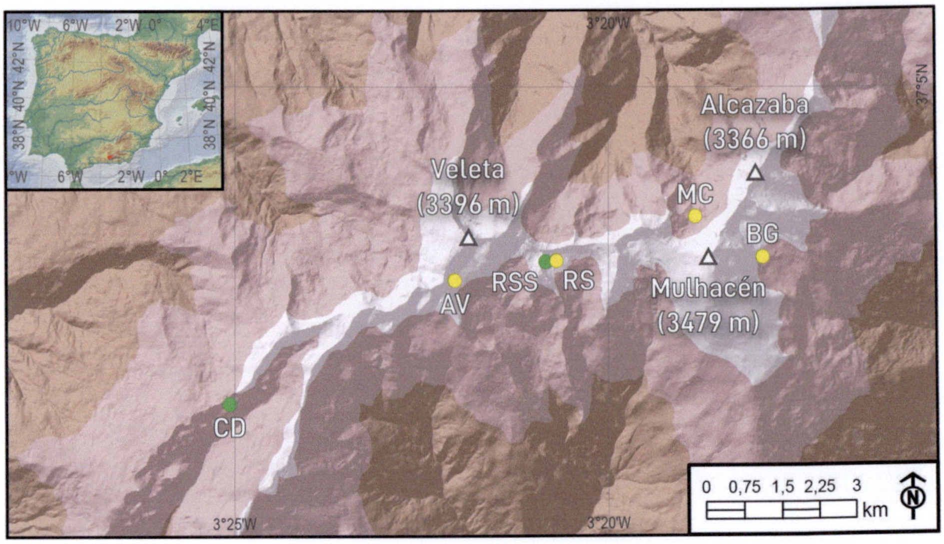


**Fig. 1** Map of the Sierra Nevada mountain range showing geographical locations of the six study lakes (circles) and highest mountain peaks (black triangles). Cuadrada (CD) and Río Seco Superior (RSS) are closed basin lakes (green circles); Aguas Verdes (AV), Río Seco (RS), Mosca (MC), and Borreguil (BG) are open basin lakes (yellow circles)

take on special importance as it is a densely populated area with severe water resource limitations that will restrict the region's capacity for development, affecting millions of people (Planton et al. 2016).

Sierra Nevada lakes are Mediterranean high-mountain lakes, located between 2800 and 3000 m asl. These high-altitude ecosystems experience intensifying temperature increases relative to lowland regions (Pepin et al. 2015). Given their Mediterranean location, they experience the intensification of summer droughts due to a rise in average summer air temperature, the reduction in annual precipitation (Nogués-Bravo et al. 2008, 2012; Sillmann et al. 2013), and a decrease in accumulated snow (Mote et al. 2005; see chapter "Snow Dynamics, Hydrology, and Erosion"). In addition, the Sierra Nevada region and the South East of Spain are significantly affected by atmospheric deposition of Saharan dust, which is distinguished by an exceptionally high phosphorus, calcium, and alkalinity content (Rogora et al. 2004; see chapter "Atmospheric Inputs and Biogeochemical Consequences in High-Mountain Lakes"). Saharan dust export to the atmosphere has increased exponentially in recent decades as a result of North African droughts (Prospero and Lamb 2003), human-induced desertification (Moulin and Chiapello 2006), and the development of commercial agriculture in the Sahel region (Mulitza et al. 2010). The effect of Saharan dust on Sierra Nevada aquatic ecosystems on a short time scale has been shown in several studies (see chapter "Atmospheric Inputs and Biogeochemical Consequences in High-Mountain Lakes").

To understand the magnitude and nature of the transformations occurring in ecosystems due to global change, long-term studies are needed. One approach to the study of the long-term effects of climate change in lakes is through the lake sedimentary record, which allows the reconstruction of past conditions of lakes and their watersheds (Smol 2008). To that end, it is common to analyze a suite of different indicators preserved in dated lake sediments, since ecosystem responses to climate and global change are diverse and complex, and thus multidisciplinary approaches produce the most robust interpretations. The use of several independent indicators allows for a better understanding of past processes in response to environmental changes.

To examine the changes that have occurred over the past two centuries in response to global change, we analyzed several paleolimnological variables preserved in sediment cores obtained between 2008 and 2011 from six Sierra Nevada lakes: Rio Seco (RS), Rio Seco Superior (RSS), Aguas Verdes (AV), Borreguil (BG), Mosca (MC), and Cuadrada (CD) (Fig. 1). These lakes are located between ~2800 and ~3000 m asl and are permanent, shallow, littoral-dominated systems with no clear differentiation between littoral and profundal zones. Some lakes are surrounded by high mountain wet grasslands and others are not. Sierra Nevada lakes are clear and fishless, with low primary production and low alkalinity values (see chapter "High Mountain Lakes as Remote Sensors of Global Change"). The lakes are remote and have experienced minimal anthropogenic disturbance that is limited to livestock grazing and mountaineers during the summer. Because of their remote, high-altitude location, and their low primary production and low alkalinity, the footprint of climate change and Saharan input may be clearly expressed, making them excellent sites for studying the effects of long-term global change.

Sediment cores were retrieved in the deepest area of each lake and sectioned in the field into fine intervals (0.25–0.5 cm). Following extrusion, sediment samples were sealed in sterile bags and stored in a cold room at ~4 °C until analysis of the different indicators. Samples were dated by radioisotopic methods using $^{210}Pb$ and $^{137}Cs$ activities (see Jiménez et al. 2018 for details). For the last ~50 years, the mean sedimentation rates obtained ranged from 0.01 to 0.03 g $cm^{-2}$ $yr^{-1}$ so that each sediment interval corresponds to approximately 2–5 years, which permits a relatively high temporal resolution in our records. Additional analytical details can be found elsewhere for chlorophyll-*a* and cladoceran remains (Jiménez et al. 2018), chironomid remains and n-alkanes (Jiménez et al. 2019), and diatom remains (Pérez-Martínez et al. 2020b).

## 2 Synchrony of Change Among Paleolimnological Proxies and with Climatic Variables

Different paleolimnological proxies indicate that the main change in the aquatic ecosystems of Sierra Nevada over the last two centuries occurred in the 1960s–1970s and that this change coincided with trends in air temperature, precipitation, and Saharan dust deposition in the region (Figs. 2 and 3). The synchrony of change among the Sierra Nevada lakes studied indicates that this is a regional signal, although some site-specific responses were observed.

We used various climate metrics as explanatory variables for the changes observed in the lake sediment records during the last ~180 years. These included a homogenized series of instrumental temperature data (mean annual air temperature series from the Madrid station, 1869–2011) and a homogenized series of precipitation data (annual precipitation series from the San Fernando station, 1839–2011) collected from climate stations near Sierra Nevada. Given that there are no long-term series of direct measurements of Saharan dust deposition, we used two related drivers as proxies including a series of values for the Winter North Atlantic Oscillation (wNAO) index since 1864, and the Sahel precipitation index (SPI) since 1900 (Jiménez et al. 2018). The magnitude of Saharan dust emissions has been associated with both the wNAO index (Moulin et al. 1997) and the SPI (Chiapello et al. 2005; Moulin and Chiapello

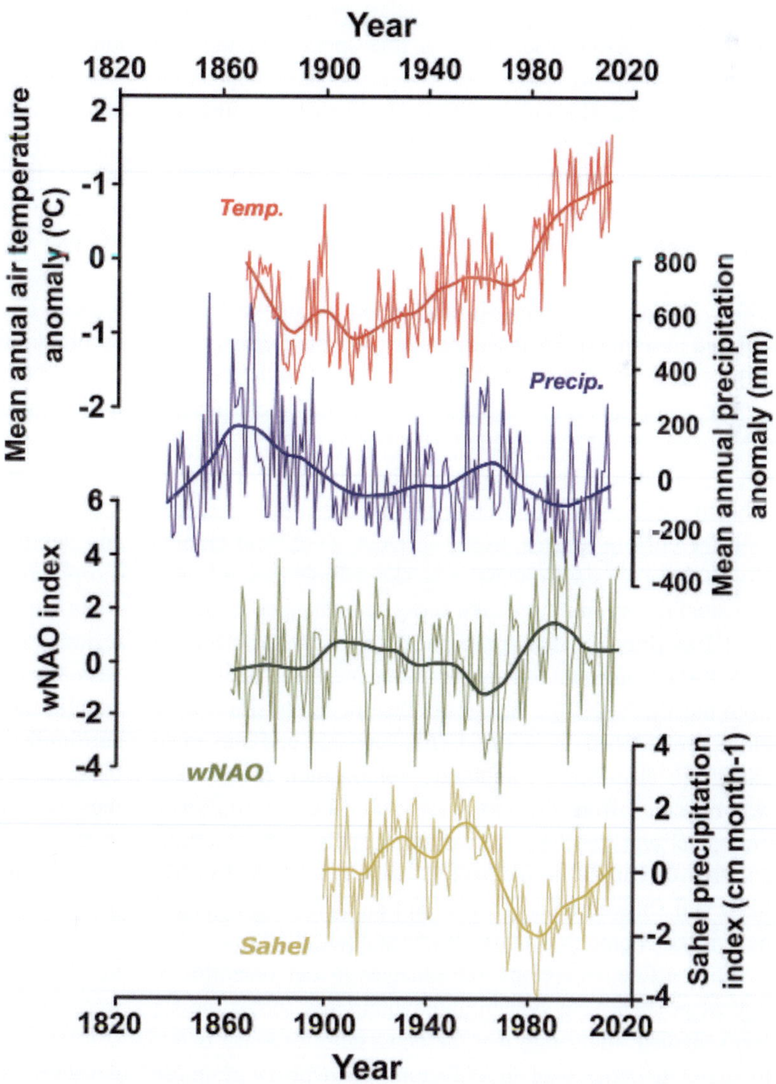


**Fig. 2** Comparison of the historical trends of the mean annual air temperature anomaly from the Madrid climate station since 1860, the annual precipitation anomaly from the San Fernando climate station since 1840, and the wNAO index and the Sahel precipitation index (SPI). Temperature anomalies are calculated from the period 1961 to 1990 and precipitation anomalies are calculated over the entire period. The anomalies of the SPI are calculated with respect to 1900 and 2013, and based on June through October averages for each year. A LOESS smoother (span = 0.2) was applied to all the variables to improve the clarity of the figure and highlight trends

2004). Both long-term series were strongly correlated to the zirconium aluminium ratio, a proxy of Saharan input (Jiménez-Espejo et al. 2014), analyzed in a sediment core from one of the Sierra Nevada lakes, and to the Saharan dust calcium (Ca) concentrations measured in an ice core obtained from a glacier in the French Alps (Preunkert and Legrand 2013). Therefore, we can consider the SPI and the wNAO index as predictors of the transport and intensity of Saharan dust events in Sierra Nevada and representative of phosphorus and calcium deposition trends in the region. For further details on climate data, see Jiménez et al. (2018).

The mean annual air temperature series from Madrid indicates a warming trend beginning at the turn of the twentieth century (Fig. 2). The temperature series also shows a breakpoint in 1971, after which the temperature values increased continuously up to the present, especially since the early 1980s. The total annual precipitation series from San Fernando (Fig. 2) indicates that the second half of the nineteenth century was wet, and then precipitation decreased from the late nineteenth century to the present, interrupted only by positive anomalies in the 1960s. The precipitation series also shows a drier period in the last 40 years, since the early 1980s. The Saharan deposition drivers (i.e., the SPI and wNAO index) indicate an increase in deposition from the 1980s onward (Fig. 2). In summary, we can say that the Sierra Nevada has experienced higher temperatures and lower precipitation as well as an increase in a Saharan deposition during the last five decades, with intensification of these tendencies in the late 1980s and 1990s. In fact, this agrees with the period of intense drought suffered in southern Spain at that time (Vicente-Serrano et al. 2017, see chapter "Climate Variability and Trends").

The main directions of change in our biological proxies were summarized using detrended correspondence analysis (DCA) for diatom assemblages and by principal component analysis (PCA) for cladoceran assemblages. In agreement with the climate records, trends in DCA, PCA, and chlorophyll-*a* from the six Sierra Nevada lake sediment records also indicated that pronounced lake ecosystem changes started in the 1970s (Fig. 3), although slight changes can be observed at $\sim$1900. Chlorophyll-*a* concentration increased in all lakes from the 1960s–1970s, changes in diatom composition were more pronounced from the 1970s onward, and the main change in cladoceran community composition in all lakes was observed from the 1980s onward (Fig. 3). Additional biological and geochemical variables, such as chironomids and leaf wax n-alkane biomarkers, were also studied in Río Seco and showed equally important changes in the second half of the twentieth century (Jiménez et al. 2019).

With a few exceptions, the changes in our paleolimnological proxies were synchronous within and across the six lakes studied, indicating that the ecosystem response is due to regional rather than local factors. Each proxy response could be associated with distinct mechanistic linkages to climate change (Fig. 4).

## 3 Environmental Changes in Sierra Nevada Lakes Linked to Climate Change

An increase in regional air temperature and a decrease in precipitation over the past few decades has triggered significant changes in Sierra Nevada summit ecosystems (Fig. 4). One of the most important climate repercussions affecting high mountain lakes is the lengthening of the ice-free period, a key variable in alpine systems (Adrian et al. 2009; Smol 2008; Moser et al. 2019) that is linked to fundamental aquatic ecosystem functioning. For example, the lengthening of the ice-free period results in a longer growing period for organisms that have a narrow temporal window to colonize the system each year through propagules (Barea-Arco et al. 2001; Pérez-Martínez et al. 2013). In addition, a longer and warmer growing season will enhance the metabolic rates of aquatic organisms and increase population numbers as well as individual growth rates. Another repercussion of longer and warmer growth periods is that the greater availability of primary producers can also affect secondary producers. Secondary producer species with longer life cycles or with higher temperature and food requirements may benefit in these situations at the expense of other more generalist species.

A decrease in precipitation in the form of snow, with or without an increase in temperature, can lead to reduced ice accumulation and a lengthening of the ice-free period. Both of these climate variables contribute to water availability in the basin as reflected by evaporation/precipitation balances (Adrian et al. 2009). In hydrologically closed lakes of the Sierra Nevada, increased evaporation from warmer and/or dryer conditions may result in a reduction in lake volume and depth, whereas in hydrologically open lakes, the same morphometrical changes may result from drying and reduced water input. The latter case may result in increased water residence time and reduced turbulent mixing, which together with warmer conditions can lead to greater water column stability in these shallow Sierra Nevada lakes. Lakes that are surrounded by wet grasslands tend to lower the availability of water in the catchment. With warmer and drier conditions, there will be a shortening of the summer waterlogging period in these meadows. In the long term, it may result in greater aridity, ultimately affecting the development of the biota associated with the meadows.

In addition to climatic impacts, Sierra Nevada lakes are also affected by Saharan dust intrusions, which are more frequent in the period from March to October (Morales-Baquero and Pérez-Martínez 2016). In recent decades, droughts in North Africa have exponentially

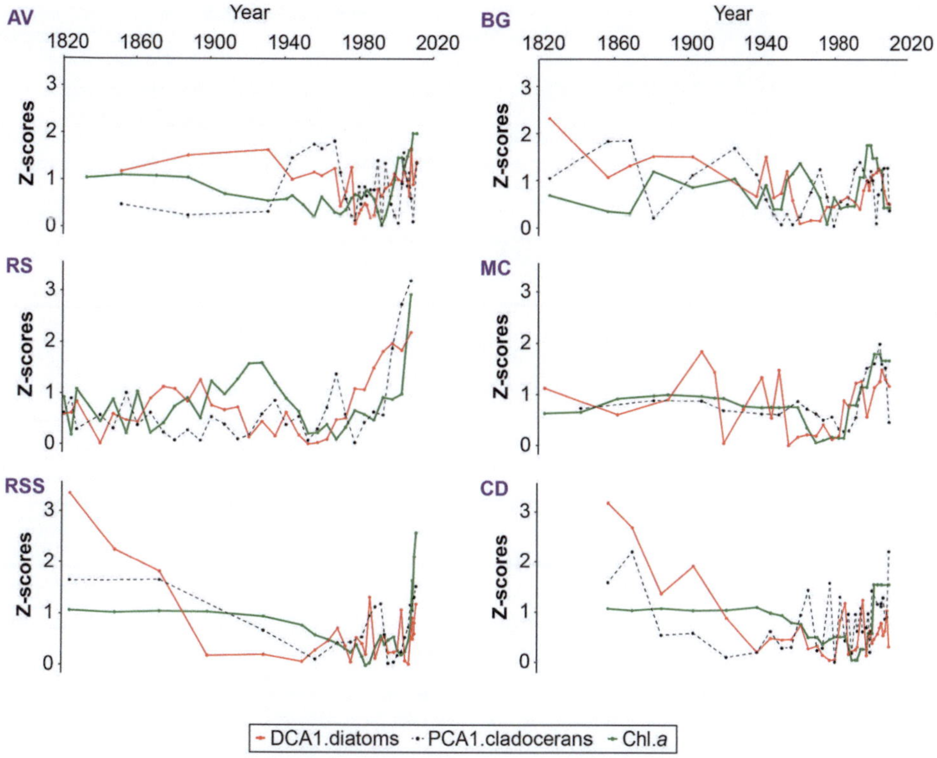


**Fig. 3** Timing of the main changes in biological proxies across the Sierra Nevada lakes. Detrended correspondence analysis (DCA) was applied to non-transformed relative abundance data as a means to summarize the main variation in the diatom assemblage data. The variation of the cladoceran assemblage data was summarized through a Principal Component Analysis (PCA). DCA axis 1 sample scores for diatom assemblages, PCA axis 1 scores for cladoceran assemblages, and chlorophyll-a (Chl.*a*) values plotted against estimated $^{210}$Pb dates for the six study lakes. Z-scores of all the variables are shown. Lake name abbreviations: Aguas Verdes (AV), Río Seco (RS), Río Seco Superior (RSS), Borreguil (BG), Mosca (MC), and Cuadrada (CD)

increased the amount of dust distributed by the atmosphere (Prospero and Lamb 2003) affecting aquatic ecosystems in the Sierra Nevada region (Morales-Baquero et al. 2006). For example, calcium and phosphorus laden dust deposited onto oligotrophic, low alkalinity Sierra Nevada lakes are considered important sources of nutrients and alkalinity, fuelling primary and secondary production (Morales-Baquero et al. 2006) and affecting base-cation availability (Pulido et al. 2006) and therefore alkalinity values.

## 3.1 Algal Biomass Increase

Trends in sedimentary chlorophyll-*a* (and its main diagenetic products), analyzed by visible reflectance spectroscopy, provide a proxy for algal biomass (Michelutti and Smol 2016). In the six lakes studied in Sierra Nevada there was a significant increasing trend in algal biomass from the 1960s–1970s to the present (Fig. 3). Regression analyzes performed with climatic variables indicate that the main predictor of trends in chlorophyll-*a* is temperature followed by the drivers of Saharan deposition, SPI, and wNAO (see Jiménez et al. 2018). The increase in air temperature during the second half of the twenieth century (Fig. 2), accompanied by longer ice-free and growing periods, provided more time for algal populations to increase, resulting in primary producers attaining higher annual accumulated biomasses. Other warming related factors that may have influenced algal growth include intensified evaporation rates and decreases in water availability in the lake catchment (Fig. 4). Indeed, southern Spain has experienced a decreasing trend in precipitation over the past 40 years, with a notable period of drought during the late-1980s and 1990s (Udelhoven et al. 2009, Vicente-Serrano et al. 2017). Collectively, these processes may contribute to increased nutrient concentrations in these non-stratified, shallow systems through evapoconcentration or through increasing rates of rock dissolution and/or snowmelt (Preston et al. 2016; Sommaruga-Wögrath et al. 1997). Nutrient recycling in these non-stratifying systems was probably also enhanced by warmer temperatures (Wilhelm and Adrian 2008). In some of the Sierra Nevada lakes, where water levels have decreased, sediment along the shore

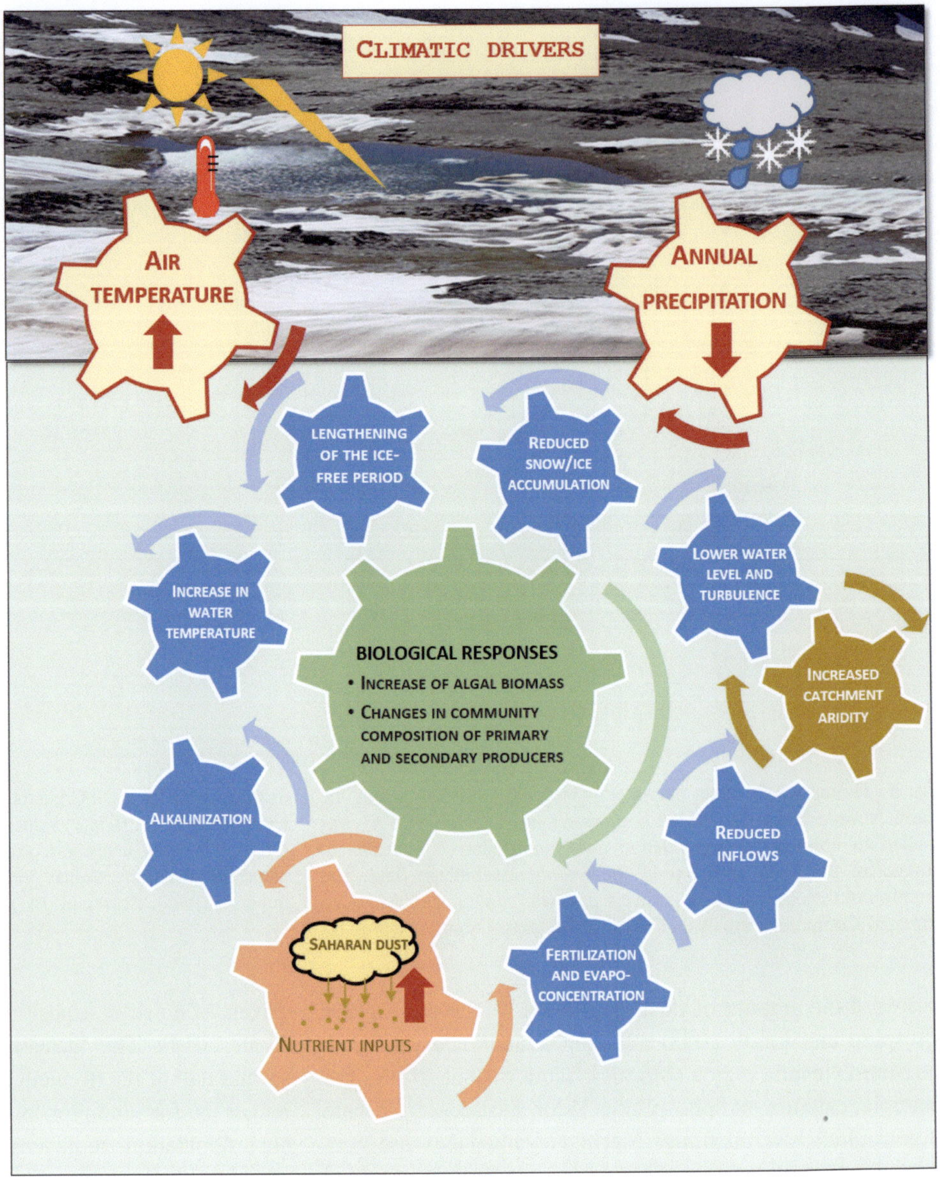


**Fig. 4** Explanatory diagram showing the effects of climatic drivers and Saharan inputs on aquatic ecosystems in the Sierra Nevada

may be exposed to the air, impeding sediment absorption of phosphorus once rehydrated, so that phosphorus remains in the water column.

The Sahel precipitation index (SPI), one of the predictors of Saharan dust transport, is a secondary explanatory variable for chlorophyll-*a* concentrations in the Sierra Nevada lakes. Phosphorus-laden atmospheric inputs probably enhanced algal growth in these nutrient-poor lakes, as has been recorded in short-term observations in a few Sierra Nevada lakes (see chapter "Atmospheric Inputs and Biogeochemical Consequences in High-Mountain Lakes"). Saharan intrusions in the area are more frequent during the ice-free season (Morales-Baquero and Pérez-Martínez 2016) and increased from the 1980s onward, when presumably the lakes experienced longer ice-free periods. In fact, an increase in the slope of our sedimentary chlorophyll-*a* data is observed in all lakes in the 1980s, concurrent with the period of accelerated temperature increase, drought, and increased Saharan deposition. The increase in algal biomass (chlorophyll-*a*) is consistent with longer and warmer periods of algal growth and greater availability of nutrients derived from both climate-driven changes and Saharan dust deposition.

In summary, the Sierra Nevada study lakes have experienced and are experiencing an increase in algal biomass since the second half of the twentieth century linked primarily to climate change and secondarily to the deposition of nutrient-laden Saharan dust. In addition, there are likely numerous interacting or cumulative effects of increased air temperature, water availability, and increased nutrients that may have contributed to the observed rise in algal biomass reported in these lakes.

## 3.2 Changes in Species Composition

The remains of freshwater aquatic organisms, such as diatoms, cladocerans, and chironomids (among others), were used as bioindicators of past environmental conditions in the six studied Sierra Nevada lakes. Detailed analyses of diatom and cladoceran remains preserved in dated sediment cores tracked changes from the late nineteenth century to the present. In one lake (Río Seco), the subfossil remains of the chironomid larval community were also analyzed. Significant compositional changes were registered in the six diatom records over the last ~180 years. This change started gradually at the turn of the twentieth century, but the most pronounced change was observed from ~1970 onward (Fig. 3). Similar to trends in sedimentary chlorophyll-*a*, diatom changes coincided with regional increases in temperature, declines in precipitation, and increases in Saharan dust deposition.

Notable compositional changes observed in the sedimentary diatom assemblages during the last ~50–60 years (Fig. 5; Pérez-Martínez et al. 2020b) include a decrease in epiphytic taxa (*Gomphonema* spp., *Pinnularia* spp., *Eunotia* spp. and *Achnanthidium minutissimum*), especially in the lakes with a greater extension of alpine meadows in their basins. In addition, a shift from diatoms that prefer more acidic conditions (e.g., *Eunotia* spp., *Brachysira brebissonii, Frustulia crassinervia*) to assemblages with higher abundances of alkaliphilous taxa (*Navicula cryptocephala, Nitzschia gracilis, Sellaphora pupula*) and a decrease and even disappearance of tychoplanktonic species (*Tabellaria flocculosa* strain IV and *Aulacoseira alpigena*) was observed (Rühland and Smol 2002; Siver and Baskette 2004; Catalan et al. 2009; Falasco and Bona 2011; Jacques et al. 2016).

The cladoceran records from the six study lakes (Fig. 5; Jiménez et al. 2018) registered a subtle change in assemblage composition ~1900, with more pronounced changes from the late 1980s to the 1990s. This response lags the changes observed in the diatoms and trends in chlorophyll-*a* but does coincide with the accelerated increase in temperature and the intense drought experienced in the region during those decades. This is indicative of a threshold-type response to climate change or a response to other climate-driven factors different from that of primary producers. The biotic change consists of a decrease in *Chydorus sphaericus* and a simultaneous increase in *Alona quadrangularis*, a species that in some lakes was absent or had only a residual presence in the deeper sediment intervals. Also notable is an increase of the *Daphnia pulex* gr. in the lakes where it is present, which is especially pronounced since the late 1980s. In Lake BG, genetic analyses revealed that the *Daphnia* taxon that became prevalent since the 1960s is the North American colonizing taxon *D.* cf. *pulex*, a different lineage than the native Eu *D.* cf. *pulicaria* from Sierra Nevada (Burillo et al. 2019; Conde-Porcuna et al. 2021).

The chironomid assemblage of Río Seco lake experienced a shift in species composition in the second half of the twentieth century (i.e., ~1950s–1960s), and can be summarized by the increase and/or new arrival of *Chironomus plumosus* type, *Heterotrissocladius marcidus* type, and *Micropsectra insignilobus* type in the uppermost section of the sediment core (Fig. 6).

The synchrony of changes in the paleolimnological proxies in all the study lakes (Figs. 3, 5, and 6; Jiménez et al. 2018; Pérez-Martínez et al. 2020b) indicates a regional response to environmental and climatic change. In particular, temperature was found to be the most important explanatory variable selected by the regression models for all taxonomic groups of organisms examined. However, aquatic biota often responds indirectly to changes in temperature, and similar to trends we observed in sedimentary chlorophyll-*a*, the increase in temperature can exacerbate changes in numerous variables to which organisms can respond (Fig. 4). In addition to temperature, Saharan dust deposition (SPI and wNAO) was identified as an important explanatory factor selected for both diatoms and cladocerans. In particular, dust deposited to oligotrophic Sierra Nevada lakes deliver both phosphorus and calcium that are important to aquatic biota in these remote lakes.

## 3.3 Changes in Lake Hydrology and More Favorable Conditions for Invertebrate Growth

In the Sierra Nevada region, a drier and warmer climate led to a reduction of snow (Bonet et al. 2016; see chapters "Climate Variability and Trends" and "Snow Dynamics, Hydrology, and Erosion"), reduced inflows to the lakes, lower water levels, and likely reduced lake water turbulent mixing. Heavily silicified tychoplanktonic diatoms such as *Aulacoseira* spp. can be used as a bioindicator of changes in thermal stability (see review by Rühland et al. 2015) as these taxa require water column mixing to maintain a planktonic existence (Kilham et al. 1996; Round et al. 1990). The dominance of *Aulacoseira* taxa during the wet and cold period of the second half of the nineteenth century and its subsequent decline to trace abundances toward the top of the Sierra Nevada cores were consistent with the onset of warmer and drier conditions ~1970 (Fig. 4). Likewise, increased water residence time and warmer temperatures potentially favor taxa with larger body size and lower growth rate, such as *A.*

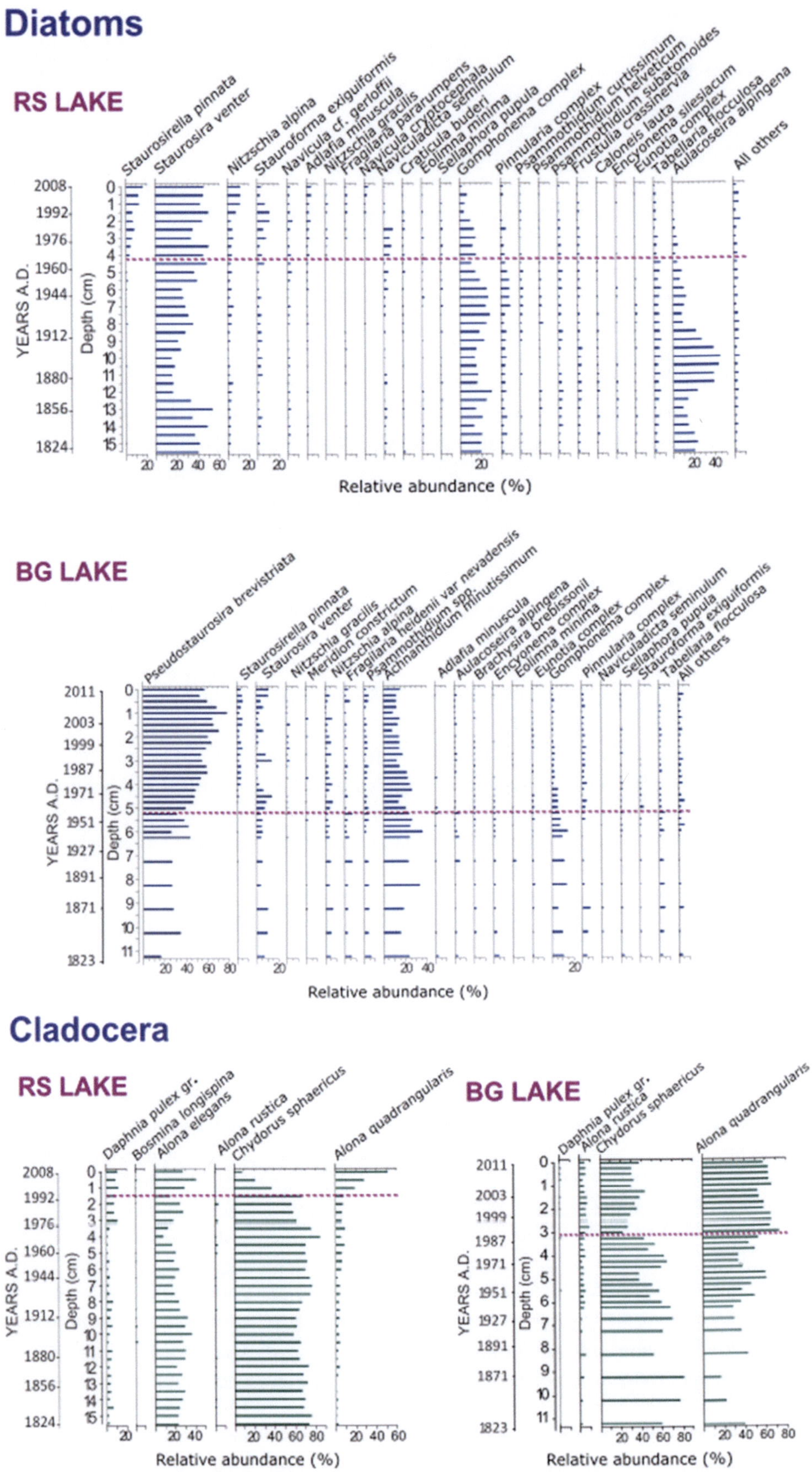


**Fig. 5** Relative abundance diagrams of the most common diatom taxa (taxa with relative abundance >1% in at least one sediment sample interval) and the most common cladoceran taxa recorded in the sediment cores from the lakes Borreguil (BG) and Río Seco (RS). The purple broken line represents the main zonation identified by the broken stick model

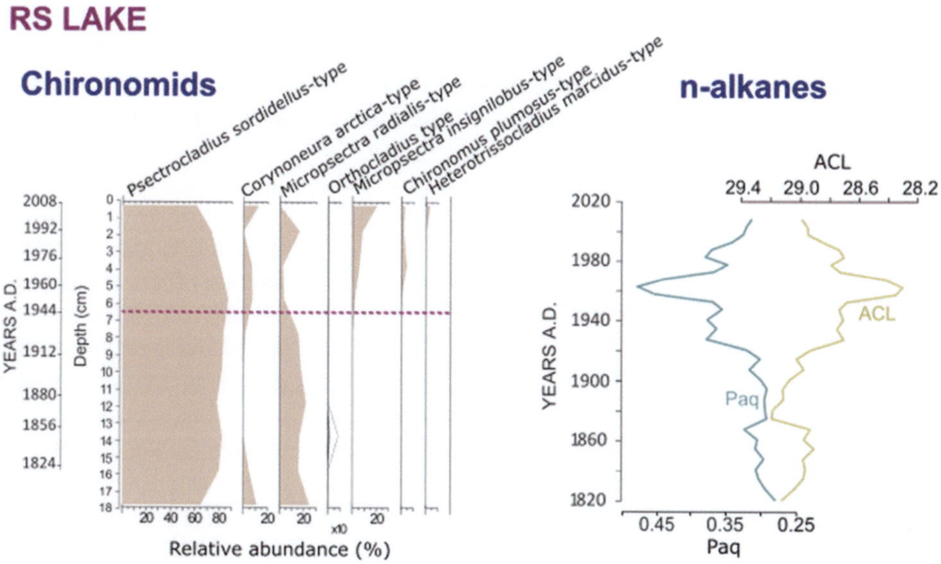


**Fig. 6** Relative abundance diagrams of the most common chironomid taxa and leaf wax n-alkane biomarkers ($P_{aq}$ and ACL) from Río Seco lake. The purple broken line represents the main zonation identified by the broken stick model

*quadrangularis* and *Daphnia pulex* gr., the latter being the largest cladoceran in the Sierra Nevada lakes. Additionally, a decrease in hydraulic flushing could favor *Daphnia* population (Jiménez et al. 2015, 2018; Pérez-Martínez et al. 2020a; Morales-Baquero et al. 2019). The importance of the advective losses of zooplankton in zooplankton abundance is well reported in different aquatic ecosystems (Bozelli 1994; Walz and Welker 1998; Rellstab et al. 2007; Beaver et al. 2013; Morales-Baquero et al. 2019).

Several biotic changes indicate a response to more favorable conditions for invertebrate growth in the lakes. One of the more notable changes among the cladocerans was a shift from *Chydorus sphaericus* to *Alona quadrangularis*. *C. sphaericus* is a species associated with colder, ultraoligotrophic waters commonly found in alpine lakes (Bigler et al. 2006; Harmsworth 1968; Lotter et al. 1997; Whiteside 1970), whereas *A. quadrangularis* is associated with warmer conditions (Bigler et al. 2006; Catalan et al. 2009; Nováková et al. 2013). Therefore, the increase (or arrival) of *A. quadrangularis* at the expense of *C. sphaericus* observed in all the lakes would be expected with the onset of more favorable environmental conditions for growth in terms of water temperature, a longer growing season and increased food availability (indicated by the increasing trend in sedimentary chlorophyll-*a*).

The chironomid species present in Río Seco (Fig. 6) before ~1950 were indicative of cold environments and oxygenated sediment (Lotter et al. 1998; Heiri et al. 2011). However, the arrival at ~1960 of *C. plumosus* type is indicative of environments richer in nutrients and organic matter and poorer in oxygen (Lotter et al. 1998; Heiri et al. 2011). This is in agreement with the increasing trend in sedimentary chlorophyll-*a* observed in the lakes at this time. The appearance and growth of warmer water taxa, such as *H. marcidus* type and *M. insignilobus* type, is consistent with an increase in lake water temperature during the period of accelerated regional warming recorded over the last ~60 years. In fact, the mean July atmospheric temperature inferred from the sedimentary chironomid record from Río Seco, based on modern chironomid assemblages examined in 274 lakes (see Heiri et al. 2011), suggests an increase of 2 °C from the 1950s onward.

Small variances in the timing of changes among our biological proxies are to be expected and can be explained by different climate response thresholds for diatoms, chironomids, and cladocerans. The most notable change in cladoceran assemblages in the Sierra Nevada sediment records occurred from the ~1980s onward and lagged the changes observed in diatoms and chlorophyll *a*. The later response of cladocerans may be explained by a threshold-like response to a combination of an accelerated rate of warming, increase in Saharan dust deposition, and a decline in water availability at this time. It is likely that primary producers have a more rapid response to temperature and resource availability than cladocerans or that the response of the latter requires a higher threshold of change in environmental factors. The coincidence of the change in cladocerans with the intense drought period in the region could indicate, however, that the response of cladocerans may be more closely linked to the effects of drought (evapoconcentration, hydrological changes) than diatoms and other primary producers.

In contrast, changes in species composition in the chironomids, a taxonomic group that has proven to be particularly useful as a temperature indicator (Walker et al. 1991), occur earlier, starting in the ~1950s. It must be considered that chironomid samples were analyzed every 2 cm and probably this coarser sampling interval makes it difficult to determine more accurately the timing of the assemblage shift, but, in any case, it coincides approximately with the increase in air temperature, confirming the sensitivity of the species of this group to temperature changes.

## 3.4 Indicators of Increasing Drought in Lake Catchments

The ice-free period on lakes in the Sierra Nevada region typically lasts from June to November; however, large interannual differences in the duration of ice cover have been reported and related to interannual variations in climatic conditions (Barea-Arco et al. 2001; Morales-Baquero et al. 2006; Pérez-Martínez et al. 2007). For example, ice-free periods ranging from 3 to 6 months have been reported in specific lakes of the Sierra Nevada. Because precipitation in Sierra Nevada is very scarce during summer, the alpine lakes experience a long period of drought during summer, which is accentuated in the driest climatological years when there is less snow and/or water accumulated in the catchment and the lake ice-free period is longer.

In our study, many of the paleolimnological proxies indicated that summer drought is becoming more pronounced in recent decades. For example, we have already mentioned the diatom taxa indicative of greater water column stability and a decrease in zooplankton advective losses (hydraulic washout) due to the lower availability of water in the catchment. Likewise, the decrease observed in the relative abundance of epiphytic diatoms, associated with vegetation and the lakeshore environment, such as *Gomphonema* spp., *Pinnularia* spp., *A. minutissimum* (Fig. 5), would indicate lower water levels and increasing drought starting in the ~1960s. The diatoms associated with this habitat are then hindered in their growth. This shift was mainly observed in those lakes with greater influence of the surrounding vegetation, i.e., lakes with a greater extension of alpine meadows in their basins and a higher ratio between the extension of these and the lake area. This assemblage change occurs mainly between the ~1960s and ~1980s, once more highlighting climate as the main driver of these changes.

In Río Seco lake, some other indicators of catchment drought, such as the n-alkane indices, enabled us to investigate the water availability for the catchment vegetation (Jiménez et al. 2019). Previous studies in Sierra Nevada (García-Alix et al. 2017) indicate that lower values of Average Chain Length (ACL, a measurement of the weighted average of the carbon chain lengths) are usually recorded in areas closer to the water pools, pointing toward higher water availability. Higher values of aquatic relative to terrestrial plant waxes ($P_{aq}$, the ratio between typical aquatic and terrestrial plant n-alkanes (Ficken et al. 2000)) are also indicative of higher water availability (Fig. 6). Hence, we recorded a period of low values of $P_{aq}$ and high values of ACL during the nineteenth century, followed by high values of $P_{aq}$ and low values of ACL afterward, indicating less water availability in the lake catchment during the growing seasons of the twentieth century. These trends match the climate records, which show a period of relatively high precipitation and low temperature until ~1920, followed by increasing temperature and decreasing precipitation afterward. This would result in intense snow/ice melting in the catchment, which is also indicated by the n-alkane indices.

## 3.5 Indicators of Lake Water Alkalinization

Although there are numerous studies focused on the effect of acid precipitation on alpine ecosystems, the Sierra Nevada region was little affected by this factor (Morales-Baquero and Pérez-Martínez 2016). However, the Sierra Nevada region does experience the effects of atmospheric deposition of Saharan dust (Jiménez et al. 2018), which is alkaline and rich in calcium and magnesium (Rogora et al. 2004). This Saharan input contributes significantly to alkalinity and base cation values of alpine lake waters in areas of low weathering rocks (Greilinger et al. 2018), such as the silicate rocks present in our Sierra Nevada study.

We observed an increase of alkaliphilous diatom species at the expense of species considered acidophilous from the ~1960s onward (Fig. 5, Pérez-Martínez et al. 2020b), indicating an alkalinization of the waters over the last few decades that can be explained by several mechanisms. A primary cause may be the increase in Saharan dust deposition since ~1980s (Fig. 2) together with a greater exposure to aerial inputs with a lengthening of the ice-free period. In addition, with warmer lake water, there is an increase in evaporation and residence time that can increase solute concentration. In our Sierra Nevada diatom records, declines in benthic fragilarioid taxa, including *Staurosira venter,* concurrent with the pronounced increase in *Staurosirella pinnata*, a taxon associated with higher levels of alkalinity (Catalan et al. 2009; Weckstrom et al. 1997), are consistent with alkalinization. In this respect, temperature, SPI, and wNAO were the main predictor variables of *S. pinnata* abundances (Pérez-Martínez et al. 2020b).

Cladocerans such as some *Daphnia* spp. can also be indicators of changes in base cations because of their calcium requirements. The calcium values in the Sierra Nevada lakes are limiting for the growth and reproduction of *Daphnia* spp. (Ashforth and Yan 2008; Hessen et al. 2000). Hence, in our Sierra Nevada sediment records, *D. pulex* gr. , a cladoceran with high calcium requirements for growth (Hessen and Rukke 2000) was likely limited by this element in our study lakes (Jeziorski et al. 2008; Jeziorski and Smol 2017).

Saharan calcium inputs, together with solute evapoconcentration, probably contributed to the observed increase of *Daphnia pulex* gr. in these lakes since the ~1970s (Fig. 5). This interpretation is corroborated by our regression analyses that indicated that Saharan deposition, together with temperature, were the main explanatory variables for this compositional shift (Jiménez et al. 2018). The combination of these conditions in the second half of the twentieth century may have favored the establishment and success of the North American colonizing taxon *D.* cf. *pulex* in BG lake (Conde-Porcuna et al. 2021; Pérez-Martínez et al. 2020a).

Paleolimnological indicators show that the long-term cumulative effects of atmospheric Ca deposition may be a critical factor in the future of Sierra Nevada aquatic ecosystems, modifying both the chemical conditions of the ecosystems and the species composition of communities.

## 3.6 Concluding Remarks

Sierra Nevada alpine lakes are influenced by a mid-latitude European and North African hot and dry climate, as well as by atmospheric deposition of Saharan dust. Intense summer drought, combined with a significant influence of Saharan dust inputs, sets the Sierra Nevada apart from other European high mountain systems. In our paleolimnological study, we recorded pronounced changes in species dominance and the appearance and disappearance of species belonging to different trophic levels, changes in chlorophyll-*a* (algal biomass) trends, changes in geochemical indicators of aridity in the catchments and in the chemical conditions of the ecosystems over the last ~60 years. We also recorded strong synchrony in changes from the six study sites and among different indicators (biotic and abiotic). The nature of the changes is consistent with a longer and warmer ice-free period, an increase in solutes by evapoconcentration, reduced turbulent mixing of the water column, and reduced water inflow and lake level. The contrast between the species composition of different groups of organisms in the early and late twentieth century allows us to appreciate the profound changes that have already taken place.

Lakes are excellent sentinels of global change (Williamson et al. 2009; Catalan et al. 2013); they provide an early warning of local environmental changes. Our findings indicate that the Sierra Nevada region is undergoing important modifications due to global change, with relatively subtle responses beginning in the nineteenth century, but then with accelerated shifts since the second half of the twentieth century. In summary, the paleolimnological indicators track an overall trend of reduced water availability in the catchment that affected the total ecosystem, including lakes and associated wet meadows, together with an increase in air and water temperatures from 1960 to 1970 onward. The changes that have thus far been reported in the Sierra Nevada region will likely be exacerbated in the future as model projections indicate further warming and drying for this region, accompanied by further increases in Saharan inputs.

**Acknowledgements** We are grateful to all our colleagues for their assistance in core collection and laboratory analyzes. We thank AEMET and San Fernando Naval Base of the Spanish Army for providing meteorological data and Dr. María Jesús Esteban-Parra from the Dept. of Applied Physics (University of Granada) for climate data. Financial support was provided by two grants to C. Pérez-Martínez (Programa Nacional de Movilidad de Recursos Humanos de Investigación Grant PR2009-0414 (MEC) and Plan Propio de Investigación y Transferencia de la Universidad de Granada Grant PPSAB2018) and two Research Projects (MMA Project 87/2007, MICINN Project CGL2011-23483) to C. P.-M. and an FPU fellowship (AP2007-00352) to L. J. from the Spanish Ministry of Education and Science.

## References

Adrian R, O'Reilly CM, Zagarese H et al (2009) Lakes as sentinels of climate change. Limnol Oceanogr 54:2283–2297

Ashforth D, Yan ND (2008) The interactive effects of calcium concentration and temperature on the survival and reproduction of Daphnia *pulex* at high and low food concentrations. Limnol Oceanogr 53:420–432

Barea-Arco J, Pérez-Martínez C, Morales-Baquero R (2001) Evidence of a mutualistic relationship between an algal epibiont and its host *Daphnia pulicaria*. Limnol Oceanogr 46:871–888

Beaver JR, Jensen DE, Casamatta DA et al (2013) Response of phytoplankton and zooplankton communities in six reservoirs of the middle Missouri River (USA) to drought conditions and a major flood event. Hydrobiologia 705:173–189

Bigler C, Heiri O, Krskova R et al (2006) Distribution of diatoms, chironomids and cladocera in surface sediments of thirty mountain lakes in south-eastern Switzerland. Aquat Sci 68:154–171

Bonet FJ, Pérez-Luque AJ, Pérez-Pérez R (2016) Trend analysis (2000–2014) of the snow cover by satellite (MODIS sensor). In: Zamora et al (eds) Global change impacts in Sierra Nevada: challenges for conservation. Consejería de Medio Ambiente y Ordenación del Territorio. Junta de Andalucía

Bozelli RL (1994) Zooplankton community density in relation to water level fluctuations and inorganic turbidity in an Amazonian lake, "Lago Batata", state of Para, Brazil. Amazoniana 13:17–32

Burillo JP, Jiménez L, Pérez-Martínez C (2019) Identifying invasive *Daphnia* species by morphological analysis of postabdominal claws in Sierra Nevada alpine lakes. J Paleolimnol 62:121–135

Catalan J, Barbieri MG, Bartumeus F et al (2009) Ecological thresholds in European alpine lakes. Freshw Biol 54:2494–2517

Catalan J, Pla-Rabes S, Wolfe AP et al (2013) Global change revealed by palaeolimnological records from remote lakes: a review. J Paleolimnol 49:513–535

Chiapello I, Moulin C, Prospero JM (2005) Understanding the long-term variability of African dust transport across the Atlantic as recorded in both Barbados surface concentrations and large-scale

Total Ozone Mapping Spectrometer (TOMS) optical thickness. J Geophys Res 110, D18S10
Ciscar JC, Feyen L, Ibarreta D et al (2018) Climate impacts in Europe the JRC PESETA III Final Report Project. JRC Sci Policy Rep JRC112769
Conde-Porcuna JM, Veiga J, Moreno EJ et al (2021) Spatiotemporal genetic structure in the *Daphnia pulex* complex from Sierra Nevada lakes (Spain): reproductive mode and first record of North American D. cf. pulex in European alpine lakes. J Plankton Res 43:380–395
Drobinski P, Da Silva N, Bastin S et al (2020) How warmer and drier will the Mediterranean region be at the end of the twenty-first century? Reg Environ Change 20:1–12
Falasco E, Bona F (2011) Diatom community biodiversity in an Alpine protected area: a study in the Maritime Alps Natural Park. J Limnol 70:157–167
Ficken KJ, Li B, Swain DL, Eglinton G (2000) An n-alkane proxy for the sedimentary input of submerged/floating freshwater aquatic macrophytes. Org Geochem 31:745–749
García-Alix A, Jiménez Espejo FJ, Toney JL et al (2017) Alpine bogs of southern Spain show human-induced environmental change superimposed on long-term natural variations. Sci Rep 7:7439
Giorgi F, Lionello P (2008) Climate change projections for the Mediterranean region. Glob Planet Change 63:90–104
Greilinger M, Schauer G, Baumann-Stanzer K et al (2018) Contribution of Saharan dust to ion deposition loads of high alpine snow packs in Austria (1987–2017). Front Earth Sci 6:126
Harmsworth RV (1968) The developmental history of Blelham Tarn (England) as shown by animal microfossils, with special reference to the Cladocera. Ecol Monogr 38:223–241
Heiri O, Brooks SJ, Birks JB, Lotter AF (2011) A 274-lake calibration data-set and inference model for chironomid-based summer air temperature reconstruction in Europe. Quat Sci Rev 30:3445–3456
Hessen DO, Rukke NA (2000) The costs of moulting in Daphnia: mineral regulation of carbon budgets. Freshw Biol 45:169–178
Hessen DO, Alstad NEW, Skardal L (2000) Calcium limitation in Daphnia magna. J Plankton Res 22:553–568
Hoerling M, Eischeid J, Perlwitz J et al (2012) On the increased frequency of Mediterranean drought. J Clim 25:2146–2161
Jacques O, Bouchard F, MacDonald LA et al (2016) Distribution and diversity of diatom assemblages in surficial sediments of shallow lakes in Wapusk National Park (Manitoba, Canada) region of the Hudson Bay Lowlands. Ecol Evol 6:4526–4540
Jeziorski A, Yan ND, Paterson AM et al (2008) The widespread threat of calcium decline in fresh waters. Science 322:1374–1377
Jeziorski AJ, Smol JP (2017) The ecological impacts of lakewater calcium decline on softwater boreal ecosystems. Environ Rev 25:245–253
Jiménez L, Conde-Porcuna JM, García-Alix A et al (2019) Ecosystem responses to climate-related changes in a Mediterranean alpine environment over the last ~180 years. Ecosystems 22:563–577
Jiménez L, Romero-Viana L, Conde-Porcuna JM, Pérez-Martínez C (2015) Sedimentary photosynthetic pigments as indicators of climate and watershed perturbations in an alpine lake in southern Spain. Limnetica 34:439–454
Jiménez L, Rühland KM, Smol JA, JP, Pérez-Martínez, C, (2018) Climate change and Saharan dust drive recent cladoceran and primary production changes in remote alpine lakes of Sierra Nevada, Spain. Glob Chang Biol 24:e139–e158
Jiménez-Espejo FJ, García-Alix A, Jiménez-Moreno G et al (2014) Saharan aeolian input and effective humidity variations over western Europe during the Holocene from a high altitude record. Chem Geol 374–375:1–12
Kelley C, Ting M, Seager R, Kushnir Y (2012) The relative contributions of radiative forcing and internal climate variability to the late 20th century winter drying of the Mediterranean region. Clim Dyn 38:2001–2015
Kilham SS, Theriot EC, Fritz SC (1996) Linking planktonic diatoms and climate change in the large lakes of the Yellowstone ecosystem using resource theory. Limnol Oceanogr 4:1052–1062
Lotter AF, Birks HJB, Hofmann W, Marchetto A (1997) Modern diatom, cladocera, chironomid, and chrysophyte cyst assemblages as quantitative indicators for the reconstruction of past environmental conditions in the Alps I. Clim J Paleolimnol 18:395–420
Lotter AF, Birks HJB, Hofmann W, Marchetto A (1998) Modern diatom, cladocera, chironomid, and chrysophyte cyst assemblages as quantitative indicators for the reconstruction of past environmental conditions in the Alps II. Nutr J Paleolimnol 19:443–463
Michelutti N, Smol JP (2016) Visible spectroscopy reliably tracks trends in paleo-production. J Paleolimnol 56:253–265
Morales-Baquero R, Carrillo P, Barea-Arco J, Villar-Argaiz P-M, M, (2006) Climate-driven changes on phytoplankton–zooplankton coupling and nutrient availability in high mountain lakes of Southern Europe. Freshw Biol 5:989–998
Morales-Baquero R, Pérez-Martínez C (2016) Saharan versus local influence on atmospheric aerosol deposition in the southern Iberian Peninsula: significance for N and P inputs. Glob Biogeochem Cycles 30:501–513
Morales-Baquero R, Pérez-Martínez C, Ramos-Rodríguez E, Sánchez-Castillo P, Villar-Argaiz M, Conde-Porcuna JM (2019) Zooplankton advective losses may affect chlorophyll-a concentrations in fishless high-mountain lakes. Limnetica 38(1):55–65
Moser KA, Baron JS, Brahney J et al (2019) Mountain lakes: Eyes on global environmental change. Glob Planet Change 178:77–95
Mote PW, Hamlet AF, Clark MP, Lettenmaier DP (2005) Declining mountain snowpack in western North America. Bull Am Meteorol Soc 86:39–49
Moulin C, Chiapello I (2004) Evidence of the control of summer atmospheric transport of African dust over the Atlantic by Sahel sources from TOMS satellites (1979–2000). Geophys Res Lett 31: L02107
Moulin C, Chiapello I (2006) Impact of human-induced desertification on the intensification of Sahel dust emission and export over the last decades. Geophys Res Lett 18:L18808
Moulin C, Lambert CE, Dulac F, Dayan U (1997) Control of atmospheric export of dust from North Africa by the North Atlantic oscillation. Nature 387:691–694
Mulitza S, Heslop D, Pittauerova D et al (2010) Increase in African dust flux at the onset of commercial agriculture in the Sahel region. Nature 466:226–228
Nogués-Bravo D, Araujo MB, Lasanta T, Lopez-Moreno JI (2008) Climate change in Mediterranean mountains during the 21st century. Ambio 37:280–285
Nogués-Bravo D, López-Moreno JI, Vicente-Serrano SM (2012) Climate change and its impact. In: Vogiatzakis IN (ed) Mediterranean mountain environments. Wiley-Blackwell, Oxford, pp 185–201
Nováková K, van Hardenbroek M, van der Knaap WO (2013) Response of subfossil Cladocera in Gerzensee (Swiss Plateau) to early Late Glacial environmental change. Palaeogeogr Palaeoclimatol Palaeoecol 391:84–89
Pepin N, Bradley RS, Diaz HF et al (2015) Elevation-dependent warming in mountain regions of the world. Nat Clim Chang 5:424–430
Pérez-Martínez C, Barea-Arco J, Conde-Porcuna JM, Morales-Baquero R (2007) Reproduction strategies of *Daphnia pulicaria* population in a high mountain lake of Southern Spain. Hydrobiologia 594:75–82
Pérez-Martínez C, Conde-Porcuna JM, Moreno E et al (2020a) Cladoceran assemblage distribution in shallow alpine lakes of

Sierra Nevada (Spain) and its relationship with environmental variables. Aquat Sci 82:4

Pérez-Martínez C, Rühland KM, Smol JP et al (2020b) Long–term ecological changes in Mediterranean mountain lakes linked to recent climate change and Saharan dust deposition revealed by diatom analyses. Sci Total Environ 727:138519

Pérez-Martínez C, Jimenez L, Moreno E, Conde-Porcuna JM (2013) Emergence pattern and hatching cues of Daphnia pulicaria (Crustacea, Cladocera) in an alpine lake. Hydrobiologia 707:47–57

Planton S, Driouech F, Rhaz KE, Lionello P (2016) The climate of the Mediterranean regions in the future climate projections. In: The Mediterranean region under climate change. Moatti JP, Thiébault S (eds) 83–91

Preston DL, Caine N, McKnight DM et al (2016) Climate regulates alpine lake ice cover phenology and aquatic ecosystem structure. Geophys Res Lett 43:5353–5360

Preunkert S, Legrand M (2013) Towards a quasi-complete reconstruction of past atmospheric aerosol load and composition (organic and inorganic) over Europe since 1920 inferred from Alpine ice cores. Clim past 9:1403–1416

Prospero JM, Lamb PJ (2003) African droughts and dust transport to the Caribbean: climate change implications. Science 302:1024–1027

Pulido-Villena E, Reche I, Morales-Baquero R (2006) Significance of atmospheric inputs of calcium over the southwestern Mediterranean region: high mountain lakes as tools for detection. Global Biogeochem Cycles 20:1–8

Rellstab C, Maurer V, Zeh M et al (2007) Temporary collapse of the Daphnia population in turbid and ultra-oligotrophic Lake Brienz. Aquat Sci 69:257–270

Rogora M, Mosello R, Marchetto A (2004) Long-term trends in the chemistry of atmospheric deposition in North western Italy: the role of increasing Saharan dust deposition. Tellus b: Chem Phys Meteorol 56:426–434

Round FE, Crawford EM, Mann DG (1990) The Diatoms. Cambridge University Press, Cambridge, Biology and Morphology of the Genera

Rühland KM, Paterson AM, Smol JP (2015) Diatom assemblage responses to warming: reviewing the evidence. J Paleolimnol 54:1–35

Rühland KM, Smol JP (2002) Freshwater diatoms from the Canadian arctic treeline and development of paleolimnological inference models. J Phycol 38:249–264

Sillmann J, Kharin VV, Zwiers FW et al (2013) Climate extremes indices in the CMIP5 multimodel ensemble: part 2. Future climate projections. J Geophys Res Atmos 118:2473–2493

Siver PA, Baskette G (2004) A morphological examination of Frustulia (Bacillariophyceae) from the Ocala National Forest, Florida, USA. Can J Bot 82:629–644

Smol JP (2008) Pollution of lakes and rivers: a paleoenvironmental perspective John Wiley & Sons

Sommaruga-Wögrath S, Koinig KA, Schmidt R et al (1997) Temperature effects on the acidity of remote alpine lakes. Nature 387:64–67

Udelhoven T, Stellmes M, del Barrio G, Hill J (2009) Assessment of rainfall and NDVI anomalies in Spain (1989–1999) using distributed lag models. Int J Remote Sens 30:1961–1976

Vicente-Serrano SM, Tomás-Burguera M, Beguería S et al (2017) A high resolution dataset of drought indices for Spain. Data 2:22

Walker IR, Smol JP, Engstrom DR, Birks HJB (1991) An assessment of Chironomidae as quantitative indicators of past climatic change. Can J Fish Aquat Sci 48:975–987

Walz N, Welker M (1998) Plankton development in a rapidly flushed lake in the River Spree system (Neuendorfer See, Northeast Germany). J Plankton Res 20:2071–2087

Weckström J, Korhola A, Blom T (1997) Diatoms as quantitative indicators of pH and water temperature in subarctic Fennoscandian lakes. Hydrobiologia 347:171–184

Williamson CE, Saros JE, Vincent W, Smol JP (2009) Lakes and reservoirs as sentinels, integrators, and regulators of climate change. Limnol Oceanogr 54:2273–2282

Wilhelm S, Adrian R (2008) Impact of summer warming on the thermal characteristics of a polymictic lake and consequences for oxygen, nutrients and phytoplankton. Freshw Biol 53:226–237

Whiteside MC (1970) Danish Cladocera: Modern ecology and core studies. Ecol Monogr 40:79–118

# Atmospheric Inputs and Biogeochemical Consequences in High-Mountain Lakes

Isabel Reche, Natalie Mladenov, Elvira Pulido-Villena, and Rafael Morales-Baquero

**Abstract**

The idiosyncrasy of the lakes of Sierra Nevada lies in the fact that glacial retreat during their formation only occurred very close to the mountaintops and their proximity to North Africa. Quaternary glaciers' retreat left a group of small lakes close to the ridgelines with small catchment areas. These lakes are close to the Sahara Desert, where atmospheric mainstream transport toward the Iberian Peninsula goes between 1500 and 4000 m above sea level. Therefore, the Sierra Nevada Mountains constitutes the main physical barrier for this atmospheric dust, and Sierra Nevada's lakes act as natural atmospheric collectors. Saharan dust intrusions and Atlantic fronts that reach the Sierra Nevada have clear seasonal, synoptic, and climatic patterns that affect the quantity and quality of atmospheric deposition. The atmospheric deposition of Saharan dust has unique chemical and biological footprints. This chapter exposes the differences in the atmospheric deposition depending on the origin (marine vs. Saharan) of air masses that reach the Sierra Nevada and their consequences for the lakes' biogeochemistry. Atmospheric deposition with Saharan dust introduces macronutrients such as phosphorus (P) and micronutrients such as calcium (Ca) and iron to the lakes. Atmospheric P inputs affect lake primary and bacterial productivity. The Ca content in the lakes and their acid-neutralizing capacity is determined mainly by atmospheric deposition. Saharan dust also introduces organic matter with a humic-like signature and bacteria into the lakes. In contrast, atmospheric deposition from marine sources introduces organic matter with an amino acid-like signature and a comparatively higher abundance of viruses. The atmospheric deposition of microorganisms has consequences for their distribution ranges and the formation of a microbial seed-bank to face future scenarios of environmental changes.

**Keywords**

Atmospheric inputs • Saharan dust • Phosphorus • Calcium • Iron • Organic matter • Airborne microorganisms • Microbial seed-banks

## 1 Introduction

### 1.1 Idiosyncrasy of Sierra Nevada Lakes

Sierra Nevada's lakes, like most high-mountain lakes, are of glacial origin. However, due to their low latitude (37° N), the glacier formation only happened in the mountaintops compared to other alpine lakes in Europe. Quaternary glaciers' retreat left a group of small lakes very close to the ridgelines at altitudes between 2800 and 3100 m. This scenario differs from lakes located at lower latitudes such as the Atlas, where lake glacier formation did not happen, despite having higher altitudes. It also differs from lakes located at higher latitudes, such as the Pyrenees or the Alps, where glacial action was more potent, generating lakes of greater size and distributed over a broader range of altitudes. Usually, the altitudes at which these lakes are found, for instance in the Alps, are lower than in the Sierra Nevada due to permanent ice. However, the rapid retreat of glaciers is currently generating new alpine lakes at higher altitudes, with consequences for planktonic life (Sommaruga 2015, Peter and Sommaruga 2016).

I. Reche (✉) · R. Morales-Baquero
Dpto Ecología e Instituto del Agua, Universidad de Granada, Granada, Spain
e-mail: ireche@ugr.es

N. Mladenov
Department of Civil, Construction, and Environmental Engineering, San Diego State University, San Diego, CA 92182, USA

E. Pulido-Villena
Mediterranean Institute of Oceanography, Marseille cedex 9, 13288 Marseille, France


R. Zamora and M. Oliva (eds.), *The Landscape of the Sierra Nevada*,
https://doi.org/10.1007/978-3-030-94219-9_18

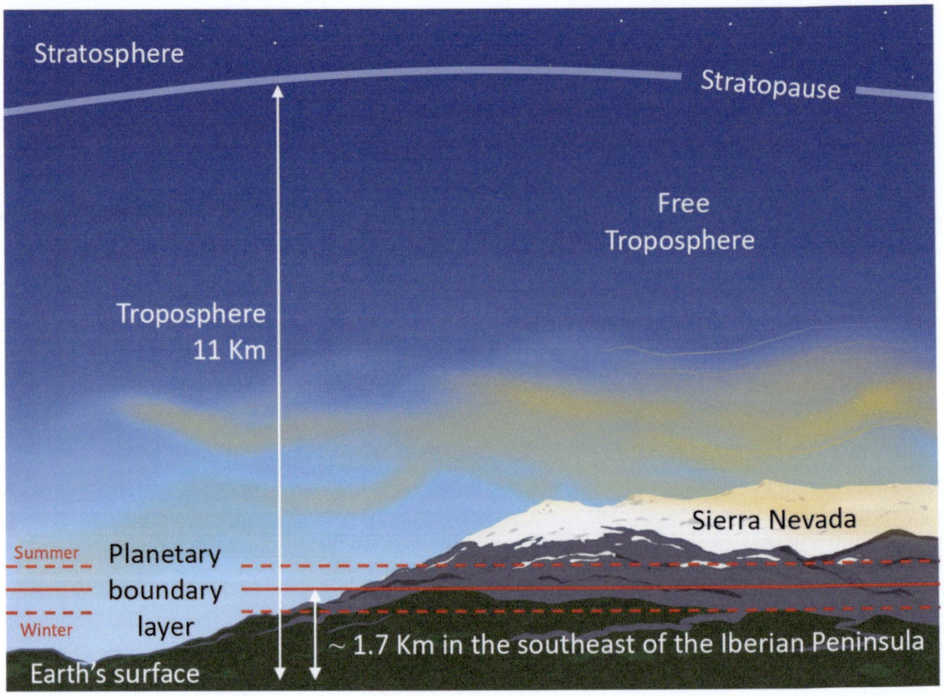


**Fig. 1** Illustration of the altitude of the planetary boundary layer (continuous red line) and its seasonal changes (dotted red lines) in relation to Sierra Nevada and the altitude of the Saharan dust intrusions

Sierra Nevada's lakes form a very confined group (the maximum distance between lakes is ca. 20 km), located above treeline near to 3000 m altitude, with small sizes (only three of them can exceed 1 ha) and headwater catchments on predominantly siliceous rocks. All lakes are subjected to a similar climatology, are limited by nutrients, and have simple food chains with a scarcity of planktonic predators and an absence of vertebrate predators. This local similarity allows us to study more precisely how regional and climatic forcing can affect the structure and function of these aquatic systems. Sierra Nevada's lakes are very responsive to atmospheric inputs due to their pristine nature, altitude, and small catchment areas. Because Sierra Nevada's lakes are located above the planetary boundary layer (1700 ± 500 m a.s.l. mean annual height, Granados-Muñoz et al. 2012), they are exposed to the deposition of materials transported by the atmospheric circulation in the free troposphere. The atmospheric microbiota collected at this altitude is similar to that collected in the free troposphere (Triadó-Margarit et al. 2019). This particular geographical location and their altitudes expose Sierra Nevada's lakes to Saharan dust intrusions whose maximum loads circulate between 1500 and 4000 m a.s.l. in the free troposphere (Talbot et al. 1986) (Fig. 1). On the other hand, in lakes larger in size than Sierra Nevada's lakes, atmospheric inputs are diluted, and therefore, dust signatures are difficult to detect. In addition, the siliceous nature of their catchments limits chemical weathering that could mask the signal of atmospheric inputs during runoff, as it occurs in lakes with more extensive and calcareous catchments. Therefore, Sierra Nevada's lakes act as "atmospheric collectors" and are very responsive to changes in long-range atmospheric inputs.

## 1.2 Seasonal, Interannual, and Recent Trends in Atmospheric Inputs on Sierra Nevada

The largest global sources of dust are located in the Northern Hemisphere, mainly in a broad "dust belt" that extends from the west coast of North Africa, over the Middle East, Central, and South Asia, to China (Prospero et al. 2002). There are significant differences in the content of dust aerosols among world localities depending on latitudinal localization inside this dust belt (Mladenov et al. 2011a). This dust belt strongly influences Sierra Nevada due to its proximity to North Africa with frequent atmospheric inputs of Saharan dust. Sierra Nevada is also under the influence of Westerlies, with fronts coming from the Atlantic Ocean that determine the rainfall patterns (Hidalgo-Muñoz et al. 2015). Therefore, the season and origin of air masses arriving at Sierra Nevada can influence atmospheric deposition and its chemical and biological footprints.

Saharan dust exports experience seasonal, interannual, and large-scale patterns in the Mediterranean region (Moulin et al. 1997; Evan et al. 2016; Sabatier et al. 2020; Cruz et al. 2021). Seasonality in dust exports is linked to changes in the Intertropical Convergence Zone (ITCZ) position. The maximum dust exports occur during late spring and summer. In the Northern Hemisphere, during the summer solstice, the

ITCZ position is located around 23° N, just over the Saharan Desert (Fig. 2a). The ascending warm air with massive dust injections rises in this zone, and high quantities of dust are exported toward the Atlantic Ocean by the predominant Easterlies winds or the Mediterranean region under particular synoptic conditions with low pressures over Portugal and high pressure over Northern Africa, channeling this dust toward the Iberian Peninsula. In contrast, during the summer solstice in the Southern Hemisphere, the ITCZ position is about 23° S far from the Sahara Desert influence, resulting in lower dust injections in the Mediterranean region (Fig. 2b).

Superimposed on this seasonal pattern are the interannual changes associated with climatic oscillations. For instance, precipitation and dust events are correlated to the North Atlantic Oscillation (NAO) index (López-Moreno et al. 2011; Hidalgo-Muñoz et al. 2015, see chapter "Climate Variability and Trends"). During the NAO negative phase, the weakness of subtropical Atlantic High promotes Atlantic fronts over the Iberian Peninsula, increasing precipitation (Fig. 2b). In contrast, the high pressure over the Iberian Peninsula blocks the Atlantic fronts during the NAO positive phase and, consequently, precipitation events are lower than during the NAO negative phase. However, dust exports from Northern Africa are more prone in this phase (Fig. 2a). More details about the influence of climatic oscillations on the rainfall patterns in the Sierra Nevada can be found in chapter "Climate Variability and Trends". At larger time scales (decades and centuries), Sahara dust exports have been related to intense droughts in the Soudano-Sahel region caused by changes in the global distribution of sea surface temperature (Prospero and Lamb 2003; Giannini et al. 2003) and the advent of commercial agriculture in the Sahel region (Mulitza et al. 2010). This last finding suggests a clear dust footprint starting with the Anthropocene.

Atmospheric deposition occurs by washing the atmosphere by the rain (termed wet deposition) or by direct

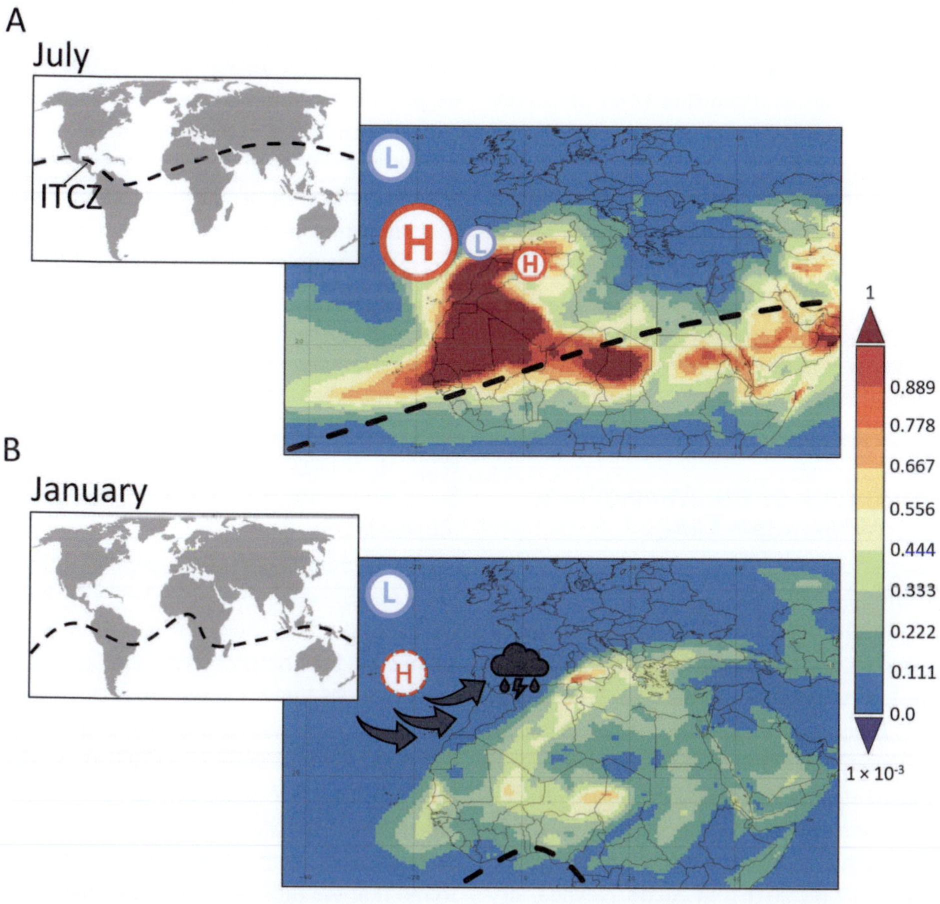


**Fig. 2** **a** Summer scenario from July 9th to 12th of 2021 with the intertropical convergence zone (ITCZ) located over the Sahara Desert (dotted black line) leading to an intense Saharan intrusion with high values of the time-averaged dust-column mass density (kg $m^{-2}$) using the second Modern-Era Retrospective analysis for Research and Applications (MERRA-2) model. Synoptic Low pressures (L) over Portugal and High pressures (H) over the North of Africa channelize the Saharan dust to the Mediterranean region in general and the Sierra Nevada in particular. **b** Winter scenario from January 9th to 12th of 2021with Atlantic fronts over the Iberian Peninsula and lower values of the time-averaged dust-column mass density (kg $m^{-2}$) using the MERRA-2 model. MERRA model is an open, online resource provided by the National Aeronautics and Space Administration (NASA) (https://giovanni.gsfc.nasa.gov/giovanni/)

sedimentation during periods without rain (termed dry deposition). The wet deposition has a chemical signature: the sum of the elements dissolved in the rain plus the aerosols it captures during the washout, some of which may be solubilized. Similarly, the chemical signature of dry deposition is composed of the fraction of water-soluble elements in atmospheric aerosols. The relative contribution of dry and wet deposition to the supply of elements from the atmosphere to Sierra Nevada depends on the rainfall regime. In the Sierra Nevada, we can find prolonged periods, longer than a month, without precipitation, during which only dry deposition is present. Saharan dust intrusions can cover thousands of square kilometers in the free troposphere, producing the deposition of particulate material that can be recorded synchronously at monitoring stations over the Peninsula (Rodríguez et al. 2001). Synchronous variables among neighbor lakes are considered a sign of climatic control at a regional scale (Baines et al. 2000; Reche et al. 2009; Morales-Baquero and Pérez-Martínez 2016). In Sierra Nevada, the deposition of particulate matter from Saharan dust intrusions occurs mainly during spring and summer associated with the south or southwest winds and the presence of cyclones in the Iberian Peninsula (Fig. 2). Every year, massive airborne plumes of Saharan dust are exported to Sierra Nevada, particularly at high altitudes (Morales-Baquero et al. 2006a). Dust is a significant source of mineral nutrients (phosphorus, calcium, and iron), organic carbon to both terrestrial (Okin et al. 2004) and aquatic ecosystems (Jickells et al. 2005; Morales-Baquero et al. 2006a; Mladenov et al. 2008), and microorganisms (Reche et al. 2009; 2018; Hervás et al. 2009; Peter et al. 2014) with chemical and biological consequences.

## 2 Chemical Footprints of the Atmospheric Inputs in Sierra Nevada's Lakes

Atmospheric deposition includes the input of gaseous (e.g., nitrogen) and lithosphere-derived (e.g., phosphorus) macronutrients as well as micronutrients (e.g., calcium, iron) to aquatic and terrestrial ecosystems (Chadwick et al. 1999; Jickells et al. 2005; Ballantyne et al. 2011; Brahney et al. 2013, 2015). In particular, Saharan dust is rich in elements of biogeochemical interest such as P, Ca, and Fe, among others, and their inputs to Sierra Nevada have been extensively studied (Morales-Baquero et al. 2006a, 2013; Pulido-Villena et al. 2006; Bhattachan et al. 2016). In Sierra Nevada, dry deposition (on average 23.6 mg $m^{-2}$ $d^{-1}$) dominates the delivery of particulate matter (PM, which is the material retained by filters of 0.7 μm pore size), being ca. three times larger than wet deposition in an annual base (Morales-Baquero et al. 2013). Likewise, dry deposition dominates the inputs of total phosphorus (TP), soluble reactive phosphorus (SRP), $Ca^{2+}$, $Mg^{2+}$, and $K^{+}$. In contrast, wet deposition dominates the inputs of $Na^{+}$, total nitrogen (TN), $NO_3^{-}$, and $SO_4^{=}$. Saharan intrusions significantly modify the chemical signature of both types of deposition. In weeks with rainfall and Saharan intrusions, dry deposition shows higher PM, TP, and $Ca^{2+}$. In contrast, in the absence of Saharan intrusions, wet deposition shows higher $Cl^{-}$ and $Na^{+}$ (Morales-Baquero et al. 2013). This chemical signature is valuable to determine, along with air mass back-trajectories and remote sensing data, the (marine vs. Saharan) origin of atmospheric deposition in Sierra Nevada.

Calcium is an element tightly linked to the carbon cycle via the carbonate-bicarbonate equilibrium affecting lake acid-neutralizing capacity (Psenner 1999) and long-term phytoplankton and zooplankton community composition (Jiménez et al. 2018; Pérez-Martínez et al. 2020). The high $CaCO_3$ content in Saharan dust significantly increases the pH of rainwater (Löye-Pilot et al. 1986). It partially counteracted acidic deposition in some alpine lakes in Europe (Psenner 1999) during the 1970s and 1980s (Rogora et al. 2004). The reported data for calcium total atmospheric deposition ranged from 13.9 to 559.8 μmol $m^{-2}$ $d^{-1}$ (on average 48 μmol $m^{-2}$ $yr^{-1}$) and showed a seasonal pattern similar to that reported for Saharan dust export to the Mediterranean region, with maxima during spring and summer (Pulido-Villena et al. 2006; Morales-Baquero et al. 2013). Calcium atmospheric inputs determine Ca concentration in Sierra Nevada's lakes (Pulido-Villena et al. 2006). The variability of this influence depends on other lake properties, such as susceptibility to evaporation and the presence/absence of outlets. For instance, the Ca concentration in the lake La Caldera (110 μmol $l^{-1}$) is about three times the Ca concentration in the lake Rio Seco (37 μmol $l^{-1}$) (Fig. 3). A larger catchment area in La Caldera than in Rio Seco and the absence of outlets can explain this lake-specific difference (Fig. 3).

Atmospheric dust also contains significant amounts of iron (Jickells et al. 2005). In Sierra Nevada, Mladenov et al. (2010) reported values of atmospheric deposition of dissolved Fe that ranged from 0.03 to 5.23 mg $m^{-2}d^{-1}$ with higher values under the influence of Saharan dust intrusions. These last authors found a robust relationship between total dissolved Fe and particulate matter in atmospheric deposition. On the other hand, during the atmospheric transit from Saharan soils to Sierra Nevada, usually around 3000 m. asl, there is enrichment in Fe solubility. Bhattachan et al. (2016) found that the dust is greatly enriched (on average 15 times) in Fe (II) in the atmospheric deposition at Sierra Nevada in comparison to the fine fraction of the parent soil collected from North African dust sources. These results are very relevant for the biological availability of this micronutrient.

Atmospheric inputs of N and P are especially relevant for Sierra Nevada lakes, as the geological characteristics of the

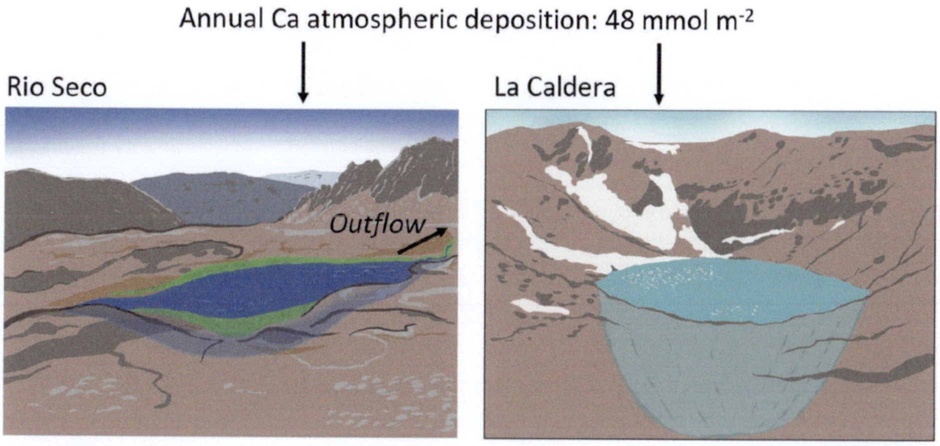


| Rio Seco | | La Caldera |
|---|---|---|
| $9.9 \times 10^4$ m$^2$ | Lake catchment area | $23.5 \times 10^4$ m$^2$ |
| 4800 mol | Catchment area deposition | 11300 mol |
| 37 μmol L$^{-1}$ | Lake Ca concentration | 110 μmol L$^{-1}$ |
| $3 \times 10^6$ L$^{-1}$ | Lake volume | $49 \times 10^6$ L$^{-1}$ |
| 111 mol | Total lake Ca content | 5400 mol |

**Fig. 3** Comparison between total atmospheric and catchment inputs of calcium and lake calcium content in two contrasting lakes like Rio Seco with a small catchment area and an outlet and La Caldera with a larger catchment area and without outlets

rocks in the catchments allow limited inputs of these macronutrients for biota development. Morales-Baquero et al. (2006a, 2013) determined the atmospheric inputs of total phosphorus (TP) and total nitrogen (TN) to Sierra Nevada. They reported values of TP from 0.1 to 10.8 μmol m$^{-2}$ d$^{-1}$ and TN from 17.2 to 533.8 μmol m$^{-2}$ d$^{-1}$. The relative contribution of dry to the total deposition of PM was approximately 80%. The TP atmospheric deposition showed consistent seasonal dynamics coupled to particulate matter (PM) deposition with maximum values during late spring and summer (Fig. 4) when the ITCZ is located over the Sahara Desert (Fig. 2). This seasonal pattern was synchronous in sites located at different altitudes from 1000 m (Lanjarón and Quéntar) to almost 3000 m above sea level (Observatory) and over the years (Fig. 4).

The lake catchment sizes generate that, despite N and P atmospheric inputs are similar for the whole area, the availability of both elements differs in each lake. Rainfall in the Sierra Nevada has a low TN: TP ratio and reaches 5.5 by weight (Morales-Baquero et al. 2006a). However, due to higher mobility of the N inorganic forms and higher P retention in unfertilized soils, the runoff has TN: TP ratios greater than 200 by weight. Consequently, N: P ratios tend to increase in lakes with larger catchments. In Sierra Nevada lakes, the N and P availability for autotrophs, measured as the ratio between dissolved inorganic nitrogen (DIN) and soluble reactive phosphorus (SRP), is lower than 5 by weight in lakes with small catchments. In contrast, this ratio increases progressively with the size of the catchments, reaching DIN: SRP ratios greater than 20 (by weight) (Morales Baquero et al. 1999). This change in the N and P availability is essential because the organisms need to incorporate both elements in a similar proportion to that which exists in their tissues. In the case of planktonic organisms, they need to assimilate about 16 atoms of N for each atom of P to grow appropriately (Redfield 1934). Therefore, in Sierra Nevada's lakes with small catchments and greater relative availability of P, plankton growth tends to be limited by N. In contrast, in lakes with larger catchment areas, P limitation is accentuated.

## 3 Footprints of Organic Matter Atmospheric Inputs in High-Mountain Lakes

Allochthonous organic matter (OM) has been introduced to high mountain environments worldwide by atmospheric deposition, and this phenomenon has significant consequences for alpine lakes. For one, chromophoric DOM (CDOM) is an essential component of atmospherically deposited organic matter. It is a key driver of lake optical properties during ice-free periods in Sierra Nevada's lakes (Reche et al. 2001; Mladenov et al. 2008, 2009), which influence light attenuation, particularly the ultraviolet radiation and associated lake biological processes.

The quantity and quality of OM atmospheric deposition have been measured using passive collectors of wet and dry deposition—including rain, snow, and dryfall (Santos et al. 2013; Oldani et al. 2017; Niu et al. 2019)—and active collectors for total suspended particulates (TSP), and coarse (PM10) and fine (PM2.5) particulate matter (Xie et al. 2016). The sources of organic matter in the atmosphere include

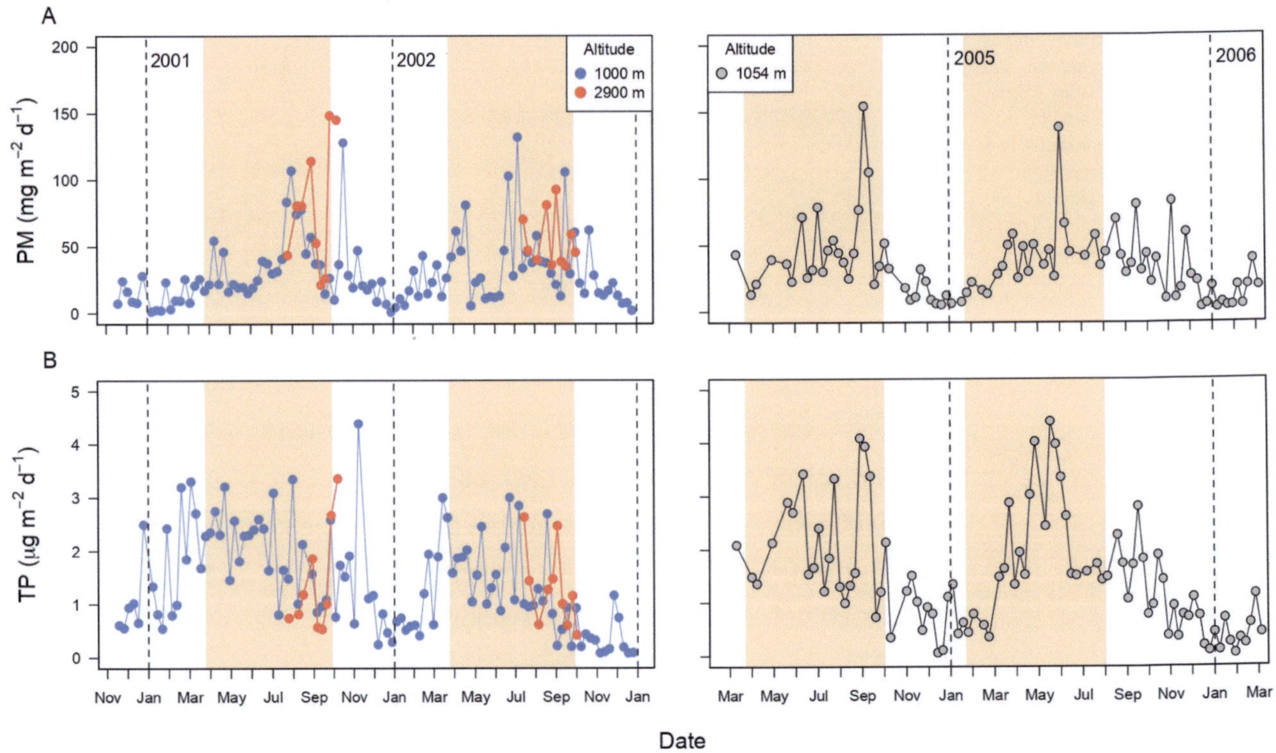


**Fig. 4** Seasonal dynamics of atmospheric deposition of particulate matter (PM) and total phosphorus (TP) at 1000 m (Lanjarón site), at 1054 m (Quéntar site), and 2900 m (Observatory of the Instituto Andaluz de Astrofísica). The maximum concentrations of PM and TP are consistently found during late spring and summer (pale brown shaded area)

primary sources, such as soils (Koulouri et al. 2008), erodible lake sediments (Washington et al. 2006), and primary biological aerosols, viruses, bacteria, fungi, and pollen (Jaenicke 2005; Bowers et al. 2009; Burrows et al. 2009; Reche et al. 2018), as well as secondary sources, such as vehicular emissions and secondary organic aerosols (Legrand et al. 2007; Xie et al. 2017). The deposition of both dissolved and particulate forms of OM (DOM and POM) are important for alpine ecosystems, and a substantial fraction of dry deposition (15% of dry deposition in Mladenov et al. 2009) is water-soluble organic carbon (WSOC) and further available for biological processing. Mladenov et al. (2009) estimated that WSOC in wet and dry deposition to a clear, alpine lake and its catchment over a one-month period represented over 70% of the total lake dissolved organic carbon (DOC) mass. Therefore, atmospheric deposition exerts a dominant influence over the distribution of organic compounds in alpine lakes.

Indeed, due to their remote position at high altitudes and the lack of surrounding vegetation, the clearest alpine lakes were shown to behave similarly to atmospheric deposition collectors and exhibit similar patterns in DOC concentration and light absorption coefficients (Mladenov et al. 2008). In Sierra Nevada, both Saharan dust-derived and marine aerosols were important sources of WSOC (Fig. 5a). Differences in the optical spectroscopic properties (UV–visible absorbance and fluorescence) between the two organic aerosol sources supported the notion that marine aerosols provide little color and comparatively more amino acid-like substances. In contrast, Saharan dust represented the primary source of CDOM with a dominance of soil fulvic and humic substances to alpine lakes (Fig. 5b). These fluorescence signatures derived from excitation-emission matrices (EEMs) of dissolved organic matter in Sierra Nevada's lakes differ substantially from the EEMs observed in boreal lakes with a more significant influence of humic and fulvic compounds from inputs of the surrounding landscape (Fig. 5c).

The influence of dust deposition on alpine lake optical properties has also been demonstrated on a global scale. For a global dataset of 86 alpine and polar lakes from the Atlas Mountains, Sierra Nevada, Pyrenees, Tyrolean Alps, Patagonian Mountains, Antarctica, and the Arctic, significant latitudinal trends were observed in lake DOC concentration, spectral slope, and spectral slope curve values (Mladenov et al. 2011a). Those geographic patterns were influenced, in part, by a site's proximity to the Saharan dust belt and other dust source regions, as demonstrated by relationships with NASA's Ozone Monitoring Instrument (OMI) aerosol index (Mladenov et al. 2011a). Similarly, kinematic trajectory analyzes implicated northern Africa as a source region for atmospheric dust carrying water-soluble organic nitrogen to

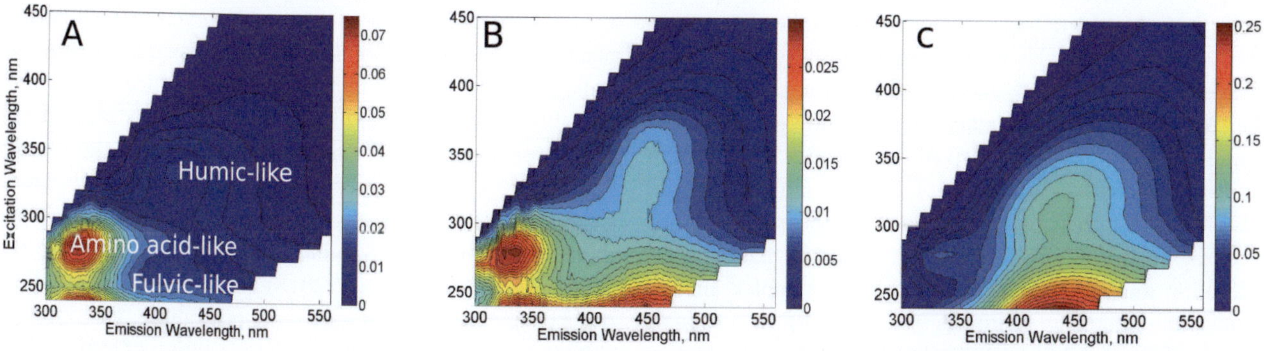


**Fig. 5** Comparison among the fluorescence spectra (Excitation-Emission Matrices, EEMs) of water-soluble organic carbon from dry deposition with Saharan dust origin on August 7th, 2008 (**a**), and dissolved organic matter from an alpine lake La Caldera (**b**) and a boreal lake Bylot 40 (**c**). The areas of amino acid-like, humic acid-like, and fulvic-acid-like peaks are shown in panel A

Mediterranean coastal regions (Mace et al. 2003). Backward trajectory analyzes also identified air masses passing through the arid west and Four Corners, USA, as dominant source areas for dry deposition of organic constituents and nutrients to Rocky Mountain lakes (Colorado, USA), especially in spring months (Mladenov et al. 2012; Oldani et al. 2017).

Multiseason and multiyear studies of DOC or WSOC in atmospheric deposition are few, potentially due to the inherent challenges of measuring wet deposition during colder months when freezing conditions and deep snowpack inhibit access to instrumentation. In the United States, the National Atmospheric Deposition Program (NADP) operates dozens of atmospheric deposition monitoring stations, primarily for tracking nutrient and base cation deposition, including one in the Colorado Rocky Mountains above treeline. Although the organic fraction is not typically analyzed in NADP collectors, there are NADP wet deposition-monitoring stations at four locations in the Colorado Rocky Mountains (one above treeline and three below treeline). These data show distinct seasonality, with peak DOC concentrations occurring during the summer months. Summer DOC deposition rates (20.4 kg C $ha^{-1}$ $yr^{-1}$) were shown to be much higher than annual averages (2.32 kg C $ha^{-1}$ $yr^{-1}$) at the Colorado Rocky Mountain site (Oldani et al. 2017). DOC in summer precipitation measured at Mt. Yulong in the Tibetan Plateau was also very high, at 19.9 kg C $ha^{-1}$ $yr^{-1}$ (Niu et al. 2019). Slightly lower DOC deposition rates were measured in summer precipitation in the Sierra Nevada Mountains, Spain, at 13.1 kg C $ha^{-1}$ $yr^{-1}$ (Mladenov et al. 2009). However, more seasonal data over the whole year are needed for more accurate comparisons.

The chemical quality of atmospheric deposition also has been shown to undergo seasonal changes that are important for biogeochemical processes in high mountain lakes. Light-absorbing organic aerosols have been categorized into black carbon (BC), from incomplete combustion of fuels, and brown carbon, from primary or secondary sources (Ram et al. 2010; Zhang et al. 2017; Laskin et al. 2018; Beres et al. 2020), and both BC and brown carbon may comprise high molecular weight humic-like substances (HULIS). Sites with WSOC derived from biomass burning in the Tibetan Plateau and Saharan dust in the Sierra Nevada (Mladenov et al. 2011b) and coast areas in Italy (Santos et al. 2013) were found to have more light-absorbing and lignaceous UV-visible absorbance and fluorescence spectral properties in the summer, which are characteristics also representative of brown carbon aerosols. By contrast, chemical and spectroscopic analyzes conducted on wet and dry deposition samples collected in the Colorado Rocky Mountains revealed that the DOM in many late spring and summer samples was lower in aromaticity, color, and polydispersity and of lower molecular weight than that of winter and fall samples (Mladenov et al. 2012). Therefore, the high DOC inputs occurring at this time were hypothesized to be labile for microorganisms and potentially able to support heterotrophic processes in water and soils, such as denitrification (Mladenov et al. 2012). The less aromatic nature of some organic aerosol deposition samples, particularly in the summer, appears to be due to enhanced solar ultraviolet (UV) radiation that has a photobleaching effect on the light-absorbing properties of organic aerosols, as suggested by Han and Kim (2017). Despite their sources and seasonality, fluorescence spectroscopic properties indicate that intense UV radiation during high altitude transport may have an important influence on most atmospherically-deposited organic matter. For example, EEMs of WSOC extracted from TSP filter samples from the Colorado Rocky Mountains were found to lack fluorescence at higher excitation wavelengths (>350 nm), most likely due to preferential photodegradation of light-absorbing WSOC at higher wavelengths (Xie et al. 2016).

The high depositional flux of organic matter to alpine lake ecosystems devoid of other carbon and nutrient sources begs the question, "how bioavailable is atmospherically deposited organic matter to alpine microorganisms?" The few studies that address this question have found alpine bacteria to be well-adapted to degrading organic matter in otherwise barren alpine environments. Using microplate bioassays on a range of snow, wet deposition, and dust deposition samples from the Colorado Rocky Mountain sites, Bigelow et al. (2020) found that all atmospheric deposition samples were able to activate metabolism by native Colorado alpine bacteria. Also, in most cases, the addition of phosphate to the bioassays did not further stimulate microbial metabolism, indicating that the atmospheric deposition samples already contained sufficient nutrients (N and P) to support biodegradation. Additional incubation experiments further demonstrated the presence of two pools (one rapid and one slowly decaying) of DOM in the atmospheric deposition and snow samples. Light-absorbing DOC in the snowpack of glacierized regions of the Tibetan Plateau, which had been primarily deposited along with mineral dust from adjacent arid regions, was studied by Yan et al. (2016). Similar to the findings for DOC in the Colorado snowpacks, much of the light-absorbing DOC (46.2%) in the Tibetan Plateau snowpack was bioavailable and could be mineralized to $CO_2$ within one month of its release. Cryoconite holes on glacier surfaces were hotspots of biodiversity on glacier surfaces that hosted metabolically active bacterial communities, including taxa that were able to degrade organic pollutants deposited by long-range transport (Pittino et al. 2018). In particular, bacteria played a significant role in degradation of chlorpyrifos and polychlorinated-biphenyls (PCBs) in these habitats. Therefore, the atmospheric deposition of organic matter in Sierra Nevada's lakes can stimulate local bacterial metabolism throughout the degradation of dissolved organic matter.

## 4 Biological Footprints of the Atmospheric Inputs in High-Mountain Lakes

Atmospheric deposition delivers macronutrients as phosphorus, micronutrients as iron and water-soluble organic matter that can boost phytoplankton and bacterioplankton growth in aquatic ecosystems (Bonnet et al. 2005; Marañón et al. 2010; Pulido-Villena et al. 2008a). In addition, atmospheric deposition can directly introduce non-native microorganisms into the recipient terrestrial and aquatic ecosystems (Hervás et al. 2009; Yamaguchi et al. 2012; Woo and Yamamoto 2020).

### 4.1 Plankton Responses to Atmospheric Deposition in Sierra Nevada's Lakes

Morales-Baquero et al. (2006a) established a seasonal link between the atmospheric deposition of total phosphorus and the chlorophyll-a concentration in two contrasting lakes as La Caldera and Rio Seco lakes. However, not all phytoplanktonic species responded uniformly to dust deposition. Pulido-Villena et al. (2008b) showed a significant growth of the chrysophyte *Chromulina nevadensis*, but phosphorus deposition did not affect other species such as the diatom *Cyclotella* sp. or green algae *Chlorella* sp. These species-specific effects also have consequences for species richness and diversity. Similarly, at a larger scale of approx. 150 years, Jiménez et al. (2018) found in dated sediment cores an increase in the concentration of inferred chlorophyll-a coupled with the intensification of atmospheric deposition in six lakes. Differences in the magnitude of the response and timing of these changes can be likely related to catchment and lake-specific differences. Pérez-Martínez et al. (2020) also found an influence of atmospheric deposition on diatom assemblages in these sediment cores (see chapter "Paleolimnological Indicators of Global Change").

Bacterioplankton is also stimulated by atmospheric deposition. Pulido-Villena et al. (2008b) found a significant correlation between the atmospheric deposition of soluble reactive phosphorus and the bacterial abundance in La Caldera Lake. In a more detailed study, Reche et al. (2009) showed that total phosphorus and water-soluble organic carbon delivered by atmospheric deposition increased bacterial abundance in oligotrophic systems of Sierra Nevada such as La Caldera Lake and Quéntar reservoir. These last authors demonstrated experimentally that Saharan dust addition had a significant and positive effect on the bacterial production and abundance but not on the species richness, diversity, or composition of the native bacterial assemblages. Both phytoplankton and bacterioplankton are food substrates for zooplankton in these alpine lakes (Cruz-Pizarro et al. 1994; Carrillo et al. 1995; Reche et al. 1997; see chapter "Snow Dynamics, Hydrology, and Erosion"). Therefore, dust inputs also have bottom-up consequences for this trophic level.

Because the pelagic trophic structure of Sierra Nevada lakes is very simple, with no vertebrate predators, zooplankton is the last trophic step where inorganic nutrients captured by autotrophs and bacteria accumulate during the growing season. The average abundance of zooplankton organisms during the thaw period can be highly variable among Sierra Nevada lakes. It can change from 101.2 ind. $L^{-1}$, in the lake with the highest density, to 0.8 ind.$L^{-1}$ in the one with the least zooplankton, and the average abundance decreases progressively as catchment size increases (Morales-Baquero and Conde-Porcuna 2000). This result suggests a progressive limitation by P as the catchment area increases. Numerous individuals of zooplankton species, such as rotifers and larval forms of copepods, are very susceptible to be P-limited (Rothhaupt 1995; Elser et al. 1996). Therefore, the atmospheric P inputs can also determine the zooplanktonic densities in the lakes.

The transfer of N and P from inorganic forms up to zooplankton requires that the development of both trophic levels be coupled. In years with particularly low temperatures, Sierra Nevada's lakes contain low zooplankton densities, high phytoplankton abundances, and high concentration of the limiting nutrient, either N or P, in each lake. These facts are interpreted as a consequence of a decoupling between the phytoplankton (with shorter generation times) and the zooplankton (with longer generation times), which hinders the transfer of inorganic nutrients up to zooplankton (Morales-Baquero et al. 2006b). This food-web climatic sensitivity shown by these small lakes confirms their value as fine sensors of climate change. It allows us to detect a progressive increase in the eutrophication of the lakes, compatible with an increase in the limitation by the scarcest nutrient in each system as global temperatures and the inputs of Saharan dust increase.

The influence of atmospheric deposition in the planktonic food webs of Sierra Nevada's lakes has been confirmed using carbon stable isotopes (Pulido-Villena et al. 2005; Morales-Baquero et al. 2006c). Irrespectively of the nature of catchment area, particulate organic matter (POM) showed more enriched $\delta^{13}C$ values in La Caldera ($\delta^{13}C = -24.5‰$) and Rio Seco ($\delta^{13}C = -26.6‰$) lakes than the phytoplankton signatures ($\delta^{13}C = -33.9‰$ and $-33.7‰$, respectively) suggesting a terrestrial vegetation influence in both lakes. This terrestrial influence, in the case of La Caldera Lake, is mostly coming from atmospheric deposition. This POM mainly was exploited by the cladoceran *Daphnia pulicaria* that showed similar $\delta^{13}C$ isotopic signatures in La Caldera ($-23.6‰$) and Rio Seco lakes ($-31.1‰$) that the corresponding POM $\delta^{13}C$ signatures. However, the copepod *Mixodiaptomus laciniatus* was $\delta^{13}C$ depleted relative to POM $\delta^{13}C$ signatures both in La Caldera ($-30.8$ ‰) and Rio Seco ($-32.1$ ‰) lakes, indicating a selective feeding on an isotopically lighter source, likely phytoplankton at least in La Caldera Lake. The results obtained show that, despite contrasting catchments, the food webs of both lakes might be partially supported by terrestrial organic carbon from atmospheric inputs and runoff.

## 4.2 Atmospheric Deposition of Microorganisms: Long-Range Transport and Seed-Bank for Future Environmental Changes

Most people are familiar with Darwin's legacy about species evolution and natural selection. However, his contribution to geology is less known, but he was an exponent of his close friend Charles Lyell's ideas about an Earth System changing slowly, gradually. That is, an evolving Earth. During his expedition in the Beagle, Darwin went through Canary and Cape Verde Islands, an area of the Atlantic Ocean that receives enormous quantities of Saharan dust. He was one of the first naturalists to report the phenomenon of dust storms over the Atlantic Ocean (Darwin 1845). He collected dust samples for sending them to his colleague Professor Ehrenberg in Berlin. Professor Ehrenberg was the first to examine the occurrence of microorganisms in the Saharan dust and is considered the founder of the science of aerobiology (Krumbein 1995). He described several protists, formerly named "Infusoria" (e.g., *Ciliophora* sp.), mostly of freshwater origin in dust samples collected over Barbados by R.H. Schomburgk and the Eastern Atlantic Ocean by C. Lyell and C. R. Darwin (Ehrenberg 1845). At the beginning of this century, Dr. Anna Gorbushina had the opportunity to explore dust subsamples of Professor Ehrenberg's collection. She found that microbes adhered to Saharan dust can live for centuries surviving the transport across the Atlantic (Gorbushina et al. 2007). This discovery opened new perspectives for the long-range transport of microorganisms and microbial storage over a long time. Microbial dormancy generates "seed" banks that allow microorganisms to dorm until environmental conditions are adequate to survive (Lennon and Jones 2011).

Aerosols in the troposphere can mobilize about $10^{18}$ cells per year (Griffin et al. 2002). These air-transported microorganisms can survive long distances suspended in dust particles following the atmospheric circulation patterns (Kellogg and Griffin 2006). In Europe, this long-range dispersal of bacteria with a Saharan origin has been reported for the Sierra Nevada, Pyrenees, and Alps mountains (Reche et al. 2009; Hervás et al. 2009; Peter et al. 2014). Above the boundary layer, high-elevation mountains are optimal sites to collect bacterial aerosols traveling through the troposphere (Triadó-Margarit et al. 2019). Airborne microorganisms can be removed from the atmosphere by rain washout (wet deposition) or by direct sedimentation during clear days (dry deposition), affecting microbial abundance, composition, and distribution in recipient ecosystems. The magnitude of microbial deposition, viability, and colonization availability of these "invaders" can affect the native microbiota, particularly in alpine lakes with high ecological value.

In Sierra Nevada, Reche et al. (2018) reported the wet and dry deposition rates of viruses and bacteria at two sites located above the atmospheric boundary layer (Fig. 1). In addition, they evaluated how the origin of air masses (marine vs. Saharan), meteorological conditions, and aerosol size can affect the deposition rates of viruses and bacteria. They quantified the total deposition rates of viruses and bacteria (Fig. 6). Virus deposition rates ranged from $0.26 \times 10^9$ to $>7 \times 10^9$ m$^{-2}$ per day (Fig. 6a). These deposition rates were not significantly different when air masses came from the Atlantic (marine source) or the Saharan Desert or under rainy or clear meteorological conditions (Fig. 6a). The deposition rates of bacteria ranged from $3 \times 10^6$ to >80

$10^6$ $m^{-2}$ per day (Fig. 6b). These rates were generally higher when air masses came from the Saharan Desert and during rainy (wet + dry deposition) than clear (only dry deposition) periods (Fig. 6b). The presence of a high abundance of dust-attached bacteria during rainy periods suggests that they might act as cloud condensation nuclei and promote precipitation (Creamean et al. 2013), or are washed out more easily from the atmosphere by rain. Bacteria from the Sahara Desert are deposited at high mountain lakes in Europe, particularly during rain events (Peter et al. 2014). The Gammaproteobacteria appear to dominate the airborne bacterial community under the influence of Saharan dust intrusions (Reche et al. 2009; Peter et al. 2014). However, the interactions among dust, bacterial identity, cloud

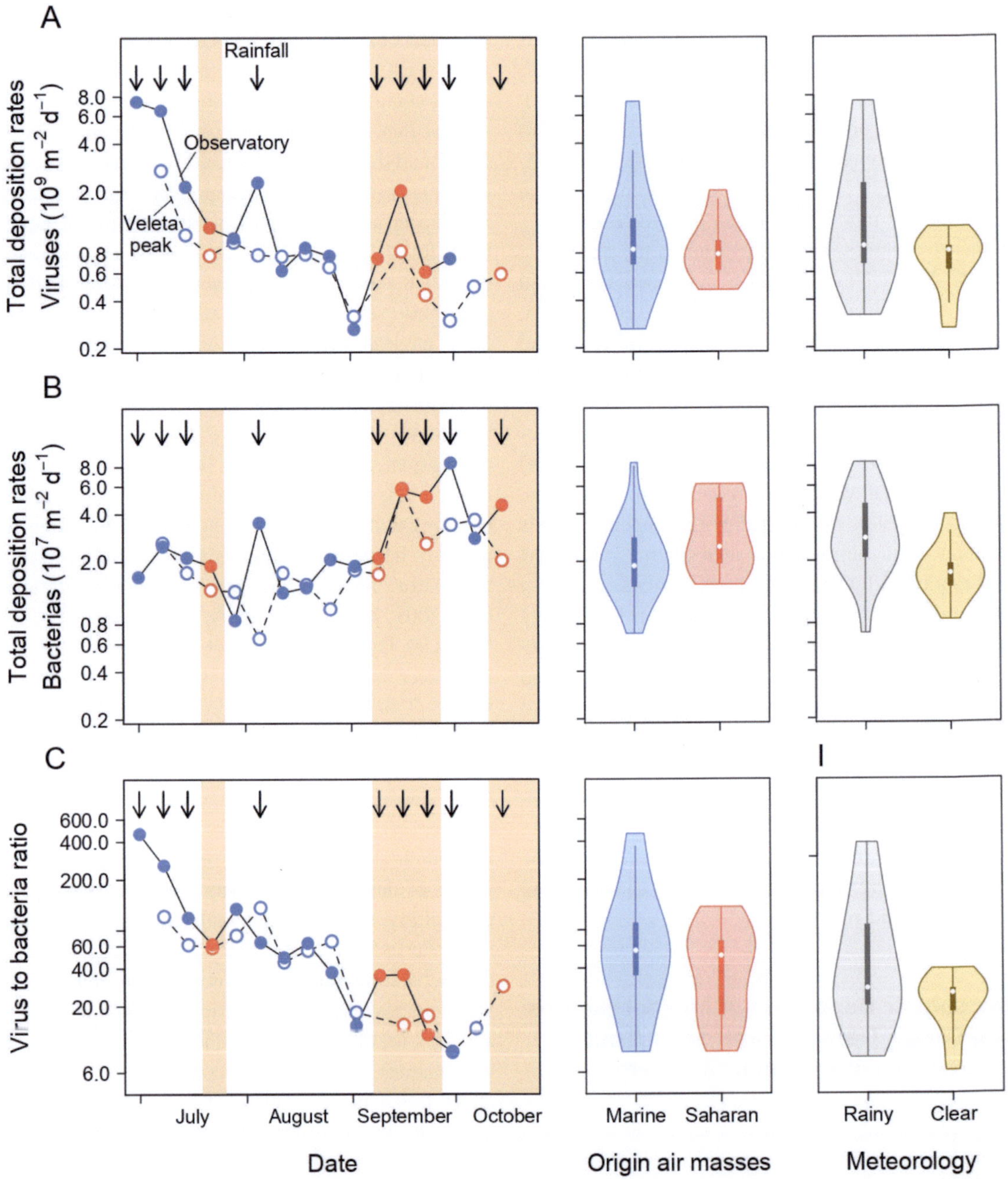


**Fig. 6** Synchronous dynamics of deposition rates of **a** viruses, **b** bacteria, and **c** virus-to-bacteria ratios at the Observatory of the Instituto Andaluz de Astrofísica (OSN) (solid circles) and Veleta Peak (VSN) (empty circles) in Sierra Nevada. Black arrows on the top indicate rain events. Samples that are predominantly of marine origin are shown in blue and samples that are predominantly Saharan are shown in orange. The median value (white dot, left panels), the 25–75% percentiles (boxes), and the non-outlier range (whiskers) in the total deposition are sorted by air-mass origin (marine vs. Saharan) and by meteorological conditions (rainy vs. clear)

formation, and precipitation remain poorly understood. These deposition rates of viruses were 9–461 times greater than the rates for bacteria (Fig. 6c). The median value of the viruses' deposition was 52-fold greater than for bacteria when air masses were predominantly of marine origin. By contrast, when the origin of air masses was predominantly Saharan, the median ratio was 28. The deposition rates of both viruses and bacteria were synchronous at the two distant sites explored (Observatory of Sierra Nevada and the Veleta peak) (Fig. 6). Synchrony among distant sites in atmospheric deposition variables is a signature of external, meteorological forcing, and long-range microbial dispersal (Reche et al. 2009; Morales-Baquero and Pérez-Martínez 2016).

Bacteria and viruses are mostly attached to particles. Therefore, it is necessary to detach these microorganisms from the dust or organic particles by washing them in a solution with a chelating agent and mechanical forces to disperse the cells and viral particles before counting their abundances by flow cytometry (Araya et al. 2019). Bioaerosols with smaller aerodynamic sizes have longer residence times in the atmosphere and are less susceptible to be removed by rain (Bowers et al. 2009; Després et al. 2012). Reche et al. (2018) obtained that ~69% of viruses and ~97% of bacteria deposited from the atmosphere were attached to particles, and proportionally more viruses were attached to the smallest airborne organic particles (<0.7 μm) than bacteria (>0.7 μm). Consequently, the atmospheric residence time of viruses appears to be longer than that of bacteria, which were associated with larger aerosols. The residence time of the microorganisms in the atmosphere depends primarily on their aerodynamic diameters and emission sources and, for example, for 3 μm bacteria-attached particles, the estimations range from 8.3 days (when the source is the desert) to 2.2 days (when the source is the sea) (Burrows et al. 2009). Many airborne bacteria from Saharan soils are dispersed in resistance forms as endospores (Fig. 7a), which can persist over time. The "ambiguous" nature of viruses about longevity (Legendre et al. 2014) makes it challenging to evaluate their viability. Viruses are deposited from the free atmosphere in billions per square meter and day (Fig. 7b), but their viability and persistence over time are still uncertain and a very exciting research challenge.

The long persistence and dispersal in the atmosphere of viruses explain observations that identical viral sequences occur at geographically distant locations and in very different environments (Short and Suttle 2005; Breibart and Rohwer 2005). This process provides a mechanism for maintaining the very high diversity of viruses and bacteria observed locally but constrained globally, consistent with a seed bank model. Long-range dispersal of viruses and bacteria can increase their distribution ranges in dormant or inactive states shaping their corresponding seed banks (Jones and Lennon 2010). The impact of the atmospheric deposition of microorganisms on the recipient ecosystems will depend on the viability of these microbes, and in the case of viruses, the occurrence of suitable hosts for replication. There is evidence that bacteria (Reche et al. 2009; Hervás et al. 2009; Peter et al. 2014) and viruses (Sharoni et al. 2015) can remain viable after atmospheric transport, which is consistent with the wide dispersal of microbes across very distant ecosystems. Hence, significant atmospheric deposition of bacteria and viruses may affect the structure and function of recipient ecosystems. Rather than being a negative consequence, this deposition provides a seed bank that should allow ecosystems to adapt to future and unpredictable environmental changes rapidly.

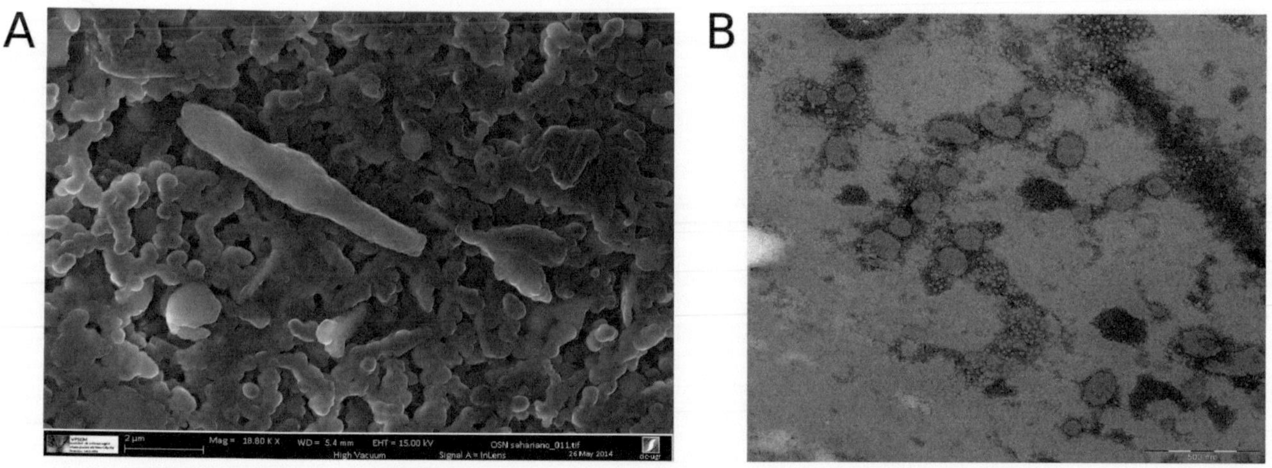


**Fig. 7** **a** Scanning electron microscopy (SEM) image of an airborne bacterium of Saharan origin. Note the spindle-shape typical for endospore formation. **b** Transmission electron microscopy (TEM) image of viruses negatively stained with uranyl acetate from aerosols of Saharan origin

## 5 Conclusions

Atmospheric deposition in the Sierra Nevada Mountains has a seasonal component with maximum dust exports during late spring and summer coinciding with the position of the intertropical convergence zone over the Saharan Desert. At larger time scales, dust exports are coupled with climatic oscillations such as North Atlantic Oscillation (NAO) with maxima during the positive NAO phase. Dust exports also have increased during the Anthropocene due to the onset of commercial agriculture in the Sahel region.

Saharan dust contains phosphorus and micronutrients like iron and bioavailable organic matter that stimulate phytoplankton and bacterial growth. Phytoplankton species respond differentially to dust deposition with consistent observations at seasonal and long-term scales. Billions of viruses and millions of bacteria per square meter and day, attached to Saharan dust particles and marine organic aggregates, are deposited above the atmospheric boundary layer in the terrestrial and aquatic ecosystems of the Sierra Nevada Mountains. This microbial atmospheric deposition expands biogeographic ranges and generates a global seed bank of microorganisms to face future environmental changes.

**Acknowledgements** We thank Dr. Ignacio Peralta-Maraver for his completely disinterested help with the illustrations and graphics and for his kindness. We also thank J. López-Ramos and R. McGrath for their assistance with sample collection, preparation, and analyzes. Funding was provided by the projects CICYT AMB99-0541, CICYT REN03-03038, ECOSENSOR (Fundación BBVA grant number BIOCON04/009), AEROGLOBAL (Junta de Andalucía grant number P06-RNM-01503), and MICROBIOGEOGRAPHY (Ministerio de Medio Ambiente, grant number 080/2007).

## References

Araya CM, Cazorla A, Reche I (2019) Detachment procedure of bacteria from atmospheric particles for flow-cytometry counting. Bio-protocol 9(12)

Baines SB, Webster KE, Kratz TK, Carpenter SR, Magnuson JJ (2000) Synchronous behavior of temperature, calcium, and chlorophyll in lakes of northern Wisconsin. Ecology 81(3):815–825

Ballantyne AP, Brahney J, Fernandez D, Lawrence CL, Saros J, Neff JC (2011) Biogeochemical response of alpine lakes to a recent increase in dust deposition in the Southwestern. US. Biogeosciences 8(9):2689–2706

Beres ND, Sengupta D, Samburova V, Khlystov AY, Moosmüller H (2020) Deposition of brown carbon onto snow: changes in snow optical and radiative properties. Atmos Chem Phys 20(10):6095–6114

Bhattachan A, Reche I, D'Odorico P (2016) Soluble ferrous iron (Fe (II)) enrichment in airborne dust. J Geophys Res Atmos 121(17):10–153

Bigelow A, Mladenov N, Lipson D, Williams M (2020) Dust deposition drives microbial metabolism in a remote, high-elevation catchment. The Holocene 30(4):589–596

Bonnet S, Guieu C, Chiaverini J, Ras J, Stock A (2005) Effect of atmospheric nutrients on the autotrophic communities in a low nutrient, low chlorophyll system. Limnol Oceanogr 50:1810–1819

Bowers RM, Lauber CL, Wiedinmyer C, Hamady M, Hallar AG, Fall R et al (2009) Characterization of airborne microbial communities at a high-elevation site and their potential to act as atmospheric ice nuclei. Appl Environ Microbiol 75:5121–5130

Brahney J, Mahowald N, Ward DS, Ballantyne AP, Neff JC (2015) Is atmospheric phosphorus pollution altering global alpine Lake stoichiometry? Global Biogeochem Cycles 29(9):1369–1383

Brahney J, Ballantyne AP, Sievers C, Neff JC (2013) Increasing $Ca^{2+}$ deposition in the western US: the role of mineral aerosols. Aeol Res 10:77–87

Breitbart M, Rohwer F (2005) Here a virus, there a virus, everywhere the same virus? Trends Microbiol 13(6):278–284

Burrows SM, Elbert W, Lawrence MG, Pöschl U (2009) Bacteria in the global atmosphere–Part 1: review and synthesis of literature data for different ecosystems. Atmos Chem Phys 9(23):9263–9280

Carrillo P, Reche I, Sánchez-Castillo P, Cruz-Pizarro L (1995) Direct and indirect effects of grazing on the phytoplankton seasonal succession in an oligotrophic lake. J Plankton Res 17(6):1363–1379

Chadwick OA, Derry LA, Vitousek PM, Huebert BJ, Hedin LO (1999) Changing sources of nutrients during four million years of ecosystem development. Nature 397:491–497

Creamean JM, Suski KJ, Rosenfeld D, Cazorla A, DeMott PJ, Sullivan RC et al (2013) Dust and biological aerosols from the Sahara and Asia influence precipitation in the Western US. Science 339:1572–1578

Cruz JA, McDermott F, Turrero MJ, Edwards RL, Martín-Chivelet J (2021) Strong links between Saharan dust fluxes, monsoon strength, and North Atlantic climate during the last 5000 years. Sci Adv 7 (26):eabe6102

Cruz-Pizarro L, Reche I, Carrillo P (1994) Plankton dynamics in a high mountain lake (Las Yeguas, Sierra Nevada). Indirect evidence of ciliates as a food source for zooplankton. Hydrobiologia 274:29–35

Darwin C (1845) An account of the fine dust which often falls on vessels in the Atlantic Ocean. Q J Geol Soc London 2:26–30

Després VR, Huffman JA, Burrows SM, Hoose C, Safatov AS, Buryak G et al (2012) Primary biological aerosol particles in the atmosphere: a review. Tellus B. 64:015598

Ehrenberg CG (1845) Neue Untersuchungen über das kleinste Leben als geologisches Moment. Mit kurzer Charakteristik von 10 neuen Gattungen und 66 neuen Arten. Berichte, Akademie Der Wissenschaften Zu Berlin 1845:53–87

Elser JJ, Dobberfuhl DR, Mackay NA, Schampel JH (1996) Organism size, life history, and N:P stoichiometry. BioScience 46(9):674–684

Evan AT, Flamant C, Gaetani M, Guichard F (2016) The past, present and future of African dust. Nature 531(7595):493–495

Giannini A, Saravanan R, Chang P (2003) Oceanic forcing of Sahel rainfall on interannual to interdecadal time scales. Science 302 (5647):1027–1030

Gorbushina AA, Kort R, Schulte A, Lazarus D, Schnetger B, Brumsack HJ et al (2007) Life in Darwin's dust: intercontinental transport and survival of microbes in the nineteenth century. Environ Microbiol 9(12):2911–2922

Granados-Muñoz MJ, Navas-Guzmán F, Bravo-Aranda JA et al (2012) Automatic determination of the planetary boundary layer height using lidar: one-year analysis over southeastern Spain. J Geophys Res 117:D18208

Griffin DW, Kellogg CA, Garrison VH, Shinn EA (2002) The global transport of dust: an intercontinental river of dust, microorganisms and toxic chemicals flows through the Earth's atmosphere. Am Sci 90(3):228–235

Han H, Kim G (2017) Significant seasonal change in optical properties by atmospheric humic-like substances (HULIS) in water-soluble organic carbon aerosols. Atmos Chem Phys Discuss 1–19

Hervàs A, Camarero L, Reche I, Casamayor EO (2009) Viability and potential for immigration of airborne bacteria from Africa that reach high mountain lakes in Europe. Environ Microbiol 11(6):1612–1623

Hidalgo-Muñoz JM, Gámiz-Fortis SR, Castro-Díez Y et al (2015) Long-range seasonal streamflow forecasting over the Iberian Peninsula using large-scale atmospheric and oceanic information. Water Resour Res 51:3543–3567

Jaenicke R (2005) Abundance of cellular material and proteins in the atmosphere. Science 308(5718):73–73

Jickells TD, An ZS, Andersen KK, Baker AR et al (2005) Global iron connections between desert dust, ocean biogeochemistry, and climate. Science 308(5718):67–71

Jiménez L, Rühland KM, Jeziorski A et al (2018) Climate change and Saharan dust drive recent cladoceran and primary production changes in remote alpine lakes of Sierra Nevada. Spain. Global Change Biol 24(1):e139–e158

Kellogg CA, Griffin DW (2006) Aerobiology and the global transport of desert dust. Trends Ecol Evol 21:638–644

Koulouri E, Saarikoski S, Theodosi C et al (2008) Chemical composition and sources of fine and coarse aerosol particles in the Eastern Mediterranean. Atmos Environ 42(26):6542–6550

Krumbein WE (1995) Gone with the wind—a second blow against spontaneous generation. Aerobiologia 11(3):205–211

Laskin J, Laskin A, Nizkorodov SA (2018) Mass spectrometry analysis in atmospheric chemistry. Anal Chem 90:166–189

Legendre M, Bartoli J, Shmakova L et al (2014) Thirty-thousand-year-old distant relative of giant icosahedral DNA viruses with a pandoravirus morphology. Proc Natl Acad Sci 111 (11):4274–4279

Legrand M, Preunkert S, Schock M et al (2007) Major 20th century changes of carbonaceous aerosol components (EC, WinOC, DOC, HULIS, carboxylic acids, and cellulose) derived from Alpine ice cores. J Geophys Res Atmos 112(D23)

Lennon JT, Jones SE (2011) Microbial seed banks: the ecological and evolutionary implications of dormancy. Nat Rev Microbiol 9 (2):119–130

López-Moreno JI, Vicente-Serrano SM, Morán-Tejeda E et al (2011) Effects of the North Atlantic Oscillation (NAO) on combined temperature and precipitation winter modes in the Mediterranean mountains: observed relationships and projections for the 21st century. Global Planet Change 77(1–2):62–76

Loye-Pilot MD, Martin JM, Morelli J (1986) Influence of Saharan dust on the rain acidity and atmospheric input to the Mediterranean. Nature 321(6068):427–428

Mace KA, Kubilay N, Duce RA (2003) Organic nitrogen in rain and aerosol in the eastern Mediterranean atmosphere: an association with atmospheric dust. J Geophys Res Atmos 108(D10)

Marañón E, Fernández A, Mourino-Carballido B et al (2010) Degree of oligotrophy controls the response of microbial plankton to Saharan dust. Limnol Oceanogr 55(6):2339–2352

Mladenov N, Pulido-Villena E, Morales-Baquero R et al (2008) Spatiotemporal drivers of dissolved organic matter in high alpine lakes: Role of Saharan dust inputs and bacterial activity. J Geophys Res Biogeosci 113(G2)

Mladenov N, Lopez-Ramos J, McKnight DM, Reche I (2009) Alpine lake optical properties as sentinels of dust deposition and global change. Limnol Oceanogr 54:2386–2400

Mladenov N, Reche I, Olmo FJ et al (2010) Relationships between spectroscopic properties of high-altitude organic aerosols and Sun photometry from ground-based remote sensing. J Geophys Res Biogeosci 115(G1)

Mladenov N, Sommaruga R, Morales-Baquero R et al (2011a) Dust inputs and bacteria influence dissolved organic matter in clear alpine lakes. Nat Commun 2(1):1–7

Mladenov N, Alados-Arboledas L, Olmo FJ et al (2011b) Applications of optical spectroscopy and stable isotope analyses to organic aerosol source discrimination in an urban area. Atmos Environ 45 (11):1960–1969

Mladenov N, Williams MW, Schmidt SK, Cawley K (2012) Atmospheric deposition as a source of carbon and nutrients to an alpine catchment of the Colorado Rocky Mountains. Biogeosciences 9 (8):3337–3355

Morales-Baquero R, Carrillo P, Reche I, Sánchez-Castillo P (1999) Nitrogen-phosphorus relationship in high mountain lakes: effects of the size of catchment basins. Can J Fish Aquat Sci 56(10):1809–1817

Morales-Baquero R, Conde-Porcuna JM (2000) Effect of the catchment areas on the abundance of zooplankton in high mountain lakes of the Sierra Nevada (Spain). Internationale Vereinigung Für Theoretische Und Angewandte Limnologie: Verhandlungen 27(4):1804–1808

Morales-Baquero R, Pulido-Villena E, Reche I (2006a) Atmospheric inputs of phosphorus and nitrogen to the southwest Mediterranean region: biogeochemical responses of high mountain lakes. Limnol Oceanogr 51(2):830–837

Morales-Baquero R, Carrillo P, Barea-Arco J et al (2006b) Climate-driven changes on phytoplankton–zooplankton coupling and nutrient availability in high mountain lakes of Southern Europe. Freshw Biol 51:989–998

Morales-Baquero R, Pulido-Villena E, Romera O et al (2006c) Significance of atmospheric deposition to freshwater ecosystems in the southern Iberian Peninsula. Limnetica 25(1–2):171–180

Morales-Baquero R, Pulido-Villena E, Reche I (2013) Chemical signature of Saharan dust on dry and wet atmospheric deposition in the south-western Mediterranean region. Tellus B 65:18720

Morales-Baquero R, Pérez-Martínez C (2016) Saharan versus local influence on atmospheric aerosol deposition in the southern Iberian Peninsula: Significance for N and P inputs. Global Biogeochem Cycles 30

Moulin C, Lambert CE, Dulac F, Dayan U (1997) Control of atmospheric export of dust from North Africa by the North Atlantic oscillation. Nature 387(6634):691–694

Mulitza S, Heslop D, Pittauerova D et al (2010) Increase in African dust flux at the onset of commercial agriculture in the Sahel region. Nature 466(7303):226–228

Niu H, Kang S, Shi X, Zhang G et al (2019) Dissolved organic carbon in summer precipitation and its wet deposition flux in the Mt. Yulong region, southeastern Tibetan Plateau. J Atmos Chem 76 (1):1–20

Okin GS, Mahowald N, Chadwick OA, Artaxo P (2004). Impact of desert dust on the biogeochemistry of phosphorus in terrestrial ecosystems. Global Biogeochem Cycles 18(2)

Oldani KM, Mladenov N, Williams MW et al (2017) Seasonal patterns of dry deposition at a high-elevation site in the Colorado Rocky Mountains. J Geophys Res Atmos 122(20):11–183

Pérez-Martínez C, Rühland KM, Smol JP et al (2020) Long-term ecological changes in Mediterranean mountain lakes linked to recent climate change and Saharan dust deposition revealed by diatom analyses. Sci Total Environ 727:138519

Peter H, Hörtnagl P, Reche I, Sommaruga R (2014) Bacterial diversity and composition during rain events with and without Saharan dust influence reaching a high mountain lake in the Alps. Environ Microbiol Rep 6:618–624

Peter H, Sommaruga R (2016) Shifts in diversity and function of lake bacterial communities upon glacier retreat. The ISME J 10(7):1545–1554

Pittino F, Ambrosini R, Azzoni RS et al (2018) Post-depositional biodegradation processes of pollutants on glacier surfaces. Condensed Matter 3(3):24

Pulido-Villena E, Reche I, Morales-Baquero R (2005) Food web reliance on allochthonous carbon in two high mountain lakes with contrasting catchments: a stable isotope approach. Can J Fish Aquat Sci 62(11):2640–2648

Pulido-Villena E, Reche I, Morales-Baquero R (2006) Significance of atmospheric inputs of calcium over the southwestern Mediterranean region: High mountain lakes as tools for detection. Global Biogeochem Cycles 20(2)

Pulido-Villena E, Wagener T, Guieu C (2008a) Bacterial response to dust pulses in the western Mediterranean: implications for carbon cycling in the oligotrophic ocean. Global Biogeochem Cycles 22(1)

Pulido-Villena E, Reche I, Morales-Baquero R (2008b) Evidence of an atmospheric forcing on bacterioplankton and phytoplankton dynamics in a high mountain lake. Aquat Sci 70(1):1–9

Prospero JM, Ginoux P, Torres O et al (2002) Environmental characterization of global sources of atmospheric soil dust identified with the Nimbus 7 Total Ozone Mapping Spectrometer (TOMS) absorbing aerosol product. Rev Geophys 40(1):2–1

Prospero JM, Lamb PJ (2003) African droughts and dust transport to the Caribbean: climate change implications. Science 302 (5647):1024–1027

Psenner R (1999) Living in a dusty world: Airborne dust as a key factor for Alpine Lakes. Water Air Soil Pollut 112:217–227

Ram K, Sarin MM, Hegde P (2010) Long-term record of aerosol optical properties and chemical composition from a high-altitude site (Manora Peak) in Central Himalaya. Atmos Chem Phys 10:11791–11803

Reche I, Carrillo P, Cruz-Pizarro L (1997) Influence of metazooplankton on interactions of bacteria and phytoplankton in an oligotrophic lake. J Plankton Res 19(5):631–646

Reche I, Pulido-Villena E, Conde-Porcuna JM, Carrillo P (2001) Photoreactivity of dissolved organic matter from high-mountain lakes of Sierra Nevada, Spain. Arct Antarct Alp Res 33(4):426–434

Reche I, Ortega-Retuerta E, Romera O et al (2009) Effect of Saharan dust inputs on bacterial activity and community composition in Mediterranean lakes and reservoirs. Limnol Oceanogr 54(3):869–879

Reche I, D'Orta G, Mladenov N et al (2018) Deposition rates of viruses and bacteria above the atmospheric boundary layer. The ISME J 12 (4):1154–1162

Redfield AC (1934) On the proportions of organic derivations in sea water and their relation to the composition of plankton. In: Daniel RJ (ed) James Johnstone memorial volume. University Press of Liverpool, pp 177–192

Rodríguez S, Querol X, Alastuey A et al (2001) Saharan dust contributions to PM10 and TSP levels in Southern and Eastern Spain. Atmos Environ 35:2433–2447

Rogora M, Mosello R, Marchetto A (2004) Long-term trends in the chemistry of atmospheric deposition in northwestern Italy: The role of increasing Saharan dust deposition, Tellus. Ser B 56:426–434

Rothhaupt KO (1995) Algal nutrient limitation affects rotifer growth rate but not ingestion rate. Limnol Oceanogr 40(7):1201–1208

Sabatier P, Nicolle M, Piot C et al (2020) Past African dust inputs in the western Mediterranean area controlled by the complex interaction between the Intertropical Convergence Zone, the North Atlantic Oscillation, and total solar irradiance. Climate of the past 16 (1):283–298

Santos PS, Santos EB, Duarte AC (2013) Seasonal and air mass trajectory effects on dissolved organic matter of bulk deposition at a coastal town in south-western Europe. Environ Sci Pollut Res 20 (1):227–237

Sharoni S, Trainic M, Schatz D et al (2015) Infection of phytoplankton by aerosolized marine viruses. Proc Natl Acad Sci USA 112:6643–6647

Short CM, Suttle CA (2005) Nearly identical bacteriophage structural gene sequences are widely distributed in both marine and freshwater environments. Appl Environ Microbiol 71(1):480–486

Sommaruga R (2015) When glaciers and ice sheets melt: consequences for planktonic organisms. J Plankton Res 37(3):509–518

Talbot RW, Harris RC, Browell EV et al (1986) Distribution and geochemistry of aerosols in the Tropical North Atlantic troposphere: relationship to Saharan Dust. J Geophys Res 91:5173–5182

Triadó-Margarit X, Caliz J et al (2019) High similarity in bacterial bioaerosol compositions between the free troposphere and atmospheric depositions collected at high-elevation mountains. Atmos Environ 203:79–86

Washington R, Todd MC, Lizcano G et al (2006) Links between topography, wind, deflation, lakes and dust: the case of the Bodélé Depression, Chad. Geophys Res Lett 33(9)

Woo C, Yamamoto N (2020) Falling bacterial communities from the atmosphere. Environ Microbiome 15(1):1–14

Xie M, Mladenov N, Williams MW et al (2016) Water soluble organic aerosols in the Colorado Rocky Mountains, USA: composition, sources and optical properties. Sci Rep 6(1):1–12

Xie M, Chen X, Hays MD et al (2017) Light absorption of secondary organic aerosol: composition and contribution of nitroaromatic compounds. Environ Sci Technol 51(20):11607–11616

Yamaguchi N, Ichijo T, Sakotani A et al (2012) Global dispersion of bacterial cells on Asian dust. Sci Rep 2(1):1–6

Yan F, Kang S, Li C, Zhang Y et al (2016) Concentration, sources and light absorption characteristics of dissolved organic carbon on a medium-sized valley glacier, northern Tibetan Plateau. The Cryosphere 10(6):2611–2621

Zhang Y, Xu J, Shi J, Xie C et al (2017) Light absorption by water-soluble organic carbon in atmospheric fine particles in the central Tibetan Plateau. Environ Sci Pollut Res 24:21386–21397

# Concluding Notes: Ecosystems of Sierra Nevada in the Anthropocene: A New Cocktail of Species and Ecological Interactions

**R. Zamora, J. Lorite, L. Matías, Alexandro B. Leverkus, M. Jimenez, J. M. Medina-Sánchez, C. Pérez-Martínez and I. Reche**

Mountain ecosystems, being extremely sensitive to environmental changes because of the fragile equilibrium in the biotic components of the natural system, can thus be used as sensors to detect early signs of change. The eight chapters of the Section Biotic Respones to Recent Changes review the ecological data available to assess exposure, sensitivity, and adaptive capacity of Sierra Nevada ecosystems to current and future scenarios. We have analyzed in fuller detail the biotic responses associated with the main drivers of global change detected in Sierra Nevada over the last few decades, mainly climate and land-use change and atmospheric deposition.

The Sierra Nevada massif has long been recognized as a biodiversity hotspot for both animal and plant groups, based mostly on the species richness, endemicity, and relictness. However, biodiversity includes many other facets related to functional and historical aspects of biological entities present in hotspots, from genes to communities and the whole biota. Sierra Nevada hosts 2348 species of flowering plants. With 362 "alpine" species, the Sierra Nevada exceeds by far the alpine plant species diversity of other European mountain ranges of similar size. Similar comparisons await to be explored using the rich data available for animals and those for aquatic systems in particular (lakes and rivers) as evidenced in several chapters. For instance, about 4000 species of insects have been listed for this massif, of which more than 150 are local endemics. For some groups such as ants and chrysomelid beetles, one-third of the Iberian taxa are found in Sierra Nevada. These numbers are surely just a small fraction of the actual diversity, as many groups are still understudied (e.g., Acari, Collembola, or some families of Coleoptera, Diptera, and Hymenoptera), and new species are added (and even newly described) for this massif every year. These biological data series suffer from issues of changing sampling effort and taxonomic resolution, which require backtracking and harmonizing species names. Consequently, our information on the existing biodiversity stock in Sierra Nevada and its change for major groups and microorganisms is still in its infancy.

The consequences of climate and land-use changes are typically most evident in mountain ecosystems, where expansions or contractions in species' distribution ranges along elevation gradients may occur via the migration of species to new areas. Sierra Nevada contains unique alpine flora. However, the conservation of this outstanding natural heritage is facing serious challenges, some arising from the intensification of climate change, which is becoming especially severe in Mediterranean mountains. For instance, summit vegetation has shown signals of change, with an altitudinal expansion of some species, and greater vegetation cover. Many alpine plants have become rarer in recent decades on the high mountains of the European continent, whereas plants from low elevations become more common. In Sierra Nevada, over the last years, 13 alpine species have disappeared from monitoring plots and, at the same time, five new taxa have appeared (chapter "Altitudinal Patterns and Changes in the Composition of High Mountain Plant Communities"). However, different impacts caused by

R. Zamora (✉) · A. B. Leverkus · M. Jimenez
J. M. Medina-Sánchez · C. Pérez-Martínez · I. Reche
Department of Ecology, University of Granada, Granada, Spain;
Andalusian Institute for Earth System Research (IISTA),
University of Granada, Granada, Spain
e-mail: rzamora@ugr.es

J. Lorite
Andalusian Institute for Earth System Research (IISTA), University of Granada, Granada, Spain;
Department of Botany, University of Granada, Granada, Spain

L. Matías
Department of Plant Biology and Ecology, University of Seville, Seville, Spain

M. Jimenez · J. M. Medina-Sánchez ·
C. Pérez-Martínez · I. Reche
University Institute of Water Research, University of Granada, Granada, Spain

human activities may act synergistically to exacerbate these changes. Only the adoption of measures with solid scientific basis can help to mitigate the combined effects of climatic and land-use changes. In this sense, chapter "Altitudinal Patterns and Changes in the Composition of High Mountain Plant Communities" provides a detailed roadmap to orient conservation agenda at Sierra Nevada, including both adaptive management actions and outdoor recreation activities evaluation.

Faunistic inventories conducted 30–40 years ago have been repeated in recent years. These long-term datasets can be used to evaluate spatiotemporal shifts and trends in biological communities and their relation to drivers of global change, such as land use or climate. Rising temperatures will facilitate the proliferation of forest pests, causing forest decay. Some of the pest species will presumably lengthen their activity period whereas others may extend their colonization area to new territories. For insects, an elevational migration has been confirmed for the pine processionary moth, which has shown a clear elevational displacement in Sierra Nevada and a latitudinal one in Europe, together with a phenological advance in response to global warming (chapter "Altitudinal Patterns and Changes in the Composition of High Mountain Plant Communities"). In the case of this month, the extensive and monospecific plantations of conifers have helped its survival, magnifying its defoliating impact, especially at high elevations due to warming. Thus, the processionary displays a very wide spectrum of possible responses to climatic change in interaction with land-use change.

Among vertebrates, equivalent responses have also been found. For instance, passerine birds have shown marked temporal dynamics over the last 40 years which have been strongly influenced by global change (chapter "Altitudinal Patterns and Changes in the Composition of High Mountain Plant Communities"). Some generalist mountain species are now more abundant than before in high-mountain areas, while the most typical alpine species have become progressively scarcer since the last cooling episode (Little Ice Age, thirteenth–nineteenth centuries in Sierra Nevada). Overall, montane species have declined and have been replaced by more thermophilic Mediterranean lowland species. A strong determining factor in the distribution and abundance of forest birds in Sierra Nevada is land-use change. The increase in coverage and densification of oak woodlands has favored forest species, which register a higher temporal turnover than do birds of the juniper scrublands and high summits.

Mediterranean forest ecosystems of Sierra Nevada are especially advantageous settings for studying the causes and consequences of global change. These ecosystems have a high diversity of woody species in a mosaic of ecological conditions. Ongoing changes in climate are altering ecological conditions for many woody plant species. For example, the demographic structure of both *Pinus sylvestris* and *Juniperus communis* populations differs across the elevational distribution in Sierra Nevada (chapter "Forest Dynamics Under Land-Use and Climate Change Scenarios"). Both species presented a significantly reduced proportion of young individuals in the lowermost populations, and a clear dominance of older age classes. This contrasts sharply with the pattern found at the treeline, where a higher proportion of saplings appeared and an even distribution was found across age classes. These results indicate an ongoing elevational displacement for both species in Sierra Nevada, the southernmost edge of their geographical distribution area. However, factors other than climate affect woody species growth and reproduction at the same time such as land-use change and herbivory pressure, and the result of their interactions are strongly heterogeneous across habitats, altitudinal ranges, and species.

Many forest species differ in their ability to withstand drought. Those that have a shorter life cycle can generally adapt better to changing conditions than do longer lived species. Therefore, we expect shrub species to be favored over trees. On the other hand, the future climate may be ideal for foreign species that now inhabit lower elevational levels. Changes in communities due to climate and land-use change may also be driven by alterations in disturbance regimes, and long-term monitoring is needed to assess successional trajectories across succession under multiple combinations of natural and anthropogenic disturbances. Also, fires will likely become more frequent if temperatures increase and rainfall decreases. This, together with the abandonment of forest use and the surge in recreational activities will no doubt contribute to great fire frequency and severity (chapter "Restoration of Mediterranean Forest Ecosystems After Major Disturbances: The Lanjarón Post-fire Experiment Over 15 Years of Succession"). Thus, we're not only altering the regime of fires by modifying fuels, making weather more adequate for fire propagation and igniting fires, but also by producing additional disturbances such as post-fire logging that change the baseline conditions for succession.

A comparative study of macroinvertebrate communities in the rivers of Sierra Nevada in the 1980s and today reveals substantial changes (chapter "Aquatic Animal Communities of Watercourses from Sierra Nevada"). Some species in the 1980s that were associated with the middle elevation stretch of the rivers are today found in the higher reaches. This is true of Trichoptera, which have increased their species richness in the protected area. In the case of Plecoptera, it has been found that the lower limits of the distribution of some species have contracted, while the distribution at the upper limits has remained unchanged. As a result, the diversity of species in the upper reaches of rivers is far greater than 30 years ago. Nonetheless, many macroinvertebrates will see their survival compromised under future climate-change scenarios due to the absence of strategies to cope with these new conditions. Vertebrates such as the brown trout are also in danger in Sierra Nevada due to several threats, such as water scarcity caused by water removal, poor water quality, or the presence of exotic species.

Long-term datasets can be used to evaluate spatiotemporal changes and trends in biological communities and their relation to drivers of global change. Unlike other European mountain lakes, the geographic location and geological history of Sierra Nevada cause high-mountain lakes to be exposed to several environmental stressors simultaneously: climatic anomalies (temperature and precipitation) linked to a climate-change hotspot (Mediterranean region), high UV radiation, increased atmospheric dust deposition, and allochthonous nutrient input. The interaction among these global-change drivers is altering intrinsic features shared by these diverse ecosystems, such as the high transparency, low nutrient content, and narrow temperature ranges of their waters. The great simplicity, but high sensitivity and responsivity to these drivers, of their biological communities, helps us to understand and anticipate changes induced in the state of ecosystems, such as loss of functional diversity (e.g., decline of mixotrophic microalgae) or alterations in the carbon flux through the food webs. These results qualify the high-mountain lakes of Sierra Nevada as unique sentinels of global change (chapter “High Mountain Lakes as Remote Sensors of Global Change”). Temperature was identified as the main predictor of the observed changes, whereas Saharan dust-deposition drivers were secondary explanatory variables (chapter “Paleolimnological Indicators of Global Change”). The result is a significant recent algal bloom and profound changes in the biota composition of the different groups of organisms analyzed. The nature of the change is interpreted as a result of a lengthening of the ice-free period of the lake and a rise in lake water temperature as well as reduced water availability in their catchments. The latter factor affects the volume and residence time of lake water and reflects the intensification of summer drought in the Sierra Nevada summit area over the last 50–60 years. Atmospheric deposition with Saharan dust carries macronutrients such as phosphorus and micronutrients as calcium and iron to the lakes, resulting in organic matter with a humic-like signature affecting lake UV-transparency. By contrast, atmospheric deposition from marine sources introduces organic matter with an amino acid-like signature and a comparatively higher ratio of viruses to bacteria (chapter “Atmospheric Inputs and Biogeochemical Consequences in High-Mountain Lakes”).

Overall, these results show that high mountains such as Sierra Nevada play a vital role in maintaining biodiversity in the context of global change. Species populations can seek the ideal conditions under which they can remain by moving to higher elevations according to their ecological necessities, or even taking advantage of the mosaics of ecological conditions characterizing mountain slopes. The results show a great variety of biotic responses to global change in Sierra Nevada: speciation, altitudinal displacement, niche evolution, hybridization, population persistence, and extinction. Furthermore, there are many studies in terrestrial ecosystems focused mostly on ecological interactions, including pollination, herbivory, seed dispersal, and facilitative plant interactions, which demonstrate that interactions are a fundamental part of functional biodiversity. Ecological interactions become baroque in high mountain lakes, whose biotic communities are very simple in their composition, but paradoxically, very complex in their network of interactions. In many cases, these studies bridge the gap between ecological function and evolutionary history by including phylogenetic information on the groups involved in interactions. However, today’s rapid rates of climate change may prevent many species from migrating fast enough to continue in their range of optimal conditions. In fact, biotic responses to environmental changes can have a time lag, depending on the generation time of the organism. Long-lived species, such as many woody plants, can show demographic responses with long temporal delays, while species with shorter generation times (e.g., annual plant species, insects, microorganisms) can respond almost immediately. For example, the planktonic communities of high-mountain lakes change radically in a few days in the face of the intrusion of Saharan dust. Thus, biotic time lags comprise a crucial factor in mountain ecosystems. Lagging responses to environmental changes, associated with the mosaics of ecological conditions in mountains, are two factors that powerfully determine the speed and magnitude of biotic responses to climate change. Plants, animals, and microorganisms display a wide spectrum of possible responses to global-change drivers, highlighting the need to consider all spatiotemporal scales in relation to the life cycle of an organism in order to adequately forecast the biotic responses to climate change.

These results indicate that montane ecological communities that are considered rather stable are, in fact, undergoing strong natural spatial and temporal dynamics now superimposed by changes caused by human activities, such as climate warming, water- and land-use changes, atmospheric inputs and alterations of disturbance regimes. As a result, biotic responses to these two major drivers of global change converge, threatening dominant high-mountain species while at the same time opening opportunities for other subordinate species. All of this seems to be restricting the distribution range or leading to the extinction of rare, endemic, and/or specialist species, parallel to expansions in the distribution range of generalist species. This may be increasing the similarity of taxonomic composition among communities along elevation gradients. This process can be envisaged as the first stage of taxonomic and functional homogenization processes or, in other words, a temporal decline in β-diversity. Thus, it becomes critical to determine to what degree the mosaic of ecological conditions favored by the mountain topography and ecological heterogeneity delays (or even halts) this widespread process of biotic homogenization, this being a topic of fundamental relevance for scientific investigation in the future. The maintenance of ongoing monitoring programs and associated research is crucial to test whether these changes in the composition, abundance, dominance, and altitudinal distribution will alter food webs and ecosystem functioning, as has already been demonstrated for high-mountain lakes.

**Author**: Andrés Ureña

# Looking Forward: New Approaches in Research, Management and Social Connection

**Author**: Andrés Ureña

# Remote Sensing in Sierra Nevada: From Abiotic Processes to Biodiversity and Ecosystem Functions and Services

Domingo Alcaraz-Segura, Javier Cabello, Salvador Arenas-Castro, Julio Peñas, and Ana Sofia Vaz

**Abstract**

During the last decades, remote sensing has changed the way humans observe and understand the Earth system. The repeated and increasingly detailed observations made from satellite platforms and other remote sensing procedures have revolutionized research, particularly in the atmospheric and oceanographic sciences but also in the biophysical sciences. This chapter presents a systematic literature review of the different ways in which remote sensing has been applied in Sierra Nevada, Spain. Studies ranged from basic research to how remote sensing is actually contributing to management in this mountain biosphere reserve. The chapter is structured using the ecosystem services cascade as a framework, i.e., from studies on abiotic (i.e., geophysical, atmospheric, cryospheric, and hydrological) processes to research on biodiversity and ecosystem functions and services. The number of remote sensing studies in Sierra Nevada is quickly growing but still relatively scarce (only 65 records). Most of this research was either applied or use-oriented research and found to be potentially useful to assess biodiversity conservation status and ecosystem services, indeed it frequently contained recommendations for the management of the protected area. Hence, there is an expected increase in the interdisciplinary and transdisciplinary application of remote sensing to research in Sierra Nevada.

**Keywords**

Earth observation • Satellite • Drone • Unmanned aerial vehicle • Manned aerial vehicle • Airborne sensors • Radiometer • Multispectral • Spain

## 1 Introduction

### 1.1 Remote Sensing Definition and Types

Remote sensing is described as the process of detecting, obtaining, and monitoring the physical characteristics of objects or areas by measuring their reflected and emitted radiation from distance. In general, remote-sensing data about the Earth system are typically obtained through sensors onboard ex situ platforms, such as satellites, manned aerial vehicle, and unmanned aerial vehicles or ships. Nevertheless, Earth remote sensing sensors also enable capturing in situ measurements, e.g., by using spectroradiometers and cameras portable by hand or mounted on poles or towers, that are directly taken closer from the target object or area.

The longest history of use of remote sensing data over vast distances is dominated by spaceborne and airborne systems, targeting both passive (i.e., using the reflected sunlight) and active sensors (i.e., using an artificial light source) (Chuvieco and Huete 2009). Traditionally, the most used platforms in environmental sciences have been satellite

D. Alcaraz-Segura (✉) · J. Peñas · A. S. Vaz
Department of Botany, University of Granada, Granada, Spain
e-mail: dalcaraz@ugr.es

D. Alcaraz-Segura · A. S. Vaz
Sierra Nevada Global Change Observatory. Interuniversity Institute for Earth System Research in Andalusia, University of Granada-Junta de Andalucía, Granada, Spain

D. Alcaraz-Segura · J. Cabello
Andalusian Center for the Assessment and Monitoring of Global Change, University of Almeria, Almeria, Spain

J. Cabello
Department of Biology and Geology, University of Almeria, Almeria, Spain

S. Arenas-Castro
Polytechnic Institute of Viana Do Castelo, Viana do castelo, Portugal

S. Arenas-Castro
Area of Ecology, Department of Botany, Ecology and Plant Physiology, University of Cordoba, Cordoba, Spain

A. S. Vaz
BIOPOLIS-CIBIO-InBIO, University of Porto, Porto, Portugal


R. Zamora and M. Oliva (eds.), *The Landscape of the Sierra Nevada*,
https://doi.org/10.1007/978-3-030-94219-9_19

passive sensors, such as Landsat (15–60 m/pixel), MODIS (250–1000 m/pixel), or the new Sentinel-2 (up to 10 m/pixel), but active sensors onboard satellites and aircrafts, such as LiDAR (Light Detection and Ranging) and microwave RADAR have significantly increased during the last decade. This technology, which offers excellent opportunities for monitoring all components of the Earth's surface, is attracting the interest of governments. Such is the case with the Copernicus program and its Sentinel-satellites constellation, the infrastructure of the European Space Agency, which is increasing our ability to analyze the Earth's surface through high spatial and temporal resolution imagery. In addition, since many objects and processes cannot be detected by very distant sensors, the use of lightweight unmanned aerial vehicles (UAVs) or drones, is opening a particularly cost-efficient way of collecting data at finer scales, but over smaller extents and restricted to strict flying regulations (Chen et al. 2021).

## 1.2 Applications of Remote Sensing in Environmental Sciences

Originally driven by military uses, remote sensing gradually evolved as a relevant technology in many scientific fields and for a variety of civil applications. Examples range from basic cartography for territorial planning, civil engineering, and emergency response, to meteorology, oceanography, agriculture, forestry, and natural resources assessment, among many others (Chuvieco and Huete 2009). Remote sensing data records allow continuous, updated, and cost-effective measures of the state and dynamics of many abiotic, biotic, and socioeconomic processes, hardly monitorable by other means at a global scale. For instance, satellite remote sensing data is considered the largest archive of biological information on Earth (Geller et al. 2017). Time series of surface reflectance in each band of the electromagnetic spectrum directly informs on the radiation and energy balance of each pixel around the Earth and indirectly on multiple aspects of the state and dynamics of life, such as photosynthesis (Hikosaka and Tsujimoto 2021), evapotranspiration (Chen and Liu 2020), wetness (Chandrasekar et al. 2010), land-use and land-cover (Rogan and Chen 2004), etc.

The use of remote sensing technology over local, regional, or global scales provides valuable information on the three dimensions of biodiversity (i.e., composition, structure, and function; Noss 1990) (Turner et al. 2003; Cabello et al. 2012, 2013; Nagendra et al. 2013; Pettorelli et al. 2016; Reddy 2021). Indeed, satellite remote sensing has been highlighted as a key tool to develop Essential Biodiversity Variables to inform on biodiversity status (Skidmore et al. 2021). Remote sensing can also provide inherent information to assess the macroecological drivers and trends of biodiversity patterns (Perry et al. 2009; Alcaraz-Segura et al. 2017; Arenas-Castro et al. 2018; Arenas-Castro et al. 2019; Regos et al. in press). The information provided by satellite data records on surface reflectance along the electromagnetic spectrum has not only been used to inform on biodiversity loss, a dramatically exceeded planetary boundary, whose control widely depends on the conservation of mountain environments, but also to assess the other environmental boundaries that define the safe operating space for humanity (Steffen et al. 2015). This is particularly the case for stratospheric ozone depletion, climate change, land-use change, atmospheric aerosols loading, biogeochemical flows, and freshwater use (Hughes et al. 2013).

Earth observation from remote sensors has also become essential to assess many aspects that directly affect human well-being such as those related to ecosystem functions and services (Alcaraz-Segura et al. 2013). Studies in this sense have been useful both for reporting on the ecosystem capacity to provide benefits to society (supply side), as well as the contributions of nature that are demanded by humans (demand side). Following the ecosystem services cascade as a framework (Potschin-Young et al. 2018; Czúcz et al. 2020), Earth observations have been widely applied in the basic study and characterization of geophysical and biophysical attributes, as well as of ecosystem processes and functions underlying ecosystem services provision (Alcaraz-Segura et al. 2013). From the demand side or social dimension of the ecosystem services cascade, remote sensors have also contributed to quantify the social goods, benefits, and values associated with ecosystem services (Vaz et al. 2020). In fact, remote sensing, together with geodesign, nature-based solutions, and artificial intelligence have become central in current European initiatives (e.g., COPERNICUS Program, GoGreenRoutes, LifeWatch ERIC) to address societal challenges, such as the United Nations' Sustainable Development Goals, providing environmental, social, and economic information needed to build social-ecological resilience and helping in the monitoring of the status and trends of biodiversity and ecosystem services in pursuit of sustainable solutions inspired and supported by nature.

## 1.3 Usefulness of Remote Sensing for Managing Protected Areas

Remote sensing can be a very useful tool for protected area managers (Nagendra et al. 2013; Rose et al. 2015). Managers need indicators sensitive to both long-term directional changes and near real-time processes to ensure a good understanding of ecosystem health (Lovett et al. 2007; Cabello et al. 2012). The indicators derived from remote sensing datasets allow for the monitoring and assessment of multiple responses of ecosystems to environmental changes and management actions. By using such indicators, protected area managers can adopt an adaptive management approach to evaluate the effectiveness of management actions (Westgate et al. 2013; Cabello et al. 2018). The convenience of remote sensing data to implement such an approach is due to its ability to reveal the drivers of long-term change with a consistent information update rate, while highlighting recent impacts on protected area conditions from the local to the regional scale (Cabello et al. 2016). In addition to these advantages, remote sensing has a much lower cost than repeated field measurements and field campaigns (Kennedy et al. 2009). The constant innovation in platforms and sensors enhances their potential performance, improving data acquisition resolution, signal/noise ratio, and in general, their quality. Thus, improving inputs allows us to adapt protected area monitoring programs to new ecological and societal challenges, which is of great importance considering the intensified impacts of climate and environmental change (Duan et al. 2020; Wang et al. 2020).

This chapter presents a thorough review of the multiple ways in which remote sensing has been applied in Sierra Nevada, from basic to applied research targeting abiotic, biophysical, and social-ecological attributes and processes to aid researchers, managers, and citizens in the comprehension and management of mountain environments. Despite remote sensing becoming the most powerful tool for monitoring ecosystems and biodiversity worldwide, many efforts (experimental, educational, and outreach) are still needed to actually incorporate it into environmental decision-making and management processes (Cabello et al. 2018). We consider that reviewing and ordering the advances done for Sierra Nevada according to the conceptual and regulatory frameworks followed by the protected area management teams will help to achieve this goal.

## 2 Materials and Methods

We performed a literature search of peer-reviewed publications using the whole collection of "ISI Web of Science" (ISI WOS; http://webofknowledge.com/) and Scopus (www.scopus.com) search engines during March, 2021. Our search structure, was adapted from Vaz et al. (2018), and focused on remote sensing applications in Sierra Nevada, Spain, considering the following search string:

> Title, abstract or keywords = ("Sierra Nevada") NOT ("California" OR "USA" OR "United States" OR "America" OR "colombia" OR "columbia" OR "Sierra Nevada de Santa Marta") AND ("remote sens*" OR "remote-sens*" OR "earth observation" OR "imagery" OR "UAV" OR "drone" OR "unmanned aerial" OR "aircraft*" OR "airborne" OR "air-borne" OR "spaceborne" OR "space-borne" OR "AVHRR" OR "radiomet*" OR "high-resolution" OR "high resolution" OR "very-high resolution" OR "high spatial resolution" OR "very-high spatial resolution" OR "hyper-spectral" OR "hyperspectral" OR "multispectral" OR "multi-spectral" OR "image fusion" OR "NDVI" OR "satellite*" OR "sensor*" OR "radar" OR "MODIS" OR "LiDAR" OR "sentinel*" OR "landsat*" OR "worldview*").

Our search retrieved 240 unduplicated records, which were then subjected to inclusion/exclusion criteria to eliminate irrelevant information (e.g., referred to Sierra Nevada in Colombia or in the USA or to a company named Sierra Nevada). Criteria were applied by checking the full content of each record individually, resulting in a final dataset of 65 relevant records. Each record was then fully reviewed to identify its potential use or actual ability to inform on: (1) Earth system components and types of targeted ecosystems; (2) explicit and implicit biodiversity dimensions (classified according to Noss 1990) and ecosystem services types (classified according to CICES version 5.1; Haines-Young and Potschin-Young 2018); (3) remote sensing platforms and products; and (4) types of research on and the management contributions to Sierra Nevada. To do so, we established a series of questions and categorical answers related to the former four attributes (Table 1).

## 3 Results

### 3.1 Overview on the Remote Sensing Research in Sierra Nevada

Most studies relied on satellite remote sensing (47%; Fig. 1a) and the most common products included MODIS (39%), Landsat (17%), AVHRR (7%), and Sentinel (3%). In situ remote sensing was the second most-used approach, being used alone or in combination with other approaches (16%). LIDAR was used in 11% of records. Unmanned (UAV) and manned aerial vehicles (MAV) accounted for a small proportion of records (3% and 2%, respectively), relying on orthophotography products used alone (3%) or in combination with LIDAR (7%; Fig. 1b).

Most records focused on terrestrial ecosystems only (40%) or both terrestrial and aquatic ecosystems (3%). Aquatic ecosystems alone were less addressed by remote-sensing means (3%). Most records did not explicitly focus their study neither on terrestrial nor on aquatic ecosystems (54%; Fig. 2a). We could notice a dominance of publications studying the biosphere and cryosphere (30% in each case), followed by the atmosphere (23%; Fig. 2b). The lithosphere and hydrosphere were less studied through remote sensing means (13% and 3%, respectively). A small number of records simultaneously studied interactions among several components of the Earth system. Regarding

**Table 1** Categories used to assign each record retrieved by our search on remote sensing applications in Sierra Nevada

| Questions | Categories |
|---|---|
| **1. Remote sensing platforms and products** | |
| Which type of remote sensing platform is being used in the record? | In situ |
| | Satellite |
| | Unmanned aerial platforms |
| | Manned aerial platforms |
| Which remote sensing product is being used/derived in the record? | LiDAR |
| | MODIS |
| | Landsat |
| | Others (specified in the record) |
| **2. Type, components, and critical environmental boundaries of the targeted ecosystems** | |
| Which type of ecosystems are being targeted in the record? | Terrestrial |
| | Aquatic |
| Which dimension of the Earth system is being targeted in the record? | Biosphere |
| | Atmosphere |
| | Cryosphere |
| | Hydrosphere |
| | Lithosphere |
| Which type of planetary boundary[1] does the record support? [1]Critical environmental boundaries to maintaining the Earth system within the functional space that ensures human well-being (Steffen et al. 2015) | Climate change |
| | Biodiversity loss |
| | Nitrogen cycle |
| | Phosphorus |
| | Water bodies acidification |
| | Land use |
| | Freshwater |
| | Ozone depletion |
| | Atmospheric aerosols |
| | Chemical pollution |
| **3. Ecosystem dimensions and biodiversity conservation** | |
| Is biodiversity conservation explicitly addressed in the record? | Yes |
| | No |
| Is the record potentially useful to inform on biodiversity conservation? | Yes |
| | No |
| Which ecosystem dimensions does the record focus on or is useful to assess? | Abiotic structure |
| | Biophysical structure |
| | Ecosystem function |
| | Ecosystem services |
| Are ecosystem services explicitly mentioned in the record? | Yes |
| | No |
| Is the record potentially useful to assess ecosystem services? | Yes |
| | No |
| Which ecosystem services (abiotic or biotic) does the record focus on or is useful to assess? | Specified based CICES v5.1 (Haines-Young and Potschin-Young 2018) |
| Which biodiversity dimension (i.e., composition, structure, or function) does the record focus on or is useful to assess (according to Noss 1990)? | Function |
| | Structure |
| | Composition |

(continued)

**Table 1** (continued)

| Questions | Categories |
|---|---|
| **4. Type of research and management contributions** | |
| Which type of research is shown in the record? | Applied |
| | Basic |
| | Use-oriented |
| Does the record present any explicit implications for management? | Yes |
| | No |

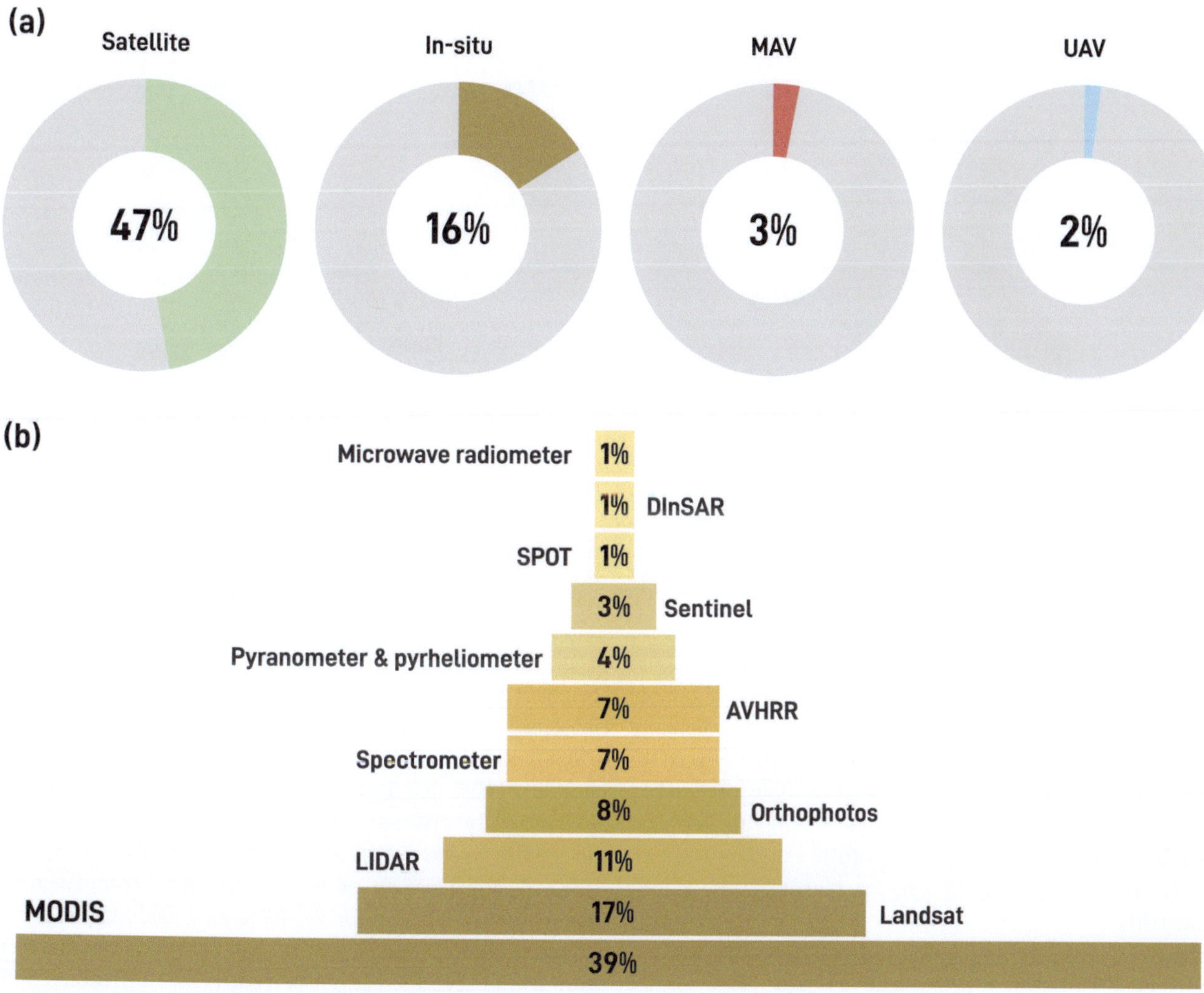


**Fig. 1** Proportion (%) of records according to different remote sensing platforms (**a**): satellite, in situ, manned (MAV), or unmanned aerial vehicle (UAV) and relying on different remote sensing products (**b**)

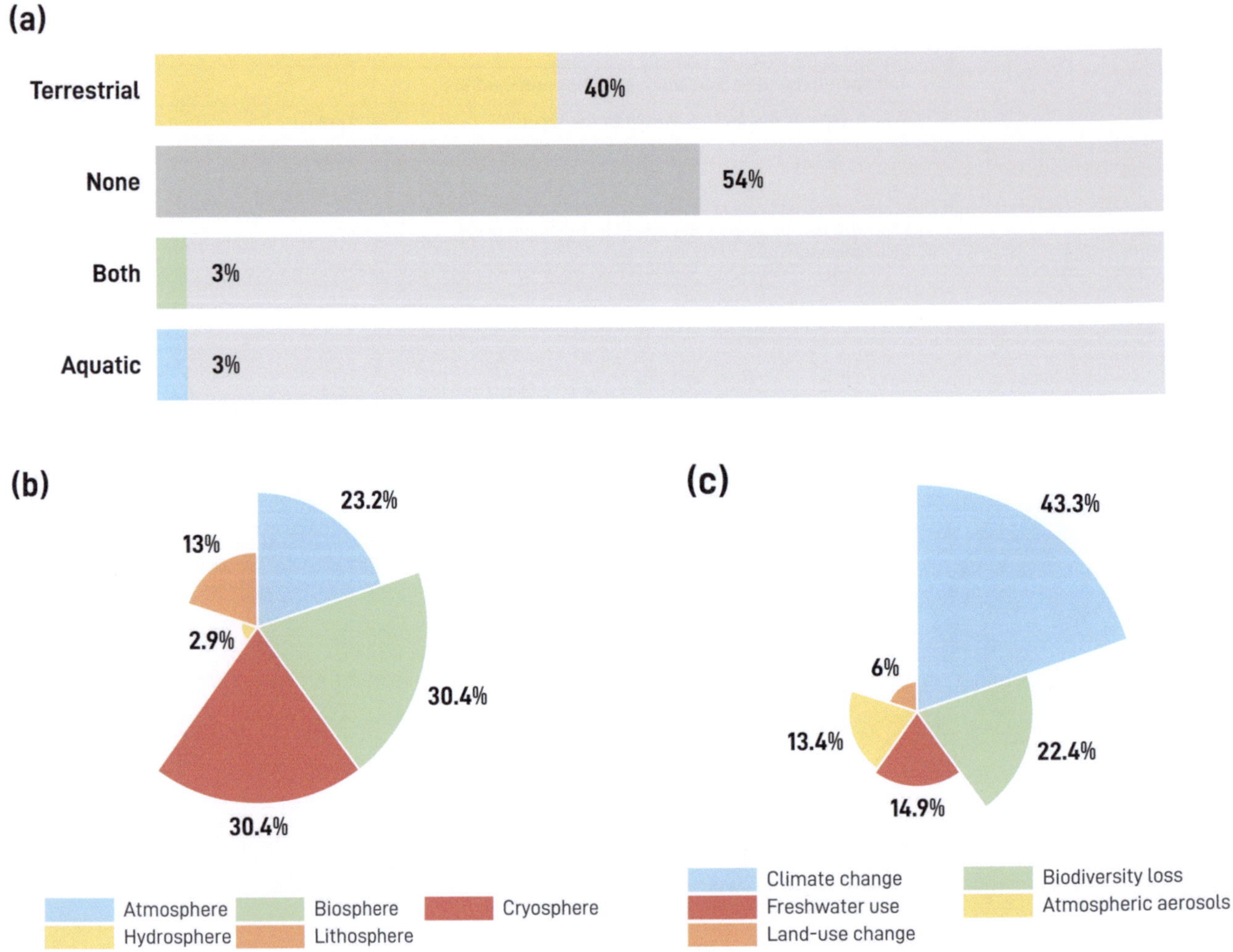


**Fig. 2** Proportion (%) of records per ecosystem type (**a**), and targeting different Earth subsystems (**b**) and planetary boundaries (Steffen et al. 2015) (**c**)

the potential of the retrieved records on the use of remote sensing in Sierra Nevada to assess the nine planetary boundaries, almost half of the records could be used to assess climate change (43%), followed by issues related to biodiversity loss (22%), freshwater use (15%), atmospheric aerosols (13%), and land-use change (6%; Fig. 2c). None of the records explicitly mentioned the planetary boundaries framework.

The vast majority of records did not explicitly address biodiversity (69%) nor ecosystem services (97% of all records) (not shown in figures). However, the findings presented in some of the retrieved records could potentially be useful for the management and conservation of biodiversity (20%), ecosystem functions (22%), or services (1%; Fig. 3 a). Focusing on those records that assessed or were found useful to infer on any dimension of biodiversity (alone or in combination with other/s dimension/s), 66% informed on biodiversity function, whereas 16% and 15% showed useful information for biodiversity structure, and composition, respectively (Fig. 3b). The great majority of records informed on the functional dimension of biodiversity alone (55%), or in combination with biodiversity composition or structure (12% and 11%, respectively). Records exclusively addressing biodiversity structure were very low (3%), with no studies exclusively addressing biodiversity composition (0%). A holistic view on biodiversity (i.e., comprehensive use of remote sensing for the integrated assessment of biodiversity composition, structure, and function) was found in only 8% of records. Eleven percent (11%) of records were not clearly found to be useful to address any biodiversity dimension. Records focused on ecosystem functioning mostly targeted the assessment of primary production and water and energy balance. The biophysical structure, and particularly, the assessment of land cover, was identified in 5% of records.

Records assessing ecosystem services or found useful to infer about them, mostly focused on abiotic ecosystem services (71%), specifically on provisioning (38%), regulating

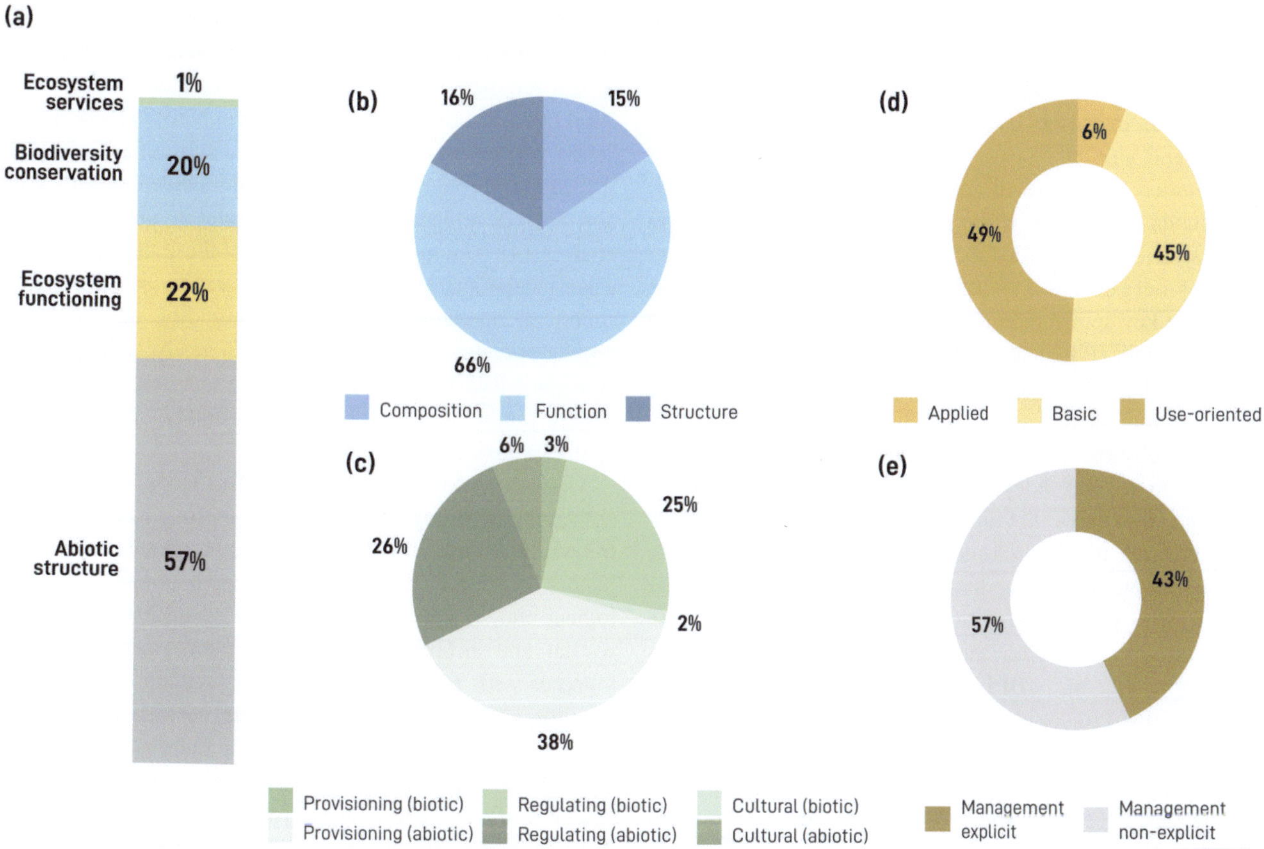


**Fig. 3** Proportion (%) of records potentially (but not explicitly) useful to address abiotic structure, ecosystem functioning, biodiversity conservation and ecosystem service (**a**), and the respective biodiversity dimensions (**b**) and types of ecosystem services (**c**). The figure also shows the proportion of records classified as use-oriented research, basic research, and applied research (**d**), as well as explicitly mentioning or not any management implications for the protected area (**e**)

(26%), and cultural services (6%; Fig. 3c). A high proportion of studies dealing with snow and water dynamics could be used to assess abiotic provisioning services related to "Surface water used for nutrition, materials, or energy" (31%), and abiotic cultural services related to "Physical and experiential interactions with natural environment." Many studies on atmospheric aerosols could be useful to assess abiotic regulating services related to "Regulation of physical, chemical, biological conditions" (23%). The production of solar energy was another abiotic provisioning service with a noticeable number of records (5%). A great proportion of papers also focused on biotic services (29%), being biotic regulating services the most abundant (25%), followed by provisioning (3%) and cultural services (2%; Fig. 3c). Mostly, biotic regulating services related to the "Regulation of physical, chemical, biological conditions," in particular with the "Maintenance of nursery populations and habitats" (20%).

Finally, most of the insights derived from remote sensing platforms and products were classified as use-oriented research (49%), or as basic research (45%). Only 6% of records were found to include applied research (Fig. 3d). Inevitably, more than half of the remote-sensing studies developed in Sierra Nevada Natural and National Park do not present or discuss any implications for managing the protected area (57%; Fig. 3e).

## 4 Discussion

### 4.1 Abiotic Processes

As expected, research on snow cover dynamics and trends was dominant in Sierra Nevada since it is a high mountain under a Mediterranean climate with a ski resort, and surrounded by irrigated agricultural fields and urban areas that frequently face snow and water scarcity (Climate Variability and Trends). Snow cover, snow depth, and snow melt have been extensively modeled in Sierra Nevada considering factors such as radiation, temperature, wind, relief, and vegetation cover and validated using both Landsat (e.g., Herrero et al. 2011; Pimentel et al. 2013) and MODIS (e.g., Pimentel et al. 2018), and Sentinel-1 and Sentinel-2 (e.g., Pratola and Navarro-Sánchez 2018) products. Snow effects on surface energy balance through albedo (Pimentel et al. 2016), on water balance through snow-water equivalent (Collados-Lara et al. 2020), on photosynthesis (Pérez-Luque et al. 2015), and on evapotranspiration, downstream irrigation use (Vivas et al. 2016), river discharge, and aquifer recharge (Jódar et al. 2018) have also been studied and modeled making use of satellite imagery. Past trends and future scenarios of snow cover in Sierra Nevada have been studied using remote sensing data (e.g., Collados-Lara et al. 2019). In situ terrestrial photography has also been key to validate the remote sensing products and model outputs (Pimentel et al. 2017; Polo et al. 2019). In summary, during the last two decades, remote sensing has been essential in developing the high level of knowledge reached on snow and water dynamics in Sierra Nevada, resulting in several independent calibrated models of snow and water dynamics that aim to be used in actual management. Overall, these studies show a decreasing trend in the persistence and extent of the snow cover area, being the precipitation regime, rather than the temperature trend, the most relevant driver on the snow regime forcing (Pérez-Palazón et al. 2015). In addition, the ongoing LACEM-LIM Project is assessing the use of satellite remote sensing together with social science (Advancing Open Science in Sierra Nevada: Current Citizen Science Campaigns) for the assessment of water level dynamics in the high mountain lakes of Sierra Nevada.

Atmospheric dust transportation (Alados-Arboledas et al. 2007; Israelevich et al. 2012) and aerosol properties, dynamics, and optical depth (e.g., Román et al. 2018) have also been extensively assessed and modeled mainly by making use of in situ remote sensing technologies, such as LIDAR and microwave radiometers (e.g., de Arruda Moreira et al. 2018), but also based on satellite imagery (e.g., Israelevich et al. 2012) and airborne sensors (Román et al. 2018). As a result, there is much-accumulated knowledge on the regional dynamics of dust transportation, aerosol properties, and their effects on air quality (de Arruda Moreira et al. 2018), ultraviolet radiation (Estellés et al. 2006), and spatio-temporal impacts on satellite vegetation indices (Reyes-Díez et al. 2015). Currently, most of this knowledge has not yet been transferred for the actual management of the Sierra Nevada protected area. However, dust intrusions and depositions have enormous effects on the composition, structure, and function of biodiversity in oligotrophic mountain lakes and rivers of Sierra Nevada. Hence, there is a huge potential for interdisciplinary and transdisciplinary research to quantify, characterize, and model the fertilization effect of aerosols, not only on aquatic, but also on the terrestrial ecosystems of Sierra Nevada. Remote sensing is expected to catalyze this collaboration under the LifeWatch SmartEcoMountains project.

Solar irradiance has been another abiotic factor widely studied in Sierra Nevada using remote sensing. A network of in situ remote sensors has been used to model solar irradiation over the complex terrain of Sierra Nevada (Bosch et al. 2008). Models of solar irradiation and subsequent effects on soil and atmospheric temperature and relative humidity have also been developed using MODIS satellite imagery (López et al. 2014). The inputs and outputs of these models, such as solar irradiation, precipitable water vapor, soil and atmospheric temperature, and relative humidity, have valuable implications for the management of Sierra Nevada protected area since they relate to direct abiotic services, such as solar energy production, and to biotic processes, such as photosynthesis and evapotranspiration.

Geomorphological studies related to landslides and paleoglacier morphology using remote sensing have also been frequent in Sierra Nevada. Landslides and floods have a long history in Sierra Nevada, showing high potential for ecological and socioeconomic (Jiménez et al. 2018). Both terrestrial and airborne LIDAR have been intensively used, together with orthophotography, to inventory, assess, and monitor multiple landslides in Sierra Nevada (e.g., Palenzuela et al. 2016). The use of satellite radar (i.e., D-InSAR) information has recently allowed the production of a landslide-risk map of Sierra Nevada (Jiménez-Perálvarez 2017), which has a direct application in the management of the protected area. In the case of glacier morphology, terrestrial LIDAR and airborne orthoimagery also allowed to characterize the glacial and postglacial evolution in Sierra Nevada (Gómez-Gutiérrez et al. 2014; Gómez-Gutiérrez et al. 2015; Gómez-Gutiérrez et al. 2016). Recently, based on the photointerpretation of very high-resolution satelite imagery (ESRI World and Google Earth imagery), a management-ready glacial and periglacial geomorphological map of Sierra Nevada has been produced (Palma et al. 2017).

## 4.2 Biodiversity Composition and Structure

The use of remote sensing to assess biodiversity and ecosystem functions and services have been restricted to satellite imagery so far. The scarcity of studies directly assessing biodiversity composition might be related to the relatively recent and expensive availability of very high-resolution imagery needed for the direct identification of species (but see Guirado et al. 2017; Blanco-Sacristan et al. under review). The severe weather conditions and abrupt relief of this high mountain have also restricted the use of UAV and MAV sensors. Nonetheless, the DETECTOR project is currently being developed (Alcaraz-Segura et al. in prep.) to fuse drone, airborne, and satellite very high-resolution imagery with deep learning neural networks to produce a wall-to-wall map of all individuals of several high-mountain shrub species in Sierra Nevada. The incorporation of satellite information into species distribution models is also opening an opportunity for the direct assessment of habitat suitability for many plants and animals at the species level (e.g., Arenas-Castro et al. 2018, 2019; Regos et al. 2019). In Sierra Nevada, we are only aware of the ongoing study of changes in the habitat suitability of *Moehringia fontqueri* Pau with the use of Landsat land surface temperature (Julio Peñas et al. in prep., personal communication).

Despite the low number of studies, remote sensing has been successfully used for the conservation status assessment of species and habitats. MODIS satellite products of vegetation indices and snow have been used to assess and model the effects of the temporal dynamics in primary productivity and snow on the demography, parasitism, forage availability, and ethology of *Capra pyrenaica* Schinz populations (Carvalho et al. 2015; Viana et al. 2018). The interannual variability and trends in MODIS vegetation indices have also been used to assess the vulnerability and conservation status of hotspots of endemic butterflies (Aragón et al. 2019) and of all southernmost remnant populations of *Quercus pyrenaica* Willd. of the Iberian Peninsula (Dionisio et al. 2012). The Global Change Observatory of Sierra Nevada has also developed an ontological system based on MODIS images to help in the assessment of the conservation status of Natura 2000 habitats (Pérez-Luque et al. 2015), intended to be used by managers for the Art. 17 Habitats Directive 6-year reporting (Schröder et al. 2013).

## 4.3 Ecosystem Processes and Functions

Ecosystem functioning has been by far the most frequent biodiversity dimension studied by remote sensing means in Sierra Nevada. All studies are restricted to the use of satellite estimators of primary production and range from the characterization of its seasonal dynamics, to the detection of long-term trends (Alcaraz-Segura et al. 2008a, 2008b; Alcaraz-Segura et al. 2009), and their local (Pérez-Luque et al. 2015; Alcaraz-Segura et al. 2016) and regional drivers (Lourenço et al. 2018). Satellite-derived descriptors of primary production dynamics have also been used to characterize ecosystem functional heterogeneity, diversity, and rarity to set geographic conservation priorities (Cabello et al. 2008, 2013; Cazorla et al. 2020). Remote-sensing estimators of primary production have dominated studies on Sierra Nevada so far. This finding may not be a surprise considering that primary production is at the bottom of the food chain and many ecological processes, and offers the most integrative response to environmental drivers (Virginia and Wall 2013). Still, many other aspects of ecosystem functioning (e.g., albedo, land surface temperature, evapotranspiration, water content, etc.) with clear and relevant ecological implications can also be assessed by remote sensing means (Arenas-Castro et al. 2019; Marcos et al. 2019, 2021). From our review, we could find only one study addressing long-term dynamics of land surface temperature using MODIS (Palade and Serrano 2014) and another one calibrating vegetation parameters to estimate evapotranspiration using Landsat (Carpintero et al. 2018). Incorporating these multiple aspects of ecosystem functioning to REMOTE, the official monitoring system of the Spanish National Parks Network Organism (Cabello et al. 2016), and ease the use of a variety of remote sensing products by scientists and managers is one of the objectives of the ongoing projects LIFE Adaptamed and LifeWatch SmartEcoMountains. At present, there is still no study explicitly addressing either biodiversity or ecosystem functioning and services in the aquatic ecosystems of Sierra Nevada using remote sensing. This gap should soon be filled by LifeWatch SmartEcoMountains project in collaboration with the ongoing LACEM-LIM project (Advancing Open Science in Sierra Nevada: Current Citizen Science Campaigns), which makes use of satellite remote sensing for the assessment of water level and chlorophyll and other pigments in the high mountain lakes of Sierra Nevada with Sentinel-2 and WorldView-3 imagery.

## 4.4 Ecosystem Services

Most of the remote sensing studies in our dataset were found to be useful for the assessment of biotic or abiotic ecosystem services in Sierra Nevada. However, only the article by Vaz et al. (2020) had as main purpose the actual assessment of final ecosystem services through the use of remote sensing means, and it was the only one addressing cultural services in Sierra Nevada. Most of the articles addressing ecosystem functions acknowledge that their study relates to intermediate provisioning or regulating ecosystem services (Potschin-Young et al. 2017), but none of them directly addressed final services as specified in the CICES classification (Haines-Young and Potschin-Young 2018). However, given the connection of intermediate services with final services via supply or production functions (Paruelo and Vallejos 2013), it is expected that much of the knowledge gained on ecosystem functions in Sierra Nevada could be soon translated to actual final ecosystem services.

## 5 Concluding Remarks and Research Gaps

Despite the vast investment on Earth Observation programs such as Copernicus in the European Union, the number of actual studies that make any use of remote sensing data or methods in Sierra Nevada is still very scarce. One of the reasons could be the high level of specialization required to use or produce remote sensing data together with the scarcity of undergraduate and postgraduate studies on remote sensing in southern Spain, particularly in the University of Granada, the research center that has produced most of the research on Sierra Nevada (see chapter "Scientific Knowledge Generated in Sierra Nevada: Bibliographic Review (1970–2021)" of this book). In addition, remote sensing experts are very much attracted to the private sector and, due to their transdisciplinary profiles, do not easily fit in the traditional disciplinary structure of university and research center departments.

One of the positive messages of our review is that most of the research conducted in Sierra Nevada via remote sensing means is potentially useful to assess biodiversity and ecosystem services, and that most of it was either applied or use-oriented research, frequently already containing recommendations for the management of the protected area. Hence, an increase in the interdisciplinary and transdisciplinary application of remote sensing to research in Sierra Nevada is expected.

**Acknowledgements** To all the researchers and naturalists that have been studying Sierra Nevada through all means and to the people that took care of its nature. This work was supported by projects LIFE Adaptamed (LIFE14349 CCA/ES/000612), DETECTOR (A-RNM-256-UGR18), EARTHCUL (PID2020-118041GB-I00), and LifeWatch SmartEcoMountains (LifeWatch-2019-10-UGR-01 Ministerio de Ciencia e Innovación-Universidad de Granada-FEDER, all co-financed by the ERDF Program. It also gained experience from the project ECOPOTENTIAL, which received funding from the European Union's Horizon 2020 Research and Innovation Program under grant agreement No. 641762; and from the NASA 2016 GEOBON Work Programme Grant # 80NSSC18K0446. This research was done within the LTSER Platform of Sierra Nevada/Granada—Spain (ES-SNE LTER_EU_ES_010) and under the agreement "Convenio de Colaboración entre la Consejería de Medio Ambiente y Ordenación del Territorio y la Universidad de Granada para el desarrollo de actividades vinculadas al Observatorio de Cambio Global de Sierra Nevada, en el marco de la Red de Observatorios de Cambio Global de Andalucía." AS Vaz was supported by Ministerio de Ciencia, Innovación y Universidades (Spain) through the 2018 Juan de la Cierva-Formación program [contract reference FJC2018-038131-I] and by the FCT—Foundation for Science and Technology through Stimulus of Scientific Employment, Individual Support (reference 2020.01175.CEECIND). SAC was funded by the PORBIOTA-Portuguese e-Infrastructure for Information and Research on Biodiversity Project (POCI-01-0145-FEDER-022127). This work contributes to the GEO BON working groups on Ecosystem Functions and Ecosystem Services.

## References

Alados-Arboledas L, Rascado JLG, Lyamani H, Navas-Guzmán F, Reyes FJO (2007) Characterization of the atmospheric aerosol by combination of LIDAR and sun-photometry. In: Proceedings of SPIE—the international society for optical engineering

Alcaraz-Segura D, Baldi G, Durante P, Garbulsky MF (2008a) Análisis de la dinámica temporal del NDVI en áreas protegidas: tres casos de estudio a distintas escalas espaciales, temporales y de gestión. Ecosistemas 17(3):108–117

Alcaraz-Segura D, Cabello J, Paruelo JM, Delibes M (2009) Use of descriptors of ecosystem functioning for monitoring a national park network: A remote sensing approach. Environ Manage 43(1):38–48

Alcaraz-Segura D, Cabello J, Paruelo JM, Delibes M (2008) Trends in the surface vegetation dynamics of the national parks of Spain as observed by satellite sensors. Appl Veg Sci 11(4):431–440

Alcaraz-Segura D, Di Bella CM, Straschnoy JV (2013) Earth observation of ecosystem services. CRC Press-Taylor and Francis Group, Boca Ratón, FL, USA, 518p

Alcaraz-Segura D, Lomba A, Sousa-Silva R, Nieto-Lugilde D, Alves P, Georges D, Vicente JR, Honrado JP (2017) Potential of satellite-derived ecosystem functional attributes to anticipate species range shifts. Int J Appl Earth Obs Geoinf 57:86–92

Alcaraz-Segura D, Reyes-Díez A, Cabello J (2016) Changes in vegetation productivity according to teledetection. In: Zamora Rodríguez RJ, Pérez-Luque AJ, Bonet FJ, Barea-Azcón JM, Aspizua R (eds) Global change impacts in Sierra Nevada: challenges for conservation. Universidad de Granada, Granada, pp 142–145

Aragón P, Sánchez-Fernández D, Hernando C (2019) Use of satellite images to characterize the spatio-temporal dynamics of primary productivity in hotspots of endemic Iberian butterflies. Ecol Indic 106:105449

Arenas-Castro S, Gonçalves J, Alves P, Alcaraz-Segura D, Honrado JP (2018) Assessing the multi-scale predictive ability of ecosystem functional attributes for species distribution modelling. PLoS One 13(6)

Arenas-Castro S, Regos A, Gonçalves JF, Alcaraz-Segura D, Honrado J (2019) Remotely sensed variables of ecosystem functioning support robust predictions of abundance patterns for rare species. Remote Sens 11(18)

Bosch JL, López G, Batlles FJ (2008) Daily solar irradiation estimation over a mountainous area using artificial neural networks. Renew Energy 33(7):1622–1628

Cabello J, Alcaraz-Segura D, Altesor A, Delibes M, Baeza S, Liras E (2008) Funcionamiento ecosistémico y evaluación de prioridades geográficas en conservación. Ecosistemas 17(3):53–63

Cabello J, Alcaraz-Segura D, Reyes A, Lourenço P, Requena JM, Bonache J, Castillo P, Valencia S, Naya J, Ramírez L, Ramírez L (2016) Serrada J (2016) System for monitoring ecosystem functioning of network of national parks of Spain with remote sensing | Sistema para el Seguimiento del funcionamiento de ecosistemas en la Red de Parques Nacionales de España mediante Teledetección. Rev Teledetec 46:119–131

Cabello J, Fernández N, Alcaraz-Segura D, Oyonarte C, Piñeiro G, Altesor A, Delibes M, Paruelo JM (2012) The ecosystem functioning dimension in conservation: insights from remote sensing. Biodivers Conserv 21(13):3287–3305

Cabello J, Lourenço P, Reyes A, Alcaraz-Segura D (2013) Ecosystem services assessment of national parks networks for functional diversity and carbon conservation strategies using remote sensing.

In: Alcaraz-Segura D, Di Bella CM, Straschnoy JV (eds) Earth observation of ecosystem services. CRC Press-Taylor and Francis Group, Boca Ratón, FL, USA, pp 179–199

Cabello J, Mairota P, Alcaraz-Segura D, Arenas-Castro S, Escribano P, Leitão PJ, Martínez-López J, Regos A, Requena-Mullor JM (2018) Satellite remote sensing of ecosystem functions: Opportunities and challenges for reporting obligations of the EU habitat directive. In: International geoscience and remote sensing symposium (IGARSS), pp 6604–6607

Carpintero E, González-Dugo MP, Jódar J, Martos-Rosillo S (2018) Use of canopy coefficients obtained from satellite data to estimate evapotranspiration over high mountain Mediterranean watersheds. In: Proceedings of the international association of hydrological sciences, pp 23–28

Carvalho J, Granados JE, López-Olvera JR, Cano-Manuel FJ, Pérez JM, Fandos P, Soriguer RC, Velarde R, Fonseca C, Ráez A, Pettorelli N, Serrano E (2015) Sarcoptic mange breaks up bottom-up regulation of body condition in a large herbivore population. Parasites Vectors 8(1)

Cazorla BP, Cabello J, Reyes A, Guirado E, Peñas J, Pérez-Luque AJ, Alcaraz-Segura D (2020) A remote sensing-based dataset to characterize the ecosystem functioning and functional diversity of a Biosphere Reserve: Sierra Nevada (SE Spain). Earth Syst Sci Data Discuss 2020:1–20

Chandrasekar K, Sai MVRS, Roy PS, Dwevedi RS (2010) Land Surface Water Index (LSWI) response to rainfall and NDVI using the MODIS Vegetation Index product 31(15):3987–4005

Chen JM, Chuvieco E, Wang M (2021) Preface, special issue of "50 years of environmental remote sensing research: 1969–2019." Remote Sens Environ 252

Chen JM, Liu J (2020) Evolution of evapotranspiration models using thermal and shortwave remote sensing data. Remote Sens Environ 237

Chuvieco E, Huete A (2009) Fundamentals of satellite remote sensing, CRC Press-Taylor and Francis Group, Boca Raton, FL, USA, 486p

Collados-Lara A-J, Pardo-Igúzquiza E, Pulido-Velazquez D (2020) Optimal design of snow stake networks to estimate snow depth in an alpine mountain range. Hydrol Process 34(1):82–95

Collados-Lara AJ, Pardo-Igúzquiza E, Pulido-Velazquez D (2019) A distributed cellular automata model to simulate potential future impacts of climate change on snow cover area. Adv Water Resour 124:106–119

Czúcz B, Haines-Young R, Kiss M, Bereczki K, Kertész M, Vári Á, Potschin-Young M, Arany I (2020) Ecosystem service indicators along the cascade: how do assessment and mapping studies position their indicators? Ecol Indic 118

de Arruda MG, Guerrero-Rascado JL, Bravo-Aranda JA, Benavent-Oltra JA, Ortiz-Amezcua P, Róman R, Bedoya-Velásquez AE, Landulfo E, Alados-Arboledas L (2018) Study of the planetary boundary layer by microwave radiometer, elastic lidar and Doppler lidar estimations in Southern Iberian Peninsula. Atmos Res 213:185–195

Dionisio MA, Alcaraz-Segura D, Cabello J (2012) Satellite-based monitoring of ecosystem functioning in protected areas: recent trends in the oak forests (*Quercus pyrenaica* Willd.) of Sierra Nevada (Spain). Int Perspect Glob Environ Chang 355:37

Duan P, Wang Y, Yin P (2020) Remote sensing applications in monitoring of protected areas: a bibliometric analysis. Remote Sens 12(5):772

Estellés V, Utrillas MP, Martínez-Lozano JA, Alcántara A, Alados-Arboledas L, Olmo FJ, Lorente J, de Cabo X, Cachorro V, Horvath H, Cañada J, García Y (2006) Intercomparison of spectroradiometers and Sun photometers for the determination of the aerosol optical depth during the VELETA-2002 field campaign. J Geophys Res Atmos 111(17)

Geller GN, Halpin PN, Helmuth B, Hestir EL, Skidmore A, Abrams MJ, Aguirre N, Blair M, Botha E, Colloff M (2017) Remote sensing for biodiversity. The GEO handbook on biodiversity observation networks. Springer, Cham, pp 187–210

Gómez-Gutiérrez Á, de Sanjosé-Blasco JJ, de Matías-Bejarano J, Berenguer-Sempere F (2014) Comparing two photo-reconstruction methods to produce high density point clouds and DEMs in the Corral del Veleta Rock Glacier (Sierra Nevada, Spain). Remote Sens 6(6):5407–5427

Gómez-Gutiérrez Á, De Sanjosé-Blasco JJ, Lozano-Parra J, Berenguer-Sempere F, De Matías-Bejarano J (2015) Does HDR Pre-processing improve the accuracy of 3D models obtained by means of two conventional SfM-MVS software packages? the case of the corral del Veleta Rock Glacier. Remote Sens 7(8):10269–10294

Gómez-Gutiérrez Á, Schnabel S, Conoscenti C, Carballo-Arias NA, Ferro V, Di Stefano C, Sanjosé Blasco JJ, De Matías J, Berenguer Sempere F (2016) Elaboración de modelos 3D de diferentes morfologías y escalas utilizando técnicas Structure-from-Motion y fotografías terrestres. Cuaternario y Geomorfol 30(1–2):23–35

Guirado E, Tabik S, Alcaraz-Segura D, Cabello J, Herrera F (2017) Deep-learning Versus OBIA for scattered shrub detection with Google Earth Imagery: Ziziphus lotus as case study. Remote Sens 9 (12)

Haines-Young R, Potschin-Young MB (2018) Revision of the common international classification for ecosystem services (CICES V5.1): a policy brief. One Ecosyst 3

Herrero J, Polo MJ, Miguel A. Losad MA (2011) Snow evolution in Sierra Nevada (Spain) from an energy balance model validated with Landsat TM data. In: Proceedings of SPIE—the international society for optical engineering

Hikosaka K, Tsujimoto K (2021) Linking remote sensing parameters to $CO_2$ assimilation rates at a leaf scale. J Plant Res 134(4):695–711

Hughes TP, Carpenter S, Rockström J, Scheffer M, Walker B (2013) Multiscale regime shifts and planetary boundaries. Trends Ecol Evol 28(7):389–395

Israelevich P, Ganor E, Alpert P, Kishcha P, Stupp A (2012) Predominant transport paths of Saharan dust over the Mediterranean Sea to Europe. J Geophys Res Atmos 117(2)

Jiménez-Perálvarez JD (2017) Landslide-risk mapping in a developing hilly area with limited information on landslide occurrence. Landslides 154 15(4):741–752

Jiménez L, Rühland KM, Jeziorski A, Smol JP, Pérez-Martínez C (2018) Climate change and Saharan dust drive recent cladoceran and primary production changes in remote alpine lakes of Sierra Nevada. Spain. Glob Chang Biol 24(1):e139–e158

Jódar J, Carpintero E, Martos-Rosillo S, Ruiz-Constán A, Marín-Lechado C, Cabrera-Arrabal JA, Navarrete-Mazariegos E, González--Ramón A, Lambán LJ, Herrera C, Herrera C, González-Dugo MP (2018) Combination of lumped hydrological and remote-sensing models to evaluate water resources in a semi-arid high altitude ungauged watershed of Sierra Nevada (Southern Spain). Sci Total Environ 625:285–300

Kennedy RE, Townsend PA, Gross JE, Cohen WB, Bolstad P, Wang YQ, Adams P (2009) Remote sensing change detection tools for natural resource managers: understanding concepts and tradeoffs in the design of landscape monitoring projects. Remote Sens Environ 113(7):1382–1396

López G, Batlles FJ, Rossi F (2014) Deriving solar radiation in a mountain area from MODIS information. In: Energy procedia, pp 1092–1099

Lourenço P, Alcaraz-Segura D, Reyes-Díez A, Requena-Mullor JM, Cabello J (2018) Trends in vegetation greenness dynamics in protected areas across borders: what are the environmental controls? Int J Remote Sens 39(14):4699–4713

Lovett GM, Burns DA, Driscoll CT, Jenkins JC, Mitchell MJ, Rustad L, Shanley JB, Likens GE, Haeuber R (2007) Who needs environmental monitoring? Front Ecol Environ 5(5):253–260

Marcos B, Gonçalves J, Alcaraz-Segura D, Cunha M, Honrado JP (2019) Improving the detection of wildfire disturbances in space and time based on indicators extracted from MODIS data: a case study in northern Portugal. Int J Appl Earth Obs Geoinf 78:77–85

Marcos B, Gonçalves J, Alcaraz-Segura D, Cunha M, Honrado JP (2021) A framework for multi-dimensional assessment of wildfire disturbance severity from remotely sensed ecosystem functioning attributes. Remote Sens 13(4):1–25

Nagendra H, Lucas R, Honrado JP, Jongman RHG, Tarantino C, Adamo M, Mairota P (2013) Remote sensing for conservation monitoring: assessing protected areas, habitat extent, habitat condition, species diversity, and threats. Ecol Indic 33:45–59

Noss RF (1990) Indicators for monitoring biodiversity: a hierarchical approach. Conserv Biol 4(4):355–364

Palade B, Serrano M (2014) Long-term high-frequency validation of MODIS LST products in a Mediterranean mountain environment. Int J Remote Sens 35(3):818–829

Palenzuela JA, Jiménez-Perálvarez JD, El Hamdouni R, Alameda-Hernández P, Chacón J, Irigaray C (2016) Integration of LiDAR data for the assessment of activity in diachronic landslides: a case study in the Betic Cordillera (Spain). Landslides 13(4):629–642

Palma P, Oliva M, García-Hernández C, Gómez Ortiz A, Ruiz-Fernández J, Salvador-Franch F, Catarineu M (2017) Spatial characterization of glacial and periglacial landforms in the highlands of Sierra Nevada (Spain). Sci Total Environ 584–585:1256–1267

Paruelo JM, Vallejos M (2013) Ecosystem services related to carbon dynamics: its evaluation using remote sensing techniques. In: Alcaraz-Segura D, Di Bella CM, Straschnoy JV (eds) Earth observation of ecosystem services. CRC Press-Taylor and Francis Group, Boca Ratón, FL, USA, pp 17–32

Pérez-Luque AJ, Pérez-Pérez R, Bonet-García FJ, Magaña PJ (2015) An ontological system based on MODIS images to assess ecosystem functioning of Natura 2000 habitats: a case study for *Quercus pyrenaica* forests. Int J Appl Earth Obs Geoinf 37:142–151

Pérez-Palazón MJ, Pimentel R, Herrero J, Aguilar C, Perales JM, Polo MJ (2015) Extreme values of snow-related variables in Mediterranean regions: Trends and long-term forecasting in Sierra Nevada (Spain). In: IAHS-AISH proceedings and reports, pp 157–162

Perry CH, Woodall CW, Liknes GC, Schoeneberger MM (2009) Filling the gap: Improving estimates of working tree resources in agricultural landscapes. Agrofor Syst 75(1):91–101

Pettorelli N, Wegmann M, Skidmore A, Mücher S, Dawson TP, Fernandez M, Lucas R, Schaepman ME, Wang T, O'Connor B, Jongman RHG, Kempeneers P, Sonnenschein R, Leidner AK, Böhm M, He KS, Nagendra H, Dubois G, Fatoyinbo T, Hansen MC, Paganini M, de Klerk HM, Asner GP, Kerr JT, Estes AB, Schmeller DS, Heiden U, Rocchini D, Pereira HM, Turak E, Fernandez N, Lausch A, Cho MA, Alcaraz Segura D, McGeoch MA, Turner W, Mueller A, St-Louis V, Penner J, Vihervaara P, Belward A, Reyers B, Geller GN (2016) Framing the concept of satellite remote sensing essential biodiversity variables: challenges and future directions. Remote Sens Ecol Conserv 2 (3):122–131

Pimentel R, Aguilar C, Herrero J, Pérez-Palazón MJ, Polo MJ (2016) Comparison between snow albedo obtained from Landsat TM, ETM+ imagery and the SPOT VEGETATION albedo product in a Mediterranean mountainous site. Hydrology 3(1)

Pimentel R, Herrero J, Polo MJ (2013) Estimating snow albedo patterns in a mediterranean site from landsat TM and ETM+ images. In: Proceedings of SPIE—the international society for optical engineering

Pimentel R, Herrero J, Polo MJ (2017) Quantifying snow cover distribution in semiarid regions combining satellite and terrestrial imagery. Remote Sens 9(10)

Pimentel R, Marín C, De Gregorio L, Callegari M, Pérez-Palazón MJ, Notarnicola C, Polo MJ (2018) Validating improved-MODIS products from spectral mixture-Landsat snow cover maps in a mountain region in southern Spain. In: Proceedings of the international association of hydrological sciences, pp 67–72

Polo MJ, Herrero J, Pimentel R, Pérez- MJ (2019) The guadalfeo monitoring network (Sierra Nevada, Spain): 14 years of measurements to understand the complexity of snow dynamics in semiarid regions. Earth Syst Sci Data 11(1):393–407

Potschin-Young M, Czúcz B, Liquete C, Maes J, Rusch GM, Haines-Young R (2017) Intermediate ecosystem services: an empty concept? Ecosyst Serv 27:124–126

Potschin-Young M, Haines-Young R, Görg C, Heink U, Jax K, Schleyer C (2018) Understanding the role of conceptual frameworks: reading the ecosystem service cascade. Ecosyst Serv 29:428–440

Pratola C, Navarro-Sánchez VD (2018) Snow cover monitoring in Hardangervidda and Sierra Nevada protected areas by using sentinel-1 time series. In: International geoscience and remote sensing symposium (IGARSS), pp 6227–6230

Reddy CS (2021) Remote sensing of biodiversity: what to measure and monitor from space to species? Biodivers Conserv 30(10):2617–2631

Regos A, Gagne L, Alcaraz-Segura D, Honrado JP, Domínguez J (2019) Effects of species traits and environmental predictors on performance and transferability of ecological niche models. Sci Rep 9(1)

Regos A, Gonçalves J, Arenas-Castro S, Alcaraz-Segura D, Guisan A and Honrado JP (in press). Mainstreaming remotely sensedecosystem functioning in ecological Niche models. Remote Sens Ecol Conserv

Reyes-Díez A, Alcaraz-Segura D (2015) Cabello-Piñar J (2015) Implications of quality filtering of Enhanced Vegetation Index (EVI) for ecosystem functioning monitoring | Implicaciones del filtrado de calidad del índice de vegetación EVI para el seguimiento funcional de ecosistemas. Rev Teledetec 43:11–29

Rogan J, Chen D (2004) Remote sensing technology for mapping and monitoring land-cover and land-use change. Prog Plann 61(4):301–325

Román R, Benavent-Oltra JA, Casquero-Vera JA, Lopatin A, Cazorla A, Lyamani H, Denjean C, Fuertes D, Pérez-Ramírez D, Torres B, Olmo FJ, Alados-Arboledas L (2018) Retrieval of aerosol profiles combining sunphotometer and ceilometer measurements in GRASP code. Atmos Res 204:161–177

Rose RA, Byler D, Eastman JR, Fleishman E, Geller G, Goetz S, Guild L, Hamilton H, Hansen M, Headley R, Wilkie D, Wilson C (2015) Ten ways remote sensing can contribute to conservation. Conserv Biol 29(2):350–359

Schröder C, Bonet Garcia FJ, Buck O, Corbane C, Deshayes M, Förster M, García-Millán V, Gil-Jimenez Y, Lang S, Ramos-Losada B, Simón-Colina A, Vanden-Borre J (2013) Validation of habitat assessment services in natura 2000 sites: streamlining the role of stakeholders from user needs to conservation status evaluation. In: GI_FORUM 2013: creating the GISociety. Verlag der Österreichischen Akademie der Wissenschaften, pp 548–551

Skidmore AK, Coops NC, Neinavaz E, Ali A, Schaepman ME, Paganini M, Kissling WD, Vihervaara P, Darvishzadeh R, Feilhauer H, Wegmann M, Wingate V (2021) Priority list of biodiversity metrics to observe from space. Nat Ecol Evol 5 (7):896–906

Steffen W, Richardson K, Rockström J, Cornell SE, Fetzer I, Bennett EM, Biggs R, Carpenter SR, De Vries W, De Wit CA, Reyers B, Sörlin S (2015) Planetary boundaries: Guiding human development on a changing planet. Science (80-)347(6223)

Turner W, Spector S, Gardiner N, Fladeland M, Sterling E, Steininger M (2003) Remote sensing for biodiversity science and conservation. Trends Ecol Evol 18(6):306–314

Vaz AS, Alcaraz-Segura D, Campos JC, Vicente JR, Honrado JP (2018) Managing plant invasions through the lens of remote sensing: A review of progress and the way forward. Sci Total Environ 642:1328–1339

Vaz AS, Moreno-Llorca RA, Gonçalves JF, Vicente JR, Méndez PF, Revilla E, Santamaria L, Bonet-García FJ, Honrado JP, Alcaraz-Segura D (2020) Digital conservation in biosphere reserves: earth observations, social media, and nature's cultural contributions to people. Conserv Lett 13(3)

Viana DS, Granados JE, Fandos P, Pérez JM, Cano-Manuel FJ, Burón D, Fandos G, Aguado MP, Figuerola J, Soriguer RC (2018) Linking seasonal home range size with habitat selection and movement in a mountain ungulate. Mov Ecol 6(1)

Virginia RA, Wall DH (2013) Ecosystem function, principles of. Encycl Biodivers Second Ed 90–95

Vivas G, Giráldez JV, Mateos L (2016) Water management in an ancestral irrigation system in southern Spain: a simulation analysis. Irrig Sci 34(5):343–360

Wang Y, Lu Z, Sheng Y, Zhou Y (2020) Remote sensing applications in monitoring of protected areas. Remote Sens 12(5):772

Westgate MJ, Likens GE, Lindenmayer DB (2013) Adaptive management of biological systems: a review. Biol Conserv 158:128–139

# Managing the Uniqueness of Sierra Nevada Ecosystems Under Global Change: The Value of in situ Scientific Research

Regino Zamora, Antonio J. Pérez-Luque, José Enrique Granados, Blanca Ramos, Javier Navarro, Javier Cano-Manuel, José Miguel Barea-Azcón, and Rut Aspízua

**Abstract**

Global change poses a challenge of great complexity for traditional systems of governance and nature management. New forms of management are required that contemplate both the adaptation of natural systems to the new conditions while trying to maximize their conservation and sustainable use in a global change scenario. In the context of decision-making and management, there is often a large amount of uncertainty, complexity, and interactions within and between human and environmental systems. Science provides society with the most objective and reliable knowledge to guide policymaking and management. Both the adaptation to new conditions and the conservation and sustainable exploitation of natural systems can be carried out if existing scientific knowledge is used. Below, we examine four case studies in which the generation of new knowledge in situ has been evident and has been applied to the improvement of environmental management, thanks to the numerous research projects of basic and applied science conducted in Sierra Nevada, but also thanks to the growing number of projects guided by the needs of environmental management. In this chapter, we summarize what has been learned to date with this philosophy of action, highlighting a series of management recommendations that, coming from collaborative experiences between scientists and managers in Sierra Nevada, can be applied in other ecosystems with similar characteristics. The selected case studies are: (1) Ibex management plan against diseases, (2) Recovery plan for high-mountain juniper scrublands, (3) Post-fire ecological restoration plan, (4) Naturalization plan for pine reforestation and development of decision-making tools.

**Keywords**

Decision-making • Usable knowledge • Scientific transference • Conservation and restoration • Adaptive management • Support-decision tools

## 1 Introduction

The environmental problems generated by global change constitute the most momentous challenge facing contemporary society. For the Mediterranean Basin, many climate models predict higher temperatures, less rainfall, and a general increase in climate variability, with more extreme temperature and precipitation events (Giorgi and Lionello 2008; see chapter "Climate Variability and Trends"). As a result, the Mediterranean Region is considered one of the most prominent climate-change hotspots in the world (Giorgi 2006; IPCC 2021). The effects of climate change are aggravated by long-term land-use change in Mediterranean ecosystems (Doblas-Miranda et al. 2017). The current challenge for environmental management is to incorporate global change as a new paradigm, both in the conceptual framework and in the concrete actions, developing new management tools to evaluate, monitor, and forecast the state of health and functioning of a given ecosystem.

In the context of decision-making and management, uncertainty, complexity and multifaceted interactions characterize the relationship between humans and environmental systems under the global-change scenario (Cash et al. 2003;

R. Zamora (✉)
Department of Ecology, University of Granada, Granada, Spain
e-mail: rzamora@ugr.es

R. Zamora · A. J. Pérez-Luque
Andalusian Institute for Earth System Research (IISTA), University of Granada, Granada, Spain

J. E. Granados · B. Ramos · J. Navarro · J. Cano-Manuel
Consejería de Agricultura, Ganadería, Pesca y Desarrollo Sostenible, Junta de Andalucía, Granada, Spain

J. M. Barea-Azcón · R. Aspízua
Agencia de Medio Ambiente y Agua, Consejería de Agricultura, Ganadería, Pesca y Desarrollo Sostenible, Granada, Spain


R. Zamora and M. Oliva (eds.), *The Landscape of the Sierra Nevada*,
https://doi.org/10.1007/978-3-030-94219-9_20

Nesshöver et al. 2017). Nevertheless, there is currently a sufficient body of ecological knowledge and technological capacity to guide effective environmental management in most ecosystems of the planet. Science provides society with objective and reliable information on which to base policy-making and managerial decisions. Scientific knowledge can also help us to predict likely future ecological responses, to design actions to maintain or even promote the ecological functions that provide key ecosystem services, and to accelerate adaptation to new conditions and thus favour ecosystem resilience.

There is ample evidence that the transformation of scientific evidence into "usable knowledge" for managers is neither automatic nor straightforward (Young et al. 2014). Traditionally, very few environmental policy decisions have been based on scientific evidence (Sutherland et al. 2004). The lack of communication between researchers and managers can cause a considerable lag between the time when a concept or theory is accepted by the scientific community and the time that the concept becomes familiar to technical and political staff. Time lags are especially troublesome when considering the global-change scenario, because most of the current managing criteria are based on outdated research but are still being used. In fact, all too many of our environmental policies and much of the street lore about the environment are still based on the science of the 1960s and 1970s (Lubchenco 1998). An additional problem is that, all too frequently, experiences in management developed in scientifically and technologically advanced countries have been transferred to other countries with different ecological conditions (climate, biotic communities, degree of human impact), as is the case of Mediterranean ecosystems. The negative results clearly indicate that most of the managing protocols derived from research in temperate and boreal ecosystems are not applicable to management and conservation in mediterranean environments (Jordano et al. 2002; Zavala et al. 2008; Zamora et al. 2016).

Adaptive management and scenario planning represent among the most promising conceptual and methodological approaches to ensure the implementation of proactive management in global-change scenarios (Williams 2011), since the objective is to create a robust, scientifically based tool for managing natural resources by maintaining all stakeholders in continuous interaction (Walters and Hilborn 1978; Holling 1978; Walters 1986). Given the lack of monitoring in traditional management, adaptive management with follow-up needs to be promoted (Lindermayer and Likens 2009). Once validated, these new techniques can be exported to the rest of the territorial matrix for immediate application. The active management of ecosystems in a global-change scenario implies developing a flexible management approach, since current forms of management may not work in the future. This makes it more necessary than ever to launch projects that assess the suitability of new and old management techniques under global change. A key element in adaptation to this change is to facilitate the self-organization capacity of ecosystems under the new environmental conditions. In this respect, we must explicitly recognize that ecosystems do not appear de novo. Instead, they emerge from within pre-existing systems that have already undergone profound changes. This fundamental idea was expressed for the first time by Odum (1962), who first used the concept of self-organization to explain how ecosystems responded to human pressure, viewing these synthetic or emerging systems as a natural outcome of the way the world was evolving with humans (Hobbs et al. 2006).

In this context, a major question to link research and management here and now is how the application of the knowledge acquired on the functioning of the ecosystems of Sierra Nevada can enable us: (1) to validate old and new management techniques in the new environmental scenarios arising from global change, (2) to develop projects aimed at promoting the mechanisms that drive the natural self-organization of ecosystems, (3) to use available environmental information as a support tool in decision-making, (4) to generate models validated with field data to forecast the dynamics of natural systems, and (5) to design the most appropriate management formulas in a changing scenario, exploring a wide spectrum of adaptation options.

Below we examine four case studies in which the generation of new knowledge in situ has been evident and has been applied to the improvement of environmental management, thanks to the numerous research projects of basic and applied science conducted in Sierra Nevada, but also thanks to the growing number of projects guided by the needs of environmental management. In this chapter, we summarize what has been learned to date with this philosophy of action, highlighting a series of management recommendations that, although coming from collaborative experiences between scientists from academic institutions (mainly from the University of Granada) and managers of the National and Natural Park of Sierra Nevada, can be applied in other ecosystems with similar characteristics. The selected case studies are: (1) Ibex management plan against diseases, (2) Conservation plan for high-mountain juniper scrublands, (3) Post-fire ecological restoration plan, (4) Naturalization plan for pine reforestation and development of decision-making tools.

## 2 Research and Management for Ibex Conservation

### 2.1 Identification of the Problem

Pathogens, viruses, bacteria, and parasites play an important role in the ecology of the hosts, regulating wild populations by acting in the selection process (Sommer 2005). The breakdown of the established balance, often triggered by changes in climatic conditions, may become detrimental to the functioning of ecosystems. In these scenarios, the emergence and/or re-emergence of the vector and zoonotic infectious diseases caused by these pathogens have a significant influence not only on wildlife but also on domestic livestock and finally on human health. Thus, there are many evident reasons to control wild faunal diseases. For example, human health may be at risk, domestic animals may also be threatened, wild species survival may be in danger, and the public opinion pushes in favour of control measures (Wobeser 2002).

Diseases are sensitive to climate, which influences the frequency and distribution and transmission vectors (fleas, ticks, mosquitoes, etc.). Wildlife is considered the most important reservoir for more than 70% of all emerging diseases that attack animals as well as humans. The interaction of these variables at a certain time and place results in epidemiological scenarios leading to the emergence and re-emergence of vector and zoonotic infectious diseases. These risks need to be better known and require new surveillance and prevention mechanisms. For this reason, the population of Sierra Nevada ibex (*Capra pyrenaica*) is being intensively monitored, not only regarding its population dynamics but also with respect to the diseases that affect it, integrating epidemiological surveillance with ecological, demographic, and reproductive considerations. In Sierra Nevada, the ibex maintains its most numerous and its most genetically diverse population in the Iberian Peninsula.

One of the main goat pathogens is *Sarcoptes scabiei*, which causes sarcoptic mange. This parasitic disease of cosmopolitan distribution afflicts a wide range of mammals, including humans, unevenly spreading within their populations (Pence and Ueckermann 2002). Sarcoidosis has been affecting several populations of ungulates in the Iberian Peninsula over the past 30 years (Moroni et al. 2021), including a large part of the Andalusian populations of ibex: Cazorla, Serranía de Ronda, and Sierra Nevada. In this latter mountain range, the earliest cases were detected in 1992 (Pérez et al. 1997), causing severe mortality. The Cazorla population suffered a mortality rate above 90% (León Vizcaino et al. 1999), and a similar trend was initially feared for the Sierra Nevada ibexes. However, no more than 200 individuals were found to be infected in a population over 10,000 (Fandos 1991), although with different dynamics.

## 2.2 The Way This Problem Had Traditionally Been Addressed

Disease management can be addressed from different strategic standpoints: prevention (impeding arrivals to a "clean" area), control (holding frequencies to tolerable levels), eradication (eliminating the disease and its causative agent from the ecosystem), or no intervention (passive management). In this context, a disease can be confronted by acting either on the etiologic agent (use of medication) or on the habitat (creating physical barriers), by managing host species (reducing population density) or modifying certain customary management measures affecting the disease (population reinforcements, re-introductions, or translocations) (Wobeser 2002).

## 2.3 Current Management Based on Research

The spatial dynamics of sarcoptic mange show sick animals occurring throughout the territory occupied by the population, so that mange-free areas do not exist. *Sarcoptes scabiei* plays a part in the ecosystem, usually being endemic, although it has epidemic cycles. Historical documents (Laza 1890) indicate that diseases influencing population dynamics have been recurring in the Sierra Nevada. Repeated contact between pathogens and hosts could have created resistance. This has been confirmed by capturing, tagging, and monitoring sick animals equipped with GPS-GSM collars (Alasaad et al. 2013; Fig. 1). Some studies involving experimental infestations have also been made, suggesting that new episodes of severe mortality by sarcoptic mange are unlikely, except in groups without previous contact with the parasite and, thus, not resistant to the disease (Castro et al. 2018; Ráez Bravo et al. in press).

Aware of this epidemiological problem, the Andalusian Regional Government implemented a protocol for the control and eradication of this disease. This started with a regional study on the status and distribution of the ibex as well as on the influence of sarcoptic mange together with other parasitic and infectious diseases that could be affecting these populations. Another measure was the construction of a fenced wild area where a number of individuals were kept in captivity in order to guarantee reproductive animals for the recovery of the population in case of severe mortality caused by this disease or any other stochastic factor. These management measures were conducted in conjunction with legislative measures ("hunting emergency" declaration), involving shooting sick animals to prevent infections and subsequent spread of disease (Ruiz Martínez et al. 1996). These measures are being applied, with specific modifications, in the wild ibex population of Sierra Nevada, and the information gathered is being used to improve disease management. This population has been monitored for the last 25 years and a number of parameters characterizing it have been identified (Granados et al. 2020).

The genetic variability of populations is key to coping with diseases. Several studies seeking to determine the variability of the Major Histocompatibility Complex

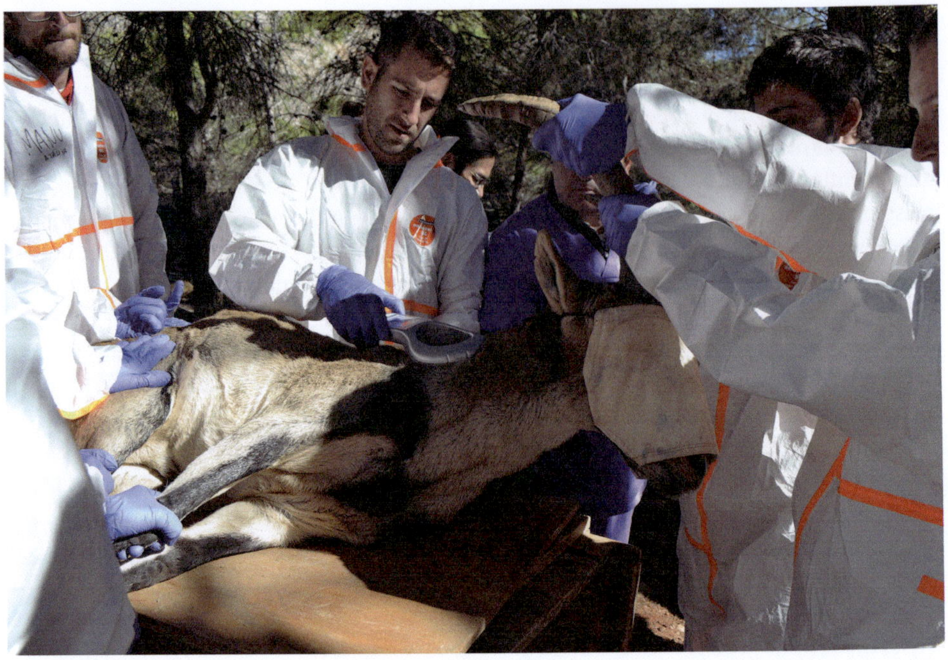

**Fig. 1** Sanitary inspection of a goat by staff of the Sierra Nevada National Park

(MHC) have been made (Amills et al. 2004; Alasaad et al. 2012a, b; Angelone-Alasaad et al. 2017, 2018). This complex gives rise to, for example, resistance to specific parasitism (Klein and Figueroa 1986; Trowsdale 1993; Bernatchez and Landry 2003). A correlation has been found between MHC diversity and the resistance to specific pathogens (Hedrick and Kim 2000). The bottlenecks undergone by several populations are responsible for their low variability (Mainguy et al. 2007). In two populations (Cazorla and Sierra Nevada), four alleles have been evidenced, three of them being shared. The results of one of the experimental infestations reveal that the alleles of the gene DRB1 of the MHC are not related to the resistance to sarcoptic mange (Ráez Bravo 2019). Studies on mitochondrial DNA highlight 25 different haplotypes occurring in the Andalusian ibex populations, 17 of these in Sierra Nevada (Márquez et al. 2020). Here, probably the higher genetic variability of this population is one of the factors explaining that this disease has had a milder effect than in the Cazorla population, which displays narrow genetic variability.

The group of captive ibex kept in the fenced-in area has been one of the mainstays of Ibex population management in Sierra Nevada. This facility has been increasingly used as a didactic resource and the centre for research on the biology and behaviour of the ibex (Espinosa et al. 2017). Thus, the fenced area with a captive population is being maintained and even enlarged, not only for conservation, but also for scientific investigation, testing new antiparasitic drugs. For example, to counteract the medicated meal, increasingly used to fight the sarcoptic mange, the farmaco-cynetics of orally supplied ivermectine has been studied. For that purpose, a number of individuals were selected to test how the concentration of ivermectine changed in blood and faeces, after supplying effective dosages. The results, even though tentative (Moroni et al. 2019), reveal concentration drops to below 20% the third day after medication delivery. This result appears to call into question the real therapeutic and prophylactic effect of orally supplied ivermectine to goats against parasites. Furthermore, these anti-parasitic drugs exert a negative effect on the environment in general and more directly on some insects. For example, the residues of the ivermectine affect the physiology of coprophagous beetles, causing high mortality both by toxicity and by compromising their defensive behaviour against predators.

Research to date has led to improved ibex management practice in Sierra Nevada. As regards the paradigm of Wobeser (2002) to implement prevention, control, eradication, and passive management, we discard the latter two measures. Instead, we adopt measures ranging from prevention to control, for instance reducing the probability of transmission and contagion of the disease by reducing densities, i.e. increasing hunting permits in the different game reserves around the Natural Park and adjacent areas. Above all, when a sick animal is encountered, individual tagging and monitoring is recommended instead of shooting if skin damage of the specimen affects less than 50% of the body surface. If injuries affect more than 50%, humanitarian shooting is preferable, since the animal may recover from the skin wounds but may suffer life-threatening damage to internal organs (Espinosa et al. 2020). New basic studies focusing primarily on disease interaction between wild and domestic herbivores, shared habitats, and prevention of more virulent strains arrive in a more and more evident scenario of global change are being undertaken.

## 3 Plan for the Conservation and Restoration of Juniper Scrublands

### 3.1 Identification of the Problem

Juniper (*Juniperus communis*) is a widespread gymnosperm in the Holarctic, but its distribution is progressively fragmented towards southern Europe, where it is located exclusively in high-mountain areas (García et al. 2000a). In Sierra Nevada, it appears mainly on north-facing slopes, between 1800 and 2800 m in elevation. Juniper scrublands have a high ecological value due to their capacity to retain soils, their abundant associated endemic flora and fauna, and their traditional use as summer grazing areas (Lorite 2002). Thus, these high-mountain scrublands are part of the Habitats (Directive 92/43/CEE).

Juniper scrublands in Sierra Nevada have been subjected to various types of anthropic disturbances for centuries, which have caused a regression in their distribution area (García et al. 1999a). The scrublands of Sierra Nevada have been used mainly for livestock, so the juniper was burned to increase the area for pasture. These fires reduce the regeneration capacity of the juniper, due to its inability to resprout. Also, junipers have been traditionally used as firewood by mountain herders and farmers. In recent decades the juniper scrublands have undergone new disturbances from the clearings for ski slopes. This type of disturbance directly affects regeneration, since it eliminates the surface layer of soil that contains organic matter and the seed bank. These clearings entail the removal of large rocks, which constitute natural landmarks in which seed-dispersing birds perch and seed rain is concentrated (García et al. 1999a; García 2001). These disturbances have caused a considerable reduction and fragmentation of the original cover of the juniper in Sierra Nevada, aggravating its regeneration problems.

### 3.2 The Way This Problem Had Traditionally Been Addressed

Juniper scrublands of Sierra Nevada have received meagre attention compared to other plant communities richer in endemic species (e.g., "Borreguiles", high-mountain meadow ecosystems; see chapter "Altitudinal Patterns and Changes in the Composition of High Mountain Plant Communities"), despite the potential importance of high-mountain scrublands as reservoirs of biodiversity and providers of multiple ecosystem services in the headwaters of river basins. Faced with this lack of information, we have conducted a series of studies in the last few decades aimed at compiling detailed scientific information in order to understand the ecological keys that help us to conserve and restore Sierra Nevada juniper scrublands.

### 3.3 Current Management Based on Research

Field investigations in the Sierra Nevada have provided crucial information for planning juniper conservation. Adult junipers in Sierra Nevada show great resistance to summer drought and snow cover as well as to browsing and trampling of herbivores (Herrero and Zamora 2014; Matías and Jump 2015). However, junipers show serious difficulties in regenerating their populations, with adult or senescent individuals dominating (García et al. 1999a, 2000a, b). This demographic pattern contrasts strongly with that of juniper populations that inhabit northern and central Europe, where juveniles may represent 50% of the entire population (García et al. 2000a). Juniper seeds have great difficulty in germinating, and their seedlings struggle to survive under the current ecological conditions in the high Mediterranean mountains (García 2001). This situation is the opposite of that occurring in more northern latitudes, with much wetter summers, where juniper populations present a high proportion of seedlings and juveniles (García et al. 2000a).

The Mediterranean high-mountains impose two fundamental restrictions on the regeneration ability of junipers: a limitation in the number of viable seeds, due to abortion and seed predation, and a limitation in the number of suitable microsites for the establishment of juniper seedlings, due to the high seedling mortality caused by the summer drought (García 2001, 2000b). Both factors determine a very low capacity for natural regeneration (García et al. 2000a) under the climatic conditions of the Mediterranean mountains. On the contrary, this juniper benefits from the mutualistic interaction that it maintains with the birds (*Turdus* ssp., see chapter "Responses of Animal Populations and Communities to Climate Change and Land-Use Shifts" of this book) that disperse its seeds on consuming the less depredated fruits (and, consequently, with a higher proportion of healthy seeds, Garcia et al. 1999b; Zamora 1990). Also, the seeds are preferentially dispersed from rocks that the birds use as perches and near water sources, which are precisely the microhabitats with the greatest recruitment success (García et al. 2000a, García 2001). At the demographic level, *J. communis* is a widely distributed species which maintains relict populations in the Mediterranean mountains thanks to great adult longevity (remnant dynamics, García and Zamora 2003).

#### 3.3.1 Lessons from Restoration Experiments

From an applied perspective, it is critical to diagnose the natural recovery potential of juniper scrublands. Under suitable ecological conditions, juniper populations can recover naturally (passive restoration). On the contrary, if the ecological conditions are unsuitable, passive restoration will not suffice, and actions will have to be taken to revitalize the process.

**(a) Passive restoration**: Junipers could potentially recolonize areas disturbed by fires or ski slopes. To test whether this potential regeneration capacity is translated into a real regeneration, we studied two areas that had undergone anthropic disturbance, one due to fire and the other due to the opening of a ski slope. To study the colonizing potential of the juniper after the fire, we chose a plot in Loma del San Juan (2200 m), on a hillside covered by a juniper scrubland that was partially burned in 1983. The plot chosen to study the recolonization of an abandoned ski slope was located in Campos de Otero, between 2200 and 2550 m in elevation. The area had been ploughed for use as a ski slope in 1978, installing a ski lift 900 m long and 35 m wide. After the two disturbances (the fire or the opening of the ski slope) no other human intervention occurred, apart from livestock grazing. Therefore, it can be considered that it has followed a dynamic of natural recolonization.

To collect data on the cover of the juniper and its demographic structure, we laid out transects within the burned area/ski slope and in the surrounding undisturbed juniper scrubland (Zamora et al. 1996). We placed 10 transects of 50 × 2 m (100 $m^2$) equally distributed in the burnt area/ski slope and 10 similar transects in the nearby undisturbed juniper scrubland to serve as a control (Zamora et al. 1996). To assess the vegetation cover, we used the intercept point method (150 points per transect), so the cover was estimated as the percentage of points occupied by a given vegetation type. Also, juniper density was calculated by counting all the individuals in each transect. The transects were carried out in the early summer of 1995 and 2021.

Despite the long time since the construction and subsequent abandonment of the ski slope (more than 40 years), this area bears hardly any junipers at present, despite the fact that it crosses the biggest juniper population of the Sierra Nevada. Juniper coverage is now 3% on the ski slope, all junipers are juveniles or small adults, while in 1995 no junipers were found. On the contrary, the coverage of *Genista versicolor* has increased remarkably since 1995, so a good part of the bare ground in 1995 is now covered by this shrub species. In the burnt area, *G. versicolor* coverage has expanded from 15% in 1995 to 73% today, while juniper coverage has gone from 0% to only 6% (Fig. 2).

Although the overall juniper coverage in undisturbed juniper scrubland remains stable or even spreads (Fig. 2), this spread occurs thanks to the growth of already established adult individuals. In 1995, juveniles represented 10% of the individuals in the population (Zamora et al. 1996) in these unaltered juniper sectors, while in 2021, practically no juveniles were found. This limitation in population recruitment corresponds closely to the very low germination and seedling establishment success found in the sowing experiments described below. In short, the regeneration of Sierra Nevada juniper has worsened both in disturbed and the undisturbed juniper scrublands. On the contrary, *G. versicolor* has steadily increased its coverage since 1995, colonizing 35% of the undisturbed juniper scrubland (see Fig. 2).

**(b) Active restoration**: The conservation and restoration of the juniper in the high-mountains of Sierra Nevada requires not only the conservation of the already established populations, but also the active restoration of the juniper populations, by sowing seeds in the appropriate microhabitats.

To evaluate seed germination and juniper seedling survival in different microhabitats, we performed sowing experiments on the northern slope of Sierra Nevada. The sowings were conducted in two different periods (1996–1998 and 2017–2019), and the germination and survival of seeds were analysed. Seeds were sown in 4 microhabitats (see García 2001): Juniper (under the canopy of junipers); Rocks (under the shadow of perching rocks); Open ground (open inter-spaces between woody vegetation), and Wet meadow (close to permanent streams). Seedling survival was analysed after summer each year. The microhabitats where juniper has the greatest recruitment success is traditional irrigation channels) and Rocks (under the shadow of perching rocks), see Fig. 3. The differences in seed germination and seedling survival observed (Fig. 3) was due to the different climatology of the years in which the sowings were made: rainy in the case of years 1996 and 1998, with more than 835 mm of precipitation per year (García 2001) and dry in years 2018, 2019, 2020, with around 410 mm of precipitation per year (with an average cumulative precipitation of 577 mm from 1941 to 2019 http://climanevada.obsnev.es). This highlights the importance of climate as a modulator of the regeneration capacity of juniper. In fact, greater aridity in recent years has collapsed recruitment, which is manifested also in the current demographic structure of the juniper, where there are hardly any juveniles. Our data lead to the conclusion that any management involving the burning or uprooting of juniper scrubland in the high mountains of the Sierra Nevada gives rise to bare soils for many years, even decades, after the disturbance. The summer aridity of the Mediterranean climate, exacerbated in recent decades, has practically collapsed the regeneration capacity of the juniper. This climatic limitation would also hinder possible attempts of active restoration. Therefore, the most effective and least expensive option for the conservation and management of juniper scrublands would be to concentrate more on efforts to maintain areas that still survive than on attempts to

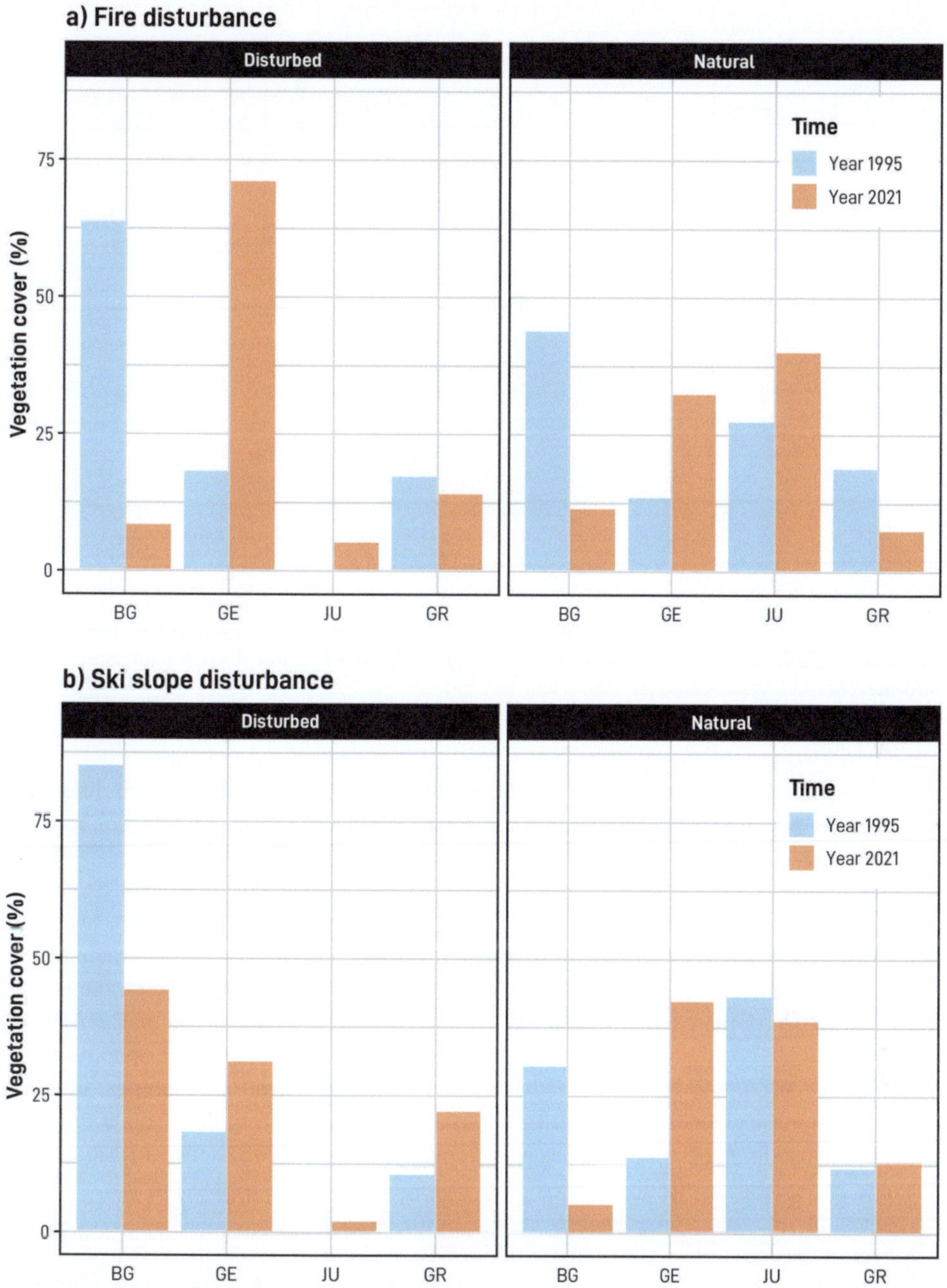


**Fig. 2** Bar plot for vegetation cover of disturbed habitat by fire and ski slope (**a** and **b**) and undisturbed (Natural) juniper scrubland as a control at two study sites (Campos de Otero and Loma del San Juan) and two periods (1995 and 2021). X-axis abbreviations: BG = Bare ground GE = *Genista versicolor*, JU = *Juniperus communis*, GR = Grasses

recover those that have already been lost. In fact, many plants now occurring as relict and endemic in the Mediterranean area are long-lived, stress-tolerant species inhabiting low disturbance habitats such as high-mountains (García and Zamora 2003). For them, persistence might be a significant strategy imposing biological inertia against extinction. Persistence of geographically peripheral populations may play a keystone role, as is the case in *J. communis* in the Mediterranean mountains, considered to be an umbrella species for herbaceous plants and invertebrate communities, many of them endemic.

Based on the knowledge acquired on the ecology of the juniper scrublands of Sierra Nevada, we offer the following recommendations for management: (1) protection of established individuals and populations against human disturbances such as clearing or burning, (2) maintenance of traditional "careo" ditches as natural wet regeneration windows (see also chapter "Singular Cultural Landscapes of the Sierra Nevada"), and (3) Conservation of dispersing bird populations, especially in the current situation where their populations have decreased a lot compared to 30–40 years ago (enhancing habitat connectivity from breeding areas and

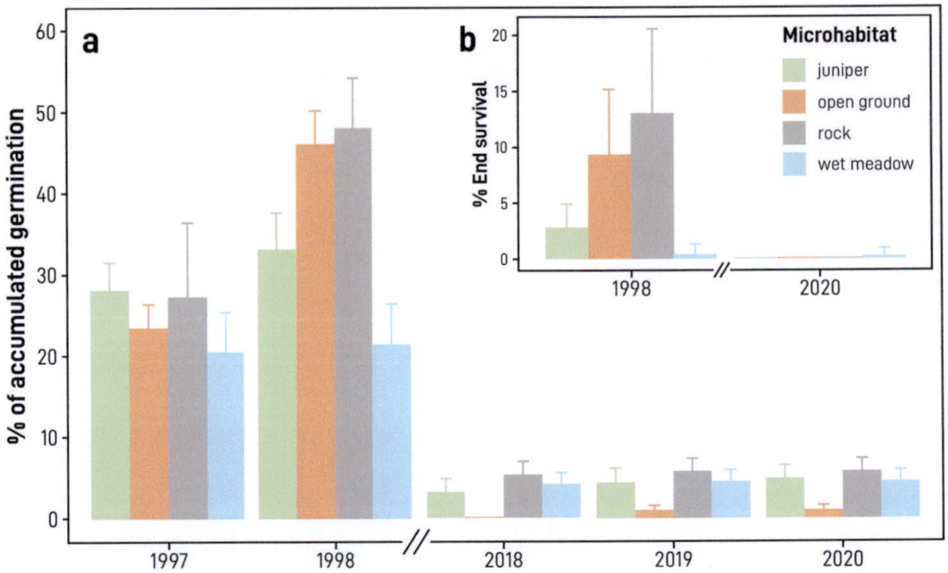


**Fig. 3** **a** Cumulative seed germination of *Juniperus communis* seeds sown in different microhabitats for each of the years studied. **b** Survival values at the end of the period (three years of monitoring) analysed for each microhabitat. Mean values are shown and bars indicate standard errors. The Cumulative losses in seed germination and seedling survival in a typical dry year are impressive: for example, from a pool of 1572 seeds dispersed by birds, there are only 51 viable seeds. From these, only 5 seeds germinated in May–June, and only one seedling survived to summer drought

minimizing hunting and harassment, see chapter "Responses of Animal Populations and Communities to Climate Change and Land-Use Shifts").

## 4 Plan for Post-Fire Ecological Restoration

### 4.1 Identification of the Problem

Wildfires are becoming unprecedented in their frequency, extent, and intensity (Seidl et al. 2017; Leverkus et al. 2021). Climatic change interacts with wildfires in multiple ways, for instance by extending the fire season and increasing the frequency of dry years (Westerlingh et al. 2006). Recent wildfires in Australia and California exemplify how drought magnifies fire propagation and intensity, leads to fire spreading to non-flammable ecosystems, increases smoke that impairs human health, and undermines the capacity of ecosystems to recover (Enright et al. 2015; Wintle et al. 2020). Scientific research has promoted policies for a healthier coexistence with fire (Turner et al. 2003; Moritz et al. 2014), including the use of prescribed fire to simulate natural processes and the creation of heterogeneous landscapes in restoration programmes to enhance regeneration in case of fire. Wildfires can provide a window of opportunity in which scientists and forest managers can take action to restore degraded ecosystems (Leverkus et al. 2019). For example, dense, monospecific pine plantations from the twentieth century, which are often unproductive because of climate and soil limitations, abound across the Mediterranean Region (Pausas and Fernández-Muñoz 2012). They were originally planted to retain soil, but they now generate many ecological challenges, such as increasing the risk of high-intensity fire and hindering the development of native vegetation (Pausas and Fernández-Muñoz 2012). Wildfires may thus release the native vegetation from the competition of the dense pine canopy (Leverkus et al. 2016) and provide management opportunities to favour more diverse and resilient communities, including species adapted to future fires and to drier conditions (Leverkus et al. 2021). In September 2005, a fire in Sierra Nevada affected nearly 3000 ha of reforested pines.

### 4.2 The Way This Problem Had Traditionally Been Addressed

Afforestation has traditionally followed a uniform plantation pattern applied to the entire stand. The scheme routinely consisted of one or a few tree species, which in terms of ecological succession has the effect of going directly from the initial stage to the last one. Given the lack of scientific criteria based on how to act, usually a new plantation is made again following similar criteria that were applied in the plantation that burned. By forestry tradition, the plantation model was aimed fundamentally at avoiding erosion and producing wood, and the species used in Spain were usually conifers (*Pinus* ssp.).

### 4.3 Current Management Based on Research

After the Lanjarón Fire, we planned a realistic ecosystem-restoration project, seeking a balance between the natural history of Sierra Nevada based on the composition, structure, and functioning of its ecosystems on the one hand, and the current and future scenarios of global change on the other hand. An ambitious post fire restoration programme was implemented by the managers of the Sierra Nevada Protected Area in collaboration with researchers from the University of Granada. The design of this project applied the most recent scientific knowledge to promote a new model of ecosystem restoration, favouring the multiple uses of the forest and promoting the diversity of ecosystem services. The restoration project itself both applied ecological principles based on existing science but also tested novel ideas and was used for research. The overall design aimed to recover an open forest, converting it to one with discontinuities, heterogeneous spatial structure, and diverse composition, in order to make it more resistant and resilient to possible disturbances such as fires, pests, and severe episodes of drought. The key points of our restoration project are: (1) To promote spatial heterogeneity, (2) To encourage phylogenetic and functional diversity, and (3) To promote the natural recovery capacity, thereby fostering ecological interactions that accelerate ecological succession (facilitation, mutualisms) and community self-organization under the new scenarios (see Fig. 4). Perhaps the most novel of these actions has been the combination of different types of treatments according to the ecological characteristics and the history of management in the area. In order to promote natural regeneration, stands have been planted with irregular shapes and different species composition.

#### 4.3.1 Promoting Spatial Heterogeneity

The basic objective of our restoration plan was to create a mosaic of vegetation that increases the biodiversity and variability of habitats in the area, resulting in a diverse landscape with a greater capacity to adapt to climatic change. Our restoration project takes advantage of natural succession dynamics. To promote natural regeneration, we planted stands containing a mixed species composition, combining shrub and tree species. In doing so, we created a mosaic of adjacent patches bearing a diverse assemblage of species belonging to different successional stages (Fig. 4).

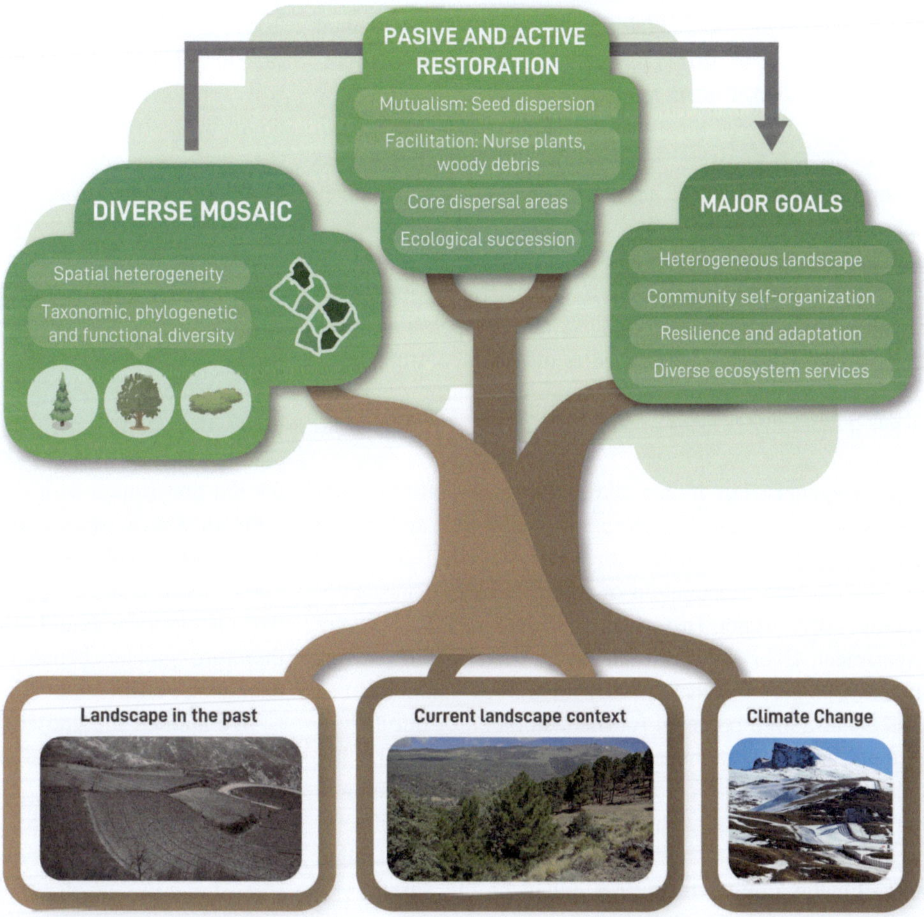


**Fig. 4** Conceptual post-fire restoration model applied in Sierra Nevada. Starting from the promotion of spatial heterogeneity and the diversity of species, the project takes advantage of the scientific knowledge that existed at the time on the possibilities and limitations of passive and active restoration in Mediterranean mountains, to generate a community of woody plants that is self-organizing, adapts better to new climatic conditions, and offers a wide range of ecosystem services in a scenario influenced by land-use change and climatic change (roots of the tree in the figure)

### 4.3.2 Encouraging Phylogenetic and Functional Diversity

It is increasingly acknowledged that, by going beyond the species concept, conservation, and restoration policies can be improved (Cadotte et al. 2011). The integration of the different dimensions of diversity (taxonomic, functional, and phylogenetic) into a multifaceted approach constitutes a nascent and promising avenue for restoration (Petchey and Gaston 2006). In our restoration project, we introduced the greatest feasible taxonomic, functional, and phylogenetic diversity into the ecosystem (Fig. 5), to encourage the plant community to self-organize, according to the balance of ecological interactions (e.g., competition, facilitation, herbivory) and the degree of adaptation of the different species to the new climatic conditions. The species used for planting belonged to different successional stages (trees and shrubs) as well as to different families and genera of woody plants common in the area, such as *Quercus ilex* subsp. *ballota Pinus pinaster* subsp. *acustiquama*, *Berberis hispanica*, *Crataegus monogyna*, *Rosa* spp., and *Rhamnus lycioides,* among many others.

### 4.3.3 Enhancing Natural Regeneration Capacity

In the project, different new restoration techniques were applied, such as using pioneer shrubs (Gomez-Aparicio et al. 2004; Castro et al. 2002, 2004; Zamora et al. 2004) and plant debris (Castro et al. 2011) as planting microsites, and creating core dispersal areas to strengthen the recovery of the most degraded areas after the fire. The restoration project applied for the first time in 2006 the concept of dispersal nuclei, which consists of an association of woody species that establish well in open sites and then grow and reproduce rapidly (e.g., *C. monogyna* and *B. hispanica*) to constitute islands of vegetation within the burnt matrix, favouring the attraction of pollinating insects and dispersing birds (Rey-Benayas et al. 2008). The birds also tend to bring seeds to the restored site from the surrounding unburnt areas and enrich it naturally (Castro et al. 2010, 2012; Leverkus et al. 2016).

**(a) Regeneration under plant debris**: A current controversial issue among restoration ecologists and forest managers is the appropriate management of dead trees after fire (see chapter "Restoration of Mediterranean Forest Ecosystems After Major Disturbances: The Lanjarón Post-fire Experiment Over 15 Years of Succession"). The use of branches, logs, or other woody debris as natural nurse objects has a high potential for the restoration of burnt sites, as they provide the benefit of a shading overstory but without underground competition. The burnt trunks and branches act as nurse structures that reduce the soil temperature and the water stress of the plants, thus producing a microclimatic improvement that enhances the survival of seedlings and saplings (Castro et al. 2011). This may reduce water deficit for seedlings (either naturally established or planted), generating a process of facilitation with high potential to benefit vegetation restoration in water-stressed ecosystems. The use of branches as ground cover also offers the advantage of facilitating tree regeneration without adding underground root competition.

Additionally, burnt wood acts as a major reservoir for nutrients that are incorporated into the soil (Marañón-Jiménez et al. 2011; Marañón-Jiménez and Castro 2013; Juan-Ovejero et al. 2021, and the structural complexity generated by trunks and branches scattered throughout the soil protects the juveniles of woody species against the trampling hooves of herbivores. Thus, branches act as nurse objects that may assist forest restoration. To explore the potential benefits of dead wood at ecosystem level, an ambitious long-term field experiment was implemented to study the effect of post-fire salvage logging in cooperation with the local forest service of the National Park (see chapter "Restoration of Mediterranean Forest Ecosystems After Major Disturbances: The Lanjarón Post-fire Experiment Over 15 Years of Succession").

**(b) Regeneration of thickets**: The first investigations aimed at using pioneer "nurse" plants for restoration purposes were conducted in Sierra Nevada at the end of the 1990s (Castro et al. 2002, 2004; Gómez-Aparicio et al. 2004; Zamora et al. 2004). It was found that, in Mediterranean settings, the environmental change (microclimate, soil fertilization, protection against herbivores) brought about by a neighbouring plant can offer benefits that exceed the costs that spatial proximity entails, thereby fostering facilitation interactions (García et al. 2000c; Gómez et al. 2001; Gómez-Aparicio et al. 2005a, b, c, 2006, 2008; Zamora et al. 2001; Castro et al. 2002; Baraza et al. 2006, 2007). In fact, many pioneer scrub species produce regeneration niches for diverse tree species in Mediterranean mountains, where summer drought and herbivory pressure are considerable (Fig. 5). To develop a reforestation technique that reproduces the natural pattern of regeneration associated with woody pioneer shrub species, we performed for the first time a series of field experiments from 1997 to 2001 in Sierra Nevada on a large scale (Gómez-Aparicio et al. 2004), with very positive results (Castro et al. 2002, 2004, 2006; Gómez-Aparicio et al. 2004). Even more positive effects were found when abiotic stress intensified, and therefore greater benefits are expected under a climate change scenario.

We applied the knowledge gained from these pioneering experimental plantations (Castro et al. 2002, 2004, 2006; Gómez-Aparicio et al. 2004; Zamora et al. 2004) in our postfire restoration project in Lanjarón. Fortunately, much of the burned area was very soon covered with pioneer shrubs (e.g., Leverkus et al. 2014). This allowed us to study the survival and growth of the tree seedlings associated with

**Fig. 5** Tree sapling (*Quercus pyrenaica*) associated with shrubs (*Berberis hispanica*) in Sierra Nevada. In Mediterranean mountains, seedlings and saplings of many tree species are facilitated by shrubs. Photograph by Francisco Bruno Navarro

shrubs. The results indicated that the presence of pioneer shrubs generally favoured the survival and growth of many seedlings that germinated after the fire, as well as many of those planted in the restoration work, although a markedly different effect was noted depending on the nurse species. For example, the thorny nurse species *Ulex parviflorus*, *G. versicolor* and *B. hispanica* facilitated restoration, improving the survival rates of all woody species (either naturally established or planted) by substantially reducing ungulate herbivory (Zamora et al. 2016). Specifically, the species that registered the highest survival rate were *Rosa canina* and *C. monogyna*, especially at low elevations. *B. hispanica* and *Q. ilex* were the species that best responded at the high elevations (Leverkus et al. 2015). In general, shrubby species survived better than did tree species (Matías et al. 2012; Zamora et al. 2016). This diversity of responses to different ecological conditions forms the basis for recovering a more resilient vegetation adapted to new climatic conditions.

Overall, the results clearly showed improved survival and growth of seedlings of many woody species associated with nurse plants across a wide range of environmental scenarios, and the role as "nurses" of different pioneer species. The results also indicate that the burnt trunks and branches act as nurse structures, reducing soil temperatures and alleviating the water stress of the plants. The burnt wood also acts as a major reservoir of nutrients that are gradually transferred to the soil. Also, the structural complexity that the trunks and branches generate by being scattered over the ground protects the juvenile woody plant species against ungulate herbivores. This translates as a higher recruitment rate and more vigorous growth of saplings, whether natural or planted. The self-organization of the woody community is evident 16 years after the Lanjarón fire, in some plots the growing arboreal species predominate while in many others the shrub species are the dominant ones. The ecological succession is different in each stand depending on the legacies, the role of interactions (facilitation, mutualism, herbivory, seed dispersal), and the actions that were carried out in the restoration project. The result is a mosaic landscape with stands of irregular size differing in species composition and successional stage. Long-term monitoring is needed to assess the landscape-scale implications of different types of management to obtain optimal landscape configurations that maximize the delivery of different ecosystem services.

## 5 Plan for Naturalizing Pine Plantations and Developing Decision-Making Tools

### 5.1 Identification of the Problem

Forest plantations are an example of widespread intensive and extensive land-use change shaping terrestrial ecosystems (Hobbs et al. 2006; Chazdon 2008). For decades, mountainous areas of Mediterranean Europe have been reforested with trees after the abandonment of croplands and pastures (Gerard et al. 2010). In Spain, 3.8 million hectares were converted to pine plantations between 1945 and 1986 to reverse the intense deforestation existing at that time (Ortuño 1990). In their current state, these dense plantations represent areas of high risk for the generation and spread of large fires and are commonly invaded by forest pests. Furthermore, these impacts are being exacerbated by current climate change that in the last term is causing massive decay processes. In this situation, it is essential to carry out a profound reassessment of the composition, structure, and functions of the reforested pine forests. Converting forest plantations into more natural forests with active regeneration, irregular structure, high biodiversity levels, and high resilience to disturbances such as pests and fires is urgently needed (Maestre and Cortina 2004; Brockerhoff et al. 2008; Pejchar et al. 2008). In this respect, interest has surged in recent decades to transform monoculture mountain afforestation to foment forest biodiversity, carbon storage, and greater resilience in the face of changing climates (Maestre and Cortina 2004; Brockerhoff et al. 2008; Pejchar et al. 2008; Leverkus et al. 2021).

### 5.2 The Way This Problem Had Traditionally Been Addressed

Pine plantations have traditionally been managed by thinning of different intensities. However, we do not know the ecological consequences of such forest treatments. For example, we do not know the optimal thinning intensity or the consequences of the structure of the forest plantation (size and internal structure of the stand) regarding the regeneration of the seed bank and the fate of the seedlings already established under the plantation canopy (González-Moreno et al. 2011). We also do not know the importance of the landscape context in which the pine reforestation is immersed, and the importance of the past land use on the legacies of *Quercus* and other woody species inside the pine plantation (Navarro-González et al. 2013). Additionally, we have a lot of uncertainty about the potential impact of climate change on the naturalization process.

### 5.3 Current Management Based on Research

The practical limitations of naturalization highlight the need to develop a conceptual model that helps us approach the problem in a planned way, based on scientific knowledge. From an applied perspective, it is critical to diagnose the natural recovery potential of a given plantation. Under suitable ecological conditions, a given plantation can recover biodiversity and heterogeneity naturally (passive restoration). On the contrary, if the ecological conditions are unsuitable, passive restoration will not suffice, and actions will have to be taken to revitalize the process (active restoration). Whether native forest species become established within forest plantations depends on both in situ land-use legacies as well as the distance to seed sources from remnant native forest fragments. In addition, other ecological factors intervene, such as the internal spatial structure of plantation patches (Utsugi et al. 2006; Gómez-Aparicio et al. 2009), vegetation type surrounding such patches (Hewitt and Kellman 2002a, b; Zamora et al. 2010), availability of seed-dispersal vectors (Zamora et al. 2010) or abiotic factors (Gómez-Aparicio et al. 2009; González-Moreno et al. 2011). The arrival of off-site propagules through organisms acting as mobile links is of special importance to plantations, where the internal resources for ecological succession are impoverished (Bengtsson et al. 2003; Lundberg and Moberg 2003; Gómez-Aparicio et al. 2009). It is crucial to place this recovery potential in a landscape context, and in a given time frame, in order to help the manager to determine the consequences of acting or not (i.e. choosing *active or passive* restoration).

Forest ecosystem services include timber and non-timber forest products (provisioning services) and regulation, habitat or support services, as well as cultural activities (TEEB 2010). The demand for regulating services such as carbon sequestration and water regulation, and for cultural services such as recreation and spiritual values, are expected to intensify because of both the burgeoning global population and rising standards of living (FAO 2010; Miura et al. 2015). Therefore, the role of planted forests as providers of diverse ecosystem services has attracted increasing attention (Brockerhoff et al. 2008, 2013; Bauhus et al. 2010; Vihervaara et al. 2012). Although the potential to enhance the ecosystem values of planted forests has been recognized for some time (Lindenmayer et al. 2015), the need to develop assessment frameworks and tools persists in order to guide informed decision-making. To comprehend how forest management affects multiple ecosystem services, we need to understand how particular forest attributes, which can be moderated by management practices, influence different ecosystem services.

Pérez-Luque and Zamora (2020) created a decision-making support system that identifies the places where active restoration is most appropriate to maximize regeneration in a pine forest as opposed to sites where passive restoration works very well and no active measures need to be implemented. **diveRpine** (*Diversification of Pine plantations* https://ajpelu.github.io/diveRpine/) is an interactive tool that simulates the way in which species diversity in pine plantations depending on the landscape, the internal structure of the plantation (past land uses, tree density), and the composition of the dispersion vectors (birds, mammals). This tool is developed in an academic context, with the ultimate goal of transferring it to decision-makers so that they can use it as a tool to support decision-making (Fig. 6a and b). **diveRpine** is parametrized with information published in scientific journals, enabling the manager to visualize different scenarios and perform simulations based on solid field data interpreted in scientific contexts. In this respect, the purpose of the application is to develop a decision-support tool that simulates the dynamics of forest ecological processes. The application projects the most likely outcomes of ecological succession in each stand based on the ecological context, and provides visual representations of the relative importance of the different ecological mechanisms involved in the process. This tool has great value to explore virtual scenarios and demonstrate the process of prioritization. The resulting simulations help the manager to identify forest stands that most need intervention (active restoration), compared to other stands where intervention is unneeded (passive restoration), depending on the ecosystem services that the manager wishes to maximize.

Thus, diveRpine enables the user to answer key management questions such as:

What variables should be taken into account to naturalize plantations?

When is it worth acting to naturalize pine plantations?

How are management actions linked to the provision of ecosystem services?

Our work provides a comprehensive view of a diverse range of ecological implications of changing from a monospecific forest plantation to a more diverse forest structure, highlighting the broad outcomes of the resulting biodiversity and ecosystem services (Baral et al. 2016; Felipe-Lucía et al. 2018).

**Box 1. How does diveRpine run?**

| |
|---|
| **1.1 Target stand** |
| **In this module, the user has to specify different aspects related to the internal structure of the pine plantation which influence the richness and diversity of species that these ecosystems can harbour** (Gómez-Aparicio et al. 2009; González-Moreno et al. 2011; Navarro-González et al. 2013) |
| **1.1.1. Tree density** |
| Tree density has a negative effect on the diversity and total species richness in pine plantations (Gómez-Aparicio et al. 2009). Greater tree density blocks light and therefore reduces the diversity of plant species. In addition, in dense pine forests, the flow of seeds is hindered for both wind-dispersed and bird-dispersed species. For example, the Eurasian jay (*Garrulus glandarius*), one of the most active dispersers in Mediterranean mountain pine-oak forests, prefers to visit less dense stands (Gómez 2003). Users can choose from among three levels of tree density: low density (<500 trees/ha), medium density (500–1500 trees/ha), and high density (>1500 trees/ha). The tree-density thresholds are based on data from Forest Inventory conducted in Sierra Nevada (Pérez-Luque et al. 2014) |
| **1.1.2. Land-use legacies** |
| Most of the afforested pines in Mediterranean mountains were planted in areas degraded by intensive human use (e.g., mountain croplands, pastures), or in shrublands or natural forests that were replaced by these conifer plantations (Pausas et al. 2004). The past land use affects the richness of species that inhabit a pine plantation (Navarro-González et al. 2013). For example, the number of recruits of *Quercus* sp. found in a pine plantation depends on the past land use, and on the distance to the seed source. There is a gradient of use intensity, in such a way that the more intense the past use before reforestation, the lower the probability of finding recruits of *Quercus* species at present (Navarro-González et al. 2013). In this module, users specify the intensity of human use before the afforestation. The gradient of human use from highest to lowest is: cultivation, pasture, scrubland, and natural forest |
| **1.2. Landscape configuration** |
| Landscape characteristics determine the functioning of certain ecological processes. The richness of species under pine plantations varies according to the number of natural vegetation patches (seed source) and the distance to them (González-Moreno et al. 2011). Thus, the greater the number of patches of natural forest, the greater the probability of native species propagules reaching the target pine |

(continued)

plantation. On the other hand, the amount and diversity of seeds provided by a seed source depend on its size (natural forest patch size) and specific composition. The distance to the seed source affects the process of seed dispersal (Hewitt and Kellman 2002a, b). Shorter distances to the natural vegetation patches boost the probability of propagules entering afforested stands (Gómez-Aparicio et al. 2009; González-Moreno et al. 2011). The degree of adjacency of the patches of natural vegetation also affects the amount of seeds that penetrate the pine plantations (Zamora et al. 2010); that is, the greater the area of adjacency, the greater the entry of seeds, and therefore the greater the probability of new species establishing themselves in these plantations

The user chooses the size and number of the natural vegetation patches, and the application randomly generates a distribution of the different vegetation patches around the target pine plantation stand. The values of species diversity existing in the seed sources (natural forest, scrublands) are established based on data from forest inventories (Pérez-Luque et al. 2014)

**1.3. Dispersal community**

We considered three groups of dispersal vectors, characterized mainly by the dispersal distance and by the quantity as well as quality of the seeds dispersed:

- *Small-sized birds*, with maximum dispersal peak of between 0 and 50 m, and rarely exceed 100 m (Jordano et al. 2007; Zamora et al. 2010)
- *Medium-sized birds*, which disperse 50% of the seeds at distances of more than 100 m. This group includes the Eurasian jay, which covers dispersion distances of up to 1000 m (Gómez 2003; Pons and Pausas 2007) and has a maximum dispersion distance of between 250 and 400 m, although this depends heavily on the landscape (Gómez 2003)
- *Mammals*, which have highly variable dispersal dynamics, but disperse over greater distances (even more than 1500 m). With maximum dispersal rates of around 650–700 m, they can also introduce non-native seeds due to their large dispersal radii (Jordano et al. 2007; Matías et al 2010)

Dispersal patterns and distances were established based on work performed mainly in Mediterranean mountain plantations (Gómez 2003; Jordano et al. 2007; Matías et al. 2010; Zamora et al. 2010). The user determines the percentage of each disperser group and a simulation is made of the number of propagules that can potentially be introduced in the target pine plantation by year

The user starts with a focal pine plantation to determine which management actions would promote the ecosystem services desired. The first step is to set the features of the target pine plantation, defining patch size, tree density, and past land use. Then, the landscape is configured by determining the number and size of patches of natural forest. With this, **diveRpine** generates a virtual landscape with the characteristics specified by the user, and it computes the initial richness value in both the natural vegetation patches and the target pine plantation. These values are expressed numerically (bottom-left side in Fig. 6b) and can also be visualized spatially (right-side in the figure). The next step is to configure the composition of the disperser community. The user specifies the percentage of small- and medium-sized bird dispersers, and the remaining percentage corresponds to mammals. Then, **diveRpine** computes a proxy of the seed input, by computing the number of propagules that can enter a pine plantation in a year, taking into account all the previous characteristics specified by the user. The result can be visualized spatially. Finally, the simulation years are specified and the final richness in the pine plantation computed

## 6 Implications of Diversifying Pine Plantations in Terms of Ecosystem Services

Several studies have demonstrated that compositional and structural diversification in forest stands bolsters stability under climate fluctuations and drought episodes (Linares et al. 2010; Choat et al. 2012; Morin et al. 2014). Similarly, greater structural heterogeneity strengthens resistance to natural disturbances such as wind and snow storms (Jactel et al. 2017). Increased compositional diversity (including genetic and phenotypic diversity) and structural diversity also enhances resilience in the forest and therefore a greater capacity to adapt to the environmental changes (Elmqvist et al. 2003; Puettmann et al. 2013). The greater the diversity of functional traits in a forest, the greater the response capacity of the forest to disturbances (Sánchez-Pinillos et al. 2016).

Managers are confronted with translating forestry activities into a panoply of ecosystem services (Fig. 7). That is, the manager can act on the composition (eliminating or planting some species instead of others) and structure (performing controlled thinning) of the plantation. These actions have functional consequences in ecological processes (e.g., thinning allows more light to enter the undergrowth, stimulating the germination of seeds and the growth of saplings). Each field-target variable on which the management is based must qualitatively and quantitatively support an ecological function and service. Then these changes, after the quantification of essential monitoring variables, must be linked to expected implications for the provision of ecosystem services, the user weighing gains against losses in possible trade-off scenarios.

Tools and assessment frameworks still need to be developed in order to guide informed decision-making (Lindenmayer et al. 2015). With **diveRpine**, we have linked science to management: first, we transformed ecological data into scientific information, and then we used this information to develop a support tool for management decision-making. This provides the manager with essential variables linked to the provision of services with which to make decisions. The novelty of our application is its conceptual and methodological approach. Conceptually, our support-decision tool is innovative by equipping managers to envisage the consequences of their decisions within a framework of assessing goods and services provided by the pine plantation. In this context, **diveRpine** is a powerful tool to support decision-making in the real world, working with key variables linked to the provision of services. In terms of methodology, it is presented in a user-friendly format, so that decision-making is based on the best possible scientific information processed with an easy-to-use tool. Although it

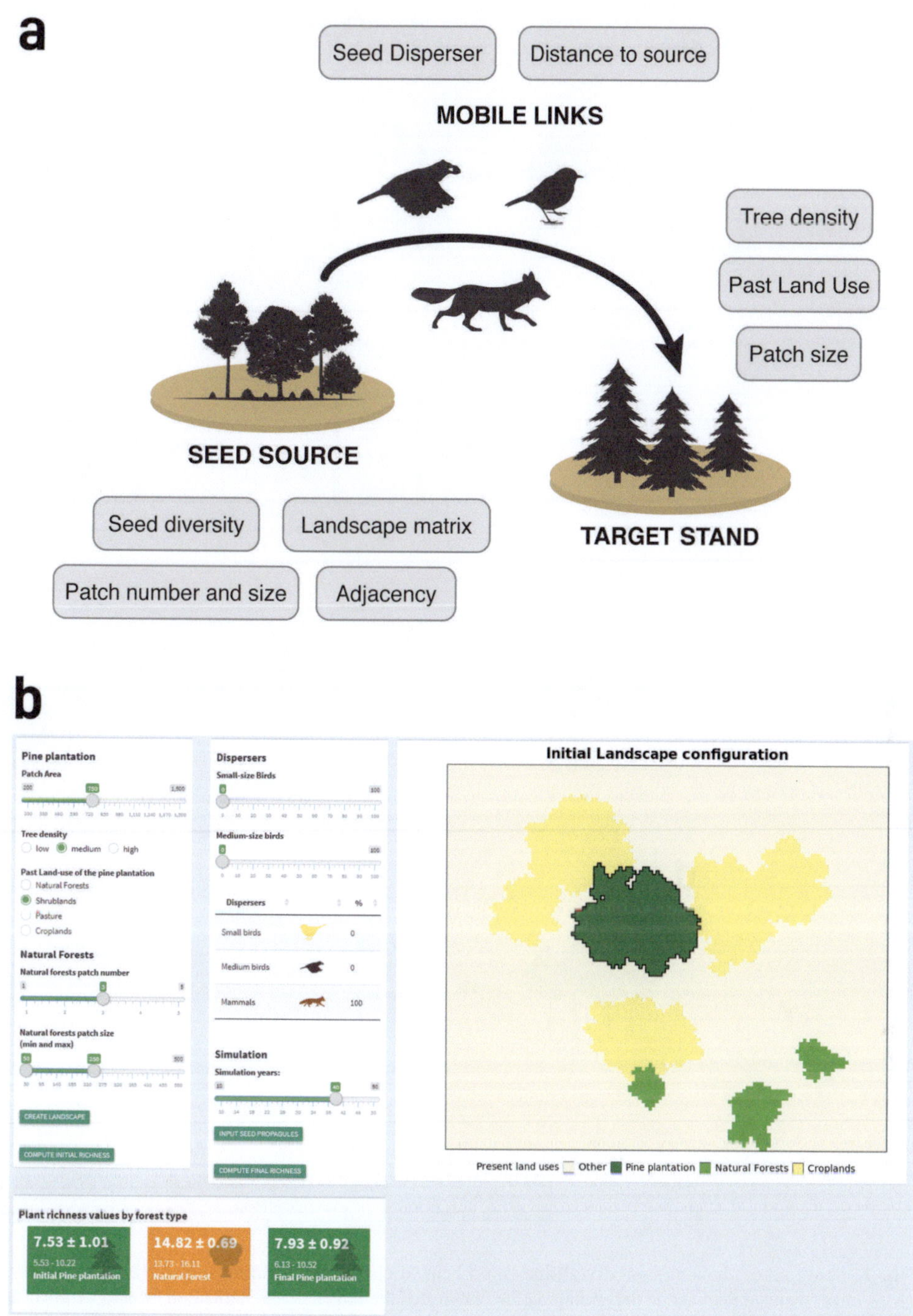


**Fig. 6** **a** The "mobile-links" conceptual framework used to develop **diveRpine** (drawing from Lundberg and Moberg 2003). **b** Screenshot of diveRpine showing the different components: modules (left side), the virtual landscape (right side), and the Richness values of pine plantations (initial values; final—after simulation—values) and natural forests (bottom-left)

has been developed with information from Mediterranean mountains, and particularly from Sierra Nevada, the conceptual and analytical approach allows **diveRpine** to be applied to any other type of plantation in any other biome. The only requirement is to have information on the relevant modules that can be applied to any ecosystem. The more complete this information is for a given plantation in a given environment, the more reliable the scenarios and subsequent simulations will be.

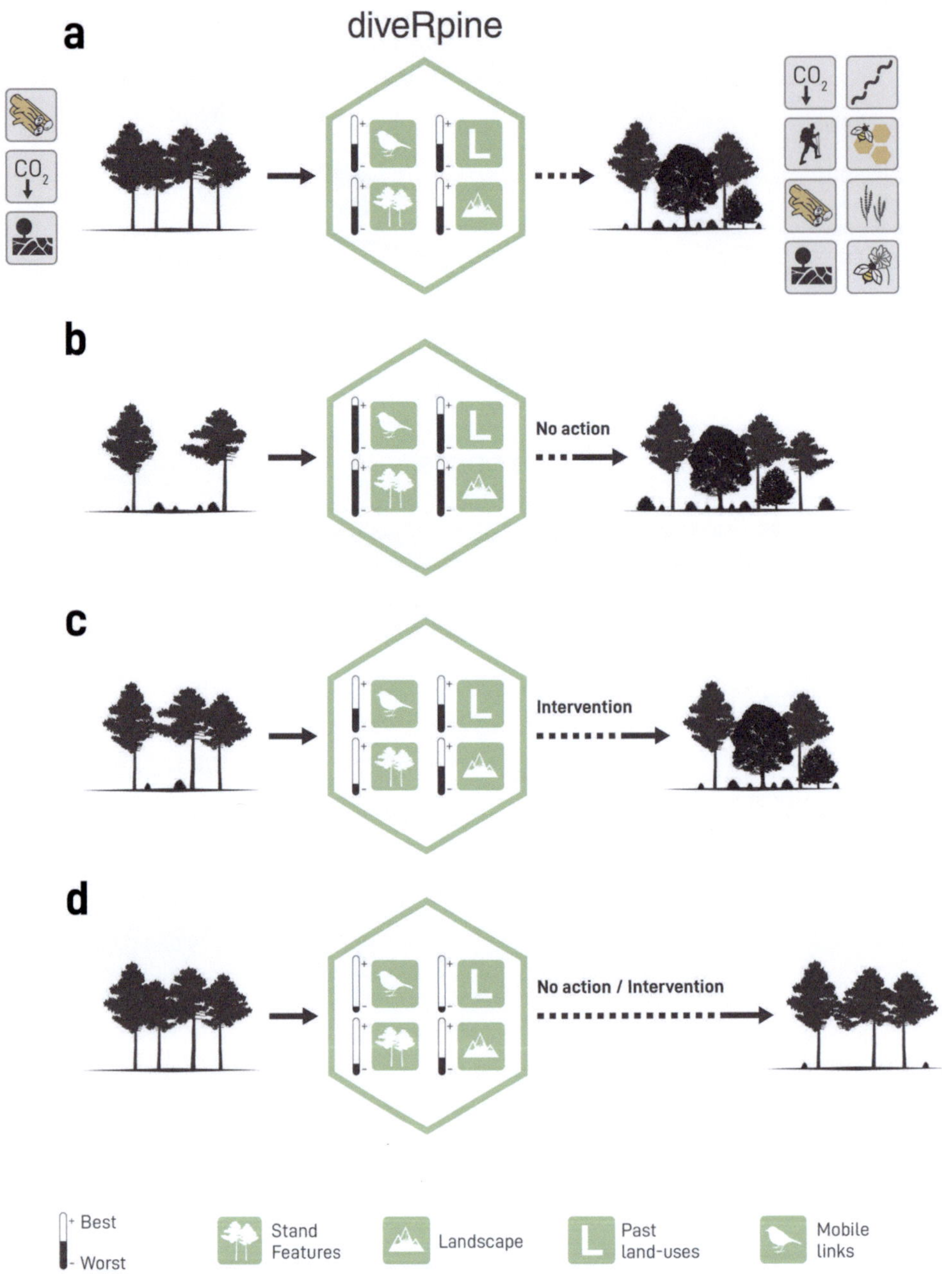


**Fig. 7** Study cases showing how **diveRpine** could help to choose a given forest management action depending on the features of the target pine plantation, landscape configuration, and disperser community. Initially **a** a dense, homogeneous and generally monospecific stand provides few ecosystem services (e.g., regulation: soil erosion, carbon sequestration; provisioning: timber production). The goal is to achieve greater diversity (composition) and heterogeneity (vertical and spatial structure) in the stand. This will increase multifunctionality and thereby provide more ecosystem services (e.g., climate regulation, pest regulation, recreational values, pollination). Under optimal conditions **b** (i.e. low tree density, heterogeneous landscape with nearby patches of natural forest, healthy populations of seeds dispersers, etc.), the diversification would follow a natural course, and thus no action (passive restoration) would be the best management option. On the contrary, under suboptimal conditions (**d**) (i.e. high tree density, isolated pine plantation, poor community of dispersers, and scarce natural sources of propagules) the manager must decide whether the action merits the investment. If the target pine plantation has some optimal conditions (**c**) (i.e. intermediate values of tree density with nearby natural forest patches providing seeds) the app could help the manager to decide whether action would help diversify the pine plantation and therefore broaden the range of ecosystem services

## 7 Concluding Comments

In the last decades, a series of long-term management projects have been developed in Sierra Nevada that address pressing environmental challenges and use scientific knowledge generated in situ to propose solutions. We have established a framework of permanent collaboration between researchers and managers of the National Park of Sierra Nevada. The results presented in this chapter highlight the benefits derived from this collaboration between researchers and managers. The focus of the four projects presented in this chapter ranges from the species to the ecosystem, and they seek solutions based on nature, enhancing the capacity for adaptation and resilience of natural systems under global change.

As opposed to the traditional management characterized by a lack of monitoring, we have opted for "adaptive management" based on follow-up of actions in order to evaluate the effects of a treatment submitted to testing. We have put into practice this philosophy in Sierra Nevada, proposing key questions from the outset, defining the goals to be pursued with the actions undertaken, and specifying the methodological and analytical details necessary to address these efforts. The implementation of adaptive management plans requires, among other things, the evaluation of management effectiveness. This is even more vital in a context of global change, since the uncertainty concerning both future conditions and the response of ecosystems to the actions is greater. This chapter exemplifies what has been learned to date under this philosophy, highlighting a series of management recommendations that may be applicable in other Mediterranean ecosystems bearing similar characteristics. The scientific knowledge generated in Sierra Nevada allows us to validate old and new management techniques in the new environmental scenarios arising from global change, to develop projects aimed at promoting the mechanisms natural self-organization of ecosystems and even to generate models validated with field data as a support tool in decision-making, exploring a wide spectrum of adaptation options.

All this previous experience has served to develop, in collaboration with the Environmental Ministry of the Regional Government of Andalusia, an ongoing ambitious LIFE governance project called ADAPTAMED (protection of key ecosystem services threatened by climate change through adaptive management of Mediterranean socioecological systems). The project focuses on implementing, monitoring, evaluating, and disseminating adaptive management measures, with an ecosystem approach. The project objectives aim to reduce the negative impact of climatic change in several Natural Protected Areas, including Sierra Nevada. As a result, an increase in the resilience of the socioecosystems concerned is expected, in such a way that their future provision of services will also be improved, in comparison to the scenario of no intervention.

**Acknowledgements** This research work was conducted in the collaborative framework of the "Sierra Nevada Global Change Observatory" monitoring programme http://obsnev.es promoted by the environmental Authority of the Regional Government of Andalusia and the University of Granada. There are many projects of the Spanish National Research Programme, the the National Parks Authonomous Body, the Junta de Andalucía, and the European Union that have financed our research in the last decades. We can highlight the following projects MIGRAME "Global Change, Migration and colonization of hábitats altitudinal gradients on Mediterranean mountains'" from the Excellence Research Group Programme of the Andalusian Government, the LIFE-ADAPTAMED (LIFE14 CCA/ES/000612): Protection of key ecosystem services by adaptive management of Climate Change endangered Mediterranean socioecosystems, the H2020 project European Long-Term Ecosystem and socioecological Research Infrastructure (eLTER) and the Thematic Center on Mountain Ecosystem & Remote sensing, Deep learning-AI e-Services University of Granada-Sierra Nevada (LifeWatch-2019-10-UGR-01) for partial funding.The information presented comes from the field works applied to conservation and restoration carried out by many researchers since the 1980s in Sierra Nevada. I especially want to highlight the work of Daniel Garcia, who provided us with the original data of his juniper seed sowings in the 90s. Jose Antonio Hódar, Jose María Gómez, Lorena Gómez Aparicio, Elena Baraza, Jorge Castro, Jose Luís Quero, Luís Matías, Alex Leverkus, Sara Marañón, Penelope Serrano, and Paulino Fandos, among others, have provided very valuable information, and have contributed to maintaining an atmosphere of permanent collaboration with the managers of the National Park. Alex Leverkus provided very useful comments to a previous draft.We thank the managers of the Sierra Nevada Natural Space for their determined support of our work in Sierra Nevada and their continued collaboration within the framework of the joint activities that we develop at the Global Change Observatory. Special thanks to both the previous managers of the National Park, Javier Sanchez and Ignacio Henares, as well as the current ones, Francisco de Asis Muñoz, and José Luis Martinez. We are also especially grateful to the park guard of the Sierra Nevada Natural Area and the technicians of Environment and Water Agency.

## References

Alasaad S, Fickel J, Rossi L et al (2012a) Applicability of major histocompatinility complex DRB1 alleles as markers to detect vertebrate hybridization: a case study from Iberian ibex x domestic goat in southern Spain. Acta Vet Scand 54(1):56

Alasaad S, Biebach I, Grossen CG et al (2012b) Microsatellite-based genotyping of MHC classII DRB1 gene in Iberian and Alpine ibex. Eur J Wildl Res 58:743–748

Alasaad S, Granados Torres JE, Fandos Paris P et al (2013) The use of radio-collars for monitoring wildlife diseases: a case study from Iberian ibex affected by *Sarcoptes scabiei* in Sierra Nevada. Spain. Parasites & Vectors 6:242

Amills M, Jiménez N, Jordana J et al (2004) Low diversity in the major histocompatibility complex class II DRB1 gene of the Spanish ibex, Capra pyrenaica. Heredity 93:266–272

Angelone-Alasaad S, Biebach I, Pérez JM et al (2017) Molecular analyses reveal unexpected genetic structure in Iberian ibex populations. PloS ONE 12 (1):e0170827

Angelone-Alasaad S, Jowers MJ, Molinar Min AR et al (2018) Hidden MHC genetic diversity in the Iberian ibex (*Capra pyrenaica*). BMC Gene 19(28)

Baral H, Keenan GMR, R (2016) A proposed framework for assessing ecosystem goods and services from planted forests. Ecosyst Serv 22:260–268

Baraza E, Zamora R, Hódar JA (2006) Conditional outcomes in plant–herbivore interactions: neighbours matter. Oikos 113(1):148–156

Baraza E, Zamora R, Hódar JA, Gómez, JM (2007) Plant–herbivore interaction: beyond a binary vision. In: Pugnaire FI, Valladares F (eds) Handbook of functional plant ecology. Marcel Dekker Inc, pp 677–718

Bauhus JP, Meer VD, Kanninen M (2010) Ecosystem goods and services from plantation forests. Earthscan, London

Bengtsson JP, Angelstam T, Elmqvist U et al (2003) Reserves, resilience and dynamic landscapes. AMBIO: J Hum Environ 32:389–396

Bernatchez L, Landry C (2003) MHC studies in nonmodel vertebrates: what have we learned about natural selection in 15 years? J Evol Biol 16:363–377

Brockerhoff EG, Jactel H, Parrotta JA et al (2008) Plantation forests and biodiversity: oxymoron or opportunity? Biodivers Conserv 17:925–951

Brockerhoff EG, Jactel H, Parrotta JA et al (2013) Role of eucalypt and other planted forests in biodiversity conservation and the provision of biodiversity-related ecosystem services. For Ecol Manage 301:43–50

Cadotte MW, Carscadden K, Mirotchnick N (2011) Beyond species: functional diversity and the maintenance of ecological processes and services. J Appl Ecol 48:1079–1087

Cash DW, Clark WC, Alcock F et al (2003) Knowledge systems for sustainable development. PNAS 100:8086–8091

Castro I, Espinosa J, Granados JE et al (2018) Characterizing the growth of *Sarcoptes scabiei* infrapopulations. Exp Appl Acarol 76:41–52

Castro J, Zamora R, Hódar JA et al (2002) Use of shrubs as nurse plants: a new technique for reforestation in Mediterranean mountains. Restor Ecol 10:297–305

Castro J, Zamora R, Hódar JA et al (2004) Benefits of using shrubs as nurse plants for reforestation in Mediterranean mountains: a 4-year study. Restor Ecol 10:297–305

Castro J, Zamora R, Hódar JA (2006) Restoring the forests of *Quercus pyrenaica* Willd using pioneer shrubs as nurse plants. Appl Veg Sci 9:137–142

Castro J, Moreno-Rueda G, Hódar JA (2010) Experimental test of postfire management in pine forests: Impact of salvage logging versus partial cutting and nonintervention on bird-species assemblages. Conserv Biol 24:810–819

Castro J, Allen CD, Molina-Morales M et al (2011) Salvage logging versus the use of burnt wood as a nurse object to promote post-fire tree seedling establishment. Restor Ecol

Castro J, Puerta-Piñero C, Leverkus ABB et al (2012) Post-fire salvage logging alters a key plant-animal interaction for forest regeneration. Ecosphere 3, art90

Chazdon RL (2008) Beyond deforestation: restoring forests and ecosystem services on degraded lands. Science 320:1458–1460

Choat B, Jansen S, Brodribb TJ et al (2012) Global convergence in the vulnerability of forests to drought. Nature 491:752–755

Doblas-Miranda E, Alonso R, Arnan X et al (2017) A review of the combination among global change factors in forests, shrublands and pastures of the Mediterranean Region: beyond drought effects. Global Planet Change 148:42–54

Elmqvist T, Folke C, Nyström M et al (2003) Response diversity, ecosystem change, and resilience. Front Ecol Environ 1:488–494

Enright NJ, Fontaine JB, Bowman DM et al (2015) Interval squeeze: altered fire regimes and demographic responses interact to threaten woody species persistence as climate changes Front. Ecol Environ 13:265–272

Espinosa J, López Olvera JR, Cano-Manuel León FJ et al (2017) Guidelines for managing captive Iberian ibex herds for conservation purposes. J Nat Conserv 40(1):24–32

Espinosa J, Pérez JM, Ráez Bravo A et al (2020) Recommendations for the management of sarcoptic mange in free-ranging Iberian ibex populations. Anim Biodivers Conserv 43(1):137–149.

Fandos P (1991) La cabra montés (*Capra pyrenaica*) en el Parque Natural de las Sierras de Cazorla. Instituto Nacional para la Conservación de la Naturaleza: Segura y Las Villas

FAO (2010) Global forest resources assessment 2010. Food Agric Organ United Nations, Rome

Felipe-Lucía MR, Soliveres S, Penone C et al (2018) Multiple forest attributes underpin the supply of multiple ecosystem services. Nat Commun 9:4839

García D (2001) Effects of seed dispersal on Juniperus communis recruitment on a Mediterranean mountain. J Veg Sci 12:839–848

García D, Zamora R, Hódar JA et al (1999a) Age structure of *Juniperus communis* L. in the Iberian Peninsula: conservation of remnant populations in Mediterranean mountains. Biol Conserv 87:215–220

García D, Zamora R, Gómez JM et al (1999b) Bird rejection of unhealthy fruits reinforces the mutualism between juniper and its avian dispersers. Oikos 84:536–544

García D, Zamora R, Gómez JM et al (2000a) Geographical variation in seed production, predation and abortion in *Juniperus communis* throughout its range in Europe. J Ecol 88:436–446

García D, Zamora R, Gómez JM et al (2000b) Do empty *Juniperus communis* seeds act as a defence against seed predation by *Apodemus sylvaticus*? Écoscience 7:214–221

García D, Zamora R, Hódar JA et al (2000c) Yew (*Taxus baccata* L.) Regeneration is facilitated by fleshy-fruited shrubs in Mediterranean environments. Biol Cons 95:31–38

García D, Zamora R (2003) Persistence and multiple demographic strategies in long-lived mediterranean plants. J Veg Sci 14:921–926

Gerard F, Petit S, Smith G et al (2010) Land cover change in Europe between 1950 and 2000 determined employing aerial photography. Progress Phys Geogr Earth Environ 34:183–205

Giorgi F (2006) Climate change hot-spots. Geophys Res Lett 33: L08707

Giorgi F, Lionello P (2008) Climate change projections for the Mediterranean region. Global Planet Change 63(2–3):90–104

Gómez JM (2003) Spatial patterns in long-distance dispersal of *Quercus ilex* acorns by jays in a heterogeneous landscape. Ecography 26:573–584

Gómez JM, Hódar JA, Zamora R et al (2001) Ungulate damage on Scots pines in Mediterranean environments: effects of association with shrubs. Can J Bot 79(6):739–746

Gómez-Aparicio L, Zamora R, Gómez JM et al (2004) Applying plant facilitation to forest restoration in Mediterranean ecosystems: a meta-analysis of the use of shrubs as nurse plants. Ecol Appl 14:1128–1138

Gómez-Aparicio L, Zamora R, Gómez JM (2005a) The regeneration status of the endangered *Acer opalus subsp. granatense* throughout its geographical distribution in the Iberian Peninsula. Biol Cons 121:195–206

Gómez-Aparicio L, Gómez JM, Zamora R et al (2005b) Canopy vs soil effects of shrubs facilitating tree seedlings in Mediterranean montane ecosystems. J Veg Sci 16:191–198

Gómez-Aparicio L, Gómez JM, Zamora R (2005c) Microhabitats shift rank in suitability for seedling establishment depending on habitat type and climate. J Ecol 93:1194–1202

Gómez-Aparicio L, Valladares F, Zamora R (2006) Differential light responses of Mediterranean tree saplings: linking ecophysiology with regeneration niche in four co-occurring species. Tree Physiol 26:947–958

Gómez-Aparicio L, Zamora R, Castro J et al (2008) Facilitation of tree saplings by nurse plants: microhabitat amelioration or protection against herbivores. J Veg Sci 19:161–172

Gómez-Aparicio L, Zavala MA, Bonet FJ, Zamora R (2009) Are pine plantations valid tools for restoring Mediterranean forests? an assessment along abiotic and biotic gradients. Ecol Appl 19:2124–2141

González-Moreno P, Quero JL, Poorter L et al (2011) Is spatial structure the key to promote plant diversity in Mediterranean forest plantations? Basic Appl Ecol 12:251–259

Granados Torres JE, Ros Candeira A, Pérez Luque AJ et al (2020) Long-term monitoring of the Iberian ibex population in the Sierra Nevada of the southeast Iberian Peninsula. Sci Data 7(203):1–10

Herrero A, Zamora R (2014) Plant responses to extreme climatic events: a field test of resilience capacity at the southern range (2014). PloS ONE 9(1):e87842

Hedrick PW, Kim TJ (2000) Genetics of complex polymorphisms: parasites and the maintenance of the major histocompatibility complex variation. In: Singh RS, Krinbas CB (eds) Evolutionary genetics: from molecules to morphology. Cambridge University Press: Cambridge, UK, pp 204–234

Hewitt N, Kellman M (2002) Tree seed dispersal among forest fragments: I. Conifer plantations as seed traps. J Biogeogr 29:337–349

Hewitt N, Kellman M (2002b) Tree seed dispersal among forest fragments: II. Dispersal abilities and biogeographical controls. J Biogeogr 29:351–363

Hobbs RJ, Arico S, Aronson J et al (2006) Novel ecosystems: theoretical and management aspects of the new ecological world order. Glob Ecol Biogeogr 15:1–7

Holling CS (ed) (1978) Adaptive environmental assessment and management. Wiley, Chichester, UK

IPCC (2021) Climate Change 2021: the physical science basis. Contribution of working Group I to the Sixth Assessment Report of the Intergovernmental Panel on Climate Change. In: Masson-Delmotte V, Zhai P, Pirani A, Connors SL, Péan C, Berger S, Caud N, Chen Y, Goldfarb L, Gomis MI, Huang M, Leitzell K, Lonnoy E, Matthews JBR, Maycock TK, Waterfield T, Yelekçi O, Yu R, Zhou B (eds) Cambridge University Press. In Press

Jactell H, Jürgen Bauhus H, Boberg J et al (2017) Tree Diversity Drives Forest Stand Resistance to Natural Disturbances. Curr Forest Rep 3 (3):223–243

Juan-Ovejero R, Molinas-González CR, Leverkus AB et al (2021) Decadal effect of post-fire management treatments on soil carbon and nutrient concentrations in a burnt Mediterranean forest. For Ecol Manage 498

Klein J, Figueroa F (1986) Evolution of the major histocompatibility complex. Crit Rev Immunol 6(4):295–386

Jordano P, Zamora R, Marañón T, Arroyo J (2002) Claves ecológicas para la restauración del bosque mediterráneo. Aspectos demográficos, ecofisiológicos y genéticos. Ecosistemas 11(1)

Jordano P, García C, Godoy JA et al (2007) Differential contribution of frugivores to complex seed dispersal patterns. Proc Natl Acad Sci 104:3278–3282

Laza E (1890) Sobre la cabra de Sierra Nevada. Actas De La Sociedad Española De Historia Natural 19:36–39

León Vizcaíno L, Ruíz de Ybáñez MR, Cubero MJ et al (1999) Sarcoptic mange in Spanish ibex from Spain. J Wildl Dis

Leverkus AB, Lorite J, Navarro FB (2014) Post-fire salvage logging alters species composition and reduces cover, richness, and diversity in Mediterranean plant communities. J Environ Manage Elsevier Ltd 133:323–331

Leverkus AB, Castro J, Delgado-Capel MJ et al (2015) Restoring for the present or restoring for the future: enhanced performance of two sympatric oaks (*Quercus ilex* and *Quercus pyrenaica*) above the current forest limit. Restor Ecol 23:936–946

Leverkus AB, Rey-Benayas JM, Castro J (2016) Shifting demographic conflicts across recruitment cohorts in a dynamic post-disturbance landscape. Ecology 97:2628–2639

Leverkus AB, Murillo PG, Doña VJ, Pausas JG (2019) Wildfires: opportunity for restoration? Science 363

Leverkus AB, Thorn S, Gustafsson L et al (2021) Environmental policies to cope with novel disturbance regimes—steps to address a world scientists' warning to humanity. Environ Res Lett 16

Linares JC, Camarero JJ, Carreira JA (2010) Competition modulates the adaptation capacity of forests to climatic stress: insights from recent growth decline and death in relict stands of the Mediterranean fir *Abies pinsapo*. J Ecol 98:592–603

Lindenmayer DB, Likens GE (2009) Adaptive monitoring: a new paradigm for long-term research and monitoring. Trends Ecol Evol 24:482–486

Lindenmayer DB, Messier C, Paquette A, Hobbs RJ (2015) Managing tree plantations as novel socioecological systems: Australian and North American perspectives. Can J for Res 45:1427–1433

Lorite J (2002) La vegetación de Sierra Nevada. In: Blanca G (ed) Flora amenazada y endémica de Sierra Nevada. Editorial Universidad de Granada, Granada, pp 23–45

Lubchenco J (1998) Entering the century of the environment: a new social contract for science. Science 279:491–497

Lundberg J, Moberg F (2003) Mobile link organisms and ecosystem functioning: implications for ecosystem resilience and management. Ecosystems 6:0087–0098

Maestre FT, Cortina J (2004) Are *Pinus halepensis* plantations useful as a restoration tool in semiarid Mediterranean areas? For Ecol Manage 198:303–317

Mainguy J, Worley K, Steeve D et al (2007) Low MHC *DRB* class II diversity in the mountain goat: past bottlenecks and possible role of pathogens and parasites. Conserv Genet 8:885–891

Marañón-Jiménez S, Castro J, Kowalski A et al (2011) Post-fire soil respiration in relation to burnt wood management in a Mediterranean mountain ecosystem. For Ecol Manage 261:1436–1447

Marañón-Jiménez S, Castro J (2013) Effect of decomposing post-fire coarse woody debris on soil fertility and nutrient availability in a Mediterranean ecosystem. Biogeochemistry 112:519–535

Marañón-Jiménez S, Castro J, Fernández-Ondoño E, Zamora R (2013) Charred wood remaining after a wildfire as a reservoir of macro- and micronutrients in a Mediterranean pine forest. Int J Wildl Fire 22:681–695

Márquez Jiménez FJ, Granados Torres J, Caruz A et al (2020) Genetic diversity of cytochrome b in Iberian ibex from Andalusia. Mammalian Biol 100:675–684

Matías L, Zamora R, Mendoza I et al (2010) Seed dispersal patterns by large frugivorous mammals in a degraded mosaic landscape. Restor Ecol 18:619–627

Matías L, Zamora R, Castro J (2012) Sporadic rainy events are more critical than increasing drought intensity for woody species recruitment in a Mediterranean community. Oecologia 169 (3):833–844

Matias L, Jump A (2015) Asymmetric changes of growth and reproductive investment herald altitudinal and latitudinal range shifts of two woody species. Glob Change Biol 21:882–896
Miura S, Amacher M, Hofer T et al (2015) Protective functions and ecosystem services of global forests in the past quarter-century. For Ecol Manage 352:35–46
Morin X, Fahse L, Mazancourt C et al (2014) Temporal stability in forest productivity increases with tree diversity due to asynchrony in species dynamics. Ecol Lett 17:1526–1535
Moritz MA, Batllori E, Bradstock RA et al (2014) Learning to coexist with wildfire. Nature 515:58–66
Moroni B, Granados Torres JE, Ráez Bravo A et al (2019) Exploring new control strategies against sarcoptic mange in Iberian ibex: a pharmacokinetic study. In: X RUSI - Reunião Ungulados Silvestres Ibéricos, 18 e 19 de Outubro de 2019. Lousã, Portugal
Moroni B, Angelone S, Pérez J M et al (2021) Sarcoptic mange in wild ruminants in Spain: solving the epidemiological enigma using microsatellite markers. Parasite Vectors 14(171)
Navarro-González I, Pérez-Luque AJ, Bonet FJ, Zamora R (2013) The weight of the past: land-use legacies and recolonization of pine plantations by oak trees. Ecol Appl 23:1267–1276
Nesshöver C, Assmuth T, Irvine KN et al (2017) The science, policy and practice of nature-based solutions: an interdisciplinary perspective. Sci Total Environ 579:1215–1227
Odum HT (1962) Ecological tools and their use. Man and the ecosystem. Proceedings of the Lockwood conference on the suburban forest and ecology. In: Waggoner PE, Ovington JD (eds) The connecticut agricultural experiment station, bulletin 652, pp 57–75
Ortuño F (1990) El plan para la repoblación forestal de España del año 1939. Análisis y comentarios. Ecología(Extra 1):373–392
Pausas JG, Bladé C, Valdecantos A et al (2004) Pines and oaks in the restoration of Mediterranean landscapes of Spain: new perspectives for an old practice: a review. Plant Ecol 171:209–220
Pausas JG, Fernández-Muñoz S (2012) Fire regime changes in the Western Mediterranean Basin: from fuel-limited to drought-driven fire regime. Clim Change 110:215–226
Pejchar L, Pringle RM, Ranganathan J et al (2008) Birds as agents of seed dispersal in a human-dominated landscape in southern Costa Rica. Biol Cons 141:536–544
Pence DB, Ueckermann E (2002) Sarcoptic mange in wildlife. Rev Sci Tech 21:385–398
Pérez JM, Ruiz-Martínez I, Granados JE, et al (1997) The dynamics of sarcoptic mange in the ibex population of Sierra Nevada in Spain. Influence of climatic factors. J Wildlife Res 2 (1):86–89
Pérez-Luque AJ, Bonet FJ, Pérez-Pérez R et al (2014) Sinfonevada: dataset of floristic diversity in Sierra Nevada forests (SE Spain). PhytoKeys 35:1–15
Pérez-Luque AJ, Bonet-García FJ, Zamora R (2019) Map of ecosystems types in Sierra Nevada mountain (southern Spain). https://doi.pangaea.de/
Pérez-Luque AJ, Zamora R (2020) diveRpine: diversification of pine plantations in mediterranean mountains. A shiny app to help forest decision makers
Petchey OL, Gaston KJ (2006) Functional diversity: back to basis and looking forward. Ecol Lett 9:741–758
Pons J, Pausas JG (2007) Acorn dispersal estimated by radio-tracking. Oecologia 153:903–911
Puettmann K, Messier C, Coates K (2013) Managing forests as complex adaptive systems. In: Messier C, Puettmann K, Coates K (eds) Managing forests as complex adaptive systems: building resilience to the challenge of global change. The earthscan forest library, New York
Ráez Bravo A (2019) Pathophysiology of sarcoptic mange in Iberian ibex. Doctoral thesis, Universidad Autónoma de Barcelona, 138 pp.
Ráez Bravo A, Granados Torres JE, Espinosa J et al (in press) Genomics reveal local skin immune reaction key to control sarcoptic mange in Iberian ibex (*Capra pyrenaica*)
Rey-Benayas JM, Bullock JM, Newton AC (2008) Creating woodland islets to reconcile ecological restoration, conservation, and agricultural land use. Front Ecol Environ 6:329–336
Ruiz Martínez I, Pérez JM, Chirosa M et al (1996) Plan de manejo y gestión de la cabra montés (*Capra pyrenaica*) ante la epizootia de sarcoptidosis en el Parque Natural de Sierra Nevada (Granada, España). 1n: Conferencia internacional sobre Sierra Nevada III: pp 47–68
Sánchez-Pinillos M, Coll L, De Cáceres M et al (2016) Assessing the persistence capacity of communities facing natural disturbances on the basis of species response traits. Ecol Ind 66:76–85
Seidl R et al (2017) Forest disturbances under climate change Nat. Clim Change 7:395–402
Sommer S (2005) The importance of immune gene variability (MHC) in evolutionary ecology and conservation. Front Zool 2:16
Sutherland WJ, Pullin AS, Dolman PM, T. M. Knight TM, (2004) The need for evidence-based conservation. Trends Ecol Evol 19:305–308
TEEB (2010) The economics of ecosystems and biodiversity: mainstreaming the economics of nature: a synthesis of the approach, conclusions and recommendations of TEEB. UNEP, Geneva
Trowsdale J (1993) Genomic structure and function in the MHC. Trends Genet 9:117–122
Turner II BL, Roger RE, Matson PA et al (2003) A framework for vulnerability analysis in sustainability science PNAS 2003 **100** (14):8074–8079
Utsugi E, Kanno H, Ueno N et al (2006) Hardwood recruitment into conifer plantations in Japan: effects of thinning and distance from neighboring hardwood forests. For Ecol Manage 237:15–28
Vihervaara P, Marjokorpi A, Kumpula T et al (2012) Ecosystem services of fast-growing tree plantations: a case study on integrating social valuations with land-use changes in Uruguay. For Policy Econ 14:58–68
Walters CJ, Hilborn R (1978) Ecological optimization and adaptive management. Annu Rev Ecol Syst 9:157–188
Walters CJ (1986) Adaptive Management of Renewable Resources. McMillan, New York, NY, USA
Westerlingh AL, Hidalgo G, Cayanand DR Swetnam TW (2006) Warming and earlier Spring increase Western U.S. Forest Wildfire Activity. Science 313(5789):940–943
Williams BK (2011) Adaptive management of natural resources-framework and issues. J Environ Manage 92:1346–1353
Wintle BA, Sarah L, Woinarski JCZ (2020) After the megafires: what next for Australian wildlife? Trends Ecol Evol 35(9):753–759
Wobeser GA (2002) Disease management strategies for wildlife. Revue Sci Et Tech (int off Epizootics) 21:159–178
Young JC, Waylen KA, Sarkki S et al (2014) Improving the science-policy dialogue to meet the challenges of biodiversity conservation: having conversations rather than talking at one-another. Biodivers Conserv 23:387–404
Zamora R (1990) The fruit diet of ring-ouzels (Turdus torquatus) wintering in Sierra Nevada (South-East Spain). Alauda 58:67–70
Zamora R, Gómez JM, García D, Hódar JA (1996) Ecología reproductiva y regeneración del matorral de la alta montaña a de Sierra Nevada: capacidad de respuesta a las perturbaciones. Sierra Nevada: Conservación y Desarrollo Sostenible, vol II, Chacón J, Rosua JL (eds), pp 407±22. University of Granada, Granada, Spain

Zamora R, Gómez JM, Hódar JA et al (2001) The effect of browsing by ungulates on Scots pine growth in a Mediterranean environment: consequences for forest regeneration. For Ecol Manage 144:33–42
Zamora R, García-Fayos P, Gómez-Aparicio L (2004) Las interacciones planta-planta y planta-animal en el contexto de la sucesión ecológica. In: Valladares F (ed) Ecología del bosque Mediterráneo en un mundo cambiante. Editorial Parques Nacionales, Madrid, ES, pp 371–393
Zamora R, Hódar JA, Matías L, Mendoza I (2010) Positive adjacency effects mediated by seed disperser birds in pine plantations. Ecol Appl 20:1053–1060
Zamora R, Pérez-Luque AJ, Bonet FJ, Barea-Azcón JM, Aspizua R (eds) (2016) Global Change Impacts in Sierra Nevada: Challenges for Conservation. Consejería de Medio Ambiente y Ordenación del Territorio. Junta de Andalucía
Zavala MA, Zamora R, Pulido F et al (2008) Nuevas perspectivas en la conservación, restauración y gestión sostenible del bosque mediterráneo. In: Valladares F (ed) Ecología del bosque mediterráneo en un mundo cambiante (Segunda edición). Páginas 511–532. Ministerio de Medio Ambiente. EGRAF, S. A., Madrid. ISBN: 978-84-8014-738-5

# Local Ecological Knowledge and the Sustainable Co-Management of Sierra Nevada's Social-Ecological System

David García-del-Amo, Cristina Gálvez-García, Irene Iniesta-Arandia, Javier Moreno-Ortiz, and Victoria Reyes-García

**Abstract**

Local ecological knowledge systems have been the basis of Sierra Nevada's social-ecological system, which has co-evolved over more than ten centuries until nowadays, based on the knowledge, practices, and innovations deriving from the relationship between people and the ecosystems on which they depend. In Sierra Nevada, this co-evolution is greatly influenced by the traditional water management system, generating a "cultural landscape." However, during the twentieth-century Sierra Nevada's social-ecological system was affected by diverse drivers of change such as climate change, rural exodus, land-use change, and conservation government policies, which are threatening its stability and the transmission of the related local ecological knowledge. Local ecological knowledge on water management, traditional agricultural systems, and knowledge related to grazing and cattle raising should be included in the co-management of the territory and representatives of this knowledge should be involved and collaborate with administration and researchers developing adaptive plants to reduce negative impacts of global change.

**Keywords**

Adaptive co-management • Drivers of change • Global environmental change • Social-ecological system • Traditional water management

## 1 The Contribution of Local Knowledge Systems to the Sustainable Co-Management of Social-Ecological Systems

During the last decades, researchers and natural resource managers have increasingly been using local knowledge and involving Indigenous Peoples and Local Communities (IPLC) in monitoring, conservation, and ecosystems' restoration efforts (Reyes-García et al. 2021). Overall, including local knowledge and promoting stakeholders' participation has resulted in better outcomes and acceptance of conservation action than when conservation is imposed by external agents. Moreover, such efforts have also had positive outcomes in the livelihood of local communities, thus being a very successful practice (Danielsen et al. 2007; Reed et al. 2016). In this line, in some research fields—such as biodiversity and ecosystem services conservation or ecosystems assessment (McElwee et al. 2020)—there have been efforts to generate new knowledge based on insights that come from different knowledge systems, i.e., science and local ecological knowledge. In fact, some global efforts, such as the Intergovernmental Science-Policy Platform on Biodiversity and Ecosystem Services (IPBES), have adopted participatory mechanisms and institutional arrangements for including IPLC and their views in the assessment (Hill et al. 2020). Overall, this trend recognizes the value of local knowledge systems for sustainable

D. García-del-Amo (✉) · I. Iniesta-Arandia · V. Reyes-García
Institut de Ciència i Tecnologia Ambientals, Universitat Autònoma de Barcelona, Barcelona, Spain
e-mail: david.garcia.delamo@uab.cat

C. Gálvez-García · J. Moreno-Ortiz
Grupo de Investigación Social y Acción Participativa GISAP, Universidad Pablo de Olavide, Sevilla, Spain

I. Iniesta-Arandia
Colectivo FRACTAL, Calle San Remigio 2, Madrid, Spain

J. Moreno-Ortiz
Cooperativa PASOS Participación y Sostenibilidad, Granada, Spain

V. Reyes-García
Institució Catalana de Recerca i Estudis Avançats (ICREA), Barcelona, Spain

V. Reyes-García
Departament d'Antropologia Social i Cultural, Universitat Autònoma de Barcelona, Barcelona, Spain


R. Zamora and M. Oliva (eds.), *The Landscape of the Sierra Nevada*,
https://doi.org/10.1007/978-3-030-94219-9_21

ecosystem management. A similar approach is gaining ground in other research topics, such as climate change research (Díaz et al. 2019; Reyes-García et al. 2019b)

We use the term "local ecological knowledge" (LEK) to refer to the worldviews, knowledge, practices, and innovations deriving from the relationship between people and nature (Berkes 2017). Other authors have used different terms to refer to these knowledge systems, like traditional ecological knowledge (TEK) (Johannes 1993; Berkes 2018), indigenous knowledge (IK) (Bicker et al. 2003; Ellen 2007), or indigenous and local knowledge (ILK) (Thaman et al. 2013), depending on the studied communities and contexts. The term captures knowledge about the natural world, techniques and technologies of resource management, local institutions governing social relations and relationships to nature, and ethical or spiritual values. LEK is situated in social-ecological contexts and is continuously evolving through the combination of written, oral, tacit, practical, and scientific knowledge obtained from various sources and validated by experimentation and direct interaction with nature (Hill et al. 2020). The application of LEK has resulted in the maintenance of the ecological integrity of many aquatic and terrestrial ecosystems (Cámara-Leret et al. 2019). Indeed, it is argued that the diversity and sophistication of LEK systems have been essential to manage the health of local and regional ecosystems and are now essential to confront societal pressures and environmental burdens (Lyver et al. 2019; Reyes-García et al. 2021).

One of the pathways through which LEK contributes to sustainable resource management is through customary governance and management practices based on formal and informal institutional and social arrangements that create and maintain biodiversity (Lepofsky 2009). Many examples exist showing that ILK-based environmental management (e.g., traditional agriculture, aquaculture, fishery, and community forestry) contribute to biodiversity maintenance in production landscapes (Aumeeruddy-Thomas et al. 2017; Zimmerer and de Haan 2019), ecosystem restoration (Reyes-García et al. 2019a), pollution buffering (Fernández-Llamazares et al. 2020), and nutrient cycling, among other positive environmental benefits. Examples of these practices include resource use agreements and collective rules governing common-pool natural resources, such as forests or waters (Ostrom 1990; Berkes 2018). Other examples of these management practices include religious beliefs and caring for intangible elements inhabiting nature and embodied in sacred forests, landscapes, lakes, springs rivers or marine areas (Berkes 2018), taboos overuse and conservation of specific species (Rafidison et al. 2020), harvesting restrictions based on spatial or temporal considerations, or selective harvesting and grazing, often to deliberately create small-scale disturbances to increase landscape heterogeneity and overall biodiversity (Bird 2015; Molnár et al. 2020). As a result of the application of management systems based on LEK, much of today's world's wild and domesticated biodiversity lies in areas traditionally owned, managed, used, and/or occupied by IPLC (Brondizio and Le Tourneau 2016). In Europe, some of these managed ecosystems are local "hotspots" of native biodiversity, including mountain hay meadows in central Europe (Babai and Molnár 2014), *dehesa* oak and cereal tree savannahs in southern Spain (Acha and Newing 2015), temperate deciduous forest in northern Spain (Guadilla-Sáez et al. 2019), or high mountain meadows (*borreguiles*) in Sierra Nevada (Blanca et al. 2001).

Beyond biodiversity-rich landscapes, management practices based on the co-evolution of natural and social systems have also resulted in complex social-ecological systems or "bio-geo-physical" units with associated social actors and institutions (Ostrom 2009). In such complex and interrelated systems, changes in the social component of the system affect the ecological component, and vice versa. Thus, changes in customary management practices can result in degradation and resource depletion, especially when local norms face the pressure of commodity and extractive industries, demands from international or urban markets for local products, and coaptation from individuals and firms pursuing resource extraction (Natcher and Brunet 2020). This is particularly the case in industrialized countries, where global dynamics, including loss of traditional livelihoods (Biró et al. 2014), landscape homogenization, new technology, or administrative regulations affect traditional biodiversity management practices (Gómez-Baggethun et al. 2013; Hernández-Morcillo et al. 2014).

In this chapter, we use the perspective of the social-ecological system to analyze the contribution of LEK to ecosystem management in Sierra Nevada. Sierra Nevada is characterized by the historical co-evolution of local communities and ecosystems, resulting in a complex social-ecological system. In that sense, Sierra Nevada can be considered a "cultural landscape" or a singular space resulting from a long and intimate relationship between peoples and their natural environment and where LEK could play a fundamental role in the future of the region, being the nexus between social and natural systems. Our chapter draws both on the literature on LEK, traditional livelihood practices, environmental research conducted on Sierra Nevada, and on our own data and personal observations while working in the area. We note that Sierra Nevada is an extensive mountain region with more than 2000 $km^2$ and more than 60 villages. This means that there may be variations in the environmental and social conditions, resulting in variations in the knowledge developed in different areas. Despite these differences, in this chapter, we consider Sierra Nevada as a unique social-ecological system. The chapter starts by exploring the historical roots of local ecosystem management, then moves to describe the most important drivers affecting Sierra Nevada's social-ecological system and local knowledge, and ends addressing the current and the possible future contributions of LEK to Sierra Nevada management and conservation.

## 2 Understanding the Historical Roots of Sierra Nevada's Social-Ecological System: Water Management as a Landscape Architect

As in other Mediterranean ecosystems (Blondel 2006), one of the most important pathways through which humans have impacted Sierra Nevada's social-ecological system has been through water management. In chapter "Singular Cultural Landscapes of the Sierra Nevada", Civantos and colleagues describe how, for centuries, human interventions have modified Sierra Nevada through the development of diverse water infrastructures like *acequias de careo*, *partidores*, *aliviaderos*, and *cimbras, minas or qanats*, to channel, guide, and harvest water running down the slopes of the mountain (Fig. 1). This complex water management system has sculpted unique cultural landscape structures around water management. Local water management systems made possible the permanent occupation of the territory because water distribution allowed the development of agriculture and livestock maintenance (Martos-Rosillo et al. 2019). Water distribution through local management systems also contributed to Sierra Nevada's singular biodiversity, as water management helps regulate the impact of the hydrological cycle in the ecosystem through the recharge of aquifers (Pulido-Bosch and Sbih 1995; Martos-Rosillo et al. 2019). Water management allows the establishment of a variety of habitats that, otherwise, would have not existed, such as chestnut clusters along high mountain acequias. The green cover that grows along the water channels also helps retain soil and prevents erosion and water filtrations feed water sources for animal and human uses (Jódar et al. 2017; Martos Rosillo et al. 2018). In the words of an *acequiero* (person in each irrigation community in charge of controlling, managing, and distributing water through the acequias) of one of the irrigation communities in Órgiva:

> *Sierra Nevada is a sponge that catches the water up high, in the upper parts in winter and spring, to prevent it from going quickly to the sea. It then releases it, little by little, in the lower areas, with water infiltrating and re-emerging on several occasions, thus giving life for several months, especially in summer, to the areas where it passes.*

Beyond their economic and ecological importance, traditional water management and harvesting techniques have also been crucial for structuring Sierra Nevada's social system (Guzmán Álvarez et al. 2010). Like other societies which have developed complex irrigation systems, in Sierra Nevada, local knowledge on water management is considered part of the local identity and cultural heritage. Moreover, water management rules determine social interactions (Fernald et al. 2015). Water in Sierra Nevada is managed through a complex system organized around irrigation communities formed by the owners of irrigated lands which have centuries of family history in common (Espinar Moreno 1989). The irrigation communities of the Sierra Nevada villages have been the basis of the structure of their social organization until a few decades ago, since irrigation communities determined the amount of water that corresponded to each member, which is directly related to the irrigated land surface of each member, and therefore largely determined a family's yield and earnings. Many irrigation communities were formally established during the twentieth century, although their existence is intrinsically linked to the creation of water infrastructures (Cressier 1995), and there are references to some irrigation communities dating back to the thirteenth century, which indicates the long-term importance of this social structure (Espinar Moreno 1989).

To manage water, irrigation communities engage in collective action, including a multiplicity of activities that involve social interactions as much as interactions with the environment, such as cleaning, maintaining, and restoring water channels (Fig. 1b). All these interactions raise conflicts related to irrigation water and the consequent emergence of rules for water distribution (Gálvez-García 2015). These rules, refined over time, are usually very precise and consider both biophysical factors (e.g., water evapotranspiration at different times of the year, or the time that takes the water to arrive from one plot to the next one, which is related to the characteristics of the soil of each portion of terrain), and social factors (e.g., watering order and weekly need of water of each community member), to have an equitable distribution of water. These rules also regulate the possibility to lend, borrow, exchange, or even sell water between members (Guzmán Álvarez 2010b).

Overall, local knowledge on water management has allowed the intense modification of the landscape, contributing to creating a particular social-ecological system that proved to be extremely resilient over centuries (Cressier 1995). Water management allowed the development of agricultural and livestock activities in the area, which together formed a biodiversity-rich productive landscape. Traditional agricultural practices, based on heterogeneous plots of crops for self-consumption and the use of agricultural plots at different altitudes to cultivate a larger variety of crops, together with traditional grazing practices and trasterminancia movement of extensive cattle ranching

**Fig. 1** **a** *Acequia de careo* without water to carry out repair works. **b** Members of Bérchules' irrigation community cleaning an *acequia de careo*. **c** *Aliviadero* on the side. **d** *Aliviadero* overflow and flow limiter. **e** *Partidor* of *acequia de careo* of the Poqueira ravine. **f** Water catchment for an *acequia de careo*. **g** *Acequia de careo* of Bérchules

favoring the dispersion of seeds, played a fundamental role in the high levels of Sierra Nevada's biodiversity (Martin Civantos 2011; Ramos Font et al. 2015; Martos Rosillo et al. 2018). Having to conduct repair and maintenance activities in the entire network of *acequias*, from the top of the mountain to the middle and bottom of the valleys, where communities are located, has made local communities excellent sentinels of their territory. Local inhabitants in Sierra Nevada are able to quickly identify changes in the system and react to them, thus maintaining operational multifunctional cultural landscapes (Vahí Serrano and Prados Velasco 2011). Thus, through social interactions oriented to operationalize management rules to successfully maintain water infrastructures, irrigation communities have also contributed to maintaining the social structure, as well as favored the transmission of local knowledge on water management (Ruiz-Ballesteros and Gálvez-García 2014).

## 3 The Twentieth and Twenty-First Centuries: Big Transformations Driving Social-Ecological System Decay and LEK Erosion

Since the mid-twentieth century, major transformations have led to a complete reconfiguration of Sierra Nevada territory and social-ecological system. Important drivers of change affecting Sierra Nevada in the twentieth century include technological and social change driven by market pressures, population dynamics, government policies, and climate change. These changes have led to the overall reconfiguration of the social-ecological system, resulting in the loss of wild and cultivated biodiversity, but also deterioration and abandonment of traditional water management systems, with an overall impact on Sierra Nevada ecosystems (Zamora et al. 2016) and the erosion of local knowledge systems (Iniesta-Arandia et al. 2014).

A major driver of change in Sierra Nevada's social-ecological system has been ***technological and social changes driven by market pressures***. After the Spanish Civil War (1936–39), inequalities in access to land intensified in the region, driving peasants to self-exploitation (Soto Fernández et al. 2007). According to oral testimonies, in La Taha, for example, the rent paid for the land was frequently two-thirds of the harvest, sometimes even three-fourths. Thus, post-war times were very hard in this region, with most families conducting subsistence activities, often being plunged into poverty. In this scenario, the local economic system relied more on the exchange of products and work than on money circulation. For example, during fieldwork, local people told us about the "*tornapeón*," a type of community work, through which peasants helped each other by alternatively working on the fields that needed more work (Gálvez-García 2015). The opening of the Spanish economy to international markets in the 1960s gave way to a more market-oriented economy. The modernization of agriculture made traditional mountain agricultural systems non-profitable economically, as the mechanization of agricultural processes in mountain areas, where small plots with difficult access predominate, is costly or not possible (Fig. 2).

While the drift of farmers toward commercial agriculture, based on the logic of the optimization of economic benefits and cost reduction, has not been unilateral in Sierra Nevada, it has gradually gained terrain over the years, with peasants adopting the new agricultural practices and technologies required for the production standards of agro-food industries and discontinuing the practice of traditional activities (Fig. 3a). As in many areas of the world (Martin et al. 2019; Labeyrie et al. 2021), the shift from traditional livelihoods has resulted in the abandonment of local livestock breeds, and many traditional landraces better adapted to local conditions, particularly rainfed crops grown in high mountain areas such as varieties of legumes, cereals, and potatoes. In some cases, local varieties have been substituted by varieties with a market value, but most often, the cultivation of such crops has been abandoned altogether, as farmers in the low areas have shifted to the monoculture of almond and olives trees, crops that required low workforce. Nowadays, local landraces such as green beans *martillosa, agostiza* or *colorá*, remain only in the memory of the elders. Only in some areas, a few varieties are kept for domestic use. These include *uva de barco,* a grape variety that was highly valued at the beginning of the twentieth century, becoming listed on Wall Street (Alonso et al. 2006), which still can be found in small familiar plots in Ohanes and other villages around, *patatas de la sierra* (Fig. 3b), a potato landrace typical in villages whose territory includes the highest areas of the mountain like Güejar Sierra, Monachil, Jérez del Marquesado, or Capilleira and Trevelez in the Alpujarras area, *frigüelos* (beans) from the Alpujarras, local varieties of tomatoes, or lettuce *pico de pájaro*. Similarly, the *pajuna,* a local cow breed, declined dramatically, and currently only a few ranchers have this breed, although efforts are underway to recover it (Horcada-Ibáñez et al. 2016).

Other activities traditionally conducted by people living in Sierra Nevada have also been replaced or abandoned by the adoption of new technologies, further contributing to eroding local knowledge systems. For example, vegetable fibers traditionally used to craft agricultural accessories, such as esparto grass, wicker, or straw, are not used anymore as they have been substituted by plastic fibers (Fig. 3d). Similarly, the extraction of essential oils, like thyme, has also been abandoned, and handmade wool looms have disappeared. In some cases, this has caused the decrease of abundance of some of these vegetal species which in turn are

**Fig. 2** Terraced slopes divided into small cultivation plots

important nurse plants for other species (Padilla and Pugnaire 2006). In other cases, the abandonment of the practices has resulted in other environmental impacts, such as shepherds throwing wool in the forest, due to its lack of economic value (Personal observation).

A second driver of change in Sierra Nevada's social-ecological system is ***population dynamic***s linked to technological changes occurring in the primary sector. In areas like Sierra Nevada, the disarticulation of traditional economic activities and a loss of competitiveness in the global markets has resulted in an intense rural exodus and the masculinization of agriculture (Rodríguez Martínez 2001). During the second mid of the twentieth century many people moved away from rural areas in general, and mountain areas in particular, where their work was neither recognized nor valued. There was also a change in people's social expectations, particularly among young people, who saw new economic possibilities in cities. All in all, and except for some municipalities near the city of Granada such as Gójar, Monachil, La Zubia, Dúrcal, and Padul, from the second half of the twentieth century, the municipalities of Sierra Nevada have experienced a population decrease (Prados Velasco and Valle Ramos 2010). Overall, the population in the area went from more than 133,000 inhabitants in 1940 to less than 87,000 in 2000, the population also experiencing an important aging process (Fig. 3 e). These population dynamics have greatly affected LEK systems, as LEK transmission channels broke with the young generations moving to cities.

It should be noted that, despite the general trend of population decrease, nowadays, some areas of the Alpujarras, the Marquesado del Zenete, or the Andaráx Valley show a population growth due to the arrival of the new population (Prados Velasco and Valle Ramos 2010). However, this new migration process does not necessarily reinforce the conservation and maintenance of the LEK, since the new inhabitants do not share the same worldview as local populations. In some areas (like Andarax Valley), the newly arrived population has imported intensive agricultural techniques, thus further contributing to the erosion of LEK systems.

A third driver of change affecting Sierra Nevada social-ecological system and associated LEK refers to ***government policies*** that have largely contributed to the transformation of the territory through land-use change, but which have also impacted LEK use and transmission.

**Fig. 3** **a** Shepherd milking in a traditional way, **b** Farmer with *patatas de la sierra*, **c** Shepherd with his flock, **d** Traditional braiding with esparto, **e** Old farmer using a traditional tool

Government policies affecting land use include the increase of forest extension resulting from *pinus* reforestations campaigns and the reduction of agro-pastoral activities associated with the creation first of the Natural Park (1989) and then of the National Park (1999). In the 1950's, when the area was still suffering the effects of the Spanish Civil War and experiencing a high migration rate, the state bought large extensions of land to councils and private owners to implement reforestation plans with the goals to reduce soil erosion and produce wood. Since 1956, government reforestation plans have resulted in a change in the vegetation cover in 42% of the territory of Sierra Nevada (Zamora et al. 2016). Decades later, other government policies, i.e., the declaration of Natural Park (1989) and National park (1999), generated additional changes by modifying and restricting the local population's access to several areas and resources

(Mena et al. 2014). Restricted activities included the collection of aromatic and medicinal plants, grazing in the areas of high peaks, as well as hunting of wild game species. These restrictions have affected the transmission of LEK related to such resources. For example, before such restrictions, it was common that families to have two residences, one in the village, which they occupied during the winter, and another in the high mountain area, sometimes a simple cabin, which they occupied during the summer months to take advantage of the pastures for cattle and agriculture of high mountain crops. The restriction of cultivating and grazing in high mountain areas, together with the changes in the professional interests of the new generations, has caused farmers, herders, and ranchers to change their habits and ways of working, including sharing tasks that facilitate knowledge transmission (Personal communication from local elders). Thus, families have ceased to cultivate in the high areas far from the nuclei of the villages, and groups of herders have ceased to share long periods of time in high mountain areas. Nowadays, lonely shepherds carry out daily grazing trips around the village (Fig. 3c) or leave the cattle to graze alone in the areas of high peaks, visiting it weekly or fortnightly.

Government policies also affect LEK through policies that put LEK-based traditional resource management systems in an administrative limbo. On one hand, legislation oriented to impose certain commercial standards in products has made it almost impossible to maintain a livelihood based on traditional practices. For example, the certification of seeds, the commercial preferences of crops and landraces that large trading companies buy, and the reduction in production prices due to agricultural and livestock technification, threaten the continuity of small producers who maintain traditional practices and associated knowledge.

On the other side, the multiplicity of rules imposed by different branches of the administration puts farmers between a rock and a hard place. Traditional resource management systems typically have a multifunctional nature (i.e., they provide multiple socioeconomic and environmental benefits), which hinders their classification in compartmentalized administrations. For example, irrigation communities are under the jurisdiction of the Andalusian Environment and Water Agency, which is part of the Regional Ministry of Agriculture, Livestock, Fisheries, and Sustainable Development. This Ministry has historically been oriented to modern and intensive agriculture, deploying plans for the technification of irrigation. But a good extension of the area in Sierra Nevada has been declared as a National Park, thus falling under the National Parks Autonomous Agency, which has an environmental focus. The implementation of policies from the two agencies translates into local communities receiving subsidies for opposite purposes. Thus, in recent years, the National Park has invested a large proportion of its budget in repairs of different stretches of the traditional irrigation system. At the same time, the Regional Ministry of Agriculture Livestock, Fisheries, and Sustainable Development offered financial aid to irrigation communities of Sierra Nevada to modernize agricultural areas by implementing drip irrigation systems. Traditional water management infrastructures and drip irrigation systems encapsulate opposing mindsets about water management: one is configured to redistribute water in a system where water is not an exclusively human good and the other is used with a purely instrumentalist mindset trying to maximize the benefits of a scarce resource. Such contradictory policies affect the social-ecological networks where irrigation systems are embedded and, as a consequence, irrigation communities become weak and fragmented, some of them being absorbed by local and regional governments losing part of their agency (Ruiz-Ballesteros and Gálvez--García 2014). The loss of norms and local institutions for self-governance, like irrigation communities, diminishes the local capacity of response against global challenges such as water scarcity, biodiversity loss, fires, or plagues, and reduces the local capacity for collective action.

Another driver of change in the Sierra Nevada's social-ecological system has been ***climate change***, which is affecting mountain regions with great intensity around the world (Zamora et al. 2017). In the Mediterranean region, the increase of the minimum and maximum temperatures and reduction of the amount of rainfall and snowfalls are reducing water availability, impacting ecosystems in multiple ways, and affecting negatively traditional livelihoods (Jiménez Olivencia 2010; Morales et al. 2019; Ruiz-Morales et al. 2020). Recent research in Sierra Nevada shows that local populations are aware of a great diversity of climate change impacts affecting the region, showing differences between the different geographical zones of Sierra Nevada. Climate change impacts related to temperatures, precipitation and droughts, rivers' flow, snow cover and snowfields, the abundance of wild fauna and crop's diseases and pests were the most frequently perceived, although the perception of impacts changes according to geographical zones and informants' characteristics (García-del-Amo et al. 2021). Moreover, Sierra Nevada inhabitants perceive climate change as the main driver of change acting in the region, particularly affecting their socioeconomic system through its cascading effects from the climatic, physical, and biological systems. For example, local inhabitants argue that changes in water availability, changes in pasture and crop productivity, and changes in crop diseases and pests have a high repercussion on Sierra Nevada's resilience. In addition, part of this LEK has had no time to adapt to the new uncertain context. For instance, making predictions about weather or water availability becomes much more difficult for peasants and herders.

Changes in the social-ecological system have made reliance on LEK less necessary for daily survival, affecting the *social recognition* of this knowledge system by local communities, including knowledge holders themselves. LEK is, by definition, a dynamic knowledge system that needs shared action contexts to stay alive (Berkes 2018), for which the lack of practical use has hampered LEK intergenerational transmission and renovation. In Sierra Nevada, the interruption of LEK transmission is often promoted by LEK knowledge holders who argue that they have worked hard "to give a better future to their children," offering them the possibility to study and engage in economic activities in the industry or service sectors (Guzmán Álvarez 2010a). These changes in values have also impacted the local identity and the prestige and consideration given to specific activities. For example, herders, who were traditionally considered important LEK-holders, have lost their social recognition in comparison to the "entrepreneurs"' of stabled cattle farms. A paradigmatic example can be found in the "*cabañuelas*," a traditional local weather predictions system. In some villages of Sierra Nevada, people who performed *cabañuelas* went from being one of the most respected and admired figures in the community to being socially downplayed, as the practice lost its social recognition and was considered outdated.

Altogether, the drivers of change described above threaten the continuity of the social-ecological system and the transmission of associated LEK, making the twenty-first-century scenario profoundly challenging for mountain communities.

## 4 LEK Role in Preserving the Current Social-Ecological System of Sierra Nevada

Despite the many changes described above, LEK systems continue to play a role in preserving Sierra Nevada's social-ecological system. In this section, we examine how LEK continues contributing to the preservation of the Sierra Nevada's social-ecological system through traditional water management, agriculture, and extensive cattle raising.

***Local knowledge on water management*** is essential for the proper functioning of Sierra Nevada's social-ecological system. As mentioned, the network of traditional water infrastructure favors aquifer recharge and helps conserve soil's phreatic level (Jódar et al. 2017; Martos-Rosillo et al. 2019) providing water for agriculture and human consumption for the local communities located on the slopes of the mountain. But water infrastructures are not operational without the associated local knowledge and enough members to maintain them. Water management is also at the basis of the local fabric and, even today, local knowledge on water management continues to help solve water-related conflicts between different social actors (peasants, herders, local population, and touristic properties) particularly during periods of water scarcity. Irrigation communities can predict the availability of water during the dry season, and measure and distribute water most fairly and efficiently. In addition, their practices and norms are flexible enough to allow the adaptation to unexpected events. For example, during the drought of 2012, the irrigation community of *La Taha*, in the High Alpujarra, managed water scarcity modifying the established irrigation turns, but also introducing temporary changes in the distribution between villages to take advantage of the available water (Gálvez-García 2015). Moreover, in several communities in Sierra Nevada, irrigation communities still have many active members and can create social pressure in defence of rights. This was seen, for example, when the regional administration announced a plan to transfer the water from Trevélez River to the coast for touristic and agricultural uses. The opposition of irrigation communities of Trevélez, Busquístar, Pórtugos, and La Taha, supported by local herders and neighbors' associations put a halt to the government plan. On the contrary, the lack of active irrigation communities in some villages of Sierra Nevada has generated important social conflicts. In those villages, the high percentage of abandoned fields and the reduced number of peasants have resulted in the appropriation of the local irrigation system by the local administration, which always prioritizes human consumption, particularly during the dry summer months (Personal observation in La Taha and Ohanes). This decision goes to the detriment of farmers, who maintain the water management system throughout the year, but are not entitled to manage water in the most critical period of the agricultural cycle.

The maintenance of ***traditional agricultural systems***, and particularly local landraces and associated knowledge, contributes to preserving the resilience of the social-ecological systems against future threats (Martin et al. 2019; Labeyrie et al. 2021). Although agriculture is no longer the main economic engine of Sierra Nevada, it continues to play an important role in the maintenance of its social and ecological fabric, particularly subsistence agriculture (Fig. 4). Nowadays, local landraces are not usually found in commercial channels, but some small farmers still cultivate local varieties, preserving part of the agricultural genetic pool that existed in the region. In the testimonies, we have mentioned the bigger resistance of local varieties to pests, as the example of cherry trees or apple trees in the High Alpujarra. The altitudinal gradient of the area allowed to intercrop different varieties and types of crops, which have fewer pests than in lower areas. However, nowadays, due to climate change effects, agricultural seasons are longer, extending the cropping seasons. However, higher temperatures are

**Fig. 4** Local farmers applying traditional irrigation techniques, *riego a manta*

allowing the survival of some pests even at high elevations, adding an extra difficulty to Sierra Nevada farmers.

LEK associated with ***grazing and extensive cattle raising*** also contributes to maintaining the stability of the social-ecological system of Sierra Nevada. Researchers have shown that the practice of grazing and extensive cattle raising favors the maintenance of high mountain pastures in Sierra Nevada, as well as the provision of the ecosystem services generated (Varela and Robles-Cruz 2016). Similarly, sustainable grazing pressure prevents soil erosion, maintaining its fertility, and controlling the biomass amount, acting as traditional wildfire prevention (Ruiz-Mirazo and Robles 2012; Varela et al. 2018). Traditional grazing practices have played a fundamental role in preserving the biodiversity of Sierra Nevada, favoring the dispersal of endozoochory seeds of high mountain pastures (Ramos Font et al. 2015), without negatively affecting endangered protected species (Robles et al. 2016). Moreover, the practice of grazing the Mediterranean scrub spp. is even beneficial for human health, as it improves the quality of goat milk (Gutiérrez-Peña et al. 2013). Currently, the number of shepherds and cattle in the area is much lower than it was several decades ago (Ruiz-Morales et al. 2020), which impacts the general configuration of the social-ecological system. Shepherds, during grazing practices, interact with different elements of the ecosystems, contributing to the preservation of the multifunctionality of heterogeneous cultural landscapes (Mena et al. 2014). Like traditional water management, grazing practices rely on LEK and the close and continuous relationship of shepherds with the different elements of the ecosystems. For example, shepherds know which natural springs have water in the different seasons; they also know the order in which livestock should be taken to the different pastures, so the feeder lasts longer; the date and the area in which the different pastures are available; the wild species with high nutritional value like *mierga, albejana, jaramargo, trigillos, or ballicos*; the toxic species in the region, like the *gayomba*; the plant species that have veterinary properties but that can harm animals when eaten in excess, like the *ajedrea* or clovers, or even the wild edible plants they can consume when traveling like *hinojos, collejas*, mushrooms species, and wild berries. As with agricultural varieties, there are still people who have local livestock breeds such as the *pajuna* cow, which is adapted to the scarce local resources, and better support the inclemency of the extreme climates of the region (Ruiz-Morales et al. 2020).

In recent years, the importance of LEK systems in the maintenance of Sierra Nevada's social-ecological system has been increasingly acknowledged at several levels. Within academia, a research group from the University of Granada, has co-led since 2014 an interdisciplinary European FP-7 project, Mediterranean Mountainous Landscapes (MeMoLa) to enhance the importance of the multifunctionality of cultural mountain landscapes. Participants in the project include several research groups, the Natural Park administration, local and national civil associations, and some irrigation communities. This project studied the importance of LEK in the recognition of soil quality to improve the distribution of crops and landraces, conducted hydrological research to study the ecological benefits of traditional water management systems, carried out educational activities, and proposed modifications in the hydrological policies, giving visibility to the problems of the irrigation communities of Sierra Nevada. For example, one of the outcomes of this project has been a proposal to reconsider the high mountain irrigation communities using *acequias de careo* as an exceptional case by the Hydrographic Confederation, due to their ecological contribution. They have also promoted the involvement of university students in the annual cleaning and reparation of irrigation ditches, thus favoring the exchange of knowledge and the approach of the students to the irrigation communities' reality (Fig. 5).

At the level of the administration, there have also been efforts to promote the maintenance of traditional practices and associated knowledge. A paradigmatic example of this effort has been the creation of a shepherds' school (www.escueladepastoresdeandalucia.es), currently in its tenth edition. This initiative encourages the maintenance of traditional practices by offering training to young people interested in extensive livestock farming. Training includes issues such as the ecological importance of extensive livestock farming and its impact on ecosystems (Ruiz-Morales et al. 2020). The shepherds' school also aims to revalue the social recognition of this profession, which has serious problems of generational renewal. The administration has also recognized as autochthonous the "*pajuna*" breed, conducting analysis to study its quality (Horcada-Ibáñez et al. 2016) and promoting its commercialization and production. Also at the administration level, managers of Sierra Nevada Natural and National park are conducting an adaptive management project (Adaptamed) for the development of adaptation measures for climate change impacts, in which they recognize the ecological value of historical irrigation systems and associated LEK. This project, which will include local irrigation communities, ranchers, and shepherds in workshops and meetings for decision-making, will be an exceptional opportunity to integrate local communities and their LEK in the co-management of the territory, increasing their active role in governance. These actions help to revitalize LEK in Sierra Nevada in a practical way, by recovering its daily use.

Finally, LEK is also experiencing some revitalization through the influence of the arrival of a new population to the area. In the last decades, the neo-rural population, also known as amenity migrants, has brought to the area new practices and needs that have generated different types of impacts. Despite their diversity, this movement has been able to organize itself to improve the commercialization of local products. This is the case of the cooperative *Las Torcas* in the Alpujarra, or *Valle y Vega*, linked to the valley floor of Lecrín Valley and the valley floor of Granada at the foot of the mountain. This new population has introduced new practices and knowledge (i.e., the use of *consuelda* or *cola de caballo* to combat pests, biodynamic agriculture, sprinkler irrigation), but they have also contributed to the maintenance of local knowledge systems. For example, the neo-rural community is responsible for the reintroduction of seeds that were locally lost (e.g., the *pico de pájaro* lettuce in High Alpujarra) and there are examples where they participated in the local work to clean annually the acequias, becoming new repositories of local water management

**Fig. 5** Group of students helping to clean and repair an *acequia de careo* (MeMoLa project)

knowledge. Many of the amenity migrants of the mountain villages start their gardens using the knowledge they acquire from locals, enriching it with their own background. For instance, the Hortigas Agroecological Cooperative, formed by urban migrants, without agricultural background, learned from the local residents of Dúrcal, including the elders. These exchanges favor the transmission of their LEK to new populations, who modify it with their own knowledge systems, in this case, agro-ecological knowledge. This cooperative, with more than 100 members, has been operating for 15 years, linking people from the city of Granada, usually students, with Dúrcal farmers. Since the members of the cooperative have to work the gardens at least once a month, this dynamic favors the transmission of knowledge. Overall, the interactions of the neo-rural population with the local communities have in many cases contributed to the valorization of LEK and the development of other agroecological initiatives.

## 5 The Future of LEK in the Context of Global Environmental Change

LEK systems are not ethereal, so LEK cannot be collected and conserved for future adaptation and implementation. Rather LEK is embodied in knowledge holders, as part of their heritage and culture. LEK is only alive when used by local communities to continue to live in a specific environment. Therefore, although LEK is always changing and facing new challenges, to survive it must preserve its usefulness for the local population—not only in an economic way, but also performing symbolic, identity, and social roles.

Traditional agricultural and livestock farming practices, and traditional irrigation infrastructures and their associated knowledge have proven to be fundamental for the maintenance of the resilience of Sierra Nevada's social-ecological system, and therefore could also be beneficial for the conservation of their ecosystems against current impacts and future threats. Therefore local and regional administrations need to take additional steps on the process of integrating this knowledge system in the management of the territory. Indeed, co-management, i.e., management involving local communities and administration, is becoming a common practice growingly implemented in protected areas and indigenous territories (Krupnik et al. 2010; Armitage et al. 2011; Danielsen et al. 2014; Reed et al. 2016). The co-management approach would imply that the local population of Sierra Nevada should be a fundamental stakeholder, involved in all the decisions about the future of their territory becoming more involved in the management of the territory and natural resources. In particular, since the traditional water management system is the backbone of Sierra Nevada, and the LEK associated is fundamental to preserve the ecological stability of the region (Jódar et al. 2017; Martos-Rosillo et al. 2019), this centuries-old knowledge on how, where, when, or how much water needs to be redistributed along the slopes of Sierra Nevada to have correct functioning of the ecosystems of different bioclimatic zones should be integrated with the co-management of Sierra Nevada. For example, during our ethnographic work, several people mentioned local priorities such as investment in infrastructures of water-supply for livestock on the high-lands during the summer or the maintenance of irrigation systems, beyond the current focus on the *acequias de careo* located on the top of the mountain but also in the lower areas of irrigation systems and hydraulic infrastructures (*cimbras*, *minas* or *qanats*, *partidores*, *albercas*, or fountains), with higher repercussion on the irrigation communities. A co-management system should be able to consider such local priorities at the same level as the priorities determined by the administration. Irrigation communities should be key actors when assessing the management of hydric resources. Moreover, their inclusion through a collaborative network would favor communication between the different irrigating communities and the exchange of knowledge, strengthening the ties between them and enriching their collective LEK.

In the same way, shepherds and ranchers should be effectively integrated and take part in adaptive management plans of high-mountain meadows of Sierra Nevada and wildfire prevention systems (Ruiz-Mirazo and Robles 2012; Varela et al. 2018). Livestock can prevent the rewilding of the territory, which can happen in national parks after removing traditional livestock practices (Guadilla-Sáez et al. 2019), and plays a fundamental role in preserving the biodiversity of Sierra Nevada, hence shepherds and ranchers should be engaged in biodiversity conservation and seed dispersion plans (Ramos Font et al. 2015; Robles et al. 2016). The Andalusian Shepherds School presents a great opportunity to create this change of vision in the management of the territory, involving the local population, adding value, and revitalizing the LEK of their communities, also creating a change in the social perception of these traditional activities (Ruiz-Morales et al. 2020). A similar role can be played by other initiatives led by extensive livestock farming women, who are trying to give visibility and revalue the historical role of women in livestock contexts and their current contributions to this sector (Fernández-Giménez et al. 2019).

To ensure the maintenance of these traditional practices and associated LEK, local communities need to continue benefiting, both economically and socially, from their knowledge (Foley and McCay 2014; Oosterveer et al. 2014). However, maintaining the benefits generated by LEK will be difficult if there is no social recognition and appreciation of the value of LEK and its bearers. The high physical effort of products derived from traditional practices and their low profitability results in a lack of generational uptake, as the new generations do not prioritize traditional livelihoods (Fig. 6). Beyond the theoretical recognition of LEK and its holders, substantial support from administrations that add economic value to products derived from traditional practices is needed. For example, the current legislation has different quality standards (Protected Designation of Origin, Protected Geographical Indication, Traditional Specialty Guaranteed, and Artisanal Food certificate) which currently only include a few products from Sierra Nevada. This legislation could be modified to revalue processed and raw agricultural and livestock products coming from traditional practices and local varieties and breeds of local communities and small producers or rural areas (Ruiz Morales et al. 2012). The Artisanal Food certificate recognizes the traditional artisanal exceptionality, allowing it to carry out product elaboration following different procedures from the rest of the production companies in the same sector, as is the case of artisan cheeses (Decree 352/2011). Similarly, being within a protected area could be recognized with the quality recognition of natural parks of Andalusia, which encourages the sustainable development of local populations around protected areas.

Such changes in legislation could be accompanied by other actions, to enable small farmers to commercialize the local agricultural varieties they still have. The promotion of the Andalusian seed network and the recognition and cataloging of the autochthonous fruit and vegetable varieties of Sierra Nevada, which currently includes 18 local varieties, will favor their social demand and conservation. Due to the proximity between producer and consumer, local consumption groups and short consumption channels have always been seen as good ways to commercialize local products (Vara-Sanchez et al. 2021), so legally favoring this type of channel is essential. In the same way, information and educational campaigns explaining the added value of artisanal products, such as the ecological and social benefits of consuming local agricultural varieties, could increase consumers' awareness. Cooperativism among small producers is essential to create alternatives in the primary and secondary sectors, which are highly industrialized and technified. In this line, different cooperatives of women shepherds are trying to revitalize the Merino wool sector in Spain, with historical worldwide recognition. These cooperatives

**Fig. 6** Ranchers of different generations grouping cows

highlight the importance of women's LEK in this sector, from the breeding and production of the animals to the processing, transformation, and final distribution of wool (Herrero 2020), which should be another alternative to explore in Sierra Nevada.

In general, public administrations could support the maintenance and regeneration of LEK by encouraging short marketing channels and supporting the organization of small independent farmers in truly democratic cooperatives. The maintenance of LEK associated with local landraces and autochthonous livestock breeds, as well as the traditional processing of primary products, can increase the food sovereignty of the Sierra Nevada communities (Pimbert 2006). The use of LEK can also increase the economic possibilities for the population and open new opportunities for young and women in rural regions, creating social and economic resilience, and reducing their dependence on other sectors such as tourism or the demands of the global market that do not usually favor the added value of the associated LEK.

Finally, academia is also in a process of increasingly valuing LEK, in some cases promoting the co-production of new actionable knowledge that brings together scientific knowledge and LEK to address current sustainability challenges (Tengö et al. 2017; Norström et al. 2020). Currently, researchers are working on the co-production of knowledge in fields like agroecology, biodiversity conservation, or natural resources management (Altieri 2002; Armitage et al. 2011). Nowadays, there is also a great concern about the need to also include the LEK of IPLC in climate change research (Krupnik et al. 2010; Berkes 2017; Díaz et al. 2019). In that sense, there is extensive literature showing the variety of climate change impacts in the climatic, physical, biological, and human systems perceived by local communities, and the level of detail that those communities are able to detect describing those impacts (Reyes-García et al. 2019a). Moreover, a recent study has shown similar results with local communities of Sierra Nevada, who are direct witnesses of climate change impacts on their social-ecological system, for which their knowledge would help to substantially improve the current databases of climate change impact at a local level, contributing to the development of local mitigation and adaptation measures (García-del-Amo 2021). Integrating local communities of Sierra Nevada in the monitoring of environmental changes happening in the region would contribute not only to improve the monitoring system, but also to design more realistic management plans in the context of climate change. Global change is transforming the socio-ecological systems of Sierra Nevada in many different ways, and the creation of interdisciplinary working groups including decision-makers, researchers, technicians, members of the irrigation community, shepherds, ranchers, and beekeepers might be the best way to obtain a correct assessment of the biophysical and sociocultural changes happening in Sierra Nevada (Garcia-del-Amo et al. 2021). These interdisciplinary working groups, in their participatory process, must be focused and designed so that they can have a real impact on the policies and norms that regulate the management of the territory.

## 6 Conclusions

In this chapter, we have analyzed the importance and contributions of LEK to maintain the proper functioning of Sierra Nevada's social-ecological system. Throughout the twentieth and twenty-first centuries, the different drivers of global environmental change have greatly eroded the proper functioning of this social-ecological system and endangered the transmission of LEK to the new generations.

The numerous challenges derived from global changes require considering all possible sources of knowledge in addressing them. A way to better assess LEK would be to engage local communities in the co-governance of the territory through the co-management and co-production of knowledge. Promotion of local varieties and breeds and favoring the continuity of small local producers and cooperatives will be necessary to ensure the future of local communities and the continuity of these traditional activities. Interdisciplinary and inclusive methodologies can help to avoid power imbalances and to ensure mutual understanding between both sources of knowledge to achieve true participation of the local population, empowering them in the management and conservation of the territory.

If economic and legislative restrictions block local communities to maintain their traditional activities; if the pressure upon their economies and ways of managing environmental resources becomes excessive; if there is not a legitimation of their ways of life; if new formulas for agropastoralism activities are not provided, it will be hard for LEK to survive and become an effective tool for dealing with the new challenges. More than protecting LEK, it is important to protect the needs of those who hold it.

**Acknowledgements** We thank the collaboration of people from Sierra Nevada who shared with us their valuable, but so often underrated, knowledge. We acknowledge financial support from the Spanish Ministry of Economy and Competitiveness, through the María de Maeztu Programme for Units of Excellence in R&D (MdM-2015-0552), the Spanish Ministry of Science, Innovation, and Universities through the Juan de la Cierva Incorporación Programme (IJCI-2017-33405), the project grant CSO2014-59704-P, and funding from the European Research Council (ERC) under grant agreement No 771056-LICCI-ERC-2017-COG. This work contributes to the "María de Maeztu" Programme for Units of Excellence (CEX2019-000940-M).

## References


Acha A, Newing HS (2015) Cork oak landscapes, promised or compromised lands? A case study of a traditional cultural landscape in Southern Spain. Hum Ecol 43:601–611

Alonso F, Gonz C, Jos J, et al (2006) La uva de Almería. Dos siglos de cultivo e historia de la variedad Ohanes, 1st edn. El Ejido (ALmería)

Altieri MA (2002) Agroecology: the science of natural resource management for poor farmers in marginal environments. Agric Ecosyst Environ 93:1–24

Armitage D, Berkes F, Dale A et al (2011) Co-management and the co-production of knowledge: learning to adapt in Canada's Arctic. Glob Environ Chang 21:995–1004

Aumeeruddy-Thomas Y, Moukhli A, Haouane H, Khadari B (2017) Ongoing domestication and diversification in grafted olive–oleaster agroecosystems in Northern Morocco. Reg Environ Chang 17:1315–1328

Babai D, Molnár Z (2014) Small-scale traditional management of highly species-rich grasslands in the Carpathians. Agric Ecosyst Environ 182:123–130

Berkes F (2017) Environmental governance for the anthropocene? Social-ecological systems, resilience, and collaborative learning. Sustain

Berkes F (2018) Sacred ecology, 4th edn. Routledge

Bicker A, Ellen R, Parkes P (2003) Indigenous enviromental knowledge and its transformations. Routledge

Bird RB (2015) Disturbance, complexity, scale: new approaches to the study of human-environment interactions. Annu Rev Anthropol 44:241–257

Biró É, Babai D, Bódis J, Molnár Z (2014) Lack of knowledge or loss of knowledge ? Traditional ecological knowledge of population dynamics of threatened plant species in East-Central Europe. J Nat Conserv 22:318–325

Blanca G, López-Onieva MR, Lorite J, et al (2001) Flora amenazada y endémica de Sierra Nevada

Blondel J (2006) The "design" of Mediterranean landscapes: a millennial story of humans and ecological systems during the historic period. Hum Ecol 34:713–729

Brondizio ES, Le Tourneau FM (2016) Environmental governance for all. Science 352(80-):1272–1273

Cámara-Leret R, Fortuna MA, Bascompte J (2019) Indigenous knowledge networks in the face of global change. Proc Natl Acad Sci U S A 116:9913–9918

Cressier P (1995) Hidráulica rural tradicional de origen medieval en Andalucía y Marruecos. Elementos de análisis práctico. In: González Alcantud J, Malpica Cuello A (eds) El agua: mitos, ritos y realidades. Anthropos Editorial, Barcelona

Danielsen F, Mendoza MM, Tagtag A et al (2007) Increasing conservation management action by involving local people in natural resource monitoring. Ambio 36:566–570

Danielsen F, Topp-Jørgensen E, Levermann N et al (2014) Counting what counts: using local knowledge to improve Arctic resource management. Polar Geogr 37:69–91

Díaz S, Settele J, Brondízio ES, et al (2019) Pervasive human-driven decline of life on Earth points to the need for transformative change. Science 366(80-)

Ellen R (2007) Local science vs. global science: approaches to indigenous knowledge in international development. Berghahn books

Espinar Moreno M (1989) Estudio sobre la propiedad particular de las aguas de la acequia de Jarales (1267–1528). Problemas de abastecimiento urbano y regadíos de tierras entre las alquerías de Abrucena y Abla. I Coloq. Hist. y Medio Físico. El agua en Zo. áridas 249–266

Fernald A, Guldan S, Boykin K et al (2015) Linked hydrologic and social systems that support resilience of traditional irrigation communities. Hydrol Earth Syst Sci 19:293–307

Fernández-Giménez ME, Oteros-Rozas E, Ravera F (2019) Co-creando conocimientos para la acción con mujeres del entorno pastoralista en España. Cabezón de la Sal, Cantabria, España

Fernández-Llamazares Á, Garteizgogeascoa M, Basu N, et al (2020) A state-of-the-art review of indigenous peoples and environmental pollution. Integr Environ Assess Manag 16:324–341

Foley P, McCay B (2014) Certifying the commons: eco-certification, privatization, and collective action. Ecol Soc

Gálvez-García C (2015) Saberes locales en el mundo global: Huertas, agua y conocimiento agroecológico en la Alpujarra Alta Occidental. Univ Pablo Olavide 442

García-del-Amo D, Mortyn PG, Reyes-García V (2021) Local perception of climate change impacts in Sierra Nevada, Spain. Analyzing sociodemographic and geographical patterns. Reg Environ Change (under revision)

Gómez-Baggethun E, Corbera E, Reyes-García V (2013) Traditional ecological knowledge and global environmental change : research findings and policy implications. Ecol Soc 18

Guadilla-Sáez S, Pardo-de-Santayana M, Reyes-García V (2019) The role of traditional management practices in shaping a diverse habitat mosaic in a mountain region of Northern Spain. Land Use Policy 89:104235

Gutiérrez-Peña R, Delgado-Pertíñez M, Fernández-Cabanás V, et al (2013) Composición en ácidos grasos y contenidos de vitaminas A y E de la leche de cabra de la raza payoya en sistemas de pastoreo arbustivo-mediterráneo. Pastos 43:36–43

Guzmán Álvarez JR, Cerrillo, Navarro RM, (Coords.) (2010) El agua domesticada, los paisajes de los regadíos de montaña en Andalucía. Agencia Andaluza del Agua. Consejería de Medio Ambiente, Junta de Andalucía, Sevilla

Guzmán Álvarez JR (2010a) Las acequias de careo de Sierra Nevada. In: El agua domesticada. Los paisajes de los regadíos de montaña en Andalucía. Agencia Andaluza del Agua, Consejería de Medio Ambiente, Junta de Andalucía, pp 284–291

Guzmán Álvarez JR (2010b) Usos y costumbres para el reparto del agua. In: El agua domesticada. Los paisajes de los regadíos de montaña en Andalucía. Agencia Andaluza del Agua, Consejería de Medio Ambiente, Junta de Andalucía, pp 167–189

Hernández-Morcillo M, Hoberg J, Oteros-Rozas E, et al (2014) Traditional ecological knowledge in Europe: Status Quo and insights for the environmental policy agenda. Environment 56:3–17

Herrero MT (2020) Women and wool in Dehesas: gender overview of the wool supply chain in Spain cultural landscapes and biodiversity in the Mediterranean Basin. DC, USA, Washington

Hill R, Adem Ç, Alangui WV et al (2020) Working with indigenous, local and scientific knowledge in assessments of nature and nature's linkages with people. Curr Opin Environ Sustain 43:8–20

Horcada-Ibáñez A, Polvillo-Polo O, Lafuente-García A, et al (2016) Beef quality of native Pajuna breed calves in two production systems. Agrociencia (Montecillo) 50:167–182

Iniesta-Arandia I, Garcia-del-Amo D, García-Nieto AP, et al (2014) Factors influencing local ecological knowledge maintenance in Mediterranean watersheds: insights for environmental policies. Ambio 44:285–296

Jiménez Olivencia Y (2010) Consecuencias del abandono del regadío en la montaña mediterránea. In: Agencia Andaluza del Agua de la Consejería de Medio Ambiente de la Junta de Andalucía y la Universidad de Córdoba (ed) El agua domesticada. Los paisajes de los regadíos de montaña en Andalucía. Granada, pp 508–513

Jódar J, Cabrera JA, Martos-Rosillo S et al (2017) Groundwater discharge in high-mountain watersheds: a valuable resource for downstream semi-arid zones. The case of the Bérchules River in

Sierra Nevada (Southern Spain). Sci Total Environ 593–594:760–772

Johannes RE (1993) Integrating traditional ecological knowledge and management with environmental impact assessment. In: Inglis JT (ed) Traditional ecological knowledge. Concepts and cases. International Program on Traditional Ecological Knowledge International Development Research Centre, Ottawa, Ontario

Krupnik I, Aporta C, Laidler GJ, et al (2010) SIKU: Knowing our ice: documenting Inuit sea ice knowledge and use

Labeyrie V, Renard D, Aumeeruddy-Thomas Y, et al (2021) The role of crop diversity in climate change adaptation: insights from local observations to inform decision making in agriculture. Curr Opin Environ Sustain 51:15–23

Lepofsky D (2009) The past, present, and future of traditional resource and environmental management. J Ethnobiol 29:161–166

Lyver POB, Timoti P, Davis T, Tylianakis JM (2019) Biocultural hysteresis inhibits adaptation to environmental change. Trends Ecol Evol 34:771–780

Martin AR, Cadotte MW, Isaac ME et al (2019) Regional and global shifts in crop diversity through the Anthropocene. PLoS ONE 14:1–18

Martin Civantos JM (2011) Working in landscape archaeology: the social and territorial significance of the agricultural revolution in al-Andalus. Early Mediev Eur 19:385–410

Martos Rosillo S, Ruiz Constán A, Martín Civantos JM, et al (2018) Careos : siembra y cosecha de agua en la cuenca del río Bérchules (Sierra Nevada, Granada). Instituto Geológico y Minero de España (IGME)

Martos-Rosillo S, Ruiz-Constán A, González-Ramón A, et al (2019) The oldest managed aquifer recharge system in Europe: New insights from the Espino recharge channel (Sierra Nevada, southern Spain). J Hydrol 578

McElwee P, Fernández-Llamazares Á, Aumeeruddy-Thomas Y, et al (2020) Working with indigenous and local knowledge (ILK) in large-scale ecological assessments: reviewing the experience of the IPBES Global Assessment. J Appl Ecol 57:1666–1676

Mena Y, Ruiz FA, Gutiérrez R et al (2014) La función de la ganadería en el mantenimiento y el aprovechamiento de los montes mediterráneos. Ambient La Rev Del Minist Medio Ambient 108:52–67

Molnár Z, Kelemen A, Kun R et al (2020) Knowledge co-production with traditional herders on cattle grazing behaviour for better management of species-rich grasslands. J Appl Ecol 57:1677–1687

Morales FDAR, Genís JMC, Guerrero YM (2019) Current status, challenges and the way forward for dairy goat production in Europe. Asian-Australasian J Anim Sci 32:1256

Natcher DC, Brunet ND (2020) Extractive resource industries and indigenous community-based monitoring: cooperation or cooptation? Extr Ind Soc 7:1279–1282

Norström AV, Cvitanovic C, Löf MF et al (2020) Principles for knowledge co-production in sustainability research. Nat Sustain 3:182–190

Oosterveer P, Adjei BE, Vellema S, Slingerland M (2014) Global sustainability standards and food security: exploring unintended effects of voluntary certification in palm oil. Glob Food Sec 3:220–226

Ostrom E (1990) Governing the commons. Cambridge University Press, Cambridge, United Kingdom, The evolution of institutions for collective action

Ostrom E (2009) A general framework for analyzing sustainability of social-ecological systems. Science 325(80-):419–422

Padilla FM, Pugnaire FI (2006) The role of nurse plants in the restoration of degraded environments. Front Ecol Environ 4:196–202

Pimbert M (2006) Transforming knowledge and ways of knowing for food sovereignty. Lied

Prados Velasco MJ, Valle Ramos C del (2010) Naturbanización y cambios en la población de los espacios naturales de Doñana y Sierra Nevada. Doc d'anàlisi geogràfica 56:435–460

Pulido-Bosch A, Sbih Y Ben (1995) Centuries of artificial recharge on the southern edge of the Sierra Nevada (Granada,Spain). Environ Geol 26:57–63

Rafidison VM, Rakouth B, Carrière SM et al (2020) Multiple values of isolated and clusters of Ficus tree species protected by Betsileo farmers in rural landscapes in Madagascar: implications for biodiversity conservation. Springer, Netherlands

Ramos Font ME, González Rebollar JL, Robles Cruz AB (2015) Endozoochorous dispersal of wild legumes: from seed recovery to field establishment. Ecosistemas 24:14–21

Reed J, Van Vianen J, Deakin EL et al (2016) Integrated landscape approaches to managing social and environmental issues in the tropics: learning from the past to guide the future. Glob Chang Biol 22:2540–2554

Reyes-García V, Fernández-Llamazares Á, McElwee P, et al (2019a) The contributions of indigenous peoples and local communities to ecological restoration. Restor Ecol 27:3–8

Reyes-García V, García-del-Amo D, Benyei P, et al (2019b) A collaborative approach to bring insights from local indicators of climate change impacts into global climate change research. Curr Opin Environ Sustain 1–8

Reyes-García V, Fernández-Llamazares Á, Aumeeruddy-Thomas Y, et al (2021) Recognizing indigenous peoples' and local communities' rights and agency in the post-2020 Biodiversity Agenda. Ambio

Robles AB, Ramos ME, Salazar C, González Rebollar JL (2016) Response of vegetation to exclusion and grazing in Mediterranean high-mountain wet pastures (Sierra Nevada, Granada, Spain). Options Méditerranéennes Ser A Mediterr Semin 245:241–245

Rodriguez Martínez F (2001) Las montañas andaluzas en la encrucijada del desarrollo rural. Ería 54–55:129–139

Ruiz Morales FA, García LN, Sayadi S, Castel Genis, José María Mena Guerrero Y (2012) Comercialización de los quesos tradicionales andaluces. Instituto de Investigación y Formación Agraria y Pesquera Junta de Andalucía. Consejería de Agricultura, Pesca y Medio Ambiente Este, Sevilla

Ruiz-Ballesteros E, Gálvez-García C (2014) Community, common-pool resources and socio-ecological systems: Water management and community building in southern Spain. Hum Ecol 42:847–856

Ruiz-Mirazo J, Robles AB (2012) Impact of targeted sheep grazing on herbage and holm oak saplings in a silvopastoral wildfire prevention system in south-eastern Spain. Agrofor Syst 86:477–491

Ruiz-Morales FA, Vázquez M, Camúñez JA et al (2020) Characterization and challenges of livestock farming in Mediterranean protected mountain areas (Sierra Nevada, Spain). Spanish J Agric Res 18:1–14

Soto Fernández D, Herrera González de Molina A, González de Molina Navarro M, Ortega Santos A (2007) La protesta campesina como protesta ambiental, siglos XVIII–XX. Hist Agrar Rev Agric e Hist Rural 277–302

Tengö M, Hill R, Malmer P et al (2017) Weaving knowledge systems in IPBES, CBD and beyond—lessons learned for sustainability. Curr Opin Environ Sustain 26–27:17–25

Thaman R, Lyver PO, Mpande R, et al (2013) The contribution of indigenous and local knowledge systems to IPBES. Thaman R, Lyver P, Mpande R, Perez E, Cariño J, Takeuchi K (eds) The contribution of indigenous and local knowledge systems to IPBES: building synergies with science. IPBES Expert Meeting Report, UNESCO/UNU. Paris, UNESCO, p 49

Vahí Serrano A, Prados Velasco MJ (2011) Acequias de careo en el Parque Nacional de Sierra Nevada (España). Interacciones entre patrimonio territorial y paisajes culturales. II Congr Iberoam y X Jorn "Técnicas Restauración y Conserv del Patrimonio"

Vara-Sánchez I, Gallar-Hernández D, García-García L, et al (2021) The co-production of urban food policies: Exploring the emergence of new governance spaces in three Spanish cities. Food Policy 103

Varela E, Robles-Cruz AB (2016) Ecosystem services and socio-economic benefits of Mediterranean grasslands. Options Méditerranéennes Série A Séminaires Méditerranéens 13–27

Varela E, Górriz-Mifsud E, Ruiz-Mirazo J, López-i-Gelats F (2018) Payment for targeted grazing: integrating local shepherds intowildfire prevention. Forests 9

Zamora R, Pérez-Luque AJ, Bonet FJ, et al (2016) Global Change Impacts in Sierra Nevada: Challenges for Conservation. Consejería de Medio Ambiente y Ordenación del Territorio, Junta de Andalucía, Granada, Spain., Granada

Zamora R, Antonio JP, Bonet FJ (2017) Monitoring global change in high mountains. In: High mountain conservation in a changing world, pp 385–413

Zimmerer KS, de Haan S (2019) Agrobiodiversity. MIT press, Integrating knowledge for a sustainable future

# Data Model, E-Infrastructure Services, and the Virtual Research Environment (VRE)

Pablo David Guerrero Alonso, Manuel Merino Ceballos, Ricardo Moreno Llorca, Andrea Ros Candeira, and Regino Zamora

**Abstract**

Sierra Nevada is a unique setting in which a broad and extensive research tradition converges in both time and space with intensive monitoring of the environment. Our aim is to establish a large data set together with an information network based on a powerful e-Infrastructure of communications, supercomputing, and distributed "cloud." This e-infrastructure will be made available to the scientific community, environmental managers, and the general public. It will allow, among many other things, access to large volumes of data including biotic components (fauna and flora) as well as abiotic ones (atmospheric, terrestrial, and freshwater elements). This will allow the user to work with analytical tools in a number of virtual research environments (VREs) such as virtual laboratories for researchers, support tools for environmental managers, or social science data for the general public. To answer the questions and provide the information required by the different users of the system, we propose the creation of the following VLABs: Ecosystems, Species, and Climate. These laboratories will have a specific part (associated with their subject matter) and another transversal part that can be extended to any laboratory of the same type. Laboratories will emerge from the scientific proposal resulting from the use of e-infrastructure. The following have been identified as examples: (i) Climate: Meteorology in Sierra Nevada environment—VRE ClimaNevada; (ii) Species: Iberian ibex (*Capra pyrenaica*)—VRE Capra; (iii) Ecosystems: Sierra Nevada high-mountain lakes—VRE MountainLakes. All laboratories will have tools that can be structured in the following interdependent hierarchical levels: Access to data, Statistics, GIS & Artificial Intelligence, Modelling and Simulation. Currently, a multitude of initiatives around the world promote the creation and use of VREs. These are web-based, community-oriented, flexible, and secure collaborative working environments designed to meet the premises of Open Science. In the present study, we identify and analyze the main challenges to be solved (starting with the data) to fully achieve the proposed vision.

**Keywords**

Sierra Nevada • Virtual Research Environment (VRE) • Virtual Labs (VLABs) • FAIR principles • Data management plan • Open science • Metadata

## 1 Introduction

Broad-scale, interdisciplinary scientific research, such as that conducted in Sierra Nevada, address multifaceted goals and problems that require the integration of data from multiple sources and the application of complex analytical procedures (Rüegg et al. 2014). The monitoring program implemented by the Sierra Nevada Global Change Observatory is designed to use all the information compiled in long-term research related to Sierra Nevada pertaining to different disciplines over the last several decades. Biodiversity, Ecosystems, and Global Change research areas present extremely diverse data sources (e.g. biological collections, measurements, observational data, meteorological projections, satellite images, genetic analyzes, etc.). A growing demand for more detailed, high-quality data and information concerning natural resources and ecosystem functions calls for trained personnel (e.g., resource specialists, data managers, researchers) working in collaboration to manage data and broad information assets (Cook and Lineback 2008).

P. D. Guerrero Alonso (✉) · M. Merino Ceballos · R. Moreno Llorca · A. Ros Candeira · R. Zamora
Andalusian Institute for Earth System Research (IISTA), University of Granada, Granada, Spain

R. Zamora
Department of Ecology, Faculty of Sciences, University of Granada, Granada, Spain


R. Zamora and M. Oliva (eds.), *The Landscape of the Sierra Nevada*,
https://doi.org/10.1007/978-3-030-94219-9_22

A fundamental pillar of our research proposal revolves around the pressing need to improve the reproducibility of ecological research. Modern ecology, like many other scientific disciplines, relies on computation as a crucial component of the research process. In recent years, however, it has become clear that failure to follow appropriate statistical and programming procedures has led to a "reproducibility crisis" in many scientific disciplines, where many results fail to be reproduced or replicated due to flawed statistical reasoning or inadequate computational practice (Stodden et al. 2016). The practice of modern and future science thus requires combining substantive expertise on a subject speciality (ecology in our case) with the necessary statistical and computational skills to be able to appropriately manage large datasets, integrate and assimilate them into reproducible, open databases, and build up robust statistical models of biodiversity patterns and ecological processes. The demand for reproducible research from funders, journals, and society is growing fast, and we must adapt ensuring we follow appropriate data science techniques and reproducible workflows (Peng 2011; Stodden et al. 2016).

With regard to data access, two challenges must be overcome: technological and social (Reichman et al. 2011). The first is related to the inherent complexity of ecological information, while the second reflects the need to remedy certain administrative and social reluctance to share data. The latter is especially paradoxical when it comes to research or monitoring which is financed with public funds, since the subsequent dissemination of data managed with public funds is limited by discrepancies in the competences of the managers of such data (Serrano 2018). In this context, in relation to the "Open Science" framework, which includes open access to data (Hampton et al. 2015), a number of good practices were needed for the management of scientific data that were clearly specified, and widely shared and applied. Some relevant examples of these practices can be found in previous studies (Corti et al. 2020; Goodman et al. 2014).

Virtual Research Environments (VREs) refer to innovative working systems that seek to enhance cooperation between researchers in all modern research settings (Barker et al. 2019). They promote novel approaches, the integration of all types of information, the elimination of organizational barriers, and the incorporation of new as well as emerging technologies that enhance information processing and analysis capabilities. Activities that promote the creation and development of VREs, in addition to being supported by the scientific community, are expanding worldwide (Lawrence et al. 2015). A multitude of organizations, both national and international, are promoting programs that facilitate such activities. Some examples outside the European Union are Science Gateways Community Institute (SGCI) in the USA, The National eResearch Collaboration Tools and Resources project (Nectar) in Australia, Canada's National Research and Education Network (CANARIE) in Canada. Within the European Union, the Horizon 2020 program has supported a large number of European VRE projects, such as Building research environments fostering Innovation, Decision-making, Governance and Education for Blue Growth (BlueBridge), Scientific research in the Earth Sciences (Ever-Est), A Europe-wide Interoperable Virtual Research Environment to Empower Multidisciplinary Research Communities and Accelerate Innovation and Collaboration (VRE4EIC), World-wide E-infrastructure for structural biology (WEST-Life), VREs for regional interdisciplinary communities in south-eastern Europe and the eastern Mediterranean (VI-SEEM) and Multi-Scale Complex Genomics (MUG). All these platforms, known as VREs, aim to share information and provide tools for generating new knowledge.

Our main goal is to follow common criteria that allow us to aggregate and integrate scientific data in an Open Science" framework for the Sierra Nevada Global-Change Observatory (Rüegg et al. 2014). The information system, called Linaria (https://linaria.obsnev.es), acts as a repository to store raw data gathered by the monitoring program as well as information generated through the processing of such data, allowing its integration and analysis. To achieve this objective, we developed a data-management approach intrinsically connected with the FAIR philosophy and a Data Management Plan as a tool to take into account the overall data life cycle. For this task to be completed, the common information management policy is necessary (this being known as FAIR principles) and then the tools that allow new information to be generated by interacting with data will be built or developed (VRE).

## 2 The FAIR Principles

The need for the efficient management of research data, in order to maximize the opportunities for reuse of such information has existed for decades. The FAIR initiative principles can be viewed as a consolidation of these earlier efforts, having emerged from a multi-stakeholder vision of an infrastructure supporting machine-actionable data reuse (Wilkinson et al. 2016). The FAIR Principles offer a set of guidelines that a data publication should comply with in order to make the data Findable, Accessible, Interoperable, and Reusable, as detailed below:

- **Findability**: (Meta)data will be deposited in recognized free- and open-access repositories. The choice of one repository or another will depend on the nature of the data. For example, sampling event data from species monitoring can be deposited in the Global Biodiversity Information Facility (GBIF), while environmental data can be published in more general repositories such as PANGAEA. Whatever the case, the repository chosen must meet these three requirements related to the FAIR principles: (1) their registry services must provide a globally unique and persistent identifier; (2) they must offer long-term maintenance of data storage as well as availability to license the data. The amount of metadata to be completed in the repository platforms is extremely varied. For this reason, to ensure rich metadata, we will develop a metadata catalogue giving free and open access to metadata and linked data.

- **Accessibility**: Data accessibility will be guaranteed by the repository, offering the possibility of retrieval from their identifiers without specialized or proprietary tools or communication methods. In the case of sensitive or restricted data, the contact protocol will be clear and explicit in the metadata catalogue, which will constantly be available.
- **Interoperability**: We envisage interoperability at two levels: internally, between data integrated into our model, and externally, with the integration of (meta)data in international networks (e.g., LTER, GBIF, etc.). The interoperability of datasets will be ensured by the use of (meta)data standards (e.g., Ecological Metadata Language, Darwin Core) and also the use of controlled vocabularies, ontologies, thesauri (e.g., EnvThes).
- **Reusability**: Depositing data in a repository does not guarantee its reusability. It is necessary to provide completed metadata concerning the collection, provenance, and use conditions (license), among other aspects. This information will be available at all times for potential users.

## 2.1 Data Management Plan (DMP)

A Data Management Plan (DMP) describes the data life cycle that will be produced during research. In broad terms, the data management strategy for any project should address the following key issues, which should be reflected in a DMP: data collection (type, source, volume, and format), data organization and management, metadata, quality assurance/quality control, data storage and preservation, data policies, data dissemination, roles and responsibilities regarding data management tasks, and finally budget (Michener 2015). In particular, Michener and Jones (2012) identified the essential components that a DMP should include. Below, we highlight the aspects that we consider key to the development of these concepts as follows.

*What quantities and types of data will be collected?*

Firstly, data generated by a project will be clearly distinguished from those reused from other projects. This should be made clear with the identification by DOI and a permanent URL for the metadata of the dataset. Information on data types, sources, volume, and format is indispensable in order to plan data organization and integration. Data formats will be those widely used by the scientific community (e.g., databases, raster and vectorial spatial formats, CSV). Quality assurance and quality control (QA/QC) procedures should be present before, during, and after data gathering (Michener and Jones 2012). The standardization of data collection following the OGC O&M standard specifications or specific data models can be useful to prevent errors and facilitate data integration.

*How will data be organized and managed?*

Dealing with data integration requires both logical consistency and flexibility (Rüegg et al. 2014), especially when complex data is generated and requires database storage. Such cases include research areas within Biodiversity, Ecosystems, and Climate Change, which include aspects ranging from high-frequency sensor data to human observations of species, all with different temporal and spatial resolution. Some of the tools used for data organization include relational database management systems or geographic information systems (Michener and Jones 2012). The temporal and spatial resolution will depend on the specific sampling design and the scientific questions defined in the project. The integration of datasets that differ in resolution and accuracy provides a data management challenge, e.g., for the purpose of modeling (Rüegg et al. 2014).

*Metadata: the five Ws*

What does well-documented data mean? For some researchers, e.g., (Michener and Jones 2012), a Data Management Plan (DPM) should provide information often called the 5 Ws, these being Who, What, When, Where, and Why. These usually end in "How" (which has the same consonant sound but is not written with the "W"):

- Who created, collected, and managed the data.
- What the content and format of the data is.
- When the data were collected.
- Where the data were collected and stored.
- Why the data were generated (project context).
- How the data were generated, processed, assured, and analyzed.

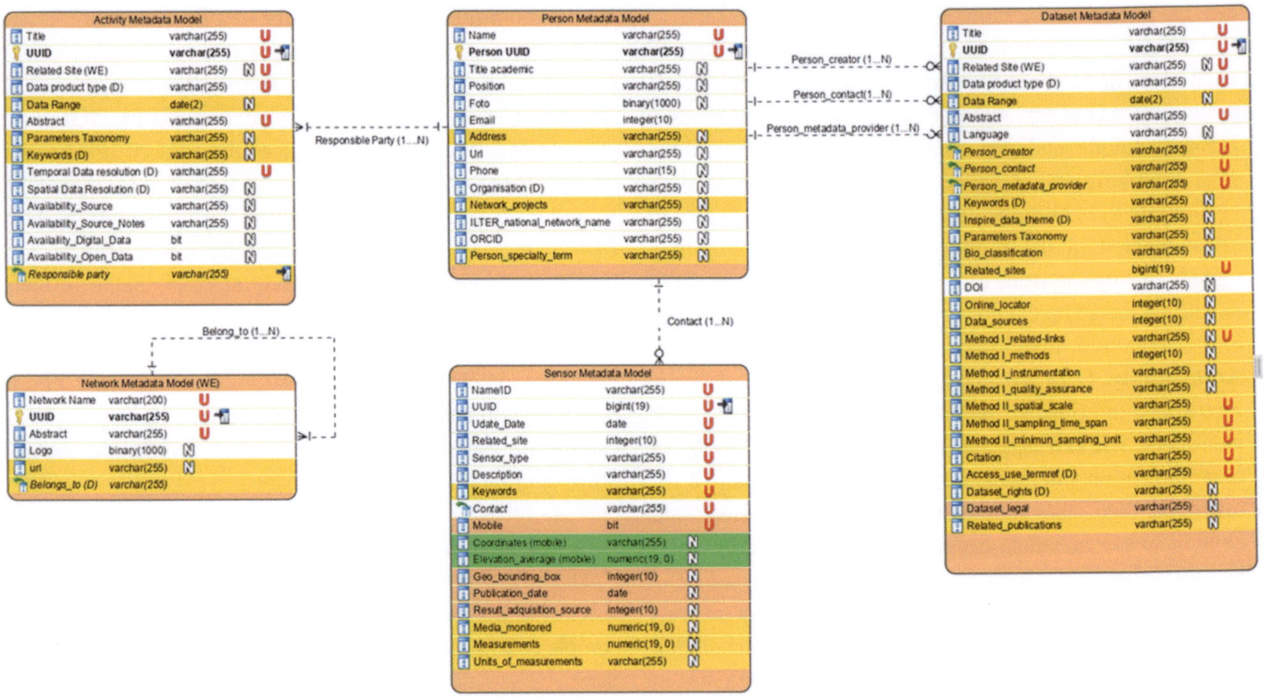


**Fig. 1** Metadata model according to FAIR principles

This is relevant not only to interpret data, but also for reusability by future users, such as: researchers, managers of territorial resources (environmental administration, territory administration, etc.), non-governmental organizations, and the general public.

*Data storage and preservation*

To avoid data loss and ensure its long-term preservation, data should be deposited in a data repository. A basic recommendation is to store at least three copies in at least two different locations (Michener 2015). Ideally, institutional repositories (e.g., university) can be combined with internationally recognized repositories (e.g., GBIF).

*Data policies*

A DMP must include explicit policy statements (Michener 2015) on data access, sharing, and reuse. These include possible restrictions due to sensitive data, how the data should be cited, use licensing, etc.

*Data dissemination*

In addition to the depositing of data in open repositories, other ways of disseminating data and adding value to it include the publishing of data and metadata in a data paper (Chavan and Penev 2011). Possibly, data deposition in international repositories will entail data transformation (e.g., Darwin Core format in GBIF). Making data products available to the scientific community and general public has multiple benefits, one being the progress of science with new research ("Data sharing and the future of science" 2018).

Compliance with the FAIR principles and the existence of a DMP allow the user to normalize data description and its management, as well as meet organizational or funder requirements. The FAIR principles do not touch on controversial issues such as technology or the approach used in the implementation. This means that they have already been accepted by several policymakers and funding organizations for research projects. The value of applying these principles is reflected in their incorporation into the European Union's Horizon 2020 Program of Research and Innovation projects or in European e-infrastructures such as Lifewatch ERIC. Similarly, the development of a DMP is a requirement in any Horizon 2020 project. Both FAIR principles and DMP are connected, and a DMP should address these principles in order to achieve a data management procedure in accordance with Open Science.

## 2.2 Metadata and Data Model

All data will be open access according to the current trend of Open Science. They will be deposited in recognized institutional repositories and replicated in other specific networks (e.g., GBIF). A metadata catalogue will provide access to all data sets. Data will be shared by using accepted standards for data exchange to ensure interoperability. We propose (Fig. 1) the following model of metadata to describe data that is compatible with accepted international standards and FAIR principles (Wohner et al. 2019).

## 3 Virtual Research Environment (VRE)

Traditionally, a virtual research environment (VRE) is a set of online systems, tools, and processes interoperating to facilitate or enhance the research process. Horizon 2020, the European Commission's research and innovation framework program, suggested that a VREs "should integrate resources across all layers of the e-infrastructure (networking, computing, data, software, user interfaces), should foster cross-disciplinary data interoperability and should provide functions that allow data citation and promote data sharing and trust" (Barker et al. 2019).

The Open Data model adopted for VREs, as mentioned elsewhere (Carusi and Reimer 2010) does the following:

- Encourages collaboration between researchers, regardless of the research center or the country in which they work.
- Increases the opportunities for synergy, so that efforts are joined to achieve the objectives in less time.
- Optimizes resources that facilitate results more efficiently than before.

We define a VRE with tools structured in the following interdependent hierarchical levels, and with AI and GIS as transversal elements (although we mention it in the hierarchical breakdown): Access to data/metadata, Statistics, GIS & AI, Modeling, and Simulation.

This VRE expands the capabilities of traditional VREs (oriented to the scientific community) with services specifically aimed at two new types of users: the general public and environmental managers (if the focus is placed on the field of Biodiversity).

The different layers shown in the figure would allow users to work together in a real-time environment that facilitates the exchange of information resources, overcoming the limitations of traditional working practices. This makes the following possible: to collaboratively discover, access, evaluate, and process heterogeneous datasets in real time; to initialize and force a model by applying the different tools in the system; and to share data, models, algorithms, and scientific results (including the traceability of workflows and processes that would facilitate reproducibility of modeling and simulations).

The minimum functionality required in each of these layers is explained in Box 1.

**Box 1. Layers of the Virtual Research Environment (VRE)**

**Layers "Interoperability" and "Search and Discovery"**

Interoperability represents the ability to run processes seamlessly across system boundaries without losing context and meaning. This is achieved as follows:

- Understanding how processes can be interconnected.
- Specifying the semantics of messages within these processes so that requirements and context can be agreed upon by all parties involved.
- Developing standards to support these processes effectively, so that messages can be exchanged between different entities in a scalable way.
- Providing implementation guidelines on how semantics are transformed into syntactically equivalent messages that can be understood and processed automatically by disparate systems.

In the field, interoperability and the possibility of finding available resources imply answering questions such as:

- Which semantic resources were used or are being used and in which scientific domain? Vocabularies, thesauri, and ontologies.
- Which metadata standards do were followed? ISO 19115, ISO 19139, Dublin Core, Darwin Core, INSPIRE ISO 19139, OGC standards, SSN, SensorML.
- Which metadata language, ontologies, and controlled vocabularies were used? EnvThes, DCAT, SSN, OBOE, re3data, Nature.

The starting point to answer these questions is the adoption of the FAIR principles and the use of a DMP. The existence of these elements together with the definition of specific semantics for the treatment and exchange of information constitute the starting point for the construction of this layer, ensuring that all the content generated within our system can be operated and found from all the applications included in it. The common strategy to achieve this objective, as we have already hinted, is to exploit semantic resources such as metadata, vocabularies, and ontologies in order to bolster interoperability between data integrated into the platform.

It is desirable for our system to be able to discover and aggregate semantic resources from all scientific

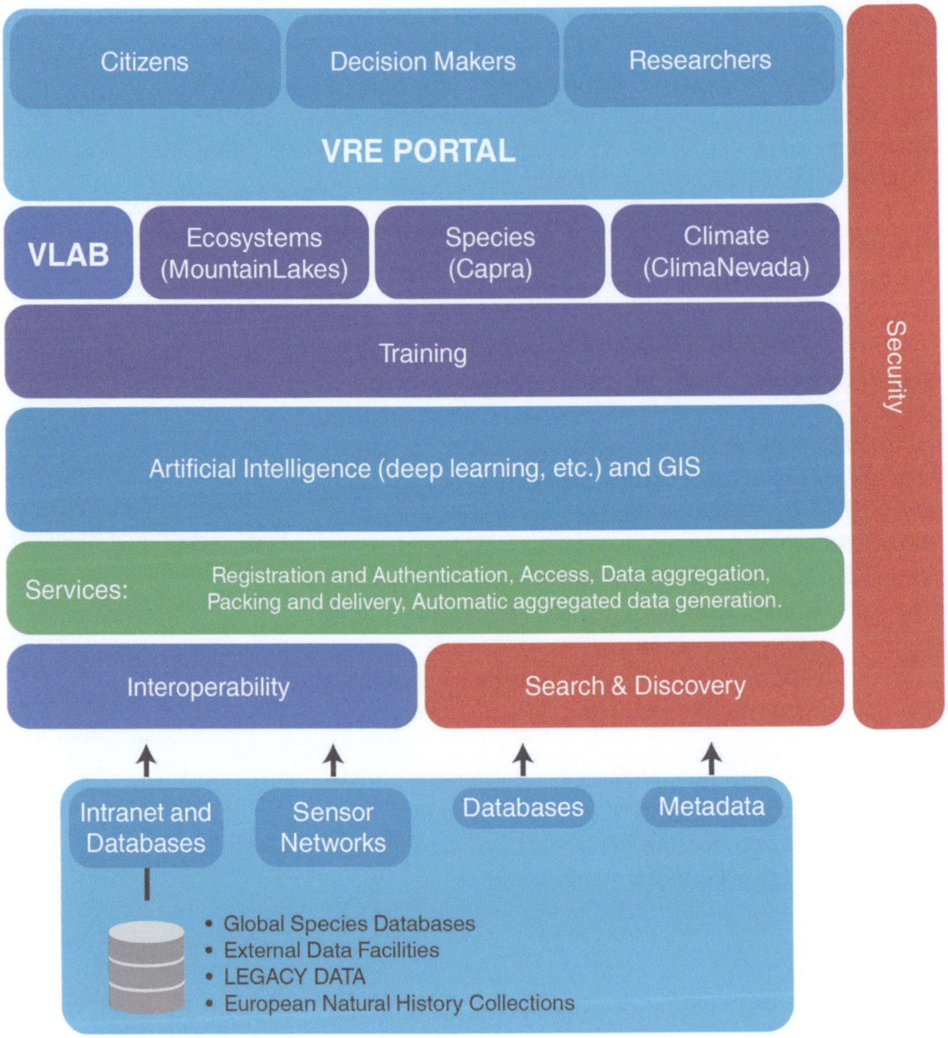


**Fig. 2** Functional layers of the Virtual Research Environment (interaction with data, Basic Services, specific tools (GIS & AI), security, training, laboratories, and access portal)

domains within a catalogue and generate a multidisciplinary semantic index of concepts/terms. However, the great diversity of existing semantics (each with its own specific application programming interface, API) makes it infeasible to scale a solution in order to collect a large amount of semantics from existing repositories. Therefore, this layer defines the semantics integrated into the system that allows the following:

- Compatibility of all the elements that can be shared with the rest of the internal elements of the platform.
- Protocol and the labeling of elements that identify the levels and sublevels (depending on the axes that we will see below) of the VRE from which the different data packages, models, algorithms, etc., are accessible.
- The means to find and access any element within the VRE.
- The manner to incorporate new semantics that allows sharing and finding accessible non-indexed content within the platform.

It should be noted that internally it is essential to provide the system with a specific API that allows access to the encapsulated content at different hierarchical levels depending on the final application that uses such encapsulated content.

**Layer "Services"**

This layer manages access to the existing service catalogue in our VRE. Different functionalities are offered in the form of services to be used from any part of the VRE according to the semantics defined in the System, or from outside the VRE by other systems. This system will offer two levels of services:

- Internal services are accessible from within the VRE. These are services with specific encapsulations that adhere to the internal semantics of the platform. The fact that they are accessible only from within the VRE is for two markedly different reasons:
  - Services are not open or are exclusively for internal use. Its own definition indicates its exclusivity.
  - Newly created services contain packaged content that can be used only by existing services within the VRE. These services can be used in the future from outside the VRE as long as applications capable of using them are built.
- Universal services that will be open and accessible online.

The semantics of the existing services in the VRE must take into account the profiles of the users accessing the VRE, since they will act on the information at different levels.

- The scientific community will have access to raw data and all existing services in the VRE.
- The general public will have pre-filtered data of public interest and services that are considered of public interest.
- Decision-makers will have access to information already filtered and validated and assessed by the scientific community and will be able to use specific services adapted to their needs.

Some examples of services that should exist in the VRE are the following:

- Registration and authentication service. This service will identify the different types of users, allowing access to VRE.
- Data Access service. The fundamental characteristic of this service is that all the data will be available from the same point, but different collections of data will be displayed depending on the type of user:
  - Scientific Community: Access to all raw data stored in the system using filters customized at any time by the user.
  - Natural resource managers: Access to information that has already been filtered and processed in order to facilitate decisions.
  - General public: Pre-filtered and processed data will be available to speed up access (maximum, average, and minimum values taken at certain intervals, statistical data on species, etc.)

This service will be provided through a portal, but also as a software service to be integrated into other systems or applications.

- Data aggregation service. This provides interconnection with the system according to the standards set by new data sources.
- Packing and data delivery service. It is most common for the scientific community to request and consult raw data for further processing or local study. This will require a system that allows the secure packaging and delivery of data to the user.
- Automatic aggregated data generation. Service for the general public, environmental managers, and the scientific community. This service will periodically generate data that must be available for the two types of users described and that require predefined processing.

**Layer "GIS and Artificial Intelligence"**

This layer combines two groups of applications that can be applied or used transversely throughout the VRE in many different ways. We will refer to these two groups of applications as "GIS" and "Artificial Intelligence."

Today, most of the information that we have is georeferenced; that is, the information has a geographic position and related information of value. Geographic Information Systems (GIS) are tools that allow the user to integrate, visualize, and relate all these large amounts of georeferenced data, thus facilitating the analysis of the different problems and aiding decision making in a more efficient way. Therefore, this GIS layer is essential in our VRE, which in this layer will contain tools that allow the following:

- Reading, editing, storing, and managing spatial data.
- Data analysis, from simple queries to the building of complex models.
- Representation of results such as maps, reports, or graphs.

GIS systems enable a visual exploration of the data, helping the user to identify and define problems to address, observe behaviors or trends, etc. In addition, these systems make it possible to superimpose layers, and the superposition can be visualized in a GIS related to the association between different spatial variables.

### Sierra Nevada Study Cases

*Case 1*: In the Sierra Nevada area, the following problems to be solved can be identified as study cases, the specific details of which depend both on the validation of the hypotheses by researchers as well as on the data available, for which the aforementioned algorithms will help achieve the desired results:

Species-level research (*Capra hispanica*—Population trend): A good estimate of the number of individuals is essential for the analysis of its environmental and socioeconomic impact (in the case of large populations) and to activate the appropriate protection mechanisms for the species (in the case of small populations).

Climate-level research (Prediction of climate scenarios): Certain climate scenarios, especially the most extreme ones (at the level of rainfall or temperatures) have a strong impact on the level of biodiversity, which also entails an impact at the socioeconomic level. With the help of meteorologists, a forecasting system for extreme weather scenarios can be designed, based on weather data from previous years.

Ecosystem-level research (Mountain Lakes):

- Prediction of the size of the lakes: The area covered by the lakes is an important indicator of the behavior of the lake ecosystem. The climate data and the size reached in previous years (at different times of the year) would allow an estimation of the size prior to the thaw.
- Prediction of the level of eutrophication: the degree of eutrophication of lakes determines a large part of the ecosystem. These data from previous years together with the data that characterize eutrophication (climatology, aerosol deposition, etc.) would allow the design of a prediction model.

*Case 2*: In the Sierra Nevada ecosystems, several problems can be identified as study cases. The first step is to study the viability of hypotheses with biodiversity researchers and, from available images (data), to apply Deep Learning algorithms with the following objectives:

Species-level research (*Capra hispanica* population location): The geographical location of the different populations of ibex, especially in summer, exerts a major effect on biodiversity as well as on the economy of the area (agriculture and mountain livestock, sighting of populations, etc.). The areas occupied by ibex populations can be estimated from distribution images from previous years together with influencing factors such as climatic data, population size, and herbaceous surface area.

Ecosystem level research (Mountain Lakes):

- Prediction of the level of the ice sheets: The surface and duration of the ice surface area in the lakes is a factor that significantly affects the lake ecosystem. Estimates can be based on photographs, with a model learned from photographs tagged with the size of the ice sheet.
- Prediction of lake size: In this case, unlike the previous section (where the aim is to predict the size before the summer period), the size of the lake would be estimated at the time of taking a photograph with a set of images of the lake labeled with its surface area.

Land-use maps: Land use has a decisive influence on the ecosystem and therefore on biodiversity. Effective land-use maps can be established using a model with images labeled according to different uses by zones. In some cases, these maps can serve as feedback for previously built models (e.g. the location of ibex populations), to try to improve the performance of these models.

*Case 3*: Automatic classification of biological collections: From images of digitized biological collections, researchers could be interested in certain specific characteristics of these individuals, especially the changes that take place over the years. This involves manual viewing and labeling of a considerable number of images. Using the analysis of common inquiries that researchers make, a sample of representative images can be selected to design a prediction model that provides automatic classification of a group of individuals. An example would be counting the number of flowers (or fruits) on a plant.

### Layer Training

This layer must include manuals and interactive user guides that allow new users of the VRE to learn how to operate all the tools included in the VRE. It should be kept in mind that the VRE will be different for each user depending on the person's profile. Different users will not see the same environment or access the same tools. The environment will adjust to each type of profile (scientists, managers, and the general public).

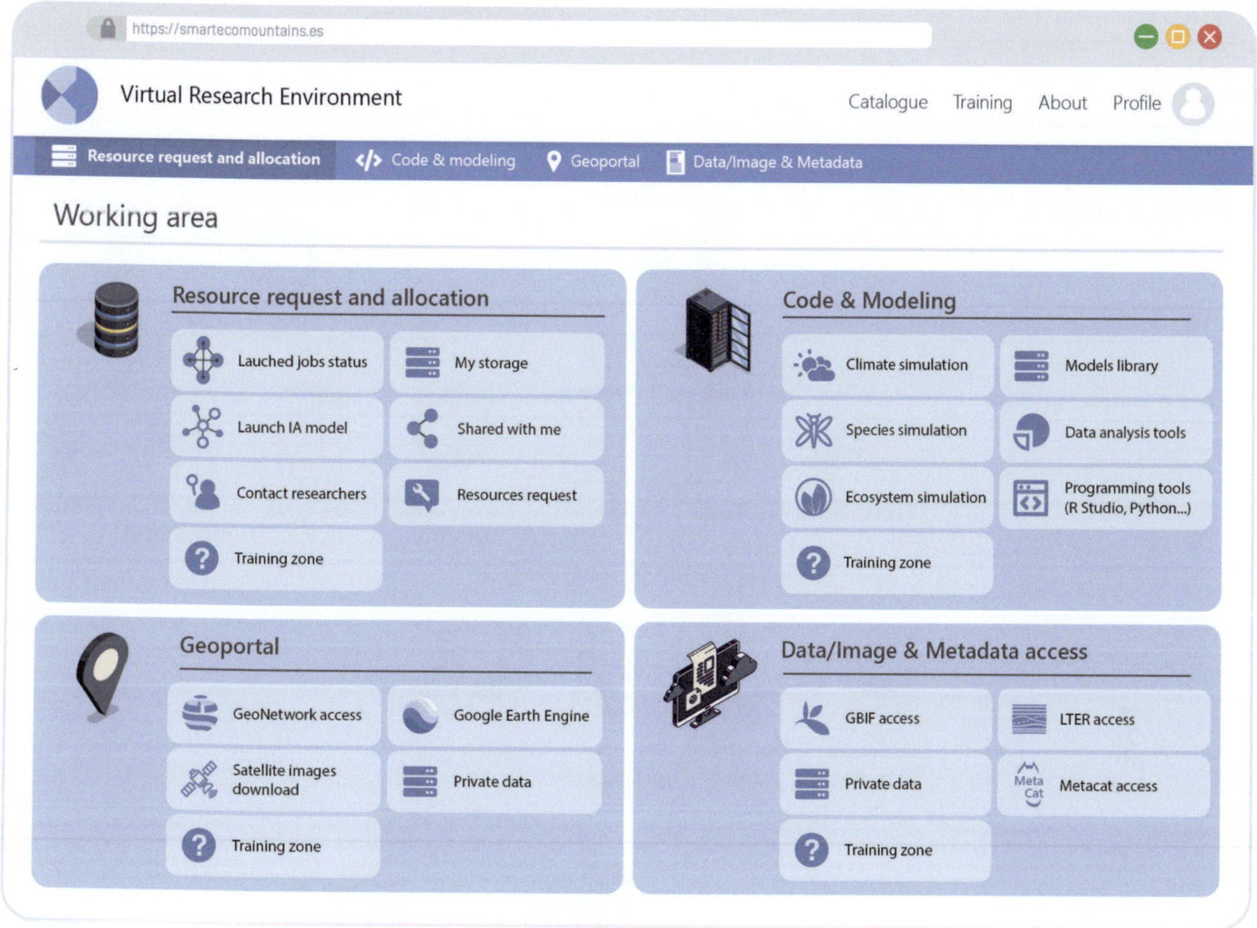


**Fig. 3** Visualization of the access portal for the scientific profile. Resource management area, programming area, information geolocation area, and access to information sources area

The VRE working environment must be user-friendly and fully adaptable to the profile of each user, so that the tools requested by the user are incorporated into the system. It is true that there is some difficulty or reluctance in the scientific community to shift from traditional and consolidated research practices and facilities to the innovative ones promoted by VRE (Carusi and Reimer 2010), mainly due to the complexity and large amount of resources that these environments offer and that in most cases involve an unaffordable technological change. In order to overcome this difficulty, it is essential for the system to have assistants and specific training in the use of VRE to enable its smooth adoption by the user community.

One of the applications that could form part of the Training layer (Box 1 and Figs. 2, 3) is a game/university zone (Fig. 4) aimed at facilitating the use of the VRE at the university. This would allow future researchers to learn how to handle this type of environment while at the same time allowing them to learn how to use all the tools related to this type of environment.

## 4 VRE Target Profiles: Researchers, Managers, and the General Public

The VRE needs to ensure the usability of the information by the target audience: scientists, managers (decision-makers), and the general public. For this, the information collected must be accessible to different actors in society, in formats appropriate to each user profile and it should be easily transformed into knowledge useful for them (Jacobson et al. 2006). Accessibility to information should be ensured by the Virtual Laboratories (VLABs—top layer Fig. 1). This is especially relevant in the case of biodiversity data (our target field), where the implementation of research continues to be under-utilized in decision making (Spierenhurg 2012). This multi-user model is needed because decision-making is complex, iterative, and often selective in terms of the information used (Young et al. 2014). Therefore, the VREs should provide access panels adapted to each user profile,

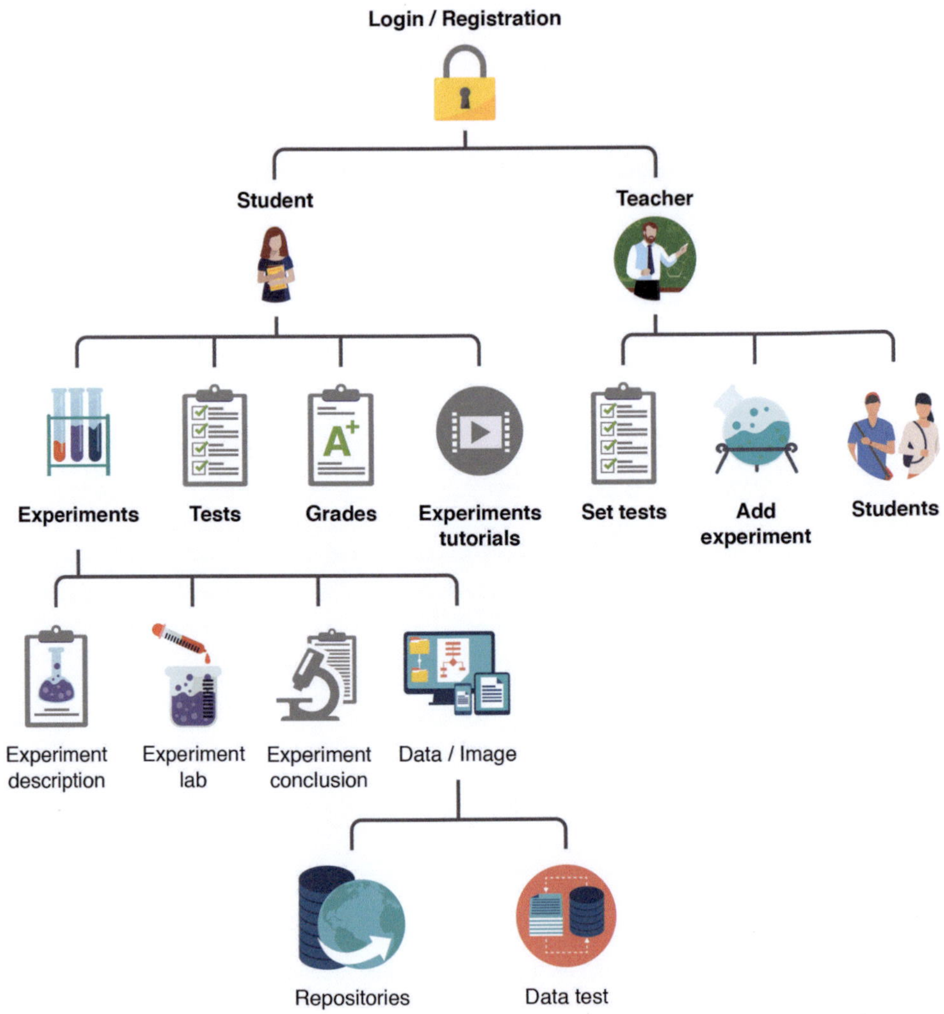


**Fig. 4** Game/University zone Visualization of the training area to students in this supporting tool, allowing their learning and evaluation, as well as access to the full power of the VRE through experiments monitored by a teacher

including tools that allow interactive dialogue between users.

It is therefore necessary to establish methods that foment long-term interactive dialogue between the various stakeholders. Here, we present a methodological approach to identify the type of information, its format, and how it will be transformed, according to the profile of the end user (Researchers, Managers, or the General Public—Fig. 5) in a VRE. The method is presented as a case study focusing on the Sierra Nevada environment with the University of Granada as a research center.

As a result of this interaction, an initial definition of the needs and objectives of each of the actors described can be fulfilled, while specific tools and processes for this can be designed. For the case of Sierra Nevada, and according to the user's profile, the following strategies were proposed:

- **Identification of scientific needs**. The needs of scientists were identified based on two fundamental sources: firstly, an updated database of the research already completed, based on a literature review made using several repositories (Web of Science, Scopus, and Dialnet, among others); and secondly, information gathered by the scientific institutions regarding the study site. In the case of UGR-Sierra Nevada, the Global Change Observatory of Sierra Nevada structured all the knowledge compiled by different thematic areas (Zamora et al. 2016), such as climatology, biodiversity changes, ecosystem services, socioeconomic issues, and paleo-perspectives. Once "Who Researches" and "What is Researched" were audited, taking into account the scope of the VRE, a series of surveys of the scientific community were conducted. The objective was for researchers to define the degree of interest that each had in accessing data from the different topic areas in the development of their research activity. For each thematic area, a list of potentially interested researchers was compiled in order to hold thematic workshops. Therefore, for each topic area, the conditions of users were determined by defining the following aspects: data required, distribution format,

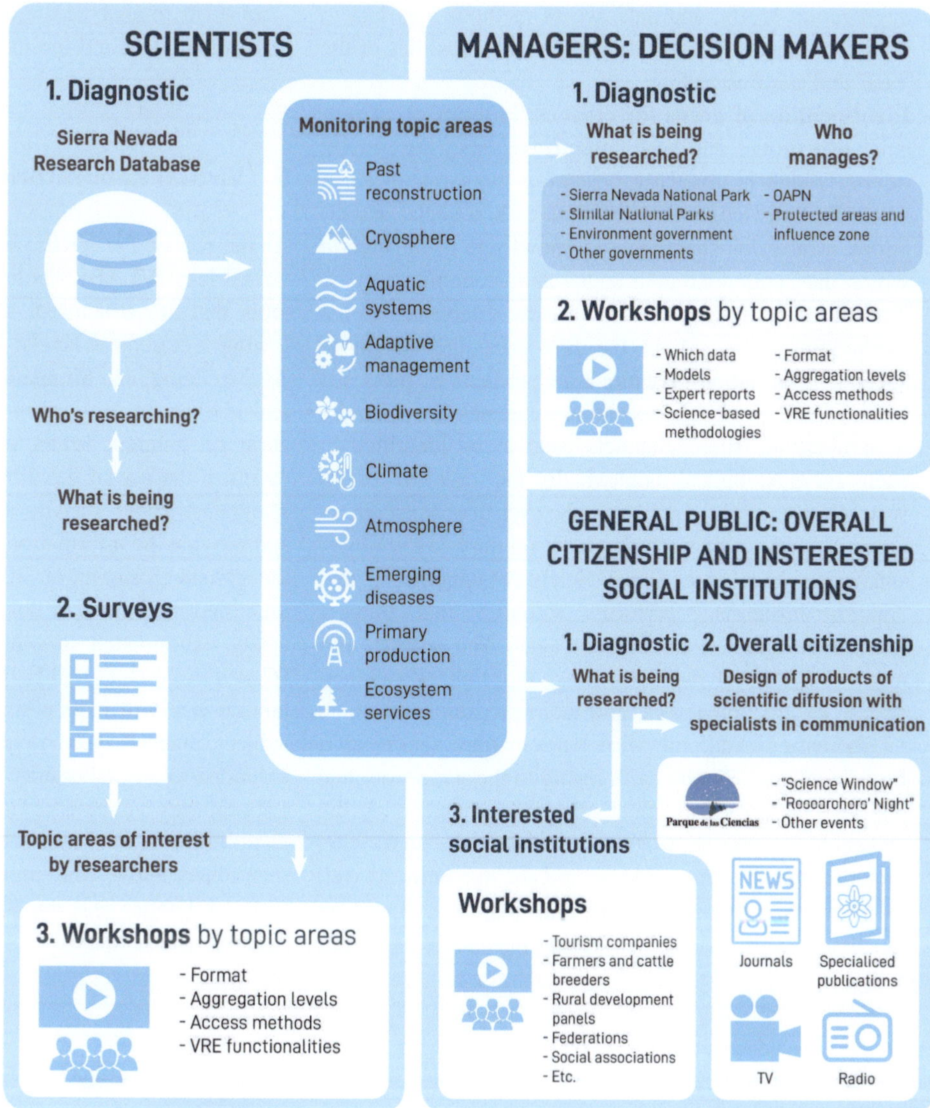


**Fig. 5** Methodological framework for identifying information. This conceptual scheme shows us a methodology to provide the VRE with contents of interest according to the user's access profile

desirable aggregation level(s), and access methods. All this information served as the basis for defining the functionalities of the VRE services and the e-infrastructure for the scientist user profile.

- **Identification of needs for managers in decision-making**. There is ample evidence that the transformation of scientific evidence into "usable knowledge" for managers is neither automatic nor straightforward (Young et al. 2014). As pointed out elsewhere (Vogel et al. 2007), the reality is that too often "scientific production is more likely not to meet the needs of users, i.e. not what professionals need." Communication barriers between scientists and managers have been studied within the epistemological boundaries of translational ecology (Enquist et al. 2017), examining temporal rhythms in their tasks, scale, etc. To eliminate these boundaries, the solution increasingly accepted in the scientific community is that the creation of this useful knowledge requires co-production between scientists and managers (Meadow et al. 2015). Therefore, the methodology is based on the inclusion of managers from the early stages in the definition of VRE tools. Based on the diagnosis of what is being researched, all management stakeholders who wish to participate in the definition of VRE products in Sierra Nevada environment will be defined. Subsequently, a series of workshops should be carried out by thematic areas, to define, within each of them, the following aspects: Data, models, scientific-technical reports, or scientifically based methodologies for decision-making. The format, aggregation levels, and data access methods also had to be determined in order to define the functionalities of the VRE and e-infrastructure.

- **Identification of needs for citizens**. In terms of citizenship, two groups can be differentiated. First, the general public, which is gradually becoming more interested in scientific knowledge. Second, entities and social associations, which demand access to knowledge of the area in which they undertake their social and economic activity. Two different approaches are therefore established in the methodology (Fig. 5). In the first case, the design of specific information-dissemination products is proposed in coordination with the specialized channels for scientific dissemination. These channels need to be identified in each socio-ecological context. In the case of Sierra Nevada, the main actor was the Science Museum at Granada city which, through temporary exhibitions, activities such as the "Science window," and "micro-exhibitions," performs activities that transfer knowledge adapted to different ages. The science and nature sections of the daily press (specialized publications), TV documentaries, and radio programs, in which UGR-Sierra Nevada has wide participation, also reach a broad audience. Finally, with regard to the socioeconomic entities concerned, the first step is to diagnose all the stakeholders who show great interest in the study site, including those in the business and social realms, as well as in sports, tourism, and territorial planning. Based on this diagnosis, a series of workshops need to be organized in order to analyze the information requirements in each thematic area and the means for their dissemination.

## 5 Virtual Laboratories (VLABs)

At the top of the VREs, below the user layer, appear the virtual labs (VLAB). These are environments that integrate tools that are structured, as mentioned above, in the following hierarchical levels: Data access, Statistics, GIS and AI, Modeling, and Simulation. The organization of the tools and transversal services in these categories allows the creation of different levels of abstraction that simplify and optimize the use of the different technologies that the VRE integrates. These VLABs are used to answer the questions and provide the information required by the different users of the system (Sáenz et al. 2017). In Sierra Nevada, we have identified three lines of development for VLABs, which are easily extrapolated to other mountains: Climate, Species (*Capra hispanica*), and Ecosystems (mountain lakes). These laboratories will have a specific part (associated with their subject matter) and another transversal part that can be extended to any laboratory of the same type. These Virtual Laboratories will emerge from the scientific proposal, taking a multidimensional approach composed of the layers/tools mentioned above, the users (Researchers, Managers, and General Public) and the thematic laboratory (Ecosystems, Species, and Climate).

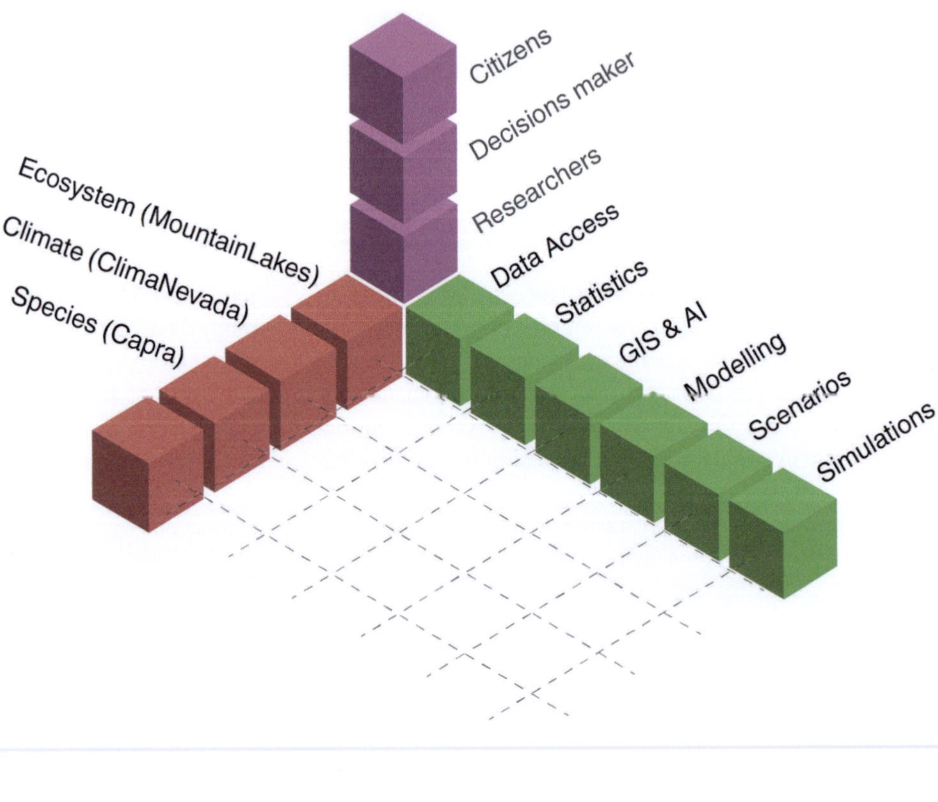


**Fig. 6** VRE multidimensional approach. This way of structuring tools, users, and thematic laboratories allows us to optimize the flow of information while defining the interrelation between the different functional layers of the VRE

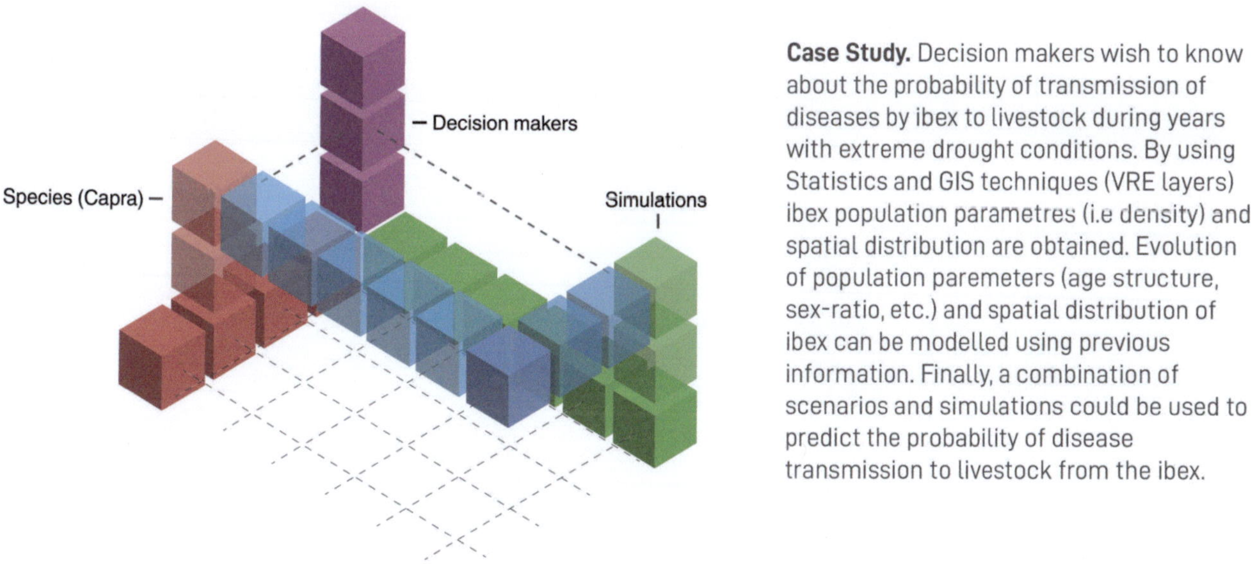


**Fig. 7** Interaction between VRE, users, and thematic laboratories. This example shows the great capacity of the multilevel structure described above to access different types of information and tools

In this multidimensional approach, the relation can be represented between the VRE (composed of the layers/tools mentioned above), the interlocutors (users), and the thematic laboratory (Sáenz et al. 2017). This approach answers questions posed by users using tools provided by VRE (Fig. 6).

This three-dimensional division maximizes the use of existing information and allows all the different kinds of elements to be exchanged within the same level. For example, it could be verified whether distribution models of a certain species also explain the distribution of other species that share the environment. This approach enables the resolution of doubts posed by users using tools provided by VRE (Fig. 7). For instance, a key uncertainty might concern the spatial distribution of the Iberian ibex during a drought year. This general question could be formulated in a more specific way, depending on the user:

- Researcher: What is the spatio-temporal dynamics of the species? What are the ecological reasons explaining his distribution? How does this distribution influence the eutrophication of lakes affected by the concentration of goats (consumption of endemic species, accumulation of excrements) near the lagoons in very dry years?
- Manager: to what extent can the Iberian ibex affect mountain crops? Will it also affect lakes (eutrophication, protected plant species, etc.)? Are hunting licenses needed for species control?
- Citizen: What mountain crops will be affected? Is there interaction with livestock? Where it is more likely to observe herds of ibex in a tourism context?

In the next section, we illustrate the power that this new three-dimensional conception of the VRE has to generate new knowledge not only for researchers, but also for managers and the general public.

## 6 Generation of Knowledge Supported by the Use of Virtual Laboratories (VLABs)

Ecosystems are complex, intermeshed systems in which many of the elements are so intertwined that it is difficult to untangle them and analyze them individually. In this way, an integral study of all the actors involved requires the design of a work methodology that brings together all the elements to allow conclusions based on the aggregation of all the factors. Following the line of examples outlined above, there is a clear example of a study that integrates the three virtual laboratories initially planned. Drought years have a significant impact (both ecological and socioeconomic) on the environment of the Sierra Nevada. During dry years, grassland areas, which provide grazing for Iberian ibex, significantly diminish in availability, with several consequences:

- Some ibex populations may move to higher altitudes in search of food, feeding on the only pastures that remain green during the summer thanks to the proximity of small lakes and streams, affecting some pasture areas and having an impact on the endemic flora located at the summits of Sierra Nevada.

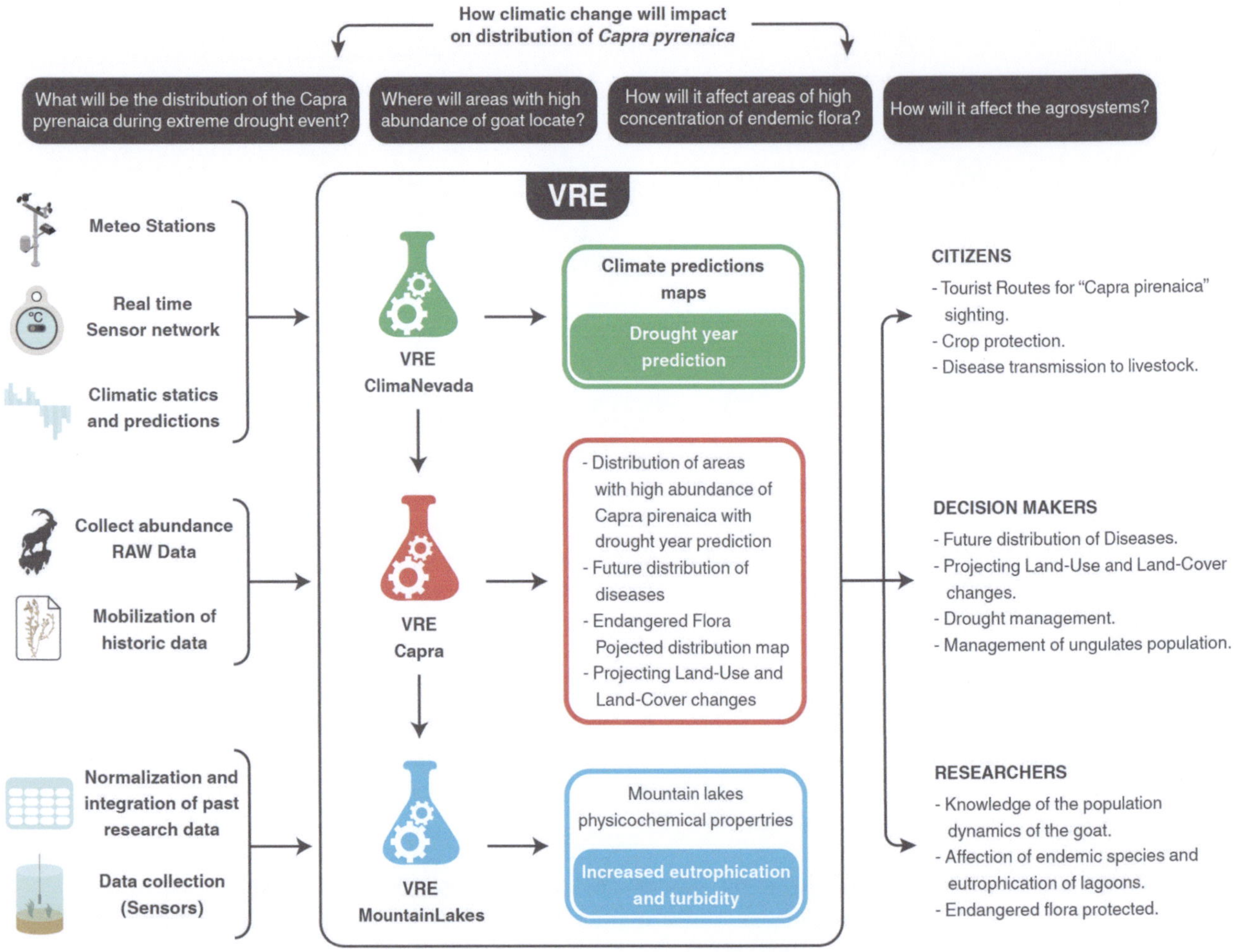


**Fig. 8** Example questions and flow of information. The figure shows the flow of information that enters each of the thematic laboratories, how information can be crossed between them, and finally obtain the answers we are looking for according to the profile of the target user

- Other ibex populations may move to lower altitudes in search of food and may affect grazing or mountain farming areas, with consequent economic damage.
- Ibex populations could become concentrated more than usual in the surroundings of mountain lakes, significantly affecting the level of eutrophication of these lakes, and thereby altering their ecosystem dynamics (Fig. 8).
- All previous efforts to construct the aforementioned models on the location and population trend of ibex, size and level of eutrophication of lakes, land use, etc., would serve to design an integral prediction model for dry years, which would allow the appropriate actions to be taken in order to reduce the numerous harmful effects of these drought periods. The following diagram shows the flow of information when the different VLABs integrated into the VRE are used to answer particular questions (Fig. 9).

The following figure exemplifies how the same question has different interpretations depending on the target user profile and the tools used to analyze the information available.

## 7 Conclusions

Every day terabytes/petabytes of data with information on multiple disciplines are generated and stored. This huge capacity of information generation raises several unresolved issues such as the technological limitation to fully exploit, share and preserve this data, the possibility of using this information for research while respecting at all times the authorship of such data and preserving its proper use, and finally, the possibility of allowing the general public access

| Scope | Questions | User Profiles | | | Methods of analysis | | | | | |
|---|---|---|---|---|---|---|---|---|---|---|
| | | **Scientist (S)** | **Manager (M)** | **Citizen (C)** | Data Access | Statistics | GIS | Modelling | Scenarios | Simulations |
| | | | | | A.I. (Deep learning, blockchain, etc.) | | | | | |
| VRE Climate | What are the climate scenarios in SN? | Analyze the implications on water availability. Its influence on the distribution species. | Knowing the availability of water: impact on biodiversity and agrosystems. | Enjoy the landscape and economic report of tourism activities. | S M | S M | S M | S M | S M | S |
| VRE Ecosystem High-mountain lakes | How will the eutrophication of water develop in lakes? | Analyze the extent to which atmospheric deposition or ibex concentration are responsible foreutrophication. | Management measures: management of ungulates. | Visual quality of the lakes. | S M | S M C | S M C | | | |
| VRE Species Iberian wild goats | What is the spatial distribution of wild goats in drought years? | Knowledge of the spatial dynamics of the wild goat<br>- How will they affect crops?<br>- How will they affect endangered species? | Species management, hunting licences, livestock interactions | Impact on crops, interaction with livestock.<br>Enjoyment of wildlife observation. | S M | S M C | S M C | S M | | |
| | What is the incidence of diseases in drought years? | Health status of the species and population control. | Population control, isolation of livestock, treatment campaigns. | Farmers and population (possibility of infection). | S M | S M C | S M C | S M | S M | S M |

**Fig. 9** Interaction between VRE, users, thematic laboratories, and methodologies. In this table, we intend to expand on the information in Fig. 8, with more questions that can be solved with the way we have proposed to structure the information

to the results of scientific research and to the knowledge that can be extracted from all this information in a friendly and intuitive way (general public and managers). The technological solution that attempts to resolve all these issues is Virtual Research Environments (VREs).

In the last decade, a multitude of international initiatives has arisen that are committed to the creation and use of virtual research environments as a means of responding to current demands. The "Horizon 2020" program of the European Union had already defined the need to develop Virtual Research Environments in different areas of work, including the environment, and therefore provided funding to make progress on these issues. In the environmental field, the European Union already in its Seventh Framework Programme (completed in 2011) began to detect these needs and began to pave the way by preparing the document "ESFRI Roadmap" (European Strategy Forum on Research infrastructures) among which was what is now Lifewatch ERIC.

Lifewatch ERIC is a European e-Science Infrastructure for Biodiversity and Ecosystem Research. It aims to strengthen scientific, technological, and innovation capacity in the field of biodiversity. In particular, it aims to establish a large data, information, and knowledge network based on a powerful e-Infrastructure of communications, supercomputing, and distributed cloud. The users of this e-Infrastructure are the scientific community, environmental managers, and citizens. It will allow, among many other things, access to large volumes of data related to biodiversity (biotic: fauna, flora, abiotic: climatic, terrestrial, oceanic, and fluvial/freshwater) and will allow working on them with analysis tools, available in a series of virtual research environments such as virtual laboratories (ERV), or with decision support tools for environmental managers, as well as access to data and social science tools by the general public.

At the University of Granada we are developing precisely these new conceptual and instrumental frameworks with existing information on biodiversity and mountain ecosystems through our Smart Ecomountains Thematic Center. Through this initiative, the University of Granada is directly involved in the development of the European Lifewatch ERIC infrastructure, focusing on mountain ecosystems biodiversity and ecosystem services with the application of Remote Sensing and Artificial Intelligence.

**Acknowledgements** This work has been carried out within the framework of the collaboration agreement between the Ministry of Environment and Territorial Planning of the Regional Government of Andalusia and the University of Granada for the implementation of activities related to the Sierra Nevada Global Change Observatory. We are also grateful for the partial support received from the LIFE-ADAPTAMED (LIFE14 CCA/ES/ 000612) projects: Protection of key ecosystem services threatened by climate change through adaptive management of Mediterranean socio-ecosystems, from the H2020 European Long-Term Ecosystem and socio-ecological Research Infrastructure (eLTER) project and from the Smart Ecomountain

Lifewatch ERIC Thematic Center (LifeWatch-2019-10-UGR-01). We are grateful to the management and technical team of the Sierra Nevada Natural Area for their continued collaboration in the framework of the joint activities that we develop in the Global Change Observatory. Juan Miguel González Aranda, Antonio José Sáenz Albanés, and Antonio Pérez Luque have contributed to the development of the ideas expressed in this chapter.

## References

Barker M, Olabarriaga SD, Wilkins-Diehr N et al (2019) The global impact of science gateways, virtual research environments and virtual laboratories. Future Gener Comput Syst 95:240–248

Carusi A, Reimer T (2010) Virtual research environment collaborative landscape study. JISC Report

Chavan V, Penev L (2011) The data paper: a mechanism to incentivize data publishing in biodiversity science. BMC Bioinform 12(S15):S2

Cook R, Lineback P (2008) Sierra nevada network data management plan

Corti L, Van den Eynden V, Bishop L, et al (2020) Managing and sharing research data: a guide to good practice (second). SAGE Publications Ltd.

Enquist CA, Jackson F, Garfin ST et al (2017) Foundations of translational ecology. Front Ecol Environ 15(10):541–550

Goodman A, Pepe A, Blocker AW et al (2014) Ten simple rules for the care and feeding of scientific data. PLoS Comput Biol 10(4): e1003542

Hampton SE, Anderson SS, Bagby SC, et al (2015) The Tao of open science for ecology. Ecosphere 6(7):art120

Jacobson SK, Morris JK, Sanders JS, et al (2006) Understanding barriers to implementation of an adaptive land management program. Conserv Biol 20(5):1516–1527. http://www.jstor.org/stable/3879143

Lawrence KA, Zentner M, Wilkins-Diehr N et al (2015) Science gateways today and tomorrow: positive perspectives of nearly 5000 members of the research community. Concurrency Comput Pract Exp 27(16):4252–4268

Meadow AM, Ferguson DB, Guido Z et al (2015) Moving toward the deliberate coproduction of climate science knowledge. Weather, Clim Soc 7(2):179–191

Michener WK (2015) Ten simple rules for creating a good data management plan. PLoS Comput Biol 11(10):e1004525

Michener WK, Jones MB (2012) Ecoinformatics: supporting ecology as a data-intensive science. Trends Ecol Evol 27(2):85–93

Peng RD (2011) Reproducible research in computational science. Science 334(6060):1226

Reichman OJ, Jones MB, Schildhauer MP (2011) Challenges and opportunities of open data in ecology. Science 331(6018):703

Rüegg J, Gries C, Bond-Lamberty B et al (2014) Completing the data life cycle: using information management in macrosystems ecology research. Front Ecol Environ 12:24–30

Sáenz AJ, Alcácer C, Rodríguez S (2017) Entornos Virtuales de Investigación para la toma de decisiones en el áambito del Medio Ambiente Urbano. WPS Review International on Sustainable Housing and Urban Renewal 1(5)

Serrano JC (2018) Repositorios públicos frente a la mercantilización de la Ciencia: apostando por la ciencia abierta y la evaluación cualitativa. MÉI: Métodos de Información 9(17):74–101. https://dialnet.unirioja.es/servlet/articulo?codigo=6840541

Spierenhurg M (2012) Getting the message across: biodiversity science and policy interfaces: a review. GAIA 21(2):125–134. http://www.ingentaconnect.com/content/oekom/gaia

Stodden V, McNutt M, Bailey DH et al (2016) Enhancing reproducibility for computational methods. Science 354(6317):1240

Vogel C, Moser SC, Kasperson RE, Dabelko GD (2007) Linking vulnerability, adaptation, and resilience science to practice: pathways, players, and partnerships. Glob Environ Change 17(3):349–364

Wilkinson MD, Dumontier M, Aalbersberg IJ et al (2016) The FAIR guiding principles for scientific data management and stewardship. Sci Data 3(1):160018

Wohner C, Peterseil J, Poursanidis D (2019) DEIMS-SDR—a web portal to document research sites and their associated data. Ecol Inform 51:15–24

Young JC, Waylen KA, Sarkki S et al (2014) Improving the science-policy dialogue to meet the challenges of biodiversity conservation: having conversations rather than talking at one-another. Biodivers Conserv 23(2):387–404

Zamora R, Pérez Luque AJ, Bonet FJ et al (2016) Global change impacts in Sierra Nevada: challenges for conservation. Consejería de Medio Ambiente y Ordenación del Territorio. Junta de Andalucía. http://hdl.handle.net/10481/5468.

# Advancing Open Science in Sierra Nevada: Current Citizen Science Campaigns

Manuel Villar-Argaiz, Eulogio Corral Arredondo,
María del Carmen Fajardo-Merlo, and José Miguel Barea-Azcón

**Abstract**

Citizen science is a very valuable complement to scientific efforts of tracking environmental change. The contribution of volunteers in scientific research is growing in Sierra Nevada, and we here report three pilot experiences that include research on high mountain glacial lakes, butterflies, and stream macroinvertebrates. So far, participation has mainly led to the involvement of citizens into routine monitoring programs, alleviating personnel and resource shortcomings. A necessary step forward in open science is to develop a roadmap to extend dissemination by storytelling and outreach activities that reach civil society. By doing so, we expect to raise a new alliance, one that closes the gap between the necessary rapprochement among science, citizens, and environment managers.

**Keywords**

Citizen science • Global change • Monitoring programs • Lakes • Butterflies • Macroinvertebrates

## 1 Open Science in the Era of a Changing Earth

There is no simple definition for what Open Science (onwards OS) is. OS is more than simply open access to scientific studies in open journals as it relies upon several basic pillars that warrant public access to a part or the full process of acquiring, processing, publishing, and reviewing scientific data (OECD 2015). OS is, therefore, an overarching concept aiming at reducing barriers in all aspects of science. OS should lie on a good set of good practices that warrantees transparency, reproducibility, open access, and participation and inclusiveness in all parts of the scientific process. OS has accelerated the way we think about collaborative research and should enable us to tackle new targets and goals by expanding benefits in a cost-effective manner. Collaborative research has fallen within the research that Earth and environmental scientists have always practiced. This is because understanding how ecosystems and the Biosphere as a whole works is an ultimate goal that requires multidisciplinary approaches involving numerous fields of knowledge.

Without doubt, the *sine qua non* development of environmental sciences has paved the road towards OS practices. However, it is clear that science, pushed by technological advances, has led to new ways of network collaboration and data management. For example, artificial intelligence has revolutionized these sciences. Specifically, Machine Learning in which computers learn from existing data has boosted the way we understand how the multiple dimensions of global change affect precipitation and moisture (Kolassa et al. 2018), fluxes of water, carbon, and energy (Alemohammad et al. 2017), or land cover and land use (Jin et al. 2019), among others. However, this top-down approach can be limited by lack of quantity and quality field data needed for calibration and validation of results. Data is the first and most valid primary input on which artificial intelligence and OS relies. Therefore, good science and OS, start with highly collaborative and comparative ways in which we collect data. This is a crucial step before adaptation and integration of datasets for advancing artificial intelligence to answer environmental science questions.

M. Villar-Argaiz (✉) · E. C. Arredondo
Department of Ecology, University of Granada, 18071 Granada, Spain
e-mail: mvillar@ugr.es

M. del C. Fajardo-Merlo · J. M. Barea-Azcón
Agencia de Medio Ambiente y Agua de Andalucía, Consejería de Agricultura, Ganadería Pesca y Desarrollo Sostenible, Junta de Andalucía, Seville, Spain


R. Zamora and M. Oliva (eds.), *The Landscape of the Sierra Nevada*,
https://doi.org/10.1007/978-3-030-94219-9_23

## 2 Citizen Science and the Urgency of Acquiring Data

According to National Geographic, citizen science "is the practice of public participation and collaboration in scientific research to increase scientific knowledge." Through citizen science, people participate in the various parts of the research process: collecting and providing data or analyzing data themselves. This open and participatory approach of citizens should enhance research and contribute to build a new scientific culture and better societies empowered by citizens (Gordienko 2013).

Bringing together scientists and citizens into a single forum where to analyze and debate the current scientific understanding of natural and human-driven changes to our planet is an opportunity we cannot afford to miss. It is urgent that we bring science and society closer, so that latest scientific breakthroughs help understand how natural and human changes on our planet affect lives and societies, and what the future might hold. Issues of societal importance—food system, air and water quality, geochemical cycles, or environmental services—need to be explored both from a scientific and social perspective. John Muir once said that "when we try to pick out anything by itself, we find it hitched to everything else in the Universe." We now face the daunting challenge of understanding our complex planetary system where human activities are intimately interwoven with natural processes. The new alliance fostered by citizen science is a step forward in the translation of the scientific message accurately, in a way that reaches a broad audience well beyond scientific forums and specific disciplines. Only new and creative approaches can help to translate scientific progress into real action, and the engagement of non-scientist in this process is a step forward in Earth stewardship (Dickinson et al. 2012). In doing so, citizen science could find its way to contribute to the UN's Sustainable Development Goal of halving environmental degradation by 2030.

Although citizen science has a long history in science with relevant contributions to scientific knowledge, it is still in its infancy in Sierra Nevada. In this chapter, we will review three case studies which will address cross-border research issues and programs on which citizens have been actively involved, both at the level of acquiring and processing data in Sierra Nevada.

## 3 Case Study 1: Long-Term Research of High Mountain Lakes

### 3.1 Glacial Lakes Under Threat

Global change generates complex and interactive responses in natural ecosystems that are not always easy to untangle and comprehend. For that reason alone, mountain research allows us to observe what cannot be easily discerned at lower points in the landscape. Across mountain ecosystems, lakes, and rivers are long regarded as sentinels and integrators (sensu Williamson et al. 2008) of global change, and particularly glacial lakes are the most staggering natural laboratories. As a blood test is a good indication of human health, the analysis of water provides us with detailed information on the processes affecting surrounding terrestrial and atmospheric environments. The reasons behind the use of these systems as sentinels of global change include, some of the following features: small catchments, oligotrophic nature of water with scarce nutrients, remote access and good ecological health, high altitude and extreme conditions, or relatively simple biological communities with high renewal rates (chapter "High Mountain Lakes as Remote Sensors of Global Change"). In addition, the study of lake sediments or paleolimnology provides unique witnesses of past changes both within land and atmosphere environments (chapters "Reconstruction of Past Environment and Climate Using Wetland Sediment Records from the Sierra Nevada and Paleolimnological Indicators of Global Change"). As a consequence, lakes integrate physical, chemical, and biological perturbations where scientists interpret the "footprints" of global change effects (Williamson et al. 2008).

Glacial lakes in Sierra Nevada are one of the landscape jewels and ecosystem emblems in Sierra Nevada. In addition to their extraordinary beauty and scientific value, these high mountain lakes are among the most vulnerable and threatened systems in our geography. On the one hand, the inexorable climate change with a marked decrease in rainfall is seriously altering hydrological levels and the persistence of the lagoons (Villar-Argaiz et al. 2001). On the other hand, the increasing pressure from tourists and mountaineers who access the high peaks is alarmingly straining these vulnerable ecosystems: garbage, organic pollution, disturbance of lake bed, or excessive trampling and camping in the surrounding meadows are only some of growing threats facing the conservation of these landmarks. Proper management of these sites is essential to reduce the impact of climate and anthropic change, and for this, data fuelling long-term monitoring programs are essential as they provide managers with the steering scientific information that requires effective adaptive management policies.

### 3.2 Scope and Goals

Research in Sierra Nevada high mountain lakes is not new. Lakes were long recognized as some of the most outstanding geographic features since the early travelers and naturalists in the nineteenth century pointed to their natural beauty and their glacial origin at the highest headquarters of valleys (e.g., Boissier 1839). Despite the numerous geological and geographical references, it was not until the early 1970s that surveys and scientific studies systematically reported patterns in the physico-chemical and biological characteristics of these lakes (Martínez-Silvestre 1975). Ever since, scientists have reported sporadic data. Although valuable and necessary, these sources of data fall short in several ways. First, limited budgets restrict data sets to infrequent data collection so that spatial and temporal variations cannot be accurately captured. Second, historical data report specific fields of interest, while leaving aside relevant data that could help provide a broader ecological context. The contributions of non-scientist, not only can be used to fill data gaps (Fritz et al. 2019) and overcome the ordinary resource shortcomings (Devictor et al. 2010), but are called to be a most decisive step toward OS. While, lake citizen science initiatives are widespread across countries with abundant water resources such as Canada with over 23 active programs https://livinglakescanada.ca/our-programs/), the participation of volunteers is rapidly growing in Sierra Nevada and will likely play a key stewardship and leadership role in the local communities in the near future.

### 3.3 Scientific and Practical Information

Global change threads in Sierra Nevada have never been greater, and so has the need to gather scientific information on the pressure exerted by man on high mountain ecosystems. The citizen-open and participatory approach gained a renewed impulse in Sierra Nevada thanks to a citizen science campaign that, under the name of "74 High Mountain Glacial-Lake Oases," was launched in 2018 by the Department of Ecology with the support of Sierra Nevada Global Change Observatory (SNGCO) and the National Park of Sierra Nevada. This campaign, later supported by the Spanish Foundation for Science and Technology and a LifeWatch ERIC project, strengthened citizen involvement in research activities carried out in Sierra Nevada (Fig. 1).

The engagement of citizens contributed to cover two main objectives. For researchers, it allows to strengthen Lake Long-Term Monitoring Programs by gathering scientific data that would otherwise be unapproachable by scientific means. For ordinary citizens, it offers a window of opportunity into science's complex world, one that contributes to understanding the science-based decisions that ensure the protection of high mountain ecosystems.

The project aims at engaging a group of committed volunteers in monitoring high mountain ecosystems with a particular focus on glacial lakes. Although the call is for a general audience, we encourage people who frequent uplands such as outdoors enthusiasts, hillwalkers, and kickers in general (Fig. 2). Because high mountains are not easily accessible and frequented areas, by focusing on those who spend time in these habitats we not only avoid setbacks, but increase the capacity and probability of completion the collection of data. The key to the success of this citizen science project relies on the broader motivation of volunteers that are involved at the very base of the production of data by taking part in field surveys.

Extensive monitoring campaigns are planned in advance and the activity pack is announced all year round to encourage citizens and mountaineers to join the extensive monitoring campaigns that are held during the ice-out period during spring and summer periods. Different press releases and press photocall invitations are organized to maximize publicity for events, and people subscribed for the upcoming events.

Our scientific colleagues and peer organizations contribute to publicizing the free workshops widely through social media. These workshops are preferentially held in situ using the University facilities and field stations in Sierra Nevada. Registration for attendance is open to all citizenships, although a number of seats are reserved to local and public stakeholders (mountaineer clubs, nature conservation societies, etc.). We have been pleasantly surprised, as anticipation of registrants has always been soared past by the capacity of the events.

Workshops are organized around two major activities. The first consist of a series of lectures by different field specialists that introduce attendants into the state-of-the-art knowledge of these unique high mountain ecosystems. Besides this learning experience, this first part of the workshop also aims at engaging participant's interest, and advancing competitiveness. The second self-building part of the workshop aims at implementing the skill-monitoring part to be implemented in the field. This is a hands-on demonstration where participants receive step-by-step instructions and practical hints on how to proceed with the monitoring protocols. These are training days with the purpose of creating a network of multidisciplinary research groups. These groups compose of one/two researchers, an environmental ranger, and four/five volunteers are gaining ground at the most intensive phase of

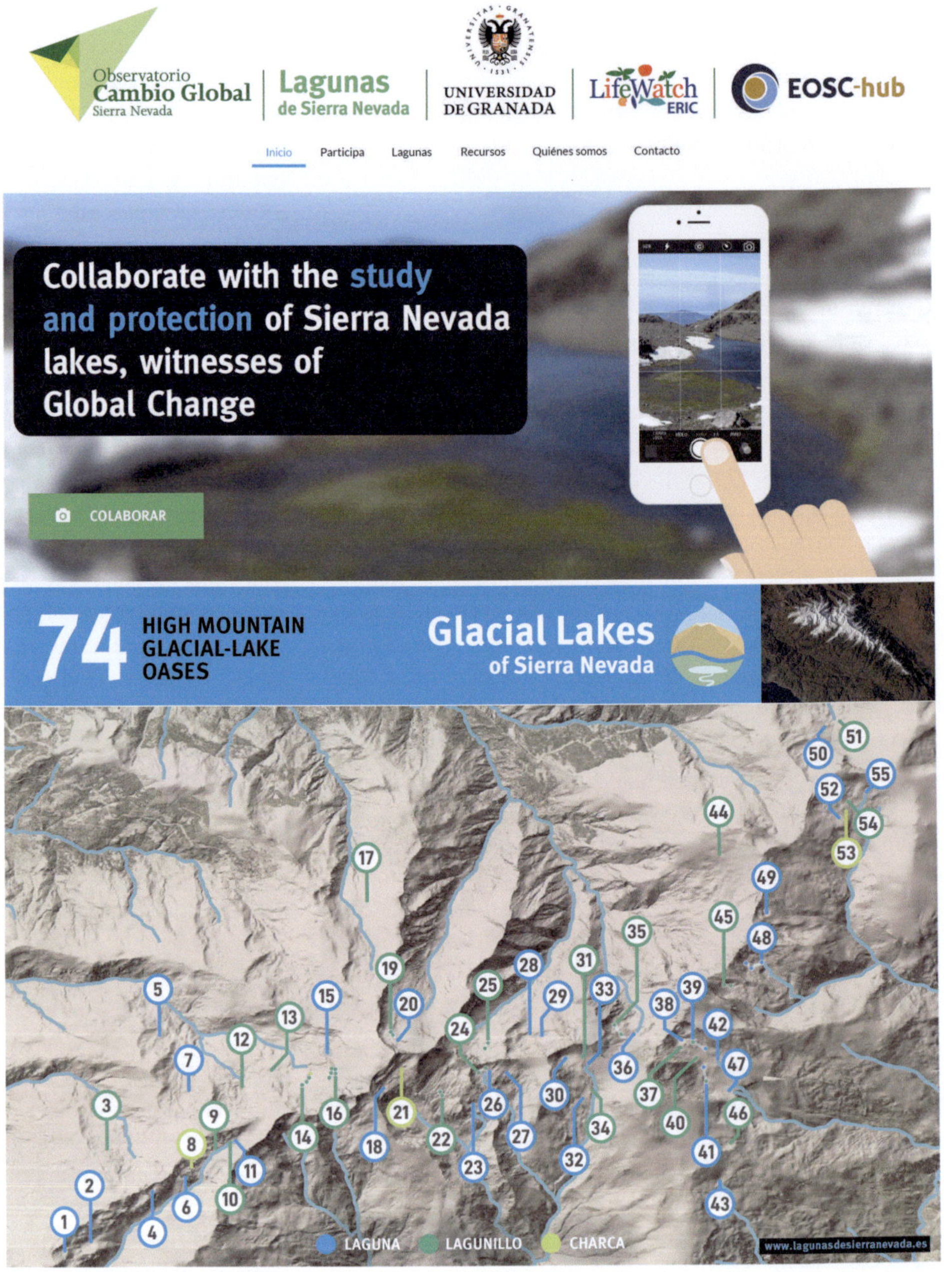


**Fig. 1** (Top panel) *Lagunas de Sierra Nevada* internet portal. (Bottom panel) Sierra Nevada map for the 74 high mountain lakes. Numbers represent the different lakes and colors refer to progressively smaller lake sizes locally known as *Laguna* (blue), *lagunillo* (green), or *charca* (yellow). Very close lakes are grouped into a single number for clarity. See full map at https://lagunasdesierranevada.es/lagunas/

the investigative process of field sampling (Fig. 2). With this traditional in-person approach, a number of citizen volunteers are incorporated in multidisciplinary teams that conduct limnological-related research alongside research mentors from across different departments of the University of Granada. One of the most valuable aspects of this consortium is learning scientific knowledge and skills in the context of applied environmental ecology.

Performance statistics and evaluation results indicate that workshops have been extraordinarily successful in the past. All participants have completed the entire workshops and are very satisfied with their experience, described the workshop as a high-quality "window" into science, and had face-to-face or via social media channels recommended the workshops to others.

The project uses a website as a major route to inviting people to join the project. The website (https://lagunasdesierranevada.es) not only serves as a port to direct people to the scientific and educational resources, but is linked to social media applications for rapid dissemination of news. Social media is the most effective method to focus on those specific groups with an existing interest in the subject, as a method of outreach to educate and engage a broader audience in a broader environmental agenda.

**Fig. 2** Photos illustrating the group of participants (top panel) and a field monitoring snapshot (bottom panel) in the "2020 Laguna de Sierra Nevada Workshop." Photos by Luís Ordoñez

## 3.4 Major Achievements

Data collected with the help of citizens has served complementary purposes, such as augmenting monitoring data, improving modeling, or even surveillance of water quality. The primary impacts of citizen science are seen in ecological studies of global change including by means of extensive and intensive monitoring campaigns.

*Extensive monitoring campaigns*. The campaign for environmental water monitoring of lakes pursues the construction of a unique and large dataset by the acquisition of water-related measurements encompassing multiple domains: the weather, the surrounding meadows, the inflow and outflows, and a myriad of physico-chemical and biological variables. Biological, chemical, and physical factors can all impact the quality of the water and are complementary means of interpreting the structure and function of water ecosystems. After training, program managers provide volunteers with a sampling protocol sheet and tick list of features where the following relevant information is registered:

- Weather conditions. Volunteers are instructed to perform a first evaluation of current weather conditions. For this, all parameters are classified into excluding categories. For example, wind categories are calm, or light, medium, or strong breeze, whereas daily weather is classified as clear, partly cloudy, cloudy, foggy, light rain, heavy rain, or snow.
- Surrounding meadows ("*borreguiles*"). Volunteers visually evaluate the length and breadth of meadow permanently or temporarily saturated with water around the lake. This information, as well as water color (transparent, grey, brown, light, or dark green) is essential to the productivity of lakes as nutrients are retained as they run into the water body.
- Aquatic plant occurrence and density. Presence of aquatic plants as a percentage of lake surface is evaluated as this is a primary influence on habitat for all biota as well as provides physical structure and influences nutrient dynamics in lakes. Assessment of cover is carried out on a quintile-scale (1 = 0%, 2 = 0–25%, 3 = 25–50%, 4 = 50–75%, and 5 = 75–100%), and the total must add up to 100%. If present, aquatic plants should be classified as free-floating or rooted plants (either emerged, submerged, or floating vegetation), and subsamples collected for professional identification.
- Lake bed substrate. Morphological structure of lake bed is registered as percentage cover of bedrock, cobble-boulders, sand, gravel, and silt/clay.
- Lake depth. Most lakes in Sierra Nevada are shallow with a mean depth of 0.8 m (Díaz-Hernández and Herrera-Martínez 2019), so volunteers can easily estimate the maximum depth of the lake.
- Alterations. Volunteers record any perturbation and unusual conditions that may affect the lake such as organic and inorganic garbage, animal excrements, etc.
- Photographic report. Several photographs are taken at a high point over the lake and at the shoreline to provide a snapshot of generic information based on expert analysis.

Visual inspection is followed by a detailed sampling using standard protocols (APHA 1985) that includes:

- Water sampling. Volunteers tour around the lake perimeter taking aleatory samples of water using pole-mounted buckets. Water is mixed in a container and subsamples are taken for quantitative assessment of:
  - Bacteria, phytoplankton, and zooplankton
  - Macronutrients (total and dissolved nutrients)
  - Major cations and anions
  - Dissolved organic carbon
  - Microplastics.
- Macroinvertebrate surveys. Macroinvertebrate are sampled over a set time period of three minutes along three aleatory-chosen transects at the littoral zone of the lake using kick net samplers.
- Netting. Zooplankton is caught in 45 μm nets and samples are taken to the laboratory for counting and measuring.
- Physico-chemical characterization. Water quality data is registered using handheld multiparametric probes. Parameters include temperature, conductivity, oxygen levels, pH, turbidity, and redox potential.

This summary of the volunteer monitoring program addresses the most relevant parameters that help understand lake structure and functioning. This includes the crucial information needed to understand lake pollution threads such as increased algal growth, nutrients, acidification, or decreased levels of dissolved oxygen, among others. Our monitoring programs aim at determining the condition of the entire watershed. This is because land-based activities as well as atmospheric phenomena affect the waters that drain from the land into the lakes.

*Intensive monitoring campaign*. Although extensive monitoring programs provide with a baseline of the state-of-the-art in the water quality of lakes, continuous interannual data is also immensely valuable if we are to detect trends in lake ecology over long term periods. For over 35 years, citizen volunteers have joined scientists in gathering critically important data on specific lakes in Sierra Nevada. As an example, Fig. 3 shows a 35-year

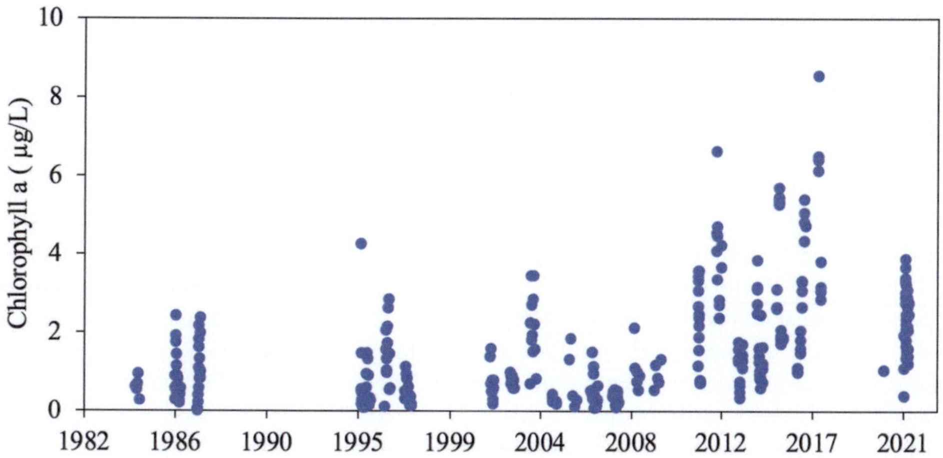

**Fig. 3** Interannual trends of chlorophyll *a* in Lake La Caldera. *Data source* Sánchez-Castillo (1986), Carrillo (1989), Villar-Argaiz (1999), González-Olalla et al. (2018)

chlorophyll-a record dating back to 1984 in Lake La Caldera. The interannual increase in chlorophyll levels is an important indicator, together with water clarity or nutrients, of a decline in water quality as it signifies the number of algae in the water with consequences for the trophic and ecological status of the lake.

*Photographic campaign*. Photographs are an incredibly important source of information to scientists. The project seeks citizens and mountaineers who, voluntarily and selflessly, contribute with photographs and field observations to build the most extensive historical records of these lagoons. Photos can be sent online (https://lagunasdesierranevada.es/) and are stored in an image database consisting of several thousand contributions some of which date back to the 1950s. By submitting photos of the lakes, volunteers help researchers to estimate historical water levels and to identify threats that can ensure law protection of our surrounding environment. Also at the interactive web platform, volunteers can, among other things, share and visualize the historical record of the photographs, find friendly access to all the scientific and disseminating information available on these lakes, or interact with the scientific teams that research these valuable ecosystems. Through social and proactive participation, we seek to strengthen the coresponsibility for the preservation of these unique sites.

## 4 Case Study 2: Monitoring Butterflies in Sierra Nevada

As highlighted throughout this chapter, citizen science is taking special importance in order to help collect relevant information on biodiversity and other key ecological processes to understand the functioning of natural systems (Jordan et al. 2015; Follett and Strezov 2015). Butterflies monitoring by volunteers is one of the finest examples of this. Throughout large areas of the world, thousands of volunteers periodically collect information on the composition and structure of butterfly communities that are subsequently stored, organized, and analyzed by professionals to offer some of the worldwide most important global change evidence. A good example of this is the recently created European system for monitoring diurnal butterflies (BMS Europe, Butterfly Monitoring Scheme), that is at this moment the greatest invertebrate monitoring network in the Planet, with near to 5.000 monitoring sites under the same methodology (Sevilleja et al. 2020a).

### 4.1 Scope and Goals

Butterflies are considered as model organisms in ecology and for this reason, the dynamics of their populations, phenology, ecology, and other attributes at the population level have usually been used as descriptors of environmental changes and are the scope of relevant studies on topics such as ecological monitoring and analysis of the effects of global change (Dennis et al. 2003; Stefanescu et al. 2003; Wallis De Vries and Van Swaay 2006; Oliver et al. 2015). In general, a good part of the arthropod species shows ecological characteristics that make them very sensitive to changes in the environment where they live (Schowalter 2016). Butterflies, in particular, usually present different requirements during their distinct life stages. This, together with their short generation times, make butterflies an ideal group for tracking the effects of global change and other drivers of environmental change. Likewise, diurnal butterflies are organisms whose study is relatively easy to approach compared to other species of invertebrates, and particularly when it comes to their adult stages. Most species are easily identifiable, and very few require additional laboratory studies to reach species-level identification. Butterfly taxonomy is nowadays relatively well studied despite the fact that recent molecular approaches to its study are providing notable novelties that can substantially modify the traditional taxonomic scenario (Dincă et al. 2015). These are some of the reasons of why butterflies are model organisms to study ecological problems and to be incorporated to a citizen science scheme. Furthermore, diurnal butterflies are a group of species that are very popular on a social level and there is an increasing number of citizens who are interested in taking part in these monitoring schemes. In fact, the volume of information available nowadays is really impressive and can only be compared with bird citizen-based knowledge regarding the study of animal biodiversity. This greater amount of knowledge available at many levels provides a large amount of information in the field, allows the improvement of citizen science networks, and establishes an information base on which to support new research. Consequently, there are more and more works that use butterflies as an object of study and in which conclusions of enormous importance are reached to understand the functioning of ecosystems and to what extent global change is affecting those functional processes that ultimately translate into ecosystem services.

## 4.2 Scientific and Practical Information

In the Sierra Nevada Protected Area, a butterfly monitoring scheme is being carried out since 2008 as part of the SNGCO (https://obsnev.es). From 2014 there are 4 transects (20% of the total) that are carried out by volunteer personnel through a citizen science program. Despite this, volunteers periodically attend to the rest of the transects to help technicians of the SNGCO. Apart from the help provided, these joint visits allow the continuation of the training process. Volunteer participation is supported by a continuous training program that was especially intensive at the beginning of the monitoring scheme. Most years a specific workshop on the butterflies of Sierra Nevada is organized. The main scope of this event is to organize the current monitoring season and to analyze main results of the previous one. The volunteers present the results obtained in their own transects and this allows the volunteer to encourage involvement in the monitoring program beyond the work of collecting information at field. Likewise, this activity allows them to assimilate the basic notions to analyze flight curves, phenological patterns, and to interpret results and to be involved in this interesting stage of the process. These sessions are also used to review in depth a taxonomic group at a Family level, paying attention to the species composition in the study area and the main characters that facilitate its identification. In the first editions of these workshops the identification sessions were oriented in a more general way to exclusive emphasize the species that could represent up to 75% of the contacts in the field. These training events have been complemented with periodic field visits, which have also allowed to homogenize methodological criteria. These training efforts have been reinforced with the publication of supporting materials such as a book on Sierra Nevada butterflies (Olivares et al. 2012) and a practical brochure that includes nearly 80% of the regional butterfly richness (Sevilleja et al. 2020b).

The social network Facebook was used as a training and collaborative tool and currently has 57 members. In this group, the members upload their photos expressing their doubts, which are attended by the group administrators. Further, the people who join this monitoring system, are incorporated initially into two specific transects where they follow a training process assisted by other volunteers with more experience. Once this training period is over, they are already considered to be able to drive and lead their own transects located at monitoring sites with greater complexity. However, and in relation to the latter, the priority to date has been to consolidate the existing network in the face of the possibility of encouraging its growth.

All the transects carried out by the volunteers are coordinated at two levels. At first, one fine scale level that implies organizing the census agenda, collecting the information generated, and even entering the data in the information management platform. This coordination is carried out by two volunteers. The second level of coordination includes the integration of the information generated in all the monitoring programs, to organize training activities, to coordinate social networks, and to avoid a networking activity with other regional, national, and European monitoring schemes. This last task is carried out by technicians of the SNGCO.

The monitoring system consists in counting butterflies at linear transect based on the Pollar and Yates (1993) method. This system consists of traversing the line of progression of the transect, counting all the adults that are detected in a band of 2.5 m on each side of the observer, five meters in front of it, and five meters above it. In short, an imaginary box with a side of five meters in which each of the butterflies that appear inside is noted. This system is internationally standardized, which allows the results to be compared at other geographical scales, providing an interest and information added to the follow-up. Butterflies are visually identified and only individuals that cannot be identified without the need for this management are captured. The capture of the specimens is carried out using an entomological sleeve and all the captured specimens are released again after identification. Taking photographs is also a tool that has proven extremely useful as an alternative to capturing. All contacts are identified and annotated to the species level whenever possible. The censuses must be carried out under specific climatic conditions, which consist of temperatures always above 14 °C, sparsely covered skies, and a wind speed below 25–30 km/h.

### 4.3 Major Achievements

Through this monitoring scheme based partially in a citizen science program a huge volume of information has been compiled, with important implications on a management strategy based in the conservation of the species and the habitats and also in the employment of these organisms as indicators of changes in some key ecological systems. At the species levels, our results have contributed to improve the knowledge of species included in the Habitat Directive and in the Regional and National Endangered Species Act (e.g., Barea-Azcón and Munguira 2019).

Furthermore, our data have allowed us to define a network of important areas for the conservation of the Sierra Nevada butterflies (hotspots), establishing seven areas based on the richness and abundance patterns inferred from a model derived from field data collected by professionals and volunteers (Fig. 4). These models have contributed to improve the existing knowledge of the environmental factors affecting the distribution of these hotspots.

Finally, the data compiled through this citizen science program has made it possible to establish the medium-term trend of the butterfly community. At the same time, interesting data have been obtained on the ecological traits of species whose trend is positive (generalist species, thermophilic, and whose Iberian distribution is associated with areas with less precipitation) or negative. Our data also allow us to infer phenological patterns.

## 5 Case Study 3: Evaluation of Stream Ecological Status Using Benthic Macroinvertebrates

The growing interest of society in environmental problems has led to the promotion of citizen science initiatives by an increasing number of organizations. While efforts to engage

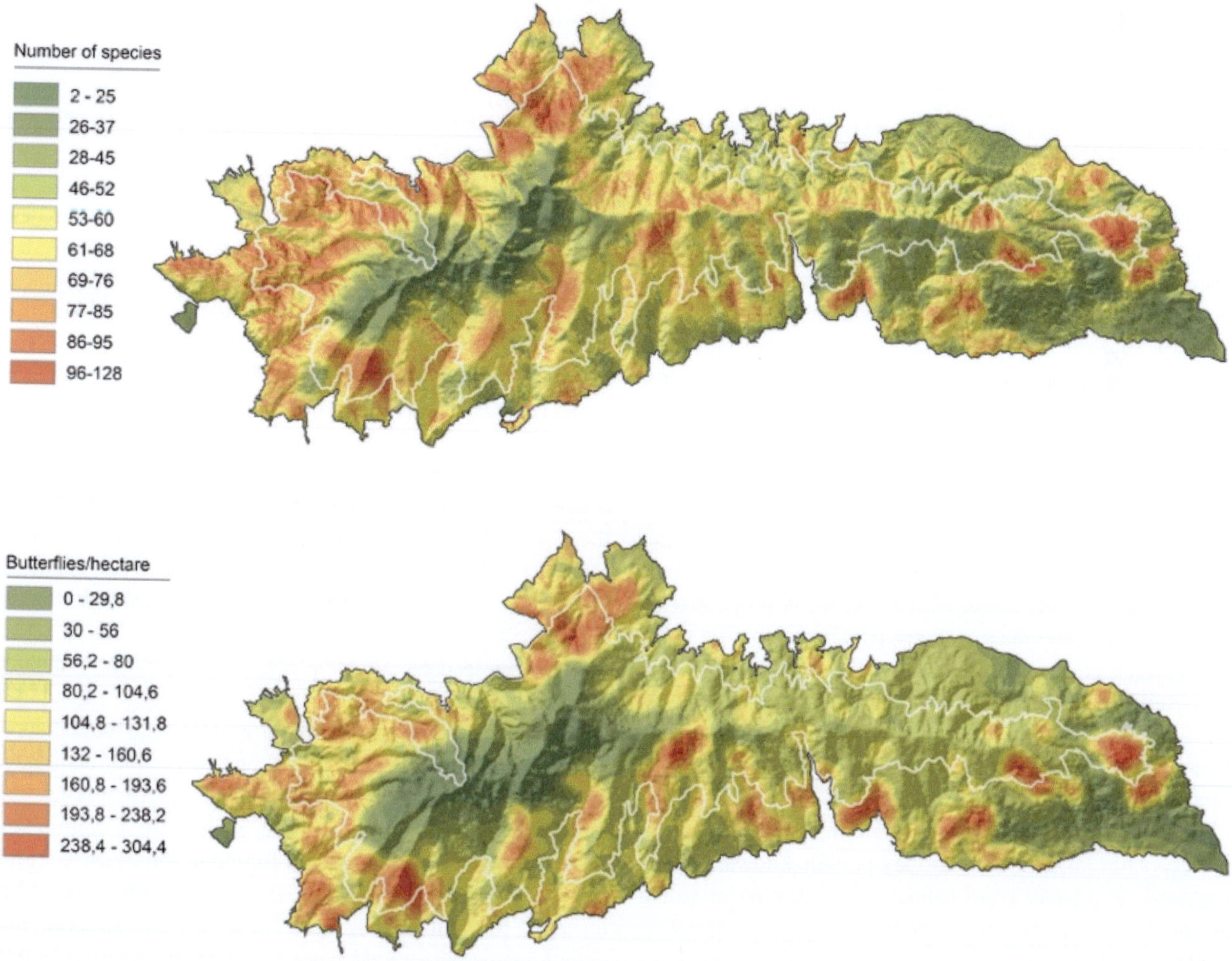


**Fig. 4** Distributional patterns of richness (top panel) and density (bottom panel) of diurnal butterflies in Sierra Nevada obtained through modeling techniques (GLM)

the public into science are more diverse than ever (Irwin 2018), aquatic macroinvertebrate communities inhabiting streams have been the focus of a large number of citizen science initiatives worldwide. For example, in 2000 the Wai Care initiative (https://waicare.org.nz/) launched a volunteer monitoring program in New Zealand's largest city which advocates citizen participation in tracking the ecological status of its river ecosystems. Likewise, the 2004 Riverfly Partnership supports a growing network of citizen science volunteers monitoring aquatic macroinvertebrates in rivers and streams across the UK (https://www.riverflies.org/). Both are examples of dynamic networks that have successfully involved the public in collaborative citizen initiatives.

## 5.1 Scope and Goals

The promotion of volunteer camp "Support for Climate Change Monitoring in Mountain River Ecosystems" by the Environmental Sciences Association (ACA) with the support of the SNGCO (https://obsnev.es) and the National Parks Autonomous Agency (OAPN), is a simple and attractive way to bring closer to society the world of management and research, especially that related with the still widely unknown world of benthic macroinvertebrates inhabiting high mountain rivers. This type of initiative allows people who do not belong to the scientific field to have a much broader perspective of the world around us and understand that our way of life as a society can have a direct impact on the composition or distribution of species that have inhabited the Sierra Nevada mountain range for thousands of years. A final goal aims at creating environmental awareness for the protection of local and regional mountain environments.

## 5.2 Scientific and Practical Information

Aquatic macroinvertebrates are widely used as an indicator of freshwater ecological status and global change. In contrast to the snapshots of chemical and physical properties of water, macroinvertebrates are long-term indicators of water quality. This is because these biological communities are highly sensitive to pollution and environmental changes over their lifetime. Therefore, the presence of specific taxa reflects the ecological status of the sampling area over a relatively large period of time.

Aquatic macroinvertebrate monitoring carried out by volunteers has large benefits as they provide knowledge and awareness of aquatic ecosystems, and contribute to filling the many gaps in macroinvertebrate data that otherwise could not be filled. In addition, data collection by volunteers allows them to extend sampling periods and contribute with information about river health and environmental change. All this information helps to inform regulatory decisions and is critical to forward public engagement in river catchment management (Moolna et al. 2019).

In addition, macroinvertebrates are relatively easy to sample using basic equipment (hand nets) and to identify using taxonomic guides. All this makes macroinvertebrates ideal bioindicators that provide an excellent resource for the assessment of a wide range of aquatic parameters. Still, one of the major drawbacks is that data collected by volunteers are often considered unreliable due to errors in sampling protocols or taxonomic identification, raising doubts about the validity of such data in scientific research programs (Vietz et al. 2016). To minimize these risks, the samples in different streams of Sierra Nevada were collected by ACA volunteers in both 2015 and 2019, applying the methodology established by the SNGCO for monitoring aquatic ecosystems (kicks), and under the initial supervision of a professional aquatic ecologist. All samples are properly fixed in 96° alcohol, and special effort is made at capturing adult specimens of Plecoptera and Tricoptera for subsequent determination at the species level with the aid of the SNGCO professional staff. All the stations sampled by volunteers correspond with some of those already established by the SNGCO, thereby reducing variability associated with the spatial and temporal variation (Fig. 5).

## 5.3 Major Achievements

The major conclusion of our initiative is that monitoring programs carried out by citizens can be a most effective means of complementing scientific monitoring programs. However, because identifying macroinvertebrates can be quite challenging, the process needs to be foresighted and validated by a professional and classification of organisms organized at a broad taxonomic level of family. In the case of the SNGCO, the data obtained by ACA volunteers were inserted with the rest of the sampling data that began in 2008, thus contributing to enlarge the time series for Sierra Nevada.

From these data, we have been able to extract valuable information about the long-term water quality of the aquatic environments and some emerging impacts of global change on aquatic macroinvertebrates. Thus less tolerant taxa, have shown changes both in their altitudinal distribution as well as flighting period. For example, imagos of *Capnioneura mitis* which used to be captured from autumn to early spring are now more frequent in late autumn (months of October through November; Fig. 6). These results suggest that *Capnioneura mitis* could have advanced its emergence period in response to rising water temperatures.

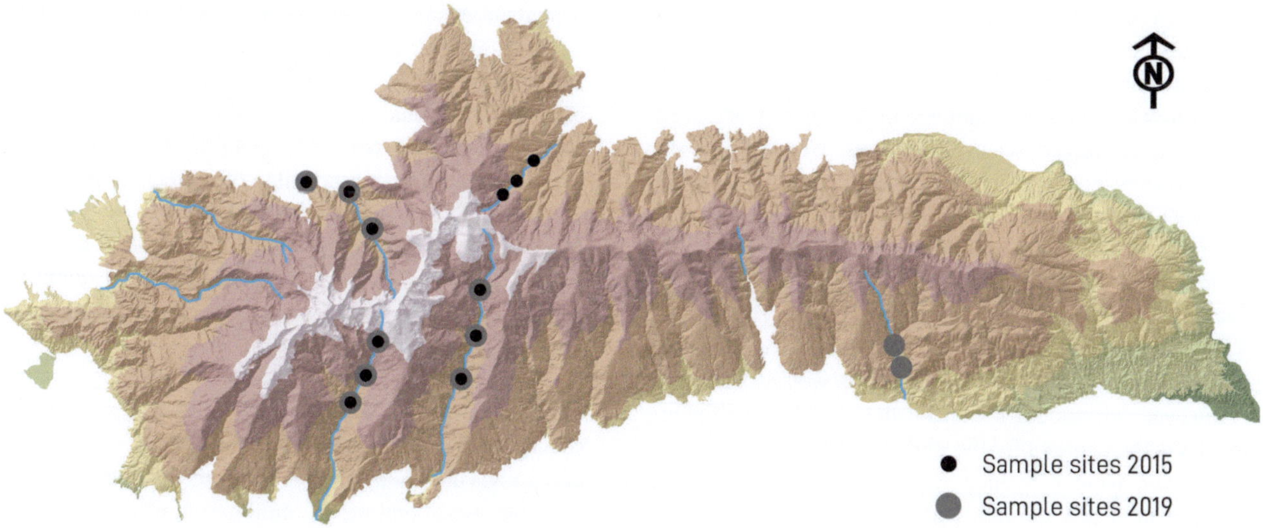


**Fig. 5** Map of monitoring sites (solid circles) in five basins of Sierra Nevada. Top northern basins (from left to right): Monachil and Genil. Bottom southern basins (from left to right): Poqueira, Trevélez, and Andaráx

**Fig. 6** Interannual dynamics of the relative contribution of *Capnioneura mitis* (two imago size classes and adults) to total macroinvertebrate abundance. Values are means for all sampling sites

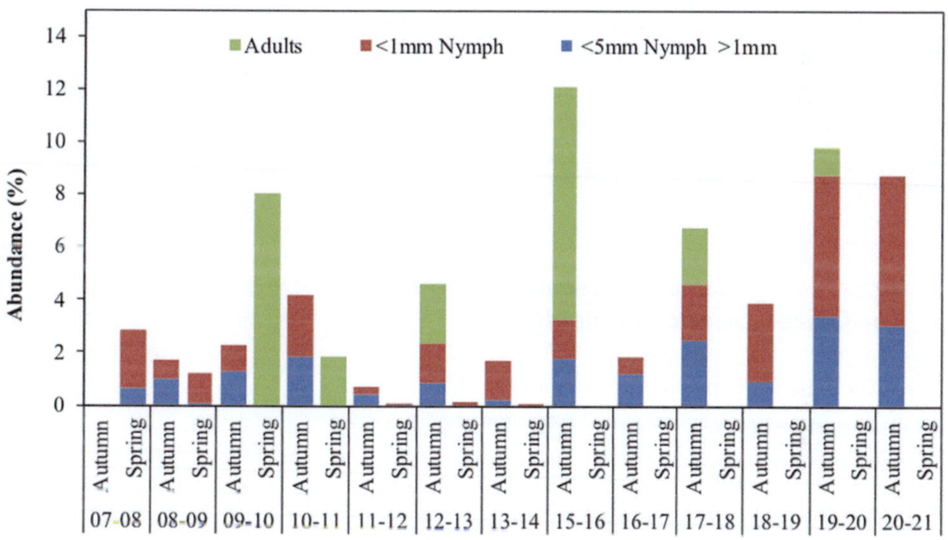


## 6 Lessons Learned from the Case Studies

Citizen science is a growing method of scientific investigation where scientists provide the training while citizens provide the data. When applied to environmental sciences, joint monitoring programs can connect scientists who study nature with citizen communities who enjoy nature. These two groups can learn from and benefit from each other, and citizens can better than anyone enhance science education through outreach and education for all ages.

Every time more environmental sciences outreach and education is developing new online tools and e-labs for increasing access and engagement with science. However, one thing that cannot change is that good science is founded on good data. This not only affirms the value of acquiring data, but underlines the foundation process of creative science that, with the help of citizens, starts with gathering the scientific data needed to solve relevant timely scientific questions. In many cases citizen science experiences revealed frailties in the National Science Programs, especially due to progressively reduced budgets and increased workloads facing scientists. Greater involvement of citizens can contribute to creatively developing solutions to overcome shortcomings and resource limitations.

We cannot backslide on Confucius' wise words "*I hear and I forget, I see and I remember, I do and I understand.*" In short, our experience with in-person learning environments showed an increasing access to an engagement with science and educational outreach. So the answer to the question of whether citizen groundwork is key for educational outreach success is "yes." Although many resources and approaches can quickly pivot to fully engage in online education, we should not renounce to in-person workshops, and prioritized citizen science programs. Nowadays, emotional appeals seem to gain ground against fact-based arguments. Perhaps, the change exists that citizen science helps revive the much-needed critical thinking (Irwin 2018). Besides all moral and educational issues, data gathered by citizens can contribute to fill the gap for missing information to crucially provide ground truth for modern technology that include continuous monitoring sensors as well as satellites.

However, we should not make the mistake of thinking that data is the panacea for the global change challenges. As Nobel prize winner Sydney Brenner once said "we are drowning in a sea of data and starving for knowledge" (Nurse 2021). Data is worthless if it does not come with new ideas that generate explanations and future actions. This is often the case in citizen science where volunteers might feel a sense of powerlessness as they find it hard to see how their work contributes to decision making in global change related topics. For this reason, we should not merely see citizen science as a means of gathering scientific information, but contribute to strengthening its role in society by fostering responsibility to stewardship action through storytelling and outreach actions (Spellman et al. 2021).

## 7 Concluding Remarks: Recommendations for Citizen Science Campaigns

Based on our own experience, we identified and prioritized needs for implementing targeted programming in citizen science campaigns. Among others:

- Professionals should supervise and provide support via training, protocols, and data calibration to ensure that data collected by citizens is scientifically robust. Such an effort is necessary to guarantee a major source of high-quality data that can serve policy-makers as well as scientists.
- When possible, citizen science campaigns should prioritize hands-in workshops where citizen volunteers receive guidance from experts in the process of scientific sampling and field/laboratory protocols.
- Publicize workshops widely through social media and other means of reaching the general public. Make workshop assistance free and prior to coursework make sure to anticipate all the necessary information and recommendations.
- If demand exceeds organization capacity, priority should be given to those citizens showing a greater contribution and commitment to the campaign in the past. Ideally, participants should belong to the local communities providing people with an opportunity to appreciate, enjoy, and understand the value of protecting the natural resources of the environment they live in.
- Workshops should be held face-to-face, and preferentially on-site and close to the study sites. By doing so, certain hands-on activities can be easily carried out in situ over the course of the workshop.
- If possible, explore avenues for hybrid onsite-online education. Hosted live sessions can be recorded in streaming or later made available on social media channels (YouTube, and others). Because the cost of implementing these options is minimal, translation of live experiences to online environments can improve the access of workshops as educational products. In-person workshops can then break time barriers and allow greater participation, reaching more people in a cost-effective manner and moving toward universal access to educational and scientific material.
- Explore the next generation of useful and practical citizen science tools. Develop simple, quick, and low-cost methods and technologies to independently assess water/air/soil quality to provide opportunities to increase the spatial and temporal coverage of ongoing monitoring programs. It is important to keep up citizen science with the emergence of new technologies and sensors. There is the risk of inefficient adaptation, and a major challenge arises as to integrate citizen's data with data provided by new devices and network technologies, including remote sensing.
- Citizen science programs should establish standard methods and operating procedures for the field and laboratory analyses, and these should always be free of charge and of open access to allow interoperability.
- Community science can contribute to guide the flow of information from scientific forums to decision-makers and managers by providing the essential data, awareness, and educational outreach necessary to evaluate catchment management and restoration practices.

Finally, there is a critical step where most citizen's campaigns fail and this is the feedback needed to hold the public interest on the studied issue. Regardless of the purpose for which data is collected, it is crucial that results are made public and visualized. This necessary feedback needs to connect to the initial need for data collection, so that citizens involved in the campaign don't feel their time and effort is being used wisely. Educational outreach tools can include from face-to-face annual workshops to online reports or program newsletters. Such an integrative approach may not only warrant the ownership of the whole process, but ensure the long-term sustainability of citizenship campaigns.

**Acknowledgements** We would like to thank the work of the volunteers and colleagues involved in data collection. In particular, we are extremely grateful to Enriqueta Martos, Beatriz Quijada, Anita Beijer, Vicente Augustín, Javier Martín, Rogelio López. We are indebted to the staff of the Department of Ecology at the University of Granada who participated in all lake Workshops and led the multidisciplinary sampling teams. We are also grateful to the rangers of the Sierra Nevada Protected Area and *Asociación de Ciencias Ambientales* for supporting and contributing to field sampling. Many thanks to Blanca Ramos, Rut Aspizua, and the management team of the Sierra Nevada National Park and Natural Park, who have always supported the butterfly monitoring scheme. We are indebted to Regino Zamora, Andrea Ros, Ricardo Moreno, and all the people involved in the Sierra Nevada Global Change Observatory. This research has been financed by FEDER/Junta de Andalucía-Consejería de Transformación Económica, Industria, Conocimiento y Universidades (B-RNM-310-UGR20), (PID2020-118872RB-I00, MICIN/AEI/10.13039/501100011033), and the LifeWatch-ERIC project Smart EcoMountains (LifeWatch-2019-10-UGR-01).

## References

Alemohammad SH, Fang B, Konings AG, Aires F, Green JK et al (2017) Water, energy, and carbon with artificial neural networks (WECANN): a statistically based estimate of global surface turbulent fluxes and gross primary productivity using solar-induced fluorescence. Biogeosciences 1144(18):4101–4124

APHA (1985) Standard methods for the examination of water and wastewater. American Public Health Association, Washington, DC

Barea-Azcón JM, Munguira ML (2019) *Polyommatus (Plebicula) golgus*. In: Jubete F et al (ed) Bases técnicas para la conservación de los lepidópteros amenazados en España. Asociación de Naturalistas Palentinos, pp. 122–136

Boissier PE (1839–1845) Voyage botanique dans le midi de l'Espagne pendant l'année 1837. In: Gide et Cie., librairies-éditeurs (eds), vol 1–2

Carrillo P (1989) Análisis de las interacciones tróficas en el plancton de un sistema oligotrófico. Ph.D Thesis, University of Granada, Spain

Dennis RLH, Shreeve TG, Van Dyck H (2003) Towards a functional resource-based concept for habitat: a butterfly biology viewpoint. Oikos 102:417–426

Devictor V, Whittaker RJ, Beltrame C (2010) Beyond scarcity: citizen Science programmes as useful tools for conservation biogeography. Divers Distrib 16:354–362

Díaz-Hernandez JL, Herrera-Martínez AJ (2019) Hydrological characteristics and paradoxes of Mediterranean high-mountain water-bodies of the Sierra Nevada, SE Spain. Hydrology 6:59

Dickinson JL, Shirk J, Bonter D, Bonney R et al (2012) The current state of citizen science as a tool for ecological research and public engagement. Front Ecol Environ 10:291–297

Dincă V, Montagud S, Talavera G, Hernández-Roldán J et al (2015) DNA barcode reference library for Iberian butterflies enables a continental-scale preview of potential cryptic diversity. Sci Rep 5:12395

Follett R, Strezov V (2015) An analysis of citizen science based research: usage and publication patterns. PLoS ONE 10:e0143687

Fritz S et al (2019) Citizen science and the United Nations sustainable development goals. Nat Sustain 2:922–930

González-Olalla JM, Medina-Sánchez JM, Lozano IL, Villar-Argaiz M et al (2018) Climate-driven shifts in algal bacterial interaction of high mountain lakes in two years spanning a decade. Sci Rep 8:10278

Gordienko YG (2013) Green paper on citizen science. European Commission, Socientize EU Project

Irwin A (2018) Citizen science: comes of age. Nature 562:480–482

Jin Z, Azzari G, You C, Di Tommaso S, Aston S et al (2019) Smallholder maize area and yield mapping at national scales with Google Earth Engine. Remote Sens Environ 228:115–128

Jordan R, Crall A, Gray S, Phillips T et al (2015) Citizen science as a distinct field of inquiry. Bioscience 65:208–211

Kolassa J, Reichle RH, Liu Q, Alemohammad SH et al (2018) Estimating surface soil moisture from SMAP observations using a neural network technique. Remote Sens Environ 204:43–59

Martínez-Silvestre R (1975) First report on the limnology of the alpine lake La Caldera, in the Penibetic Mountains (Sierra Nevada, Granada, Spain). SIL Proceedings 19:1133–1139

Moolna A, Duddy M, Fitch B, While K (2019) Citizen science and aquatic macroinvertebrates: public engagement for catchment-scale pollution vigilance. Ecoscience 27:303–317

Nurse P (2021) Biology must generate ideas as well as data. Nature 597:305

OECD (2015) Making open science a reality. OECD Science, Technology and Industry Policy Papers, No. 25, OECD Publishing, Paris

Olivares FJ, Barea-Azcón JM, Pérez-López FJ, Tinaut JA et al (2012) Mariposas diurnas de Sierra Nevada. Consejería de Medio Ambiente, Junta de Andalucía

Oliver TH, Marshall HH, Morecroft MD, Brereton T et al (2015) Interacting effects of climate change and habitat fragmentation on drought-sensitive butterflies. Nat Clim Change 5:941–945

Pollar E, Yates TJ (1993) Monitoring butterflies for ecology and conservation. Champan & Hall, London

Sánchez-Castillo P (1986) Estudio de las comunidades fitoplanctónicas de las lagunas de alta montaña de Sierra Nevada. Ph.D. Thesis, University of Granada, Spain

Schowalter TD (2016) Insect ecology: an ecosystem approach. Elsevier Academic (4th ed). San Diego, CA, USA

Sevilleja CG, Barea-Azcón JM, Fernández-Zamudio R, Munguira ML (2020b) Field guide for butterfly identification. Andalusia. Able (Assessing Butterflies in Europe) project. https://butterfly-monitoring.net/sites/default/files/Pdf/Field%20Guides/Field-Guide_SP_Andalucia_English%20(low).pdf

Sevilleja CG et al (2020a) Assessing butterflies in Europe – European butterfly monitoring scheme. In: Network development: Technical Report. Butterfly Conservation Europe and ABLE/eBMS (www.butterfly-monitoring.net)

Spellman KV, Cost D, Villano CP (2021) Connecting community and citizen science to stewardship action planning through scenarios storytelling. Front Ecol Evol 9:695534

Stefanescu C, Peñuelas J, Filella I (2003) Effects of climatic change on the phenology of butterflies in the northwest Mediterranean Basin. Glob Change Biol 9:1494–1506

Vietz GJ, Flatley AJ, Rutherfurd ID (2016) Comparisons of stream macroinvertebrates monitoring data from citizen scientists and an aquatic ecologist. In: Proceedings of the 8th Australian Stream Management Conference, Leura, New South Wales

Villar-Argaiz M, Medina-Sánchez JM, Cruz-Pizarro L, Carrillo P (2001) Inter- and intra-annual variability in the phytoplankton community of a high mountain lake: the influence of external (atmospheric) and internal (recycled) sources of phosphorus. Freshw Biol 46:1017–1034

Villar-Argaiz M (1999) Redes tróficas pelágicas: una perspectiva estequiométrica. Ph.D. Thesis, University of Granada, Spain

Wallis De Vries MF, Van Swaay CAM (2006) Global warming and excess nitrogen may induce butterfly decline by microclimatic cooling. Glob Change Biol 12:1620–1626

Williamson CE, Dodds W, Kratz TK, Palmet MA (2008) Front Ecol Environ 6:247–254

# Filling the Gaps in Research, Monitoring, Management and Social Connection

Regino Zamora and Ana Mellado

**Abstract**

Mountain ranges offer extraordinary opportunities to conduct research and monitor global change. Starting with a detailed analysis of the geographical setting, ecological dynamics and the history of human management, in this book, we focus on the uniqueness of the natural heritage of Sierra Nevada in both a historical and a global-change context. The 24 chapters of this book provide a full review of the diagnosis of the health status of Sierra Nevada ecosystems. All the evidence presented in this book reinforces our underlying idea of Sierra Nevada as a unique biophysical, historical and socio-economic laboratory and observatory of global change. This mountain is also of strategic importance as a provider of ecosystem services within the bounds of the National Park as well as for the surrounding region. This last chapter provides a critical review of what we have done so far, and what still needs to be done to improve research, monitoring, use of new technological tools (remote sensors, artificial intelligence, virtual research environments), user-oriented solutions and knowledge mobilization. The chapter ends with a final proposal: a mountain of all and for all, a joint journey of nature and people towards the mountain community.

**Keywords**

Anthropocene • Research and monitoring gaps • New technological tools • Knowledge mobilization • User-oriented solutions • Nature and people • Mountain community

R. Zamora (✉)
Department of Ecology, University of Granada, Granada, Spain
e-mail: rzamora@ugr.es

R. Zamora · A. Mellado
Andalusian Institute for Earth System Research (IISTA), University of Granada, Granada, Spain

## 1 Mountain Global-Change Lab

Mountains, being locally unique (Zamora et al. 2017a; Rachbek et al. 2019), are the crown jewels of our natural heritage around the world. Mountainous formations are home to a quarter of terrestrial biodiversity, a third of the world's protected areas (Körner and Ohsawa 2005; Perrigo et al. 2019), and a large number of known rare and endemic species (Antonelli et al. 2018; Enquist et al. 2019; Körner 2021). In addition, mountains store natural water reservoirs on high ground (water towers) and provide vital ecosystem services for the people who live in and around them (Kohler et al. 2014; Payne et al. 2017; Mengist et al. 2020). Thus, mountains play a central role in regional and global sustainable development. These ecosystem services have remained relatively stable over the past 10,000 years—a geologic period known as the Holocene—in which environmental change occurred naturally and Earth's regulatory capacity maintained the conditions that enabled human development. Now, human activities have reached a level that could upset the systems that maintain Earth in the desirable Holocene state (Steffen et al. 2011; Folke et al. 2021). Mountain environments are particularly threatened by the increasing pressures from human population growth, global change and land-use change, putting at risk the valuable natural resources on which a major part of today's society and economy depends. The present and future challenge of conserving the natural heritage of mountains is to determine the exposure, sensitivity and adaptability of their ecosystems in relation to global change (Hansen et al. 2014; Catalan et al. 2017; Zamora et al. 2017a; Rogora et al. 2018).

In recent decades, views of the relationship between humans and nature have changed with the increasing impacts of human activities on natural systems. Over the past 60 years, the prevailing view of conservation has changed several times, resulting in multiple framings in use today (Mace 2014). Conservation thinking moved away from species towards ecosystem's health, and a wider


R. Zamora and M. Oliva (eds.), *The Landscape of the Sierra Nevada*,
https://doi.org/10.1007/978-3-030-94219-9_24

acceptance of the notion that people are part of ecosystems emerged. Our original conception of nature conservation has evolved to incorporate territorial, social and economic aspects into the original conservation objectives. Global change is accelerating the need for a more holistic and innovative approach to their planning and management. The transfer of scientific knowledge to decision-making and the involvement of society are crucial.

Our perception of the conservation of mountains has also been changing. The conservation of mountain ecosystems has traditionally been approached as if the parameters of environmental excellence were immutable and as if mountains were cathedrals or museums instead of as what they in fact are: dynamic systems, open to natural changes and susceptible to human impacts. This reality is even more evident today because the human footprint extends throughout the planet, affecting high mountains directly and indirectly. In a global-change scenario, mountain ecosystems constitute a vital natural heritage to be preserved, and a source of natural resources providing the regulating and provisioning of cultural ecosystem services (Fig. 1). Two fundamental concepts, conservation and utilitarianism, have traditionally been considered to be antagonistic and, therefore, have been scrutinized and developed separately during the last century. In the age of human dominance over the systems of earth, known as the Anthropocene, we have the obligation and responsibility to strive for the complementarity between the two concepts (Millennium Ecosystem

**Fig. 1** Mountains provide multiple contributions to people, including (1) regulating services (e.g., clean air and water, carbon sequestration); (2) provisioning services (e.g., food and raw materials); (3) supporting services (e.g., habitat for species, maintenance of genetic diversity); (4) cultural services (e.g., historical-cultural referents and recreation); (5) reservoirs of local knowledge; (6) generators of scientific knowledge

Assessment 2005). Specifically, the conservation of mountain ecosystems both raises concerns and offers solutions to environmental and socioeconomic problems here and now. In this context, our general hypothesis is that the most diverse, heterogeneous and multifunctional mountain ecosystems offer a wider range of ecosystem services and have a greater capacity of resistance and resilience in the face of global change.

Mountain ranges offer extraordinary opportunities to conduct research and to monitor global change. In order to understand the consequences of human impact on the planet, we need systems of reference. Mountain ecosystems may represent the best-preserved reference systems in a given region, providing the opportunity to compare their dynamic behaviour with the more intense humanized ecosystems surrounding the mountains under a global-change scenario (Huber et al. 2005; Hansen et al. 2014). In this respect, mountain ecosystems are equivalent to controls in the classic experimental designs against which to compare treatments, in this case, caused by human activities (land-use changes, pollution, overexploitation of resources, etc.) in the surrounding more anthropized matrix.

Mediterranean mountain ecosystems are more prone to increase in the vulnerability of ecosystem supply due to global change than other European ecosystems. Thus, the Mediterranean mountains seem to suffer the double jeopardy of being mountains and being located near the Mediterranean basin (Nogués-Bravo et al. 2008; Doblas Miranda et al. 2017). In a scenario of global change, a high mountain such as Sierra Nevada represents an exceptional ecology field lab, offering the advantage of the enormous mosaic of ecological conditions extending along an elevational gradient. Mountains are also extremely sensitive to environmental changes because of the fragile equilibrium in the components of the natural system, and thus these ecosystems can be used as sensors to detect early signs of change. This includes processes considered more genuinely global, which can be observed better from the mountains as exceptional lookouts. In this sense, mountain ecosystems are key observatories of the atmosphere since they can be used to study remote processes including all the aspects related to climate and atmospheric dynamics such as energy balance, electromagnetic signals generated by lightning produced by the Earth's global storm activity, UV radiation, atmospheric particle transport and deposition, pollutants, greenhouse gases, pollen transport or resistant biological forms and microorganisms (Beniston 2003; Huber et al. 2005; Zamora et al. 2016; Catalan et al. 2017).

Starting with a detailed analysis of geographic, biotic and cultural singularity of mountain landscapes, we focus this book on the ecological uniqueness of Sierra Nevada as an exceptional global change laboratory. The 24 chapters of this book provide a full review of the diagnosis of the health status of ecosystems in Sierra Nevada. We first introduce the physical setting of Sierra Nevada as a landscape in continuous evolution, due to geomorphologic and climatic changes. More recently, environmental changes derived from human activities are superimposed on this natural variability, creating new global change scenarios. The ecosystems of Sierra Nevada are exposed to different drivers of global change, such as climate change, changes in land use, and allochthonous abiotic and biotic inputs. As a very brief summary, the results show the following: (1) an ancestral human footprint on the ecosystems of Sierra Nevada, particularly evident during the twentieth century; (2) a moderate climate warming, with reduced and more variable precipitation, as well as a consequent reduction in snow cover duration over the last few decades; (3) significant changes in biophysical characteristics of rivers and mountain lakes, due mainly to atmospheric inputs; and (4) shifts in the distribution, phenology, and interactions of many species of plants and animals along elevational gradients.

High mountain aquatic systems are particularly sensitive to allochthonous inputs, but terrestrial ecosystems may also be sensitive, for which there is little information (Stevens et al. 2018). The application of 'omic' technologies in connection with ecosystem processes is beginning to reveal new insights into ecosystem functioning. Additionally, new pollutants could have more incidence in the ecosystems of Sierra Nevada, such as plastics, both microplastics carried by the wind and the remains of plastic garbage abandoned in the mountains (Padha et al. 2021). Urban pollution from the city of Granada and its industrial environment might be also increasingly evident in the mountains (Battaglin et al. 2018). Otherwise, in the generalized altitudinal ascent of animal and plant species from low elevations (Pauchard et al. 2009), invasive species appear more and more frequently, particularly in aquatic systems, but also in terrestrial ecosystems, which could have their main access route through the roads that go up to the summits of Sierra Nevada. Moreover, emerging diseases using animals as transmission vectors (zoonosis) are also a growing focus of concern due to their potential impact on human health. Additionally, changes in land use and climate also bear direct consequences for socioeconomic activities; in this sense, the organization of the mountain economy in Sierra Nevada has drastically changed over the last 60 years in parallel to the development of infrastructures, tourism and industrial activity (Bonet et al. 2015, see chapter "Local Ecological Knowledge and the Sustainable Co-Management of Sierra Nevada's Social-Ecological System"). Faced with the increase in magnitude and diversity of anthropogenic impacts and their interactions, the rapid recovery of the vegetation cover after the rural abandonment is the only aspect that could be considered positive in the immersion of Sierra Nevada ecosystems in the Anthropocene.

All the evidence presented in this book reinforces our original idea of Sierra Nevada as an exceptional laboratory and observatory of global change, being also a major provider of provisioning, regulating, and cultural ecosystem services within the boundaries of the massif as well as for the surrounding dry region. Clearly, the investigations and monitoring that continue to be carried out as a continuation of those already initiated years ago will give us a measure of the health status of the Sierra Nevada ecosystems in the face of global change. Only the adoption of measures with a solid scientific basis can provide a roadmap to orient conservation and adaptive management at Sierra Nevada. These investigations and monitoring will also be the basis for the sustainable use of mountain resources.

Below, we critically review what has been done and what remains to be done in research, monitoring, use of new technological tools, knowledge mobilization, and information transfer to management and social connections in Sierra Nevada.

## 2 Research and Monitoring in Sierra Nevada

*What have we done so far?*

### 2.1 Monitoring in Sierra Nevada

The natural ecosystems of Sierra Nevada have long attracted the attention of naturalists and researchers. A global review of these investigations is presented in chapter "Scientific Knowledge Generated in Sierra Nevada: Bibliographic Review (1970–2021)". Sierra Nevada represents an exceptional ecology lab, offering the advantage of the enormous mosaic of ecological conditions existing along mountain slopes. For these reasons, a long-term monitoring programme called *Sierra Nevada Global-Change Observatory* (http://obsnev.es) promoted by the Environmental and Regional Planning Council of the Regional Government of Andalusia and the University of Granada was established some years ago to put together useful and relevant information regarding the impacts of global change on ecological systems of Sierra Nevada (Bonet et al. 2010).

Given the complexity of natural systems and the variety of factors that influence natural processes, research and monitoring programmes clearly need conceptual frameworks that help organize the information gathered. The design of the Global-Change Monitoring programme in Sierra Nevada (Zamora et al. 2016, 2017a, b) was inspired by the conceptual framework and the thematic areas proposed by the GLOCHAMORE (Global Change in Mountain Regions http://mri.scnatweb.ch/projects/glochamore, Grabherr et al. 2005) and related initiatives. Our monitoring programme is based on evaluating, through standardized protocols, the composition, structure and functioning of the ecosystems of Sierra Nevada, providing a vision of the trends of change that allow the development of an adaptive capacity. The goal was to identify the best biophysical and socioeconomic indicators of the health of our ecosystems under a global-change scenario, and to determine their ability to adapt to global change (Hansen et al. 2014; Zamora et al. 2016). The thematic areas of our research strategy are organized according to our understanding of the causes and consequences of global change, their ecological implications and the corresponding biotic and socioeconomic responses to changes (Zamora et al. 2017a, b).

Several reports have been released since the start of the monitoring in Sierra Nevada (Fig. 2). First, we identified the potential impacts of global change on the main ecosystems of Sierra Nevada and the ecosystem services affected (Bonet et al. 2010; http://sl.ugr.es/structure_obsnev_en). Then, we defined standardized methodological protocols to measure the potential impacts of the global change (Aspizua et al. 2014; http://sl.ugr.es/methods_obsnev_en). A preliminary diagnosis of the global-change impacts on the ecosystems of Sierra Nevada has been published (Zamora et al. 2016; http://sl.ugr.es/results_obsnev_en). For this report, we used information from several sources. Apart from the information generated by our monitoring programme (more than 100 variables of 48 different methodologies; Aspizua et al. 2014), we compiled information from several research groups that had been working for decades in Sierra Nevada.

The intensive monitoring of critical zones in which research and monitoring are concentrated in Sierra Nevada has been one of the pillars of the Sierra Nevada Global-Change Observatory (Bonet et al. 2010; Aspízua et al. 2014; Zamora et al. 2016). To link biotic with abiotic information at different spatial scales, we have developed the concept of Intensive Monitoring Stations (Fig. 3), which are areas of representative ecosystems, in which a high density of ecological monitoring protocols coincides spatially with abiotic measurements (Zamora et al. 2017a, 2021). An Intensive Monitoring Station can be considered a hotspot for gathering biotic and abiotic data in the same spatial location (Zamora et al. 2017a, 2021).

In short, our Sierra Nevada Global-Change Observatory (http://obsnev.es) has compiled an enormous amount of environmental information and mobilized the scientific community working in Sierra Nevada. This project has also demonstrated its ability to establish linkages with similar projects at national and international scales. The foundations have also been laid for effectively transferring scientific and monitoring information to the management field.

The continuation of the work described above requires evolving from the regional project stage to being considered as a long-term infrastructure of international relevance. Now, with the support and commitment of the public

**Fig. 2** Time course of the main reports generated by the OBSNEV (http://obsnev.es) from its starting point in 2007. The first report set the basic structure and data of the project (http://sl.ugr.es/structure_obsnev_en). A second report was published with the monitoring methodologies (http://sl.ugr.es/methods_obsnev_en), and finally a report was filed concerning the impacts of the global change in Sierra Nevada ecosystems (http://sl.ugr.es/results_obsnev_en)

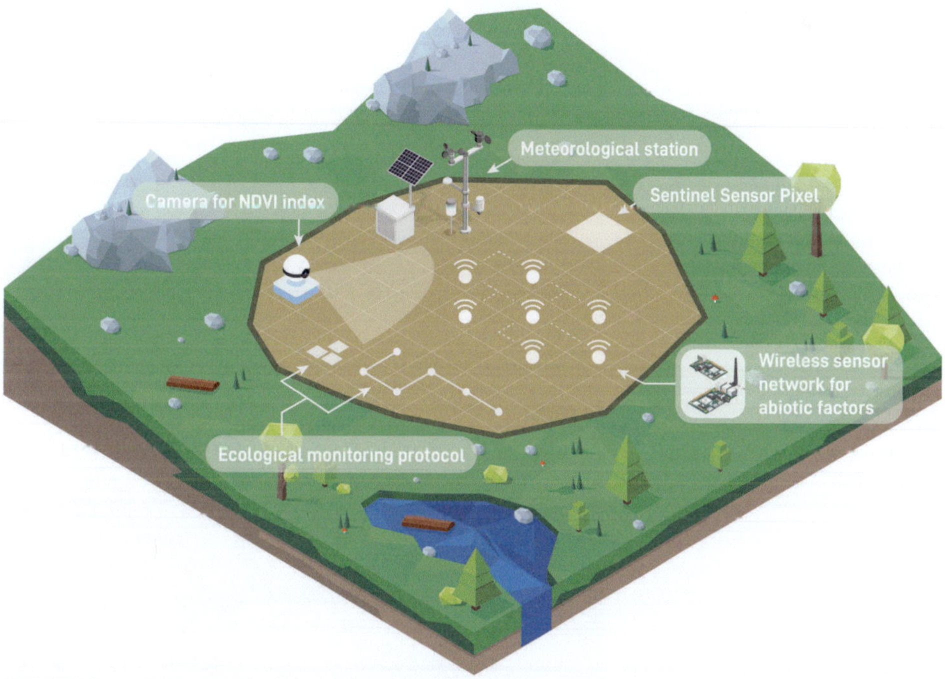


**Fig. 3** Scheme of the Intensive Monitoring Stations. These are areas with high densities of ecological monitoring protocols located around a weather station. They also include a wireless sensor network to measure abiotic variables in selected microhabitats around the station (temperature and air moisture, soil water content, irradiance, aerosol deposition, etc.) in relation to biotic responses (growth, phenology, demographic responses, etc.) Modified from Zamora et al. (2017a)

administrations, the University of Granada, and the European Research Infrastructure (ERIC) LifeWatch, it is an opportunity to turn this idea into reality. LifeWatch is a European e-science and technology infrastructure for biodiversity and ecosystem research to support the scientific community and other users. The main goal of LifeWatch-ERIC is to advance research on biodiversity and ecosystem services and to provide major contributions to addressing the big environmental challenges facing the Earth, including knowledge-based solutions to environmental managers and stakeholders for environmental preservation. To this end, LifeWatch ERIC provides access through a pan-European distributed e-Infrastructure to a multitude of data sets, services and tools enabling the construction and operation of Virtual Research Environments (Virtual Laboratories and Decision-support Applications) where specific issues related with biodiversity research are treated through the execution of integrated models.

## 2.2 Smart EcoMountains, a Thematic Center of LifeWatch-ERIC

The creation of the Mountains Thematic Center of the LifeWatch-ERIC (University of Granada-Sierra Nevada), named as Smart EcoMountains, will allow the incorporation of all existing sources of information for their analysis and interpretation in a scientific context (Fig. 4). Our proposal also includes the human dimension of biodiversity in global-change scenarios, setting realistic goals for sustainable development aligned with the "Agenda 2030".

Under the umbrella of Smart EcoMountains we will be continuously generating data on mountain-related biological, physical, chemical, climatological and socioeconomic variables studied in Sierra Nevada by in situ monitoring, automated instruments, and remote sensing, which will be compared with other mountains. All these data are freely available through open-access data repositories, enriching the information generated by international networks of mountain observatories in an effort to enhance our global understanding of mountain-related social-ecological systems (Shahgedanova et al. 2021). In addition to providing data to the global community, Smart EcoMountains combines broad expertise in multidisciplinary mountain research with a profound knowledge of ICT resources, artificial intelligence and remote sensing in order to develop tools and services, including Virtual Research Environments, that will support access to and analyses of mountain ecological and socioeconomic data (see chapter "Data Model, E-Infrastructure Services, and the Virtual Research Environment (VRE)"). Such facilities are designed to meet the needs of multiple users, including researchers, policymakers and citizens, providing users with the tools for understanding the importance of biodiversity and the effects of major environmental changes, such as climate change.

*What still needs to be done to do better?*

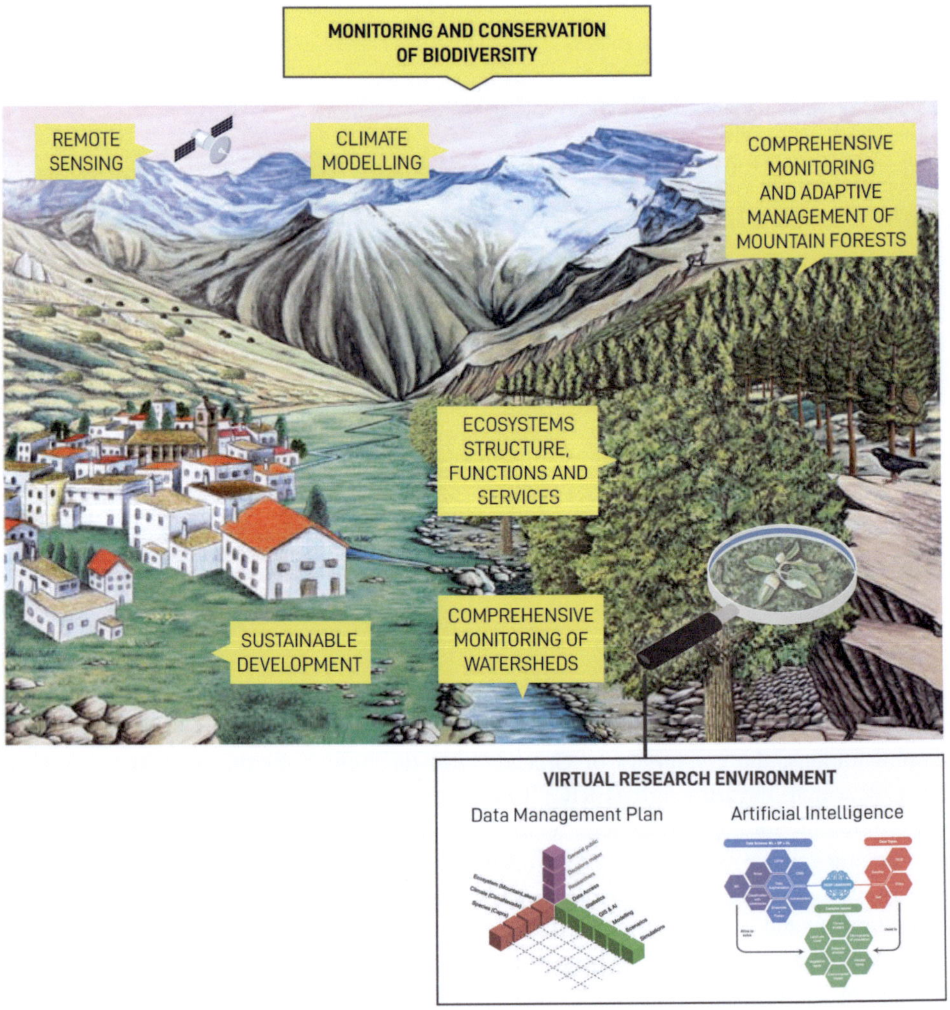


**Fig. 4** Smart EcoMountains incorporates different sources of information in the field of biodiversity, environmental sciences and global change in Sierra Nevada. It is divided into eight thematic areas: (1) monitoring and conservation of Biodiversity, (2) comprehensive monitoring of watersheds; (3) comprehensive monitoring and adaptive management of mountain forests; (4) climate monitoring and modelling; (5) ecosystems structure, functions and services; (6) Sustainable development (7) virtual research environments and (8) remote sensing

### 2.3 Back to Hubbard Brooks

Until now, research and monitoring have been conducted in isolated compartments, sectorial, monodisciplinary, reductionist, short-term, with each researcher/technician pursuing a particular interest. Faced with this lack of unity, we need a holistic, interdisciplinary and integrating conceptual framework "from local to global" governing long-term research and monitoring. The creation of Smart EcoMountains will enable the incorporation of all existing sources of information (including all scientific and technical data) in the field of biodiversity, environmental sciences and global change in Sierra Nevada for their analysis and interpretation within a global scientific interdisciplinary context. Our challenge now is to consolidate a long-term interdisciplinary physical as well as virtual European e-infrastructure for research and monitoring (including analysis, modelling, and synthesis) that will optimize the adaptive conservation and ecosystem service provisioning of mountain ecosystems.

The spatio-temporal design of the monitoring and research at our Lifewatch Thematic Center on Mountain Ecosystems is based on the successful model of Hubbard Brook (https://hubbardbrook.org) (Lindermayer and Likens 2009; Likens 2013), where, for the first time, long-term ecological experiments were performed at the basin scale from a holistic, interdisciplinary perspective, marking an inflection point in studying the impact of natural and anthropic disturbances on the natural environment. Our proposal includes an intensive monitoring programme where the hydrographic basin constitutes the main integral unit of study with the aim of evaluating the response of mountain ecosystem structure, composition, and function to disturbances (e.g. droughts, fires, atmospheric contamination, pests) and their interactions. On the other hand, a long-term perspective is needed to evaluate the impacts of these threats. For this reason, our proposal is focused on the temporal continuity of previously long-term research and monitoring from an integrated perspective, taking advantage of the long-tradition research activity of several research groups of the University of Granada and the monitoring expertise in the Sierra Nevada National Park.

Thus, the projected permanent physical and virtual research infrastructure and associated monitoring programme will work under the following philosophy: (1) carefully formulated questions; (2) maintaining/creating time series of essential variables; (3) using the new technological tools (remote sensing, IA, Virtual Research Environments, deployment of latest-generation sensors) and traditional field monitoring to answer scientific questions and to maintain essential variables.

### 2.4 New Conceptual and Technological Approaches: Exploring Fine Spatio-Temporal Heterogeneity in Mountain Ecosystems

In a climatic-change context, one of the priorities of our monitoring programme has been to gather reliable information in Sierra Nevada on climatic variables as determinants of ecosystem processes (Bonet et al. 2010; Aspízua et al. 2014). Our knowledge of ecological responses to climate change is based on data provided by weather stations, extrapolating this information to large areas using modelling tools (Lembrecht et al. 2019a, b; Zellweger et al. 2020). These macroclimatic projections can adequately represent the conditions in large, flat and homogeneous surfaces, but the realism and precision of the simulations diminish drastically when applied to mountains, where the topography generates considerable environmental heterogeneity that is often overlooked in the models (Stewart et al. 2010; Opedal et al. 2015; Lembrecht et al. 2019a, b). In fact, highly contrasting temperature mosaics have been identified in mountain environments over short distances, due mainly to topographic differences (Fridley 2009; Scherrer and Korner 2010; Korner and Hilfbrunner 2018; Garcia et al. 2020). These differences can have profound consequences for the persistence of species in a global-warming scenario (Dobrowski et al. 2009; Fridley 2009; Scherrer and Korner 2011). For this reason, in mountain ecosystems, it is essential to have in situ meteorological information with the highest spatio-temporal resolution possible in order to understand how the abiotic parameters (temperature, precipitation, wind, radiation, aerosol deposition) actually vary in the mosaic of environmental conditions related to differences in altitude, exposure, slope and types of vegetation cover (Scherrer and Korner 2010; García et al. 2020; Thornton et al. 2021).

We have recently developed the application climanevada/obsnev.es, which allows us to combine all meteorological data provided by different providers in Sierra Nevada (Pérez-Luque et al. 2021). In addition, we are currently designing and deploying sensor networks with the aim of characterizing the microclimates within the ecosystems of Sierra Nevada (Zamora et al. 2021). This network of microsensors will help us to advance in the emerging research field "Microclimate Ecology"', which integrates physical and biological aspects at different resolution scales (Zellweger et al. 2019). Our conceptual and instrumental approach will enable us to rigorously explore the cause–effect relationships between the abiotic environment and the ecological responses of organisms, as well as the possible interactions between the different spatial and temporal scales (De Frenne et al. 2013; Soranno et al. 2014). The comprehensive knowledge of these cause–effect relationships would allow us for a fine spatial mapping of biodiversity and its relationships with environmental conditions in Sierra Nevada, fostering rapid and accurate conservation actions against threats. It will also aid us in assessing the degree of decoupling between the macro- and micro-scale measurements within a given spatial and temporal context (Fridley 2009; Dobrowski et al. 2009; Zellweger et al. 2019). In this way, we will build bridges in the complex chain of causality involving Macroclimate > Microclimate > ecological response in an explicit spatial and temporal context.

## 2.5 A Long-Term and Broad-Scale Vision

The scientific approach to the study of ecological processes is not only based on solid theoretical frameworks and good experimental designs aimed at testing hypotheses but also requires time series derived from long-term studies (Likens 1989; Carpenter 1998). Long-term ecological and environmental studies have contributed profoundly to the development of a multitude of foundational advances in ecology across a diversity of natural ecosystems (Hughes et al. 2017). For example, the inextricable relationship between temperate forest and stream ecosystems emerged from long-term forest manipulations and monitoring at the Hubbard Brook Experimental Forest (Likens 1989).

The study of many natural processes requires long-term monitoring because their rate of change is very slow, because transient dynamics are likely to occur and/or because of the time lag between a given environmental change and the corresponding ecological response. In addition, long time periods are needed to be able to record the impact of extreme events, which are usually rare. Therefore, environmental assessments need a long memory to disentangle long-term from short-term changes and to reveal the full extent of ecological and biotic responses (Gonzalez et al. 2016; Hillebrand et al. 2018). For instance, while we have abiotic time series spanning tens to hundreds of years, most biodiversity time series are much shorter. In addition, these biological data series suffer from issues of changing sampling effort and taxonomic resolution, which require backtracking and harmonising species names. Consequently, our information on the degree of environmental abiotic change, as observed in the past and predicted for the future, is much more compelling than the knowledge on biodiversity change (Hillebrand et al. 2018).

Compiling long time series is also key for the appropriate parameterization and accurate validation of environmental models, such as, for example, general circulation models to forecast the climate, or models of the potential distributions of species. Long-term series will help to forecast developments in ecosystems under new scenarios, with the use of modelling tools. It is clear that long-term monitoring is a crucial part of generating accurate, solid and lasting science that will prepare us to conserve and manage ecosystems in global-change scenarios (Lovett et al. 2007; Zamora et al. 2017a). However, addressing long-term questions and phenomena cannot be effectively addressed on a project-to-project basis, with short-term funding cycles, and without maintaining a temporal coherence in seeking the same essential scientific and monitoring objectives (Likens 1989).

Long-term ecological studies are also critical for making informed decisions in many environmental management contexts, playing a vital role in environmental policy (Hughes et al. 2017). Three critical tenets underpin future long-term monitoring and management plans (Baron et al. 2009): (1) climate patterns of the past will not be the climate patterns of the future; (2) climate defines the environment and influences future trajectories of the distributions of species and their habitats; (3) specific management actions may help bolster the resilience of certain natural resources, but fundamental changes in species and their environment may be inevitable. In short, mountain landscapes of the past will not be the landscapes of the future. Science-based management will be necessary because past experience may not serve as a guide for new future conditions.

Long-term monitoring can be put into practice at different spatial scales. Understanding these spatial and temporal scales requires the creation of a "network of networks" (Peters et al. 2008) for observation, experimentation, and modelling. Thus, we have devoted effort to establishing links between our monitoring programme and other similar initiatives. First, Sierra Nevada is part of the Spanish LTER (Long-Term Ecological Research) network (http://www.lter-spain.net). Moreover, our site is connected to regional initiatives (Andalusian Network of Global Change Observatories, REDIAM, Junta de Andalucía) as well as national ones (Monitoring Global Change in National Parks, OAPN, Ministerio de Transición Ecológica) that share this philosophy of long-term studies with a broad spatial coverage. Finally, our site has participated in certain global research networks, such as the International Long-Term Ecological Research Network (ILTER) and the Program on Ecosystem Change and Society (PECS, see Holzer et al. 2018). In addition, the World Network of UNESCO Biosphere Reserves introduced the social-ecological approach into protected-area management, as well as the need to monitor changes in the biosphere resulting from human-nature interactions (Holzer et al. 2018). Despite the promising advances in long-term monitoring by these networks, one persistent challenge is the harmonization of monitoring protocols to promote cross-site comparability. This would foster more effective interoperability (Vargas et al. 2017) as well as knowledge generalization from locally driven research initiatives to broader contexts (Dick et al. 2018; Magliocca et al. 2018).

## 2.6 Looking for Essential Variables to Evaluate Exposure, Vulnerability and Adaptation to Global Changes

Due to the vital importance of healthy mountain ecosystems for the human well-being worldwide, decision-makers rely on the scientific community to deliver robust research-based understanding and accurate predictive models capable of supporting the design and implementation of appropriate management strategies that respond to global and context-specific needs. The development of such knowledge and models requires robust conceptual underpinnings and, therefore, the availability of sufficiently ample, explanatory and representative environmental data.

At present, however, the necessary long-term, homogeneous and comparable mountain observations are not easily available, due primarily to the difficult access and harsh conditions of these frequently remote sites (Shahgedanova et al. 2021). In addition, no consensus is available regarding which mountain variables should be considered observation priorities in order to detect mountain ecological change. Such variables need to take into consideration the complex topography of mountain systems and the diversity and high spatiotemporal variability of phenomena (Kulonen et al. 2020; Thornton et al. 2021). However, different global initiatives have been launched to deal with this situation. For example, the GEO-Mountains (https://www.geomountains.org/) is bringing together mountain observation networks and research institutions to facilitate the discoverability, accessibility and usability of key data and information related to the environmental and socio-economic systems across global mountain regions. The Mountain Research Initiative (MRI; https://www.mountainresearchinitiative.org/) aims to develop a global reference network of long-term environmental and socioeconomic monitoring observatories in mountain systems (Adler et al. 2020; Shahgedanova et al. 2021). This initiative regards mountain observatories as hubs for comprehensive monitoring of mountain environments, in which multidisciplinary, integrated observations of bio-physical and social environments are conducted over long periods of time according to established protocols using both in situ and remote observations (Shahgedanova et al. 2021). Otherwise, the MRI is seeking to establish the set of variables that are indispensable for better monitoring, understanding, and ultimately predicting the most important aspects of mountain biodiversity, climate, and social change at a global scale. The group is defining the Essential Mountain Variables (Kulonen et al. 2020; Thornton et al. 2021), based on the Essential Climate Variables and Essential Biodiversity Variables frameworks previously established by the Climate Observation System (GCOS; https://gcos.wmo.int/en/home) and the Biodiversity Observation Network GEO-BON (https://geobon.org/).

This standardized set of essential variables, to which Sierra Nevada will also contribute, together with appropriate data management and strong open data-sharing policies, would make it possible to compile a globally intercomparable database of diverse but consistent and useable evidence to understand and predict the dynamics of mountain social-ecological systems, guiding mitigation and adaptation measures (Fig. 5). Another front opening up is to identify specific mountain indicators considered crucial for

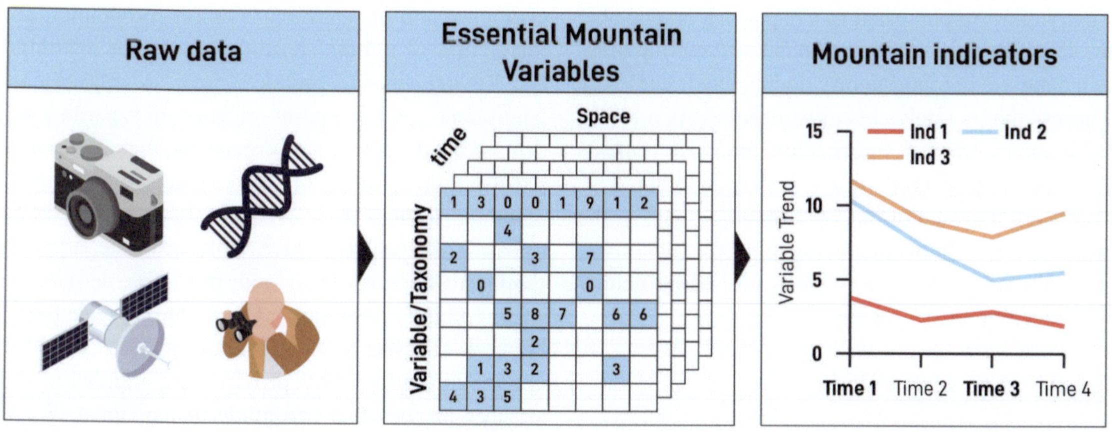


**Fig. 5** Essential Mountain Variables (EMV) are derived from raw data obtained through different observation methods, including field surveys, automated instruments, satellite remote sensing and DNA sequencing. These data are harmonized, standardized and organized as Essential Mountain Variables data products, which can be used in various ways to derive indicators that quantify spatiotemporal changes in species distributions, population abundances or other aspects of biodiversity (Figure adapted from Kissling et al. 2018)

safeguarding mountain biodiversity and ecosystem integrity (Bitsch et al. 2020). This will contribute to the Convention for Biological Diversity's vision of "Living in Harmony with Nature" by 2050 and strengthen the consistency among international frameworks and science- and evidence-based policy-making (Bitsch et al. 2020).

## 3 Use of New Technological Tools: Remote Sensors, Artificial Intelligence, Virtual Research Environments

The urgency and complexity of safeguarding mountain ecosystems in the Anthropocene require a reenvisioning of how new technological tools can best support the comprehensive assessments of biodiversity and its changes that are required for successful action. The development of monitoring design and management strategies that account for the scale, pace and complexity of anthropogenic impacts on species and ecosystems requires assessments of past and current biodiversity changes as well as robust projections of the potential future distributions of species and ecosystems (Ceballos et al. 2017; Dirzo et al. 2014; Kim et al. 2018).

*What have we done so far?*

The success of any global-change programme depends on the quality of the data it collects, manages and disseminates. The environmental information available on the ecosystems of Sierra Nevada has been compiled mainly through academic research projects and technical monitoring related to conservation and restoration. Both the research as well as the monitoring projects have been undertaken in the usual hermetic research and environmental management spaces, with little interdisciplinarity and limited capacity for transfer outside the strictly scientific or technical field. Scientific information usually ends up published in a paper that is available only by the scientific community; technical information often ends up in a database or report poorly documented, with very restricted access that is rarely consulted or used. As a result, much of the environmental information amassed remains inaccessible, and furthermore, it is also poorly announced, so potential users are unaware of it. Consequently, much of the existing information on mountain ecosystems remains invisible and thus not valued or used. What is not shared neither advances science nor serves society.

*What still needs to be done to do better?*

Science, today, is increasingly global, multi-disciplinary and networked. This trend needs innovative, dynamic and ubiquitous research supporting environments where scattered scientists can easily access data, software and processing resources managed by diverse systems in separate administration domains through their web browser. Virtual research environments (see chapter "Data Model, E-Infrastructure Services, and the Virtual Research Environment (VRE)") enable these international multi-disciplinary research-working environments and can accelerate innovation in research thanks to their integrated approach, putting technology at the service of scientific questions.

Remote sensing, together with artificial intelligence, geodesign and nature-based solutions have become central in current European initiatives (e.g., COPERNICUS Program, GoGreenRoutes, LifeWatch ERIC) to address societal challenges, such as the United Nations' Sustainable Development Goals. In Smart EcoMountains we will use these new conceptual and technological approaches, including virtual research environments for the mountain community (Fig. 6), to answer key scientific questions. We do not intend to recreate a virtual world "in the cloud", a simulated world, as the main focus of our work. On the contrary, we want to know the real world that we have in Sierra Nevada and its immediate surroundings in comparison with other mountains in Europe and the rest of the world. In that real world, in addition to biodiversity and climate change, there are researchers, managers, shepherds, and mountaineers, and, above all, citizens with a wide range of needs. This includes people who live in and from the mountains, others who simply wish to enjoy mountains for a weekend, or those who find pleasure visiting these places remotely thanks to the internet. Policymakers need to take into account the questions, concerns and interests implicit in all of these contexts.

In the face of the growing challenges brought about by human activities, predictive planning and decision-making in mountain-ecosystem conservation, restoration and sustainable development are urgently needed. A key opportunity to advance ecological modelling for predictive as well as explanatory purposes arises through a collaboration between ecologists and the earth observation community, and a close integration of remote sensing and species distribution models. Remote-sensing products have the capacity to provide continuous spatiotemporal information regarding key factors driving the distribution of organisms, thereby improving both the use and accuracy of these models for management and planning (Randin et al. 2020). In this respect, a major goal is to assess ecosystem services both from the supply side (e.g. biophysical evaluations) and from the demand side (e.g. socio-cultural evaluations). In addition, revealing the links between ecosystem structure and functions on the one hand and multiple supporting, provisioning, regulating and cultural services on the other can guide management. We will identify, evaluate and map the biophysical aspects of ecosystem services relevant to scientists, managers and public communities of Sierra Nevada. For this, we will characterize the key biophysical aspects and ecosystem services most relevant for

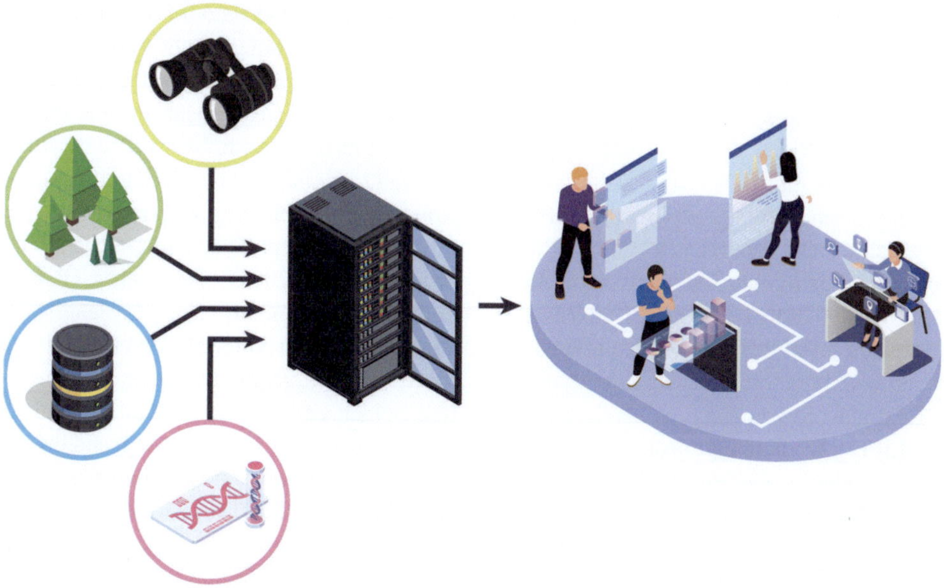

**Fig. 6** Smart EcoMountains is developing Virtual Research Environments available to the scientific community, environmental managers and the general public (chapter "Data Model, E-Infrastructure Services, and the Virtual Research Environment (VRE)"). These allow users to work together in a real-time environment that facilitates the exchange of information resources, overcoming the limitations of traditional working practices. In particular, users can collaboratively discover, access, evaluate and process heterogeneous datasets in real time and can share data, models, algorithms and scientific results. The environments are adapted to the needs of different types of users (scientists, managers, and the general public)

Sierra Nevada from a multi-actor perspective. We will also evaluate and map the biophysical aspects of ecosystem services considering their spatially explicit dimensions of ecological supply and potential demand. Finally, we will predict the supply and demand of biophysically based ecosystem services as a helpful decision-making tool in scenarios of management and environmental change. Last but not the least, always doing the necessary field validation, so that the models help us understand what happens in the real world, not in an imaginary world "in the cloud".

## 4 From Applied Science to User-Oriented Solutions and Co-Design

Scientists are dedicated to studying and understanding the composition, organization and functioning of natural systems. This knowledge enables us to identify environmental problems, and even propose solutions in the field of the scientist's specialty. It enables us to indicate what can and cannot be done, in accordance with the limitations imposed by natural laws. It also helps us to confront the uncertainty derived from the great variability and complexity inherent in natural systems. For all these reasons, scientific knowledge is necessary to make sustainable use of natural resources, because raw science is needed for applied science. However, it is not a straightforward task to reconcile the curiosity-driven approach of researchers with the problem-solving approach of managers, especially if both approaches lack long-term common goals. The challenge is for scientific information to effectively reach the policymakers, and provide feedback on their decision-making process. The challenge is also to help society in general increase its scientific awareness and become more and more involved in environmental issues.

*What have we done so far?*

Many initiatives exist to improve science-policy communication, but these largely conform to a linear model of communication in which scientific facts are transmitted to policy advisers to solve problems. The linear model of science-policy communication assumes that policymakers pose well-defined questions while scientists provide credible, legitimate, relevant and timely knowledge (Bradshaw and Borchers 2000) and that policymakers will go on to develop solutions based on this knowledge (Pielke 2007; Young et al. 2014). Following this linear model, science and policy advice/decision-making are perceived as separate domains. While this model can help start a dialogue, it is, on its own, insufficient because decision-making is complex, iterative, and often selective in the information used.

Long-term studies can be important in quantifying the effectiveness of conservation management activities, where prolonged periods might be required for systems to recover following major human disturbance. As opposed to the traditional management characterized by a lack of monitoring, we have opted for "adaptive management" based on follow-up actions in order to evaluate the effects of a treatment submitted to testing (Williams et al. 2009; Allen et al. 2011; Birgé et al. 2016). We have been pioneers putting into practice this philosophy in Sierra Nevada, proposing key questions from the outset, defining the goals to be pursued with the actions undertaken, and specifying the methodological and analytical details necessary to address these efforts (e.g. chapter "Managing the Uniqueness of Sierra Nevada Ecosystems Under Global Change: The Value of in situ Scientific Research"). All this previous experience has served to propose, in collaboration with the Environmental Ministry of the Regional Government of Andalusia, an ongoing ambitious LIFE governance project called ADAPTAMED (protection of key ecosystem services threatened by climate change through adaptive management of Mediterranean socioecological systems). The project focuses on implementing, monitoring, evaluating and disseminating adaptive management measures, with an ecosystem approach. The project objectives aim to reduce the negative impact of climatic change in several natural protected areas, including Sierra Nevada. As a result, an increase in the resilience of the ecosystems concerned is expected, in such a way that their future provision of services will also be improved, in comparison to the scenario of no intervention.

*What still needs to be done to do better?*

Management decisions should be backed by the best available science by integrating primary research and systematic reviews (Dicks et al. 2014). However, management decisions should also account for the values and preferences of stakeholders. Yet applied ecology rarely includes mechanisms to ensure that the science is framed for use and incorporated into decision-making to achieve desired management or societal outcomes (e.g. improved ecosystem function). We strongly need to connect end-users of environmental science to the field research carried out by scientists who study the basis of environmental problems. A better, more effective dialogue is needed between science and policy to underpin sustainable use and conservation. We need user-oriented solutions to environmental problems and co-design (Enquist et al. 2017; Hallet et al. 2017; Lawson et al. 2017; Wall et al. 2017). We need models that link the drivers of change to the response of ecosystems and their services, and ultimately to human well-being. Such models will not only help us to better understand how ecosystems respond to changes in the drivers but they will also allow us to quantify the differential costs and trade-offs of various policies and interventions, which is the type of information that is needed to influence policymakers to make informed decisions (Watson 2005). In short, we need an agenda for science, policy and practice (Pascual et al. 2021).

In addition to the need for a paradigm shift and a common agenda, there are bureaucratic barriers that considerably limit the transfer of scientific knowledge to the actual practice of environmental management in the real world. Scientific assessments that have influenced, to varying degrees, national and international policy decision-making have addressed global issues including long-range acid deposition, stratospheric ozone depletion, climate change and loss of biodiversity (Watson 2005). The recent COVID-19 pandemic has further highlighted the value of scientific knowledge for political decision-making. However, diverse cultures and perspectives can create different barriers to scientist and stakeholder engagement. Scientists and stakeholders also have different levels of flexibility, and this may allow them to help each other to confront challenges (Hallet et al. 2017). Flexibility has two dimensions: personal and institutional. Any group of researchers or stakeholders invariably contain people endowed with that necessary flexibility and common sense, so that, at a personal level, problems of understanding can almost always be solved. Another very different situation usually occurs at the "institutional" level of public administrations, which are characterized by mainly attending to day-to-day emergencies, which compromise their ability to plan in the medium-long term and to solve real problems. The difficulties that hinder permanent collaboration between the fields of research and management are well diagnosed (Watson 2005; Zamora et al. 2017a; Múgica de la Guerra et al. 2020), such as the compartmentalization of competencies between different administrations, the institutional coordination difficulties, the lack of a common research and monitoring program linked to management needs and based on scientific questions, the lack of vision and long-term planning, and the growing bureaucratic difficulties overwhelmed by an oversized regulatory-bureaucratic-political apparatus. This "dinosaur syndrome" of public administrations stands as the main barrier against dialogue between scientists and environmental managers being transformed into new guidelines and practices of environmental management in the Anthropocene. Fortunately, there are some promising examples, such as the Sierra Nevada Global-Change Observatory (http://obsnev.es) and the Mountains Thematic Center of the LifeWatch-ERIC, showing that this syndrome can be overcome (with the permission of the dinosaurs) by aligning different people and even institutions under the same objectives.

## 5 Knowledge Mobilization

The progress and well-being of human societies ultimately depend on the existence of functional natural systems capable of providing essential goods and services, both material and non-material. Many important mountain ecosystem services are intrinsically linked with one another via a series of complex process interactions and feedback mechanisms, in which overlapping cause–effect relationships can be disentangled only by combining different observational points (e.g. biophysical, climatological and socio-economic drivers) within and across ecosystems (Dee et al. 2017). Such an approach, known as "system thinking", urgently requires interdisciplinary perspectives and integrated observation and analysis, so that the signals of change in complex mountain social-ecological systems can be quantified, ideally in near real time.

*What have we done so far?*

Mountain scientists have historically approached the study of the physical setting, climate, aquatic and terrestrial ecosystems and socioeconomic systems in isolated ways. In Sierra Nevada, with the current progress in science and technology (see Sect. 2), we have succeeded in bringing together different perspectives from various fields of knowledge (e.g. ecology, biodiversity, climatology, sociology), which is improving our potential for adequately observing, monitoring, analysing, reporting and predicting changes in complex ecological systems and their interactions.

*What still needs to be done to do better?*

The scientific knowledge generated in Sierra Nevada still needs to be complemented with that of local inhabitants (chapter "Local Ecological Knowledge and the Sustainable Co-Management of Sierra Nevada's Social-Ecological System"). Combining both pathways of knowledge will offer the chance to identify opportunities for sustainable development and innovation, as well as to take management approaches based on systems thinking, transformation and self-organization to bring out the best in our society.

Another pending issue is to involve society in the generation of scientific knowledge. For this, however, we must first go through a communication challenge (Treise and Weigold 2002; Perales 2010). Despite the fact that science and technology are today indispensable tools for economic, educational and cultural development, the vast majority of people do not understand these tools. If we want society to adopt a critical relationship with scientific and technological knowledge and activities, we need to strengthen the channels of communication between the scientific and technological systems and the rest of society, establishing mechanisms aimed not only at disseminating scientific knowledge among the population but also at increasing confidence in science and promoting favourable attitudes towards science. In short, it requires the development of scientific and technological culture (Quintanilla 2011).

Within the framework of Smart EcoMountains, we intend to strengthen such scientific and technological culture. We propose a Learning Hub directed to researchers, decision-makers, students at all educational levels, and the general public, in order to promote knowledge sharing. A variety of educational and training courses, workshops and tools, aimed at different audiences, will be co developed by scientists, information technology specialists and experts in scientific and technological education, to ensure that messages actually reach the public. Further, a number of ongoing and upcoming citizen science projects will actively integrate society into science and innovation (Sherbinin et al. 2021). We also propose new channels of collaboration with experienced storytellers and artists, to use the power of stories and the variety of artistic forms to engage and connect with the public, allowing people to understand scientific knowledge and its impact in a deeper and more meaningful way (Dahlstrom 2014; Martinez-Conde and Macknik 2017).

## 6 Epilogue: Towards the Mountain Community

A major challenge, today and in the future, is to ensure that nature continues to provide a quality of life for all people. However, nature's benefits are unevenly distributed. Nowadays, more than half of the population of the planet has moved to the cities, and a large part of the population in developed countries has only incidental contact with nature. While urban areas may concentrate economic and political power, sparsely populated areas, such as mountain rural settlements and conservation areas, frequently have their environments degraded or their traditional ways of life constrained by rigid regulations aimed at natural conservation (Brondizio and Tourneau 2016; Sharma 2000).

Collaboration between scientists, policymakers and locals on new ways of conceiving and implementing management programmes in mountain systems could help guarantee both biodiversity conservation and a good quality of life for all (Fig. 7) (Draper 2000; Berkes 2007; Sharma et al. 2010; Brondizio and Tourneau 2016). Transformative practices aiming at a sustainable future for mountains would benefit from approaches that consider different knowledge systems and worldviews, such as nature's contribution to people (Pascual et al. 2017; Díaz et al. 2018). Local mountain communities often possess detailed knowledge on biodiversity and ecosystem trends thanks to their direct reliance on their local ecosystems, as well as their observations of change passed down through generations. In many cases, especially in hard-to-reach mountain sites, local information could enrich that of scientists, providing a valuable contribution to understand and communicate the ways in which

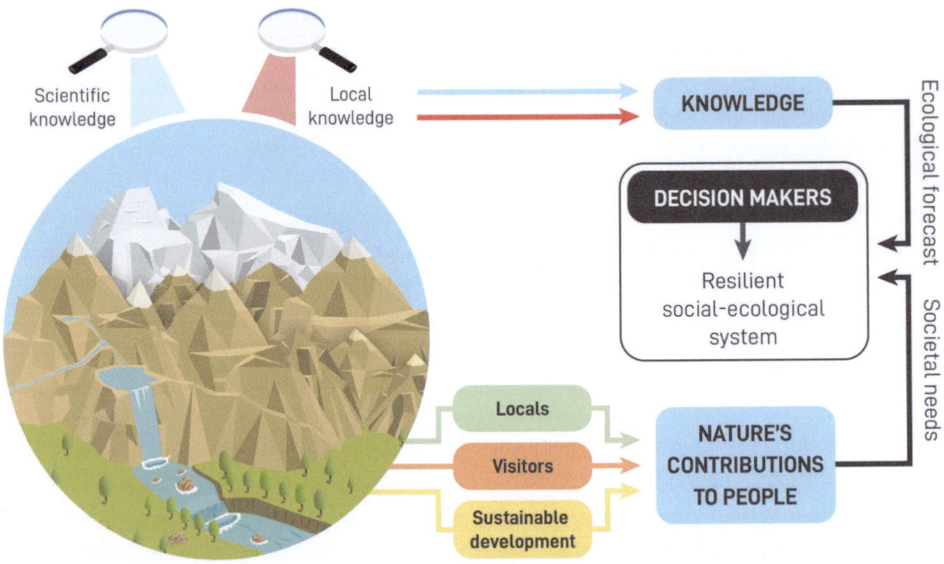


**Fig. 7** To guarantee biodiversity conservation and a good quality of life for all people, decision-makers should: (1) consider plural perspectives combining scientific knowledge with that of local communities, which together can improve our understanding of how mountain ecosystems change and how these changes affect society; and (2) consider the needs of the various types of mountain users, with the ultimate goal of maintaining resilient social-ecological mountain systems

biodiversity decline may affect the complex relationships between people and nature, and to improve our management of mountain biodiversity (Hussain 2000; Rosen et al. 2012).

Exploring new opportunities to support the viability and resilience of place-based livelihoods that sustain the resilience of mountain systems would be another way to guarantee beneficial contributions of nature to a good quality of life for all (Nepal 2002). In some situations, a problem could be turned into an opportunity. For example, the global environmental crisis is motivating the valorization of well-preserved places and traditional high-quality products that were hardly valued just a few years ago. This could trigger an opportunity for mountain dwellers in conservation areas to make a living, considering the prevailing conditions of today, while enhancing biodiversity conservation (Joshi and Dhyani 2009; Bajracharya et al. 2005).

Natural and cultural homogenization of vast planetary landscapes and societies has enhanced the aesthetic and experiential value of the dwindling numbers of well-preserved mountain regions and pockets of cultural diversity. Mountain dwellers now have the opportunity to take advantage of the valuable and unique context and culture of which they form a part. In this sense, the biodiversity crisis has recognized not only the need to protect wild species but also to recover forgotten domesticated plants and animals (Tapia 2000; Hussain and Qamar 2020). That is, attention is beginning to be focused on recovering plant varieties and animal breeds that have resulted from thousands of years of selection by farmers and pastoral communities, and that in turn have given rise to a surprising diversity of traditional products such as cheeses, breads, preserves and cured meats. In this new context, the biodiversity crisis could be an opportunity to highlight the value of the knowledge and products of local mountain communities, which could be promoted using market channels that recognize their added value in terms of ecological as well as social benefits. Existing global initiatives, such as the Mountain Partnership Products (http://www.fao.org/mountain-partnership) and Slow Food Travel (https://www.slowfood.com), try to strengthen the resilience of small-scale mountain producers by promoting new modes of consumption and new models of tourism that encourage visitors to discover ancient techniques, artisanal crafts and products, production sites, and landscapes surrounding the local producers, cooks, and hosts. In summary, we must try to understand how local communities apply their knowledge to decision-making within mountain systems, and with that understanding explore opportunities to support the viability and resilience of place-based livelihoods, which, in turn, sustain the resilience of mountain systems.

As noted in the preface, Sierra Nevada is a climatic, biogeographic, ecological and historical frontier, constituting also a mixture of the pristine and the humanized. There, the conservation of a unique natural heritage must be balanced with the need to achieve the sustainable use of the services provided by local ecosystems. Thus, the conservation of Sierra Nevada ecosystems poses not only concerns but also solutions regarding environmental and socioeconomic problems. Our proposal opens an ecosystemic approach as a line of analysis and action in the research, monitoring, and management of Sierra Nevada. We hope that the scientific information and critical views presented in this book will provide a fresh opportunity for the general public, students, managers and scientists to understand how mountain ecosystems work, how humans depend on them, and what threats we all face as a result of global change.

**Acknowledgements** This research work was conducted in the collaborative framework of the "Sierra Nevada Global Change Observatory" monitoring program http://obsnev.es, promoted by the Environmental and Regional Planning Council of the Regional Government of Andalusia and the University of Granada. We can highlight the following projects for partial funding: the LIFE-ADAPTAMED (LIFE14 CCA/ES/000612): Protection of key ecosystem services by adaptive management of Climate Change endangered Mediterranean socioecosystems, the H2020 project European Long-Term Ecosystem and socio-ecological Research Infrastructure (eLTER) and the Thematic Center on Mountain Ecosystem & Remote sensing, Deep learning-AI e-Services University of Granada-Sierra Nevada (Smart EcoMountains LifeWatch-2019-10-UGR-01). Ricardo Moreno, Pablo Guerrero, Andrea Ros and Manolo Merino have provided very valuable information and have contributed to maintaining an atmosphere of permanent collaboration in our research group. We thank the managers of the Sierra Nevada National Park for their determined support of our work in Sierra Nevada and their continued collaboration within the framework of the joint activities that we develop at the Global Change Observatory. We want to give special thanks to Blanca Ramos Losada.

## References


Adler C, Palazzi E, Kulonen A et al (2020) Monitoring mountains in a changing world: new horizons for the global network for observations and information on mountain environments (GEO-GNOME). Mt Res Dev 38:265–269

Allen CR, Fontaine JP, Kevin L et al (2011) Adaptive management for a turbulent future. J Environ Manage 92:1339–1345

Antonelli A, Kissling WD, Flantua SGA et al (2018) Geological and climatic influences on mountain biodiversity. Nat Geosci 718 (11):718–725

Aspizua R, Barea-Azcón JM, Bonet FJ, Pérez-Luque AJ, Zamora R (2014) Sierra nevada global-change observatory. Monitoring methodologies. Consejería de Medio Ambiente, Junta de Andalucía

Bajracharya SB, Furley P, Newton AC (2005) Effectiveness of community involvement in delivering conservation benefits to the Annapurna Conservation Area, Nepal. Environ Conserv 32:239–247

Baron JS, Gunderson L, Allen CA et al (2009) Options for national parks and reserves for adapting to climate change. J Environ Manage 44:1033–1042

Battaglin WA, Bradley PM, Iwanowicz L et al (2018) Pharmaceuticals, hormones, pesticides, and other bioactive contaminants in water, sediment, and tissue from Rocky Mountain National Park, 2012–2013. Sci Total Environ 643:651–673

Beniston M (2003) Climatic change in mountain regions: a review of possible impacts. Clim Variability Change High Elevation Reg: Past, Present Future 59:5–31

Berkes F (2007) Community-based conservation in a globalized world. PNAS 104:15188–15193

Birgé HE, Allen CR, Garmestani AS et al (2016) Adaptive management for ecosystem services. J Environ Manage 183:343e352

Bitsch J, Jurek M, Gjerdi L et al (2020) Policy brief: elevating mountains in the post-2020 global biodiversity framework 2.0.

Bonet FJ, Pérez-Luque AJ, Moreno-Llorca R, Zamora R (2010) Observatorio de cambio global en Sierra Nevada. Consejería de Medio Ambiente, Junta de Andalucía - Universidad de Granada, Estructura y Contenidos Básicos

Bonet-García FJ, Pérez-Luque AJ, Moreno-Llorca R et al (2015) Protected areas as elicitors of human well-being in a developed region: a new synthetic (socioeconomic) approach. Biol Conserv 187:221–229

Bradshaw GA, Borchers JG (2000) Uncertainty as information: narrowing the science–policy gap. Conserv Ecol 4(1):7

Brondizio ES, Le Tourneau FM (2016) Environmental governance for all. Science 352:1272–1273

Carpenter SR (1998) The need for large-scale experiments to assess and predict the response of ecosystems to perturbation. In: Pace ML, Groffman PM (eds) Success, limitations and frontiers in ecosystem science. Springer. New York, USA, pp 287–312

Catalan J, Ninot JM, Aniz MM (2017). High mountain conservation in a changing world. Springer International Publishing, Cham, pp 385–413

Ceballos G, Ehrlich PR, Dirzo R (2017) Biological annihilation via the ongoing sixth mass extinction signaled by vertebrate population losses and declines. PNAS 114:E6089–E6096

Dahlstrom MF (2014) Using narratives and storytelling to communicate science with non-expert audiences. PNAS 111(4):13614–13620

De Frenne P, Rodríguez-Sanchez F, Coomes DA et al (2013) Microclimate moderates plant responses to macroclimate warming. PNAS 110:18561–18565

Dee LE, Allesina S, Bonn A et al (2017) Operationalizing network theory for ecosystem service assessments. Trends Ecol Evol 32:118–130

Díaz S, Pascual U, Stenseke M et al (2018) Assessing nature's contributions to people. Science 359:270–272

Dick J, Orenstein DE, Holzer J et al (2018) What is socio-ecological research delivering? A literature survey across 25 international LTSER platforms. Sci Total Environ 622–623:1225–1240

Dicks LV, Walsh JC, Sutherland WJ (2014) Organising evidence for environmental management decisions: a '4S' hierarchy. Trends Ecol Evol 29(11):607–613

Dirzo R, Young HS, Galetti M (2014) Defaunation in the anthropocene. Science 345:401–406

Doblas-Miranda E, Alonso R, Arnan X et al (2017) A review of the combination among global change factors in forests, shrublands and pastures of the mediterranean region: beyond drought effects. Glob Planet Change 148:42–54

Dobrowski SZ, Abatzoglou JT, Greenberg JA, Schladow SG (2009) How much influence does landscape-scale physiography have on air temperature in a mountain environment? Agric Meteorol 149:1751–1758

Draper D (2000) Toward sustainable mountain communities: Balancing tourism development and environmental protection in Banff and Banff National Park, Canada. J Hum Environ Stud 29:408–415

Enquist CAF, Jackson ST, Garfin GM et al (2017) Foundations of translational ecology. Front Ecol Environ 15:541–550

Enquist BJ, Feng, X, Boyle, B et al (2019). The commonness of rarity: global and future distribution of rarity across land plants. Sci Adv 5 (11):eaaz0414

Folke C, Polasky S, Rockström J et al (2021) Our future in the anthropocene biosphere. Ambio 50:834–869

Fridley JD (2009) Downscaling climate over complex terrain: high fine scale (<1000 m) spatial variation of near-ground temperatures in a mountain forested landscape (Great Smoky Mountains). J Appl Meteorol Climatol 48:1033–1049

García MB, Domingo D, Pizarro M (2020) Rocky habitats as microclimatic refuges for biodiversity. A close up thermal approach. Environ Exp Bot 170:103886

Gonzalez A, Cardinale BJ, Allington GRH et al (2016) Estimating local biodiversity change: a critique of papers claiming no net loss of local diversity. Ecology 97:1949–1960

Grabherr G, Björnsen Gurung A, Dedieu J-P et al (2005) Long-term environmental observations in mountain biosphere reserves: recommendations from the EU GLOCHAMORE project. Mt Res Dev 25:376–382

Hallett LM, Morelli TL, Gerber LR et al (2017) Navigating translational ecology: creating opportunities for scientist participation. Front Ecol Environ 15:578–586

Hansen AJ, Piekielek N, Davis C et al (2014) Exposure of U.S. National Parks to land use and climate change 1900–2100. Ecol App 24:484–502

Hillebrand H, Blasius B, Borer ET et al (2018) Biodiversity change is uncoupled from species richness trends: consequences for conservation and monitoring. J Appl Ecol 55:169–184

Holzer JM, Adamescu MC, Bonet-García FJ et al (2018) Negotiating local versus global needs in the international long term ecological research network's socio-ecological research agenda. Environ Res Lett 13(10):105003

Huber UM, Bugmann H, Reasoner MA (eds) (2005) Global change and mountain regions: an overview of current knowledge. Springer, New York

Hughes BB, Beas-Luna R, Barner AK et al (2017) Long-term studies contribute disproportionately to ecology and policy. Bioscience 67:271–281

Hussain S (2000) Protecting the snow leopard and enhancing farmers' livelihoods. Mt Res Dev 20:226–231

Hussain A, Qamar FM (2020) Dual challenge of climate change and agrobiodiversity loss in mountain food systems in The Hindu-Kush Himalaya. One Earth 3:539–542

Joshi R, Dhyani PP (2009) Environmental sustainability and tourism–implications of trend synergies of tourism in Sikkim Himalaya Curr Sci 97(1):33–41

Kim HR, Alkemade IMD, Leadley R et al (2018) A protocol for an intercomparison of biodiversity and ecosystem services models using harmonized land-use and climate scenarios. Geosci Model Dev 11:4537–4562

Kissling WD, Ahumada JA, Bowser A et al (2018) Building essential biodiversity variables (EBVs) of species distribution and abundance at a global scale. Biol Rev 93:600–625

Kohler T, Wehrli A, Jurek M (2014) Mountains and climate change: a global concern. Sustainable Mountain Development Series. Bern, Switzerland, Centre for Development and Environment (CDE), Swiss Agency for Development and Cooperation (SDC) and Geographica Bernensia, 136 pp

Körner C (2021) Alpine plant life. Springer International Publishing, Cham

Körner C, Hiltbrunner E (2018) The 90 ways to describe plant temperature. Perspect Plant Ecol Evol Syst 30:16–21

Körner C, Ohsawa M (2005) Mountain systems. In: Hassan R, Scholes R, Ash N (eds) Ecosystems and human well-being: current state and trends. Island Press

Kulonen AL, Adler C, Palazzi E (2020) GEO-GNOME workshop "Identifying Essential Biodiversity Variables (EBVs) and Essential Societal Variables (ESVs) in Mountain Environments"

Lawson DM, Hall KR, Yung L, Enquist CAF (2017) Building translational ecology communities of practice: insights from the field. Front Ecol Environ 15:569–577

Lembrechts J, Lenoir J, Roth N et al (2019) Comparing temperature data sources for use in species distribution models: from in-situ logging to remote sensing. Glob Ecol Biogeogr 28:1578–1596

Lembrechts J, Nijs I, Lenoir J (2019) Incorporating microclimate into species distribution models. Ecography 42:1267–1279

Likens GE (2013) The hubbard brook ecosystem study: celebrating 50 years. Bull Ecol Soc Am 94:336–337

Likens GE (1989) Long-term studies in ecology: approaches and alternatives. Springer

Lindenmayer DB, Likens GE (2009) Adaptive monitoring: a new paradigm for long-term research and monitoring. Trends Ecol Evol 24:482–486

Lovett GM, Burns DA, Driscoll CT et al (2007) Who needs environmental monitoring? Front Ecol Environ 5(5):253–260

Mace GM (2014) Whose conservation? Science 345(6204):1558–1560

Magliocca NR, Ellis EC, Allington GRH et al (2018) Closing global knowledge gaps: producing generalized knowledge from case studies of social-ecological systems. Glob Environ Change 50:1–14

Martinez-Conde S, Macknik SL (2017) Opinion: finding the plot in science storytelling in hopes of enhancing science communication. PNAS 114(31):8127–8129

Mengist W, Soromessa T, Legese G (2020) Ecosystem services research in mountainous regions: a systematic literature review on current knowledge and research gaps. Sci Total Environ 702:134581

Millennium ecosystem assessment (2005) Ecosystems and human well-being: synthesis. Island Press, Washington, DC

Múgica M, Montes C, Mata Olmo R, Castell Puig C (2020) Las áreas protegidas como herramientas para reforzar las conexiones entre ciencia, gestión y sociedad. Ecosistemas 29(1):1904

Nepal SK (2002) Mountain ecotourism and sustainable development. Mt Res Dev 22(2):104–109

Nogués Bravo D, Araujo M, Lasanta T et al (2008) Climate change in mediterranean mountains during the 21st century. Ambio 37(4):28–285

Opedal ØH, Armbruster WS, Graae BJ (2015) Linking small-scale topography with microclimate, plant species diversity and intra-specific trait variation in an alpine landscape. Plant Ecol Divers 8:305–315

Padha S, Kumar R, Dhar A, Sharma P (2021) Microplastic pollution in mountain terrains and foothills: a review on source, extraction, and distribution of microplastics in remote areas. Environ Res

Pascual U, Balvanera P, Díaz S et al (2017) Valuing nature's contributions to people: the IPBES approach. Curr Opin Environ Sustain 26:7–16

Pascual U, Adams WM, Díaz S et al (2021) Biodiversity and the challenge of pluralism. Nat Sustain 4:567–572

Pauchard A, Kueffer C, Dietz H et al (2009) Ain't no mountain high enough: plant invasions reaching new elevations. Front Ecol Environ 7(9):479–486

Payne D, Spehn EM, Snethlage M, Fischer M (2017) Opportunities for research on mountain biodiversity under global change. Curr Opin Environ Sustain 29:40–47

Perales OM (2010) Problemas epistemológicos de la comunicación pública de la ciencia. Doctoral dissertation, Universidad de Salamanca

Pérez-Luque AJ, Peinó-Calero E, Rodríguez-Brito A et al (2021) ClimaNevada: Base de datos climática del Observatorio de Cambio Global de Sierra Nevada. Ecosistemas 30(1):2155

Perrigo A, Hoorn C, Antonelli A (2019) Why mountains matter for biodiversity. J Biogeogr 00:1–11

Peters DP, Groffman PM, Nadelhoffer KJ et al (2008) Living in an increasingly connected world: a framework for continental-scale environmental science. Front Ecol Environ 6:229–237

Pielke RA Jr (2007) The honest broker. Making sense of science in policy and politics. University Press, Cambridge

Quintanilla MA (2011) La ciencia y la cultura científica. ArtefaCToS, Revista De Estudios Sobre La Ciencia y La Tecnología 3(1):31–48

Rahbek C, Borregaard MK, Antonelli A et al (2019) Building mountain biodiversity: geological and evolutionary processes. Science 365:1114–1119

Randin CHF, Ashcroftd MB, Bolligere J et al (2020) Monitoring biodiversity in the anthropocene using remote sensing in species distribution models. Remote Sens Environ 239:111626

Rogora M, Frate L, Carranza ML et al (2018) Assessment of climate change effects on mountain ecosystems through a cross-site analysis in the Alps and Apennines. Sci Total Environ 624:1429–1442

Rosen T, Hussain S, Mohammad G et al (2012) Reconciling sustainable development of mountain communities with large carnivore conservation. Mt Res Dev 32:286–293

Scherrer D, Körner C (2010) Infra-red thermometry of alpine landscapes challenges climatic warming projections. Glob Change Biol 16:2602–2613

Scherrer D, Körner C (2011) Topographically controlled thermal-habitat differentiation buffers alpine plant diversity against climate warming. J Biogeogr 38:406–416

Shahgedanova M, Adler C, Gebrekirstos A et al (2021) Mountain observatories: status and prospects for enhancing and connecting a global community. Mt Res Dev 41:A1–A15

Sharma E, Chettri N, Oli KP (2010) Mountain biodiversity conservation and management: a paradigm shift in policies and practices in the Hindu Kush-Himalayas. Ecol Res 25:909–923

Sharma P (ed) (2000) Tourism as development: case studies from the Himalaya, vol 4. Himal Books

Sherbinin A, Bowser A, Chuang TR et al (2021) The critical importance of citizen science data. Front Clim 3:20

Soranno PA, Cheruvelil KS, Bissell EG et al (2014) Cross-scale interactions: quantifying multi-scaled cause–effect relationships in macrosystems. Front Ecol Environ 12:65–73

Steffen W, Persson A, Deutsch L et al (2011) The anthropocene: from global change to planetary stewardship. AMBIO 40:739–761

Stevens CJ, David TI, Storkey J (2018) Atmospheric nitrogen deposition in terrestrial ecosystems: its impact on plant communities and consequences across trophic levels. Func Ecol 32:1757–1769

Stewart JR, Lister AM, Barnes I, Dalén L (2010) Refugia revisited: individualistic responses of species in space and time. P Roy Soc B 277:661–671

Tapia ME (2000) Mountain agrobiodiversity in Peru. Mt Res Dev 20:220–225

Thornton JM, Palazzi E, Pepin NC et al (2021) Toward a definition of essential mountain climate variables. One Earth 4:805–827

Treise D, Weigold MF (2002) Advancing science communication: a survey of science communicators. Sci Commun 23(3):310–322

Vargas R, Alcaraz-Segura D, Birdsey R et al (2017) Enhancing interoperability to facilitate implementation of REDDC: case study of Mexico. Carbon Manag 8(1):57–65

Wall TU, McNie E, Garfin GM (2017) Use-inspired science: making science usable by and useful to decision makers. Front Ecol Environ 15:551–559

Watson RT (2005) Turning science into policy: challenges and experiences from the science-policy interface. Phil Trans R Soc B 360:471–477

Williams BK, Szaro, RC, Shapiro CD (2009) Adaptive management: the U.S. Department of the interior technical guide. Adaptive Management Working Group, U.S. Department of the Interior, Washington, DC

Young JC, Waylen KA, Sarkki S et al (2014) Improving the science-policy dialogue to meet the challenges of biodiversity conservation: having conversations rather than talking at one-another. Biodivers Conserv 23:387–404

Zamora R, Pérez-Luque AJ, Bonet FJ (2017) Monitoring global change in high mountains. In: Catalan J, Ninot JM, Aniz MM (eds) High mountain conservation in a changing world. Springer International Publishing, Cham, pp 385–413

Zamora R, Pérez-Luque AJ, Bonet FJ et al (2017) Global change impacts in the sierra nevada LTER-site (southern Spain). Bull Ecol Soc Am 98(2):157–164

Zamora R, Pérez-Luque AJ, Guerrero-Alonso PD et al (2021) Uniendo macro y microclima en paisajes de montaña: una aproximación conceptual e instrumental. Ecosistemas 30(1):2166

Zamora R, Pérez-Luque AJ, Bonet FJ, Barea-Azcón JM, Aspizua R (2016) Global change impacts in sierra nevada: challenges for conservation. Consejería de Medio Ambiente y Ordenación del Territorio. Junta de Andalucía

Zellweger F, De Frenne P, Lenoir J et al (2019) Advances in microclimate ecology arising from remote sensing. Trends Ecol Evol 34:327–341

Zellweger F, De Frenne P, Lenoir J et al (2020) Forest microclimate dynamics drive plant responses to warming. Science 368:772–775

**Author**: Luis Ordoñez

# Concluding Notes: Research, Adaptive Conservation and Enjoyment of Sierra Nevada in the Anthropocene

**R. Zamora, D. Alcaraz-Segura, P. Guerrero, David García-del-Amo and M. Villar-Argaiz**

Chapters of this thematic block have addressed new conceptual and technological approaches in research, management and social connection that we are developing in Sierra Nevada to assess the sensitivity and adaptive capacity of its diverse ecosystems to current and future environmental and human scenarios. Chapter "Remote Sensing in Sierra Nevada: From Abiotic Processes to Biodiversity and Ecosystem Functions and Services" introduced us to the actual use of remote sensing as a tool for monitoring ecosystem abiotic and biotic structure, functions and services throughout terrestrial, aquatic and atmospheric studies. Most of the research conducted in Sierra Nevada via remote sensing means was found potentially useful to assess biodiversity and ecosystem services. Indeed, most studies were either applied or use-oriented research, frequently already containing recommendations for the management of the protected area. Hence, an increase in the interdisciplinary and transdisciplinary application of remote sensing to research in Sierra Nevada from local to regional and global studies is expected, with subsequent implications in management.

Chapter "Managing the Uniqueness of Sierra Nevada Ecosystems Under Global Change: The Value of in situ Scientific Research" examined four case studies in which the generation of new knowledge in situ has been evident and has been applied to the improvement of environmental management, thanks to the numerous research projects of basic and applied science conducted in Sierra Nevada, but also thanks to the growing number of projects guided by the needs of environmental management. In the last decades, a series of long-term management projects have been developed in Sierra Nevada that address pressing environmental challenges and use scientific knowledge generated in situ to propose solutions. As opposed to the traditional management characterized by a lack of monitoring, we have opted for "adaptive management" based on follow-up of actions in order to evaluate the effects of a treatment submitted to testing. We have put into practice this philosophy in Sierra Nevada, proposing key questions from the outset, defining the goals to be pursued with the actions undertaken, and specifying the methodological and analytical details necessary to address these efforts. We have established a framework of permanent collaboration between researchers of academic institutions and managers of the National Park of Sierra Nevada. The results presented in chapter "Managing the Uniqueness of Sierra Nevada Ecosystems Under Global Change: The Value of in situ Scientific Research" highlight the benefits derived from this collaboration. The focus of the four projects presented in this chapter ranges from the species to the ecosystem, and they seek solutions based on nature, enhancing the capacity for adaptation and resilience of natural systems under global change.

Chapter "Local Ecological Knowledge and the Sustainable Co-Management of Sierra Nevada's Social-Ecological System" analyses the importance of local ecological knowledge to maintain the proper functioning of Sierra Nevada social-ecological system. Ecosystems of Sierra Nevada have co-evolved together with local communities during last centuries, greatly influenced by the traditional water management system, generating a "cultural landscape". Throughout the twentieth and twenty-first centuries, the different drivers of global environmental change have greatly

R. Zamora (✉)
Department of Ecology, University of Granada, Granada, Spain
e-mail: rzamora@ugr.es

D. Alcaraz-Segura
Department of Botany, University of Granada, Granada, Spain

P. Guerrero
Andalusian Institute for Earth System Research (IISTA), University of Granada, Granada, Spain

D. García-del-Amo
Institut de Ciencia i Tecnologia Ambientals, Universitat Autónoma de Barcelona, Barcelona, Spain

M. Villar-Argaiz
Department of Ecology, University of Granada, Granada, Spain

eroded the proper functioning of this social-ecological system and endangered the transmission of local ecological knowledge to the new generations. The numerous environmental challenges threatening Sierra Nevada, derived from global changes, require considering all possible sources of knowledge, including local ecological knowledge in addressing them, keeping at the same time this historical legacy alive. Bridging of local ecological knowledge and scientific knowledge for knowledge co-production would engage local communities in the co-monitoring and co-governance of the territory. At the same time, promotion of local varieties and breeds will favour the continuity of small local producers and cooperatives allowing the future of local communities and the continuity of these traditional activities. Interdisciplinary and inclusive methodologies can help to avoid power imbalances and to ensure mutual understanding between both sources of knowledge to achieve a true participation of the local population, empowering them in the co-management and conservation of the territory.

Sierra Nevada is a widely studied mountain system over a diverse range of disciplines. Every day terabytes of data with heterogeneous information in very different formats are generated. This huge capacity of information generation raises several unresolved issues such as the technological limitation to fully exploit, share and preserve this data, the possibility of using this information for research while respecting at all times the authorship of such data and preserving its proper use, and finally, the possibility of allowing the general public access to the results of scientific research and to the knowledge that can be extracted from all this information in a friendly and intuitive way. The urgency and complexity of safeguarding mountain ecosystems in the Anthropocene require a reenvisioning of how new technological tools can best support the comprehensive assessments of ecosystems health that are required for successful action. Chapter "Data Model, E-Infrastructure Services and the Virtual Research Environment (VRE)" explains precisely these new conceptual and instrumental frameworks with existing information on biodiversity and mountain ecosystems through our Smart Ecomountains initiative under the umbrella of the European Research Infrastructure Lifewatch. Through this Thematic Center, the University of Granada is directly involved in the development of the European Lifewatch ERIC infrastructure, focusing on research, monitoring and sustainable development of mountain ecosystems with the application of Remote Sensing and Artificial Intelligence.

Sierra Nevada is also an excellent place to develop citizen science projects (chapter "Advancing Open Science in Sierra Nevada: Current Citizen Science Campaigns"). In addition to the obvious interest of its ecosystems, Sierra Nevada is a very accessible high mountain, which facilitates field work. It is urgent to get a good account of what is happening and the speed at which some of the environmental changes are occurring in Sierra Nevada. Bringing together scientists and citizens into a single forum where to analyse and debate the current scientific understanding of natural and human-driven changes to our planet is an opportunity, we cannot afford to miss. New initiatives are rapidly emerging alongside social awareness for science-based decisions. Citizen science is certainly a powerful means by which citizens can participate in the future of science practice (chapter "Advancing Open Science in Sierra Nevada: Current Citizen Science Campaigns"). Perhaps the most thrilling goal is to engage citizens in building scientifically robust information. For this, professionals should supervise and provide support via training, protocols and data calibration to ensure that data collected by citizens is scientifically robust. Such an effort is necessary to guarantee a major source of high-quality data that can serve policy-makers as well as scientists. Community science can contribute to guide the flow of information from scientific forums to decision-makers and managers by providing the essential data, awareness and educational outreach necessary to evaluate conservation, management and restoration practices.

The 24 chapters of this book provide a full state-of-the-art review for the diagnosis of the health status of ecosystems in Sierra Nevada. All the evidence presented in this book reinforces our original idea of Sierra Nevada as a unique biophysical laboratory and observatory of global change. Sierra Nevada is also of strategic importance as a provider of ecosystem services within the boundaries of the massif as well as for the surrounding region. The last chapter "Filling the Gaps in Research, Monitoring, Management and Social Connection" offers a critical review of what have we done so far, and what still needs to be done to do better in research, monitoring, use of new technological tools (remote sensor, artificial intelligence, virtual research environments), knowledge mobilization, user-oriented solutions and co-design. Our proposal is that mountain ecosystems are both a natural heritage that must be preserved, and a source of natural resources providing provisioning, regulating and cultural ecosystem services. Thus, the conservation of mountain ecosystems is both a concern and a source of solutions to environmental and socioeconomic problems. Collaboration between scientists, policymakers and locals to new ways of conceiving and implementing management programmes in mountain systems could help guarantee healthy and biodiverse ecosystems and a good quality of life for all. Issues of societal importance—food system, air and water quality, geochemical cycles or environmental services—need to be explored both from a scientific and social perspective. Transformative practices aiming at a sustainable future for a mountain would benefit from approaches that consider different knowledge systems and worldviews, such as nature's contribution to people. After critically evaluating the current situation of what have we done so far, we propose a joint journey of people and nature through the mountains, pointing out what we should avoid, and what we should do.

# Index




R. Zamora and M. Oliva (eds.), *The Landscape of the Sierra Nevada*,
https://doi.org/10.1007/978-3-030-94219-9

FSC
www.fsc.org
MIX
Papier aus verantwortungsvollen Quellen
Paper from responsible sources
FSC® C105338

If you have any concerns about our products,
you can contact us on
**ProductSafety@springernature.com**

In case Publisher is established outside the EU,
the EU authorized representative is:
**Springer Nature Customer Service Center GmbH**
**Europaplatz 3, 69115 Heidelberg, Germany**

Printed by Libri Plureos GmbH
in Hamburg, Germany